Encyclopaedia of VECTOR CALCULUS

Encyclopaedia of VECTOR CALCULUS

Volume 1

Dr. Rakesh Kumar Pandey

ANMOL PUBLICATIONS PVT. LTD.
NEW DELHI - 110 002 (INDIA)

ANMOL PUBLICATIONS PVT. LTD.
Regd. Office: 4360/4, Ansari Road, Daryaganj,
New Delhi-110 002 (India)
Ph.: 23278000, 23261597
Branch Office: No. 1015, Ist Main Road, BSK IIIrd Stage
IIIrd Phase, IIIrd Block,
Bangalore-560 085 (India)
Tel.: 080-41723429
Visit us at: www.anmolpublications.com

Encyclopaedia of Vector Calculus

First Edition, 2009
ISBN 978-81-261-4110-4 (Set)

PRINTED IN INDIA

Printed at Mehra Offset Press, Delhi.

Contents

Preface

Encyclopaedia of Vector Calculus is the foundation stone on which a vast amount of Applied Mathematics is based. Topics such as Fluid Dynamics, Solid Mechanics and Electromagnetism depend heavily on the Calculus of Vector Quantities in Three Dimensions. This Encyclopaedia covers the material in a comprehensive but concise manner, combining mathematical rigour with Physical Insight. There are many diagrams to illustrate the physical meaning of the Mathematical concepts, which is essential for a full understanding of the subject. There are worked examples that cover all of the material. The a complete chapter is devoted to most important applications of Vector Calculus including *Heat, Mechanics and Electromagnetism*.

Author

List of Symbols

$\pm$	$\not\subset$	$\subset$	$\supset$	$\subseteq$	$\in$	$\notin$	∂
$\neq$	$\equiv$	$\approx$	$\angle$	∇	Σ	Π	Ω
ϕ	α	β	χ	ε	γ	η	λ
μ	ν	$\perp$	π	θ	ρ	σ	τ
$\Re$	ω	ξ	ψ	ζ	Υ	$\exists$	$\int$
$\iint$	$\iiint$	$\oint$	$\times$	$\sqrt{\ }$	$\therefore$	$\because$	Δ
$\geq$	$\leq$	$\Rightarrow$	$\Downarrow$	$\Leftrightarrow$	∞	$\propto$	$\mathbb{R}$
$\mathbb{Z}$	$\mathbb{N}$	$\mathbb{C}$	$\mathbb{Q}$				

Chapter 1

Introduction to Vector Calculus

Vector calculus (also called *vector analysis*) is a field of mathematics concerned with multivariable real analysis of vectors in an inner product space of two or more dimensions (some results — those that involve the cross product — can only be applied to three dimensions). It consists of a suite of formulae and problem solving techniques very useful for engineering and physics. Vector analysis has its origin in quaternion analysis, and was formulated by the American engineer and scientist J. Willard Gibbs and the British engineer Oliver Heaviside.

Vector calculus is concerned with scalar fields, which associate a scalar to every point in space, and vector fields, which associate a vector to every point in space. For example, the temperature of a swimming pool is a scalar field: to each point we associate a scalar value of temperature. The water flow in the same pool is a vector field: to each point we associate a velocity vector.

VECTOR OPERATIONS

Vector calculus studies various differential operators defined on scalar or vector fields, which are typically expressed in terms of the del operator (∇). The four most important operations in vector calculus are:

Operation	*Notation*	*Description*	*Domain/Range*
Gradient	$\text{grad}(f) = \Delta f$	Measures the rate and direction of change in a scalar field.	Maps scalar fields to vector fields.
Curl	$\text{curl}(F) = \Delta \times f$	Measures the tendency to rotate about a point in a vector field.	Maps vector fields to vector fields.
Divergence	$\text{div}(F) = \Delta \cdot F$	Measures the magnitude of a source or sink at a given point in a vector field.	Maps vector fields to scalar fields.
Laplacian	$\Delta f = \Delta^2 f = \Delta \cdot \Delta f$	A composition of the divergence and gradient operations.	Maps scalar fields to scalar fields.

A quantity called the Jacobian is useful for studying functions when both the domain and range of the function are multivariable, such as a change of variables during integration.

THEOREMS

Likewise, there are several important theorems related to these operators which generalize the fundamental theorem of calculus to higher dimensions:

Theorem	*Statement*	*Description*
Gradient theorem	$\varphi(q)-\varphi(p)=\int_L \nabla\varphi \cdot dr.$	The line integral through a gradient (vector) field equals the difference in its scalar field at the endpoints of the curve.
Green's theorem	$\int_L Ldx + Mdy = \iint_D\left(\frac{\partial M}{\partial x}-\frac{\partial L}{\partial y}\right)dA$	The integral of the scalar curl of a vector field over some region in the plane equals the line integral of the vector field over the curve bounding the region.
Stokes' theorem	$\int_C \nabla\times F\cdot d\Sigma = \oint_{\partial\Sigma} F\cdot dr,$	The integral of the curl of a vector field over a surface equals the line integral of the vector field over the curve bounding the surface.
Divergence theorem	$\iiint_V (\nabla\cdot F)dV = \iint_{\partial V} F\cdot dS,$	The integral of the divergence of a vector field over some solid equals the integral of the flux through the surface bounding the solid.

The use of vector calculus may require the handedness of the coordinate system to be taken into account. Most of the analytic results are easily understood, in a more general form, using the machinery of differential geometry, of which vector calculus forms a subset.

INTRODUCTION TO VECTORS

VECTORS

We have three types of physical quantities:

1. *Scalars:* Magnitude Only. e.g. Mass, Length and Time.
2. *Vectors:* Magnitude and Direction. e.g. Velocity and Force.
3. *Tensors:* Not covered here.

The simplest way to differentiate vectors from scalars is to take Speed vs. Velocity.

In the path shown above, assume that the particle has a fixed speed. While it's speed does not change, it does have an acceleration, as the direction changes constanly.

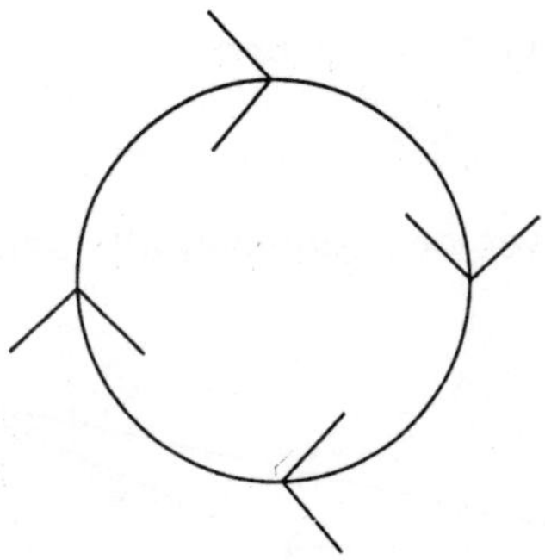

Free Vectors vs. Bound Vectors

In this course we will be dealing with free vectors. Free vectors are deemed to be equal if they have the same magnitude and direction, regardless of their location. The same does not hold true for bound vectors.

Free Vectors

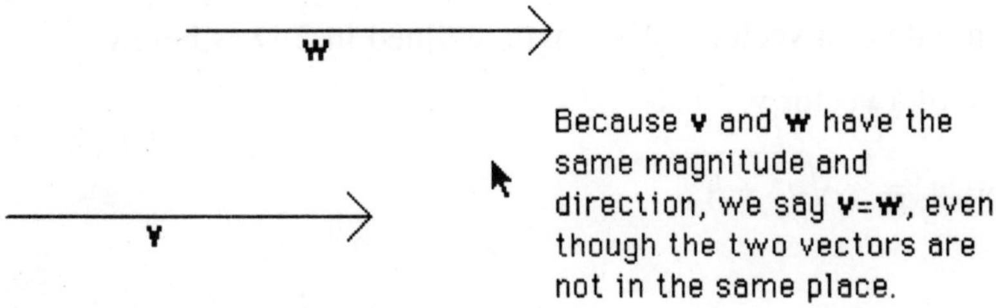

Bound Vectors

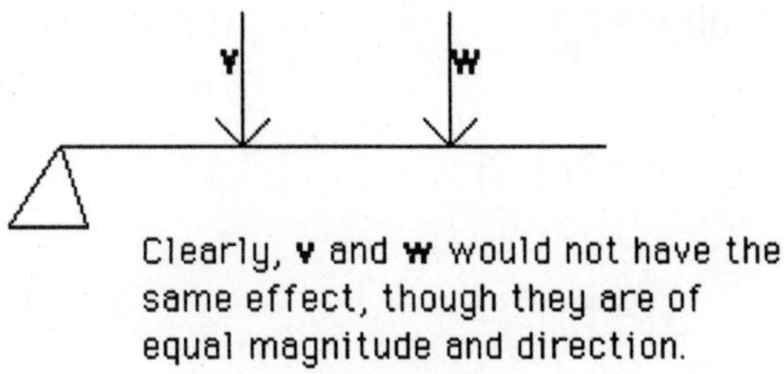

Unit Vectors

Def: A unit vector is a vector of length (magnitude) one.

The 2-D space with which we are very familiar has two particular unit vectors reffered to as a "Basis." They are the i and j vectors. Because they are a Basis every vector can be written in terms of these two vectors. For example v = 3i + 2j. Thus, one can determine the components of v. It is important; however, to differentiate between the scalar and vector components.

Scalar Components: 3,2 Vector Components: 3i, 2j

Laws of Addition

There are two ways to add vectors, geometrically and algebraically.

Geometric Addition

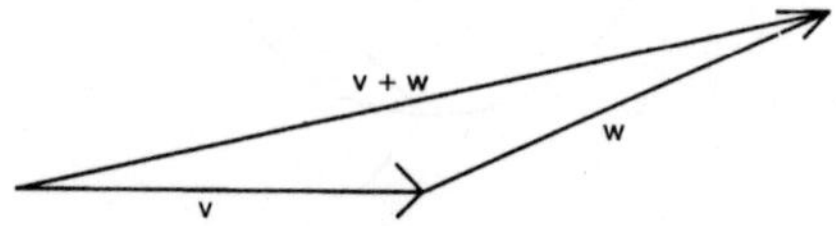

The parrallelogram method of vector addition

Algebraic Addition

v = 3i – 2j, w = i + j

v + w = 4i – j

Magnitude

The Magnitude of a vector is it's length, defined in 2-D as below.

Magnitude of a vector v:

$$|v| = |ai + bj| = +\sqrt{a^2 + b^2}$$

Direction

Def: The direction of v is a unit vector parallel to v

$$\text{dir v} = \frac{v}{|v|}$$

$$v = 2i + 2j$$

$$|v| = 5$$

$$\text{dir v} = \frac{3}{5}i + \frac{4}{5}j$$

For Example: But what does this mean? We usually think of direction in terms of angles, not components of vectors. This brings us to slope.

Slope

To find θ requires using the double argument Arctangent.

Since information like slope = 1 would yield both 45° and 135° for θ.

$$v = ai + bj$$

$$r = \sqrt{a^2 + b^2}$$

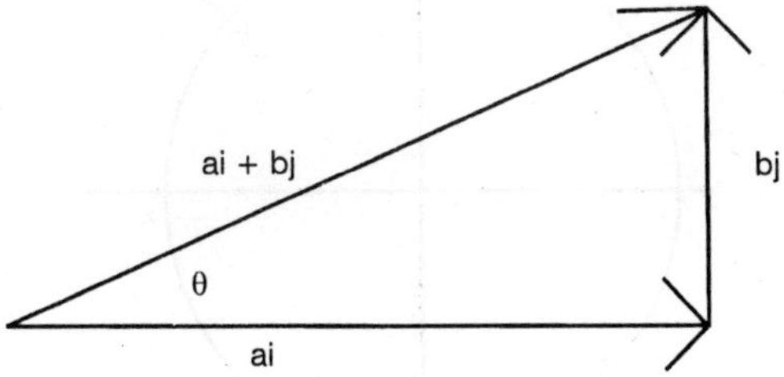

Then Slope:

$$a = r * \cos \theta$$

$$\text{slope} = b/_a = \tan \theta \; a = r * \sin \theta$$

This can be used to find unit tangents and normals to curves defined in the x-y plane. Any curve will have two unit tangents and two unit normals at any point, so it is necessary to differentiate betweent the clockwise and counterclockwise tangents and the in and out unit normals. The example below shows an inward unit normal and a clockwise unit tangent.

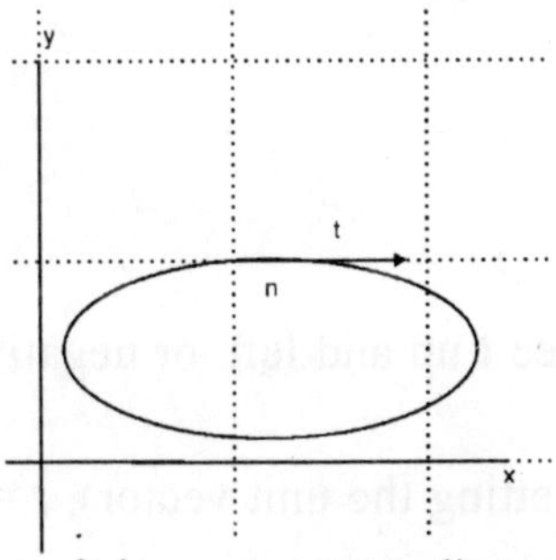

In order to find these tangents, it is necessary to "marry" calculus and slope. Since the derivitive, *dy/dx*, is equal to the slope, *b/a*, you can find the unit tangent vector simply by setting these two quantities equal to each other. Namely:

$$dy/dx = b/a$$

GEOMETRY OF VECTORS

Last lecture left off discussing the "marriage" of vectors and calculus in order to find a vector tangent to a given curve in the x-y plane. The following example illustarates exactly how this can be done.

The circle is $x^2 + y^2 = 1$.

Point *P* has coordinate $(1/\sqrt{2}, 1/\sqrt{2})$

We want the counterclockwise tangent and outward normal pictured above

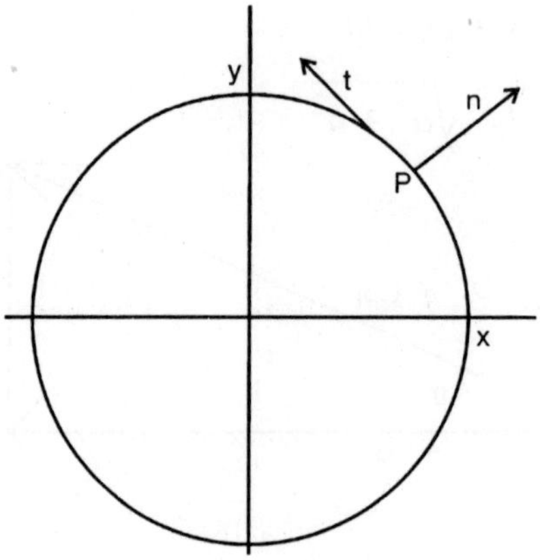

First find the slope of the tangent: $2x + 2y\dfrac{dy}{dx}$

$$= 0\frac{dy}{dx} = -\frac{x}{y}$$

$$= -1$$

Also. slope = b/a

So $b/a = -1$ and $b = -a$

Therefore $T = i - j$ or $T = j - i$

Looking at the picture we see t up and left. or negative x and positive y,

so $T = j - i$. Normalizing (getting the unit vector), $t = \dfrac{1}{\sqrt{2}}(j - i)$.

For the Normal: Slope of the normal is the neg. reciprocal of the tangent (They're ⊥.) = 1. So $N = i + j$ or $N = -i - j$. Since the normal has pos. x and y components,

We know that $N = i + j$, normallizing $n = \dfrac{1}{\sqrt{2}}(i - j)$

3-D Cartesian Coordinates

In three dimensions, the two dimentional concepts are extended, with the simple addition of a third component-the unit vector parallel to the z axis, k.

Magnitude and direction are still computed in extremely similar ways.

$$v = ai + bj + ck$$

$$|v| = +\sqrt{a^2 + b^2 + c^2}$$

$$\operatorname{dir} v = \frac{v}{|v|}$$

$$\text{Ex: } v = i - 2\,j + k$$

$$\operatorname{dir} v = \frac{1}{\sqrt{6}}\,(i - 2j + k)$$

We would still like to have angles to describe direction rather than a difficult to grasp set of components. Two Dimentions required one angle to define a direction. In three dimensions we'll need two.

Spherical Coordinates

Spherical coordinates define a vector in terms of its length (R), an angle between the vector and the z-axis (phi), and the angle between the vectors projection onto the xy plane and the x-axis (theta).

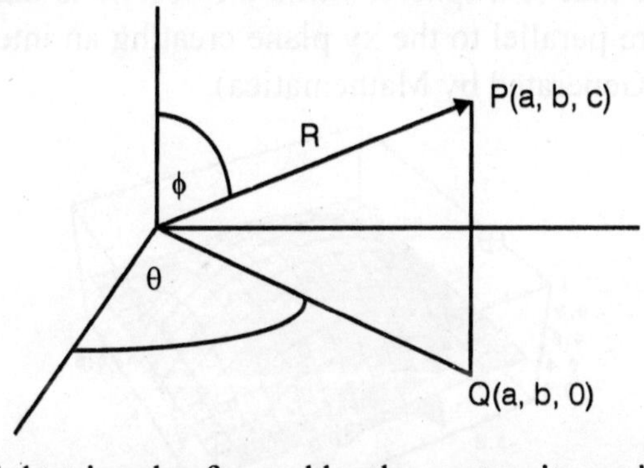

By looking at the right triangles formed by the vector, its projection, and the axes, the relationship between the Spherical coordinates and Cartesian ones can be determined. It turns out that x = R sin (phi) cos (theta); y = R sin (phi) sin (theta); and z = R cos (phi).

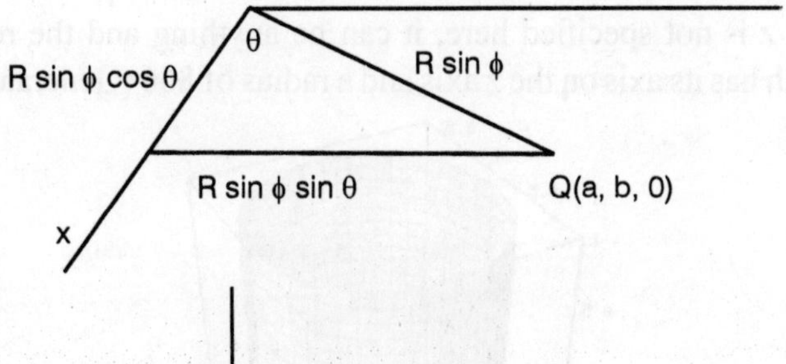

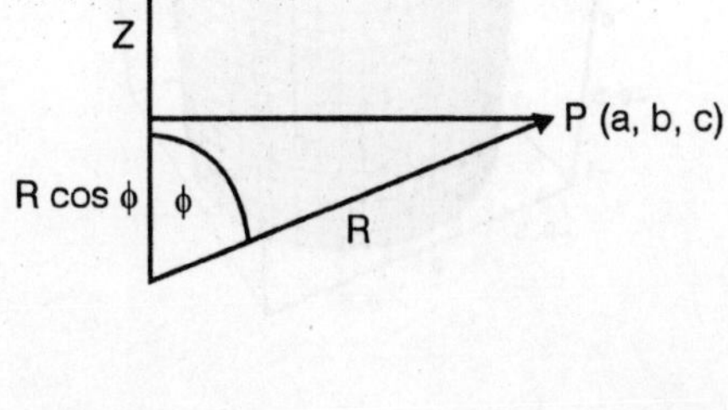

3D Geometry in three Dimentions there are four types of objects that can Appear:

1. *Surfaces:* Surfaces are of the form f(x,y,z) = 0 and are things like spherical shells or planes. Ex: x*x + y*y + z*z = 1 [sphere] or ax + by + cz = d [plane].
2. *Lines:* A line is the intersection of two surfaces and is thus defined by a system of two equations, namely, f1(x,y,z)=0; f2(x,y,z)=0 Ex: x=2; y=3.
3. *Volume:* Volumes are of the form f(x,y,z) << 0. Examples include things like a solid sphere x*x + y*y + z*z << 1.
4. *Point:* (1,2,3).

Identify the following Sets of Points (in 3D)

1. $x^2 + y^2 + z^2 = 1$

   $z = \frac{1}{2}$

The first equation is that of a sphere, while the second is that of a horizontal plane. The plane cuts the sphere parallel to the xy plane creating an intersection set of a circle. This is pictured below (Generated by Mathematica).

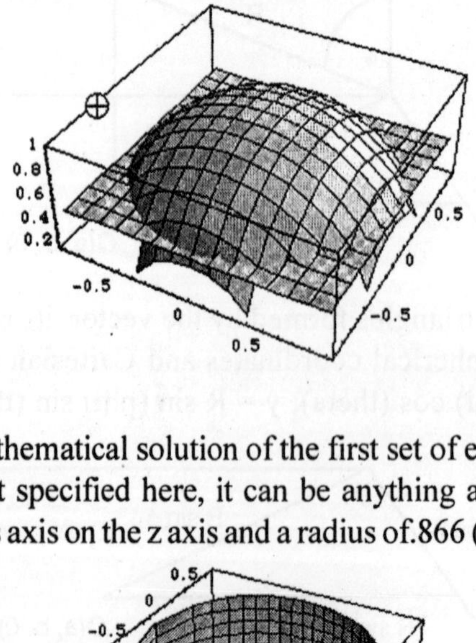

2. $x^2 + y^2 = \frac{3}{4}$

This equation is the mathematical solution of the first set of equations, but it behaves very differently. Because z is not specified here, it can be anything and the result is the cylinder pictured below, which has its axis on the z axis and a radius of.866 (Generated by Mathematica).

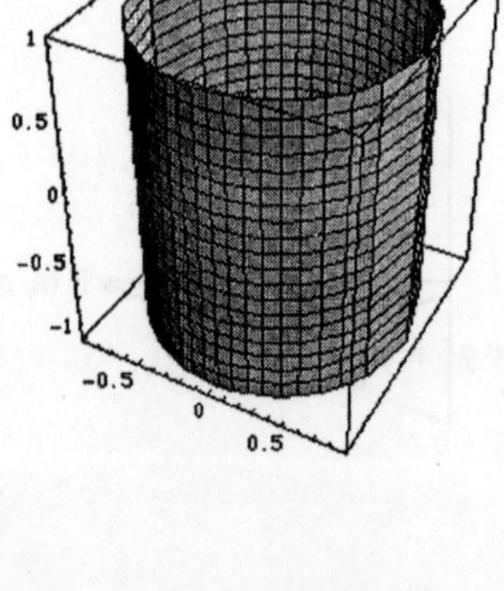

FUNCTIONS OF SEVERAL VARIABLES

Suppose we go beyond $f(x)$ and talk about $F(x,y,z)$ – *e.g.* a function of the *exact position in space*. This is just an example, of course; the abstract idea of a function of several variables can have "several" be as many as you like and "variables" be anything you choose. Another place where we encounter lots of functions of "several" variables is in THERMODYNAMICS, but for the time being we will focus our attention on the three *spatial* variables x (left-right), y (back-forth) and z (up-down).

How can we tackle *derivatives* of this function?

PARTIAL DERIVATIVES

Well, we do the obvious: we say, "Hold all the *other* variables *fixed* except [for instance] x and then treat $F(x,y,z)$ as a function only of x, with y and z as fixed *parameters*." Then we know just how to define the derivative with respect to x. The short name for this derivative is the PARTIAL DERIVATIVE *with respect to x*, written symbolically

$$\frac{\partial F}{\partial x}$$

where the fact that there are other variables being held fixed is implied by the use of the symbol ∂ instead of just d.

$$\text{Similarly } \frac{\partial F}{\partial y} \text{ for and } \frac{\partial F}{\partial z}.$$

OPERATORS

The foregoing description applies for *any* function of (x,y,z); the concept of "taking partial derivatives" is independent of what function we are taking the derivatives *of*. It is therefore practical to learn to think of $\frac{\partial}{\partial x}$ and $\frac{\partial}{\partial y}$ and $\frac{\partial}{\partial y}$ as Operators that can be applied to *any* function (like F).

Put the operator on the left of a function, perform the operation and you get a partial derivative – a new function of (x,y,z).

In general, such "operators" *change one function into another*. Physics is loaded with operators like these.

THE GRADIENT OPERATOR

The GRADIENT operator is a *vector* operator, written $\vec{\nabla}$ and called "grad" or "del." It is defined (in Cartesian coordinates x,y,z) as[1].

$$\vec{\nabla} = \hat{i}\frac{\partial}{\partial x} + \hat{j}\frac{\partial}{\partial y}\hat{k}\frac{\partial}{\partial z}$$

and can be applied directly to any *scalar* function of (x,y,z) – say, $\phi(x,y,z)$ to turn it into a *vector* function $\vec{\nabla}\phi = \hat{i}\frac{\partial}{\partial x} + \hat{j}\frac{\partial}{\partial y}\hat{k}\frac{\partial}{\partial z}$.

GRADIENTS OF SCALAR FUNCTIONS

It is instructive to work up to this "one dimension at a time." For simplicity we will stick to using ϕ as the symbol for the function of which we are taking derivatives.

The GRADIENT in One Dimension

Let the dimension be x. Then we have no "extra" variables to hold constant and the gradient of $\phi(x)$ is nothing but $\hat{i}\frac{d\phi}{\partial x}$. We can illustrate the "meaning" of $\vec{\nabla}\phi$ by an example: let $\phi(x)$ be the mass of an object times the acceleration of gravity times the height h of a hill at horizontal position x. That is $\phi(x)$, is the *gravitational potential energy* of the object when it is at horizontal position x. Then

$$\vec{\nabla}\phi = \hat{i}\frac{d\phi}{\partial x} + \hat{i}\frac{\partial}{\partial x}(mgh) = mg\left(\frac{\partial h}{\partial x}\right)i.$$

Note that $\frac{\partial h}{\partial x}$ is the *slope* of the hill and $-\vec{\nabla}\phi$ is the *horizontal component of the net force* (gravity plus the normal force from the hill's surface) on the object. That is, $-\vec{\nabla}\phi$ is the *downhill force*.

The GRADIENT in Two Dimensions

In the previous example we disregarded the fact that most hills extend in *two* horizontal directions, say x = East and y = North. [If we stick to small distances we won't notice the curvature of the Earth's surface.] In this case there are two *components* to the slope: the Eastward slope $\frac{\partial h}{\partial x}$ and the Northward slope $\frac{\partial h}{\partial y}$. The former is a measure of how steep the hill will seem if you head due East and the latter is a measure of how steep it will seem if you head due North. If you put these together to form a *vector* "steepness" (gradient)

$$\vec{\nabla}h = \hat{i}\frac{\partial h}{\partial x} + \hat{j}\frac{\partial h}{\partial y}$$

then the vector $\vec{\nabla}h$ points *uphill* – *i.e.* in the direction of the *steepest ascent*. Moreover, the gravitational potential energy $\phi = mgh$ as before [only now is a function of 2 variables, $\phi(x, y)$] so that $-\vec{\nabla}\phi$ is once again the *downhill force* on the object.

The GRADIENT in Three Dimensions

If the potential ϕ is a function of 3 variables, $\phi(x, y, z)$ [such as the three *spatial coordinates* x, y and z – in which case we can write it a little more compactly as $\phi(\vec{r})$ where $\vec{r} \equiv x\hat{i} + y\hat{j} + z\hat{k}$, the vector distance from the origin of our coordinate system to the point in space where ϕ is being evaluated], then it is a little more difficult to make up a "hill" analogy — try imagining a topographical map in the form of a 3-dimensional hologram where instead of *lines* of constant *altitude* the "equipotentials" are *surfaces* of constant ϕ. (This is just what Physicists do picture!) Fortunately the *math* extends easily to 3 dimensions (or any larger number, if that has any meaning in the context we choose).

In general, any time there is a *potential energy* function $\phi(\vec{r})$ we can immediately write down the *force* $\vec{F}$ associated with it as

$$\vec{F} \equiv -\vec{\nabla}\phi$$

A perfectly analogous expression holds for the *electric field* $\vec{E}$ [force per unit charge] in terms of the *electrostatic potential* ϕ [potential energy per unit charge]:[2]

$$\vec{E} \equiv -\vec{\nabla}\phi$$

The GRADIENT in *N* Dimensions

Although we won't be needing to go beyond 3 dimensions very often in Physics, you might want to borrow this metaphor for application in other realms of human endeavour where there are more than 3 variables of which your scalar field is a function. You could have ϕ be a measure of *happiness*, for instance [though it is hard to take reliable measurements on such a subjective quantity]; then ϕ might be a function of lots of factors, such as x_1 = freedom from violence, x_2 = freedom from hunger, x_3 = freedom from poverty, x_4 = freedom from oppression, and so on.[3] Note that with an arbitrary number of variables we get away from thinking up different names for each one and just call the ith variable "x_i."

Then we can define the GRADIENT in N dimensions as

$$\vec{\nabla}\phi = \hat{i}_1 \frac{\partial \phi}{\partial x_1} + \hat{i}_2 \frac{\partial \phi}{\partial x_2} + \ldots + \hat{i}_N \frac{\partial \phi}{\partial x_N}$$

or

$$\vec{\nabla}\phi = \sum_{i=1}^{N} \hat{i}_i \frac{\partial \phi}{\partial x_i}$$

where $\hat{i}_i$ is a UNIT VECTOR in the x_i direction.

DIVERGENCE OF A VECTOR FIELD

If we form the scalar ("dot") product of $\vec{\nabla}$ with a *vector* function $\vec{A}(x, y, z)$ we get a *scalar* result called the DIVERGENCE of $\vec{A}$:

$$\operatorname{div} \vec{A} \equiv \vec{\nabla} \cdot \vec{A} \equiv \frac{\partial A_x}{\partial x} + \frac{\partial A_y}{\partial y} + \frac{\partial A_z}{\partial z}$$

This name is actually quite mnemonic: the DIVERGENCE of a vector field is a *local* measure of its "outgoingness" — *i.e.* the extent to which there is more *exiting* an infinitesimal region of space than *entering* it. If the field is represented as "flux lines" of some indestructible "stuff" being emitted by "sources" and absorbed by "sinks," then a nonzero DIVERGENCE at some point means there must be a *source* or *sink* at that position. That is to say,

"What leaves a region is no longer in it."

For example, consider the divergence of the CURRENT DENSITY $\vec{J}$, which describes the FLUX of a CONSERVED QUANTITY such as electric charge Q. (Mass, as in the current of a river, would do just as well.)

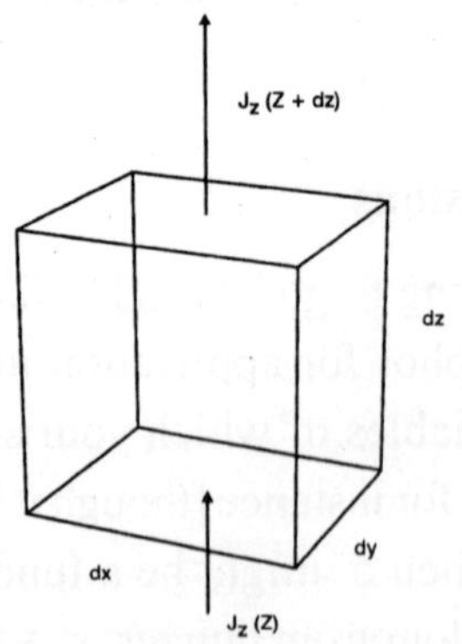

Fig. Flux into and out of a Volume Element.
$dV = dx\ dy\ dz$.

To make this as easy as possible, let's picture a *cubical* volume element $dV = dx\ dy\ dz$. In general, $\vec{J}$ will (like any vector) have three components (J_x, J_y, J_z), each of which may be a function of position (x,y,z). If we take the lower left front corner of the cube to have coordinates (x,y,z) then the upper right back corner has coordinates ($x + dx$, $y + dy$, $z + dz$). Let's concentrate first on J_z and how it depends on z.

It may not depend on z at all, of course. In this case, the amount of Q coming into the cube through the bottom surface (per unit time) will be the same as the amount of Q going out through the top surface and there will be no net gain or loss of Q in the volume-at least not due to J_z.

If J_z is bigger at the top, however, there will be a net loss of Q within the volume dV due to the "divergence" of J_z. Let's see how much: the difference between $J_z(z)$ at the bottom and $J_z(z + dz)$ at the top is, by definition, $dJ_z = \left(\frac{\partial J_z}{\partial z}\right) dz$. The flux is over the same area at top and bottom, namely $dx\, dy$, so the total *rate of loss* of Q due to the z-dependence of J_z is given by

$$\dot{Q}_z = -dx\,dy\left(\frac{\partial J_z}{\partial z}\right) dz = -\left(\frac{\partial J_z}{\partial z}\right) dx\,dy\,dz$$

or
$$\dot{Q} = -\left(\frac{\partial J_z}{\partial z}\right) dV.$$

A perfectly analogous argument holds for the x-dependence if J_x and the y-dependence of J_y, giving a total rate of change of Q

$$\dot{Q} = \left(\frac{\partial J_x}{\partial x} + \frac{\partial J_y}{\partial y} + \frac{\partial J_z}{\partial z}\right) dV$$

or
$$\dot{Q} = -\vec{\nabla} \cdot \vec{J}\, dV$$

The total amount of Q *in* our volume element dV at a given instant is just ρdV, of course, so the rate of change of the enclosed Q is just

$$\dot{Q} = \dot{\rho}\, dV$$

which means that we can write

$$\frac{\partial \rho}{\partial t} dV = -\vec{\nabla} \cdot \vec{J}\, dV$$

or, just cancelling out the common factor dV on both sides of the equation,

$$\frac{\partial \rho}{\partial t} = -\vec{\nabla} \cdot \vec{J}$$

which is the compact and elegant "differential form" of the EQUATION OF CONTINUITY.

This equation tells us that the "Q sourciness" of *each point* in space is given by the degree to which flux "lines" of $\vec{J}$ tend to radiate *away* from that point more than they converge *toward* that point – namely, the DIVERGENCE of $\vec{J}$ at the point in question. This esoteric-looking mathematical expression is, remember, just a formal way of expressing our original dumb tautology!

CURL OF A VECTOR FIELD

If we form the vector ("cross") product $\vec{\nabla}$ of with a *vector* function $\vec{A}(x,y,z)$ we get a *vector* result called the curl of $\vec{A}$:

$$\text{curl } \vec{A} \equiv \vec{\nabla} \times \vec{A} \equiv \hat{i}\left(\frac{\partial A_z}{\partial y} - \frac{\partial A_y}{\partial z}\right) + \hat{j}\left(\frac{\partial A_x}{\partial z} - \frac{\partial A_z}{\partial x}\right) + \hat{k}\left(\frac{\partial A_y}{\partial x} - \frac{\partial A_x}{\partial y}\right)$$

This is a lot harder to visualize than the DIVERGENCE, but not impossible. Suppose you are in a boat in a huge river (or Pass) where the current flows mainly in the x direction but where the speed of the current (flux of water) varies with y. Then if we call the current $\vec{J}$, we have a nonzero value for the derivative $\frac{\partial A_x}{\partial y}$, which you will recognize as one of the terms in the formula for. What does this imply? Well, if you are sitting in the boat, moving with the current, it means the current on your port side moves faster – *i.e.* forward relative to the boat – and the current on your starboard side moves slower – *i.e.* backward relative to the boat – and this implies a *circulation* of the water around the boat – *i.e.* a *whirlpool!* So $\vec{\nabla} \times \vec{J}$ is a measure of the local "swirliness" of the current $\vec{J}$, which means "curl" is not a bad name after all!

THE LAPLACIAN OPERATOR

If we form the scalar ("dot") product of $\vec{\nabla}$ with *itself* we get a scalar *second derivative* operator called the LAPLACIAN:

$$\vec{\nabla} \cdot \vec{\nabla} \equiv \nabla^2 = \frac{\partial^2}{\partial x^2} + \frac{\partial^2}{\partial y^2} + \frac{\partial^2}{\partial z^2}$$

What does the ∇^2 operator "mean?" It is the three-dimensional generalization of the one-dimensional CURVATURE operator d^2/dx^2. Consider the familiar one-dimensional function $h(x)$ where h is the height of a hill at horizontal position x. Then dh/dx is the *slope* of the hill and d^2h/dx^2 is its *curvature* (the rate of change of the slope with position). This property appears in every form of the WAVE EQUATION. In three dimensions, a nice visualization is harder (there is no extra dimension "into which to curve") but $\nabla^2\phi$ represents the equivalent property of a scalar function $\phi(x, y, z)$.

GAUSS' LAW

The EQUATION OF CONTINUITY describes the conservation of "actual physical stuff" entering or leaving an infinitesimal region of space dV. For example, $\vec{J}$ may be the *current density* (charge flow per unit time per unit area normal to the direction of flow) in which case ρ is the *charge density* (charge per unit volume); in that example the conserved "stuff" is electric charge itself. Many other examples exist, such as FLUID DYNAMICS (in which *mass* is the conserved stuff) or HEAT FLOW (in which *energy* is the conserved quantity). In ELECTROMAGNETISM, however, we deal not only with the conservation of *charge* but also with the *continuity* of abstract *vector fields* like $\vec{E}$ and $\vec{B}$. In order to visualize $\vec{E}$, we have developed the notion of "electric field lines" that cannot be broken except where they originate (from positive charges) and terminate (on negative charges). [This description only holds for *static* electric fields; when things *move* or otherwise change with time, things get a lot more complicated... and interesting!] Thus a positive charge is a "source of electric field lines" and a negative charge is a "sink" – the charges themselves stay put, but the lines of $\vec{E}$ *diverge* out of or into them.

GAUSS' LAW states that the net flux of electric field "lines" *out* of a closed surface S is proportional to the net electric charge enclosed *within* that surface. The constant of proportionality depends on which system of units one is using; in *SI* units it is $1/\varepsilon_0$. In mathematical shorthand, this reads

$$\varepsilon_o \oiint_S \vec{E} \cdot d\vec{A} = Q_{\text{cncl}}.$$

Recalling our earlier discussion of DIVERGENCE, we can think of $\vec{E}$ as being a sort of flux density of conserved "stuff" *emitted* by positive electric charges. Remember, in this case the charges themselves do not go anywhere; they simply emit (or absorb) the electric field "lines" which emerge from (or disappear into) the enclosed region. The rate of generation of this "stuff" is $Q_{\text{cncl}}/\varepsilon_0$. We can then apply GAUSS' LAW to an infinitesimal volume element using Fig. with $\varepsilon_0 \vec{E}$ in place of $\vec{J}$. Except for the "fudge factor" ε_0 and the replacement of $\dot{Q}$ by Q_{cncl}, the same arguments used to derive the EQUATION OF CONTINUITY lead in this case to a formula relating the divergence of $\vec{E}$ to the electric charge density ρ *at any point in space*, namely

$$\overrightarrow{AS} = \frac{1}{2}\overrightarrow{AB} + \lambda\overrightarrow{M_cC}.$$

This is the *differential form* of GAUSS' LAW.

POISSON AND LAPLACE

Even in its differential form, GAUSS' LAW is a little tricky to solve analytically, since it is a *vector* differential equation. Generally we have an easier time solving *scalar* differential equations, even though they may involve higher order partial derivatives. Fortunately, we can convert the former into the latter: recall that the vector electric *field* can always be obtained from the scalar electrostatic *potential* using

$$\vec{E} \equiv -\vec{\nabla}\phi$$

Thus div $\vec{E} \equiv \vec{\nabla}\cdot\vec{E} = -\vec{\nabla}\cdot\vec{\nabla}\phi$ or

$$\boxed{\vec{\nabla}\phi = -\frac{1}{\varepsilon_0}\rho}.$$

This relation is known as POISSON'S EQUATION. Its simplified cousin, LAPLACE'S EQUATION, applies in regions of space where there are *no free charges*:

$$\boxed{\vec{\nabla}^2\phi = 0}.$$

Each of these equations finds much use in real electrostatics problems. Advanced students of electromagnetism learn many types of functions that satisfy LAPLACE'S EQUATION, with different *symmetries*; since a *conductor* is always an *equipotential* (every point in a given conductor must have the same ϕ, otherwise there would be an electric field in the conductor that would cause charges to move until they cancelled out the differences in ϕ), empty regions surrounded by conductors of certain shapes must have ϕ with a spatial dependence satisfying those BOUNDARY CONDITIONS as well as LAPLACE'S EQUATION. One can often write down a complicated-looking formula for ϕ almost by inspection, using this favourite method of Physicists and Mathematicians, namely... KNOWING THE ANSWER.

GEOMETRY IN HIGHER DIMENSIONS

THE CROSS PRODUCT

We define the cross product of two three-dimensional vectors a and b by the requirements:

1. a × b is perpendicular to both a and b.
2. ||a × b|| is the area of the parallelogram spanned by a and b (i.e. the parallelogram whose adjacent sides are the vectors a and b).
3. The direction of a × b is determined by right-hand rule. (This means that if we curl the fingers of the right hand from a to b, then the thumb points in the direction of a × b.)

Recall from lecture (or your textbook) that the area of the parallelogram is ||a|| ||b|| sin, where is the angle between a and b.

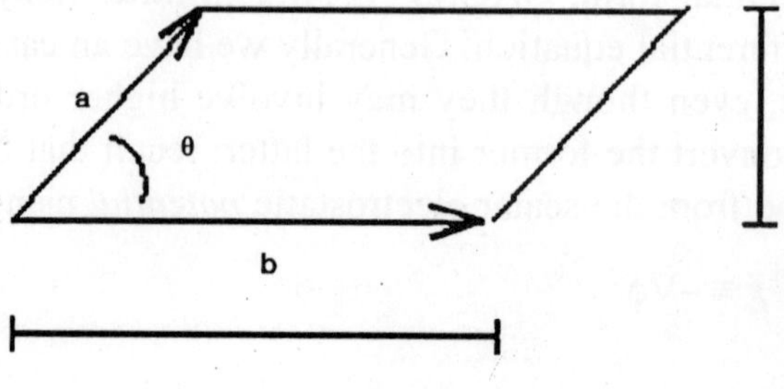

This means that the magnitude of the cross product is largest when a and b are perpendicular. On the other hand, if a and b are parallel or if either vector is the zero vector, then the cross product is the zero vector. (This is a good thing, since these latter cases are exactly when there is not a unique line perpendicular to both a and b.)

Below is a Concept-Visualization Tool (CVT) that helps illustrate how the cross product works. The vector a is shown in blue, b is green, and c is red. I've also drawn the parallelogram formed by a and b.

The Component Formula for the cross Product

Recall the geometric definition of the cross product as given in an earlier reading.

Cross Product of Unit Vectors

First, we derive the cross product of the unit vectors, i, j, and k be the standard unit vectors in R^3. (We define the cross product only in three dimensions.)

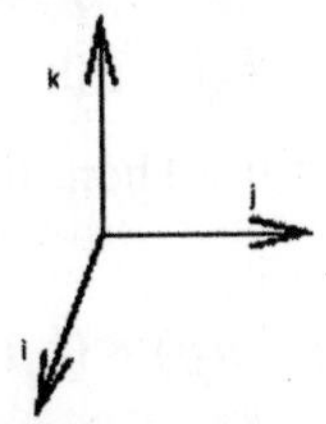

The parallelogram spanned by any two of these standard unit vectors is a unit square, which has area one. Hence, the cross product must be a unit vector. Looking at the above graph, you can use the right-hand rule to determine the following results.

$$i \times j = k$$
$$j \times k = i$$
$$k \times i = j$$

This little cycle diagram can help you remember these results.

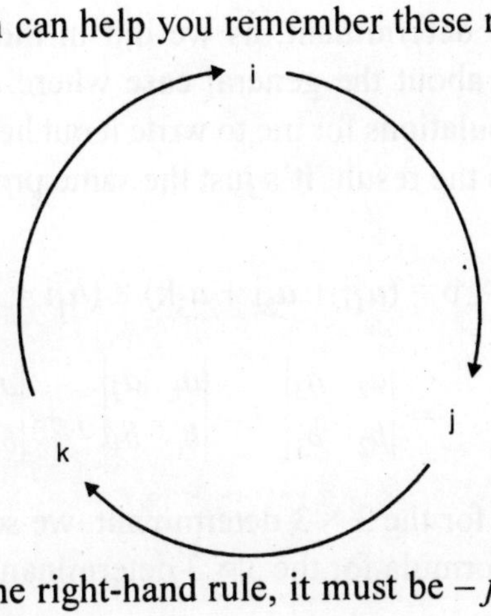

What about $i \times k$? By the right-hand rule, it must be $-j$. I think you get the idea.

General Vectors

In an earlier reading, we mentioned the special properties of the cross product: $b \times a = - a \times b$, and $a \times a = 0$. With these exceptions, the cross product behaves like regular multiplication, obeying these properties,

1. $(ya) \times b = y(a \times b) = a \times (yb)$,
2. $a \times (b + c) = a \times b + a \times c$,
3. $(b + c) \times a = b \times a + c \times a$,

where a, b, and c are vectors in R^3 and y is a scalar. We can use these properties, along with the cross product of the standard unit vectors, to write the formula for the cross product in terms of components.

We write the components of a and b as:

$$a = (a_1, a_2, a_3) = a_1\mathrm{i} + a_2\mathrm{j} + a_3\mathrm{k}$$
$$b = (b_1, b_2, b_3) = b_1\mathrm{i} + b_2\mathrm{j} + b_3\mathrm{k}$$

First, we'll assume that $a_3 = b_3 = 0$. (Then, the manipulations are much easier.) We calculate:

$$\begin{aligned} a \times b &= (a_1 i + a_2 j) \times (b_1 i + b_2 j) \\ &= a_1 b_1 (i \times i) + a_1 b_2 (i \times j) + a_2 b_1 (j \times i) + a_2 b_2 (j \times j) \end{aligned}$$

Since we know that $i \times i = 0 = j \times j$ and that $i \times j = k = -j \times i$, this quickly simplifies to

$$\begin{aligned} a \times b &= (a_1 b_2 - a_2 b_1)\mathrm{k} \\ &= \begin{vmatrix} a_1 & a_2 \\ b_1 & b_2 \end{vmatrix} k. \end{aligned}$$

Writing the result as a determinant, as we did in the last step, is a handy way to remember the result. What about the general case where a_3 and b_3 might not be zero? Well, that's too many manipulations for me to write it out here, and you probably wouldn't read it anyway. To calculate the result, it's just the same process as above. You can verify yourself that in general

$$\begin{aligned} \mathrm{a} \times \mathrm{b} &= (a_1\mathrm{i} + a_2\mathrm{j} + a_3\mathrm{k}) \times (b_1\mathrm{i} + b_2\mathrm{j} + b_3\mathrm{k}) \\ &= \begin{vmatrix} a_2 & a_3 \\ b_2 & b_3 \end{vmatrix} i - \begin{vmatrix} a_1 & a_3 \\ b_1 & b_3 \end{vmatrix} j + \begin{vmatrix} a_1 & a_2 \\ b_1 & b_2 \end{vmatrix} k. \end{aligned}$$

Looking at the formula for the 3×3 determinant, we see that the formula for a cross product looks a lot like the formula for the 3×3 determinant. If we allow a matrix to have the vector i, j, and k as entries, the 3 × 3 determinant gives a handy mnemonic to remember the cross product:

$$a \times b = \begin{vmatrix} i & j & k \\ a_1 & a_2 & a_3 \\ b_1 & b_2 & b_3 \end{vmatrix}$$

This is the way I remember how to compute the cross product.

COORDINATES CARTESIAN COORDINATE SYSTEM IN THE PLANE AND IN SPACE

A Cartesian coordinate system in the space consists of number lines (axes) perpendicularly intersecting in one point O called the origin. The orientation of the axes is chosen according to the„Right-Hand-Rule“ as shown in the figure.

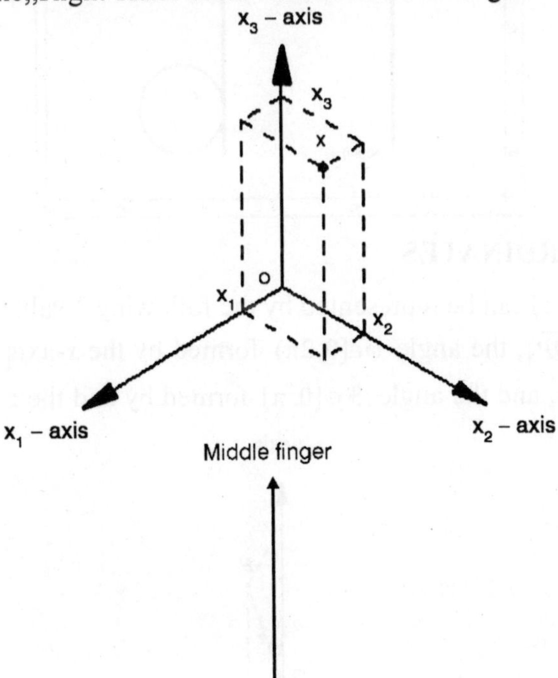

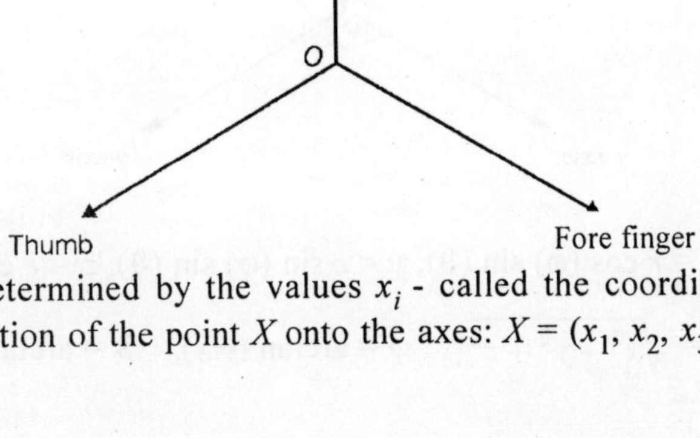

A point X is determined by the values x_i - called the coordinates - obtained by perpendicular projection of the point X onto the axes: $X = (x_1, x_2, x_3)$. Using a notation without subscripts, the coordinates are usually symbolized by (x, y, z) and the number axes are called x –, y – and z-axis. A Cartesian coordinate system in the plane is defined in an analogous manner.

PIXEL COORDINATES

When drawing to a screen, the depicted objects are often given by pixel coordinates. The depicted rectangle corresponds to a window with 90 × 120 pixels and the coordinates of the coloured pixels refer to the upper left corner of the window.

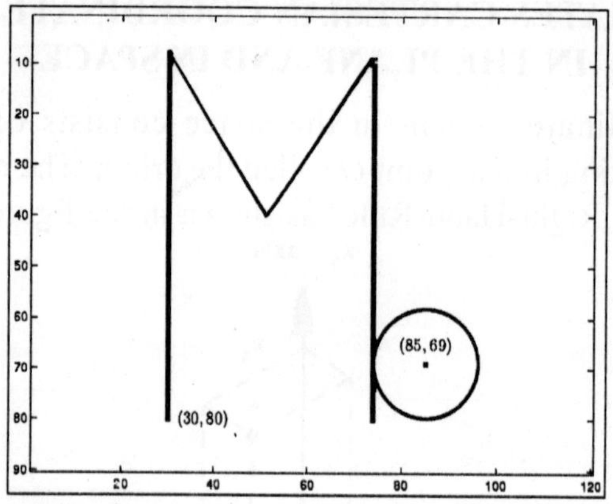

SPHERICAL COORDINATES

A point $P = (x, y, z)$ can be represented by the following 3 values: the point's distance from the origin $r = |\overline{OP}|$, the angle $\varphi \in [0, 2\pi)$ formed by the x-axis and the projection of $\overline{OP}$ onto the xy-plane, and the angle $\vartheta \in [0, \pi]$ formed by and the z-axis.

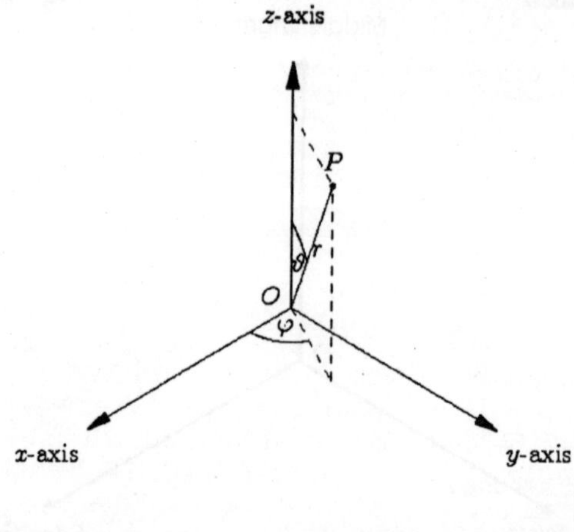

We have

$$x = r \cos(\varphi) \sin(\vartheta),\ y = r \sin(\varphi) \sin(\vartheta),\ z = r \cos(\vartheta)$$

and
$$r = \sqrt{x^2 + y^2 + z^2}, \quad \varphi = \arctan(y/x), \quad \vartheta = \arctan\sqrt{x^2 + y^2 / z^2}),$$

but you have got to select the appropriate branch of depending on the signs of and:

$x \geq 0 \wedge y \geq 0$	$\varphi \in [0, \pi/2)$
$x \leq 0$	$\varphi \in [\pi/2, 3\pi/2)$
$x \geq 0 \wedge y < 0$	$\varphi \in [3\pi/2, 2\pi)$
$z \leq 0$	$\vartheta \in [\pi/2, \pi)$

In Cartesian coordinates the points

$$P_1 = (1, 1, 1), \quad P_2 = (1, -1, -1), \quad P_3 = (-1, 1, -1), \quad P_4 = (-1, -1, 1),$$

form the vertices of a regular tetrahedron with edge length $2\sqrt{2}$ and a barycentre which lies in the origin O.

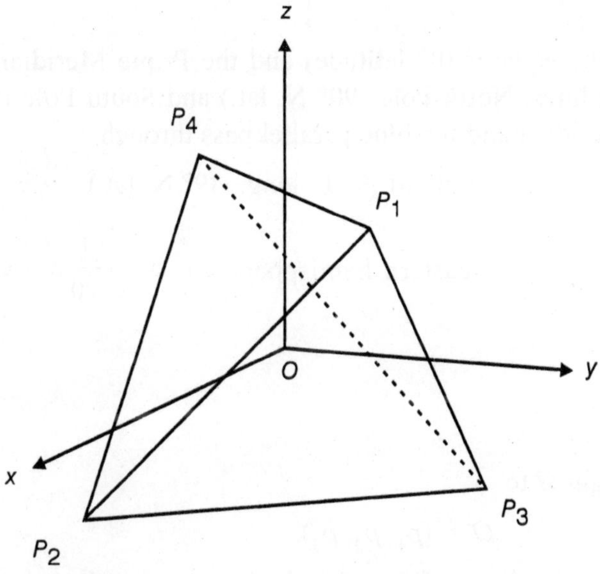

Represented by spherical coordinates (r, ϑ, φ) the vertices are given by

$$P_1' = (\sqrt{3}, \psi, \pi/4), \qquad P_1' = (\sqrt{3}, \pi - \psi, \pi/4),$$

$$P_3' = (\sqrt{3}, \pi, -\psi, 3\pi/4), \qquad P_4' = (\sqrt{3}, \psi, -3\pi/4),$$

with $\psi = \arccos(1/\sqrt{3})$ respectively $\psi = \arctan\sqrt{2}$.

LONGITUDE AND LATITUDE

If we use appropriate angle ranges, then longitudes and latitudes on the earth's surface correspond to spherical coordinates:

Eastern longitude	0° – 180°	$\varphi = 0 \ldots \pi$
Western longitude	0° – 180°	$\varphi = 0 \ldots -\pi$
Northern latitude	0° – 90°	$\vartheta = \pi/2 \ldots 0$

Southern latitude 0° – 90° $\vartheta = \pi/2 \ldots \pi$

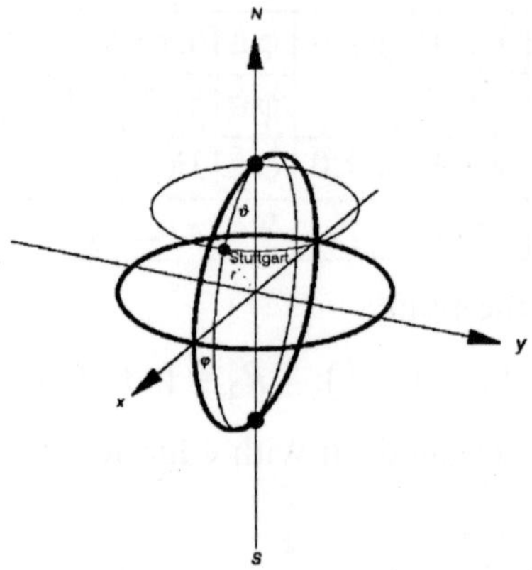

In the picture the equator (0° latitude) and the Prime Meridian (0° longitude) are represented by bold lines. North Pole (90° N. lat.) and South Pole (90° S. lat.) are also plotted. The blue meridian and the blue parallel pass through

Stuttgart (9° E. long., 49° N. lat.)

and correspond to $\varphi = \frac{1}{20}\pi$ (eastern hemisphere) or $\varphi = -\frac{1}{20}\pi$ (western hemisphere) and $\vartheta = \frac{41}{180}\pi$.

TRANSLATION

Moving the origin O to

$$O' = (p_1, p_2, p_3),$$

without changing the directions of the axes, the coordinates change as follows:

$$X' = (x_1 - p_1, x_2 - p_2, x_3 - p_3)$$

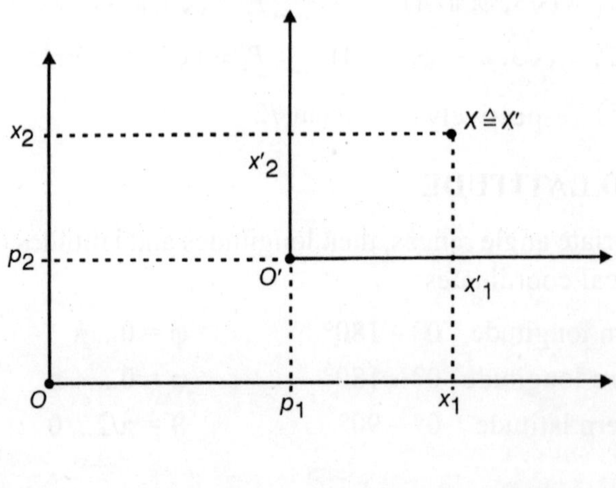

FRAME OF REFERENCE IN MOTION

The uniform motion of an object in the plane can be described in terms of a given Cartesian coordinate system by

$$(x(t), y(t) = p + \alpha t, q + \beta t), t \geq 0.$$

Now, let us observe this motion sitting in a train moving along the -axis at constant velocity v in direction of increasing numbers. If we attach another Cartesian coordinate system to the train that coincides with the system given above at time $t = 0$, then the coordinates of the object in our moving system are given by

$$x' = x - vt, y' = y.$$

Thus, the object's velocity that we perceive generally differs from its actual velocity.

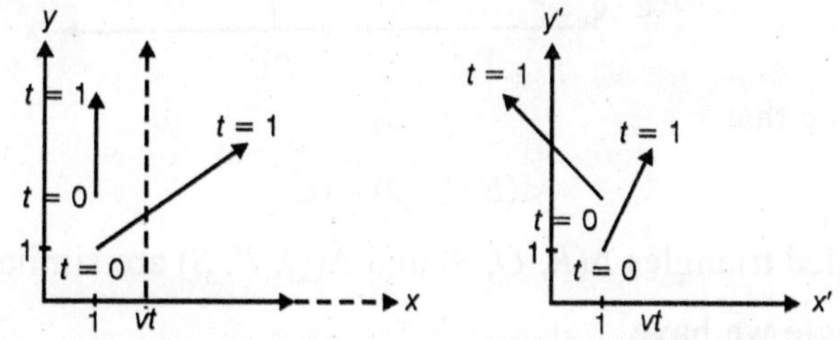

For example, if $v = 2$, then we have for an object moving from $P = (1, 1)$ to Q (4, 3) (red arrow) in a given time unit

$$P' = (1, 1), Q' = (4 - 2, 3 - 0) = (2, 3).$$

Correspondingly, for the actual and the observed velocities we obtain

$$v_a = \sqrt{(4-1)^2 + (3-1)^2} = \sqrt{13}, v_o = \sqrt{5};$$

that is, the motion, observed from the train, seems to be decelerated.

ROTATION

Rotating the xy-plane by the angle about the z-axis the coordinates of a point $P = (p_1, p_2, p_3)$ transform as follows:

$$x' = \cos(\alpha)x + \sin(\alpha)y, \quad y' = -\sin(\alpha)x + \cos(\alpha)y, \quad z' = z.$$

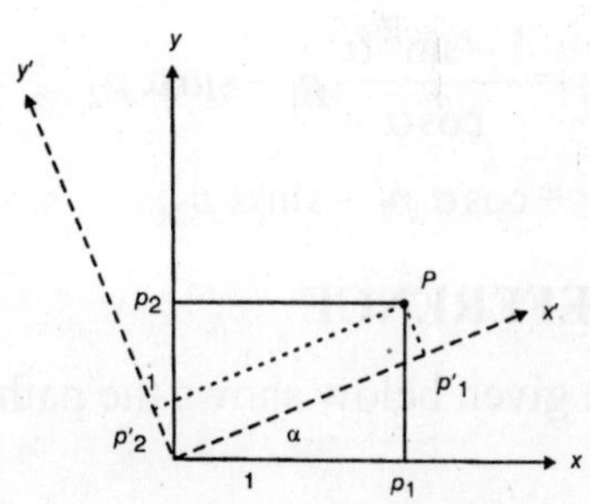

Analogous formulas are obtained by rotations about the *yz*-and the *zx*-plane.

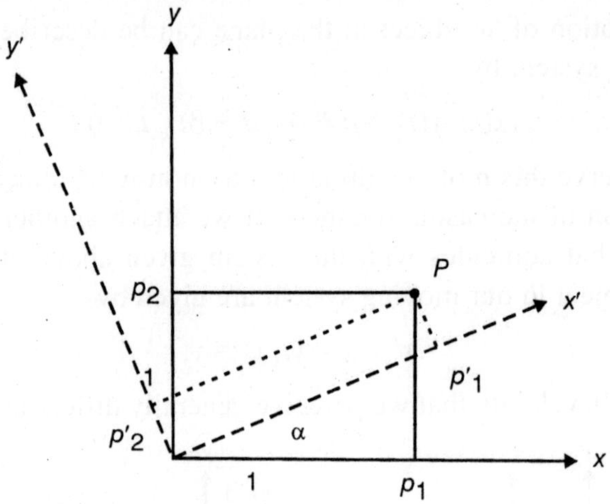

First of all we see that

$$\sphericalangle(S, P, Q) = \alpha,$$

that is, the right-angled triangles $\Delta(R, O, S)$ and $\Delta(Q, P, S)$ are similar.

In the first triangle we have

$$|\overline{OS}| = p_1/\cos\alpha$$

and

$$|\overline{RS}| = p_1 \tan\alpha.$$

From the second triangle we obtain

$$p'_2 = \cos\alpha\ |\overline{PS}| = \cos\alpha(p_2 - |\overline{RS}|)$$

$$= \cos\alpha\ p_2 - \sin\alpha\ p_1$$

and

$$p'_1 = |\overline{OS}| + |\overline{SQ}| = p_1/\cos\alpha + \sin\alpha(p_2 - |\overline{RS}|)$$

$$= \frac{1-\sin^2\alpha}{\cos\alpha} p_1 + \sin\alpha\, p_2$$

$$= \cos\alpha\, p_1 + \sin\alpha\, p_2.$$

ROTATING FRAME OF REFERENCE

The left side of the picture given below shows the paths of a rectilineal and a circular motion:

$$G: (x, y) = (1, 2 + t/(2\pi)),$$

$$G: (x, y) = (1 + \cos(t), 2 - \sin(t).$$

On the right the two paths are pictured in a frame of reference rotating at an angular velocity $\omega = 1$. Hence, they are pictured as they would appear to an observer having a ride on a roundabout.

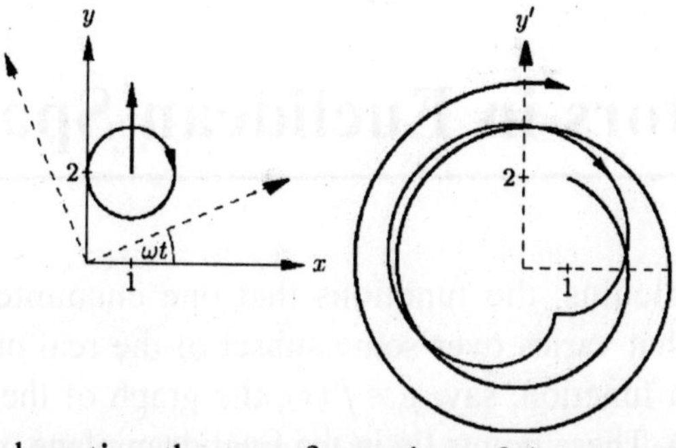

For the rectilineal motion the transformation of the coordinates yields

$$x' = c + (2 + t/(2\pi))s, \quad y' = -s + (2 + t/(2\pi))c$$

with

$$c = \cos(\omega t), \quad s = \sin(\omega t).$$

Since the factor within the brackets increases if the parameter does, we obtain a spiral. The pictured windings correspond to the time interval $0 \le t \le 4\pi$.

The path's shape of the circular motion in the frame rotating at $\omega = 1$ cannot directly be seen from the transformed coordinates

$$x' = c(1 + c) + s(2 - s), \quad y' = -s(1 + c) + c(2 - s).$$

As our example shows, the direction of the observed motion can change abruptly. Even bends can appear on the observed path. In such a singular point the observed velocity equals zeros with respect to the rotating frame.

Chapter 2

Vectors in Euclidean Space

In single-variable calculus, the functions that one encounters are functions of a variable (usually x or t) that varies over some subset of the real number line (which we denote by $\mathbb{R}$.). For such a function, say, $y = f(x)$, the graph of the function f consists of the points $(x, y) = (x, f(x))$. These points lie in the Euclidean plane, which, in the Cartesian or rectangular coordinate system, consists of all ordered pairs of real numbers (a, b). We use the word "Euclidean" to denote a system in which all the usual rules of Euclidean geometry hold. We denote the Euclidean plane by $\mathbb{R}^2$; the "2" represents the number of *dimensions* of the plane. The Euclidean plane has two perpendicular coordinate axes: the x-axis and the y-axis.

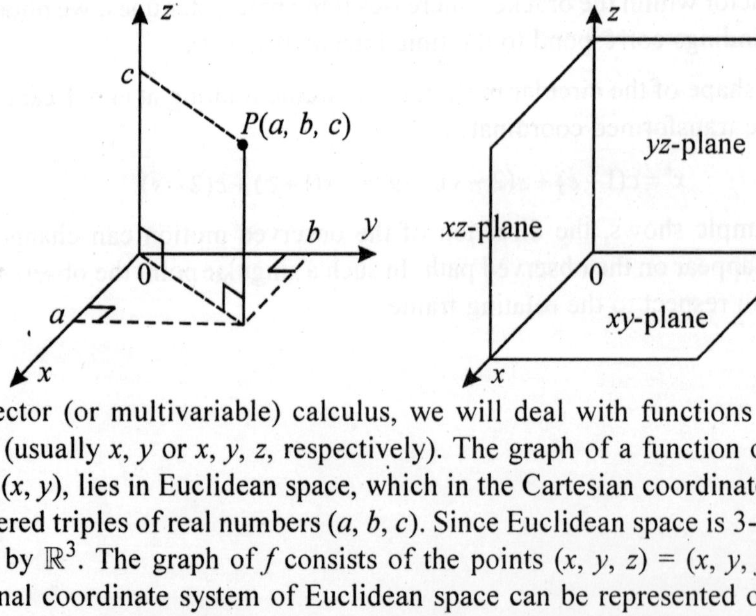

In vector (or multivariable) calculus, we will deal with functions of two or three variables (usually x, y or x, y, z, respectively). The graph of a function of two variables, say, $z = f(x, y)$, lies in Euclidean space, which in the Cartesian coordinate systemconsists of all ordered triples of real numbers (a, b, c). Since Euclidean space is 3-dimensional, we denote it by $\mathbb{R}^3$. The graph of f consists of the points $(x, y, z) = (x, y, f(x, y))$. The 3-dimensional coordinate system of Euclidean space can be represented on a flat surface, such as this page or a blackboard, only by giving the illusion of three dimensions. Euclidean space has three mutually perpendicular coordinate axes (x, y and z), and three mutually perpendicular coordinate planes: the xy-plane, yz-plane and xz-plane.

The coordinate system shown in Figure is known as a right-handed coordinate system, because it is possible, using the right hand, to point the index finger in the positive direction of the x-axis, the middle finger in the positive direction of the y-axis, and the thumb in the positive direction of the z-axis.

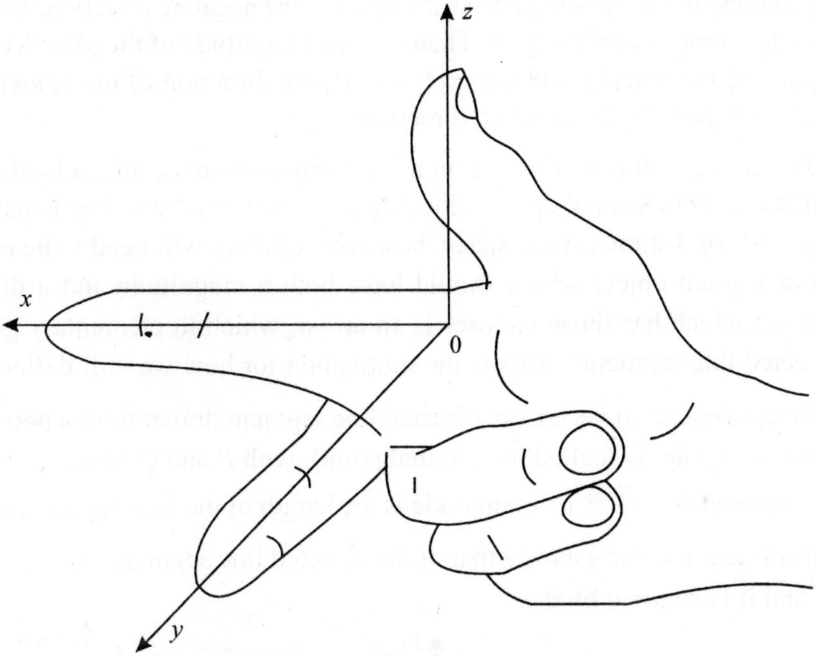

Fig. Right-handed Coordinate System

An equivalent way of defining a right-handed system is if you can point your thumb upwards in the positive z-axis direction while using the remaining four fingers to rotate the x-axis towards the y-axis. Doing the same thing with the left hand is what defines a left-handed coordinate system. Notice that switching the x- and y-axes in a right-handed system results in a left-handed system, and that rotating either type of system does not change its "handedness". Throughout the book we will use a right-handed system.

For functions of three variables, the graphs exist in 4-dimensional space (i.e. R^4), which we can not see in our 3-dimensional space, let alone simulate in 2-dimensional space. So we can only think of 4-dimensional space abstractly.

So far, we have discussed the *position* of an object in 2-dimensional or 3-dimensional space. But what about something such as the velocity of the object, or its acceleration? Or the gravitational force acting on the object? These phenomena all seem to involve motion and *direction* in some way. This is where the idea of a *vector* comes in.

You have already dealt with velocity and acceleration in single-variable calculus. For example, for motion along a straight line, if $y = f(t)$ gives the displacement of an object after time t, then $dy/dt = f'(t)$ is the velocity of the object at time t. The derivative $f'(t)$ is just a number, which is positive if the object is moving in an agreedupon "positive" direction, and negative if it moves in the opposite of that direction. So you can think of that number, which was called the velocity of the object, as having two components: a *magnitude*, indicated by a nonnegative number, preceded by a *direction*, indicated by a plus or minus symbol (representing motion in the positive direction or the negative direction, respectively), i.e. $f'(t) = \pm a$ for some number $a \geq 0$. Then a is the magnitude of the velocity (normally called the *speed* of the object), and the $\pm$ represents the direction of the velocity (though the + is usually omitted for the positive direction).

For motion along a straight line, i.e. in a 1-dimensional space, the velocities are also contained in that 1-dimensional space, since they are just numbers. For general motion along a curve in 2- or 3-dimensional space, however, velocity will need to be represented by a multidimensional object which should have both a magnitude and a direction. A geometric object which has those features is an arrow, which in elementary geometry is called a "directed line segment". This is the motivation for how we will define a vector.

Definition: A (nonzero) vector is a directed line segment drawn from a point P (called its initial point) to a point Q (called its terminal point), with P and Q being distinct points. The vector is denoted by $\overrightarrow{PQ}$ ℝ Its magnitude is the length of the line segment, denoted by $\left\|\overrightarrow{PQ}\right\|$, and its direction is the same as that of the directed line segment. The zero vector is just a point, and it is denoted by 0.

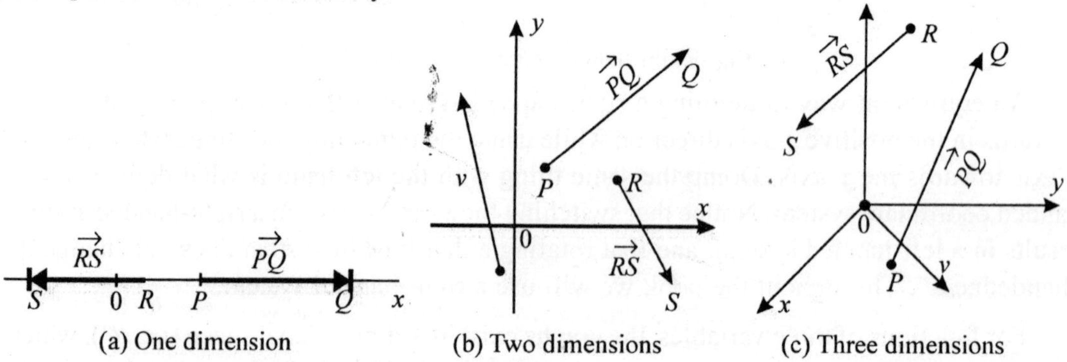

Fig. Vectors in Different Dimensions

To indicate the direction of a vector, we draw an arrow from its initial point to its terminal point. We will often denote a vector by a single bold-faced letter (e.g. v) and use the terms "magnitude" and "length" interchangeably. Note that our definition could apply to systems with any number of dimensions.

A few things need to be noted about the zero vector. Our motivation for what a vector is included the notions of magnitude and direction. What is the magnitude of the zero vector? We define it to be zero, i.e. $\| 0 \| = 0$. This agrees with the definition of the zero vector as just a point, which has zero length. What about the direction of the zero vector?

A single point really has no well-defined direction. Notice that we were careful to only define the direction of a *nonzero* vector, which is well-defined since the initial and terminal points are distinct. Not everyone agrees on the direction of the zero vector. Some contend that the zero vector has *arbitrary* direction (i.e. can take any direction), some say that it has *indeterminate* direction (i.e. the direction can not be determined), while others say that it has *no* direction. Our definition of the zero vector, however, does not require it to have a direction, and we will leave it at that.

Now that we know what a vector is, we need a way of determining when two vectors are equal. This leads us to the following definition.

Definition: Two nonzero vectors are equal if they have the same magnitude and the same direction. Any vector with zero magnitude is equal to the zero vector.

By this definition, vectors with the same magnitude and direction but with different initial points would be equal. For example, the vectors u, v and w all have the same magnitude $\sqrt{5}$ (by the Pythagorean Theorem). And we see that u and w are parallel, since they lie on lines having the same slope $\frac{1}{2}$, and they point in the same direction. So $u = w$, even though they have different initial points. We also see that v is parallel to u but points in the opposite direction. So u , v.

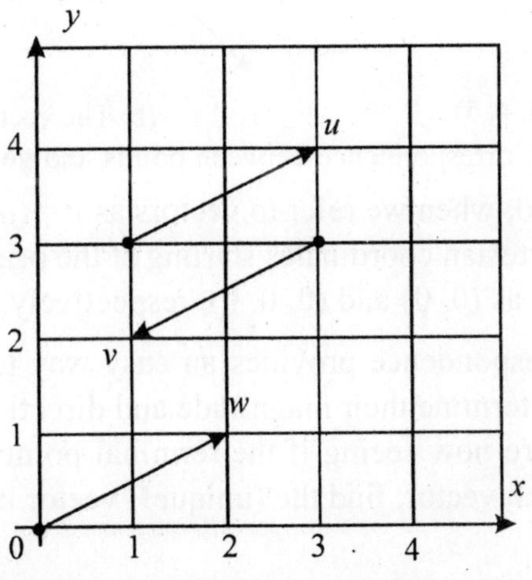

So we can see that there are an infinite number of vectors for a given magnitude and direction, those vectors all being equal and differing only by their initial and terminal points. Is there a single vector which we can choose to represent all those equal vectors?

Unless otherwise indicated, when speaking of "the vector" with a given magnitude and direction, we will mean the one whose initial point is at the origin of the coordinate system.

Thinking of vectors as starting from the origin provides a way of dealing with vectors in a standard way, since every coordinate system has an origin. But there will be times when it is convenient to consider a different initial point for a vector (for example, when adding vectors, which we will do in the next section).

Another advantage of using the origin as the initial point is that it provides an easy correspondence between a vector and its terminal point.

Example: Let v be the vector in $\mathbb{R}^3$ whose initial point is at the origin and whose terminal point is (3, 4, 5). Though the *point* (3, 4, 5) and the vector v are different objects, it is convenient to write $v = (3, 4, 5)$. When doing this, it is understood that the initial point of v is at the origin (0, 0, 0) and the terminal point is (3, 4, 5).

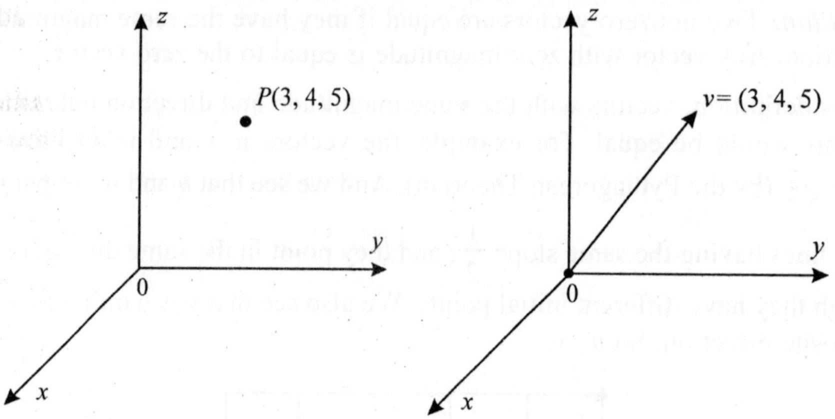

(a) The point (3, 4, 5) (b) The vector (3, 4, 5)

Fig. Correspondence between points and vectors

Unless otherwise stated, when we refer to vectors as $v = (a, b)$ in $\mathbb{R}^2$ or $v = (a, b, c)$ in $\mathbb{R}^3$, we mean vectors in Cartesian coordinates starting at the origin. Also, we will write the zero vector 0 in $\mathbb{R}^2$ and $\mathbb{R}^3$ as (0, 0) and (0, 0, 0), respectively.

The point-vector correspondence provides an easy way to check if two vectors are equal, without having to determine their magnitude and direction. Similar to seeing if two points are the same, you are now seeing if the terminal points of vectors starting at the origin are the same. For each vector, find the (unique!) vector it equals whose initial point is the origin.

Then compare the coordinates of the terminal points of these "new" vectors: if those coordinates are the same, then the original vectors are equal. To get the "new" vectors starting at the origin, you *translate* each vector to start at the origin by subtracting the coordinates of the original initial point from the original terminal point. The resulting point will be the terminal point of the "new" vector whose initial point is the origin. Do this for each original vector then compare.

Example: Consider the vectors $\overrightarrow{PQ}$ and $\overrightarrow{RS}$ in $\mathbb{R}_3$, where $P = (2, 1, 5)$, $Q = (3, 5, 7)$, $R = (1, -3, -2)$ and $S = (2, 1, 0)$. Does $\overrightarrow{PQ} = \overrightarrow{RS}$?

Solution: The vector $\overrightarrow{PQ}$ is equal to the vector v with initial point (0, 0, 0) and terminal point $Q - P = (3, 5, 7) - (2, 1, 5) = (3 - 2, 5 - 1, 7 - 5) = (1, 4, 2)$.

Similarly, $\overrightarrow{RS}$ is equal to the vector w with initial point (0, 0, 0) and terminal point $S - R = (2, 1, 0) - (1, -3, -2) = (2 - 1, 1 - (-3), 0 - (-2)) = (1, 4, 2)$.

So $\overrightarrow{PQ} = v = (1, 4, 2)$ and $\overrightarrow{RS} = w = (1, 4, 2)$.

$\therefore \quad \overrightarrow{PQ} = \overrightarrow{RS}$

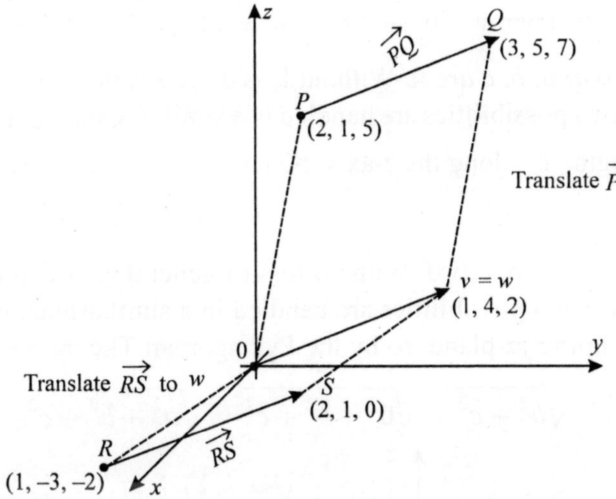

Recall the distance formula for points in the Euclidean plane:

For points $P = (x_1, y_1)$, $Q = (x_2, y_2)$ in $\mathbb{R}^2$, the distance d between P and Q is:

$$d = \sqrt{(x_2 - x_1)^2 + (y_2 - y_1)^2}$$

By this formula, we have the following result: For a vector $\overrightarrow{PQ}$ in $\mathbb{R}^2$ with initial point $P = (x_1, y_1)$ and terminal point $Q = (x_2, y_2)$, the magnitude of $\overrightarrow{PQ}$ is:

$$\| \overrightarrow{PQ} \| = \sqrt{(x_2 - x_1)^2 + (y_2 - y_1)^2}$$

Finding the magnitude of a vector $v = (a, b)$ in $\mathbb{R}^2$ is a special case of formula with $P = (0, 0)$ and $Q = (a, b)$:

For a vector $v = (a, b)$ in $\mathbb{R}^2$, the magnitude of v is:

$$\| v \| = \sqrt{a^2 + b^2}$$

To calculate the magnitude of vectors in $\mathbb{R}^3$, we need a distance formula for points in Euclidean space:

Theorem: The distance d between points $P = (x_1, y_1, z_1)$ and $Q = (x_2, y_2, z_2)$ in $\mathbb{R}^3$ is:

$$d = \sqrt{(x_2 - x_1)^2 + (y_2 - y_1)^2 (x_2 - z_1)^2}$$

The proof will use the following result:

Theorem: For a vector $v = (a, b, c)$ in R^3, the magnitude of v is:

$$\| v \| = \sqrt{a^2 + b^2 + c^2}$$

Proof: There are four cases to consider:

Case 1: $a = b = c = 0$. Then $v = 0$, so $\| v \| = 0 = \sqrt{0^2 + 0^2 + 0^2} = \sqrt{a^2 + b^2 + c^2}$.

Case 2: Exactly two of a, b, c *are* 0. Without loss of generality, we assume that $a = b = 0$ and $c \neq 0$ (the other two possibilities are handled in a similar manner). Then $v = (0, 0, c)$, which is a vector of length $| c |$ along the z-axis. So $\| v \| = | c | = \sqrt{c^2} = \sqrt{0^2 + 0^2 + c^2} = \sqrt{a^2 + b^2 + c^2}$.

Case 3: Exactly one of a, b, c *is* 0. Without loss of generality, we assume that $a = 0$, $b \neq 0$ and $c \neq 0$ (the other two possibilities are handled in a similar manner). Then $v = (0, b, c)$, which is a vector in the yz-plane, so by the Pythagorean Theorem we have

$$\| v \| = \sqrt{b^2 + c^2} = \sqrt{0^2 + b^2 + c^2} = \sqrt{a^2 + b^2 + c^2}$$

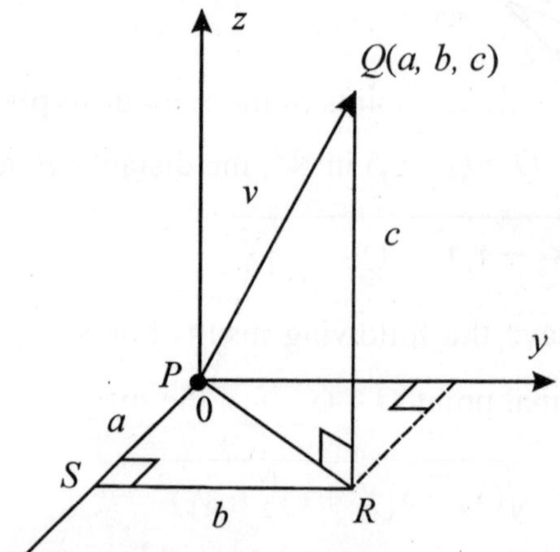

Case 4: *none of* a, b, c *are* 0. Without loss of generality, we can assume that a, b, c are all positive (the other seven possibilities are handled in a similar manner). Consider the points $P = (0, 0, 0)$, $Q = (a, b, c)$, $R = (a, b, 0)$, and $S = (a, 0, 0)$. Applying the Pythagorean Theorem to the right triangle $\Delta PS\,R$ gives $| PR |^2 = a^2 + b^2$. A second application of the Pythagorean Theorem, this time to the right triangle ΔPQR, gives $\| v \| = | PQ | = \sqrt{| PR |^2 + | QR |^2} = \sqrt{a^2 + b^2 + c^2}$. This proves the theorem.

Example: Calculate the following:

1. The magnitude of the vector $\overrightarrow{PQ}$ in ¡^{2} with $P = (-1, 2)$ and $Q = (5, 5)$.

Solution: By formula, $\|\overrightarrow{PQ}\| = \sqrt{(5-(-1))^2+(5-2)^2} = \sqrt{36+9} = \sqrt{45} = 3\sqrt{5}$.

2. The magnitude of the vector $v = (8, 3)$ in $\mathbb{R}^2$.

Solution: By formula, $\| v \| = \sqrt{8^2+3^2} = \sqrt{73}$.

3. The distance between the points $P = (2, -1, 4)$ and $Q = (4, 2, -3)$ in $\mathbb{R}^2$.

Solution: By formula, the distance $d = \sqrt{(4-2)^2+(2-(-1))^2+(-3-4)^2} = \sqrt{4+9+49} = \sqrt{62}$.

4. The magnitude of the vector $v = (5, 8, -2)$ in $\mathbb{R}^3$.

Solution: By formula, $\| v \| = \sqrt{5^2+8^2+(-2)^2} = \sqrt{25+64+4} = \sqrt{93}$.

EXERCISES

1. Calculate the magnitudes of the following vectors:

   (a) $v = (2, -1)$ (b) $v = (2, -1, 0)$ (c) $v = (3, 2, -2)$ (d) $v = (0, 0, 1)$ (e) $v = (6, 4, -4)$

2. For the points $P = (1, -1, 1)$, $Q = (2, -2, 2)$, $R = (2, 0, 1)$, $S = (3, -1, 2)$, does $\overrightarrow{PQ} = \overrightarrow{RS}$?

3. For the points $P = (0, 0, 0)$, $Q = (1, 3, 2)$, $R = (1, 0, 1)$, $S = (2, 3, 4)$, does $\overrightarrow{PQ} = \overrightarrow{RS}$?

4. Let $v = (1, 0, 0)$ and $w = (a, 0, 0)$ be vectors in $\mathbb{R}^3$. Show that $\| w \| = | a | \| v \|$.

5. Let $v = (a, b, c)$ and $w = (3a, 3b, 3c)$ be vectors in $\mathbb{R}^3$. Show that $\| w \| = 3 \| v \|$.

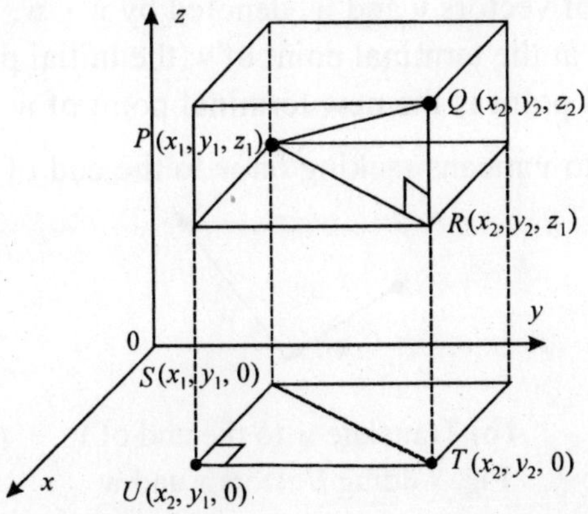

Vector Algebra

Now that we know what vectors are, we can start to perform some of the usual algebraic operations on them (e.g. addition, subtraction). Before doing that, we will introduce the notion of a *scalar*.

Definition: A scalar is a quantity that can be represented by a single number.

For our purposes, scalars will always be real numbers. Examples of scalar quantities are mass, electric charge, and speed (not velocity).4 We can now define *scalar multiplication* of a vector.

Definition: For a scalar k and a nonzero vector v, the scalar multiple of v by k, denoted by kv, is the vector whose magnitude is $|k| \| v \|$, points in the same direction as v if $k > 0$, points in the opposite direction as v if $k < 0$, and is the zero vector 0 if $k = 0$. For the zero vector 0, we define $k0 = 0$ for any scalar k.

Two vectors v and w are parallel (denoted by $v \| w$) if one is a scalar multiple of the other. You can think of scalar multiplication of a vector as stretching or shrinking the vector, and as flipping the vector in the opposite direction if the scalar is a negative number.

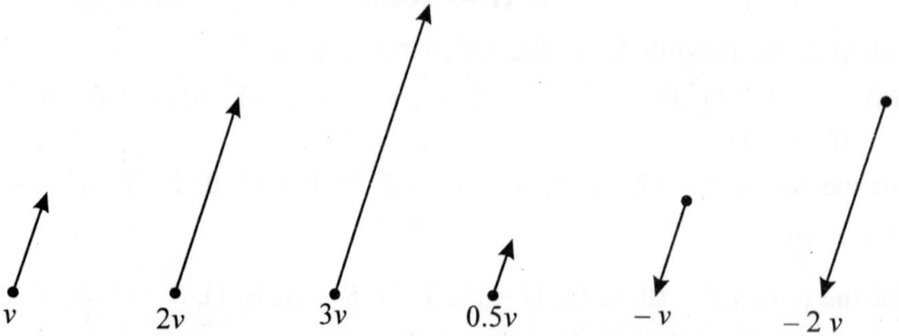

Recall that translating a nonzero vector means that the initial point of the vector is changed but the magnitude and direction are preserved. We are now ready to define the *sum* of two vectors.

Definition: The sum of vectors v and w, denoted by $v + w$, is obtained by translating w so that its initial point is at the terminal point of v; the initial point of $v + w$ is the initial point of v, and its terminal point is the new terminal point of w.

Intuitively, adding w to v means tacking on w to the end of v.

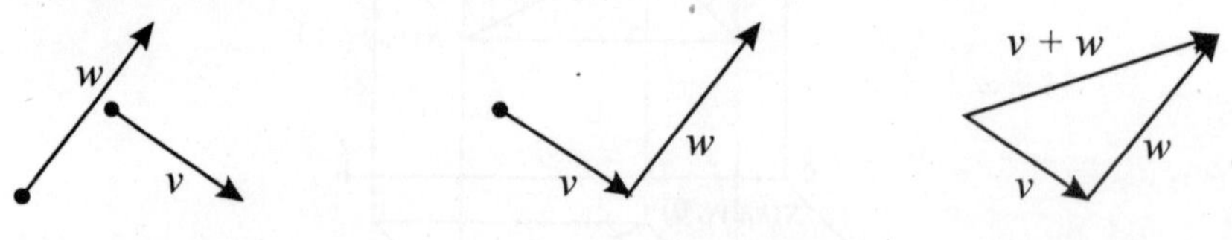

(a) Vectors v and w (b) Translate w to the end of v (c) The sum $v + w$

Fig. Adding Vectors v and w

Notice that our definition is valid for the zero vector (which is just a point, and hence can be translated), and so we see that $v + 0 = v = 0 + v$ for any vector v. In particular, $0 + 0 = 0$. Also, it is easy to see that $v + (-v) = 0$, as we would expect. In general, since the scalar multiple $-v = -1\ v$ is a well-defined vector, we can define vector subtraction as follows: $v - w = v + (-w)$.

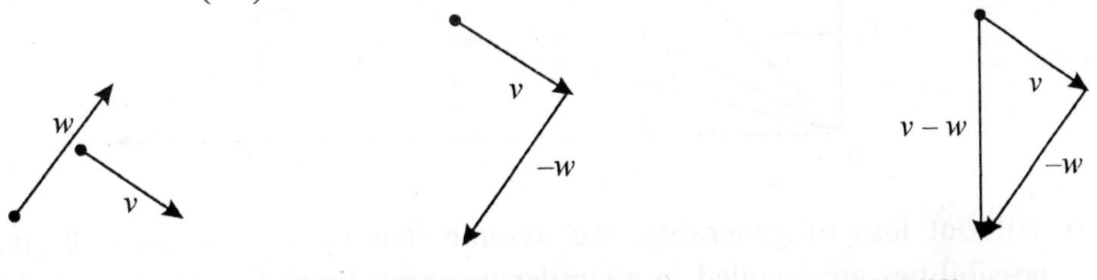

(a) Vectors v and w (b) Translate $-w$ to the end of v (c) The difference $v - w$

Fig. Subtracting vectors v and w

Figure shows the use of "geometric proofs" of various laws of vector algebra, that is, it uses laws from elementary geometry to prove statements about vectors. For example, (a) shows that $v + w = w + v$ for any vectors v, w. And (c) shows how you can think of $v - w$ as the vector that is tacked on to the end of w to add up to v.

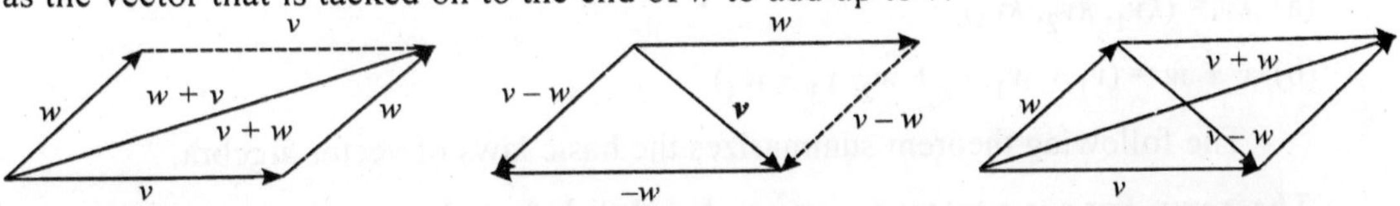

(a) Add vectors (b) Subtract vectors (c) Combined add/subtract

Fig. Geometric" Vector Algebra

Notice that we have temporarily abandoned the practice of starting vectors at the origin. In fact, we have not even mentioned coordinates in this section so far. Since we will deal mostly with Cartesian coordinates in this book, the following two theorems are useful for performing vector algebra on vectors in $\mathbb{R}^2$ and $\mathbb{R}^3$ starting at the origin.

Theorem: Let $v = (v_1, v_2)$, $w = (w_1, w_2)$ be vectors in $\mathbb{R}^2$, and let k be a scalar. Then

(a) $kv = (kv_1, kv_2)$

(b) $v + w = (v_1 + w_1, v_2 + w_2)$

Proof: (a) Without loss of generality, we assume that $v_1, v_2 > 0$ (the other possibilities are handled in a similar manner). If $k = 0$ then $kv = 0v = 0 = (0, 0) = (0v_1, 0v_2) = (kv_1, kv_2)$, which is what we needed to show. If $k \neq 0$, then (kv_1, kv_2) lies on a line with slope $\dfrac{kv_2}{kv_1} = \dfrac{v_2}{v_1}$, which is the same as the slope of the line on which v (and hence kv) lies, and (kv_1, kv_2) points in the same direction on that line as kv. Also, by formula the magnitude of (kv_1, kv_2) is p(kv_1) $\sqrt{(kv_1)^2 + (kv_2)^2} = \sqrt{k^2v_1^2 + k^2v_2^2} = \sqrt{k^2(v_1^2 + v_2^2)} = |k|\sqrt{v_1^2 + v_2^2} = |k|\,\|v\|$. So kv and (kv_1, kv_2) have the same magnitude and direction. This proves (a).

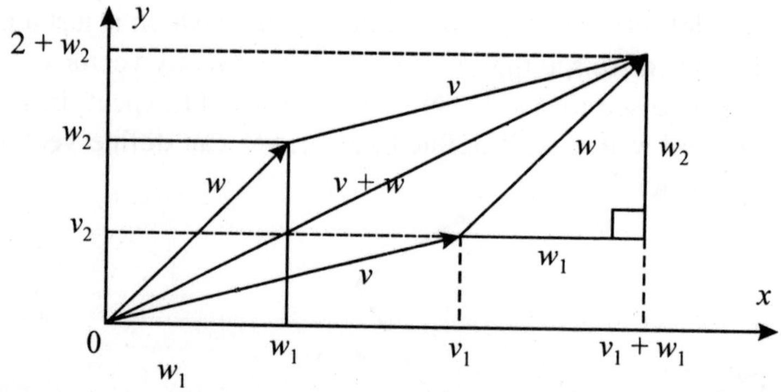

(b) Without loss of generality, we assume that v_1, v_2, w_1, $w_2 > 0$ (the other possibilities are handled in a similar manner). From Figure, we see that when translating w to start at the end of v, the new terminal point of w is $(v_1 + w_1, v_2 + w_2)$, so by the definition of $v + w$ this must be the terminal point of $v + w$. This proves (b).

Theorem: Let $v = (v_1, v_2, v_3)$, $w = (w_1, w_2, w_3)$ be vectors in $\mathbb{R}^3$, let k be a scalar. Then

(a) $kv = (kv_1, kv_2, kv_3)$

(b) $v + w = (v_1 + w_1, v_2 + w_2, v_3 + w_3)$

The following theorem summarizes the basic laws of vector algebra.

Theorem: For any vectors u, v, w, and scalars k, l, we have

(a) $v + w = w + v$ Commutative Law

(b) $u + (v + w) = (u + v) + w$ Associative Law

(c) $v + 0 = v = 0 + v$ Additive Identity

(d) $v + (-v) = 0$ Additive Inverse

(e) $k(lv) = (kl)v$ Associative Law

(f) $(v + w) = kv + kw$ Distributive Law

(g) $(k + l)v = kv + lv$ Distributive Law

Proof: (a) We already presented a geometric proof of this in Figure.

(b) To illustrate the difference between analytic proofs and geometric proofs in vector algebra, we will present both types here. For the analytic proof, we will use vectors in $\mathbb{R}^3$ (the proof for $\mathbb{R}^2$ is similar).

Let $u = (u_1, u_2, u_3)$, $v = (v_1, v_2, v_3)$, $w = (w_1, w_2, w_3)$ be vectors in $\mathbb{R}^3$. Then

$u + (v + w) = (u_1, u_2, u_3) + ((v_1, v_2, v_3) + (w_1, w_2, w_3))$

$= (u_1, u_2, u_3) + (v_1 + w_1, v_2 + w_2, v_3 + w_3)$ by Theorem

$= (u_1 + (v_1 + w_1), u_2 + (v_2 + w_2), u_3 + (v_3 + w_3))$ by Theorem

$= ((u_1 + v_1) + w_1, (u_2 + v_2) + w_2, (u_3 + v_3) + w_3)$ by properties of real numbers

$= (u_1 + v_1, u_2 + v_2, u_3 + v_3) + (w_1, w_2, w_3)$ by Theorem

$= (u + v) + w$

This completes the analytic proof of (b). Figure provides the geometric proof.

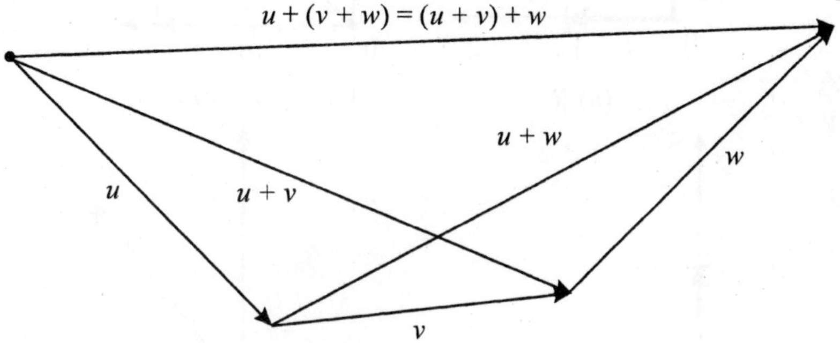

Fig. Associative Law for vector addition

(c) We already discussed.

(d) We already discussed.

(e) We will prove this for a vector $v = (v_1, v_2, v_3)$ in $\mathbb{R}^3$ (the proof for $\mathbb{R}^2$ is similar):

$k(lv) = k(lv_1, lv_2, lv_3)$ by Theorem

$= (klv_1, klv_2, klv_3)$ by Theorem

$= (kl)(v_1, v_2, v_3)$ by Theorem

$= (kl)v$

(*f*) and (*g*): Left as exercises for the reader.

A unit vector is a vector with magnitude 1. Notice that for any nonzero vector *v*, the vector $\frac{v}{\|v\|}$ is a unit vector which points in the same direction as *v*, since $\frac{1}{\|v\|} > 0$ and $\left\|\frac{v}{\|v\|}\right\| = \frac{\|v\|}{\|v\|} = 1$. Dividing a nonzero vector *v* by $\|v\|$ is often called *normalizing v*.

There are specific unit vectors which we will often use, called the basis vectors: $i = (1, 0, 0)$, $j = (0, 1, 0)$, and $k = (0, 0, 1)$ in $\mathbb{R}^3$; $i = (1, 0)$ and $j = (0, 1)$ in $\mathbb{R}^2$.

These are useful for several reasons: they are mutually perpendicular, since they lie on distinct coordinate axes; they are all unit vectors: $\|i\| = \|j\| = \|k\| = 1$; every vector can be written as a unique scalar combination of the basis vectors: $v = (a, b) = a\,i + b\,j$ in $\mathbb{R}^2$, $v = (a, b, c) = a\,i + b\,j + ck$ in $\mathbb{R}^3$.

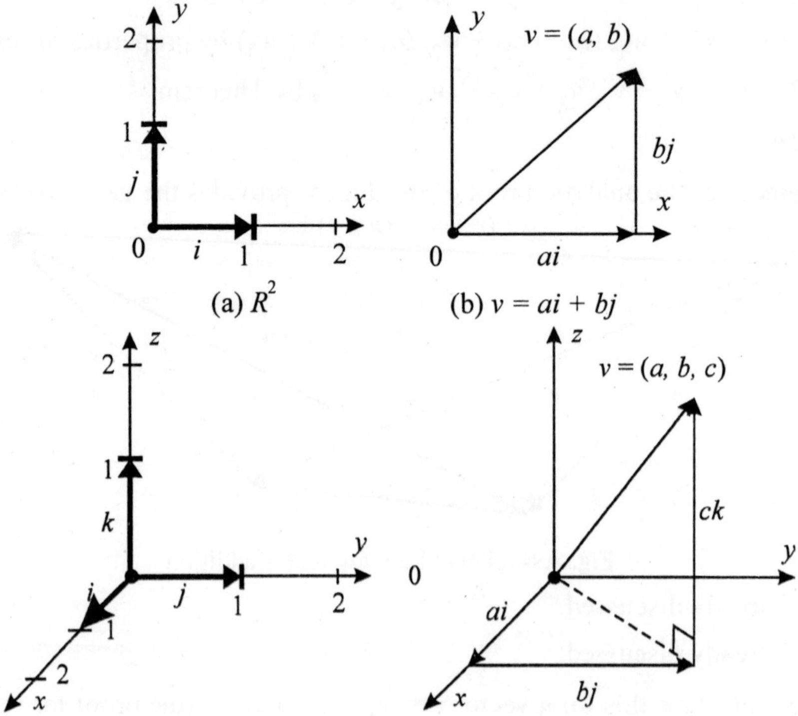

(a) R^2 (b) $v = ai + bj$ (c) R^3 (d) $= v = ai + bj + ck$

Fig. Basis Vectors in Different Dimensions

When a vector $v = (a, b, c)$ is written as $v = a\,i + b\,j + ck$, we say that v is in *component form*, and that a, b, and c are the *i, j, and k components*, respectively, of v. We have:

$$v = v_1\,i + v_2 j + v_3\,k,\ k \text{ a scalar} \Rightarrow kv = kv_1\,i + kv_2 j + kv_3\,k$$

$$v = v_1\,i + v_2 j + v_3\,k,\ w = w_1\,i + w_2 j + w_3\,k \Rightarrow v + w$$

$$= (v_1 + w_1)i + (v2 + w_2)j + (v_3 + w_3)k$$

$$v = v_1\,i + v_2 j + v_3\,k \Rightarrow \|v\| = \sqrt{v_1^2 + v_2^2 + v_3^2}$$

Example: Let $v = (2, 1, -1)$ and $w = (3, -4, 2)$ in $\mathbb{R}^3$.

(a) Find $v - w$.

Solution: $v - w = (2 - 3, 1 - (-4) - 1 - 2) = (-1, 5, -3)$

(b) Find $3v + 2w$.

Solution: $3v + 2w = (6, 3, -3) + (6, -8, 4) = (12, -5, 1)$

(c) Write v and w in component form.

Solution: $v = 2i + j - k,\ w = 3i - 4j + 2k$

(d) Find the vector u such that $u + v = w$.

Solution: By Theorem, $u = w - v = -(v - w) = -(-1, 5, -3) = (1, -5, 3)$.

(e) Find the vector u such that $u + v + w = 0$.

Solution: By Theorem, $u = -w - v = -(3, -4, 2) - (2, 1, -1) = (-5, 3, -1)$.

(f) Find the vector u such that $2u + i - 2j = k$.

Solution: $2u = -i + 2j + k \Rightarrow u = -\frac{1}{2}i + j + \frac{1}{2}k$

(g) Find the unit vector $\frac{v}{\|v\|}$.

Solution: $\frac{v}{\|v\|} = \frac{1}{\sqrt{2^2 + 1^2 + (-1)^2}}(2,1,-1) = \left(\frac{2}{\sqrt{6}}, \frac{1}{\sqrt{6}}, \frac{-1}{\sqrt{6}}\right)$

We can now easily prove Theorem from the previous section. The distance d between two points $P = (x_1, y_1, z_1)$ and $Q = (x_2, y_2, z_2)$ in $\mathbb{R}^3$ is the same as the length of the vector $w - v$, where the vectors v and w are defined as $v = (x_1, y_1, z_1)$ and $w = (x_2, y_2, z_2)$. So since $w - v = (x_2 - x_1, y_2 - y_1, z_2 - z_1)$, then $d = \| w - v \| = \sqrt{(x_2 - x_1)^2 + (y_2 - y_1)^2 + (z_2 - z_1)^2}$ by Theorem.

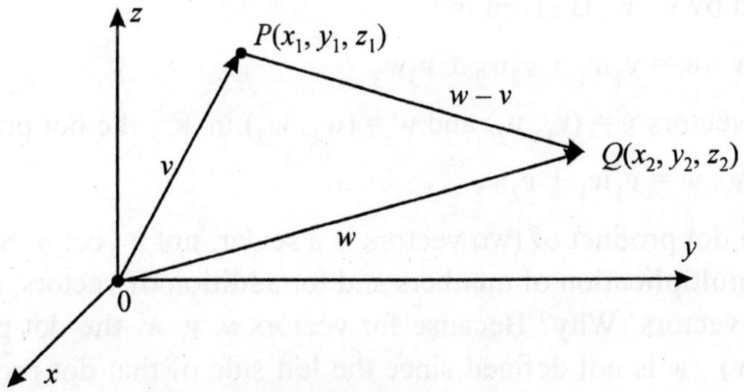

Fig. Proof of Theorem : $d = \|w - v\|$

EXERCISES

1. Let $v = (-1, 5, -2)$ and $w = (3, 1, 1)$.

   (a) Find $v - w$.

   (b) Find $v + w$.

   (c) Find $\frac{v}{\|v\|}$.

(d) Find $\left\|\frac{1}{2}(v-w)\right\|$.

(e) Find $\left\|\frac{1}{2}(v+w)\right\|$.

(f) Find $-2\,v + 4w$.

(g) Find $v - 2w$.

(h) Find the vector u such that $u + v + w = i$.

(i) Find the vector u such that $u + v + w = 2\,j + k$.

(j) Is there a scalar m such that $m(v + 2w) = k$? If so, find it.

1. We know that every vector in $\mathbb{R}^3$ can be written as a scalar combination of the vectors i, j, and k. Can every vector in $\mathbb{R}^3$ be written as a scalar combination of just i and j, i.e. for any vector v in $\mathbb{R}^3$, are there scalars m, n such that $v = mi + n$? Justify your answer.

Dot Product

You may have noticed that while we did define multiplication of a vector by a scalar in the previous section on vector algebra, we did not define multiplication of a vector by a vector. We will now see one type of multiplication of vectors, called the *dot product*.

Definition: Let $v = (v_1, v_2, v_3)$ and $w = (w_1, w_2, w_3)$ be vectors in $\mathbb{R}^3$. The dot product of v and w, denoted by $v \,.\, w$, is given by:

$$v \,.\, w = v_1w_1 + v_2w_2 + v_3w_3$$

Similarly, for vectors $v = (v_1, v_2)$ and $w = (w_1, w_2)$ in $\mathbb{R}^2$, the dot product is:

$$v \,.\, w = v_1w_1 + v_2w_2$$

Notice that the dot product of two vectors is a scalar, not a vector. So the associative law that holds for multiplication of numbers and for addition of vectors, does *not* hold for the dot product of vectors. Why? Because for vectors u, v, w, the dot product $u \,.\, v$ is a scalar, and so $(u \,.\, v) \,.\, w$ is not defined since the left side of that dot product (the part in parentheses) is a scalar and not a vector.

For vectors $v = v_1\, i + v_2 j + v_3\, k$ and $w = w_1\, i + w_2 j + w_3\, k$ in component form, the dot product is still $v \,.\, w = v_1w_1 + v_2w_2 + v_3w_3$.

Also notice that we defined the dot product in an analytic way, i.e. by referencing vector coordinates. There is a geometric way of defining the dot product, which we will now develop as a consequence of the analytic definition.

Definition: The angle between two nonzero vectors with the same initial point is the smallest angle between them.

We do not define the angle between the zero vector and any other vector. Any two nonzero vectors with the same initial point have two angles between them: θ and 360° – θ. We will always choose the smallest nonnegative angle θ between them, so that $0° \le \theta \le 180°$.

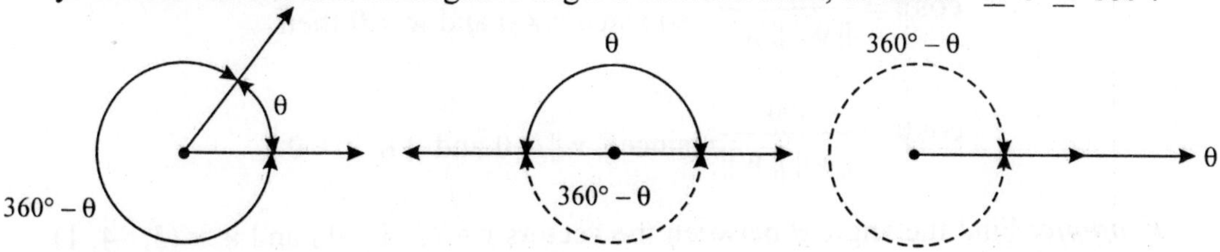

(a) 0° < θ < 180° (b) θ = 180° (c) θ = 0°

Fig. Angle between Vectors

We can now take a more geometric view of the dot product by establishing a relationship between the dot product of two vectors and the angle between them.

Theorem: Let *v*, *w* be nonzero vectors, and let θ be the angle between them. Then

$$\cos\theta = \frac{v.w}{\|v\|\,\|w\|}$$

Proof: We will prove the theorem for vectors in $\mathbb{R}^3$ (the proof for $\mathbb{R}^2$ is similar). Let $v = (v_1, v_2, v_3)$ and $w = (w_1, w_2, w_3)$. By the Law of Cosines, we have

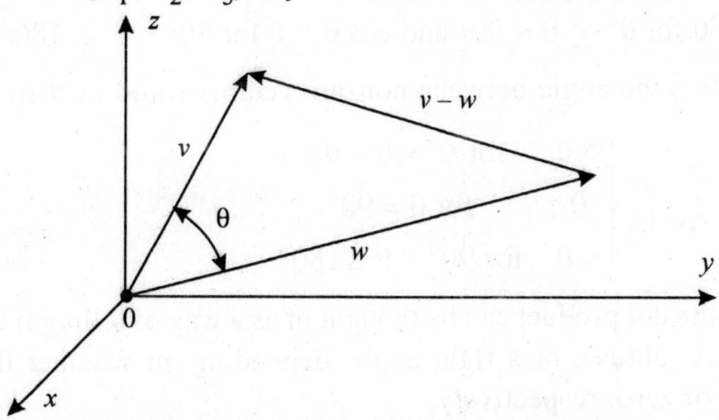

$$\|v - w\|^2 = \|v\|^2 + \|w\|^2 - 2\,\|v\|\,\|w\|\cos\theta$$

(note that equation holds even for the "degenerate" cases θ = 0° and 180°).

Since $v - w = (v_1 - w_1, v_2 - w_2, v_3 - w_3)$, expanding $\|v - w\|^2$ in equation gives $\|v\|^2 + \|w\|^2 - 2\,\|v\|\,\|w\|\cos\theta = (v_1 - w_1)^2 + (v_2 - w_2)^2 + (v_3 - w_3)^2$

$$= \left(v_1^2 - 2v_1w_1 + w_1^2\right) + \left(v_2^2 - 2v_2w_2 + w_2^2\right) + \left(v_3^2 - 2v_3w_3 + w_3^2\right)$$

$$= \left(v_1^2 - v_2^2 + v_3^2\right) + \left(w_1^2 + w_2^2 + w_3^2\right) - 2\left(v_1w_1 + v_2w_2 + v_2w_2 + v_3w_3\right)$$

$$= \|v\|^2\|w\|^2 - 2(v.w)\text{, so}$$

$$-2\,\|v\|\,\|w\|\cos\theta = -2(v.w),\text{ so since } v \neq 0 \text{ and } w \neq 0 \text{ then}$$

$$\cos\theta = \frac{v.w}{\|v\|\,\|w\|},\text{ so since } v \neq 0 \text{ and } w \neq 0 \text{ then}$$

$$\cos\theta = \frac{v.w}{\|v\|\,\|w\|},\text{ since } \|v\| > 0 \text{ and } \|w\| > 0.$$

Example: Find the angle θ between the vectors $v = (2, 1, -1)$ and $w = (3, -4, 1)$.

Solution: Since $v \,.\, w = (2)(3) + (1)(-4) + (-1)(1) = 1$, $\|v\| = \sqrt{6}$, and $\|w\| = \sqrt{26}$, then

$$\cos\theta = \frac{v.w}{\|v\|\,\|w\|} = \frac{1}{\sqrt{6}\sqrt{26}} = \frac{1}{2\sqrt{39}} \approx 0.08 \Rightarrow \theta = 85.41^\circ$$

Two nonzero vectors are perpendicular if the angle between them is 90°. Since cos 90° = 0, we have the following important corollary to Theorem:

Corollary: Two nonzero vectors v and w are perpendicular if and only if $v \,.\, w = 0$.

We will write $v \perp w$ to indicate that v and w are perpendicular.

Since $\cos\theta > 0$ for $0^\circ \leq \theta < 90^\circ$ and $\cos\theta < 0$ for $90^\circ < \theta \leq 180^\circ$, we also have:

Corollary: If θ is the angle between nonzero vectors v and w, then

$$v.w \text{ is } \begin{cases} > 0 & \text{for } 0^\circ \leq \theta < 90^\circ \\ 0 & \text{for } \theta = 90^\circ \\ < 0 & \text{for } 90^\circ < \theta \leq 180^\circ \end{cases}$$

By Corollary, the dot product can be thought of as a way of telling if the angle between two vectors is acute, obtuse, or a right angle, depending on whether the dot product is positive, negative, or zero, respectively.

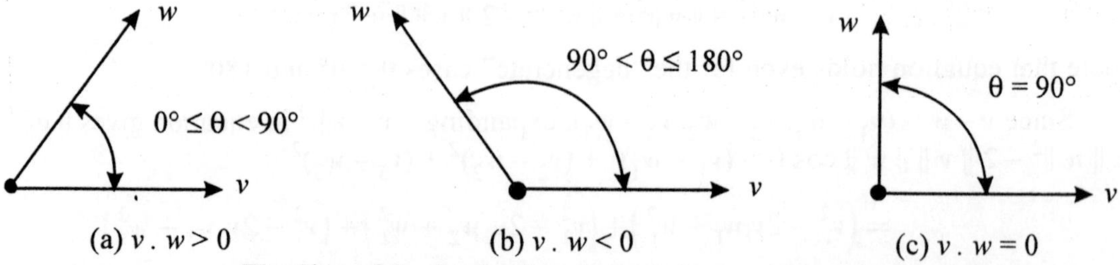

Fig. Sign of the dot Product & Angle between Vectors

Example: Are the vectors $v = (-1, 5, -2)$ and $w = (3, 1, 1)$ perpendicular?

Solution: Yes, $v \perp w$ since $v \,.\, w = (-1)(3) + (5)(1) + (-2)(1) = 0$.

The following theorem summarizes the basic properties of the dot product.

Theorem: For any vectors u, v, w, and scalar k, we have

(a) $v \,.\, w = w \,.\, v$ Commutative Law

(b) $(kv) \,.\, w = v \,.\, (kw) = k(v \,.\, w)$ Associative Law

(c) $v \,.\, 0 = 0 = 0 \,.\, v$

(d) $u \,.\, (v + w) = u \,.\, v + u \,.\, w$ Distributive Law

(e) $(u + v) \,.\, w = u \,.\, w + v \,.\, w$ Distributive Law

(f) $| v \,.\, w | \leq \| v \| \| w \|$ Cauchy-Schwarz Inequality[5]

Proof: The proofs of parts (a)-(e) are straightforward applications of the definition of the dot product, and are left to the reader as exercises. We will prove part (f). (f) If either $v = 0$ or $w = 0$, then $v \,.\, w = 0$ by part (c), and so the inequality holds trivially. So assume that v and w are nonzero vectors. Then by Theorem,

$$v \,.\, w = \cos\theta \, \| v \| \, \| w \|, \text{ so}$$

$$| v \,.\, w | = | \cos\theta | \, \| v \| \, \| w \|, \text{ so}$$

$$| v \,.\, w | \leq \| v \| \, \| w \| \text{ since } | \cos\theta | \leq 1.$$

Using Theorem we see that if $u \,.\, v = 0$ and $u \,.\, w = 0$, then $u \,.\, (kv + lw) = k(u \,.\, v) + l(u \,.\, w) = k(0) + l(0) = 0$ for all scalars k, l. Thus, we have the following fact:

If $u \perp v$ and $u \perp w$, then $u \perp (kv + lw)$ for all scalars k, l.

For vectors v and w, the collection of all scalar combinations $kv + lw$ is called the span of v and w. If nonzero vectors v and w are parallel, then their span is a line; if they are not parallel, then their span is a plane. So what we showed above is that a vector which is perpendicular to two other vectors is also perpendicular to their span.

The dot product can be used to derive properties of the magnitudes of vectors, the most important of which is the *Triangle Inequality*, as given in the following theorem:

Theorem: For any vectors v, w, we have

(a) $\| v \|^2 = v \,.\, v$

(b) $\| v + w \| \leq \| v \| + \| w \|$ Triangle Inequality

(c) $\| v - w \| \geq \| v \| - \| w \|$

Proof: (a) Left as an exercise for the reader.

(b) By part (a) and Theorem, we have

$$\| v + w \|^2 = (v + w) \,.\, (v + w) = v \,.\, v + v \,.\, w + w \,.\, v + w \,.\, w$$

$= \| v \|^2 + 2(v \,.\, w) + \| w \|^2$, so since $a \leq | a |$ for any real number a, we have

$\leq \| v \|^2 + 2 \mid v \, . \, w \mid + \| w \|^2$, so by Theorem we have

$\leq \| v \|^2 + 2 \| v \| \| w \| + \| w \|^2 = (\| v \| + \| w \|)^2$ and so

$\| v + w \| \leq \| v \| + \| w \|$ after taking square roots of both sides, which proves (b).

(c) Since $v = w + (v - w)$, then $\| v \| = \| w + (v - w) \| \leq \| w \| + \| v - w \|$ by the Triangle Inequality, so subtracting $\| w \|$ from both sides gives $\| v \| - \| w \| \leq \| v - w \|$.

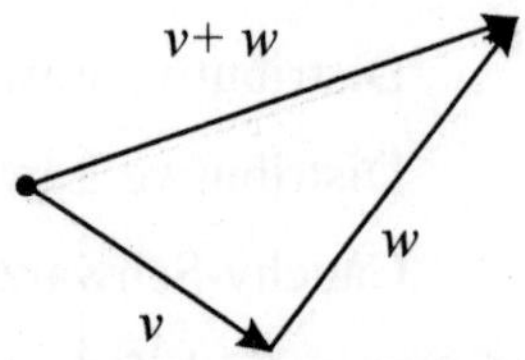

The Triangle Inequality gets its name from the fact that in any triangle, no one side is longer than the sum of the lengths of the other two sides. Another way of saying this is with the familiar statement "the shortest distance between two points is a straight line."

EXERCISES

1. Let $v = -3i - 2j - k$ and $w = 6i + 4j + 2k$. Calculate $v \, . \, w$.
2. Let $v = (5, 1, -2)$ and $w = (4, -4, 3)$. Calculate $v \, . \, w$.

   For Exercises 3-8, find the angle θ between the vectors v and w.
3. $v = (2, 1, 4), w = (1, -2, 0)$
4. $v = (4, 2, -1), w = (8, 4, -2)$
5. $v = -i + 2j + k, w = -3i + 6j + 3k$
6. $v = i, w = 3i + 2j + 4k$
7. $v = (5, 1, -2), w = (4, -4, 3)$
8. $v = (7, 2, -10), w = (2, 6, 4)$
9. Let $v = (8, 4, 3)$ and $w = (-2, 1, 4)$. Is $v \perp w$? Justify your answer.
10. Let $v = (6, 0, 4)$ and $w = (0, 2, -1)$. Is $v \perp w$? Justify your answer.
11. Prove or give a counterexample: If $u \, . \, v = u \, . \, w$, then $v = w$.
12. Prove or give a counterexample: If $v \, . \, w = 0$ for all v, then $w = 0$.
13. Prove or give a counterexample: If $u \, . \, v = u \, . \, w$ for all u, then $v = w$.
14. Prove that $\| v \| - \| w \| \leq \| v - w \|$ for all v, w.
15. Let α, β, and γ be the angles between a nonzero vector v in $\mathbb{R}^3$ and the vectors i, j, and k, respectively. Show that $\cos^2 \alpha + \cos^2 \beta + \cos^2 \gamma = 1$.

(Note: α, β, γ are often called the *direction angles* of v, and $\cos \alpha$, $\cos \beta$, $\cos \gamma$ are called the *direction cosines*.)

Cross Product

We have defined the dot product, which gave a way ofmultiplying two vectors. The resulting product, however, was a scalar, not a vector. In this section we will define a product of two vectors that does result in another vector. This product, called the *cross product*, is only defined for vectors in $\mathbb{R}^3$. The definition may appear strange and lacking motivation, but we will see the geometric basis for it shortly.

Definition: Let $v = (v_1, v_2, v_3)$ and $w = (w_1, w_2, w_3)$ be vectors in $\mathbb{R}^3$. The cross product of v and w, denoted by $v \,.\, w$, is the vector in $\mathbb{R}^3$ given by:

$$v \,.\, w = (v_2w_3 - v_3w_2, v_3w_1 - v_1w_3, v_1w_2 - v_2w_1)$$

Example: Find $i \times j$.

Solution: Since $i = (1, 0, 0)$ and $j = (0, 1, 0)$, then

$$\begin{aligned} i \times j &= ((0)(0) - (0)(1), (0)(0) - (1)(0), (1)(1) - (0)(0)) \\ &= (0, 0, 1) \\ &= k \end{aligned}$$

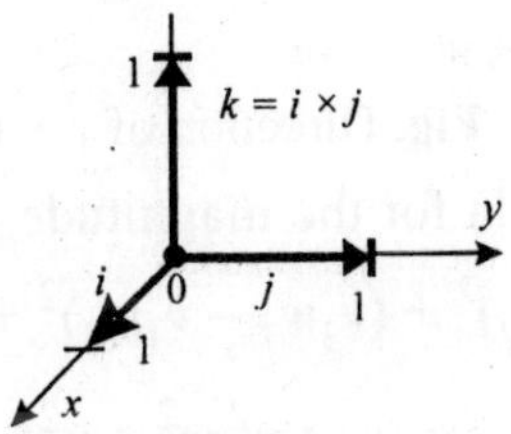

Similarly it can be shown that $j \,.\, k = i$ and $k \,.\, i = j$.

In the above example, the cross product of the given vectors was perpendicular to both those vectors. It turns out that this will always be the case.

Theorem: If the cross product $v \,.\, w$ of two nonzero vectors v and w is also a nonzero vector, then it is perpendicular to both v and w.

Proof: We will show that $(v \,.\, w) \,.\, v = 0$:

$$\begin{aligned} (v \,.\, w) \,.\, v &= (v_2w_3 - v_3w_2, v_3w_1 - v_1w_3, v_1w_2 - v_2w_1) \,.\, (v_1, v_2, v_3) \\ &= v_2w_3v_1 - v_3w_2v_1 + v_3w_1v_2 - v_1w_3v_2 + v_1w_2v_3 - v_2w_1v_3 \\ &= v_1v_2w_3 - v_1v_2w_3 + w_1v_2v_3 - w_1v_2v_3 + v_1w_2v3 - v_1w_2v_3 \\ &= 0, \text{ after rearranging the terms.} \end{aligned}$$

$\therefore \quad v \times w \perp v$ by Corollary.

The proof that $v \times w \perp w$ is similar.

Corollary: If the cross product $v \times w$ of two nonzero vectors v and w is also a nonzero vector, then it is perpendicular to the span of v and w.

The span of any two nonzero, nonparallel vectors *v*, *w* in $\mathbb{R}^3$ is a plane *P*, so the above corollary shows that *v* . *w* is perpendicular to that plane. As shown in Figure, there are two possible directions for $v \times w$, one the opposite of the other. It turns out that the direction of $v \times w$ is given by the *right-hand rule*, that is, the vectors *v*, *w*, $v \times w$ form a right-handed system. Recall that this means that you can point your thumb upwards in the direction of $v \times w$ while rotating *v* towards *w* with the remaining four fingers.

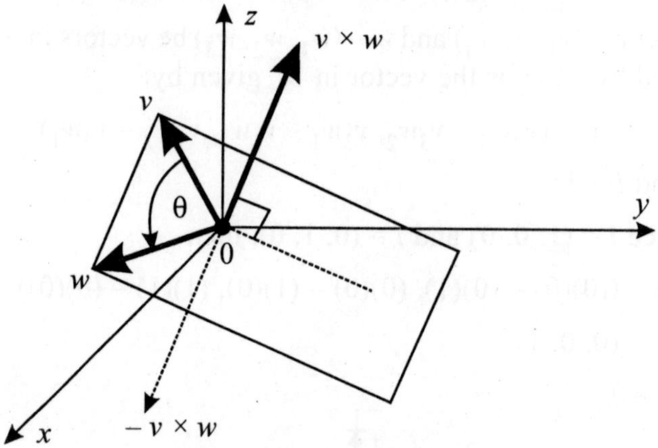

Fig. Direction of $v \times w$

We will now derive a formula for the magnitude of $v \times w$, for nonzero vectors *v*, *w*:

$$\| v \times w \|^2 = (v_2w_3 - v_3w_2)^2 + (v_3w_1 - v_1w_3)^2 + (v_1w_2 - v_2w_1)^2$$

$$= v_2^2w_3^2 - 2v_2w_2v_3w_3 + v_3^2w_2^2 + v_3^2w_1^2$$

$$-2v_1w_1v_3w_3 + v_1^2w_3^2 + v_1^2w_2^2 - 2v_1w_1v_2w_2 + v_2^2w_1^2$$

$$= v_1^2\left(w_2^2 + w_3^2\right) + w_2^2\left(v_1^2 + w_3^2\right) + v_3^2\left(w_1^2 + w_2^2\right)$$

$$-2(v_1w_1v_2w_2 + v_1w_1v_3w_3 + v_2w_2v_3w_3)$$

and now adding and subtracting $v_1^2w_1^2, v_2^2w_2^2$, and $v_3^2w_3^2$ on the right side gives

$$= v_1^2\left(w_1^2 + w_2^2 + w_3^2\right) + v_2^2\left(w_1^2 + w_2^2 + w_3^2\right) + v_3^2\left(w_1^2 + w_2^2 + w_3^2\right)$$

$$-\left(v_1^2w_1^2 + v_2^2w_2^2 + v_3^2w_3^2 + 2(v_1w_1v_2w_2 + v_1w_1v_3w_3 + v_2w_2v_2w_3)\right)$$

$$= \left(v_1^2 + v_2^2 + v_3^2\right)\left(w_1^2 + w_2^2 + w_3^2\right) \; -((v_1w_1)^2 + (v_2w_2)^2 + (v_3w_3)^2$$

$$+ 2(v_1w_1)(v_2w_2) + 2(v_1w_1)(v_3w_3) + 2(v_2w_2)(v_3w_3))$$

so using $(a + b + c)^2 = a^2 + b^2 + c^2 + 2ab + 2ac + 2bc$ for the subtracted term gives

$$= \left(v_1^2 + v_2^2 + v_3^2\right)\left(w_1^2 + w_2^2 + w_3^2\right) - (v_1w_1 + v_2w_2 + v_3w_3)^2$$

$$= \| v \|^2 \| w \|^2 - (v \,.\, w)^2$$

$$= \| v \|^2 \| w \|^2 \left(1 - \frac{(v.w)^2}{\| v \|^2 \| w \|^2}\right), \text{ since } \| v \| > 0 \text{ and } \| w \| > 0, \text{ so by Theorem}$$

$$= \| v \|^2 \| w \|^2 (1 - \cos^2 \theta), \text{ where } \theta \text{ is the angle between } v \text{ and } w, \text{ so}$$

$\| v \,.\, w\|^2 = \| v \|^2 \| w \|^2 \sin^2 \theta$, and since $0° \leq \theta \leq 180°$, then $\sin \theta \geq 0$, so we have:

If θ is the angle between nonzero vectors v and w in $\mathbb{R}^3$, then

$\| v \,.\, w \| = \| v \| \, \| w \| \sin \theta$

It may seem strange to bother with the above formula, when the magnitude of the cross product can be calculated directly, like for any other vector. The formula is more useful for its applications in geometry, as in the following example.

Example: Let ΔPQR and $PQRS$ be a triangle and parallelogram, respectively.

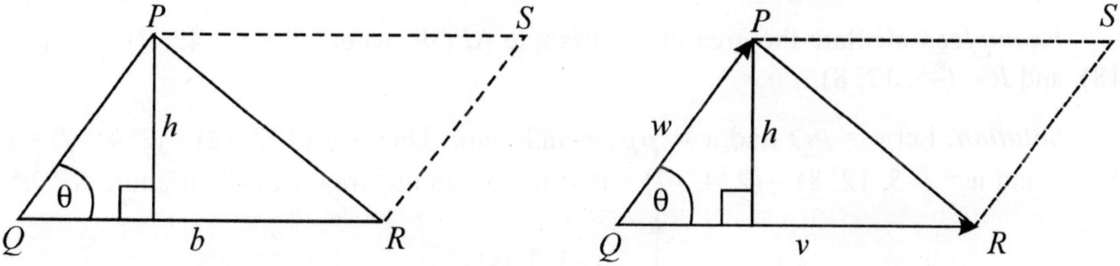

Think of the triangle as existing in $\mathbb{R}^3$, and identify the sides QR and QP with vectors v and w, respectively, in $\mathbb{R}^3$. Let θ be the angle between v and w. The area A_{PQR} of ΔPQR is $\frac{1}{2}bh$, where b is the base of the triangle and h is the height. So we see that

$$b = \| v \| \text{ and } h = \| w \| \sin \theta$$

$$A_{PQR} = \frac{1}{2} \| v \| \, \| w \| \sin \theta$$

$$= \frac{1}{2} \| v \times w \|$$

So since the area A_{PQRS} of the parallelogram $PQRS$ is twice the area of the triangle ΔPQR, then

$$A_{PQRS} = \| v \| \, \| w \| \sin \theta$$

By the discussion in Example, we have proved the following theorem:

Theorem: Area of triangles and parallelograms

(a) The area A of a triangle with adjacent sides v, w (as vectors in $\mathbb{R}^3$) is:

$$A = \frac{1}{2} \| v \times w \|$$

(b) The area A of a parallelogram with adjacent sides v, w (as vectors in $\mathbb{R}^3$) is:

$$A = \| v \times w \|$$

It may seem at first glance that since the formulas derived in Example were for the adjacent sides QP and QR only, then the more general statements in Theorem that the formulas hold for *any* adjacent sides are not justified. We would get a different *formula* for the area if we had picked PQ and PR as the adjacent sides, but it can be shown that the different formulas would yield the same value, so the choice of adjacent sides indeed does not matter, and Theorem is valid.

Theorem makes it simpler to calculate the area of a triangle in 3-dimensional space than by using traditional geometric methods.

Example: Calculate the area of the triangle ΔPQR, where $P = (2, 4, -7)$, $Q = (3, 7, 18)$, and $R = (-5, 12, 8)$.

Solution: Let $v = \overrightarrow{PQ}$ and $w = \overrightarrow{PR}$, as in Figure. Then $v = (3, 7, 18) - (2, 4, -7) = (1, 3, 25)$ and $w = (-5, 12, 8) - (2, 4, -7) = (-7, 8, 15)$, so the area A of the triangle ΔPQR is

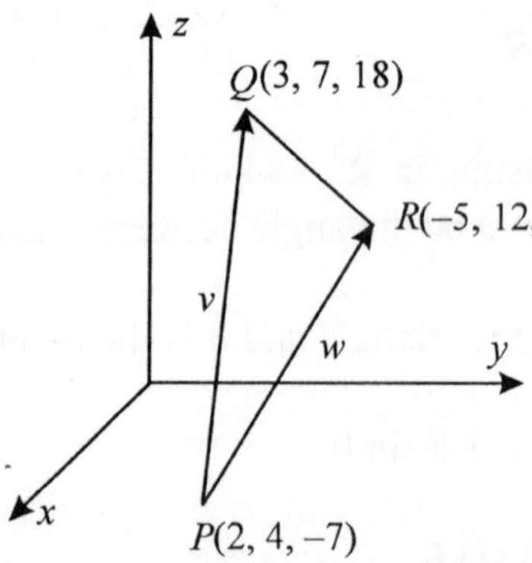

$$A = \frac{1}{2} \| v \times w \| = \frac{1}{2} \| (1, 3, 25) \times (-7 \times 8 \times 15) \|$$

$$= \frac{1}{2} \| ((3)(15) - (25)(8), (28)(-7) - (1)(15), (1)(8) - (3)(-7) \|$$

$$= \frac{1}{2} \| (-155, -190, 29) \|$$

$$= \frac{1}{2}\sqrt{(-155)^2 + (-190)^2 + 29^2} = \frac{1}{2}\sqrt{60966}$$

$A \approx 123.46$

Example: Calculate the area of the parallelogram $PQRS$, where $P = (1, 1)$, $Q = (2, 3)$, $R = (5, 4)$, and $S = (4, 2)$.

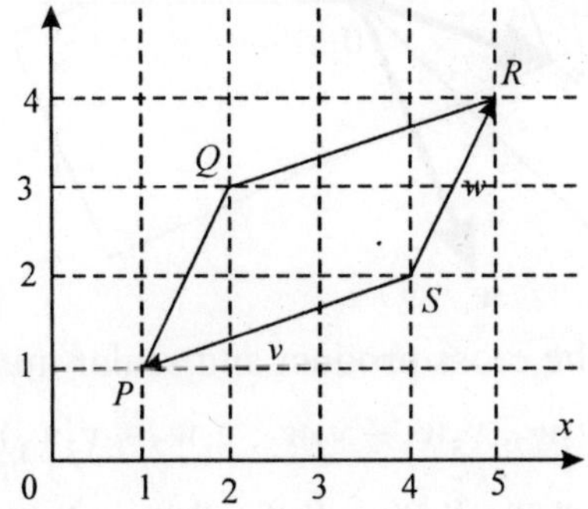

Solution: Let $v = \overrightarrow{SP}$ and $w = \overrightarrow{SR}$, as in Figure. Then $v = (1, 1) - (4, 2) = (-3, -1)$ and $w = (5, 4) - (4, 2) = (1, 2)$. But these are vectors in $\mathbb{R}^2$, and the cross product is only defined for vectors in $\mathbb{R}^3$. However, $\mathbb{R}^2$ can be thought of as the subset of $\mathbb{R}^3$ such that the z-coordinate is always 0. So we can write $v = (-3, -1, 0)$ and $w = (1, 2, 0)$. Then the area A of $PQRS$ is

$$A = \| v \times w \|$$

$$= \| (-3, -1, 0) \times (1, 2, 0) = ((-1)(0) - (0)(2), (0)(1) - (-3)(0), (-3)(2) - (-1)(1)) \|$$

$$= \| (0, 0, -5) \|$$

$$A = 5$$

The following theorem summarizes the basic properties of the cross product.

Theorem: For any vectors u, v, w in $\mathbb{R}^3$, and scalar k, we have

(a) $v \,.\, w = -w \,.\, v$ Anticommutative Law

(b) $u \,.\, (v + w) = u \,.\, v + u \,.\, w$ Distributive Law

(c) $(u + v) \,.\, w = u \,.\, w + v \,.\, w$ Distributive Law

(d) $(kv) \,.\, w = v \,.\, (kw) = k(v \,.\, w)$ Associative Law

(e) $v \,.\, 0 = 0 = 0 \,.\, v$

(f) $v \,.\, v = 0$

(g) $v \,.\, w = 0$ if and only if $v \parallel w$

Proof: The proofs of properties (b)-(f) are straightforward. We will prove parts (a) and (g) and leave the rest to the reader as exercises.

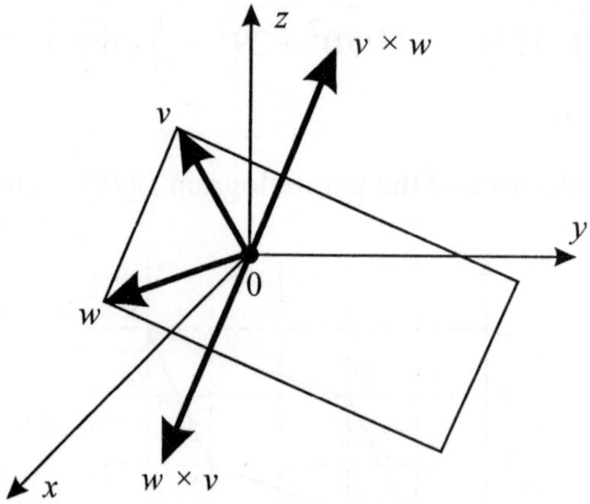

(a) By the definition of the cross product and scalar multiplication, we have:

$$\begin{aligned} v \times w &= (v_2w_3 - v_3w_2, v_3w_1 - v_1w_3, v_1w_2 - v_2w_1) \\ &= -(v_3w_2 - v_2w_3, v_1w_3 - v_3w_1, v_2w_1 - v_1w_2) \\ &= -(w_2v_3 - w_3v_2, w_3v_1 - w_1v_3, w_1v_2 - w_2v_1) \\ &= -w \times v \end{aligned}$$

Note that this says that $v \times w$ and $w \times v$ have the same magnitude but opposite direction.

(g) If either v or w is 0 then $v \times w = 0$ by part (e), and either $v = 0 = 0w$ or $w = 0 = 0v$, so v and w are scalar multiples, i.e. they are parallel.

If both v and w are nonzero, and θ is the angle between them, then by formula (1.11), $v \times w = 0$ if and only if $\| v \| \| w \| \sin \theta = 0$, which is true if and only if $\sin \theta = 0$ (since $\| v \| > 0$ and $\| w \| > 0$). So since $0° \leq \theta \leq 180°$, then $\sin \theta = 0$ if and only if $\theta = 0°$ or $180°$. But the angle between v and w is 0° or 180° if and only if $v \parallel w$.

Example: Adding to Example, we have

$$i \times j = k \qquad j \times k = i \qquad k \times i = j$$

$$j \times i = -k \qquad k \times j = -i \qquad i \times k = -j$$

$$i \times i = j \times j = k \times k = 0$$

Example: *Volume of a parallelepiped:* Let the vectors u, v, w in $\mathbb{R}^3$ represent adjacent sides of a parallelepiped P, as in Figure. Show that the volume of P is the *scalar triple product* $u \cdot (v \times w)$.

Solution: Recall that the volume of a parallelepiped is the area A of the base parallelogram times the height h. By Theorem, the area A of the base parallelogram is $\| v \times w \|$. And we can see that since $v \times w$ is perpendicular to the base parallelogram determined by v and w, then the height h is $\| u \| \cos \theta$, where θ is the angle between u and $v \times w$. By Theorem we know that

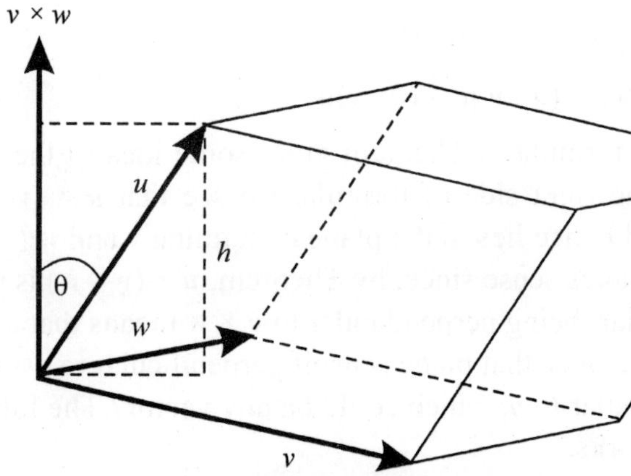

Fig. Parallelepiped *P*

$$\cos\theta = \frac{u.(v \times w)}{\| u \| \, \| v \times w \|}. \text{ Hence,}$$

$$\begin{aligned} vol(P) &= Ah \\ &= \| v \times w \| \frac{\| u \| u.(v \times w)}{\| u \| \, \| v \times w \|} \\ &= u \, . \, (v \times w) \end{aligned}$$

In Example the height *h* of the parallelepiped is $\| u \| \cos \theta$, and not "$\| u \| \cos \theta$, because the vector *u* is on the same side of the base parallelogram's plane as the vector $v \times w$ (so that $\cos \theta > 0$). Since the volume is the same no matter which base and height we use, then repeating the same steps using the base determined by *u* and *v* (since *w* is on the same side of that base's plane as $u \times v$), the volume is $w \, . \, (u \times v)$. Repeating this with the base determined by *w* and *u*, we have the following result:

For any vectors *u, v, w* in $\mathbb{R}^3$,

$$u \, . \, (v \times w) = w \, . \, (u \times v) = v \, . \, (w \times u)$$

(Note that the equalities hold trivially if any of the vectors are 0.)

Since $v \times w = -w \times v$ for any vectors *v, w* in $\mathbb{R}^3$, then picking the wrong order for the three adjacent sides in the scalar triple product in formula will give you the negative of the volume of the parallelepiped. So taking the absolute value of the scalar triple product for *any* order of the three adjacent sides will *always* give the volume:

Theorem: If vectors *u, v, w* in $\mathbb{R}^3$ represent any three adjacent sides of a parallelepiped, then the volume of the parallelepiped is $| \, u \, . \, (v \times w) \, |$.

Another type of triple product is the *vector triple product* $u \times (v \times w)$. The proof of the following theorem is left as an exercise for the reader:

Theorem: For any vectors u, v, w in $\mathbb{R}^3$,

$$u \times (v \times w) = (u \cdot w)v - (u \cdot v)w$$

An examination of the formula in Theorem gives some idea of the geometry of the vector triple product. By the right side of formula, we see that $u \times (v \times w)$ is a scalar combination of v and w, and hence lies in the plane containing v and w (i.e. $u \times (v \times w)$, v and w are coplanar). This makes sense since, by Theorem, $u \times (v \times w)$ is perpendicular to both u and $v \times w$. In particular, being perpendicular to $v \times w$ means that $u \times (v \times w)$ lies in the plane containing v and w, since that plane is itself perpendicular to $v \times w$. But then how is $u \times (v \times w)$ also perpendicular to u, which could be any vector? The following example may help to see how this works.

Example. Find $u \times (v \times w)$ for $u = (1, 2, 4)$, $v = (2, 2, 0)$, $w = (1, 3, 0)$.

Solution: Since $u \cdot v = 6$ and $u \cdot w = 7$, then

$$\begin{aligned} u \times (v \times w) &= (u \times w)v - (u \cdot v)w \\ &= 7\,(2, 2, 0) - 6\,(1, 3, 0) = (14, 14, 0) - (6, 18, 0) \\ &= (8, -4, 0) \end{aligned}$$

Note that v and w lie in the xy-plane, and that $u \times (v \times w)$ also lies in that plane. Also, $u \times (v \times w)$ is perpendicular to both u and $v \times w = (0, 0, 4)$.

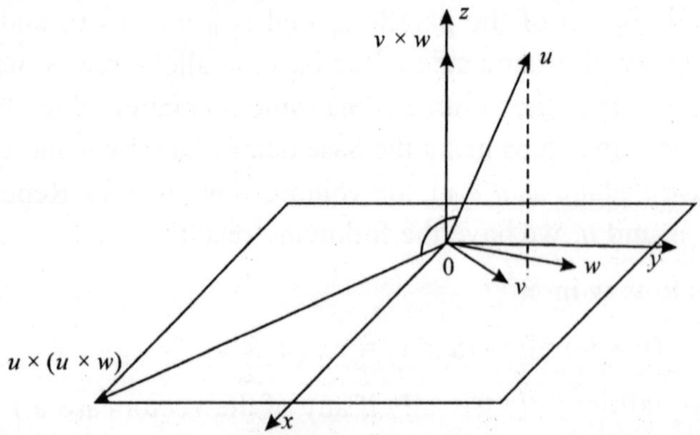

For vectors $v = v_1\, i + v_2\, j + v_3\, k$ and $w = w_1\, i + w_2\, j + w_3\, k$ in component form, the cross product is written as: $v \times w = (v_2w_3 - v_3w_2)i + (v_3w_1 - v_1w_3)j + (v_1w_2 - v_2w_1)k$. It is often easier to use the component form for the cross product, because it can be represented as a *determinant*. We will not go too deeply into the theory of determinants7; we will just cover what is essential for our purposes.

A 2 × 2 matrix is an array of two rows and two columns of scalars, written as

$$\begin{bmatrix} a & b \\ c & d \end{bmatrix} \text{ or } \begin{pmatrix} a & b \\ c & d \end{pmatrix}$$

where a, b, c, d are scalars. The determinant of such a matrix, written as

$$\begin{vmatrix} a & b \\ c & d \end{vmatrix} \text{ or } \det \begin{bmatrix} a & b \\ c & d \end{bmatrix},$$

is the scalar defined by the following formula:

$$\begin{vmatrix} a & b \\ c & d \end{vmatrix} = ad - bc.$$

It may help to remember this formula as being the product of the scalars on the downward diagonal minus the product of the scalars on the upward diagonal.

Example:

$$\begin{vmatrix} 1 & 2 \\ 3 & 4 \end{vmatrix} = (1)(4) - (2)(3) = 4 - 6 = -2$$

A 3×3 matrix is an array of three rows and three columns of scalars, written as

$$\begin{bmatrix} a_1 & a_2 & a_3 \\ b_1 & b_2 & b_3 \\ c_1 & c_2 & c_3 \end{bmatrix} \text{ or } \begin{pmatrix} a_1 & a_2 & a_3 \\ b_1 & b_2 & b_3 \\ c_1 & c_2 & c_3 \end{pmatrix}$$

and its determinant is given by the formula:

$$\begin{vmatrix} a_1 & a_2 & a_3 \\ b_1 & b_2 & b_3 \\ c_1 & c_2 & c_3 \end{vmatrix} = a_1 \begin{vmatrix} b_2 & b_3 \\ c_2 & c_3 \end{vmatrix} - a_2 \begin{vmatrix} b_1 & b_3 \\ c_1 & c_3 \end{vmatrix} + a_3 \begin{vmatrix} b_1 & b_2 \\ c_1 & c_2 \end{vmatrix}$$

One way to remember the above formula is the following: multiply each scalar in the first row by the determinant of the 2×2 matrix that remains after removing the row and column that contain that scalar, then sum those products up, putting alternating plus and minus signs in front of each (starting with a plus).

Example:

$$\begin{vmatrix} 1 & 0 & 2 \\ 4 & -1 & 3 \\ 1 & 0 & 2 \end{vmatrix} = 1 \begin{vmatrix} -1 & 3 \\ 0 & 2 \end{vmatrix} - 0 \begin{vmatrix} 4 & 3 \\ 1 & 2 \end{vmatrix} + 2 \begin{vmatrix} 4 & -1 \\ 1 & 0 \end{vmatrix}$$

$$= 1(-2 - 0) - 0(8 - 3) + 2(0 + 1) = 0$$

We defined the determinant as a scalar, derived from algebraic operations on scalar entries in a matrix. However, if we put three *vectors* in the first row of a 3×3 matrix, then the definition still makes sense, since we would be performing scalar multiplication on those three vectors (they would be multiplied by the 2 × 2 scalar determinants as before). This gives us a determinant that is now a vector, and lets us write the cross product of $v = v_1\, i + v_2 j + v_3\, k$ and $w = w_1\, i + w_2 j + w_3\, k$ as a determinant:

$$v \times w = \begin{vmatrix} i & j & k \\ v_1 & v_2 & v_3 \\ w_1 & w_2 & w_3 \end{vmatrix} = \begin{vmatrix} v_2 & v_3 \\ w_2 & w_3 \end{vmatrix} i - \begin{vmatrix} v_1 & v_3 \\ w_1 & w_3 \end{vmatrix} j + \begin{vmatrix} v_1 & v_2 \\ w_1 & w_2 \end{vmatrix} k$$

$$= (v_2w_3 - v_3w_2)i + (v_3w_1 - v_1w_3)j + (v_1w_2 - v_2w_1)k$$

Example: Let $v = 4\, i - j + 3k$ and $w = i + 2k$. Then

$$v \times w = \begin{vmatrix} i & j & k \\ 4 & -1 & 3 \\ 1 & 0 & 2 \end{vmatrix} = \begin{vmatrix} -1 & 3 \\ 0 & 2 \end{vmatrix} i - \begin{vmatrix} 4 & 3 \\ 1 & 2 \end{vmatrix} j + \begin{vmatrix} 4 & -1 \\ 1 & 0 \end{vmatrix} k = -2\, i - 5\, j + k$$

The scalar triple product can also be written as a determinant. In fact, by Example, the following theorem provides an alternate definition of the determinant of a 3 × 3 matrix as the volume (or negative volume) of a parallelepiped whose adjacent sides are the rows of the matrix. The proof is left as an exercise for the reader.

Theorem: For any vectors $u = (u_1, u_2, u_3)$, $v = (v_1, v_2, v_3)$, $w = (w_1, w_2, w_3)$ in $\mathbb{R}^3$:

$$u.(v \times w) = \begin{vmatrix} u_1 & u_2 & u_3 \\ v_1 & v_2 & v_3 \\ w_1 & w_2 & w_3 \end{vmatrix}$$

Example: Find the volume of the parallelepiped with adjacent sides $u = (2, 1, 3)$, $v = (-1, 3, 2)$, $w = (1, 1, -2)$.

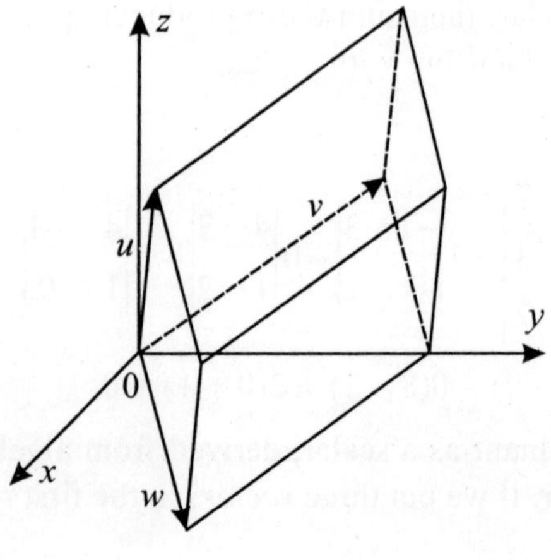

Solution: By Theorem, the volume of the parallelepiped P is the absolute value of the scalar triple product of the three adjacent sides (in any order). By Theorem,

$$u.(v \times w) = \begin{vmatrix} 2 & 1 & 3 \\ -1 & 3 & 2 \\ 1 & 1 & -2 \end{vmatrix}$$

$$= \begin{vmatrix} 3 & 2 \\ 1 & -2 \end{vmatrix} - 1\begin{vmatrix} -1 & 2 \\ 1 & -2 \end{vmatrix} + 3\begin{vmatrix} -1 & 3 \\ 1 & 1 \end{vmatrix}$$

$= 2(-8) - 1(0) + 3(-4) = -28$, so

$vol\,(P) = |-28| = 28.$

Interchanging the dot and cross products can be useful in proving vector identities:

Example: Prove: $(u \times v) . (w \times z) = \begin{vmatrix} u.w & u.z \\ v.w & v.z \end{vmatrix}$ for all vectors u, v, w, z in $\mathbb{R}^3$.

Solution: Let $x = u \times v$. Then

$(u \times v) . (w \times z) = x . (w \times z)$

$= w . (z \times x)$ (by formula)

$= w . (z \times (u \times v))$

$= w . ((z . v)u - (z . u)v)$ (by Theorem)

$= (z . v)(w . u) - (z . u)(w . v)$

$= (u . w)(v . z) - (u . z)(v . w)$ (by commutativity of the dot product).

$$= \begin{vmatrix} u.w & u.z \\ v.w & v.z \end{vmatrix}$$

EXERCISES

For Exercises 1-6, calculate $v \times w$.

1. $v = -i + 2j + k$, $w = -3i + 6j + 3k$
2. $v = i$, $w = 3i + 2j + 4k$
3. $v = (5, 1, -2)$, $w = (4, -4, 3)$
4. $v = (7, 2, -10)$, $w = (2, 6, 4)$
5. $v = (2, 1, 4)$, $w = (1, -2, 0)$
6. $v = (1, 3, 2)$, $w = (7, 2, -10)$

For Exercises 7-8, calculate the area of the triangle ΔPQR.

8. $P = (4, 0, 2)$, $Q = (2, 1, 5)$, $R = (-1, 0, -1)$
7. $P = (5, 1, -2)$, $Q = (4, -4, 3)$, $R = (2, 4, 0$

For Exercises 9-10, calculate the area of the parallelogram *PQRS*.

9. $P = (2, 1, 3)$, $Q = (1, 4, 5)$, $R = (2, 5, 3)$, $S = (3, 2, 1)$
10. $P = (-2, -2)$, $Q = (1, 4)$, $R = (6, 6)$, $S = (3, 0)$

For Exercises 11-12, find the volume of the parallelepiped with adjacent sides *u, v, w*.

11. $u = (1, 1, 3)$, $v = (2, 1, 4)$, $w = (5, 1, -2)$
12. $u = (1, 3, 2)$, $v = (7, 2, -10)$, $w = (1, 0, 1)$

For Exercises 13-14, calculate $u \cdot (v \times w)$ and $u \times (v \times w)$.

13. $u = (1, 1, 1)$, $v = (3, 0, 2)$, $w = (2, 2, 2)$
14. $u = (1, 0, 2)$, $v = (-1, 0, 3)$, $w = (2, 0, -2)$
15. Calculate $(u \times v) \cdot (w \times z)$ for $u = (1, 1, 1)$, $v = (3, 0, 2)$, $w = (2, 2, 2)$, $z = (2, 1, 4)$.
16. If v and w are unit vectors in $\mathbb{R}^3$, under what condition(s) would $v \times w$ also be a unit vector in $\mathbb{R}^3$? Justify your answer.
17. Show that if $v \times w = 0$ for all w in $\mathbb{R}^3$, then $v = 0$.
18. Prove the following for all vectors v, w in R^3:
    (a) $\| v \times w \|^2 + | v \cdot w |^2 = \| v \|^2 \| w \|^2$
    (b) If $v \cdot w = 0$ and $v \times w = 0$, then $v = 0$ or $w = 0$.
19. Consider the vector equation $a \times x = b$ in $\mathbb{R}^3$, where $a \neq 0$. Show that:
    (a) $a \cdot b = 0$
    (b) $x = \dfrac{b \times a}{\| a \|^2} + ka$ is a solution to the equation, for any scalar k
20. Prove the *Jacobi identity*: $u \times (v \times w) + v \times (w \times u) + w \times (u \times v) = 0$
21. Show that u, v, w lie in the same plane in $\mathbb{R}^3$ if and only if $u \cdot (v \times w) = 0$.
22. For all vectors u, v, w, z in $\mathbb{R}^3$, show that

$$(u \times v) \times (w \times z) = (z \cdot (u \times v))w - (w \cdot (u \times v))z$$

and that

$$(u \times v) \times (w \times z) = (u \cdot (w \times z))v - (v \cdot (w \times z))u$$

Why do both equations make sense geometrically?

Lines and Planes

Now that we know how to perform some operations on vectors, we can start to deal with some familiar geometric objects, like lines and planes, in the language of vectors.

The reason for doing this is simple: using vectors makes it easier to study objects in 3-dimensional Euclidean space. We will first consider lines.

Line through a point, Parallel to a Vector

Let $P = (x_0, y_0, z_0)$ be a point in $\mathbb{R}^3$, let $v = (a, b, c)$ be a nonzero vector, and let L be the line through P which is parallel to v .

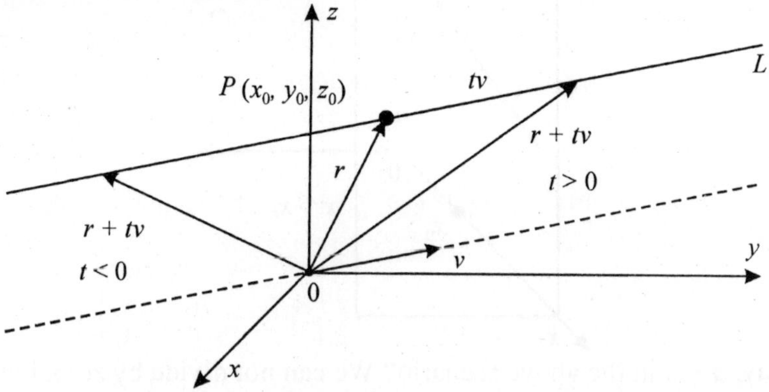

Let $r = (x_0, y_0, z_0)$ be the *vector* pointing from the origin to P. Since multiplying the vector v by a scalar t lengthens or shrinks v while preserving its direction if $t > 0$, and reversing its direction if $t < 0$, then we see from Figure that every point on the line L can be obtained by adding the vector tv to the vector r for some scalar t. That is, as t varies over all real numbers, the vector $r + tv$ will point to every point on L. We can summarize the *vector representation of L* as follows:

For a point $P = (x_0, y_0, z_0)$ and nonzero vector v in $\mathbb{R}^3$, the line L through P parallel to v is given by

$$r + tv, \text{ for } -\infty < t < \infty$$

where $r = (x_0, y_0, z_0)$ is the vector pointing to P.

Note that we used the correspondence between a vector and its terminal point. Since $v = (a, b, c)$, then the terminal point of the vector $r + tv$ is $(x_0 + at, y_0 + bt, z_0 + ct)$. We then get the *parametric representation of L* with the *parameter t*:

For a point $P = (x_0, y_0, z_0)$ and nonzero vector $v = (a, b, c)$ in $\mathbb{R}^3$, the line L through P parallel to v consists of all points (x, y, z) given by

$$x = x_0 + at,\ y = y_0 + bt,\ z = z_0 + ct, \text{ for } -\infty < t < \infty$$

Note that in both representations we get the point P on L by letting $t = 0$.

In formula, if $a \neq 0$, then we can solve for the parameter $t : t = (x - x_0)/a$. We can also solve for t in terms of y and in terms of z if neither b nor c, respectively, is zero: $t = (y - y_0) / b$ and $t = (z - z_0) / c$. These three values all equal the same value t, so we can write the following system of equalities, called the *symmetric representation of L*:

For a point $P = (x_0, y_0, z_0)$ and vector $v = (a, b, c)$ in $\mathbb{R}^3$ with a, b and c all nonzero, the line L through P parallel to v consists of all points (x, y, z) given by the equations

$$\frac{x - x_0}{a} = \frac{y - y_0}{b} = \frac{z - z_0}{c}$$

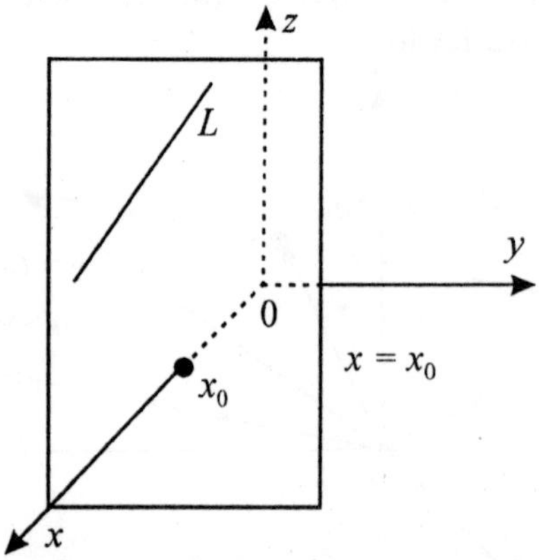

What if, say, $a = 0$ in the above scenario? We can not divide by zero, but we do know that $x = x_0 + at$, and so $x = x_0 + 0t = x_0$. Then the symmetric representation of L would be:

$$x = x_0, \frac{y - y_0}{b} = \frac{z - z_0}{c}$$

Note that this says that the line L lies in the *plane* $x = x_0$, which is parallel to the yz-plane. Similar equations can be derived for the cases when $b = 0$ or $c = 0$. You may have noticed that the vector representation of L in formula is more compact than the parametric and symmetric formulas. That is an advantage of using vector notation. Technically, though, the vector representation gives us the *vectors* whose terminal points make up the line L, not just L itself. So you have to remember to identify the vectors $r + tv$ with their terminal points. On the other hand, the parametric representation *always* gives just the points on L and nothing else.

Example : Write the line L through the point $P = (2, 3, 5)$ and parallel to the vector $v = (4, -1, 6)$, in the following forms: (a) vector, (b) parametric, (c) symmetric. Lastly: (d) find two points on L distinct from P.

Solution: (a) Let $r = (2, 3, 5)$. Then by formula (1.16), L is given by:

$$r + tv = (2, 3, 5) + t\,(4, -1, 6), \text{ for } -\infty < t < \infty$$

(b) L consists of the points (x, y, z) such that

$$x = 2 + 4t,\ y = 3 - t,\ z = 5 + 6t, \text{ for } -\infty < t < \infty$$

(c) L consists of the points (x, y, z) such that

$$\frac{x - 2}{4} = \frac{y - 3}{-1} = \frac{z - 5}{6}$$

(d) Letting $t = 1$ and $t = 2$ in part (b) yields the poi[illegible]6, 2, 11) and (10, 1, 17) on L.

Line through two Points

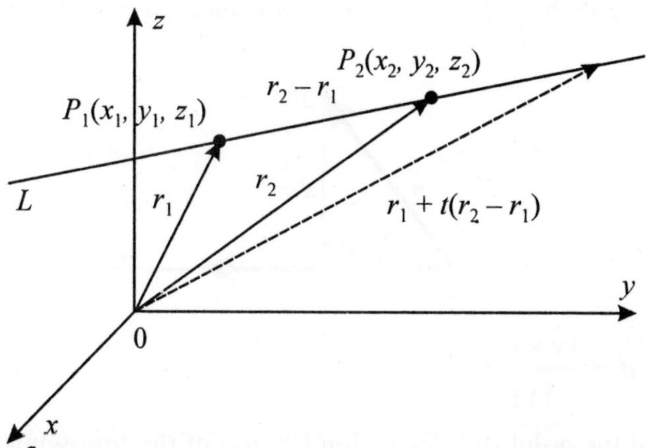

Let $P_1 = (x_1, y_1, z_1)$ and $P_2 = (x_2, y_2, z_2)$ be distinct points in $\mathbb{R}^3$, and let L be the line through P_1 and P_2. Let $r_1 = (x_1, y_1, z_1)$ and $r_2 = (x_2, y_2, z_2)$ be the vectors pointing to P_1 and P_2, respectively. Then as we can see from Figure, $r_2 - r_1$ is the vector from P_1 to P_2. So if we multiply the vector $r_2 - r_1$ by a scalar t and add it to the vector r_1, we will get the entire line L as t varies over all real numbers. The following is a summary of the vector, parametric, and symmetric forms for the line L:

Let $P_1 = (x_1, y_1, z_1)$, $P_2 = (x_2, y_2, z_2)$ be distinct points in $\mathbb{R}^3$, and let $r_1 = (x_1, y_1, z_1)$, $r_2 = (x_2, y_2, z_2)$. Then the line L through P_1 and P_2 has the following representations:

Vector:

$$r_1 + t(r_2 - r_1)\text{, for } -\infty < t < \infty$$

Parametric:

$$x = x_1 + (x_2 - x_1)t,\ y = y_1 + (y_2 - y_1)t,\ z = z_1 + (z_2 - z_1)t,\ \text{for } -\infty < t < \infty$$

Symmetric:

$$\frac{x - x_1}{x_2 - x_1} = \frac{y - y_1}{y_2 - y_1} = \frac{z - z_1}{z_2 - z_1} \quad (\text{if } x_1 \neq x_2, y_1 \neq y_2, \text{ and } z_1 \neq z_2)$$

Example: Write the line L through the points $P_1 = (-3, 1, -4)$ and $P_2 = (4, 4, -6)$ in parametric form.

Solution: By formula, L consists of the points (x, y, z) such that

$$x = -3 + 7t,\ y = 1 + 3t,\ z = -4 - 2t,\ \text{for } -\infty < t < \infty$$

Distance between a Point and a Line

Let L be a line in $\mathbb{R}^3$ in vector form as $r + tv$ (for $-\infty < t < \infty$), and let P be a point not on L. The distance d from P to L is the length of the line segment from P to L which is perpendicular to L. Pick a point Q on L, and let w be the vector from Q to P. If θ is the angle between w and v, then $d = \| w \| \sin \theta$. So since $\| v \times w \| = \| v \| \, \| w \| \sin \theta$ and $v \neq 0$, then:

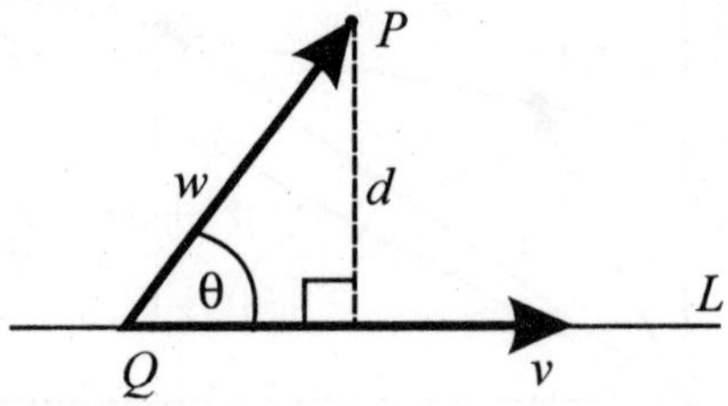

$$d = \frac{\| v \times w \|}{\| v \|}$$

Example: Find the point of intersection (if any) of the following lines:

$$\frac{x+1}{3} = \frac{y-2}{2} = \frac{z-1}{-1} \text{ and } x+3 = \frac{y-8}{-3} = \frac{z+3}{2}$$

Solution: First we write the lines in parametric form, with parameters s and t:

$$x = -1 + 3s,\ y = 2 + 2s,\ z = 1 - s \text{ and } x = -3 + t,\ y = 8 - 3t,\ z = -3 + 2t$$

The lines intersect when $(-1 + 3s, 2 + 2s, 1 - s) = (-3 + t, 8 - 3t, -3 + 2t)$ for some s, t:

$$-1 + 3s = -3 + t : \Rightarrow t = 2 + 3s$$

$$2 + 2s = 8 - 3t : \Rightarrow 2 + 2s = 8 - 3(2 + 3s) = 2 - 9s$$

$$\Rightarrow 2s = -9s \Rightarrow s = 0 \Rightarrow t = 2 + 3(0) = 2$$

$1 - s = -3 + 2t : 1 - 0 = -3 + 2(2) \Rightarrow 1 = 1$ (Note that we had to check this.)

Letting $s = 0$ in the equations for the first line, or letting $t = 2$ in the equations for the second line, gives the point of intersection $(-1, 2, 1)$. We will now consider planes in 3-dimensional Euclidean space.

Plane through a Point, Perpendicular to a Vector

Let P be a plane in $\mathbb{R}^3$, and suppose it contains a point $P_0 = (x_0, y_0, z_0)$. Let $n = (a, b, c)$ be a nonzero vector which is perpendicular to the plane P. Such a vector is called a normal vector (or just a *normal*) to the plane.

Now let (x, y, z) be any point in the plane P. Then the vector $r = (x - x_0, y - y_0, z - z_0)$ lies in the plane P. So if $r \neq 0$, then $r \perp n$ and hence $n \cdot r = 0$. And if $r = 0$ then we still have $n \cdot r = 0$.

Conversely, if (x, y, z) is any point in $\mathbb{R}^3$ such that $r = (x - x_0, y - y_0, z - z_0) \neq 0$ and $n \cdot r = 0$, then $r \perp n$ and so (x, y, z) lies in P. This proves the following theorem:

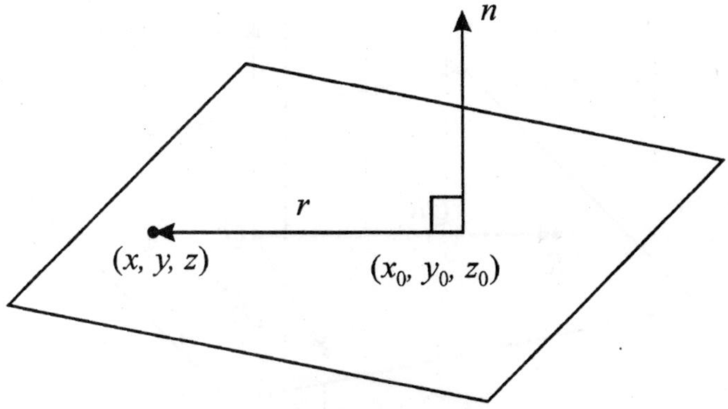

Fig. The Plane P

Theorem: Let P be a plane in $\mathbb{R}^3$, let (x_0, y_0, z_0) be a point in P, and let n = (a, b, c) be a nonzero vector which is perpendicular to P. Then P consists of the points (x, y, z) satisfying the vector equation:

$$n \cdot r = 0$$

where $r = (x - x_0, y - y_0, z - z_0)$, or equivalently:

$$a(x - x_0) + b(y - y_0) + c(z - z_0) = 0$$

The above equation is called the point-normal form of the plane P.

Example: Find the equation of the plane P containing the point (–3, 1, 3) and perpendicular to the vector $n = (2, 4, 8)$.

Solution: By formula (1.25), the plane P consists of all points (x, y, z) such that:

$$2(x + 3) + 4(y - 1) + 8(z - 3) = 0$$

If we multiply out the terms in formula and combine the constant terms, we get an equation of the plane in normal form:

$$ax + by + cz + d = 0$$

For example, the normal form of the plane in Example is $2x + 4y + 8z - 22 = 0$.

Plane Containing three Noncollinear Points

In 2-dimensional and 3-dimensional space, two points determine a line. Two points do not determine a plane in $\mathbb{R}^3$. In fact, three *collinear* points (i.e. all on the same line) do not determine a plane; an infinite number of planes would contain the line on which those three points lie. However, three *noncollinear* points do determine a plane. For if Q, R and S are noncollinear points in $\mathbb{R}^3$, then $\overrightarrow{QR}$ and $\overrightarrow{QS}$ are nonzero vectors which are not parallel (by noncollinearity), and so their cross product $\overrightarrow{QR} \times \overrightarrow{QS}$ is perpendicular to both $\overrightarrow{QR}$ and $\overrightarrow{QS}$ So $\overrightarrow{QR}$ and $\overrightarrow{QS}$ (and hence Q, R and S) lie in the plane through the point Q with normal vector $n = \overrightarrow{QR} \times \overrightarrow{QS}$.

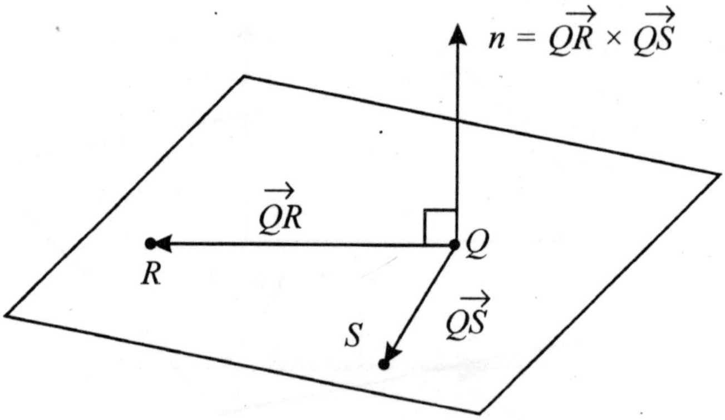

Fig. Noncollinear points Q, R, S

Example: Find the equation of the plane P containing the points (2, 1, 3), (1, –1, 2) and (3, 2, 1).

Solution: Let $Q = (2, 1, 3)$, $R = (1, -1, 2)$ and $S = (3, 2, 1)$. Then for the vectors $\overrightarrow{QR} = (-1, -2, -1)$ and $\overrightarrow{QS} = (1, 1, -2)$, the plane P has a normal vector

$$n = \overrightarrow{QR} \times \overrightarrow{QS} = (-1, -2, -1) \times (1, 1, -2) = (5, -3, 1)$$

So using formula with the point Q (we could also use R or S), the plane P consists of all points (x, y, z) such that:

$$5(x-2) - 3(y-1) + (z-3) = 0$$

or in normal form,

$$5x - 3y + z - 10 = 0$$

We mentioned earlier that skew lines in $\mathbb{R}^3$ lie on separate, parallel planes. So two skew lines do not determine a plane. But two (nonidentical) lines which either intersect or are parallel do determine a plane. In both cases, to find the equation of the plane that contains those two lines, simply pick from the two lines a total of three noncollinear points (i.e. one point from one line and two points from the other), then use the technique above, to write the equation. We will leave examples of this as exercises for the reader.

Distance between a Point and a Plane

The distance between a point in $\mathbb{R}^3$ and a plane is the length of the line segment from that point to the plane which is perpendicular to the plane. The following theorem gives a formula for that distance.

Theorem: Let $Q = (x_0, y_0, z_0)$ be a point in $\mathbb{R}^3$, and let P be a plane with normal form $ax + by + cz + d = 0$ that does not contain Q. Then the distance D from Q to P is:

$$D = \frac{|ax_0 + by_0 + cz_0 + d|}{\sqrt{a^2 + b^2 + c^2}}$$

Proof: Let $R = (x, y, z)$ be any point in the plane P (so that $ax + by + cz + d = 0$) and let $r = \overrightarrow{RQ} = (x_0 - x, y_0 - y, z_0 - z)$. Then $r \neq 0$ since Q does not lie in P. From the normal form equation for P, we know that $n = (a, b, c)$ is a normal vector for P. Now, any plane divides $\mathbb{R}^3$ into two disjoint parts. Assume that n points toward the side of P where the point Q is located. Place n so that its initial point is at R, and let θ be the angle between r and n. Then $0° < \theta < 90°$, so $\cos\theta > 0$. Thus, the distance D is $\cos\theta \, \| r \| = | \cos\theta | \, \| r \|$.

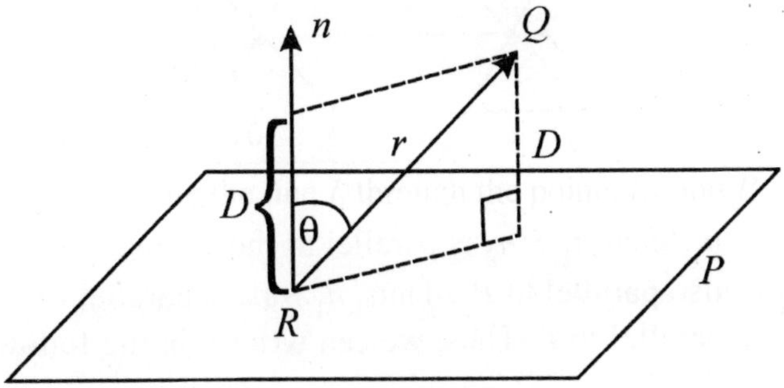

By Theorem, we know that $\cos\theta = \dfrac{n.r}{\| n \| \, \| r \|}$, so

$$D = |\cos\theta| \, \| r \| = \frac{| n.r |}{\| n \| \, \| r \|} \| r \| = \frac{| n.r |}{\| n \|}$$

$$= \frac{| a(x_0 - x) + b(y_0 - y) + c(z_0 - z) |}{\sqrt{a^2 + b^2 + c^2}}$$

$$= \frac{| ax_0 + by_0 + cz_0 - (ax + by + cz) |}{\sqrt{a^2 + b^2 + c^2}}$$

$$= \frac{| ax_0 + by_0 + cz_0 - (-d) |}{\sqrt{a^2 + b^2 + c^2}} = \frac{| ax_0 + by_0 + cz_0 + d |}{\sqrt{a^2 + b^2 + c^2}}$$

If n points away from the side of P where the point Q is located, then $90° < \theta < 180°$ and so $\cos\theta < 0$. The distance D is then $| \cos\theta | \, \| r \|$, and thus repeating the same argument as above still gives the same result.

Example: Find the distance D from (2, 4, –5) to the plane from Example.

Solution: Recall that the plane is given by $5x - 3y + z - 10 = 0$. So

$$D = \frac{| 5(2) - 3(4) + 1(-5) - 10 |}{\sqrt{5^2 + (-3)^2 + 1^2}} = \frac{| -17 |}{\sqrt{35}} = \frac{17}{\sqrt{35}} \approx 2.87$$

Line of Intersection of two Planes

Note that two planes are parallel if they have normal vectors that are parallel, and the planes are perpendicular if their normal vectors are perpendicular. If two planes do intersect, they do so in a line. Suppose that two planes P_1 and P_2 with normal vectors n_1 and n_2, respectively, intersect in a line L.

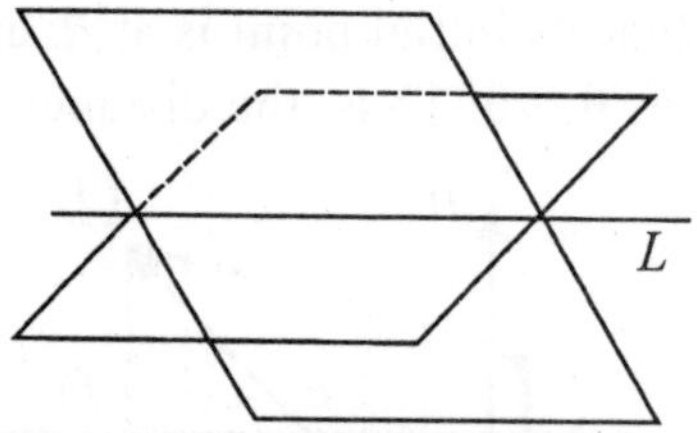

Figure

Since $n_1 \times n_2 \perp n_1$, then $n_1 \times n_2$ is parallel to the plane P_1. Likewise, $n_1 \times n_2 \perp n_2$ means that $n_1 \times n_2$ is also parallel to P_2. Thus, $n_1 \times n_2$ is parallel to the intersection of P_1 and P_2, i.e. $n_1 \times n_2$ is parallel to L. Thus, we can write L in the following vector form:

$$L : r + t(n_1 \times n_2), \text{ for } -\infty < t < \infty$$

where r is any vector pointing to a point belonging to both planes. To find a point in both planes, find a common solution (x, y, z) to the two normal form equations of the planes. This can often be made easier by setting one of the coordinate variables to zero, which leaves you to solve two equations in just two unknowns.

Example: Find the line of intersection L of the planes $5x - 3y + z - 10 = 0$ and $2x + 4y - z + 3 = 0$.

Solution: The plane $5x - 3y + z$ “ $10 = 0$ has normal vector $n_1 = (5, -3, 1)$ and the plane $2x + 4y - z + 3 = 0$ has normal vector $n_2 = (2, 4, -1)$. Since n_1 and n_2 are not scalar multiples, then the two planes are not parallel and hence will intersect. A point (x, y, z) on both planes will satisfy the following system of two equations in three unknowns:

$$5x - 3y + z - 10 = 0$$

$$2x + 4y - z + 3 = 0$$

Set $x = 0$ (why is that a good choice?). Then the above equations are reduced to:

$$-3y + z - 10 = 0$$

$$4y - z + 3 = 0$$

The second equation gives $z = 4y + 3$, substituting that into the first equation gives $y = 7$. Then $z = 31$, and so the point $(0, 7, 31)$ is on L. Since $n_1 \times n_2 = (-1, 7, 26)$, then L is given by:

$$\underline{r} + t(n_1 \times n_2) = (0, 7, 31) + t(-1, 7, 26), \text{ for } -\infty < t < \infty$$

or in parametric form:

$$x = -t,\ y = 7 + 7t,\ z = 31 + 26t, \text{ for } -\infty < t < \infty$$

EXERCISES

For Exercises 1-4, write the line L through the point P and parallel to the vector v in the following forms: (a) vector, (b) parametric, and (c) symmetric.

1. $P = (2, 3, -2)$, $v = (5, 4, -3)$
2. $P = (3, -1, 2)$, $v = (2, 8, 1)$
3. $P = (2, 1, 3)$, $v = (1, 0, 1)$
4. $P = (0, 0, 0)$, $v = (7, 2, -10)$

For Exercises 5-6, write the line L through the points P_1 and P_2 in parametric form.

5. $P_1 = (1, -2, -3)$, $P_2 = (3, 5, 5)$
6. $P_1 = (4, 1, 5)$, $P_2 = (-2, 1, 3)$

For Exercises 7-8, find the distance d from the point P to the line L.

7. $P = (1, -1, -1)$, $L : x = -2 - 2t, y = 4t, z = 7 + t$
8. $P = (0, 0, 0)$, $L : x = 3 + 2t, y = 4 + 3t, z = 5 + 4t$

For Exercises 9-10, find the point of intersection (if any) of the given lines.

9. $x = 7 + 3s, y = -4 - 3s, z = -7 - 5s$ and $x = 1 + 6t, y = 2 + t, z = 3 - 2t$
10. $\dfrac{x-6}{4} = y + 3 = z$ and $\dfrac{x-11}{3} = \dfrac{y-14}{-6} = \dfrac{z+9}{2}$.

For Exercises 11-12, write the normal form of the plane P containing the point Q and perpendicular to the vector n.

11. $Q = (5, 1, -2)$, $n = (4, -4, 3)$
12. $Q = (6, -2, 0)$, $n = (2, 6, 4)$

For Exercises 13-14, write the normal form of the plane containing the given points.

13. (1, 0, 3), (1, 2, –1), (6, 1, 6)
14. (–3, 1, –3), (4, –4, 3), (0, 0, 1)

For Exercises 17-18, find the distance D from the point Q to the plane P.

15. $Q = (4, 1, 2)$, $P : 3x - y - 5z + 8 = 0$
16. $Q = (0, 2, 0)$, $P : -5x + 2y - 7z + 1 = 0$

For Exercises 19-20, find the line of intersection (if any) of the given planes.

17. $x + 3y + 2z - 6 = 0$, $2x - y + z + 2 = 0$
18. $3x + y - 5z = 0$, $x + 2y + z + 4 = 0$
19. Find the point(s) of intersection (if any) of the line $\dfrac{x-6}{4} = y + 3 = z$ with the plane.

Surfaces

In the previous section we discussed planes in Euclidean space. A plane is an example of a *surface*, which we will define informally as the solution set of the equation $F(x, y, z) = 0$ in $\mathbb{R}^3$, for some real-valued function F.

For example, a plane given by $ax + by + cz + d = 0$ is the solution set of $F(x, y, z) = 0$ for the function $F(x, y, z) = ax + by + cz + d$. Surfaces are 2-dimensional. The plane is the simplest surface, since it is "flat". In this section we will look at some surfaces that are more complex, the most important of which are the sphere and the cylinder.

Definition: A sphere S is the set of all points (x, y, z) in $\mathbb{R}^3$ which are a fixed distance r (called the radius) from a fixed point $P_0 = (x_0, y_0, z_0)$ (called the center of the sphere):

$$S = \{(x, y, z) : (x - x_0)^2 + (y - y_0)^2 + (z - z_0)^2 = r^2\}$$

Using vector notation, this can be written in the equivalent form:

$$S = \{x : \| x - x_0 \| = r\}$$

where $x = (x, y, z)$ and $x_0 = (x_0, y_0, z_0)$ are vectors.

Figure illustrates the vectorial approach to spheres.

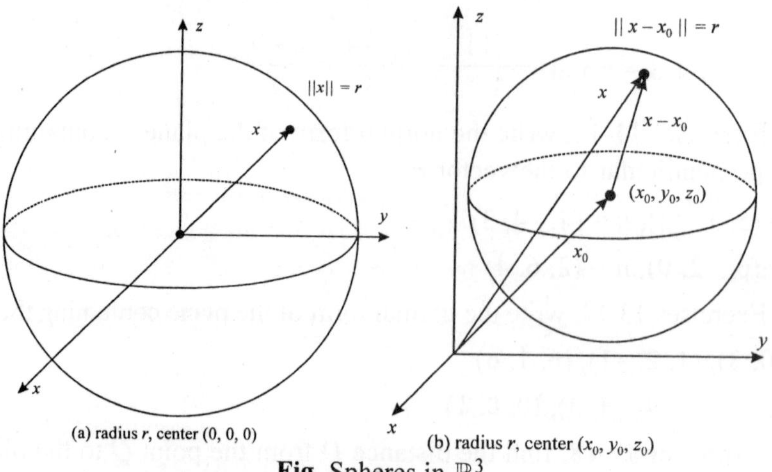

Fig. Spheres in $\mathbb{R}^3$

Note in Figure that the intersection of the sphere with the xy-plane is a circle of radius r (i.e. a *great circle*, given by $x^2 + y^2 = r^2$ as a subset of $\mathbb{R}^2$). Similarly for the intersections with the xz-plane and the yz-plane. In general, a plane intersects a sphere either at a single point or in a circle.

Example : Find the intersection of the sphere $x^2 + y^2 + z^2 = 169$ with the plane $z = 12$.

Solution: The sphere is centered at the origin and has radius $13 = \sqrt{169}$, so it does intersect the plane $z = 12$.

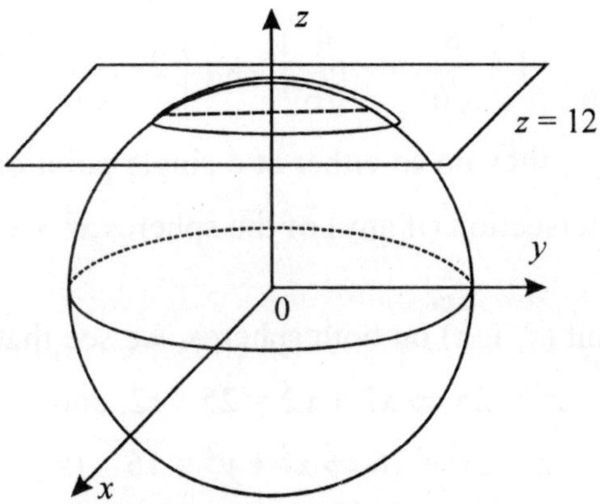

Putting $z = 12$ into the equation of the sphere gives

$$x^2 + y^2 + 122 = 169$$

$$x^2 + y^2 = 169 - 144 = 25 = 5^2$$

which is a circle of radius 5 centered at (0, 0, 12), parallel to the *xy*-plane.

If the equation in formula is multiplied out, we get an equation of the form:

$$x^2 + y^2 + z^2 + ax + by + cz + d = 0$$

for some constants *a*, *b*, *c* and *d*. Conversely, an equation of this form *may* describe a sphere, which can be determined by completing the square for the *x*, *y* and *z* variables.

Example: Is $2x^2 + 2y^2 + 2z^2 - 8x + 4y - 16z + 10 = 0$ the equation of a sphere?

Solution: Dividing both sides of the equation by 2 gives

$$x^2 + y^2 + z^2 - 4x + 2y - 8z + 5 = 0$$

$$(x^2 - 4x + 4) + (y^2 + 2y + 1) + (z^2 - 8z + 16) + 5 - 4 - 1 - 16 = 0$$

$$(x - 2)^2 + (y + 1)2 + (z - 4)^2 = 16$$

which is a sphere of radius 4 centered at (2, –1, 4).

Example: Find the points(s) of intersection (if any) of the sphere from Example and the line $x = 3 + t$, $y = 1 + 2t$, $z = 3 - t$.

Solution: Put the equations of the line into the equation of the sphere, which was $(x - 2)^2 + (y + 1)^2 + (z - 4)^2 = 16$, and solve for *t*:

$$(3 + t - 2)^2 + (1 + 2t + 1)^2 + (3 - t - 4)^2 = 16$$

$$(t + 1)^2 + (2t + 2)^2 + (-t - 1)^2 = 16$$

$$6t^2 + 12t - 10 = 0$$

The quadratic formula gives the solutions $t = -1 \pm \dfrac{4}{\sqrt{6}}$. Putting those two values into the equations of the line gives the following two points of intersection:

$$\left(2+\frac{4}{\sqrt{6}},-1+\frac{8}{\sqrt{6}},4-\frac{4}{\sqrt{6}}\right) \text{ and } \left(2-\frac{4}{\sqrt{6}},-1-\frac{8}{\sqrt{6}},4+\frac{4}{\sqrt{6}}\right)$$

If two spheres intersect, they do so either at a single point or in a circle.

Example: Find the intersection (if any) of the spheres $x^2 + y^2 + z^2 = 25$ and $x^2 + y^2 + (z - 2)^2 = 16$.

Solution: For any point (x, y, z) on both spheres, we see that

$x^2 + y^2 + z^2 = 25 \Rightarrow x2 + y2 = 25$ " $z2$, and

$x^2 + y^2 + (z - 2)^2 = 16 \Rightarrow x^2 + y^2 = 16 - (z - 2)^2$, so

$16 - (z - 2)^2 = 25 - z^2 \Rightarrow 4z - 4 = 9 \Rightarrow z = 13/4$

$\Rightarrow x^2 + y^2 = 25 - (13/4)^2 = 231/16$

$\therefore$ The intersection is the circle $x^2 + y^2 = \dfrac{231}{16}$ of radius $\dfrac{\sqrt{231}}{4} \approx 3.8$ centered at $\left(0, 0, \dfrac{13}{4}\right)$.

The cylinders that we will consider are *right circular cylinders*. These are cylinders obtained by moving a line L along a circle C in $\mathbb{R}^3$ in a way so that L is always perpendicular to the plane containing C. We will only consider the cases where the plane containing C is parallel to one of the three coordinate planes.

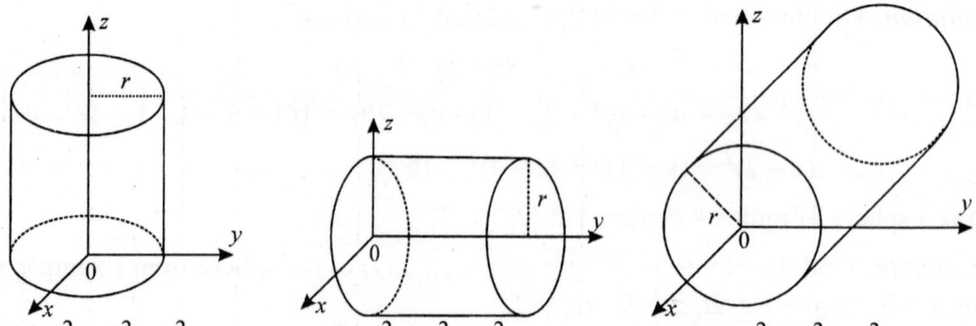

(a) $x^2 + y^2 = r^2$, any z (b) $x^2 + z^2 = r^2$, any y (c) $y^2 + z^2 = r^2$, any x

Fig. Cylinders in $\mathbb{R}^3$

For example, the equation of a cylinder whose base circle C lies in the xy-plane and is centered at $(a, b, 0)$ and has radius r is

$$(x - a)^2 + (y - b)^2 = r^2,$$

where the value of the z coordinate is unrestricted. Similar equations can be written when the base circle lies in one of the other coordinate planes. A plane intersects a right circular cylinder in a circle, ellipse, or one or two lines, depending on whether that plane is parallel, oblique, or perpendicular, respectively, to the plane containing C. The intersection of a surface with a plane is called the trace of the surface.

The equations of spheres and cylinders are examples of *second-degree equations* in $\mathbb{R}^3$, i.e. equations of the form

$$Ax^2 + By^2 + Cz^2 + Dxy + Exz + Fyz + Gx + Hy + Iz + J = 0$$

for some constants $A, B, ..., J$. If the above equation is not that of a sphere, cylinder, plane, line or point, then the resulting surface is called a quadric surface.

One type of quadric surface is the ellipsoid, given by an equation of the form:

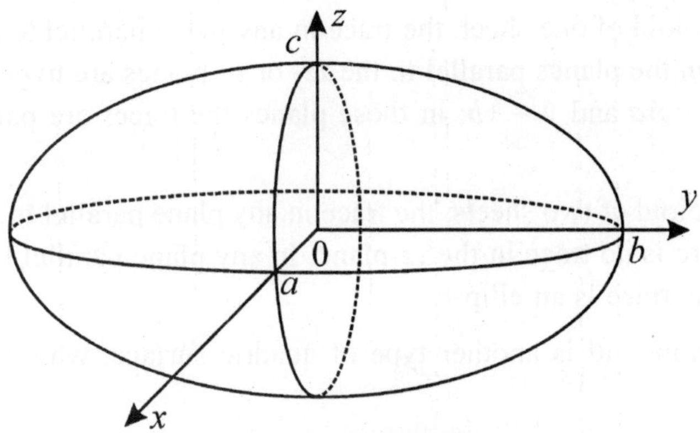

Fig. Ellipsoid

$$\frac{x^2}{a^2} + \frac{y^2}{b^2} + \frac{z^2}{c^2} = 1$$

In the case where $a = b = c$, this is just a sphere. In general, an ellipsoid is egg-shaped (think of an ellipse rotated around its major axis). Its traces in the coordinate planes are ellipses.

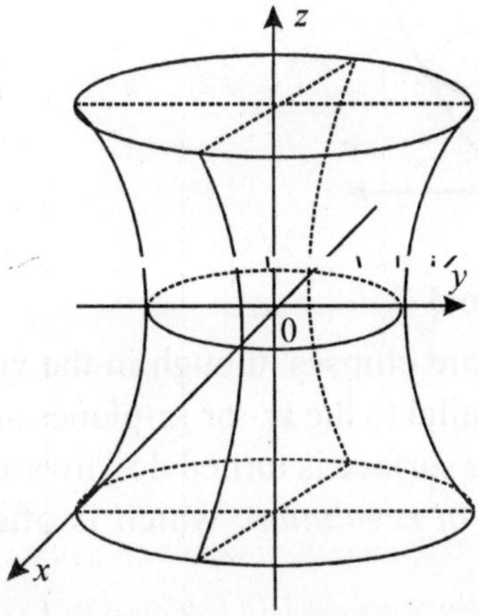

Fig. Hyperboloid of one sheet

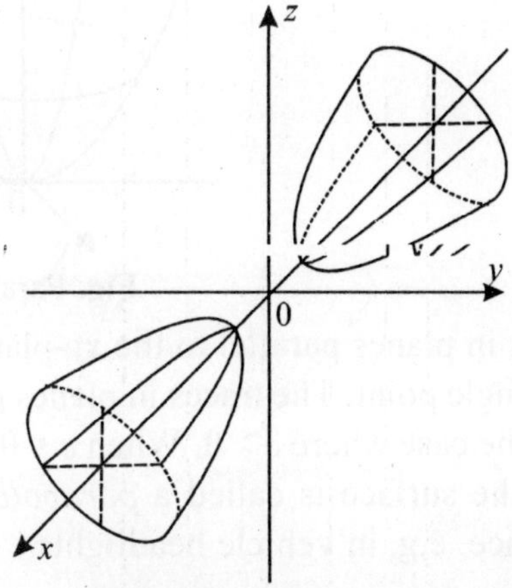

Fig. Hyperboloid of two sheets

Two other types of quadric surfaces are the hyperboloid of one sheet, given by an equation of the form:

$$\frac{x^2}{a^2}+\frac{y^2}{b^2}-\frac{z^2}{c^2}=1$$

and the hyperboloid of two sheets, whose equation has the form:

$$\frac{x^2}{a^2}-\frac{y^2}{b^2}-\frac{z^2}{c^2}=1$$

For the hyperboloid of one sheet, the trace in any plane parallel to the *xy*-plane is an ellipse. The traces in the planes parallel to the *xz*- or *yz*-planes are hyperbolas, except for the special cases $x = \pm a$ and $y = \pm b$; in those planes the traces are pairs of intersecting lines.

For the hyperboloid of two sheets, the trace in any plane parallel to the *xy*- or *xz*plane is a hyperbola. There is no trace in the *yz*-plane. In any plane parallel to the *yz*-plane for which $|x| > |a|$, the trace is an ellipse.

The elliptic paraboloid is another type of quadric surface, whose equation has the form:

$$\frac{x^2}{a^2}+\frac{y^2}{b^2}=\frac{z}{c}$$

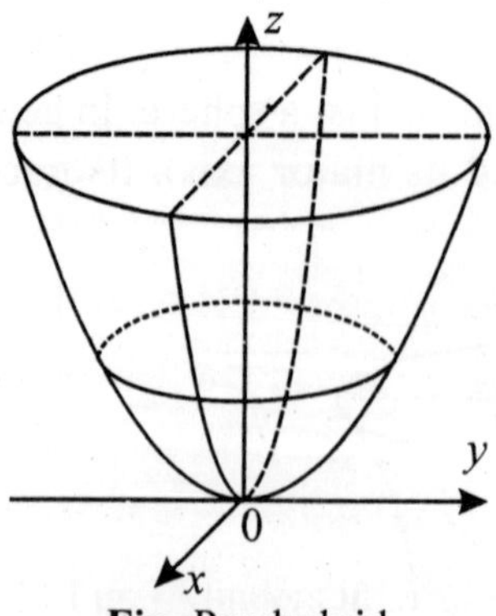

Fig. Paraboloid

The traces in planes parallel to the *xy*-plane are ellipses, though in the *xy*-plane itself the trace is a single point. The traces in planes parallel to the *xz*- or *yz*-planes are parabolas. Figure shows the case where $c > 0$. When $c < 0$ the surface is turned downward. In the case where $a = b$, the surface is called a *paraboloid of revolution*, which is often used as a reflecting surface, e.g. in vehicle headlights.

A more complicated quadric surface is the hyperbolic paraboloid, given by:

$$\frac{x^2}{a^2} - \frac{y^2}{b^2} = \frac{z}{c}$$

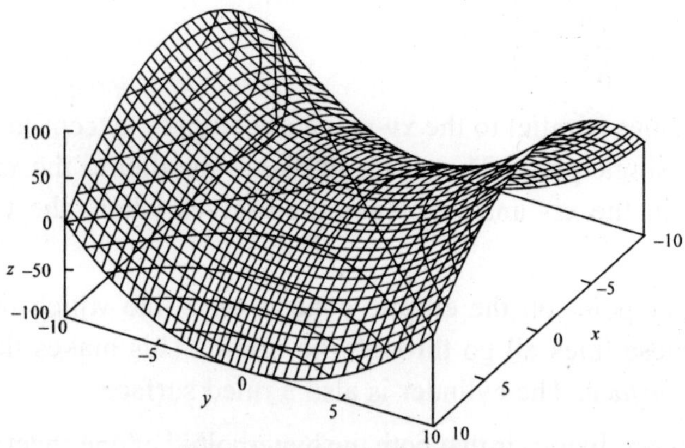

Fig. Hyperbolic Paraboloid

The hyperbolic paraboloid can be tricky to draw; using graphing software on a computer can make it easier. For example, Figure was created using the free Gnuplot package. It shows the graph of the hyperbolic paraboloid $z = y^2 - x^2$, which is the special case where $a = b = 1$ and $c = -1$ in equation. The mesh lines on the surface are the traces in planes parallel to the coordinate planes. So we see that the traces in planes parallel to the xz-plane are parabolas pointing upward, while the traces in planes parallel to the yz-plane are parabolas pointing downward.

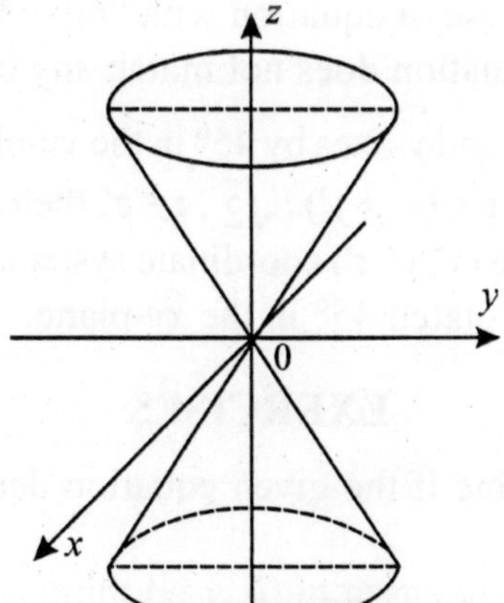

Fig. Elliptic Cone

Also, notice that the traces in planes parallel to the xy-plane are hyperbolas, though in the xy-plane itself the trace is a pair of intersecting lines through the origin. This is true in general when $c < 0$ in equation. When $c > 0$, the surface would be similar to that in Figure, only rotated 90° around the z-axis and the nature of the traces in planes parallel to the xz- or yz-planes would be reversed.

The last type of quadric surface that we will consider is the elliptic cone, which has an equation of the form:

$$\frac{x^2}{a^2}+\frac{y^2}{b^2}-\frac{z^2}{c^2}=0$$

The traces in planes parallel to the *xy*-plane are ellipses, except in the *xy*-plane itself where the trace is a single point. The traces in planes parallel to the *xz*- or *yz*-planes are hyperbolas, except in the *xz*- and *yz*-planes themselves where the traces are pairs of intersecting lines.

Notice that every point on the elliptic cone is on a line which lies entirely on the surface; in Figure these lines all go through the origin. This makes the elliptic cone an example of a *ruled surface*. The cylinder is also a ruled surface.

What may not be as obvious is that both the hyperboloid of one sheet and the hyperbolic paraboloid are ruled surfaces. In fact, on both surfaces there are *two* lines through each point on the surface. Such surfaces are called *doubly ruled surfaces*, and the pairs of lines are called a *regulus*.

It is clear that for each of the six types of quadric surfaces that we discussed, the surface can be translated away from the origin (e.g. by replacing x^2 by $(x - x_0)^2$ in its equation). It can be proved11 that *every* quadric surface can be translated and/or rotated so that its equation matches one of the six types that we described.

For example, $z = 2xy$ is a case of equation with "mixed" variables, e.g. with $D \neq 0$ so that we get an *xy* term. This equation does not match any of the types we considered.

However, by rotating the *x*- and *y*-axes by 45° in the *xy*-plane by means of the coordinate transformation $x = (x'-y')/\sqrt{2}$, $y = (x'+y')/\sqrt{2}$, $z = z'$, then $z = 2xy$ becomes the hyperbolic paraboloid $z' = (x')^2 - (y')^2$ in the (x', y', z') coordinate system. That is, $z = 2xy$ is a hyperbolic paraboloid as in equation, but rotated 45° in the *xy*-plane.

EXERCISES

For Exercises 1-4, determine if the given equation describes a sphere. If so, find its radius and center.

1. $x^2 + y^2 + z^2 - 4x - 6y - 10z + 37 = 0$
2. $x^2 + y^2 + z^2 + 2x - 2y - 8z + 19 = 0$
3. $2x^2 + 2y^2 + 2z^2 + 4x + 4y + 4z - 44 = 0$
4. $x^2 + y^2 - z^2 + 12x + 2y - 4z + 32 = 0$
5. Find the point(s) of intersection of the sphere $(x - 3)^2 + (y + 1)^2 + (z - 3)^2 = 9$ and the line $x = -1 + 2t$, $y = -2 - 3t$, $z = 3 + t$.

6. Find the intersection of the spheres $x^2 + y^2 + z^2 = 9$ and $(x - 4)^2 + (y + 2)^2 + (z - 4)^2 = 9$.
7. Find the intersection of the sphere $x^2 + y^2 + z^2 = 9$ and the cylinder $x^2 + y^2 = 4$.
8. Find the trace of the hyperboloid of one sheet $\frac{x^2}{a^2}+\frac{y^2}{b^2}-\frac{z^2}{c^2}=1$ in the plane $x = a$, and the trace in the plane $y = b$.
9. Find the trace of the hyperbolic paraboloid $\frac{x^2}{a^2}-\frac{y^2}{b^2}=\frac{z}{c}$ in the xy-plane.
10. It can be shown that any four *noncoplanar* points (i.e. points that do not lie in the same plane) determine a sphere. Find the equation of the sphere that passes through the points (0, 0, 0), (0, 0, 2), (1, –4, 3) and (0, –1, 3).
11. Show that the hyperboloid of one sheet is a doubly ruled surface, i.e. each point on the surface is on two lines lying entirely on the surface. as $\frac{x^2}{a^2}-\frac{z^2}{c^2}=1-\frac{y^2}{b^2}$, factor each side. Recall that two planes intersect in a line.)
12. Show that the hyperbolic paraboloid is a doubly ruled surface.

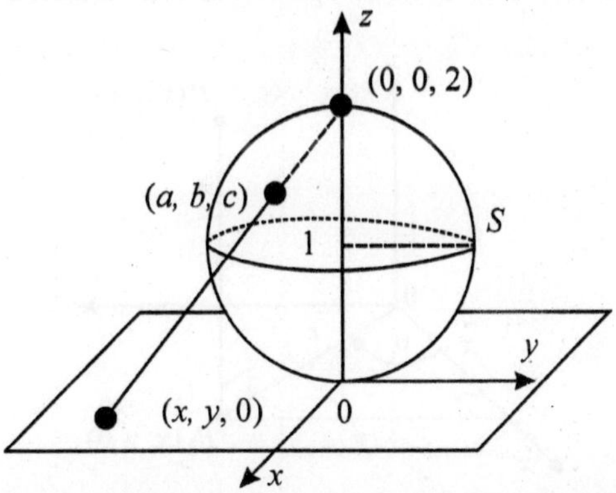

13. Let S be the sphere with radius 1 centered at (0, 0, 1), and let S^* be S without the "north pole" point (0, 0, 2). Let (a, b, c) be an arbitrary point on S^*. Then the line passing through (0, 0, 2) and (a, b, c) intersects the xy-plane at some point $(x, y, 0)$. Find this point $(x, y, 0)$ in terms of a, b and c.

(Note: Every point in the xy-plane can be matched with a point on S^*, and vice versa, in this manner. This method is called *stereographic projection,* which essentially identifies all of $\mathbb{R}^2$ with a "punctured" sphere.)

Curvilinear Coordinates

The Cartesian coordinates of a point (x, y, z) are determined by following a family of straight paths from the origin: first along the x-axis, then parallel to the y-axis, then parallel to the z-axis, as in Figure. In *curvilinear coordinate systems*, these paths can be curved. The two types of curvilinear coordinates which we will consider are cylindrical and spherical coordinates. Instead of referencing a point in terms of sides of a rectangular parallelepiped, as with Cartesian coordinates, we will think of the point as lying on a cylinder or sphere. Cylindrical coordinates are often used when there is symmetry around the z-axis, while spherical coordinates are useful when there is symmetry about the origin.

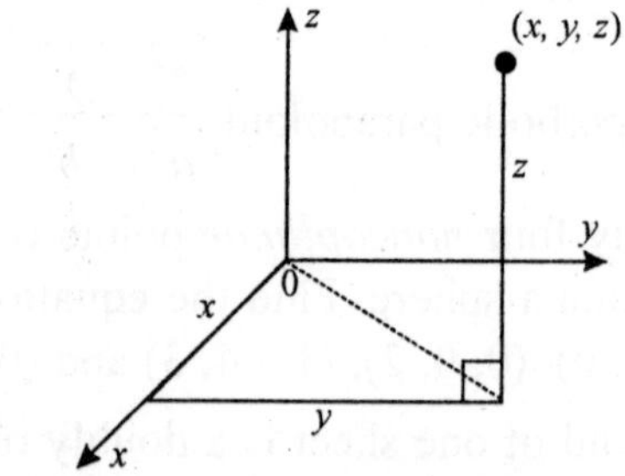

Let $P = (x, y, z)$ be a point in Cartesian coordinates in $\mathbb{R}^3$, and let $P_0 = (x, y, 0)$ be the projection of P upon the xy-plane. Treating (x, y) as a point in $\mathbb{R}^2$, let (r, θ) be its polar coordinates. Let ρ be the length of the line segment from the origin to P, and let ϕ be the angle between that line segment and the positive z-axis. ϕ is called the *zenith angle*. Then the cylindrical coordinates (r, θ, z) and the spherical coordinates (ρ, θ, z) of $P(x, y, z)$ are defined as follows:

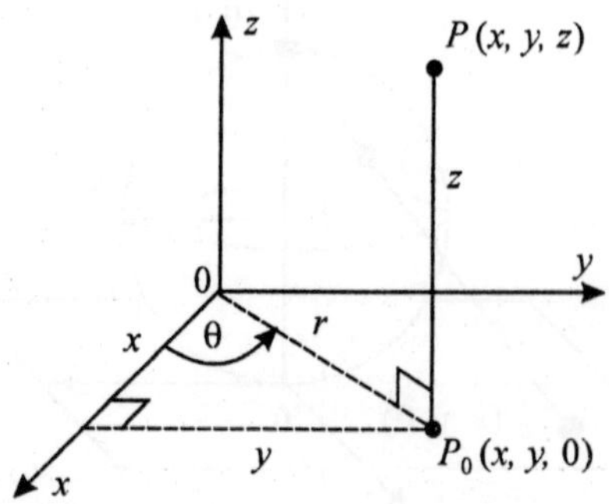

CYLINDRICAL COORDINATES

Cylindrical coordinates (r, θ, z):

$$x = r\cos\theta \qquad r = \sqrt{x^2 + y^2}$$

$$y = r\sin\theta \qquad \theta = \tan^{-1}\left(\frac{y}{x}\right)$$

$$z = z \qquad z = z$$

where $0 \le \theta \le \pi$ if $y \ge 0$ and $\pi < \theta < 2\pi$ if $y < 0$

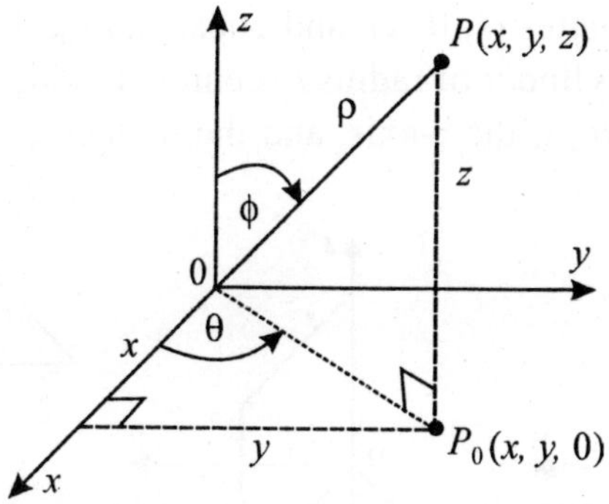

SPHERICAL COORDINATES

Spherical Coordinates (ρ, θ, ϕ):

$$x = \rho \sin \phi \cos \theta \qquad \rho = \sqrt{x^2 + y^2 + z^2}$$

$$y = \rho \sin \phi \sin \theta \qquad \theta = \tan^{-1}\left(\frac{y}{x}\right)$$

$$z = \rho \cos \phi \qquad \phi = \cos^{-1}\left(\frac{z}{\sqrt{x^2 + y^2 + z^2}}\right)$$

where $0 \le \theta \le \pi$ if $y \ge 0$ and $\pi < \theta < 2\pi$ if $y < 0$

Both θ and ϕ aremeasured in radians. Note that $r \ge 0$, $0 \le \theta < 2\pi$, $\rho \ge 0$ and $0 \le \phi \le \pi$. Also, θ is undefined when $(x, y) = (0, 0)$, and ϕ is undefined when $(x, y, z) = (0, 0, 0)$.

Example: Convert the point (–2, –2, 1) from Cartesian coordinates to (a) cylindrical and (b) spherical coordinates.

Solution: (a) $r = \sqrt{(-2)^2 + (-2)^2} = 2\sqrt{2}, \theta = \tan^{-1}\left(\frac{-2}{-2}\right)$

$$\tan^{-1}(1) = \frac{5\pi}{4},$$

since $y = -2 < 0$.

$$\therefore \quad (r, \theta, z) = \left(2\sqrt{2}, \frac{5\pi}{4}, 1\right)$$

(b) $\rho = \sqrt{(-2)^2 + (-2)^2 + 1^2} = \sqrt{9} = 3,$

$$\phi = \cos^{-1}\left(\frac{1}{3}\right) \approx 1.23 \text{ radians.}$$

$$\therefore (\rho, \theta, \phi,) = \left(3, \frac{5\pi}{4}, 1.23\right)$$

For cylindrical coordinates (r, θ, z), and constants r_0, θ_0 and z_0, we see from Figure that the surface $r = r_0$ is a cylinder of radius r_0 centered along the z-axis, the surface $\theta = \theta_0$ is a half-plane emanating from the z-axis, and the surface $z = z_0$ is a plane parallel to the xy-plane.

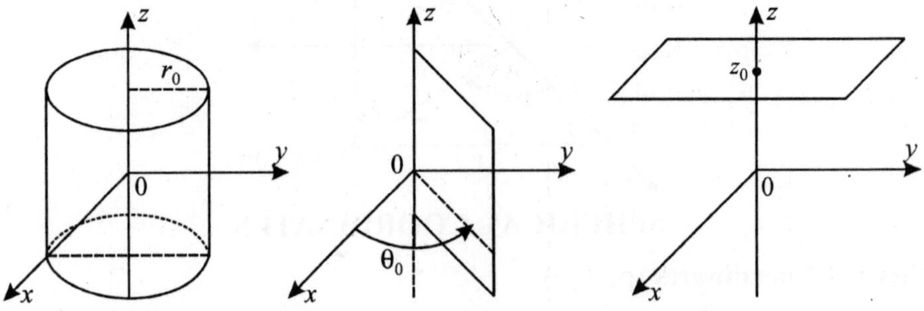

(a) $r = r_0$ (b) $\theta = \theta_0$ (c) $z = z_0$

Fig. Cylindrical Coordinate Surfaces

For spherical coordinates (ρ, θ, ϕ), and constants ρ_0, θ_0 and ϕ_0, we see from Figure that the surface $\rho = \rho_0$ is a sphere of radius ρ_0 centered at the origin, the surface $\theta = \theta_0$ is a half-plane emanating from the z-axis, and the surface $\phi = \phi_0$ is a circular cone whose vertex is at the origin.

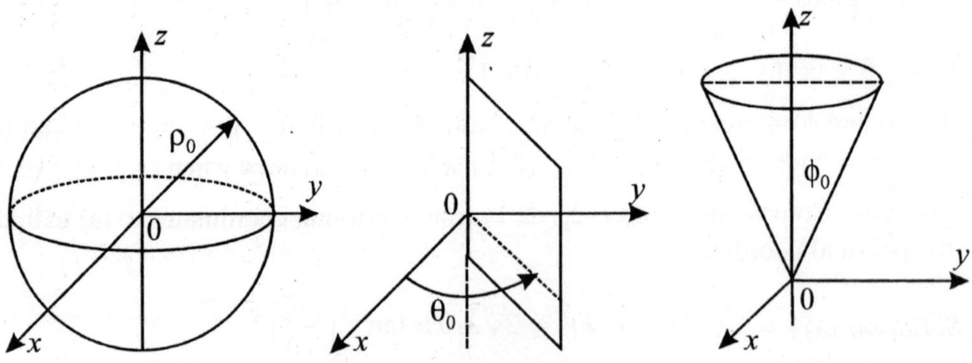

(a) $\rho = \rho_0$ (b) $\theta = \theta_0$ (c) $\phi = \phi_0$

Fig. Spherical Coordinate Surfaces

Figures show how these coordinate systems got their names. Sometimes the equation of a surface in Cartesian coordinates can be transformed into a simpler equation in some other coordinate system, as in the following example.

Example: Write the equation of the cylinder $x^2 + y^2 = 4$ in cylindrical coordinates.

Solution: Since $r = \sqrt{x^2 + y^2}$, then the equation in cylindrical coordinates is $r = 2$.

Using spherical coordinates to write the equation of a sphere does not necessarily make the equation simpler, if the sphere is not centered at the origin.

Example: Write the equation $(x - 2)^2 + (y - 1)^2 + z^2 = 9$ in spherical coordinates.

Solution: Multiplying the equation out gives

$$x^2 + y^2 + z^2 - 4x - 2y + 5 = 9, \text{ so we get}$$

$$\rho^2 - 4\rho \sin\phi \cos\theta - 2\rho \sin\phi \sin\theta - 4 = 0, \text{ or}$$

$$\rho^2 - 2 \sin\phi\, (2\cos\theta - \sin\theta)\rho - 4 = 0$$

after combining terms. Note that this actually makes it more difficult to figure out what the surface is, as opposed to the Cartesian equation where you could immediately identify the surface as a sphere of radius 3 centered at (2, 1, 0).

Example: Describe the surface given by $\theta = z$ in cylindrical coordinates.

Solution: This surface is called a *helicoid.* As the (vertical) z coordinate increases, so does the angle θ, while the radius r is unrestricted. So this sweeps out a (ruled!) surface shaped like a spiral staircase, where the spiral has an infinite radius. Figure shows a section of this surface restricted to $0 \le z \le 4\pi$ and $0 \le r \le 2$.

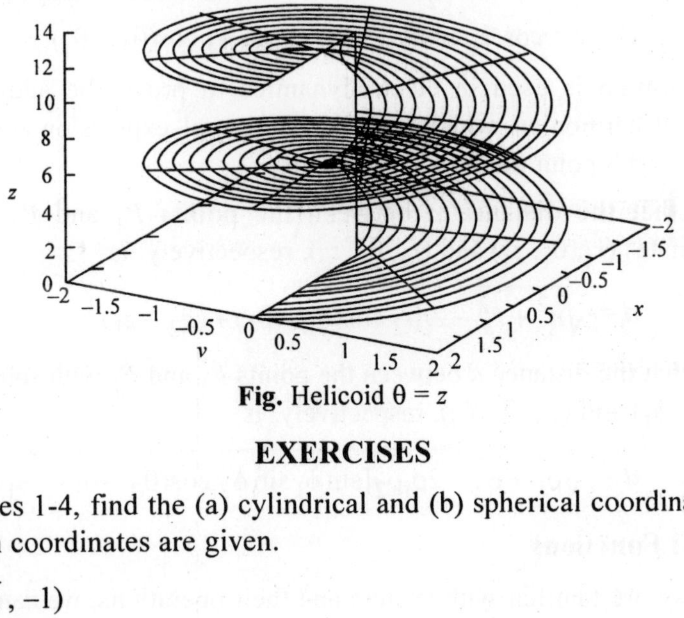

Fig. Helicoid $\theta = z$

EXERCISES

For Exercises 1-4, find the (a) cylindrical and (b) spherical coordinates of the point whose Cartesian coordinates are given.

1. $(2, 2\sqrt{3}, -1)$
2. $(-5, 5, 6)$
3. $(\sqrt{21}, -\sqrt{7}, 0)$
4. $(0, \sqrt{2}, 2)$

For Exercises 5-7, write the given equation in (a) cylindrical and (b) spherical coordinates.

5. $x^2 + y^2 + z^2 = 25$
6. $x^2 + y^2 = 2y$

7. $x^2 + y^2 + 9z^2 = 36$
8. Describe the intersection of the surfaces whose equations in spherical coordinates are $\theta = \frac{\pi}{2}$ and $\phi = \frac{\pi}{4}$.
9. Show that for $a \neq 0$, the equation $\rho = 2a \sin \phi \cos \theta$ in spherical coordinates describes a sphere centered at $(a, 0, 0)$ with radius $|a|$.
10. Let $P = (a, \theta, \phi)$ be a point in spherical coordinates, with $a > 0$ and $0 < \phi < \pi$. Then P lies on the sphere $\rho = a$. Since $0 < \phi < \pi$, the line segment from the origin to P can be extended to intersect the cylinder given by $r = a$ (in cylindrical coordinates). Find the cylindrical coordinates of that point of intersection.
11. Let P_1 and P_2 be points whose spherical coordinates are $(\rho_1, \theta_1, \phi_1)$ and $(\rho_2, \theta_2, \phi_2)$, respectively. Let v_1 be the vector from the origin to P_1, and let v_2 be the vector from the origin to P_2. For the angle ^{3} between v_1 and v_2, show that

$$\cos \gamma = \cos \phi_1 \cos \phi_2 + \sin \phi_1 \sin \phi_2 \cos(\theta_2 - \theta_1).$$

This formula is used in electrodynamics to prove the addition theorem for spherical harmonics, which provides a general expression for the electrostatic potential at a point due to a unit charge.

12. Show that the distance d between the points P_1 and P_2 with cylindrical coordinates (r_1, θ_1, z_1) and (r_2, θ_2, z_2), respectively, is

$$d = \sqrt{r_1^2 + r_2^2 - 2r_1r_2 \cos(\theta_2 - \theta_1) + (z_2 - z_1)^2}$$

13. Show that the distance d between the points P_1 and P_2 with spherical coordinates $(\rho_1, \theta_1, \phi_1)$ and $(\rho_2, \theta_2, \phi_2)$, respectively, is

$$d = \sqrt{\rho_1^2 + \rho_2^2 - 2\rho_1\rho_2[\sin\phi_1 \sin\phi_2 \cos(\theta_2 - \theta_1) + \phi_1 \cos\phi_2;]}$$

Vector-Valued Functions

Now that we are familiar with vectors and their operations, we can begin discussing functions whose values are vectors.

Definition: A vector-valued function of a real variable is a rule that associates a vector $f(t)$ with a real number t, where t is in some subset D of $\mathbb{R}^1$ (called the domain of f). We write $f : D \to \mathbb{R}^3$ to denote that f is a mapping of D into $\mathbb{R}^3$.

For example, $f(t) = ti + t^2j + t^3k$ is a vector-valued function in $\mathbb{R}^3$, defined for all real numbers t. We would write $f : \mathbb{R} \to \mathbb{R}^3$. At $t = 1$ the value of the function is the vector $i + j + k$, which in Cartesian coordinates has the terminal point (1, 1, 1).

A vector-valued function of a real variable can be written in component form as

$$f(t) = f_1(t)i + f_2(t)j + f_3(t)k$$

or in the form

$$f(t) = (f_1(t), f_2(t), f_3(t))$$

for some real-valued functions $f_1(t)$, $f2(t)$, $f3(t)$, called the *component functions* of f. The first form is often used when emphasizing that $f(t)$ is a vector, and the second form is useful when considering just the terminal points of the vectors. By identifying vectors with their terminal points, a curve in space can be written as a vector-valued function.

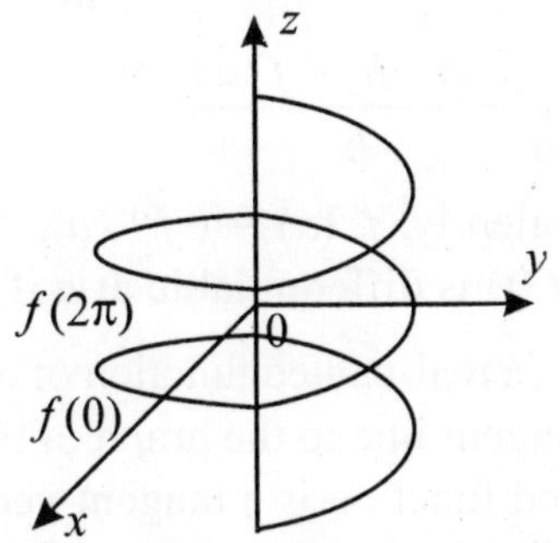

Example: Define $f: \mathbb{R} \to \mathbb{R}^3$ by $f(t) = (\cos t, \sin t, t)$. This is the equation of a *helix*. As the value of t increases, the terminal points of $f(t)$ trace out a curve spiraling upward. For each t, the x- and y-coordinates of $f(t)$ are $x = \cos t$ and $y = \sin t$, so

$$x^2 + y^2 = \cos^2 t + \sin^2 t = 1.$$

Thus, the curve lies on the surface of the right circular cylinder $x^2 + y^2 = 1$.

It may help to think of vector-valued functions of a real variable in $\mathbb{R}^3$ as a generalization of the parametric functions in $\mathbb{R}^2$ which you learned about in single-variable calculus. Much of the theory of real-valued functions of a single real variable can be applied to vector-valued functions of a real variable. Since each of the three component functions are real-valued, it will sometimes be the case that results from singlevariable calculus can simply be applied to each of the component functions to yield a similar result for the vector-valued function. However, there are times when such generalizations do not hold. The concept of a limit, though, can be extended naturally to vector-valued functions, as in the following definition.

Definition: Let $f(t)$ be a vector-valued function, let a be a real number and let c be a vector. Then we say that the limit of $f(t)$ as t approaches a equals c, written as $\lim_{t \to a} f(t) = c$, if $\lim_{t \to a} \| f(t) - c \| = 0$. If $f(t) = (f_1(t), f_2(t), f_3(t))$, then

$$\lim_{t \to a} f(t) = \left(\lim_{t \to a} f_1(t), \lim_{t \to a} f_2(t), \lim_{t \to a} f_3(t) \right)$$

provided that all three limits on the right side exist.

The above definition shows that continuity and the derivative of vector-valued functions can also be defined in terms of its component functions.

Definition: Let $f(t) = (f_1(t), f_2(t), f_3(t))$ be a vector-valued function, and let a be a real number in its domain. Then $f(t)$ is continuous at a if $\lim_{t \to a} f(t) = f(a)$. Equivalently, $f(t)$ is continuous at a if and only if $f_1(t)$, $f_2(t)$, and $f_3(t)$ are continuous at a.

The derivative of $f(t)$ at a, denoted by $f'(a)$ or $\frac{df}{dt}(a)$, is the limit

$$f'(a) = \lim_{h \to 0} \frac{f(a+h) - f(a)}{h}$$

if that limit exists. Equivalently, $f'(a) = (f_1'(a), f_2'(a), f_3'(a))$, if the component derivatives exist. We say that $f(t)$ is differentiable at a if $f'(a)$ exists.

Recall that the derivative of a real-valued function of a single variable is a real number, representing the slope of the tangent line to the graph of the function at a point. Similarly, the derivative of a vector-valued function is a tangent vector to the curve in space which the function represents, and it lies on the *tangent line* to the curve.

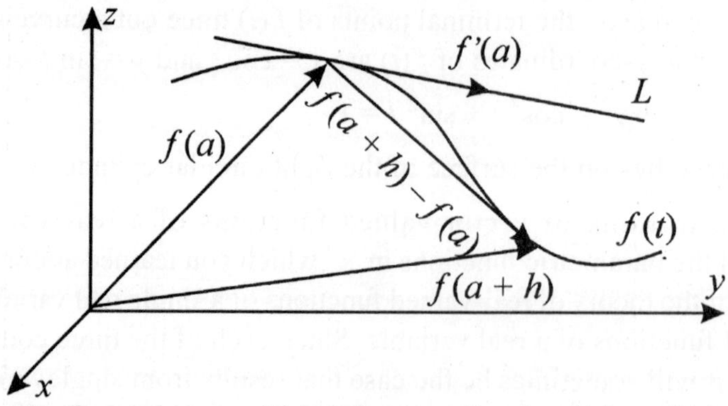

Fig. Tangent vector $f'(a)$ and tangent line $L = f(a) + sf'(a)$

Example: Let $f(t) = (\cos t, \sin t, t)$. Then $f'(t) = (-\sin t, \cos t, 1)$ for all t. The tangent line L to the curve at $f(2\pi) = (1, 0, 2\pi)$ is $L = f'(2\pi) + s f'(2\pi) = (1, 0, 2\pi) + s(0, 1, 1)$, or in parametric form: $x = 1$, $y = s$, $z = 2\pi + s$ for $-\infty < s < \infty$.

A scalar function is a real-valued function. Note that if $u(t)$ is a scalar function and $f(t)$ is a vector-valued function, then their product, defined by $(u f)(t) = u(t) f(t)$ for all t, is a vector-valued function (since the product of a scalar with a vector is a vector).

The basic properties of derivatives of vector-valued functions are summarized in the following theorem.

Theorem: Let $f(t)$ and $g(t)$ be differentiable vector-valued functions, let $u(t)$ be a differentiable scalar function, let k be a scalar, and let c be a constant vector. Then

(a) $\frac{d}{dt}(c) = 0$

(b) $\frac{d}{dt}(kf) = k\frac{df}{dt}$

(c) $\frac{d}{dt}(f + g) = \frac{df}{dt} + \frac{dg}{dt}$

(d) $\frac{d}{dt}(f - g) = \frac{df}{dt} - \frac{dg}{dt}$

(e) $\frac{d}{dt}(uf) = \frac{du}{dt}f + u\frac{df}{dt}$

(f) $\frac{d}{dt}(f.g) = \frac{df}{dt}.g + f.\frac{dg}{dt}$

(g) $\frac{d}{dt}(f \times g) = \frac{df}{dt} \times g + f \times \frac{dg}{dt}$

Proof: The proofs of parts (a)-(e) follow easily by differentiating the component functions and using the rules for derivatives from single-variable calculus. We will prove part (f), and leave the proof of part (g) as an exercise for the reader.

(f) Write $f(t) = (f_1(t), f_2(t), f_3(t))$ and $g(t) = (g_1(t), g_2(t), g_3(t))$, where the component functions $f_1(t), f_2(t), f_3(t), g_1(t), g_2(t), g_3(t)$ are all differentiable real-valued functions. Then

$$\frac{d}{dt}(f(t).g(t)) = \frac{d}{dt}(f_1(t)\, g_1(t) + f_2(t)\, g_2(t) + f_3(t)\, g_3(t))$$

$$= \frac{d}{dt}(f_1(t)g_1(t)) + \frac{d}{dt}(f_2(t)g_2(t) + \frac{d}{dt}(f_3(t)g_3(t))$$

$$= \frac{df_1}{dt}(t)g_1(t) + f_1(t)\frac{dg_1}{dt}(t) + \frac{df_2}{dt}(t)g_2(t) + f_2(t)$$

$$\frac{dg_2}{dt} + \frac{df_3}{dt}(t)g_3(t) + f_3(t)\frac{dg_3}{dt}(t)$$

$$= \left(\frac{df_1}{dt}(t), \frac{df_1}{dt}(t), \frac{df_1}{dt}(t)\right).(g_1(t), g_2(t), g_3(t))$$

$$+ (f_1(t), f_2(t), f_3(t)).\left(\frac{dg_1}{dt}(t), \frac{dg_1}{dt}(t), \frac{dg_1}{dt}(t)\right)$$

$$= \frac{df}{dt}(t).g(t) + f(t).\frac{dg}{dt}(t) \text{ for all } t.$$

Example: Suppose $f(t)$ is differentiable. Find the derivative of $\| f(t) \|$.

Solution: Since $\| f(t) \|$ is a real-valued function of t, then by the Chain Rule for real valued functions, we know that $\frac{d}{dt}\| f(t)\|^2 = 2\| f(t)\| \frac{d}{dt}\| f(t)\|$.

But $\| f(t) \|^2 = f(t).f(t)$, so $\frac{d}{dt}\| f(t)\|^2 = \frac{d}{dt}(f(t).f(t))$. Hence, we have

$2\| f(t)\| \frac{d}{dt}\| f(t)\| = \frac{d}{dt}(f(t).f(t)) = f'(t).f(t) + f(t).f'(t)$ by Theorem, so

$= 2f'(t).f(t)$, so if $\| f(t) \| \neq 0$ then

$$\frac{d}{dt}\| f(t)\| = \frac{f'(t).f(t)}{\| f(t)\|}.$$

We know that $\| f(t) \|$ is constant if and only if $\frac{d}{dt}\| f(t)\| = 0$ for all t. Also, $f(t) \perp f'(t)$ if and only if $f'(t).f(t) = 0$. Thus, the above example shows this important fact:

If $\| f(t) \| \neq 0$, then $\| f(t) \|$ is constant if and only if $f(t) \perp f'(t)$ for all t.

This means that if a curve lies completely on a sphere (or circle) centered at the origin, then the tangent vector $f'(t)$ is always perpendicular to the *position vector* $f(t)$.

Example: The *spherical spiral* $f(t) = \left(\frac{\cos t}{\sqrt{1+a^2t^2}}, \frac{\sin t}{\sqrt{1+a^2t^2}}, \frac{-at}{\sqrt{1+a^2t^2}}\right)$, for $a \neq 0$

Figure shows the graph of the curve when $a = 0.2$. In the exercises, the reader will be asked to show that this curve lies on the sphere $x^2 + y^2 + z^2 = 1$ and to verify directly that $f'(t).f(t) = 0$ for all t.

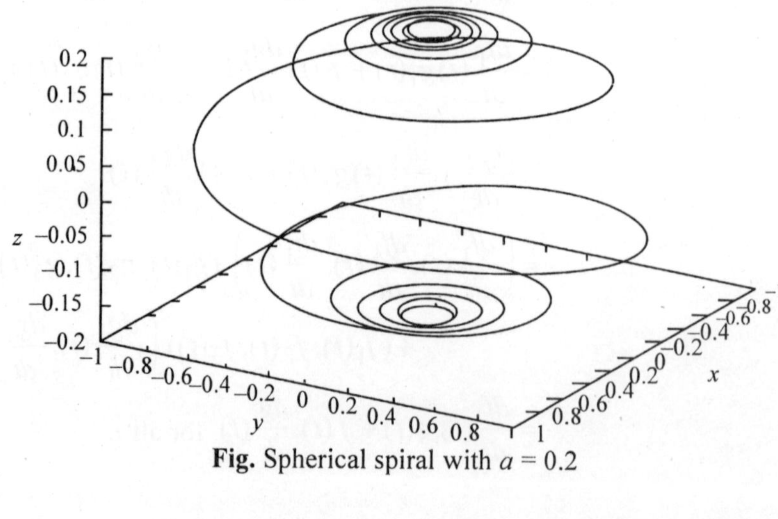

Fig. Spherical spiral with $a = 0.2$

Just as in single-variable calculus, higher-order derivatives of vector-valued functions are obtained by repeatedly differentiating the (first) derivative of the function:

$$f''(t) = \frac{d}{dt} f'(t), f'''(t) = \frac{d}{dt} f''(t), \ldots, \frac{d^n f}{dt^n} = \frac{d}{dt}\left(\frac{d^{n-1} f}{dt^{n-1}}\right)$$

(for $n = 2, 3, 4 \ldots$)

We can use vector-valued functions to represent physical quantities, such as velocity, acceleration, force, momentum, etc. For example, let the real variable t represent time elapsed from some initial time ($t = 0$), and suppose that an object of constant mass m is subjected to some force so that it moves in space, with its position (x, y, z) at time t a function of t. That is, $x = x(t), y = y(t), z = z(t)$ for some real-valued functions $x(t), y(t), z(t)$. Call $r(t) = (x(t), y(t), z(t))$ the position vector of the object. We can define various physical quantities associated with the object as follows:

Position: $r(t) = (x(t), y(t), z(t))$

Velocity: $v(t) = \dot{r}(t) = r'(t) = \dfrac{dr}{dt}$

$= (x'(t), y'(t), z'(t))$

Acceleration: $a(t) = \dot{v}(t) = v'(t) = \dfrac{dv}{dt}$

$= \ddot{r}(t) = r''(t) = \dfrac{d^2 r}{dt^2}$

$= (x''(t), y''(t), z''(t))$

Momentum: $p(t) = mv(t)$

Force: $F(t) = p(t) = p'(t) = \dfrac{dp}{dt}$ (Newton's Second Law of Motion)

The magnitude $\| v(t) \|$ of the velocity vector is called the *speed* of the object. Note that since the mass m is a constant, the force equation becomes the familiar $F(t) = ma(t)$.

Example: Let $r(t) = (5 \cos t, 3 \sin t, 4 \sin t)$ be the position vector of an object at time $t \geq 0$. Find its (a) velocity and (b) acceleration vectors.

Solution: (a) $v(t) = r(t) = (-5 \sin t, 3 \cos t, 4 \cos t)$

(b) $a(t) = v(t) = (-5 \cos t, -3 \sin t, -4 \sin t)$

Note that $\| r(t) \| = \sqrt{25\cos^2 t + 25\sin^2 t} = 5$ for all t, so by Example we know that $r(t) \cdot \dot{r}(t) = 0$ for all t (which we can verify from part (a)). In fact, $\| v(t) \| = 5$ for all t also.

And not only does $r(t)$ lie on the sphere of radius 5 centered at the origin, but perhaps not so obvious is that it lies completely within a *circle* of radius 5 centered at the origin. Also, note that $a(t) = -r(t)$. It turns out that whenever an object moves in a circle with constant speed, the acceleration vector will point in the opposite direction of the position vector (i.e. towards the center of the circle).

Recall that if r_1, r_2 are position vectors to distinct points then $r_1 + t(r_2 - r_1)$ represents a line through those two points as t varies over all real numbers. That vector sum can be written as $(1 - t)r_1 + tr_2$. So the function $l(t) = (1 - t)r_1 + tr_2$ is a line through the terminal points of r_1 and r_2, and when t is restricted to the interval [0, 1] it is the line segment between the points, with $l(0) = r_1$ and $l(1) = r_2$.

In general, a function of the form $f(t) = (a_1t + b_1, a_2t + b_2, a_3t + b_3)$ represents a line in $\mathbb{R}^3$. A function of the form $f(t) = (a_1t^2 + b_1t + c_1, a_2t^2 + b_2t + c_2, a_3t^2 + b_3t + c_3)$ represents a (possibly degenerate) parabola in $\mathbb{R}^3$.

Example: *Bezier curves* are used in Computer Aided Design (CAD) to approximate the shape of a polygonal path in space (called the *Bézier polygon* or *control polygon*). For instance, given three points (or position vectors) b_0, b_1, b_2 in $\mathbb{R}^3$, define

$$b_0^1(t) = (1-t)b_0 + tb_1$$

$$b_1^1(t) = (1-t)b_1 + tb_2$$

$$b_0^2(t) = (1-t)b_0^1(t) + tb_1^1(t)$$

$$= (1 - t)^2 b_0 + 2r(1 - t)b_1 + t^2 b_2$$

for all real t. For t in the interval [0, 1], we see that $b_0^1(t)$ is the line segment between b_0 and b_1, and $b_1^1(t)$ is the line segment between b_1 and b_2. The function $b_0^2(t)$ is the Bezier curve for the points b_0, b_1, b_2. Note from the last formula that the curve is a parabola that goes through b_0 (when $t = 0$) and b_2 (when $t = 1$).

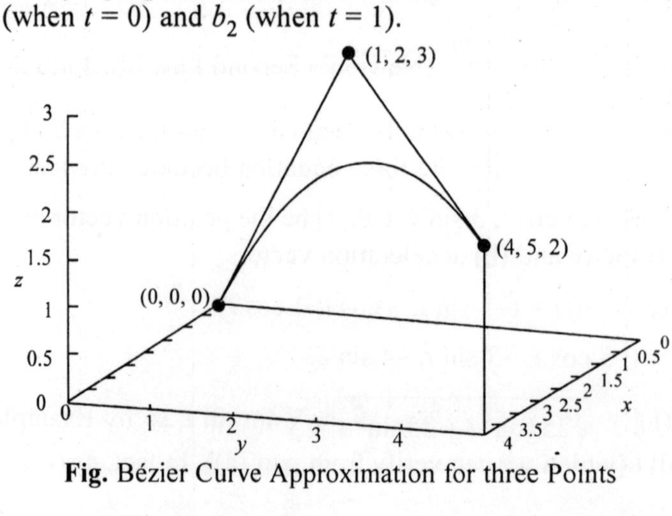

Fig. Bézier Curve Approximation for three Points

As an example, let $b_0 = (0, 0, 0)$, $b_1 = (1, 2, 3)$, and $b_2 = (4, 5, 2)$. Then the explicit formula for the Bezier curve is $b_0^2(t) = (2t + 2t^2, 4t + t^2, 6t - 4t^2)$, as shown in Figure, where the line segments are $b_0^1(t)$ and $b_1^1(t)$, and the curve is $b_0^2(t)$.

In general, the polygonal path determined by $n \geq 3$ noncollinear points in $\mathbb{R}^3$ can be used to define the Bézier curve recursively by a process called *repeated linear interpolation*. This curve will be a vector-valued function whose components are polynomials of degree $n - 1$, and its formula is given by *de Casteljau's algorithm*. In the exercises, the reader will be given the algorithm for the case of $n = 4$ points and asked to write the explicit formula for the Bézier curve for the four points shown in Figure.

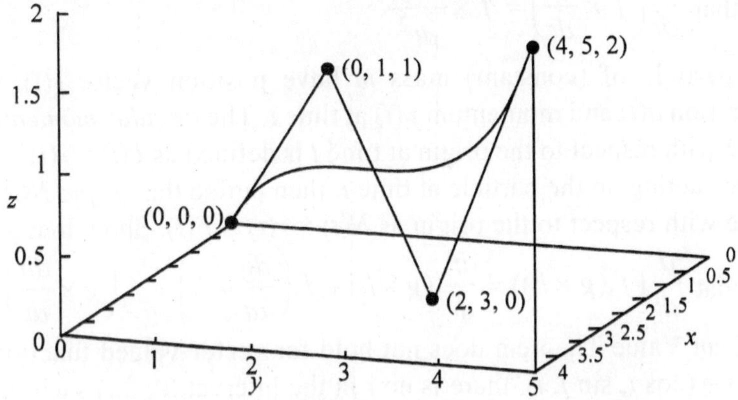

Fig. Bézier Curve Approximation for four Points

EXERCISES

For Exercises 1-4, calculate $f'(t)$ and find the tangent line at $f(0)$.

1. $f(t) = (t + 1, t^2 + 1, t^3 + 1)$
2. $f(t) = (e^t + 1, e^{2t} + 1, e^{t^2} + 1)$
3. $f(t) = (\cos 2t, \sin 2t, t)$
4. $f(t) = (\sin 2t, 2\sin^2 t, 2\cos t)$

For Exercises 5-6, find the velocity $v(t)$ and acceleration $a(t)$ of an object with the given position vector $r(t)$.

5. $r(t) = (t, t - \sin t, 1 - \cos t)$
6. $r(t) = (3\cos t, 2\sin t, 1)$
7. Let $f(t) = \left(\dfrac{\cos t}{\sqrt{1+a^2t^2}}, \dfrac{\sin t}{\sqrt{1+a^2t^2}}, \dfrac{-at}{\sqrt{1+a^2t^2}}\right)$, with $a \neq 0$.

   (a) Show that $\| f(t) \| = 1$ for all t.

(b) Show directly that $f'(t) \cdot f(t) = 0$ for all t.

8. If $f'(t) = 0$ for all t in some interval (a, b), show that $f(t)$ is a constant vector in (a, b).

9. For a constant vector $c \neq 0$, the function $f(t) = tc$ represents a line parallel to c.
   (a) What kind of curve does $g(t) = t^3 c$ represent? Explain.
   (b) What kind of curve does $h(t) = e^t c$ represent? Explain.
   (c) Compare $f'(0)$ and $g'(0)$. Given your answer to part (a), how do you explain the difference in the two derivatives?

10. Show that $\frac{d}{dt}\left(f \times \frac{df}{dt}\right) = f \times \frac{d^2 f}{dt^2}$.

11. Let a particle of (constant) mass m have position vector $r(t)$, velocity $v(t)$, acceleration $a(t)$ and momentum $p(t)$ at time t. The *angular momentum* $L(t)$ of the particle with respect to the origin at time t is defined as $L(t) = r(t) \times p(t)$. If $F(t)$ is the force acting on the particle at time t, then define the *torque* $N(t)$ acting on the particle with respect to the origin as $N(t) = r(t) \times F(t)$. Show that $L'(t) = N(t)$.

12. Show that $\frac{d}{dt}(f.(g \times h)) = \frac{df}{dt}.(g \times h) + f.\left(\frac{dg}{dt} \times h\right) + f.\left(g \times \frac{dh}{dt}\right)$.

13. The Mean Value Theorem does not hold for vector-valued functions: Show that for $f(t) = (\cos t, \sin t, t)$, there is no t in the interval $(0, 2\pi)$ such that

$$f'(t) = \frac{f(2\pi) - f(0)}{2\pi - 0}.$$

14. The Bézier curve $b_0^3(t)$ for four noncollinear points b_0, b_1, b_2, b_3 in $\mathbb{R}^3$ is defined by the following algorithm (going from the left column to the right):

$$b_0^1(t) = (1-t)b_0 + tb_1 \qquad b_0^2(t) = (1-t)b_0^1(t) + tb_1^1(t)$$

$$b_0^3(t) = (1-t)b_0^2(t) + tb_1^2(t) \qquad b_1^1(t) = (1-t)b_1 + tb_2$$

$$b_1^2(t) = (1-t)b_1^1(t) + tb_2^1(t) \qquad b_2^1(t) = (1-t)b_2(t) + tb_3$$

   (a) Show that $b_0^3(t) = (1-t)b_0 + 3t(1-t)^2 b_1 + 3t^2(1-t)b_2 + t^3 b_3$.
   (b) Write the explicit formula for the Bezier curve for the points $b_0 = (0, 0, 0)$, $b_1 = (0, 1, 1)$, $b_2 = (2, 3, 0)$, $b_3 = (4, 5, 2)$.

15. Let $r(t)$ be the position vector for a particle moving in ¡3. Show that

$$\frac{d}{dt}(r \times (v \times r)) = \| r \|^2 a + (r \,.\, v)v - (\| v \|^2 + r \,.\, a)r.$$

16. Let r(t) be the position vector in $\mathbb{R}^3$ for a particle that moves with constant speed $c > 0$ in a circle of radius $a > 0$ in the xy-plane. Show that a(t) points in the opposite direction as $r(t)$ for all t.

Arc Length

Let $r(t) = (x(t), y(t), z(t))$ be the position vector of an object moving in $\mathbb{R}^3$. Since $\| v(t) \|$ is the speed of the object at time t, it seems natural to define the distance s traveled by the object from time $t = a$ to $t = b$ as the definite integral

$$s = \int_a^b \| v(t) \| \, dt = \int_a^b \sqrt{x'(t)^2 + y'(t)^2 + z'(t)^2 \, dt},$$

which is analogous to the case from single-variable calculus for parametric functions in $\mathbb{R}^2$. This is indeed how we will define the distance traveled and, in general, the arc length of a curve in $\mathbb{R}^3$.

Definition: Let $f(t) = (x(t), y(t), z(t))$ be a curve in $\mathbb{R}^3$ whose domain includes the interval $[a, b]$. Suppose that in the interval (a, b) the first derivative of each component function $x(t)$, $y(t)$ and $z(t)$ exists and is continuous, and that no section of the curve is repeated. Then the arc length L of the curve from $t = a$ to $t = b$ is

$$s = \int_a^b \| f'(t) \| \, dt = \int_a^b \sqrt{x'(t)^2 + y'(t)^2 + z'(t)^2 \, dt}$$

A real-valued function whose first derivative is continuous is called *continuously differentiable* (or a C^1 function), and a function whose derivatives of all orders are continuous is called *smooth* (or a C^∞ function). All the functions we will consider will be smooth. A *smooth curve* $f(t)$ is one whose derivative $f'(t)$ is never the zero vector and whose component functions are all smooth.

Note that we did not *prove* that the formula in the above definition actually gives the length of a section of a curve. A rigorous proof requires dealing with some subtleties, normally glossed over in calculus texts, which are beyond the scope of this book.

Example: Find the length L of the helix $f(t) = (\cos t, \sin t, t)$ from $t = 0$ to $t = 2\pi$.

Solution: By formula, we have

$$L = \int_0^{2\pi} \sqrt{(-\sin t)^2 + (\cos t)^2 + 1^2 \, dt} = \int_0^{2\pi} \sqrt{\sin^2 t + \cos^2 + 1 dt}$$

$$\int_0^{2\pi} \sqrt{2} dt = \sqrt{2}(2\pi - 0) = 2\sqrt{2}\pi$$

Similar to the case in $\mathbb{R}^2$, if there are values of t in the interval $[a, b]$ where the derivative of a component function is not continuous then it is often possible to partition $[a, b]$ into subintervals where all the component functions are continuously differentiable (except at the endpoints, which can be ignored). The sum of the arc lengths over the subintervals will be the arc length over $[a, b]$.

Notice that the curve traced out by the function $f(t) = (\cos t, \sin t, t)$ from Example is also traced out by the function $g(t) = (\cos 2t, \sin 2t, 2t)$.

For example, over the interval $[0, \pi]$, $g(t)$ traces out the same section of the curve as $f(t)$ does over the interval $[0, 2\pi]$. Intuitively, this says that $g(t)$ traces the curve twice as fast as $f(t)$. This makes sense since, viewing the functions as position vectors and their derivatives as velocity vectors, the speeds of $f(t)$ and $g(t)$ are $\| f'(t) \| = \sqrt{2}$ and $\| g'(t) \| = 2\sqrt{2}$, respectively. We say that $g(t)$ and $f(t)$ are different *parametrizations* of the same curve.

Definition: Let C be a smooth curve in $\mathbb{R}^3$ represented by a function $f(t)$ defined on an interval $[a, b]$, and let $\alpha : [c, d] \to [a, b]$ be a smooth one-to-one mapping of an interval $[c, d]$ onto $[a, b]$. Then the function $g : [c, d] \to \mathbb{R}^3$ defined by $g(s) = f(\alpha(s))$ is a parametrization of C with parameter s. If α is strictly increasing on $[c, d]$ then we say that $g(s)$ is *equivalent* to $f(t)$.

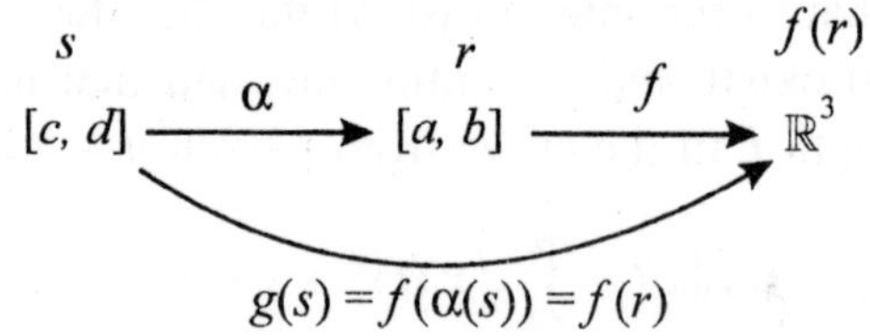

Note that the differentiability of $g(s)$ follows from a version of the Chain Rule for vector-valued functions.

Theorem Chain Rule: If $f(t)$ is a differentiable vector-valued function of t, and $t = \alpha(s)$ is a differentiable scalar function of s, then $f(s) = f(\alpha(s))$ is a differentiable vector-valued function of s, and

$$\frac{df}{ds} = \frac{df}{dt}\frac{dt}{ds}$$

for any s where the composite function $f(\alpha(s))$ is defined.

Example: The following are all equivalent parametrizations of the same curve:

$$f(t) = (\cos t, \sin t, t) \text{ for } t \text{ in } [0, 2\pi]$$

$$g(s) = (\cos 2s, \sin 2s, 2s) \text{ for } s \text{ in } [0, \pi]$$

$$h(s) = (\cos 2\pi s, \sin 2\pi s, 2\pi s) \text{ for } s \text{ in } [0, 1]$$

To see that $g(s)$ is equivalent to $f(t)$, define $\alpha : [0, \pi] \to [0, 2\pi]$ by $\alpha(s) = 2s$. Then α is smooth, one-to-one, maps $[0, \pi]$ onto $[0, 2\pi]$, and is strictly increasing (since $\alpha'(s) = 2 > 0$ for all s). Likewise, defining $\alpha : [0, 1] \to [0, 2\pi]$ by $\alpha(s) = 2\pi s$ shows that $h(s)$ is equivalent to $f(t)$.

A curve can have many parametrizations, with different speeds, so which one is the best to use? In some situations the arc length parametrization can be useful. The idea behind this is to replace the parameter t, for any given smooth parametrization $f(t)$ defined on $[a, b]$, by the parameter s given by

$$s = s(t) = \int_a^t \| f'(u) \| \, du.$$

In terms of motion along a curve, s is the distance traveled along the curve after time t has elapsed. So the new parameter will be distance instead of time. There is a natural correspondence between s and t: from a starting point on the curve, the distance traveled along the curve (in one direction) is uniquely determined by the amount of time elapsed, and vice versa.

Since s is the arc length of the curve over the interval $[a, t]$ for each t in $[a, b]$, then it is a function of t. By the Fundamental Theorem of Calculus, its derivative is

$$s'(t) = \frac{ds}{dt} = \frac{d}{st} \int_a^t \| f'(u) \| \, du = \| f'(t) \| \text{ for all } t \text{ in } [a, b].$$

Since $f(t)$ is smooth, then $\| f'(t) \| > 0$ for all t in $[a, b]$. Thus $s'(t) > 0$ and hence $s(t)$ is strictly increasing on the interval $[a, b]$. Recall that this means that s is a one-to-one mapping of the interval $[a, b]$ onto the interval $[s(a), s(b)]$. But we see that

$$s(a) = \int_a^a \| f'(u) \| \, du = 0 \text{ and } s(b) = \int_a^b \| f'(u) \| \, du = L$$

$= \text{arc length from } t = a \text{ to } t = b$

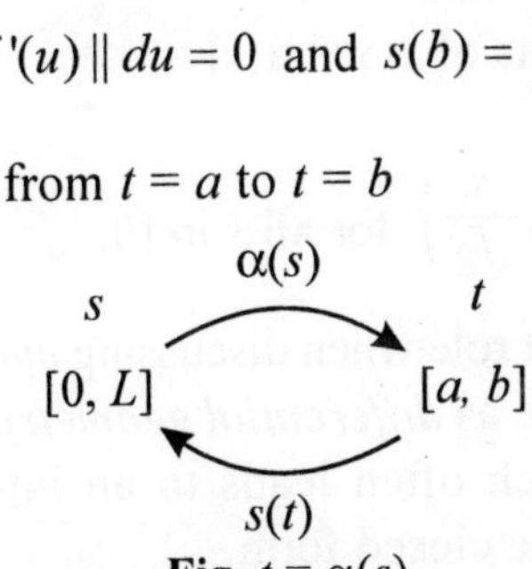

Fig. $t = \alpha(s)$

So the function $s : [a, b] \to [0, L]$ is a one-to-one, differentiable mapping onto the interval $[0, L]$. From single-variable calculus, we know that this means that there exists an inverse function $\alpha : [0, L] \to [a, b]$ that is differentiable and the inverse of $s : [a, b] \to [0, L]$. That is, for each t in $[a, b]$ there is a unique s in $[0, L]$ such that $s = s(t)$ and $t = (s)$. And we know that the derivative of α is

$$\alpha'(s) = \frac{1}{s'(\alpha(s))} = \frac{1}{\| f'(\alpha(s)) \|}$$

So define the arc length parametrization $f : [0, L] \to \mathbb{R}^3$ by

$$f(s) = f(\alpha(s)) \text{ for all } s \text{ in } [0, L].$$

Then $f(s)$ is smooth, by the Chain Rule. In fact, $f(s)$ has *unit speed*:

$f'(s) = f'(\alpha(s))\,\alpha'(s)$ by the Chain Rule, so

$$= f'(\alpha(s))\frac{1}{\| f'(\alpha(s)) \|}, \text{ so}$$

$\| f'(\alpha(s)) \| = 1$ for all s in $[0, L]$.

So the arc length parametrization traverses the curve at a "normal" rate.

In practice, parametrizing a curve $f(t)$ by arc length requires you to evaluate the integral $s = \int_a^t \| f'(u) \|\, du$ in some closed form (as a function of t) so that you could then solve for t in terms of s.

If that can be done, you would then substitute the expression for t in terms of s (which we called $\alpha(s)$) into the formula for $f(t)$ to get $f(s)$.

Example: Parametrize the helix $f(t) = (\cos t, \sin t, t)$, for t in $[0, 2\pi]$, by arc length.

Solution: We have

$$s = \int_0^t \| f'(u) \|\, du = \int_0^t \sqrt{2}\, du = \sqrt{2}t \text{ for all } t \text{ in } [0, 2\pi].$$

So we can solve for t in terms of s: $t = \alpha(s) = \dfrac{s}{\sqrt{2}}$.

$$\therefore \quad f(s) = \left(\cos\frac{s}{\sqrt{2}}, \sin\frac{s}{\sqrt{2}}, \frac{s}{\sqrt{2}}\right) \text{ for all } s \text{ in } [0, \sqrt{2}\pi]. \text{ Note that } \| f'(s) \| = 1.$$

Arc length plays an important role when discussing *curvature* and *moving frame fields*, in the field of mathematics known as *differential geometry*. The methods involve using an arc length parametrization, which often leads to an integral that is either difficult or impossible to evaluate in a simple closed form.

The simple integral in Example is the exception, not the norm. In general, arc length parametrizations are more useful for theoretical purposes than for practical computations. Curvature and moving frame fields can be defined without using arc length, which makes their computation much easier, and these definitions can be shown to be equivalent to those using arc length. We will leave this to the exercises.

The arc length for curves given in other coordinate systems can also be calculated:

Theorem: Suppose that $r = r(t)$, $\theta = \theta(t)$ and $z = z(t)$ are the cylindrical coordinates of a curve $f(t)$, for t in $[a, b]$. Then the arc length L of the curve over $[a, b]$ is

$$L = \int_a^b \sqrt{r'(t)^2 + r(t)^2\theta'(t)^2 + z'(t)^2}\, dt$$

Proof: The Cartesian coordinates $(x(t), y(t), z(t))$ of a point on the curve are given by

$$x(t) = r(t) \cos \theta(t), y(t) = r(t) \sin \theta(t), z(t) = z(t)$$

so differentiating the above expressions for $x(t)$ and $y(t)$ with respect to t gives

$$x'(t) = r'(t) \cos \theta(t) - r(t) \theta'(t) \sin \theta(t), y'(t)$$
$$= r'(t) \sin \theta(t) + r(t)\theta'(t) \cos \theta(t)$$

and so

$$\begin{aligned} x'(t)^2 + y'(t)^2 &= (r'(t) \cos \theta(t) - r(t)\theta'(t) \sin \theta(t))^2 \\ &\quad + (r'(t) \sin \theta(t) + r(t)\theta'(t) \cos \theta(t))^2 \\ &= r'(t)^2(\cos^2 \theta + \sin^2 \theta) + r(t)^2 \theta'(t)^2(\cos^2 \theta + \sin^2 \theta) \\ &\quad - 2r'(t)r(t)\theta'(t) \cos \theta \sin \theta + 2r'(t)r(t)\theta'(t) \cos \theta \sin \theta \\ &= r'(t)^2 + r(t)^2 \theta'(t)^2, \text{ and so} \end{aligned}$$

$$\begin{aligned} L &= \int_a^b \sqrt{x'(t)^2 + y'(t)^2 + z'(t)^2}\, dt \\ &= \int_a^b \sqrt{r'(t)^2 + r(t)^2\theta'(t)^2 + z'(t)^2}\, dt \end{aligned}$$

Example: Find the arc length L of the curve whose cylindrical coordinates are $r = e^t$, $\theta = t$ and $z = e^t$, for t over the interval $[0, 1]$.

Solution: Since $r'(t) = et$, $\theta'(t) = 1$ and $z'(t) = e^t$, then

$$\begin{aligned} L &= \int_0^1 \sqrt{r'(t)^2 + r(t)^2\theta'(t)^2 + z'(t)^2}\, dt \\ &= \int_0^1 \sqrt{e^{2t} + e^{2t}(1) + e^{2t}}\, dt \\ &= \int_0^1 e^t \sqrt{3} dt = \sqrt{3}(e-1) \end{aligned}$$

EXERCISES

For Exercises 1-3, calculate the arc length of $f(t)$ over the given interval.

1. $f(t) = (3 \cos 2t, 3 \sin 2t, 3t)$ on $[0, \pi / 2]$
2. $f(t) = ((t^2 + 1) \cos t, (t^2 + 1) \sin t, 2\sqrt{2}\, t)$ on $[0, 1]$
3. $f(t) = (2 \cos 3t, 2 \sin 3t, 2t^{3/2})$ on $[0, 1]$
4. Let $f(t)$ be a differentiable curve such that $f(t) \neq 0$ for all t. Show that

$$\frac{d}{dt}\left(\frac{f(t)}{\| f(t) \|}\right) = \frac{f(t) \times (f'(t) \times f(t))}{\| f(t) \|^3}$$

5. Let $f(t)$ be a smooth curve such that $f'(t) \neq 0$ for all t. Then we can define the *unit tangent vector* T by

$$T(t) = \frac{f'(t)}{\| f'(t) \|}$$

Show that

$$T'(t) = \frac{f'(t) \times (f''(t) \times f'(t))}{\| f'(t) \|^3}$$

6. Find T, N, B and κ at each point of the helix $f(t) = (\cos t, \sin t, t)$.
7. Show that the arc length L of a curve whose spherical coordinates are $\rho = \rho(t)$, $\theta = \phi(t)$ and $\phi = \phi(t)$ for t in an interval $[a, b]$ is

$$L = \int_a^b \sqrt{\rho'(t)^2 + (\rho(t)^2 \sin^2 \phi(t))\theta'(t)^2 + \rho(t)^2 \phi'(t)^2}\, dt.$$

Chapter 3

Algebra of Vector

We know what vectors are, we can start to perform some of the usual algebraic operations on them (e.g. addition, subtraction). Before doing that, we will introduce the notion of a *scalar*.

Definition: A scalar is a quantity that can be represented by a single number.

For our purposes, scalars will always be real numbers. Examples of scalar quantities are mass, electric charge, and speed (not velocity).4 We can now define *scalar multiplication* of a vector.

Definition: For a scalar k and a nonzero vector v, the scalar multiple of v by k, denoted by kv, is the vector whose magnitude is $|k| \, \| v \|$, points in the same direction as v if $k > 0$, points in the opposite direction as v if $k < 0$, and is the zero vector 0 if $k = 0$. For the zero vector 0, we define $k0 = 0$ for any scalar k.

Two vectors v and w are parallel (denoted by $v \parallel w$) if one is a scalar multiple of the other. You can think of scalar multiplication of a vector as stretching or shrinking the vector, and as flipping the vector in the opposite direction if the scalar is a negative number.

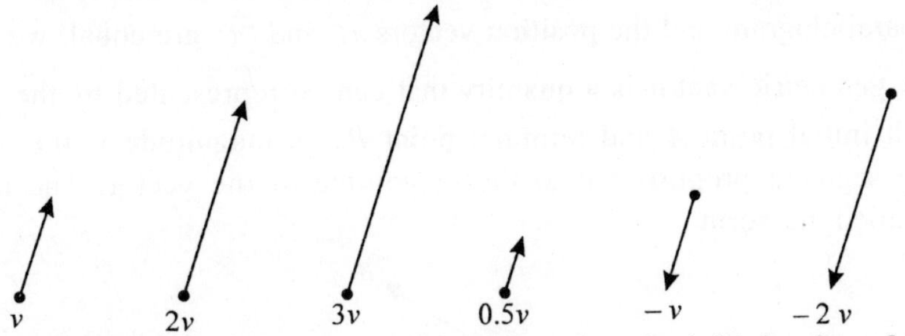

Recall that translating a nonzero vector means that the initial point of the vector is changed but the magnitude and direction are preserved. We are now ready to define the *sum* of two vectors.

Vectors as Directed line Segments

A vector is represented by a directed line segment. The length of the segment measuring the magnitude of the vector and the arrowhead indicating the direction of the vector.The direction of the vector is indicated by an arrow pointing from the tail to the head. If the tail is at point A and the head is at point B, the vector from A to B is written $\overrightarrow{AB}$. The arrowhead is usually placed at one end (or near the end) of the line segment; this end is called the head of the vector and the other end is called the tail of the vector. Such vectors are called free vectors because they are distinguished only by their length and direction, and not by their position in space.

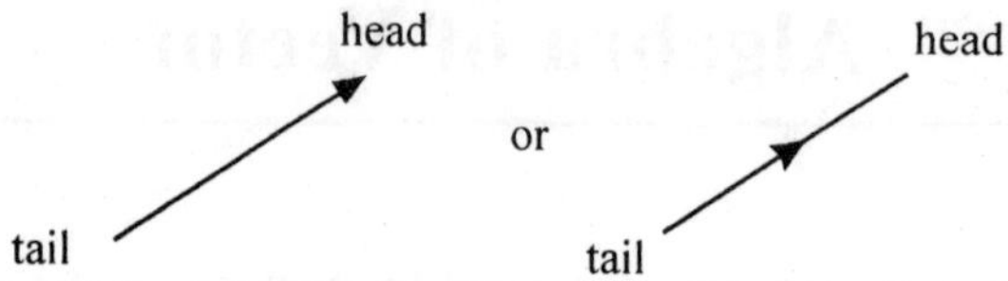

The vector with the same magnitude as u but opposite direction, the negative of u, is labelled $-u$.

Representation of a Vector

Given points A and B, we may think of the position vector $v = \overrightarrow{AB}$ as the vector that acts on the point A to get B.

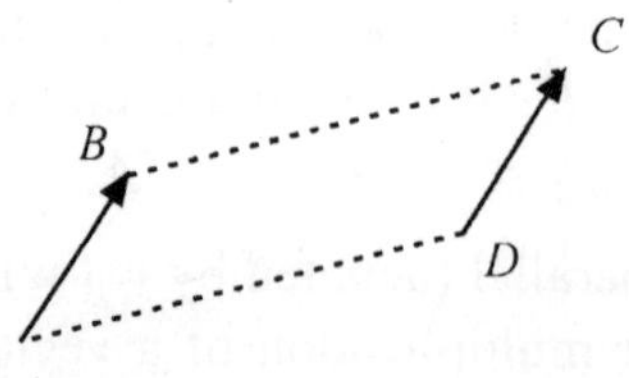

If the line segments AB and DC have the same direction and the same length, then $ABCD$ is a parallelogram and the position vectors $\overrightarrow{AB}$ and $\overrightarrow{DC}$ are equal; we write this as $\overrightarrow{AB} = \overrightarrow{DC}$. A geometric vector is a quantity that can be represented by the directed line segment with initial point A and terminal point B. Its magnitude is the length of the directed line segment proportional to the magnitude of the vector. The magnitude is sometimes called the norm .

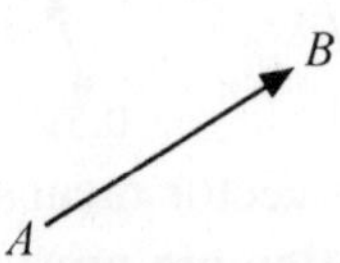

Analytically a vector is represented by a symbol $\overrightarrow{AB}$ or a bold face letter U. The magnitude of $\overrightarrow{AB}$AB is written as $|\overrightarrow{AB}|$ or $|AB|$ or simply AB. The magnitude of U is written as $|U|$ or simply U.

Zero or Null Vector

All vectors have length and direction, there is one exception, the zero vector. The zero vector $\hat{0}$ is the vector from a point A to the same point A.

We can write $0 = \overrightarrow{AA}$ no matter where the point A is. The zero vector does not have a direction but it does have a length; not surprisingly, its length is 0. Thus we can write $|0| = 0$, i.e., it is a vector of magnitude zero . It is written as $\hat{0}$.

Graphically it is a point. If $\overrightarrow{AB} = 0$, then $|\overrightarrow{AB}| = 0$ and A and B coincides.

Equality of Vectorss

Two vectors a and b whose magnitudes are equal i.e. $a = b$ and which are in the same direction regardless of their initial points, are called equal vectors.

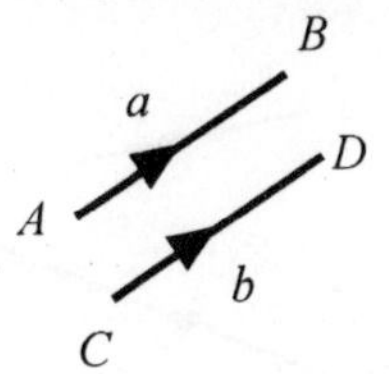

For any three vectors a, b and c,

$r = a + b + c = (a + b) + c = a + (b + c) + (a + c) + b$

This result can be extended to any number of vectors.

Free Vector

A vector which can be shifted along its line of action or parallel to itself and whose initial point can be anywhere, is called a free vector.

Negative of a Vector

The negative of a vector u is the vector –u which has the same direction and inclination but is of the opposite sense (of direction). The negative of the vector from A to B is the vector from B to A. It has the same length as $\overrightarrow{AB}$ but the opposite direction. Thus the negative of the vector $\overrightarrow{AB}$ is $\overrightarrow{BA}$ and we also write $\overrightarrow{BA} = -\overrightarrow{AB}$.

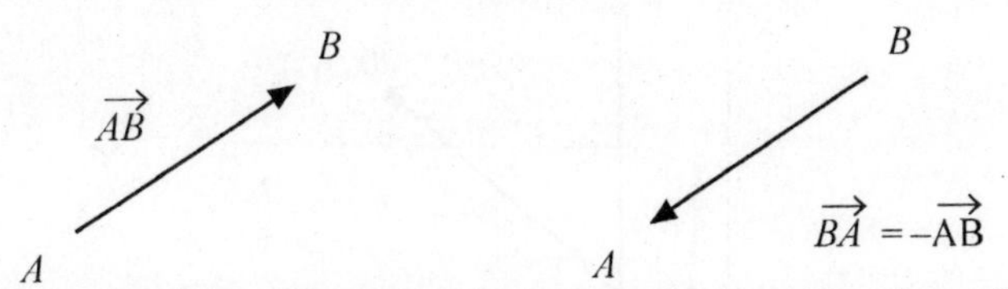

Fig. The Resultant of a System of Vectors is the least Number of Vectors that will Replace the given System.

Unit Vector

A vector in a given direction having unit magnitude is called a unit vector in that direction . It is usually denoted by $\hat{n}, \hat{e}, \hat{a}, \hat{b}$ etc. If $\vec{U}$ is given vector then the unit vector in the direction (or along or parallel) to $\vec{U}$ is given by

$$\hat{U} = \frac{\vec{U}}{|\vec{U}|}$$

in Cartesian coordinates it is typical to use $\hat{i}, \hat{j}$ and $\hat{k}$ to represent unit vectors along $\overrightarrow{OX}, \overrightarrow{OY}$ and $\overrightarrow{OZ}$ respectively. A vector which specifies a position in space with respect to the origin of the coordinate system could then be written.

Position Vector of a Point

If A and B are any two points then $\overrightarrow{AB}$ is called the position vector of point B with reference to point A.

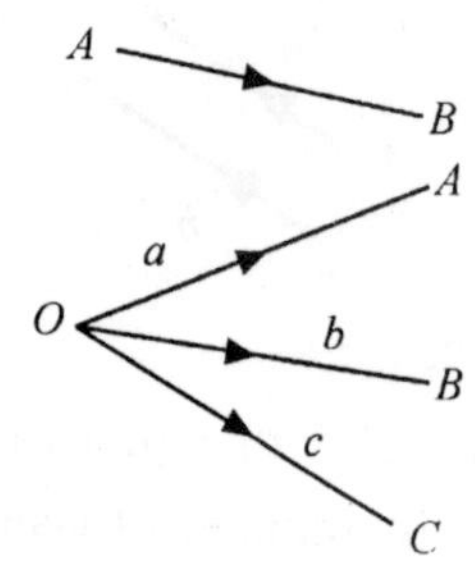

Position vector $\vec{r} = x\hat{i} + y\hat{j} + z\hat{k}$

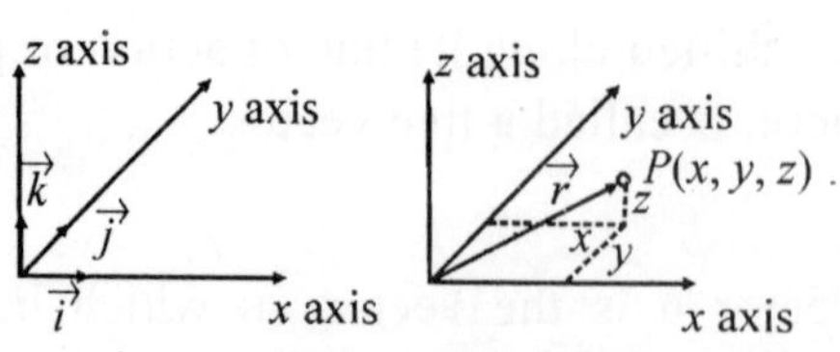

Position vector is a convenient mathematical construct to encapsulate the twin ideas of magnitude and direction of the position, occupied by an object. Position vector is a vector that extends from the reference point to the position of the particle.

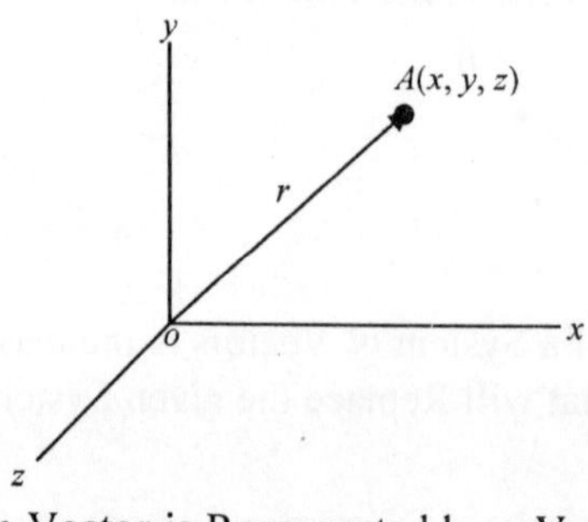

Fig. Position Vector is Represented by a Vector, Joining Origin to the Position of point object.

It is easy to realize that vector representation of position is appropriate, where directional properties of the motion are investigated. As a matter of fact, three important directional attributes of motion, namely displacement, velocity and acceleration are expressed in terms of position vectors.

Thus, all directional attributes of motion is based on the processing of position vectors.

Position Vector in Component form

One of the important characteristics of position vector is that it is rooted to the origin of the coordinate system. We shall find that most vectors associated with physical quantities, having directional properties, are floating vectors and not rooted to a point of the coordinate system. Recall that scalar components are graphically obtained by dropping two perpendiculars from the ends of the vector to the axes. In the case of position vector, one of the end is the origin itself. As position vector is rooted to the origin, the scalar components of position vectors in three mutually perpendicular directions of the coordinate system are equal to the coordinates themselves. Also, scalar components of position vector, r, by definition in the designated directions of the axes are :

$$x = r\cos\alpha$$

$$y = r\cos\beta$$

$$z = r\cos\gamma$$

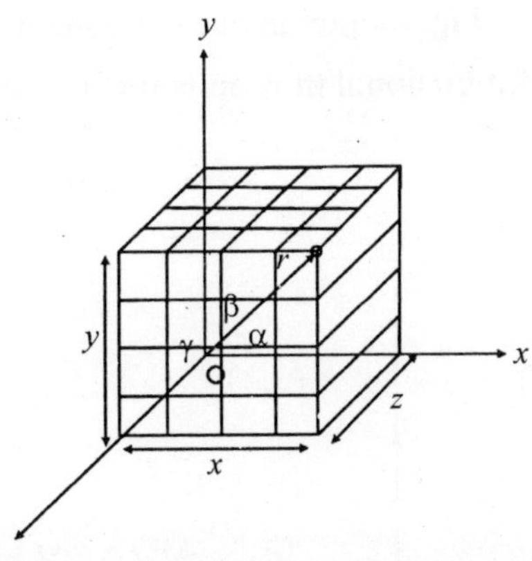

Fig. Scalar components are equal to coordiantes of the position

and position vector in terms of components (coordinates) is :

$$r = xi + yj + zk$$

where i , j and k are unit vectors in x, y and z directions.

The magnitude of position vector is given by :

$$r = |r| = \sqrt{x^2 + y^2 + z^2}$$

MOTION TYPES

In two dimensional motion, two of the three coordinates change with time. The remaining third coordinate is constant. By appropriately choosing the coordinate system, we can eliminate the need of specifying the third coordinate.

In one dimensional motion, only one of the three coordinates is changing with time. Other two coordinates are constant through out the motion. particle with the values of changing coordinate

A motion along x –axis or parallel to x – axis is, thus, described by x - component of the position vector i.e. x – coordinate of the position as shown in the figure.

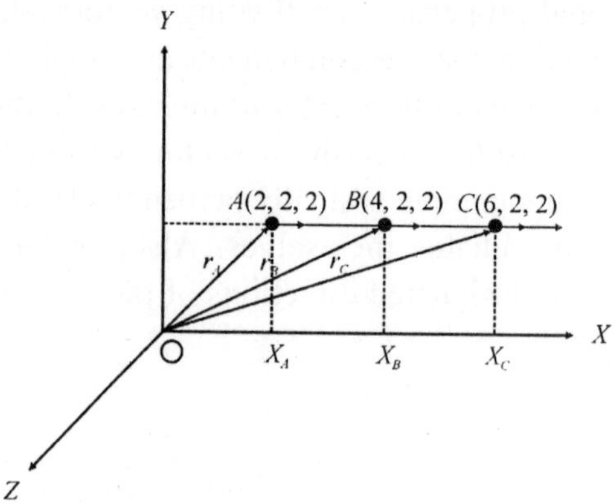

Fig. Motion in one Dimesnion

The description of one dimensional motion is further simplified by shifting axis to the path of motion as shown below.

x, y = 0, z = 0

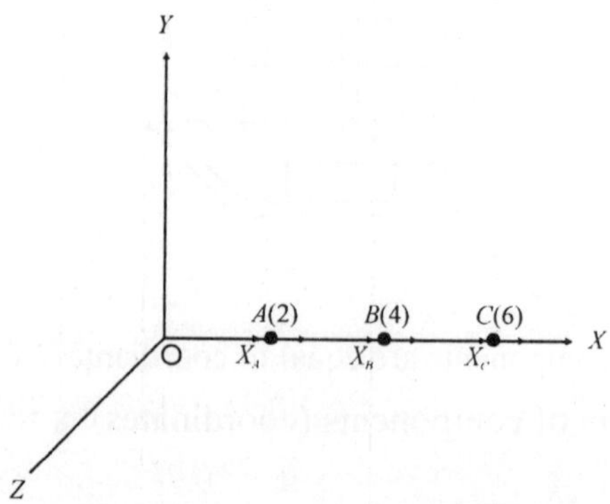

Fig. Motion in one Dimesnion

In this case, position vector itself is along x – axis and, therefore, its magnitude is equal to x – coordinate. We shall use a, b, c etc. to denote position vectors of points A, B, C etc. respectively, relative to origin O.

REPRESENTATION OF A VECTOR IN A PLANE

Let $P\ (x, y)$ be any point on the coordinate plane. With $\overrightarrow{OX}$ and $\overrightarrow{OY}$ as coordinate axes and O as the origin. Then $\overrightarrow{OP}$ stands for the position vector of point P relative to origin. Let $P = \overrightarrow{OP}$.

Any such vector can be broken down into two components, a horizontal (along $\overrightarrow{OX}$) and a vertical (along $\overrightarrow{OY}$) component.

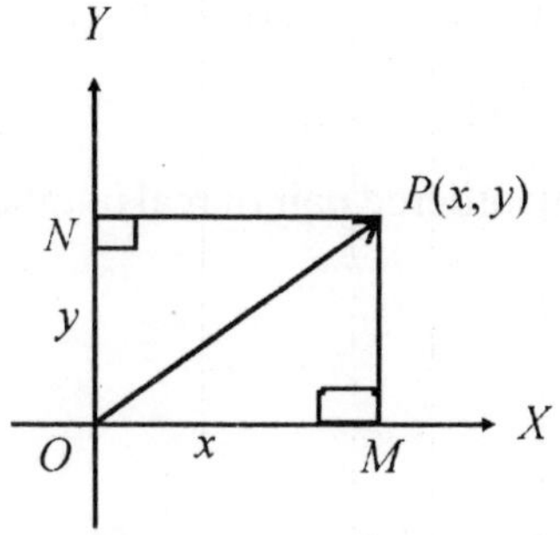

Let $\overrightarrow{OM}$ and $\overrightarrow{ON}$ be the horizontal and vertical component of $\overrightarrow{OP}$ i.e., P. Now $\overrightarrow{OM}$ and $\overrightarrow{ON}$ are non-collinear vectors but they are collinear with $\hat{i}$ and $\hat{j}$ respectively. So $\overrightarrow{OM} = x\ \hat{i}$ and $\overrightarrow{ON} = y\ \hat{j}$. If we use the tail up rule we have, $\overrightarrow{OP} = \overrightarrow{OM} + \overrightarrow{MP}$. But $\overrightarrow{MP} = \overrightarrow{ON} = y\ \hat{j}$.

$\therefore \quad \overrightarrow{OP} = x\hat{i} + y\ \hat{j}$ i.e., $p = x\hat{i} + y\ \hat{j}$

Here x and y are components of p along x-axis and y-axis respectively. This ordered pair $(x\ , y)$ associated with p is known as the rectangular components of p.

Let $\angle POX = \theta$ then

$$\cos\theta = \frac{|x|}{|p|} \text{ and } \sin\theta = \frac{|y|}{|p|} \text{ so that } p = \sqrt{x^2 + y^2}$$

which is the magnitude of p.

This concept can be extended to higher dimensions such as three dimension.

Also, note that any vector $\overrightarrow{AB}$ can be shifted so that its initial point A is set at 'O', then $\overrightarrow{AB}$ is in 'standard position'. Now $\overrightarrow{AB} = \overrightarrow{OP}$ then $\overrightarrow{OP}$ replaces $\overrightarrow{AB}$.

Thus $\overrightarrow{OP}$ is the standard vector in the plane for all vectors in the plane of the same magnitude and direction as $\overrightarrow{OP}$.

$$\therefore \quad OP = OM + ON$$

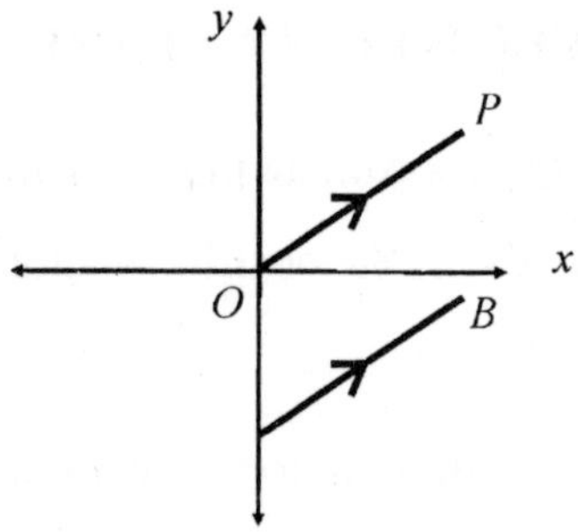

If coordinates of A are (xa, ya) and that of B are (xb, yb) then shifting A to O and making $B = P$ we have coordinates of P are $(xb - xa, yb - ya)$

$\therefore \overrightarrow{AB}$ or $\overrightarrow{OP} \equiv (xb - xa, yb - ya)$

Thus an algebraic vector is an ordered pair of real numbers that corresponds to a standard geometric vector $\overrightarrow{OP}$.

ALGEBRA OF VECTORS

A. Addition of Vectors

Vector addition represents the net effect of the directional quantities

Vectors operate with other scalar or vector quantities in a particular manner. Unlike scalar algebraic operation, vector operation draws on graphical representation to incorporate directional aspect. Basically, vector addition is dealt in two conceptually equivalent ways:

1.Graphical and

2.Algebraic (vector)

The resulting vector after addition is termed as sum or resultant vector. The resultant vector corresponds to the "resultant" or "net" effect of a physical quantity having directional attribute. The effect of a force system on a body, is determined by the resultant force acting on it. The idea of resultant force, in this case, conveys that the resulting force (vector) has the same effect as that of the forces (vectors) being added.

$$R = F_1 + F_2 + F_3 + F_4$$

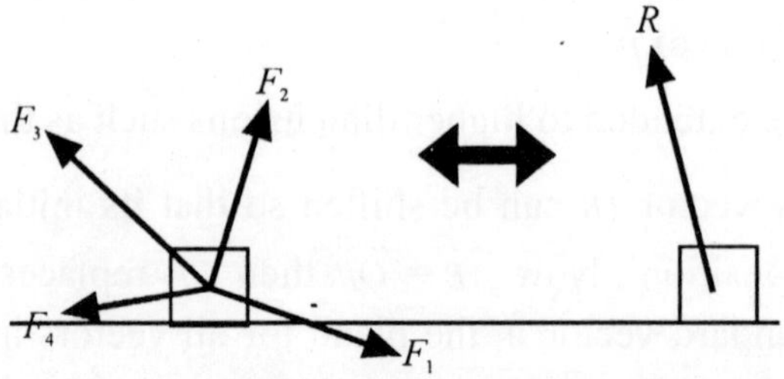

Fig. Resultant Force

It is important to emphasize here that vector rules do not distinguish between vector types (whether displacement or acceleration vector). This means that the rules are general as far as vectors are concerned. Vector addition rule restricts that only like vectors can be added. It is expected also. The requirement is similar to scalar algebra where 2 plus 3 is always 5, but we need to add similar quantity like 2 meters plus 3 meters is 5 meters. But, we can not add, for example, distance and temperature.

Vector Addition: Graphical Method

Let us examine the example of displacement of a person in two different directions. The two displacement vectors, perpendicular to each other, are added to give the "resultant" vector. In this case, the closing side of the right triangle represents the sum (i.e. resultant) of individual displacements *AB* and *BC*.

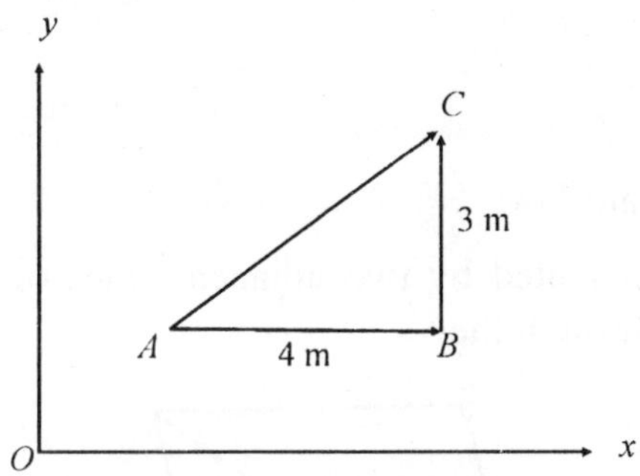

Fig. Displacement $AC = AB + BC$

The method used to determine the sum in this particular case (in which, the closing side of the triangle represents the sum of the vectors in both magnitude and direction) forms the basic consideration for various rules dedicated to implement vector addition.

Triangle Law

Definition: Triangle law of Vector Addition. If two vectors are represented by two sides of a triangle in sequence, then third closing side of the triangle, in the opposite direction of the sequence, represents the sum (or resultant) of the two vectors in both magnitude and direction.

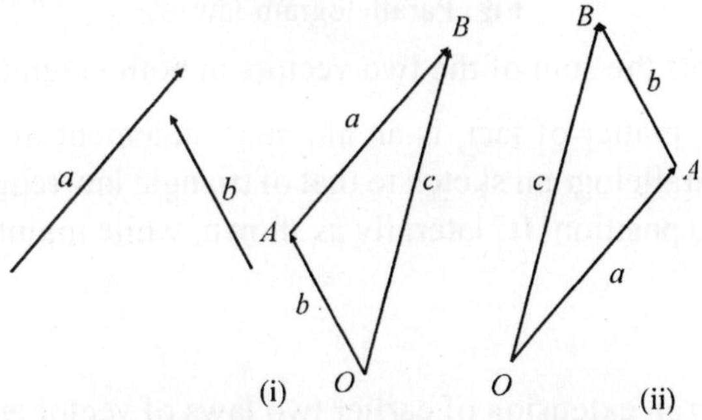

Fig. Triangle law of Vector Addition

Here, the term "sequence" means that the vectors are placed such that tail of a vector begins at the end of arrow head of the vector placed before it.

The triangle law does not restrict where to start i.e. with which vector to start. Also, it does not put conditions with regard to any specific direction for the sequence of vectors, like clockwise or anti-clockwise, to be maintained. In figure, the law is applied starting with vector,b; whereas the law is applied starting with vector, a, in figure. In either case, the resultant vector, c, is same in magnitude and direction.

This is an important result as it conveys that vector addition is commutative in nature i.e. the process of vector addition is independent of the order of addition. Mathematically speaking :

$$a + b = b + a$$

Parallelogram Law

Definition: Parallelogram law.

If two vectors are represented by two adjacent sides of a parallelogram, then the diagonal of parallelogram through the

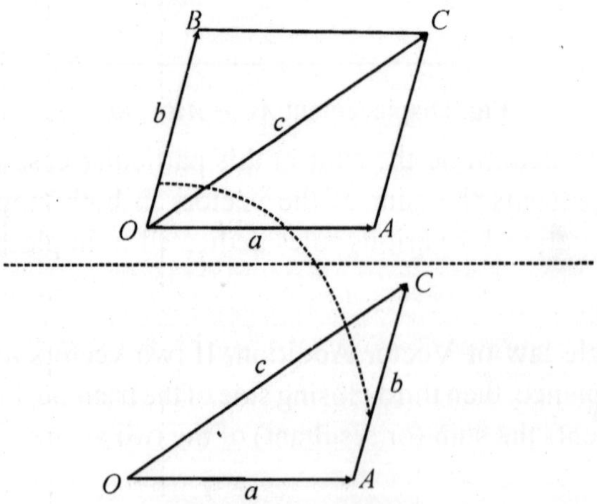

Fig. Parallelogram law

common point represents the sum of the two vectors in both magnitude and direction.

Parallelogram law, matter of fact, is an alternate statement of triangle law of vector addition. Converting parallelogram sketch to that of triangle law requires shifting vector, *b*, from the position *OB* to position *AC* laterally as shown, while maintaining magnitude and direction.

Polygon Law

The polygon law is an extension of earlier two laws of vector addition. It is ,matter of fact, successive application of triangle law to more than two vectors. A pair of vectors (*a*, *b*) is added in accordance with triangle law. The intermediate resultant vector $(a + b)$ is then added to third vector (c) again, successively till all vectors to be added have been exhausted.

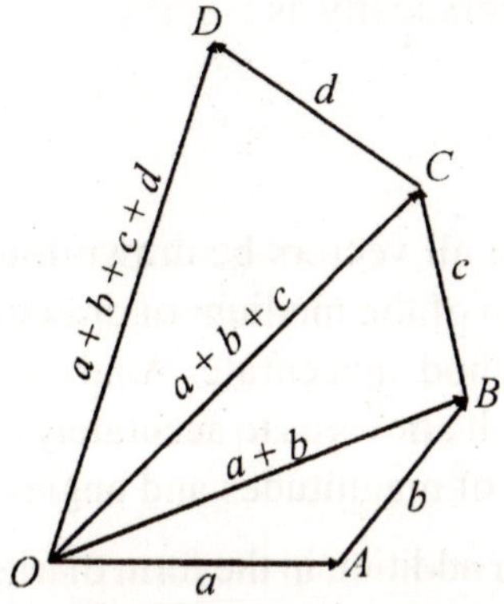

Fig. Successive Application of Triangle Law

Definition: Polygon law.

Polygon law of vector addition : If (n–1) numbers of vectors are represented by (n-1) sides of a polygon in sequence, then n^{th} side, closing the polygon in the opposite direction, represents the sum of the vectors in both magnitude and direction.

In the figure shown below, four vectors namely a, b, c and d are combined to give their sum. Starting with any vector, we add vectors in a manner that the subsequent vector begins at the arrow end of the preceding vector.

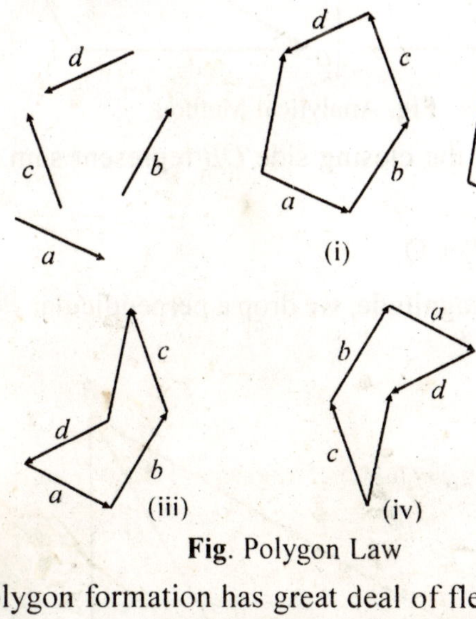

Fig. Polygon Law

Matter of fact, polygon formation has great deal of flexibility. It may appear that we should elect vectors in increasing or decreasing order of direction. But, this is not so. This point is demonstrated in figure, in which the vectors b and c have simple been exchanged in their positions in the sequence without affecting the end result.

It means that the order of grouping of vectors for addition has no consequence on the result. This characteristic of vector addition is known as "associative" property of vector addition and is expressed mathematically as :

$$(a + b) + c = a + (b + c)$$

Analytical Method

Vector method requires that all vectors be drawn true to the scale of magnitude and direction. The inherent limitation of the medium of drawing and measurement techniques, however, renders graphical method inaccurate. Analytical method, based on geometry, provides a solution in this regard. It allows us to accurately determine the sum or the resultant of the addition, provided velues of magnitudes and angles are supplied.

Here, we shall analyze vector addition in the form of triangle law to obtain the magnitude of the sum of the two vectors. Let P and Q be the two vectors to be added, which make an angle, with each other. We arrange the vectors in such a manner that two adjacent sides OA and AB of the triangle OAB, represent two vectors P and Q respectively.

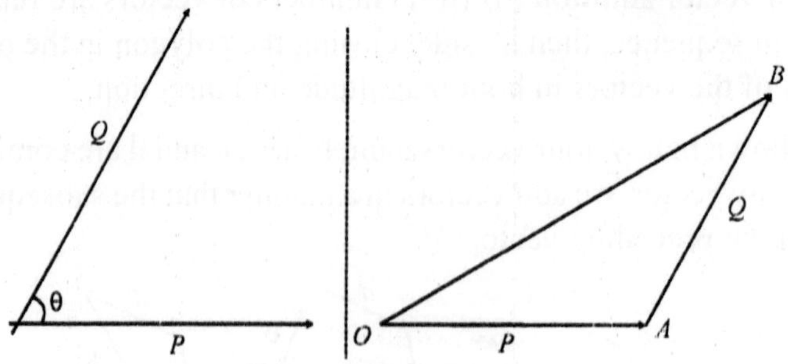

Fig. Analytical Method

According to triangle law, the closing side OB represent sum of the vectors in both magnitude and direction.

$$OB = OA + AB = P + Q$$

In order to determine the magnitude, we drop a perpendicular BC on the extended line OC.

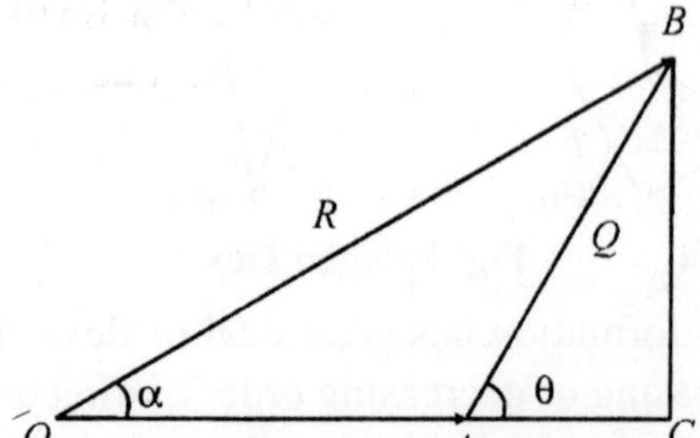

Fig. Analytical MSethod

In Δ ACB,

$$AC = AB\cos\theta = Q\cos\theta$$

$$BC = AB\sin\theta = Q\sin\theta$$

In right Δ *OCB*, we have:

$$OB = \sqrt{OC^2 + BC^2} = \sqrt{(OA + AC)_2 + BC^2}$$

Substituting for AC and BC,

$$|OB| = \sqrt{(P + Q\cos\theta)^2 + (Q\sin\theta)^2}$$

$$P = \sqrt{P^2 + Q^2\cos^2\theta + 2PQ\cos\theta + Q^2\sin^2\theta}$$

$$R = OB = \sqrt{P^2 + Q^2 + 2PQ\cos\theta}$$

Let "α" be the angle that line *OA* makes with *OC*, then

$$\tan a = \frac{BC}{OC} = \frac{Q\sin\theta}{P + Q\cos\theta}$$

The equations give the magnitude and direction of the sum of the vectors. The above equation reduces to a simpler form, when two vectors are perpendicular to each other. In that case, $\theta = 90°$; sinq = sin90° = 1; cos θ = cos90° = 0

and,

$$OB = \sqrt{P^2 + Q^2}$$

$$\tan\alpha = \frac{Q}{P}$$

These results for vectors at right angle are exactly same as determined, using Pythagoras theorem.

Vector addition: Algebraic Method

Graphical method is meticulous and tedious as it involves drawing of vectors on a scale and measurement of angles. More importantly, it does not allow algebraic operations that otherwise would give a simple solution. We can, however, extend algebraic techniques to vectors, *provided vectors are represented on a rectangular coordinate system*. The representation of a vector on coordinate system uses the concept of unit vector and component.

Components of a Vector

A scalar component, also known as projection, of a vector *AB* in the positive direction of a straight line *C'C* is defined as:

$$AC = |AB| \cos\theta = AB \cos\theta$$

where θ, is the angle that vector AB makes with the specified direction C'C as shown in the figure.

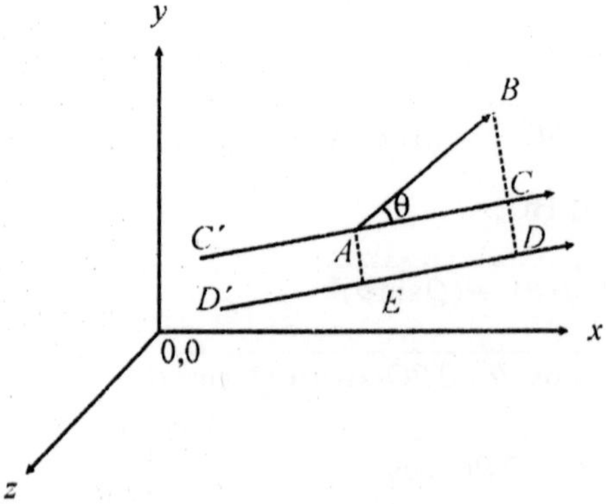

It should be understood that the component so determined is not rooted to the common point "*A*" or any specific straight line like *C'C* in a given direction. Matter of fact, component of vector *AB* drawn on any straight lines like *D'D* parallel to *C'C* are same. Note that the component i.e. projection of the vector is graphically obtained by drawing two perpendicular lines from the ends of the vector on the straight line in the specified direction.

$$ED = AC = AB \cos\theta,$$

Following the above definition of component of vector (specifically known as scalar component) , we can resolve a given vector in three components in three mutually perpendicular directions of rectangular coordinate system.

Note that we measure angle with respect to parallel lines to the axes. By convention, we denote components by using the non-bold type face of the vector symbol with a suffix representing direction (*x* or *y* or *z*).

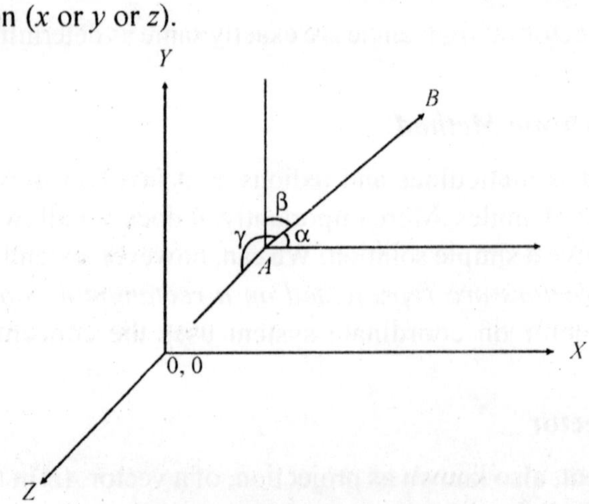

Fig. Components of a Vector

$$AB_x = \text{AB} \cos\alpha$$

$$AB_y = \text{AB} \cos\beta$$

$$AB_y = \text{AB} \cos\gamma$$

where α, β and γ are the angles that vector AB makes with the positive directions of x, y and z directions respectively.

As a vector can be resolved in a set of components, the reverse processing of components in a vector is also expected. A vector is composed from vector components in three directions along the axes of rectangular coordinate system by combining three component vectors. The vector component is the vector counterpart of the scalar component, which is obtained by multiplying the scalar component with the unit vector in axial direction. The vector components of a vector a, are $a_x i$, $a_y j$ and $a_z k$.

A vector a, is equal to the vector sum of component vectors in three mutually perpendicular directions of rectangular coordinate system.

$a = normal\ a_x i + normal\ a_y j + normal\ a_z k$

where i , j and k are unit vectors in the respective directions.

Magnitude of the sum of two vectors (a + b) in x and y direction, which are perpendicular to each other, is :

$$|\vec{a}| = \sqrt{a_x^2 + a_y^2} = a_x^2 + a_y^2$$

and the magnitude of the sum of vectors (a + b) and third vector c, which are perpendicular to each other, is :

$$|\vec{a}| = \sqrt{a_x^2 + a_y^2} = \left(a_x^2 + a_y^2\right)^{1/2}$$

$$|\vec{c}| = \sqrt{a_x^2 + a_y^2 + a_z^2}$$

Vector Addition using Component of Vector

We are now equipped to represent a vector as a combination of component vectors. Also, the stage is, now, set to design a frame work, which allows vector addition with algebraic methods. The frame work for vector addition draws on two important concepts. The first concept is that a vector can be equivalently expressed in terms of three component vectors.

$a = a_x i + a_y j + a_z k$ and $b = b_x i + b_y j + b_z k$

The component vector form has important significance. It ensures that component vectors to be added are restricted to three known directions only. This paradigm eliminates the possibility of unknown direction. The second concept is that vectors along a direction can be treated algebraically. If two vectors are along the same line, then resultant is given as:

When , $\theta = 0°$, $\cos\theta = \cos 0° = 1$ and,

$$|\vec{R}| = \sqrt{P^2 + 2PQ + Q^2} = \sqrt{(P+Q)^2} = P + Q$$

When θ , $= 180°$, $\cos\theta = \cos 180° = -1$

and, $$|\vec{R}| = \sqrt{P^2 - 2PQ + Q^2} = \sqrt{(P-Q)^2} = P - Q$$

Thus, we see that the magnitude of the resultant is equal to algebraic sum of the magnitudes of the two vectors. Using these two concepts, the addition of vectors is affected as outlined here:

1. Represent the vectors (a and b) to be added in terms of components:
   $a = a_x i + a_y j + a_z k$ and $b = b_x i + b_y j + b_z k$
2. Group components in a given direction:
   $a + b = a_x i + a_y j + a_z k + b_x i + b_y j + b_z k$
   $a+b = (a_x+b_x)\, i + (a_y+b_y) j + (a_z+b_z)\, k$
3. Find the magnitude and direction of the sum, using analytical method

Properties of Vector Addition

1. Vector addition is commutative i.e. $a + b = b + a$ Extension to sums of two or more vectors is immediate.
2. Vector addition is associative. Observe the adjoining fig. we have $\overrightarrow{OA} = a$, $\overrightarrow{AB} = b$, $\overrightarrow{BC} = c$ and $\overrightarrow{OC} = r$ using tail-tip rule, for triangle OAB, $\overrightarrow{OB} = a + b$, for triangle ABC, $\overrightarrow{AC} = b + c$ for triangle OBC, c i.e. $r = (a + b) + c$. Also, for triangle OAC, $\overrightarrow{OC} = (a + c) + c$ i.e $r = a + (b + c)$.

   Thus, $r = (a + b) + c = a + (b + c)$.

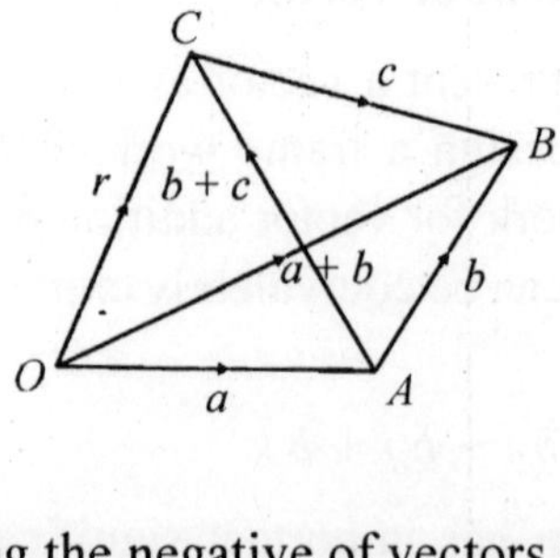

Subtraction of a Vector

It is accomplished by adding the negative of vectors i.e.
$a - b = a + (-b)$. Note also that $-(a + b) = -a - b$.

Subtraction is considered an addition process with one modification that the second vector (to be subtracted) after reversing its direction is added to the first vector (from which the second vector is subtracted). Let us consider the problem of subtracting vector, b, from , a. Using graphical techniques, we first reverse the direction of vector, b, and obtain the sum applying triangle or parallelogram law.

Symbolically,

$$a - b = a + (-b)$$

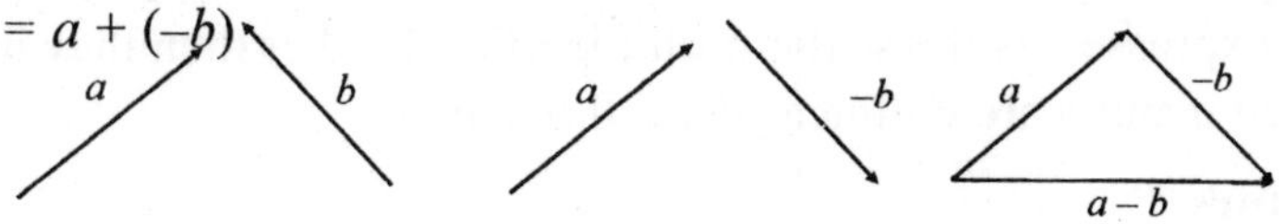

(i) (ii) (iii)

Fig. Subtraction of Vector

Composition of Vectors

It is the process of determining the resultant of a system of vectors. For this we have to draw a vector polygon, placing tail of each vector at the tip of the preceding vector. Then draw a vector from the tail of the first vector to tip of the last vector in the system. Later on we will show that not all vector systems reduce to a single vector. Also note that the order of vectors are drawn is immaterial.

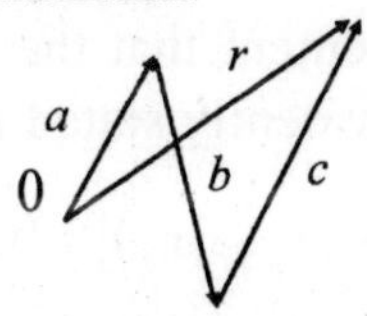

Multiplication of Vectors by Scalars

(A) Let m be a scalar and a be a vector then the product ma whose magnitude is $|m|$ times of $|a|$ and which is similarly or oppositely directed to a depending upon whether m is positive or negative.

Then

1. for $m > 0$, m a is a vector of magnitude m $|a|$ and is in the direction of a .
2. for $m = 0$, ma is a null vector.
3. for $m < 0$, ma is the vector of magnitude $-m$ $|a|$ and is exactly opposite to the vector a

From this, it follows that $(-1)\, a = -a$ and $0\, a = 0$

(B) Other operations with scalars m and n are

$(m + n)\, a = m\, a + n\, a$

$m\, (a + b) = m\, a + m\, b$

$mn\, (a) = n\, (ma) = (m\, n\, a)$

Linear Combination of Vectors

Let a and b be two vectors and x and y be any real numbers. Then the vector $r = x\,a + y\,b$ is called the linear combination of vectors a and b.

Collinear Vectors

Two vectors a and b are said to be collinear, if they are parallel to one or on the same line. This clearly indicates that 1) they must be coincident or (2) they must be parallel. If two vectors are expressed as the scalar multiple of each other then they must be collinear. i.e. if $a = x\,b$ then a and b are collinear if $x \neq 0$ scalar.

Coplanar Vectors

Geometric objects lying in a common plane are said to be coplanar. Three non-collinear points determine a plane and so are trivially coplanar. Four points are coplanar iff the volume of the tetrahedron defined by them is 0,

$$\begin{vmatrix} x_1 & y_1 & z_1 & 1 \\ x_2 & y_2 & z_2 & 1 \\ x_3 & y_3 & z_3 & 1 \\ x_4 & y_4 & z_4 & 1 \end{vmatrix} = 0$$

co-planarity is equivalent to the statement that the pair of lines determined by the four points are not skew, and can be equivalently stated in vector form as

$$(x_3 - x_1)\cdot[(x_2 - x_1)\times(x_4 - x_3)] = 0.$$

An arbitrary number of n points $x_1, x_2, \ldots, x_n$ can be tested for co-planarity by finding the point-plane distances of the points $x_1, \ldots, x_n$ from the plane determined by (x_1, x_2, x_3) and checking if they are all zero. If so, the points are all coplanar. A set of n vectors V is coplanar if the nullity of the linear mapping defined by V has dimension 1, the matrix rank of V (or equivalently, the number of its singular values) is $(n - 1)$.

If vectors are parallel to a given plane, then they are coplanar. Note that if a and b be two non-collinear vectors which can be uniquely expressed as $r = x\,a + y\,b, x, y \neq 0$ scalars, then they must be coplanar.

Parallel lines in three-dimensional space are coplanar, but skew lines are not.

MULTIPLICATION

Vector multiplication provides concise and accurate representation of natural laws, which involve vectors.

The vector multiplication, however, is not an unique mathematical construct like scalar multiplication.

The rules of vector multiplication have been formulated to encapsulate physical processes in their completeness. Since we know that the direction of change of velocity is the direction of acceleration, we draw the vector diagram and find out the direction of acceleration. We can see that the direction of acceleration turns out to act in vertically downward direction.

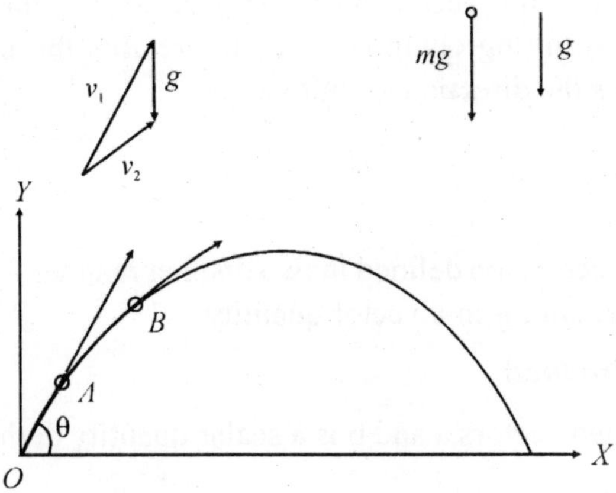

Fig. Direction of Acceleration

The conceptualization of physical laws in vector form, however, provides us with powerful means to arrive at the result in relatively simpler manner. If we look at the flight of particle in parabolic motion, then we observe that the motion of particle is under the force of gravity, which is acting vertically downward. There is no other force (neglecting air resistance). Now, from second law of motion, we know that :

$$F_{Resultant} = ma$$

This equation reveals that the direction of acceleration is same as that of the resultant force acting on the particle. Thus, acceleration of the particle in parabolic motion is acting vertically downward.

Multiplication of vector with Scalar

Multiplication of a vector *A*, with another scalar quantity, a, results in another vector, *B*. The magnitude of the resulting vector is equal to the product of the magnitude of vector with the scalar quantity. The direction of the resulting vector, however, is same as that of the original vector.

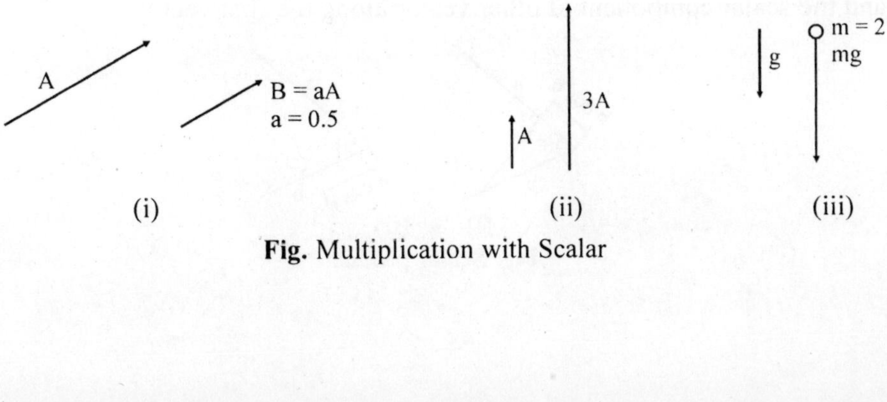

Fig. Multiplication with Scalar

We have already made use of this type of multiplication intuitively in expressing a vector in component form.

$$A = A_x i + A_y j + A_z k$$

In this vector representation, each component vector is obtained by multiplying the scalar component with the unit vector. As the unit vector has the magnitude of 1 with a specific direction, the resulting component vector retains the magnitude of the scalar component, but acquires the direction of unit vector.

$$A_x = A_x i$$

Products of Vectors

Thus, products of vectors are defined in two distinct manner – one resulting in a scalar quantity and the other resulting in a vector quantity.

Scalar Product (dot Product)

Scalar product of two vectors a and b is a scalar quantity defined as :

$$a.b = ab \cos\theta,$$

where "a" and "b" are the magnitudes of two vectors and "θ" is the angle between the direction of two vectors. It is important to note that vectors have two angles θ and 2π – θ. We can use either of them as cosine of both "θ" and "2π – θ" are same.

The notation "a.b" is important and should be mentally noted to represent a scalar quantity – even though it involves bold faced vectors. It should be noted that all quantities on the right hand side are scalars.

Meaning of scalar Product

We can read the definition of scalar product in either of the following manners :

$$a.b = a\,(b \cos\theta)$$
$$a.b = b\,(a \cos\theta)$$

Recall that "*b* cos θ" is the scalar component of vector b along the direction of vector a and "*a* cos θ" is the scalar component of vector a along the direction of vector *b*. Thus, we may consider the scalar product of vectors *a* and *b* as the product of the magnitude of one vector and the scalar component of other vector along the first vector.

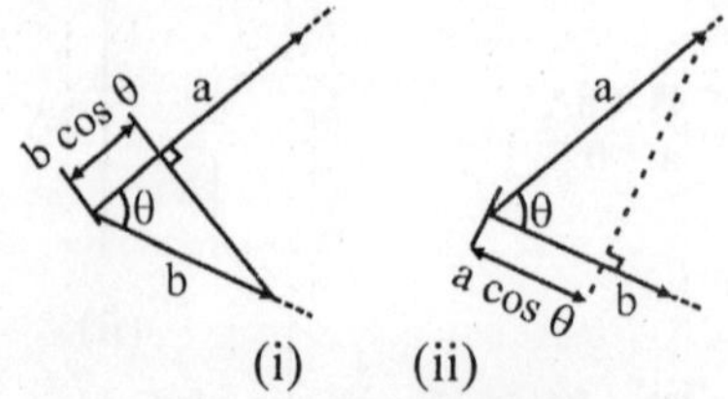

Fig. Scalar product

The figure below shows drawing of scalar components. The scalar component of vector b in figure is obtained by drawing perpendicular from the tip of the vector, b, on the direction of vector, a. Similarly, the scalar component of vector a in figure is obtained by drawing perpendicular from the tip of the vector, a, on the direction of vector, *b*.

The two alternate ways of evaluating dot product of two vectors indicate that the product is commutative i.e. independent of the order of two vectors :

$$a \,.\, b = b \,.\, a$$

Values of Scalar Product

The value of dot product is maximum for the maximum value of $\cos\theta$. Now, the maximum value of cosine is $\cos 0° = 1$. For this value, dot product simply evaluates to the product of the magnitudes of two vectors.

$$(a.b)_{\max} = ab$$

For $\theta = 180°, \cos 180° = -1$ and

$$a.b = -ab$$

Thus, we see that dot product can evaluate to negative value as well. This is a significant result as many scalar quantities in physics are given negative value. The work done, for example, can be negative, when the work is done on the system; instead of work being done by the system (positive).

The scalar product evaluates to zero for $\theta = 90^0$ and 270^0 as cosine of these angles are zero. These results have important implication for unit vectors. The dot products of same unit vector evaluates to 1.

$$i.i = j.j = k.k = 1$$

The dot products of combination of different unit vectors evaluate to zero.

$$i.j = j.k = k.i = 0$$

Example: A block of mass "m" moves from point *A* to *B* along a smooth plane surface under the action of force as shown in the figure. Find the work done if it is defined as:

$$W = F.\Delta x$$

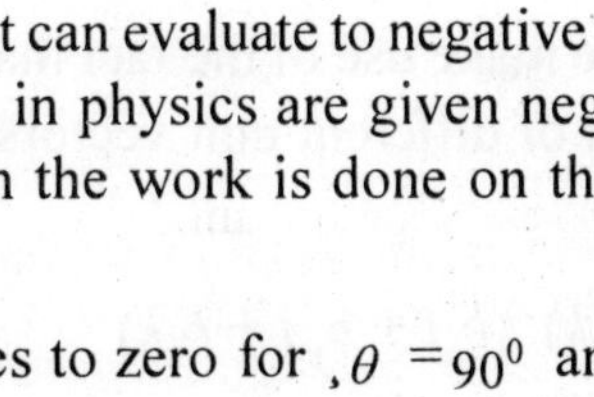

Fig. Work Done

where *F* and Δ x are force and displacement vectors.

Solution: Expanding the expression of work, we have:

Work done

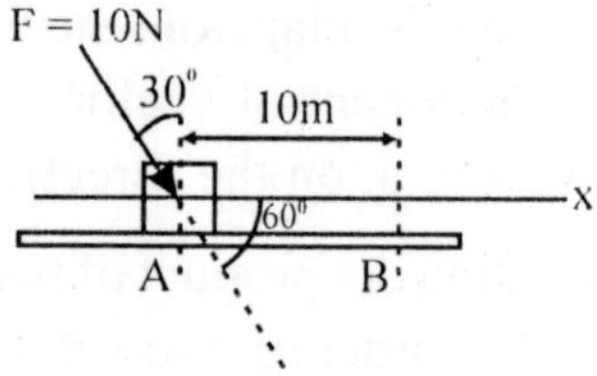

Fig. Work Done

$$W = F.\Delta x = F\Delta x\cos,$$

Here, $F = 10\ N$, $\Delta x = 3$ m and $\cos\theta = \cos 60° = 0.5$.

$$W = 10 \times 10 \times 0.5 = 50\ J$$

Scalar Product in Component form

Two vectors in component forms are written as :

$$a = a_x i + a_y j + a_z k$$

$$b = b_x i + b_y j + b_z k$$

In evaluating the product, we make use of the fact that multiplication of the same unit vectors is 1, while multiplication of different unit vectors is zero. Finally, the dot product evaluates to scalar terms:

$$\vec{a} \cdot \vec{b} = (a_x i + a_y j + a_z k)\ .(b_x i + b_y j + b_z k)$$

$$\vec{a} \cdot \vec{b} = a_x i.b_x i + a_y j.b_y j + a_z k.b_z k$$

$$\vec{a} \cdot \vec{b} = a_x b_x + a_y b_y + a_z b_z$$

Importance of dot Product

An important use of the dot product is to test whether or not two vectors are orthogonal. Two vectors are orthogonal if the angle between them is 90 degrees.

For Example: <2, –3, 3> and <3, 3, 1> are orthogonal since the dot product is

$$2(3) + (-3)(3) + 3(1) = 0.$$

Projections of a Vector

One important use of dot products is in projections. The scalar projection of b onto a is the *length* of the segment AB shown in the figure below. The vector projection of b onto a is the *vector* with this length that begins at the point A points in the same direction (or opposite direction if the scalar projection is negative) as a.

Thus, mathematically, the scalar projection of b onto a is $|b| \cos \theta$ (where θ is the angle between a and b) which from (*) is given by

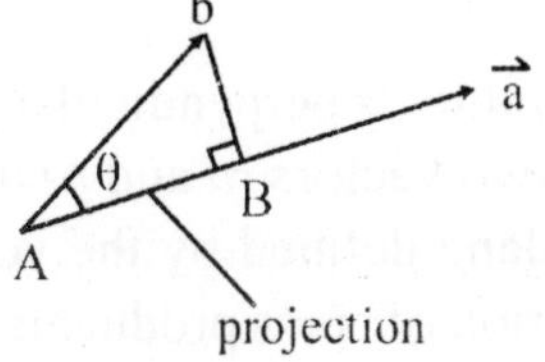

Projection of b on $a = \dfrac{\vec{a}\cdot\vec{b}}{|\vec{a}|}$

This quantity is also called the component of b in the a direction (hence the notation comp). And, the vector projection is merely the unit vector $a/|a|$ times the scalar projection of b onto a:

$$proj_a \vec{b} = \frac{\vec{a}\cdot\vec{b}}{|\vec{a}|}\frac{\vec{a}}{|\vec{a}|} = \frac{\vec{a}\cdot\vec{b}}{|\vec{a}|^2}\vec{a}$$

Thus, the scalar projection of b onto a is the magnitude of the vector projection of b onto a.

Example: Suppose you wish to find the work W done in moving a particle from one point to another. From physics we know $W = Fd$ where F is the magnitude of the force moving the particle and d is the distance between the two points. However, this relation is only valid when the force acts in the direction the particle moves. Suppose this is not the case. Let the force vector be $F = <5, 9, 7>$ and the displacement vector be $d = <4, 5, 3>$. In this case, the work is the product of the distance moved (the magnitude of the displacement vector) and the magnitude of the component of the force that acts in the direction of displacement (the scalar projection of F onto d):

$$W = |\vec{d}||\vec{F}|\cos\theta = \vec{F}\cdot\vec{d}$$

Thus, the work done by the force to displace the particle from say the origin to the point (5,9,7) is $W = \vec{F}\cdot\vec{d} = 4(5)+9(5)+7(3) = 20+45+21 = 86$ Joules

Vector Product (cross product)

The cross product of two vectors a and b is a third vector. The magnitude of the vector product is given by the following expression:

$$|c| = |a \times b| = ab\sin\theta$$

where, is the smaller of the angles between the two vectors. It is important to note that vectors have two angles θ and 2π – θ. We should use the smaller of the angles as sine of 0 and 2π – θ are different.

Direction of Vector Product

The direction of the vector product is perpendicular to the plane containing vectors, *a*, and , *b*. We must understand that two vectors (*a* and *b*) define an unique plane. The vector product is perpendicular to this plane defined by the vectors as shown in the figure. The most important aspect of the direction of cross product is that it is independent of the angle, θ, enclosed by the vectors. The enclosed angle , θ, only impacts the magnitude of the cross product and not its direction.

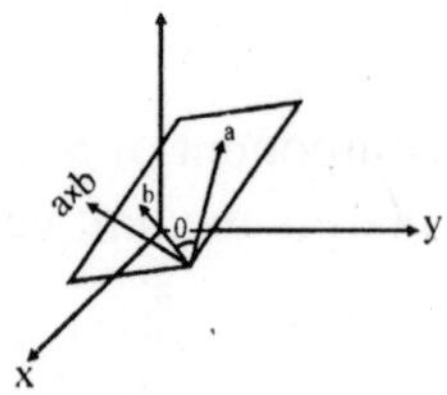

Fig. Direction of Vector Product

Incidentally, the requirement for determining direction suits extremely well with rectangular coordinate system. We know that rectangular coordinate system comprises of three planes, which are at right angles to each other. It is, therefore, suggested that we must orient our coordinate system in such a manner that vectors lie in one of the three planes defined by the rectangular coordinate system. This arrangement renders the cross product oriented in the direction of the remaining axis as shown in the figure.

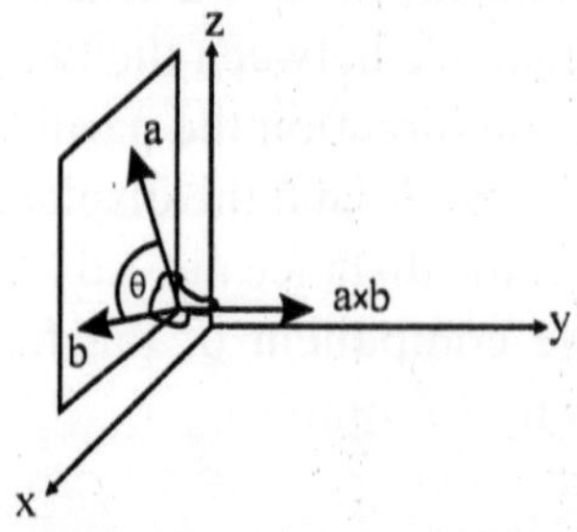

Fig. Direction of Vector Product

Matter of fact, direction of the vector product is not yet actually determined. We can draw the vector product perpendicular to the plane on either of the two sides.

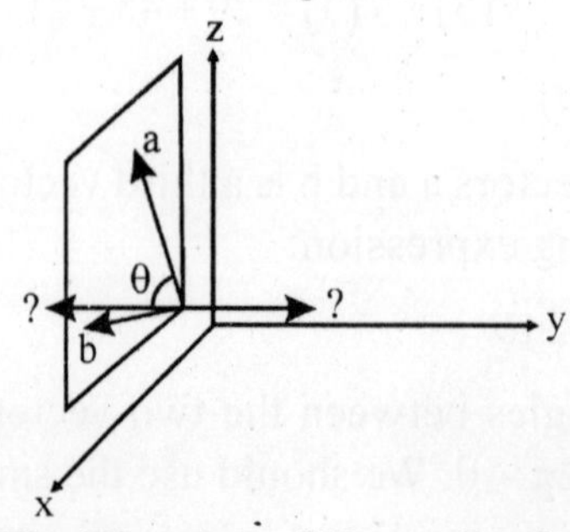

Fig. Direction of Vector Product

For example, the product can be drawn either along the positive direction of y – axis or along the negative direction of y – axis direction .

The direction of the vector product, including which side of the plane, is determined by right hand rule for vector products. According to this rule, we place right fist such that the curl of the fist follows as we proceed from the first vector, *a*, to the second vector, *b*. The stretched thumb, then gives the direction of vector product.

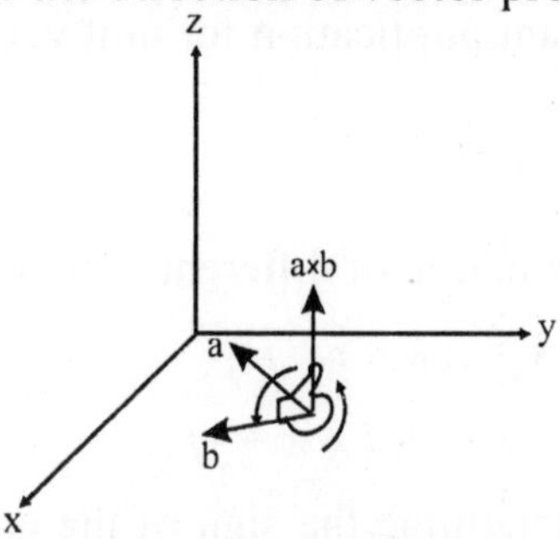

Fig. Direction of Vector Product

When we apply this rule to the case considered before, we find that the vector product is in the positive y – direction as shown below:

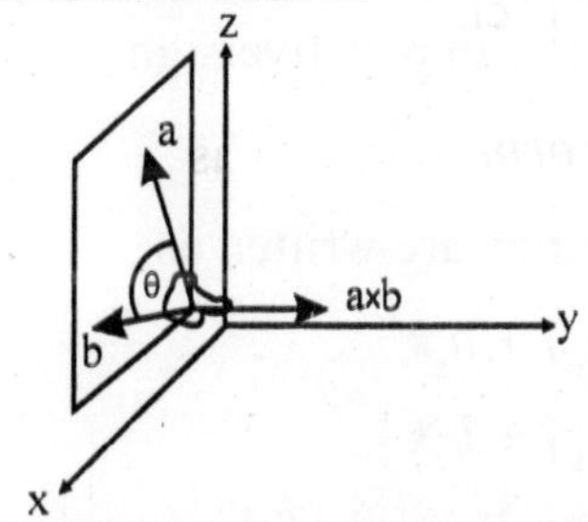

Fig. Direction of Vector Product

Here, we notice that we move in the anti-clockwise direction as we move from vector, *a*, to vector, *b*, while looking at the plane formed by the vectors. This fact can also be used to determine the direction of the vector product. If the orientation is anticlockwise, then the vector product is directed towards us; otherwise the vector product is directed away on the other side of the plane. Further, it is important to note that the direction of cross product can be on a particular side of the plane, depending upon whether we take the product from *a* to *b* or from *b* to *a*. Implying:

$$a \times b = -b \times a$$

Thus, vector product is not commutative.

Values of Cross Product

The value of vector product is maximum for the maximum value of sinθ. Now, the maximum value of $\sin\theta$ is $\sin 90° = 1$. For this value, the vector product evaluates to the product of the magnitude of two vectors multiplied.

Thus, we see that cross product evaluates to negative value, which corresponds to the negative direction of the vector product i.e. other side of the planar formed by the vectors.

$$|a \times b|_{max} = ab$$

and $$|a \,'\, b|_{max} = -ab$$

The vector product evaluates to zero for $\theta = 0°$ and $180°$ as sine of these angles are zero. These results have important implication for unit vectors. The dot products of same unit vector evaluates to 0.

$$i \times i = j \times j = k \times k$$

The cross products of combination of different unit vectors evaluate as :

$$i \times j = k; j \times k = i; k \times i = j$$

$$j \times i = -k; k \times j = -i; i \times k = -j$$

There is a simple rule to determine the sign of the cross product. We write the unit vectors in sequence *i,j,k*. Now, we form pair of vectors as we move from left to right like $i \times j$, $j \times k$ and right to left at the end like $k \times i$ in cyclic manner. The cross products of these pairs result in the remaining unit vector with positive sign. Cross products of other pairs result in the remaining unit vector with negative sign.

Cross product in Component Form

Two vectors in component forms are written as:

$$a = a_x i + a_y j + a_z k$$

$$b = b_x i + b_y j + b_z k$$

In evaluating the product, we make use of the fact that multiplication of the same unit vectors gives the value of 0, while multiplication of two different unit vectors result in remaining vector with appropriate sign. Finally, the vector product evaluates to vector terms:

$$a \times b = (a_x i + a_y j + a_z k) \times (b_x i + b_y j + b_z k)$$

$$a \times b = a_x i \times b_y j + a_x i \times b_z k + a_y j \times b_x i + a_y j \times b_z k + a_z k \times b_x i + a_z k \times b_y j$$

$$a \times b = a_x b_x k - a_x b_z j \times a_y b_x k + a_y b_z i + a_z b_x j - a_z b_y i$$

$$a \times b = (a_y b_z - a_z b_y) i + (a_z b_x - a_x b_z) j + (a_x b_y - a_y b_x) k$$

Evidently, it is difficult to remember above expression. If we know expanding matrix, then we can write above expression in the matrix form, which is easy to remember.

$$\vec{a} \times \vec{b} = \begin{vmatrix} i & j & k \\ a_x & a_y & c_x \\ b_x & b_y & c_y \end{vmatrix}$$

Properties of the Cross Product:

1. The length of the cross product of two vectors is $|a \times b| = ab \sin\theta$
2. The length of the cross product of two vectors is equal to the area of the parallelogram determined by the two vectors.
3. Anticommutativity: $\vec{a} \times \vec{b} = -\vec{b} \times \vec{a}$
4. Multiplication by scalars: $(c\vec{a}) \times \vec{b} = c(\vec{a} \times \vec{b}) = \vec{a} \times (c\vec{b})$
5. Distributivity: $\vec{a} \times (\vec{b} \times \vec{c}) = \vec{a} \times \vec{b} + \vec{a} \times \vec{c}$

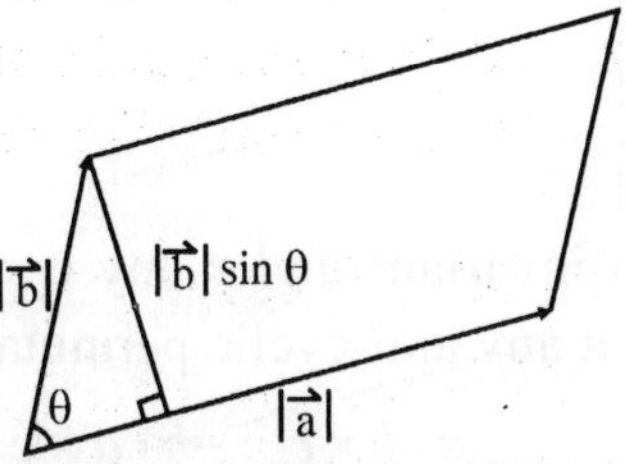

Note that the result for the length of the cross product leads directly to the fact that two vectors are parallel if and only if their cross product is the zero vector. This is true since two vectors are parallel if and only if the angle between them is 0 degrees (or 180 degrees).

Product of three Vectors

Scalar Triple Product

The scalar triple product of three vectors *a*, *b*, and *c* is $(a \times b) \cdot c$. It is a scalar product because it evaluates to a single number (unlike the cross product). The scalar triple product is important because its

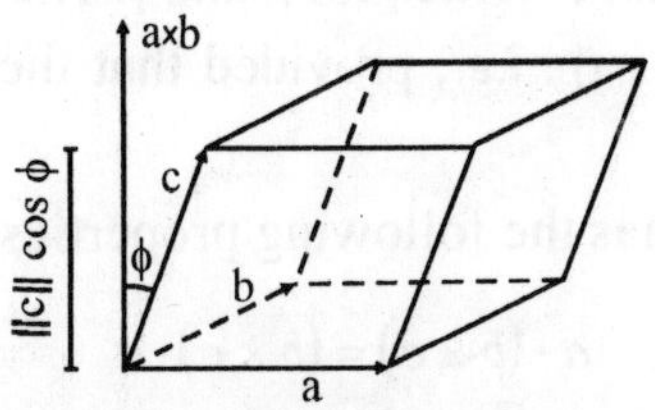

absolute value $|(a \times b) \cdot c|$ is the volume of the parallelepiped spanned by *a*, b, and *c* (i.e., the parallelepiped whose adjacent sides are the vectors *a*, *b*, and *c*).

This results from simple geometry. The volume is the area of the base times the height. We already know the area of the parallelogram base: $\|a \times b\|$. The height is the component of *c* in the direction normal to the base, i.e., in the direction of $a \times b$. Hence the height is $\left|\|c\| \cos\phi\right|$, where ϕ is the angle between *c* and $a \times b$. (Why do we need the absolute value? $\cos\phi$ would be negative if $\phi > \pi/2$.)

The volume of the parallelepiped is therefore

$$\text{Volume} = \left\| \vec{a} \times \vec{b} \right\| \left\| \vec{c} \right\| \cos\phi \Big| = \left| \left(\vec{a} \times \vec{b} \right) \cdot \vec{c} \right|$$

Using the formula for the cross product in components form, we can write the scalar triple product in component form as Write scalar triple product in component form as

$$\left(\vec{a} \times \vec{b} \right) \cdot \vec{c} = \begin{vmatrix} a_2 & a_3 \\ b_2 & b_3 \end{vmatrix} c_1 - \begin{vmatrix} a_1 & a_3 \\ b_1 & b_3 \end{vmatrix} c_2 + \begin{vmatrix} a_1 & a_2 \\ b_1 & b_2 \end{vmatrix} c_3$$

$$= \begin{vmatrix} c_1 & c_2 & c_3 \\ a_1 & a_2 & a_3 \\ b_1 & b_2 & b_3 \end{vmatrix}$$

The triple product is also invariant under any cyclic permutation of *a*, *b*, and *c*, $a \cdot b \times c = b \cdot c \times a = c \cdot a \times b$ but any anti-cyclic permutation causes it to change sign,

$$a \cdot b \times c = -b \cdot a \times c$$

The scalar triple product is zero if any two of *a*, *b*, and *c* are parallel, or if *a*, *b*, and *c* are co-planar. If *a*, *b*, and care non-coplanar, then any vector *r* can be written in terms of them:

$$r = \alpha a + \beta b + \gamma c$$

Forming the dot product of this equation with *b* × *c*, we then obtain

$$r \cdot b \times c = \alpha a \cdot b \times c$$

so

$$\alpha = \frac{r \cdot b \times c}{a \cdot b \times c}$$

Analogous expressions can be written for β and γ. The parameters α, and γ are uniquely determined provided $a \cdot b \times c \neq 0$: *i.e.*, provided that the three basis vectors are not co-planar

The scalar triple product has the following properties

$$a \cdot (b \times c) = (b \times c) \cdot a$$

$$a \cdot (b \times c) = c \cdot (a \times b) = (b \times c) \cdot a$$

$$a \cdot (b \times c) = -a \cdot (c \times b)$$

Vector triple Product

The vector triple product is defined as the cross product of one of the vectors with the cross product of the other two. This is known as the *vector triple product* because it results in a vector.

$$a \times (b \times c) = b(a \cdot c) - c(a \cdot b)$$

For three vectors *a*, *b*, and *c*, the vector triple product is defined $a \times (b \times c)$

1. The brackets are important because $a \times (b \times c) \neq (a \times b) \times c$

2. In fact, it can be demonstrated that

$$a \times (b \times c) = (a \cdot c)b - (a \cdot b)c$$

and

$$(a \times b) \times c = (a \cdot c) - (b \cdot c)a$$

Let us try to prove the first of the above theorems. The left-hand side and the right-hand side are both proper vectors, so if we can prove this result in one particular coordinate system then it must be true in general.

Let us take convenient axes such that the *x*-axis lies along *b*, and *c* lies in the *x*-*y* plane. It follows that $b = (b_x, 0, 0)$, $c = (c_x, c_y, 0)$ and $a = (a_x, a_y, a_z)$.

The vector $b \times c$ is directed along the *z*-axis: $b \times c = (0, 0, b_x c_y)$.

It follows that $a \times (b \times c)$ lies in the *x*-*y* plane:

$$a \times (b \times c) = (a_y b_x c_y, - a_x b_x c_y, 0).$$

To evaluate the right-hand side, of the above equation we need

$$a \cdot c = a_x c_x + a_y c_y \text{ and } a \cdot b = a_x b_x.$$

It follows that the right-hand side is

$$\text{R.H.S} = ([a_x c_y + a_y c_y]b_x, 0) - (a_x b_x c_y, a_x b_x c_y, 0)$$

$$= (a_x c_y b_x, -a_x b_x c_y, 0) = \text{LH.S}$$

which proves the theorem.

The vector triple product has the properties

$$(a \times b) \times c = -c \times (a \times b) = c \times (b \times a)$$
$$a \times b(b \times c) = -a \cdot (c \times b)$$
$$a \times (b \times c) = (a \cdot c)b - (a \cdot b)c$$

ILLUSTRATIVE EXAMPLES

1. The three corners of a triangular slab have co-ordinates of (8, 12, 10), (8, 12, 0) and (1, 2, 1).

   Calculate:

   (a) The length of each of the three sides of the slab;

   (b) The three internal angles;

   (c) The area;

   (d) The equation of the plane in which the slab lies.

   Express this in vector and Cartesian form.

Solution:

Vectors: $a = (8,12,10)$ $\quad b = (8,12.0)$ $\quad c = (1,2,1)$

$$A\bar{B} = b = a = (0,0,-10)$$

$$\overline{BA} = a - b = (0,0,10)$$

$$|\overline{AB}| = |\overline{BA}| = \sqrt{100 = 10}$$

$$\overline{BC} = c - b = (-7,-10,1)$$

$$\overline{CB} = b - c = (7,10,-1)$$

$$|\overline{BC}| = |\overline{CB}| = \sqrt{49+100+1} = \sqrt{150} = 12.25$$

Side lengths = 10, 10.68 and 15.17 units.

$$\overline{AB} \cdot \overline{AC} = |\overline{AB}||\overline{AC}|\cos\theta$$

$$90 = 10 \times 15.17 \cos\theta$$

$$\cos\theta = 0.593$$

$$\theta = 53.6°$$

$$B\bar{A} \cdot BC = |\overline{BA}||\overline{BC}|\cos\theta$$

$$10 = 10 \times 12.25 \cos\theta$$

$$\cos\theta = 0.0816$$

$$\theta = 85.32°$$

$$C\bar{A} \cdot C\bar{B} = |CA||CB|\cos\theta$$

$$140 = 15.17 \times 12.25 \cos\theta$$

$$\cos\theta = 0.753$$

$$\theta = 41.11°$$

$$Area = \frac{1}{2}|\overline{AB} \times \overline{AC}|$$

$$A\bar{B} \times A\dot{C} = \begin{vmatrix} i & j & k \\ 0 & 0 & -10 \\ -7 & -10 & -9 \end{vmatrix} = i\begin{vmatrix} 0 & -10 \\ -10 & -9 \end{vmatrix} - i\begin{vmatrix} 0 & -10 \\ -7 & -9 \end{vmatrix} + i\begin{vmatrix} 0 & 0 \\ -7 & -10 \end{vmatrix}$$

$$= 0 - 100, -(0-70), 0$$

$$= (-100, 70, 0)$$

$$Area = \frac{1}{2}\sqrt{1000+4900+0} = \frac{122.06}{2} = 61.03 units^2$$

Equation of the plane of the triangular slab.

General equation of plane

$$r \cdot n = a \cdot n$$

Where n is a normal vector and a is the position vector of a point in the plane.

A normal vector is given by the cross product of two vector which lie in the plane so we can take n as

$$n = A\bar{B} \times A\bar{C} = -100, 70, 0$$

And if we take the corner point (8,12,10) as the point in the plane then the vector equation of the plane becomes

$$r \cdot (-100, 70, 0) = (8,12,10)(-100, 70, 0) = 4$$

In Cartesian form

$$(x,y,z) \cdot (-100,70,0) = 40 -100x + 70y = 40$$

This can be checked for the other two corner points·

$$b = (8,12,0)$$

$$-800 + 840 + 0 = 40$$

$$c = (1,2,1)$$

$$-100+140+0=40$$

2. The three corners of a triangular slab have co-ordinates of (8, 12, 10), (8, 12, 0) and (1, 2, 1). Calculate:
   (a) The length of each of the three sides of the slab;
   (b) The three internal angles;
   (c) The area;
   (d) The equation of the plane in which the slab lies.
   (f) Express this in vector and Cartesian form.

Solution: a = (8,12,10) b=(8,12,0) c=(1,2,1)

$$AB = b - a = (0,0,-10)$$

$$\overline{AB} = a - b = (0,0,10)$$

$$\left|\overline{AB}\right| = \left|\overline{AB}\right| = \sqrt{100} = 10$$

$$\overline{BC} = b - c = (-7,-10,1)$$

$$CB = b - c = (7,10,-1)$$

$$\left|\overline{BC}\right| = \left|\overline{CB}\right| = \sqrt{49+100+1} = \sqrt{150} = 12.25$$

$$\overline{CA} = a - c = (7,10,9)$$

$$\overline{AC} = c - a = (-7,10,-9)$$

$$\left|\overline{CA}\right| = \left|\overline{AC}\right| = \sqrt{49+100+81} = \sqrt{230} = 15.17$$

Side lengths = 10, 10.68 and 15.17 units.

$$\overline{AB}\cdot\overline{AC}=\left|\overline{AB}\right|\left|\overline{AC}\right|\cos\theta$$

$$90=10\times15.17\cos\theta$$

$$\cos\theta=0.593$$

$$\theta=53.6^{o}$$

$$B\bar{A}\ \cdot B\bar{C}\ =\left|BA\right|\left|BC\right|\cos\theta$$

$$10=10\times12.25\cos\theta$$

$$\cos\theta=0.0816$$

$$\theta=85.32^{\circ}$$

$$C\bar{A}\ \cdot C\bar{B}\ =\left|CA\right|\left|CB\right|\cos\theta$$

$$140=15.17\times12.25\cos\theta$$

$$\cos\theta=0.753$$

$$\theta=41.11^{\circ}$$

The area of the surface is given by

$$Area=\frac{1}{2}\left|\overline{AB}\times\overline{AC}\right|$$

$$A\bar{B}\ \times A\bar{C}\ =\begin{vmatrix} i & j & k \\ 0 & 0 & -10 \\ -7 & -10 & -9 \end{vmatrix}=i\begin{vmatrix} 0 & -10 \\ -10 & -9 \end{vmatrix}-i\begin{vmatrix} 0 & -10 \\ -7 & -9 \end{vmatrix}+i\begin{vmatrix} 0 & 0 \\ -7 & -10 \end{vmatrix}$$

$$=0-100,-(0-70),0$$

$$=(-100,70,0)$$

$$Area=\frac{1}{2}\sqrt{1000+4900+0}=\frac{122.06}{2}=61.03\,\text{units}^2$$

Equation of the plane of the triangular slab.

General equation of plane

$$\mathrm{r}\cdot\mathrm{n}=\mathrm{a}\cdot\mathrm{n}$$

Where n is a normal vector and a is the position vector of a point in the plane. A normal vector is given by the cross product of two vector which lie in the plane so we can take n as

$$\mathrm{n}=A\bar{B}\times A\bar{C}=-100,70,0$$

And if we take the corner point (8,12,10) as the point in the plane then the vector equation of the plane becomes

$$\mathbf{r}\cdot(-100,70,0)=(8,12,10)\,(-100,70,0)=4$$

In Cartesian form

$$(x, y, z)\cdot(-100, 70, 0) = 4C$$

$$-100x + 70y = 4C$$

This can be checked for the other two corner points

b = (8,12,0)

–800 + 840 + 0 = 40

c = (1,2,1)

–100 + 140 + 0 = 40

3. A, B and C are three points with co-ordinates (4, 0, 8) , (2, 6,10) and (0,1,1) (in meters) respectively. Two forces act through A, one of 13.266 kN in the direction AB and a second given (in units of kN) by the vector (8, 1, 2).

Calculate:

(a) The resultant force through A

(b) The resultant force through A in the direction of the vector (1, 2, 2).

(c) The moment due to each force about the point C.

(d) The total moment about the point C.

Solution: A = (4,0,8) B = (2,6,10) C = (0,1,1)

(a) *AB* = b – a = (2 – 4, 6 – 0, 10 – 8) = (–2, 6, 2)

$$\text{Unit vector} = \frac{AB}{|AB|} = \frac{(-2,6,2}{\sqrt{2^2+6^2+2^2}} = \frac{1}{\sqrt{44}}(-2,6,2) = \frac{1}{6.633}(-2,6,2)$$

$$\text{Force vector} = 13.266\frac{1}{6.633}(-2,6,2) = (-4,12,4)$$

Resultant force = (-4,12,4)+(8,1,2) = (4,13,6)

$$\text{(b) Unit vector in direction (2,1,2)} = \frac{(1,2,2)}{\sqrt{1^2+2^2+2^2}} = \frac{(1,2,2)}{3}$$

$$\text{Component in direction (1,2,2)} = \left(\frac{1,2,2}{3}\right)\cdot(4,13,6) = \frac{4+26+12}{3} = \frac{42}{3} = 14$$

(c) Moment = r × F

r = a – c = (4, 0, 1) – (0, 1, 1)

= (4, –1,7)

$$M_1 = (4,1,7)\times(-4,12,4)$$

$$= \begin{vmatrix} i & j & k \\ 4 & -1 & 7 \\ -4 & 12 & 4 \end{vmatrix}$$

$$= i\begin{vmatrix} -1 & 7 \\ 2 & 4 \end{vmatrix} - j\begin{vmatrix} 4 & 7 \\ -4 & 4 \end{vmatrix} + k\begin{vmatrix} 4 & -1 \\ -4 & 12 \end{vmatrix}$$

$$= (-88,-44,44)$$

$$M_2 = (4,1,7)\times(-4,13,6)$$

$$= \begin{vmatrix} i & j & k \\ 4 & -1 & 7 \\ 8 & 1 & 2 \end{vmatrix}$$

$$= i\begin{vmatrix} -1 & 7 \\ 1 & 2 \end{vmatrix} - j\begin{vmatrix} 4 & 7 \\ 8 & 2 \end{vmatrix} + k\begin{vmatrix} 4 & -1 \\ 8 & 1 \end{vmatrix}$$

$$= (-9,48,12)$$

(d) Total moment $\boldsymbol{M_{tot}}=\boldsymbol{M_1} + \boldsymbol{M_2}$ =(-97,4,56)

Alternatively, using the resultant force to get the total moment

$$M_{tot} = (4,1,7)\times(-4,13,6)$$

$$= \begin{vmatrix} i & j & k \\ 4 & -1 & 7 \\ 4 & 13 & 6 \end{vmatrix}$$

$$= i\begin{vmatrix} -1 & 7 \\ 13 & 6 \end{vmatrix} - j\begin{vmatrix} 4 & 7 \\ 4 & 6 \end{vmatrix} + k\begin{vmatrix} 4 & -1 \\ 4 & 13 \end{vmatrix}$$

$$= (-97,4,56)$$

4. A force of 5 units acts through the point (1, 4, 3) in the direction given by the vector (0, 2, 2). Two other forces, given by the vectors (2, 4, 0) and (2, 4, 6) act through the points (2, 2, 2) and (0, 0, 2) respectively.

Find:

1. The sum of the components from all forces in the direction of the vector (2, 1, 2).
2. The moment due to each force about the point (1, 4, 3).
3. The total moment about the point (1, 4, 3)

Hint: Convert the first force to a vector form. Calculate the unit vector of the direction vector given .

Solution: Modulus of the vector $= |0,2,2| = \sqrt{0+4+4} = \sqrt{8}$

Unit vector $= \dfrac{1}{\sqrt{8}}(0,2,2)$

Force vector $= \dfrac{4}{\sqrt{8}}(0,2,2) = \left(0,\sqrt{8},\sqrt{8}\right) = (0, 2.82, 2.828)$

Calculate the unit vector in the direction of the required components

Modulus $= |2,1,2| = \sqrt{4,1,4} = \sqrt{9}$

Force 1 component $= \dfrac{1}{\sqrt{9}}(2,1,2)\cdot\left(0,\sqrt{8},\sqrt{8}\right) = 3.495$

Force 2 component $= \dfrac{1}{\sqrt{9}}(2,1,2)\cdot(2,4,0) = \dfrac{8}{3} = 2.667$

Force 3 component $= \dfrac{1}{\sqrt{9}}(2,1,2)\cdot(2,4,6) = \dfrac{20}{3} = 6.667$

Moments about the point (1,4,3)

Moment

$$M = r \times F$$

where r is the position vector of a point on the line of action of the force relative to the point where the moment is calculated.

Moment due to Force 1

$$r = (1,4,3)-(1,4,3)=0$$

$$M_1 = 0$$

Moment due to Force 2

$$r = (2,2,2) - (1,4,3) = (1, -2, -1)$$

$$M_1 = (1, -2, -1) \times (2, 4, 0)$$

$$= \begin{vmatrix} i & j & k \\ 1 & -2 & -1 \\ 2 & 4 & 0 \end{vmatrix} = i\begin{vmatrix} -2 & -1 \\ 4 & 0 \end{vmatrix} - j\begin{vmatrix} 1 & -1 \\ 2 & 0 \end{vmatrix} + k\begin{vmatrix} 1 & -2 \\ 2 & 4 \end{vmatrix}$$

$$= (0 + 4, -(0 + 2), 4 + 2)$$

$$= (4, -2, 6)$$

Moment due to Force 3

$$r = (0,0,2) - (1, 4, 3) = (-1, -4, -1)$$

$$M_3 = (-1, -4, -1) \times (2, 4, 6)$$

$$= \begin{vmatrix} i & j & k \\ -1 & -4 & -1 \\ 2 & 4 & 6 \end{vmatrix} = i\begin{vmatrix} -4 & -1 \\ 4 & 6 \end{vmatrix} - j\begin{vmatrix} -1 & -1 \\ 2 & 6 \end{vmatrix} + k\begin{vmatrix} -1 & -4 \\ 2 & 4 \end{vmatrix}$$

$$= (-24 + 4, -(0 + 2), -4 + 8)$$

$$= (-20, -2, 4)$$

Total moment

$$M = M_1 + M_2 + M_3 = (-16, -4, 10)$$

EXERCISE

1. Show that $(a \times b) \times c = a \times (b \times c)$ if and only if $(c \times a) \times b = 0$.
2. Find a set of vectors reciprocal to the vectors $2i + 3j - k$, $i - j - 2k$ and $-i + 2j + 2k$.
3. Two side of the triangle are formed of the vectors a $= 3i + 6j - 2k$ and b $= 4i - j + 3k$. Find the angles of the triangle.
4. Find all possible vectors a and b such that $a \times b = 2i + 2j - k$, $a + b = i - 3j + 4k$.
5. Find the value of l for which the give vectors $\lambda i + j + 2k$, $i + \lambda j - k$ and $2i - j + \lambda k$ are coplanar.
6. Let be the three a $= 2i - j + k$, b $= i + 2j - k$ and c $= i + j - 2k$ vectors . Find the vector in the plane of b and c whose projection on a is of magnitude $\sqrt{\frac{2}{3}}$.
7. If the vectors $ai + j + k$, $i + bj + k$ and $i + j + ck$ $(a \neq b \neq c \neq 1)$ are coplanar vectors> find the value of $\frac{1}{1-a} + \frac{1}{1-b} + \frac{1}{1-c}$.
8. If a $= i + j - k$, b $= i - j + k$ and c are unit vectors perpendicular to the vector a and coplanar with a nd b. Find the unit vector d perpendicular to both a and c.
9. If $d = \lambda(a \times b) + \mu(b \times c) + v(c \times a)$ and $[a\ b\ c] = \frac{1}{8}$. Find the value of $\lambda + \mu + v$.

Chapter 4

Definition of a Limit

While a table of numbers can certainly suggest that a limit has a certain value, it cannot definitely prove that the limit has that value. For instance, look at the function we looked at earlier.

$$f(x) = \frac{x^3 - 4x^2 + x + 6}{x^2 - 5x + 6}$$

The table of values we found earlier for x nearing 2 was:

$f(1.5)$	=	2.5	$f(2.5)$	=	3.5
$f(1.7)$	=	2.7	$f(2.3)$	=	3.3
$f(1.8)$	=	2.8	$f(2.2)$	=	3.2
$f(1.9)$	=	2.9	$f(2.1)$	=	3.1
$f(1.99)$	=	2.99	$f(2.01)$	=	3.01
$f(1.999)$	=	2.999	$f(2.001)$	=	3.001

The values certainly look like they are approaching 3, but how do we know for certain? Perhaps they are approaching 3.000075 or 2.999996. What we need is a precise definition of a limit, which will tell us when we are exactly correct.

$$\lim_{x \to a} f(x) = L$$

if for every number $\in > 0$ there is a corresponding number $\delta > 0$ such that

$$| d(x) - L | < \in$$

whenever

$$0 < | x - a| < \delta.$$

Using our example function, $f(x)$ is as above, $a = 2$, and we think that $L = 3$. Let be any positive number. Then we can choose δ to be equal to $\in$. In which case, if $0 < | x - 2 | < \delta$,

$$| f(x) - L | = \left| \frac{x^3 - 4x^2 + x + 6}{x^2 - 5x + 6} - 3 \right|$$

$$= \left| \frac{x^3 - 4x^2 + x + 6}{x^2 - 5x + 6} - \frac{3x^2 + 15x + 18}{x^2 - 5x + 6} \right|$$

$$= \left| \frac{x^3 - 7x^2 + 16x - 12}{x^2 - 5x + 6} \right| = \left| \frac{(x-3)(x-2)(x-2)}{(x-3)(x+2)} \right|.$$

We can cancel the $(x - 3)$ term from the top and bottom of the fraction, as well as one of the $(x - 2)$ terms from the top with the same on the bottom, to get

$$| f(x) - L | = | x - 2 | < \delta = \in .$$

So, since the definition fits exactly, we can state with certainty that the limit as x approaches 2 of $f(x)$ is 3. Also, we can notice that if we used any value other than 3 for L, we wouldn't get the cancellation we did in the last step, so we would not have been able to fit the definition.

Below are the definitions of several related limit concepts.

LEFT-HAND [OR RIGHT-HAND] LIMIT

Let f be a function defined on some open interval (b,a) [or (a,b)]. We say *the left-hand [or right-hand] limit of f(x) as x approaches a is L*, (or *the limit of f(x) as x approaches a from the left [or right] is L*) and we write

$$\lim_{x \to a^-} f(x) = L \left[\lim_{x \to a^+} f(x) = L \right]$$

if for every number $\in > 0$ there is a corresponding number $\delta > 0$ such that

$$| f(x) - L | < \in$$

whenever

$$a - \delta < x < a \; [a < x < a + \delta].$$

[NEGATIVE] INFINITE LIMIT

Let f be a function defined on some open interval that contains the number a, except possibly at a itself. We say *the limit of f(x) as x approaches a is [negative] infinity*, and we write

$$\lim_{x \to a} f(x) = [-]\infty$$

if for every number $N > 0$ $[N < 0]$ there is a corresponding number $\delta > 0$ such that

$$f(x) > N \; [f(x) < N]$$

whenever

$$0 < |x - a| < \delta.$$

LIMIT AT [NEGATIVE] INFINITY

Let f be a function defined on some open interval from a to infinity [from negative infinity to a]. Then we say *the limit of f(x) as x approaches [negative] infinity is L*, and we write

$$\lim_{x \to [-]\infty} f(x) = L$$

if for every number $\in > 0$ there is a corresponding number N such that

$$|f(x) - L| < \in$$

whenever

$$x > N \ [x < N].$$

[NEGATIVE] INFINITE LIMIT AT {NEGATIVE} INFINITY

Let f be a function defined on some open interval from a to infinity {from negative infinity to a}. We say *the limit of f(x) as x approaches {negative} infinity is [negative] infinity*, and we write

$$\lim_{x \to \{-\}\infty} f(x) = [-]\infty$$

if for every number $M > 0$ there is a corresponding number N such that

$$f(x) > M \ [f(x) < M]$$

whenever

$$x > N \ \{x < N\}.$$

Similar definitions can be made for infinite-valued left- or right-hand limits.

BASIC LIMIT LAWS

Listed here are a couple of basic limits and the standard limit laws which, when used in conjunction, can find most limits. They are listed for standard, two-sided limits, but they work for all forms of limits. However, note that if a limit is infinite, then the limit does *not* exist.

Basic Limits

If c is a constant, then $\lim_{x \to a} c = c.$.

$$\lim_{x \to a} x = a.$$

Limit Laws

Addition Law

If the limits $\lim_{x \to a} f(x)$ and $\lim_{x \to a} g(x)$ both exist, then

$$\lim_{x \to a} f(x) + g(x) = \lim_{x \to a} f(x) + \lim_{x \to a} g(x).$$

Subtraction Law

If the limits $\lim_{x \to a} f(x)$ and $\lim_{x \to a} g(x)$ both exist, then

$$\lim_{x \to a} f(x) - g(x) = \lim_{x \to a} f(x) - \lim_{x \to a} g(x).$$

Constant Law

If c is a constant, and the limit $\lim_{x \to a} f(x)$ exists, then

$$\lim_{x \to a} c \cdot f(x) = c \cdot \lim_{x \to a} f(x).$$

Multiplication Law

If the limits

$$\lim_{x \to a} f(x)$$

and

$$\lim_{x \to a} g(x)$$

both exist, then.

$$\lim_{x \to a} f(x) \cdot g(x) = \lim_{x \to a} f(x) \cdot \lim_{x \to a} g(x).$$

Division Law

If the limits $\lim_{x \to a} f(x)$ and $\lim_{x \to a} g(x)$ both exist, and $\lim_{x \to a} g(x) \neq 0$, then

$$\lim_{x \to a} \frac{f(x)}{g(x)} = \frac{\lim_{x \to a} f(x)}{\lim_{x \to a} g(x)}.$$

Power Law

If n is an integer, and the limit $\lim_{x \to a} f(x)$ exists, then

$$\lim_{x \to a} (f(x))^n = \left(\lim_{x \to a} f(x) \right)^n.$$

Root Law

If n is an integer, the limit $\lim_{x \to a} f(x)$ exists, and that limit is positive if n is even, then

$$\lim_{x \to a} \sqrt[n]{f(x)} = \sqrt[n]{\lim_{x \to a} f(x)}.$$

Squeeze Law

If $f(x) \le g(x) \le h(x)$ for all x in an open interval that contains a, except possibly at a itself, and $\lim_{x\to a} f(x) = \lim_{x\to a} h(x) = L$, then

$$\lim_{x\to a} g(x) = L.$$

Composition Law

If f is continuous at b and $\lim_{x\to a} g(x) = b$, then

$$\lim_{x\to a} f(g(x)) = f(b) = f\left(\lim_{x\to a} g(x)\right).$$

A QUICK NOTE ON SEQUENCES

A *sequence* is usually thought of as a list of numbers, such as the sequence

1, 4, 9, 16, 25, ...

and we often want to know whether the sequence converges or diverges, and if it converges, what its limit value is.

All of the discussion about limits on these pages will refer to functions of real variables. However, these results can be fully adapted to sequences, due to the following result:

If $\lim_{x\to a} f(x) = L$ and $f(n) = a_n$ when n is an integer, then

$$\lim_{x\to\infty} a_n = L.$$

From this, we see that all of the limit laws, and l'Hôpital's Rule, hold for sequences as well as functions of real variables.

ZERO OVER ZERO

This is the standard case for l'Hôpital's Rule. If you are finding a limit of a fraction, where the limits of both the numerator and the denominator are zero, then l'Hôpital's Rule says that *the limit of the fraction is the same as the limit of the fraction of the derivatives.* For example,

$$\lim_{x\to 2} \frac{x^2 - 4}{x^2 - 3x + 2} = \frac{2x}{2x - 3}.$$

Both the top and bottom limits of the first fraction are zero, so we can use l'Hôpital's Rule and take derivatives. Note that $2x$ is the derivative of $x^2 - 4$, and $2x - 3$ is the derivative of $x^2 - 3x + 2$. We can then continue to find that

$$\lim_{x\to 2}\frac{2x}{2x-3}=\frac{1}{4}$$

by using the Division Limit Law.

Consider this more complicated example.

$$\lim_{x\to 0}\frac{\sin x - x}{x^3} = \lim_{x\to 0}\frac{\cos x - 1}{3x^2}$$

As x goes to zero, the limits of $\cos x - 1$ and $3x^2$ are both (still) zero, so we can apply l'Hôpital's Rule again.

$$\lim_{x\to 0}\frac{\cos x - 1}{3x^2} = \lim_{x\to 0}\frac{-\sin x}{6x^2} = \lim_{x\to 0}\frac{-\cos x}{6} = -\frac{1}{6}$$

Note that we used l'Hôpital's Rule twice more in that last line. As long as the limits of the numerator and denominator are still both zero (or both infinte), l'Hôpital's Rule can be used.

INFINITY OVER INFINITY

This is another common use of l'Hôpital's Rule. If you are finding a limit of a fraction, where the limits of both the numerator and the denominator are infinite, then l'Hôpital's Rule says that *the limit of the fraction is the same as the limit of the fraction of the derivatives*. For example,

$$\lim_{x\to\infty}\frac{x}{e^x} = \lim_{x\to\infty}\frac{x}{e^x} = 0.$$

Note that both x and e^x approach infinity as x approaches infinity, so we can use l'Hôpital's Rule. Also, the derivative of x is 1, and the derivative of e^x is (still) e^x.

Here is another example.

$$\lim_{x\to\infty}\frac{x^2-4}{x^2-3x+2} = \lim_{x\to\infty}\frac{2x}{2x-3}$$

Note $2x$ is the derivative of $x^2 - 4$, and $2x - 3$ is the derivative of $x^2 - 3x + 2$. Now, the limits of both $2x$ and $2x - 3$ are still infinite, so we can use l'Hôpital's Rule again:

$$\lim_{x\to\infty}\frac{2x}{2x-3} = \lim_{x\to\infty}\frac{2}{2} = 1.$$

Here is a more complicated example:

$$\lim_{x\to 0^+}\frac{\ln x}{\cot x} = \lim_{x\to 0^+}\frac{\frac{1}{x}}{-\csc^2 x} = \lim_{x\to 0^+}\frac{-\sin^2 x}{x}.$$

Here we use the fact that csc(x) = 1/sin(x) to simplify the fraction. Note that for the last fraction, the limit of the numerator and denominator are both zero, which is another case where we can apply l'Hôpital's Rule.

$$\lim_{x\to 0^+} \frac{-\sin^2 x}{x} = \lim_{x\to 0^+} \frac{-2\sin x.\cos x}{1} = \frac{-2\cdot 0\cdot 1}{1} = 0$$

INFINITY TIMES ZERO

Consider

$$\lim_{x\to 0^+} \ln x \cdot \sin x.$$

In this case, there is no fraction in the limit. Since the limit of ln(x) is negative infinity, we cannot use the Multiplication Limit Law to find this limit. We can convert the product ln(x)*sin(x) into a fraction:

$$\lim_{x\to 0^+} \ln x \cdot \sin x = \lim_{x\to 0^+} \frac{\ln x}{\frac{1}{\sin x}} = \lim_{x\to 0^+} \frac{\ln x}{\csc x}.$$

Now, we have a fraction where the limits of both the numerator and denominator are infinite. Thus, we can apply l'Hôpital's Rule:

$$\lim_{x\to 0^+} \frac{\ln x}{\csc x} = \lim_{x\to 0^+} \frac{\frac{1}{x}}{-\csc x \cdot \cot x} = \lim_{x\to 0^+} \frac{\frac{1}{x}}{-\frac{1}{\sin x}\frac{1}{\tan x}}$$

$$= \lim_{x\to 0^+} \frac{-\sin x \cdot \tan x}{x}.$$

Remember that the derivative of ln(x) is $1/x$, and the derivative of csc(x) is -csc(x)cot(x).

We can now use l'Hôpital's Rule again, as the limits of both the top and the bottom are zero, using the Product Rule to take the derivative of the numerator.

$$\lim_{x\to 0^+} -\frac{\sin x \cdot \tan x}{x} = \lim_{x\to 0^+} -\frac{\cos x \cdot \tan x + \sin x \cdot \sec^2 x}{1}$$

$$\lim_{x\to 0^+} -\left(\sin x + \frac{\tan x}{\cos x}\right) = -\left(0 + \frac{0}{1}\right) = 0$$

INFINITY MINUS INFINITY

Often, particularly with fractions, l'Hôpital's Rule can help in cases where one term with infinite limit is subtracted from another term with infinite limit. This is generally done by finding common denominators. For example:

$$\lim_{x\to 1}\frac{x}{x-1}-\frac{1}{\ln x} = \lim_{x\to 1}\frac{x\ln x - x + 1}{(x-1)\ln x}.$$

Now, we have just one fraction, and both the numerator and denominator have a limit of zero. So, we can apply l'Hôpital's Rule.

$$\lim_{x\to 1}\frac{x\ln x - x + 1}{(x-1)\ln x} = \lim_{x\to 1}\frac{1.\ln x + x\cdot\frac{1}{x} - 1}{1\cdot\ln x + \frac{x-1}{x}} = \lim_{x\to 1}\frac{\ln x + 1 - 1}{\ln x + \frac{x-1}{x}}$$

$$= \lim_{x\to 1}\frac{\ln x}{\frac{x\ln x + (x-1)}{x}} = \lim_{x\to 1}\frac{x\ln x}{x\ln x + (x-1)}$$

We can now use l'Hôpital's Rule again, since the limits of the numerator and denominator are again zero, using the Product Rule carefully on the top and bottom of the fraction.

$$\lim_{x\to 1}\frac{x\ln x}{x\ln x + (x-1)} = \lim_{x\to 1}\frac{1\cdot\ln x + x\cdot\frac{1}{x}}{1\cdot\ln x + x\cdot\frac{1}{x} + 1} = \lim_{x\to 1}\frac{\ln x + 1}{\ln x + 2}$$

$$= \frac{\ln 1 + 1}{\ln 1 + 2} = \frac{0+1}{0+2} = \frac{1}{2}$$

LIMITS OF POWERS

All of the different forms of powers of limits are handled in the same way. Since these functions don't have any obvious fractions in them, it doesn't look like l'Hôpital's Rule will apply at all to them. However, with the use of the exponential function, we can put a function into fractional form. Recall that

$$a^b = \exp(b \ln a)$$

and use that to find the limit. And, since the exponential function is continuous, you can use the Composition Limit Law to bring the limit inside the exponential function. In general, handle exponential limits in this way:

$$\lim f(x)^{g(x)} = \lim \exp(g(x)\cdot\ln f(x)) = \text{lm} \exp\left(\frac{\ln f(x)}{\frac{1}{g(x)}}\right)$$

$$= \exp\left(\lim\frac{\ln f(x)}{\frac{1}{g(x)}}\right)$$

Consider the following limit:

$$\lim_{x\to 0^+} x^x.$$

Using the regular limit laws, we cannot find this limit. Use the exponential function to change the form of your limit.

$$\lim_{x\to 0^+} x^x = \lim_{x\to 0^+} \exp(x\ln x)$$

Now we can use the Composition Limit Law to bring the limit "inside" the exponential function:

$$\lim_{x\to 0^+} \exp(x\ln x) = \exp\left(\lim_{x\to 0^+} x\ln x\right)$$

but only if we can find a limit for the inside function. Now, we can rewrite the limit as follows:

$$\lim_{x\to 0^+} x\ln x = \lim_{x\to 0^+} \frac{\ln x}{\frac{1}{x}}$$

and since both ln(x) and $1/x$ have infinite limit, we can use l'Hôpital's Rule on the limit.

$$\lim_{x\to 0^+} \frac{\ln x}{\frac{1}{x}} = \lim_{x\to 0^+} \frac{\frac{1}{x}}{\frac{1}{x^2}} = \lim_{x\to 0^+} \frac{-x^2}{x} = \lim_{x\to 0^+} -x = 0$$

So, $$\lim_{x\to 0^+} x^x = e^0 = 1.$$

Here is another example of how this method can work.

$$\lim_{x\to 0^+} (\cos x)^{\frac{1}{x^2}} = \lim_{x\to 0} \exp\left(\frac{\ln\cos x}{x^2}\right) = \exp\left(\lim_{x\to 0} \frac{\ln\cos x}{x^2}\right)$$

Now, we can use l'Hôpital's Rule on the fraction, since both the numerator and denominator have limit zero,

$$\lim_{x\to 0} \frac{\ln\cos x}{x^2} = \lim_{x\to 0} \frac{\frac{1}{\cos x}\cdot(-\sin x)}{2x} = \lim_{x\to 0} -\frac{\tan x}{2x}$$

and then use it again to find the limit.

$$\lim_{x\to 0} -\frac{\tan x}{2x} = \lim_{x\to 0} -\frac{\sec^2 x}{2} = -\frac{\sec^2 0}{2} = -\frac{1^2}{2} = -\frac{1}{2}$$

Now we can state that

$$\lim_{x\to 0} (\cos x)^{\frac{1}{x^2}} = e^{-\frac{1}{2}} = \frac{1}{\sqrt{2}}.$$

Here is a third example.

$$\lim_{x\to 1^+}\left(\frac{1}{x-1}\right)^{\ln x} = \lim_{x\to 1^+} \exp(\ln x \cdot - \ln(x-1)) = \exp\left(\lim_{x\to 1^+} -\frac{\ln x-1)}{\frac{1}{\ln x}}\right)$$

Now apply l'Hôpital's Rule to the fraction, as both the top and the bottom of the fraction have infinite limits.

$$\lim_{x\to 1^+} -\frac{\ln x-1)}{\frac{1}{\ln x}} = \lim_{x\to 1^+} -\frac{\frac{1}{x-1}}{-\frac{1}{\ln^2 x}\cdot\frac{1}{x}} = \lim_{x\to 1^+} \frac{x\ln^2 x}{x-1}$$

We can use l'Hôpital's Rule again here, this time with both limits being zero.

$$\lim_{x\to 1^+} \frac{x\ln^2 x}{x-1} = \lim_{x\to 1^+} \frac{1\cdot \ln^2 x + 2x\ln x\cdot \frac{1}{x}}{1}$$

$$= \lim_{x\to 1^+} \ln^2 x + 2\ln x = 0^2 + 2{\cdot}0 = 0$$

So, we can put the limit back in to find the final answer.

$$\lim_{x\to 1^+}\left(\frac{1}{x-1}\right)^{\ln x} = e^0 = 1$$

SERIES, CONVERGENCE, AND SERIES TESTS

You may recall, from back when you first started studying integration, that you approximated the area under a curve by adding up a bunch of rectangles.

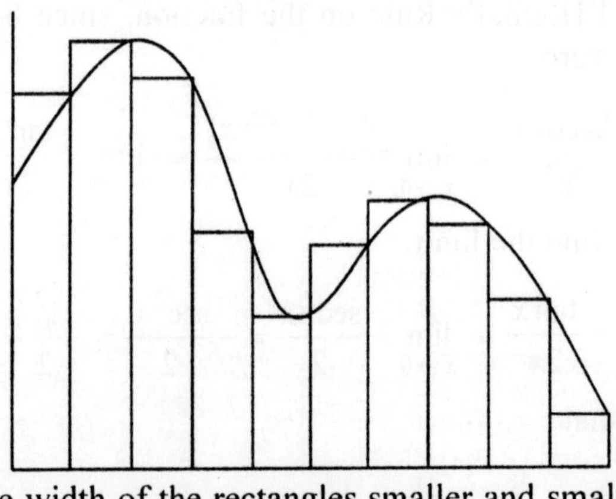

You then made the width of the rectangles smaller and smaller, finding the limit of the area as the width went to zero, to find the area. In a sense, you were adding up an infinite number of rectangle areas. Series are a way to formally add up an an infinite amount of numbers, in a way that makes sense.

However, determining whether or not the series you are looking at converges, that is, whether or not it has a real sum, can be tricky. In addition to some basic information on series, these pages should be able to help you determine whether or not your series converges.

In this set of pages, the series of interest will always by symbolized as the sum, as n goes from 1 to infinity, of $a[n]$. In addition, any auxilliary sequence will be symbolized as the sum, as n goes from 1 to infinity, of $b[n]$. Symbolically, that is

$$\sum_{n=1}^{\infty} a_n \text{ and } \sum_{n=1}^{\infty} b_n .$$

THE DEFINITION OF A SERIES

Let

$$a_1, a_2, a_3, \ldots$$

be a sequence. We call the sum

$$a_1 + a_2 + a_3 + \ldots + a_n + \ldots$$

an *infinite series* (or just a *series*) and denote it as .

$$\sum_{n=1}^{\infty} a_n .$$

We define a second sequence, $s[n]$, called the *partial sums*, by ,

$$\delta_1 = a_1,$$
$$\delta_2 = a_1 + a_2,$$
$$\delta_3 = a_1 + a_2 + a_3$$

or, in general,

$$\delta_n = \sum_{i=1}^{n} a_i .$$

We then define convergence as follows:

DEFINITION OF CONVERGENCE, OR THE LIMIT OF A SERIES

Given a series

$$\sum_{n=1}^{\infty} a_n = a_1 + a_2 + a_3 + \ldots + a_n + \ldots$$

let $s[n]$ denote its nth partial sum:

$$\delta_n = \sum_{n=1}^{\infty} a_n = a_1 + a_2 + a_3 + \ldots + a_n.$$

If the sequence $s[n]$ has a limit, that is, if there is some s such that for all > 0 there exists some $N > 0$ such that $|s[n] - s| < \in$, then the series is called *convergent*, and we say the series *converges*. We write

$$a_1 + a_2 + a_3 + \ldots + a_n + \ldots = \delta$$

or

$$\sum_{n=1}^{n} a_n = \delta.$$

The number s is called the *sum of the series*. If the series does not converge, the series is called *divergent*, and we say the series *diverges*.

LIST OF COMMON SERIES

These pages list several series which are important for comparison purposes. They are useful for the comparison tests: the 'regular' Comparison Test and the Limit Comparison Test. Click on the name of the series to get more information on the series.

Series List

Geometric Series

$a + ar + ar^2 + \ldots$ The term of the series is ar^n. This series converges if $|r| < 1$ and diverges otherwise. If it converges, it converges to $a/(1\text{-}r)$.

p-series

The term of the p-series is $1/n^p$. This series converges if $p>1$ and diverges otherwise. If $p=1$, this is the *harmonic series*.

List of Series Tests

The series of interest will always by symbolized as the sum, as n goes from 1 to infinity, of $a[n]$. In addition, any auxilliary sequence will be symbolized as the sum, as n goes from 1 to infinity, of $b[n]$. Or, symbolically,

$$\sum_{n=1}^{n} a_n \text{ and } \sum_{n=1}^{n} b_n .$$

Click on the name of the test to get more information on the test.

The Common Series Tests

Divergence Test

If the limit of $a[n]$ is not zero, or does not exist, then the sum diverges.

$$\lim_{n\to\infty} a_n \text{ does not exist, or}$$

$$\sum_{n=1}^{n} a_n \neq 0 \Rightarrow \lim_{n\to\infty} a_n \text{ diverges}$$

Integral Test

If you can define f so that it is a continuous, positive, decreasing function from 1 to infinity (including 1) such that $a[n]=f(n)$, then the sum will converge if and only if the integral of f from 1 to infinity converges.

f continuous, positive, decreasing on $[1, \infty)$

such that $\quad a_n = f(n)$

$$\sum_{n=1}^{\infty} a_n \text{converges} \Leftrightarrow \int_1^{\infty} f(x)\,dx \text{ converges}$$

Please note that this does *not* mean that the sum of the series is that same as the value of the integral. In most cases, the two will be quite different.

Comparison Test

Let $b[n]$ be a second series. Require that all $a[n]$ and $b[n]$ are positive. If $b[n]$ converges, and $a[n] <= b[n]$ for all n, then $a[n]$ also converges. If the sum of $b[n]$ diverges, and $a[n] >= b[n]$ for all n, then the sum of $a[n]$ also diverges.

$$a_n, b_n > 0 \text{ for all } n$$

$$\sum_{n=1}^{\infty} b_n \text{converges, } a_n \leq b_n \text{ for all } n \Rightarrow \sum_{n=1}^{\infty} a_n \text{converges}$$

$$\sum_{n=1}^{\infty} a_n \text{diverges, } a_n \geq b_n \text{ for all } n \Rightarrow \sum_{n=1}^{\infty} a_n \text{diverges}$$

Limit Comparison Test

Let $b[n]$ be a second series. Require that all $a[n]$ and $b[n]$ are positive.

1. If the limit of $a[n]/b[n]$ is positive, then the sum of $a[n]$ converges if and only if the sum of $b[n]$ converges.

2. If the limit of $a[n]/b[n]$ is zero, and the sum of $b[n]$ converges, then the sum of $a[n]$ also converges.
3. If the limit of $a[n]/b[n]$ is infinite, and the sum of $b[n]$ diverges, then the sum of $a[n]$ also diverges.

$$a_n, b_n > 0 \text{ for all } n$$

$$\lim_{n\to\infty} \frac{a_n}{b_n} > 0 \Rightarrow \left(\sum_{n=1}^{\infty} a_n \text{converges}, \Leftrightarrow \sum_{n=1}^{\infty} b_n \text{converges} \right)$$

$$\lim_{n\to\infty} \frac{a_n}{b_n} = 0 \text{ and } \sum_{n=1}^{\infty} b_n \text{converges}, \Rightarrow \sum_{n=1}^{\infty} a_n \text{converges}$$

$$\lim_{n\to\infty} \frac{a_n}{b_n} = \infty \text{ and } \sum_{n=1}^{\infty} b_n \text{diverges} \Rightarrow \sum_{n=1}^{\infty} a_n \text{diverges}$$

Alternating Series Test

If $a[n] = (-1)^{(n+1)}b[n]$, where $b[n]$ is positive, decreasing, and converging to zero, then the sum of $a[n]$ converges.

$$a_n = (-1)n{+}1.\ b_n,\ b_n > 0 \text{ for all } n,\ b_n \text{ decreasing}$$

Absolute Convergence Test

If the sum of $|a[n]|$ converges, then the sum of $a[n]$ converges.

$$\sum_{n=1}^{\infty} |\, a_n| \text{ converges} \Rightarrow \sum_{n=1}^{\infty} a_n \text{ converges}$$

$$\lim_{n\to\infty} \left| \frac{a_{n+1}}{a_n} \right| < 1 \Rightarrow \sum_{n=1}^{\infty} a_n \text{ converges}$$

Ratio Test

If the limit of $|a[n+1]/a[n]|$ is less than 1, then the series (absolutely) converges. If the limit is larger than one, or infinite, then the series diverges.

$$\lim_{n\to\infty} \left| \frac{a_{n+1}}{a_n} \right| < 1 \Rightarrow \sum_{n=1}^{\infty} a_n \text{ diverges}$$

Root Test

If the limit of $|a[n]|^{(1/n)}$ is less than one, then the series (absolutely) converges. If the limit is larger than one, or infinite, then the series diverges.

$$\lim_{n\to\infty} \sqrt[n]{|a_n|} < 1 \Rightarrow \sum_{n=1}^{\infty} a_n \text{ converges}$$

$$\lim_{n\to\infty} \sqrt[n]{|a_n|} < 1 \Rightarrow \sum_{n=1}^{\infty} a_n \text{ diverges}$$

Start of Tests

Look at your series. Treat the individual terms of the series as a sequence instead. Does this sequence converge to zero? If not, then the series does not converge. This is called the Divergence Test. If the limit is zero, then continue with the tests. If it's difficult to tell, then continue with the tests. For example,

$$\sum_{n=1}^{\infty} n^2 \text{ diverges since } \lim_{n\to\infty} n^2 \neq 0$$

TRY TO FIND THE FORM OF THE SERIES TERMS

Try to determine the form of the terms of the series. Pick the type that best seems to fit your series.

1. Type A: n never appears in a power.
2. Type B: n only appears in a power.
3. Type C: n appears in both a power and a base.
4. Type D: $n!$ appears.

Some examples:

Type A $\dfrac{n^3 + \sqrt{n}}{2n^2 \sqrt[R]{3n + 2n^2}}$

Type B $\dfrac{2^{n+2}}{3^{n+1}}$

Type C $\left(\dfrac{3n^3 + 2n - 1}{2n^3 + \dfrac{1}{n}} \right)^n$

Type D $\dfrac{n!}{n^n}$

LIST OF COMMON MACLAURIN SERIES

$$\frac{1}{1-x} = \sum_{n=0}^{\infty} x^n = 1 + x + x^2 + x^3 + \ldots \text{ for } x \text{ in } (-1, 1)$$

$$e^x = \sum_{n=0}^{\infty} \frac{x^n}{n!} = 1 + x + \frac{x^2}{2} + \frac{x^3}{6} + \ldots \text{ for } x \text{ all } x$$

$$\sin(x) = \sum_{n=0}^{\infty} (-1)^n \frac{x^{2n+1}}{(2n)!} = x - \frac{x^3}{6} + \frac{x^5}{120} - \frac{x^7}{5040} + \ldots \text{ for } x \text{ all } x$$

$$\cos(x) = \sum_{n=0}^{\infty} (-1)^n \frac{x^{2n+1}}{(2n)!} = 1 - \frac{x^3}{2} + \frac{x^5}{24} - \frac{x^6}{720} + \ldots \text{ for all } x$$

$$\arctan(x) = \sum_{n=0}^{\infty} (-1)^n \frac{x^{2n+1}}{2n+1} = x - \frac{x^3}{3} + \frac{x^5}{5} - \frac{x^7}{7} + \ldots \text{ for } x \text{ in} [-1, 1]$$

EXERCISES

For Exercises 1-6, state the domain and range of the given function.

1. $f(x, y) = x^2 + y^2 - 1$
2. $f(x, y) = \dfrac{1}{x^2 + y^2}$
3. $f(x, y) = \sqrt{x^2 + y^2 - 4}$
4. $f(x, y) = \dfrac{x^2 + 1}{y}$
5. $f(x, y, z) = \sin(xyz)$
6. $f(x, y, z) = \sqrt{(x-1)(yz-1)}$

For Exercises 7-18, evaluate the given limit.

7. $\lim\limits_{(x,y)\to(0,0)} \cos(xy)$
8. $\lim\limits_{(x,y)\to(0,0)} e^{xy}$
9. $\lim\limits_{(x,y)\to(0,0)} \dfrac{x^2 - y^2}{x^2 + y^2}$

10. $\displaystyle\lim_{(x,y)\to(0,0)} \frac{xy^2}{x^2+y^2}$

11. $\displaystyle\lim_{(x,y)\to(1,-1)} \frac{x^2-2xy+y^2}{x-y}$

12. $\displaystyle\lim_{(x,y)\to(0,0)} \frac{xy^2}{x^2+y^2}$

13. $\displaystyle\lim_{(x,y)\to(1,1)} \frac{x^2-y^2}{x-y}$

14. $\displaystyle\lim_{(x,y)\to(0,0)} \frac{x^2-2xy+y^2}{x-y}$

15. $\displaystyle\lim_{(x,y)\to(0,0)} \frac{y^4\sin(xy)}{x-y}$

16. $\displaystyle\lim_{(x,y)\to(0,0)} (x^2+y^2)\cos\left(\frac{1}{xy}\right)$

17. $\displaystyle\lim_{(x,y)\to(0,0)} \frac{x}{y}$

18. $\displaystyle\lim_{(x,y)\to(0,0)} \cos\left(\frac{1}{xy}\right)$

19. Show that $f(x, y) = \dfrac{1}{2\pi\sigma^2}e^{-(x^2+y^2)/2\sigma^2}$, for $\sigma > 0$, is constant on the circle of radius $r > 0$ centered at the origin. This function is called a *Gaussian blur*, and is used as a filter in image processing software to produce a "blurred" effect.
20. Suppose that $f(x, y) \le f(y, x)$ for all (x, y) in $\mathbb{R}^2$. Show that $f(x, y) = f(y, x)$ for all (x, y) in $\mathbb{R}^2$.
21. Use the substitution $r = \sqrt{x^2+y^2}$ to show that

$$\lim_{(x,y)\to(0,0)} \frac{\sin\sqrt{x^2+y^2}}{\sqrt{x^2+y^2}} = 1.$$

Chapter 5

Calculus of Vectors

In this chapter we learn about Vector Calculus i.e. Differentiation and Integration of vectors and their application.

LIMIT AND CONTINUITY

Let a vector function *f* be defined for all values of t in some neighborhood about a point t_0 possibly at t0. Then, a vector a is called the limit of f as $t \to t0$ if given any $\varepsilon > 0$, however small, there exists a real number $\delta > 0$ such that

$|f(t) - a| < \varepsilon$, whenever $0 < |t - t_0| < \delta$

and we write $\lim_{t \to t_0} f(t) = a$ or $\underset{t \to t_0}{Lt} f(t) = a$.

The same definition holds for the limit of a scalar function if *f* is replaced by a scalar function *f*, and vector *a* by a scalar *a*.

We state below some important theorems on limits of a vector function.

Theorem: If $f = f_1 i + f_2 j + f_3 k$ and $a = a_1 i + a_2 j + a_3 k$, then

$\lim_{t \to t_0} f(t) = a$ if and only if $\lim_{t \to t_0} f_i(t) = a_i, i = 1, 2, 3$.

Theorem: Let *f* and g be vector functions of a scalar variable t. Let f be a scalar function of t.

Then,

1. $\underset{t \to t_0}{Lt}[f(t) + g(t)] = \underset{t \to t_0}{Lt} f(t) + \underset{t \to t_0}{Lt} g(t)$

2. $\underset{t \to t_0}{Lt}[f(t) - g(t)] = \underset{t \to t_0}{Lt} f(t) - \underset{t \to t_0}{Lt} g(t)$

3. $\underset{t\to t_0}{Lt}\left[f(t)\cdot g(t)\right]=\left\{\underset{t\to t_0}{Lt} f(t)\right\}\cdot\left\{\underset{t\to t_0}{Lt} g(t)\right\}$

4. $\underset{t\to t_0}{Lt}\left[f(t)\times g(t)\right]=\left\{\underset{t\to t_0}{Lt} f(t)\right\}\times\left\{\underset{t\to t_0}{Lt} g(t)\right\}$

5. $\underset{t\to t_0}{Lt}\left[\phi(t)\times g(t)\right]=\left\{\underset{t\to t_0}{Lt} \phi(t)\right\}\times\left\{\underset{t\to t_0}{Lt} g(t)\right\}$

A vector function *f* is said to be continuous at $t = t_0$ if it is defined in some neighborhood of t_0 and $\underset{t\to t_0}{Lt} f(t)=\underset{t\to t_0}{Lt} f(t)$

Theorem: If $f = f_1 i + f_2 j + f_3 k$ is a function of t, then *f* is continuous at t_0 if and only if f_1, f_2 and f_3 are all continuous at t_0.

DIÛERENTIATION OF VECTOR-VALUED FUNCTIONS

A curve C is deûned by r = r(t), a vector-valued function of one (scalar) variable. Let us imagine that C is the path taken by a particle and t is time. The vector r(t) is the position vector of the particle at time t and r(t + h) is the position vector at a later time t + h. The average velocity of the particle in the time interval [t, t + h] is then

$$\frac{\text{displacement vector}}{\text{lengh of time interval}}=\frac{r(t+h)-r(t)}{h}$$

In terms of the components of r this is

$$\left(\frac{x(t+h)-x(t)}{h},\ \frac{y(t+h)-y(t)}{h},\ \frac{z(t+h)-z(t)}{h}\right)$$

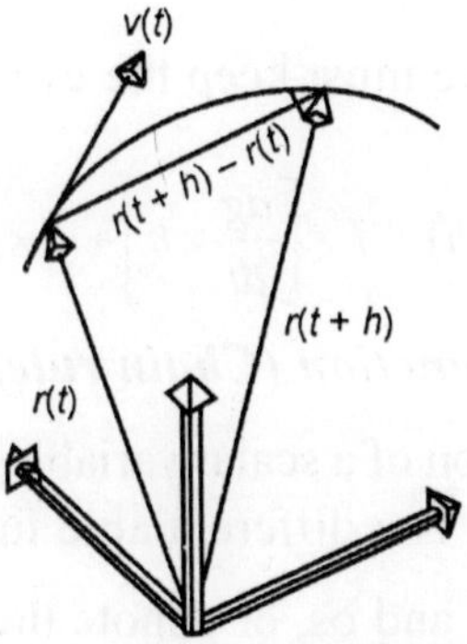

If each of the scalar function x, y and z are diûerentiable, then this vector has a limit

$$\left(\frac{dx}{dt},\frac{dy}{dt},\frac{dz}{dt}\right)=(\dot{x},\dot{y},\dot{z})$$

which is the instantaneous velocity of the particle v(t). This means that (if the motion is smooth) then

$$v(t) = \lim_{h \to} \frac{r(t+h) - r(t)}{h} = \frac{d}{dt} r(t)$$

Rules for Differentiation. To differentiate the vector product of vector function, the order of factors must be preserved because the vector product is not commutative.. If f, g and h are differentiable vector functions and ϕ is a scalar function of a scalar variable t, then

1. $\frac{d}{dt}(f \pm g) = \frac{df}{dt} \pm \frac{dg}{dt}$

2. $\frac{d}{dt}(f \cdot g) = f \cdot \frac{dg}{dt} + g \cdot \frac{df}{dt}$

   [Since f.g = g.f,\in evaluating $\frac{d}{dt}(f \cdot g)$, we do not care about the order of f and g]

3. $\frac{d}{dt}(f \times g) = f \times \frac{dg}{dt} + \frac{df}{dt} \times g$

   [Since f × g = –g′f,\ in evaluating $\frac{d}{dt}(f \times g)$, we do not care about the order of f and g]

4. $\frac{d}{dt}(\phi f) = \phi \frac{df}{dt} + f \frac{d\phi}{dt}$

5. $\frac{d}{dt}[fgh] = \left[\frac{df}{dt} gh\right] + \left[f \frac{dg}{dt} h\right] + \left[fg \frac{dh}{dt}\right]$

   Where [fgh] denotes the scalar triple product of f, g and h.

   [In evaluating $\frac{d}{dt}[fgh]$, we must keep the cyclic order of f, g and h in tact]

6. $\frac{d}{dt}(f \times (g \times h)) = \frac{df}{dt} \times (g \times h) + f \times \left[\frac{dg}{dt} \times h\right] + f \times \left[g \times \frac{dh}{dt}\right]$

Derivative of a Function of a Function (Chain rule)

Let r be a differentiable function of a scalar variables s and s be a differentiable function of another scalar variable t, then r is a differentiable function of t.

Let δt be a small change in t and δs, δr denote the corresponding changes in s and r respectively.

Now, $$\frac{\delta r}{\delta t} = \frac{\delta r}{\delta s} \cdot \frac{\delta s}{\delta t}$$

Taking limits as $\delta t \to 0$ (and hence $\delta s \to 0$), we have

$$\underset{\delta t \to 0}{Lt} \frac{\delta r}{\delta t} = \left(\underset{\delta t \to 0}{Lt} \frac{\delta r}{\delta s} \right) \cdot \left(\underset{\delta t \to 0}{Lt} \frac{\delta s}{\delta t} \right) \text{ or } \frac{dr}{dt} = \frac{dr}{ds} \cdot \frac{ds}{dt}$$

Constant Vectors

A vector is said to be constant if both its magnitude and direction do not change.

Partial Derivatives of Vector Functions of more than one Variable

So far, we have considered vector functions of a single scalar variable t. But, we come across with many vector functions of more than one independent scalar variable.

If f is a vector function of independent variables x, y and z, we write f = f (x, y, z).

Let δx be a small change in x, while other variables remain unchanged. Let f be the corresponding change in f,

Then, $\delta f = f(x + \delta x, y, z) - f(x, y, z)$

Dividing by δx both sides we get

$$\frac{\delta f}{\delta x} = \frac{f(x+\delta x, y, z) - f(x, y, z)}{\delta x}$$

Taking limits $\delta x \to 0$, we have

$$\underset{\delta x \to 0}{Lt} \frac{\delta f}{\delta x} = \underset{\delta x \to 0}{Lt} \frac{f(x+\delta x, y, z) - f(x, y, z)}{\delta x}$$

If this limit exists, we denote it by $\frac{\partial f}{\partial x}$ or f_x and is called the partial derivative of f w. r. t x i.e.,

$$\frac{\partial f}{\partial x} = \underset{\delta x \to 0}{Lt} \frac{\delta f}{\delta x} = \underset{\delta x \to 0}{Lt} \frac{f(x+\delta x, y, z) - f(x, y, z)}{\delta x}$$

Similarly we define the partial derivatives

$$\frac{\partial f}{\partial y} = \underset{\delta y \to 0}{Lt} \frac{\delta f}{\delta y} = \underset{\delta y \to 0}{Lt} \frac{f(x, y+\delta y, z) - f(x, y, z)}{\delta y}$$

and $$\frac{\partial f}{\partial z} = \underset{\delta z \to 0}{Lt} \frac{\delta f}{\delta z} = \underset{\delta z \to 0}{Lt} \frac{f(x, y, z+\delta z) - f(x, y, z)}{\delta z}$$

Higher –order partial derivatives may also be defined in the same way. For example,

$$f_{xx} = \frac{\partial^2 f}{\partial x^2} = \frac{\partial}{\partial x}\left(\frac{\partial f}{\partial x}\right) \qquad f_{yy} = \frac{\partial^2 f}{\partial y^2} = \frac{\partial}{\partial y}\left(\frac{\partial f}{\partial y}\right)$$

$$f_{xy} = \frac{\partial}{\partial y}\left(\frac{\partial f}{\partial x}\right) = \frac{\partial^2 f}{\partial y \partial x}, \qquad f_{yx} = \frac{\partial}{\partial x}\left(\frac{\partial f}{\partial y}\right) = \frac{\partial^2 f}{\partial x \partial y}$$

If f has continuous partial derivatives of second order, then

$$\frac{\partial^2 f}{\partial x \partial y} = \frac{\partial^2 f}{\partial y \partial x}.$$

Rules for Partial Differentiation

Rules for partial differentiation of vectors are similar to those used in elementary calculus for scalar functions. Thus, if f and g are differentiable vector functions of x, y, z and ϕ is a differentiable scalar function of x, y, z then (using x as an example)

1. $\frac{\partial}{\partial x}(f \pm g) = \frac{\partial f}{\partial x} \pm \frac{\partial g}{\partial x}$

2. $\frac{\partial}{\partial x}(f \cdot g) = f \cdot \frac{\partial g}{\partial x} + g \cdot \frac{\partial f}{\partial x}$

3. $\frac{\partial}{\partial x}(f \times g) = f \times \frac{\partial g}{\partial x} + \frac{\partial f}{\partial x} \times g$

4. $\frac{\partial}{\partial x}(\phi f) = \phi \frac{\partial f}{\partial x} + f \frac{\partial \phi}{\partial x}$

5. $\frac{\partial^2}{\partial y \partial x}(f \cdot g) = \frac{\partial}{\partial y}\left\{\frac{\partial}{\partial x}(f \cdot g)\right\} = \frac{\partial}{\partial y}\left\{f \cdot \frac{\partial g}{\partial x} + g \cdot \frac{\partial f}{\partial x}\right\} = \frac{\partial}{\partial y}\left(f \cdot \frac{\partial g}{\partial x}\right) + \frac{\partial}{\partial y}\left(g \cdot \frac{\partial f}{\partial x}\right)$

$$= f \cdot \frac{\partial^2 g}{\partial y \partial x} + \frac{\partial f}{\partial x} \cdot \frac{\partial g}{\partial x} + \frac{\partial f}{\partial x} \cdot \frac{\partial g}{\partial x} + \frac{\partial^2 f}{\partial y \partial x} \cdot g$$

6. $\frac{\partial^2}{\partial x \partial y}(f \times g) = \frac{\partial}{\partial x}\left\{\frac{\partial}{\partial y}(f \times g)\right\} = \frac{\partial}{\partial x}\left\{f \times \frac{\partial g}{\partial y} + \frac{\partial f}{\partial y} \times g\right\} = \frac{\partial}{\partial x}\left(f \times \frac{\partial g}{\partial y}\right) + \frac{\partial}{\partial x}\left(\frac{\partial f}{\partial y} \times g\right)$

$$= f \times \frac{\partial^2 g}{\partial x \partial y} + \frac{\partial f}{\partial x} \times \frac{\partial g}{\partial y} + \frac{\partial f}{\partial y} \times \frac{\partial g}{\partial x} + \frac{\partial^2 f}{\partial x \partial y} \times g$$

If $f(x,y,z) = f_1(x,y,z)i + f_2(x,y,z)j + f_3(x,y,z)k$, then

$$\frac{\partial f}{\partial x} = \frac{\partial f_1}{\partial x} i + \frac{\partial f_2}{\partial x} j + \frac{\partial f_3}{\partial x} k \text{ etc}$$

If a vector function f is a function of x, y, z and the variables x, y, z are expressed in terms of new variables s, t, then the function will be a function of s and t. Then, we can write

$$\frac{\partial f}{\partial t}=\frac{\partial f}{\partial x}\cdot\frac{\partial x}{\partial t}+\frac{\partial f}{\partial y}\cdot\frac{\partial y}{\partial t}+\frac{\partial f}{\partial z}\cdot\frac{\partial z}{\partial t} \text{ etc.}$$

Total Differentials

Differentials of vectors follow rules similar to those of elementary calculus. For example,

1. If $f(x,y,z)=f_1(x,y,z)i+f_2(x,y,z)j+f_3(x,y,z)k$, then

   $df(x,y,z)=df_1(x,y,z)i+df_2(x,y,z)j+df_3(x,y,z)k$

2. $d(f\cdot g)=f\cdot dg+g\cdot df$
3. $d(f\times g)=f\times dg+df\times g$
4. If f = f(x, y, z), then $df=\frac{\partial f}{\partial x}dx+\frac{\partial f}{\partial y}dy+\frac{\partial f}{\partial z}dz$, etc

Space Curve

If r = f(t) denotes a vector function of a scalar variable t and r is the position vector of a point P(x, y, z), in space, then as t varies, so does r and therefore the end point P or r describes a continuous curve. Therefore, a curve C can always be represented by expressing the position vector r of any point on it as a function of some scalar variable t.

If a $\leq$ t $\leq$ b, the end point of the vector r(a) is called initial point of the curve C and the end point of the vector r(b) is called the end point of C. if r (a) = r (b), then curve C is a closed curve. That is, a curve is closed if its initial and end point coincide. If there exist a < t_1 and t_2 < b such that r (t_1) = r (t_2) = r_0 (say), then r is called a multiple point. Geometrically, a multiple point occurs where curve C crosses itself or is tangent to itself. A curve C is called a simple curve if it has no multiple points. A curve which is both simple and closed, is called a simple closed curve.

Let P be a fixed point on the curve having position vector r corresponding to some t. Let Q be the point corresponding to t+δ t and have position vector r + δr. Then,

$$\underset{\delta t\to 0}{Lt}\frac{PQ}{\delta t}=\underset{\delta t\to 0}{Lt}\frac{\delta r}{\delta s}=\frac{dr}{dt}$$

is a vector tangent to the space curve C at P. This vector is called the tangent vector to C at P. A point t on a space curve C is called singular or non-singular according as the value of $\frac{dr}{dt}$ is zero or non-zero at t. The direction of the space curve C at a non-singular point P is the same as that of the tangent vector to C at P.

A vector function which has a continuous derivative and no singular points is called a smooth vector function.

If the parameter t is the time, then represents $\frac{dr}{dt}$ the velocity v, with which the end point of r describes the curve and $\frac{dv}{dt}=\frac{d^2r}{dt^2}$ represents its acceleration a along the curve.

ILUSTRATIVE EXAMPLES

1. $If r = t^2 i + \sin tj + (1-t)k \; find \frac{dr}{dt} and \frac{d^2r}{dt^2}$

Solution: $\frac{dr}{dt} = 2ti + \cos ij - k$ and $\frac{d^2r}{dt^2} = 2i - \sin tj$

2. Show that if *a, b, c are constant vectors when* $r = at^2 + bt + c$ *is the path of a point moving with constant acceleration.*

Solution: The velocity of the point $=\frac{dr}{dt} = 2at = b$ and the acceleration $\pm \frac{d^2r}{dt^2} = 2a =$ constant vector. Thus the point whose path is r = at² = dt + c is moving with constant acceleration.

3. If $r = a\cos ti + a\sin tj + tk, find \frac{dr}{dt}, \frac{d^2r}{dt^2} and \left|\frac{d^2r}{dt^2}\right|$

$$\frac{dr}{dt} = \frac{d(a\cos t)}{dt}i + \frac{d}{dt}(a\sin t)j + \frac{d}{dt}(t)k = -a\sin ti + a\cos tj + k$$

$$\frac{d^2r}{dt^2} = \frac{d}{dt}(a - \sin t)i + \frac{d(a\cos t)j}{dt} + \frac{d}{dt}(l)k = -a\cos ti - a\sin tj$$

$$\therefore \left|\frac{d^2r}{dt^2}\right| = \sqrt{[(-a\cos t)^2 + (-a\sin t)^2]} = a\sqrt{[\cos t^2 + \sin t^2]} = a$$

4. If $r = a\cos ti + a\sin tj + at\tan ak$ *find* $\left|\frac{dr}{dt}\times\frac{d^2r}{dr^2}\right|$ *and* $\left|\frac{dr}{dt}\,\frac{d^2r}{dt^2}\,\frac{d^3r}{dt^3}\right|$.

Solution: $\frac{dr}{dt} = -a\sin tj + a\tan ak$ $\frac{d^2r}{dt^2} = -a$ cos it-a sin tj+0 as $a \tan \alpha$ =constant.

$$\frac{d^3r}{dt^3} = a\sin ti - a\cos tj.$$

$$\therefore \frac{dr}{dt}\frac{d^2r}{dt^2} = \begin{vmatrix} i & j & k \\ -a\sin t & a\cos t & a\tan\alpha \\ -a\cos t & -a\sin t & 0 \end{vmatrix}$$

$$= i(-a^2 \sin t \tan\alpha) + j(-a^2 \cos t \tan\alpha) \ +k(a^2 \sin^2 t + a^2 \cos^2 t)$$

$$\therefore \left|\frac{dr}{dt} \times \frac{d^2r}{dt^2}\right| = \sqrt{[(a^2 \sin t \tan a^2 + (a^2 \cos t \tan\alpha)^2 + a^4]}$$

$$= \sqrt{[a^4 \tan^2 \alpha(\sin^2 t + \cos^2 t) + a^4]} = a^2 \sec\alpha.$$

Again, we know that $[abc] = \begin{vmatrix} a_1 & a_2 & a_3 \\ b_1 & b_2 & b_3 \\ c_1 & c_2 & c_3 \end{vmatrix}$ •

Therefore $\left[\frac{dr}{dt}\frac{d^2r}{dt^2}\frac{d^3r}{dt^3}\right] = \begin{vmatrix} -a\sin t & a\cos t & a\tan\alpha \\ -a\cos t & -a\sin t & 0 \\ a\sin t & -a\cos t & 0 \end{vmatrix} = a \tan\alpha\,(a^2 \cos^2 t + a^2 \sin 2t)$

$= a^3 \tan\alpha$

5. *Find the unit tangent vector at any point on the curve $x = t^2 + 2$, $y = 4t - 5$, $z = 2t^2 - 6t$ where t is any variable. Also determine the unit tangent vector at the point $t=2$.*

Solution: Let r be the position vector of any point (x, y, z) on the given curve, then we have $r = xi + yj + zk = (t^2 + 2)i + (4t - 5)j + (2t^2 - 6t)k$

$$\Rightarrow \frac{dr}{dr} = 2ti + 4j + (4t - 6)k \text{ and } a = \frac{d^2r}{dt^2} = \frac{d^2x}{dt^2}i + \frac{d^2y}{dt^2}j$$

Then multiply (5) scalarly by $\hat{t} = \dfrac{\frac{dr}{dt}}{\left|\frac{dr}{dt}\right|} = \dfrac{2ti + 4j + (4t-6)k}{2\sqrt{5t^2 - 12t + 13}} = \dfrac{ti + 2j + (2t-3)k}{\sqrt{(5t^2 - 12t + 13)}}$

Then the unit tangent vector at the point $t=2$ is given by

$$\frac{2i + 2j + (2\times 2 - 3)k}{\sqrt{(5\times 4 - 12\times 2 + 13)}} = \frac{1}{3}(2i + 2j + k).$$

6. *Find the angle between the tangents to the curve* $r = t^2 i - 2tj + t^3k$ *at the points* $t = 1$ *and* $t = 2$.

Solution: $r = t^2 i = 2tj + t^3k \quad \Rightarrow \dfrac{dr}{dt} = 2ti - 2j + 3t^2k$

Let T_1 and T_2 be the vectors along the tangents at $t=1$ and $t=2$ respectively, then we have

$$T_1 = \left(\frac{dr}{dt}\right)_{t=1} = 2i - 2j + 3k; \quad \text{and} \quad T_1 = \left(\frac{dr}{dt}\right)_{t=2} = 4i - 2j + 12k.$$

Let q be the required angle between the tangents, then we have $T_1 \cdot T_2 = |T_1||T_2| \cos q$

$$\Rightarrow (2)(4) + (-2)(-2) + (3)(12) = \sqrt{[2^2 + (-2)^2 + 3^2]} = \sqrt{[4^2 + (-2)^2 + 12^2]}\cos\theta$$

$$\Rightarrow 48 = \sqrt{(17)}\sqrt{(164)} \cos q \Rightarrow \cos q$$

$$= \frac{48}{\sqrt{(17)}\sqrt{(164)}} \Rightarrow \theta = \cos^{-1}\left\{\frac{48}{\sqrt{(17)}\sqrt{(164)}}\right\}$$

7. *Find the curvature and torsion of the curve* $x = a \cos t$, $y = a \sin t$, $z = bt$.

Note. Any curve drawn on a circular cylinder and cutting its generators at a constant angle is known as a circular helix.

The vector equation of the curve is given by $r = a\cos ti + a\sin tj + btk \therefore (dr/dt) = -a\sin ti + a\cos tj + bk$

Its arc length from $P_0(t=0)$ to any point $P(t)$ is

$$= \int_0^t |dr/dt|\, dt = \sqrt{(a^2 + b^2)t} \Rightarrow \frac{ds}{dt} = \sqrt{(a^2 + b^2)}$$

Also $\hat{t} = \dfrac{dr}{dt} = \dfrac{dr}{dt} / \dfrac{ds}{dt} = \dfrac{-a\sin ti + a\cos tj + bk}{\sqrt{(a^2 b^2)}}$ and $= \dfrac{dt}{ds} = \dfrac{d\hat{t}}{dt} / \dfrac{ds}{dt} = \dfrac{-a(\cos ti + \sin tj)}{a^2 + b^2}$

$$\Rightarrow \quad k = \left|\frac{d\hat{t}}{ds}\right| = \frac{a}{a^2 + b^2}$$

Also $\hat{n} = -(\cos ti + \sin tj)$ and $\hat{b} = \hat{t} \times \hat{n} = (b\sin ti - b\cos tj + ak)/\sqrt{(a^2 + b^2)}$

$$\Rightarrow \quad \frac{d\hat{b}}{ds} = \frac{db}{dt} \Big/ \frac{ds}{dt} = b(\cos ti + \sin tj)/(a^2 + b^2) = -\tau\hat{n} = \tau(\cos ti = \sin tj)$$

$$\Rightarrow \quad \tau = \frac{b}{a^2 + b^2}$$

EXERCISE

1. Given $R = t^m A + t^n B$, where A, B are constant vectors, show that, if R and d^2R/dt^2 are parallel vectors, then m + n = 1, unless m = n.

2. If $P = 5t^2 I + t^3 J - tK$ and $Q = 2I \sin t - J \cos t + 5tK$, find (i) $\frac{d}{dt}(P.Q)$; (ii) $\frac{d}{dt}(P \times Q)$.

3. If $\frac{dU}{dt} = W \times U$ and $\frac{dV}{dt} = W \times V$, prove that $\frac{d}{dt}(U \times V) = W \times (U \times V)$.1.
   If r = a cos wt + b sin wt

   Show that $$r \times \frac{dr}{dt} = \omega a \times b.$$

   and $$\frac{d^2 r}{dt^2} = -\omega^2 r$$

   a, b, w being constants.

4. If $r = a \cos ti + a \sin tj$ + at tan ak.

   find $\left|\frac{dr}{dt} \times \frac{d^2 r}{dt^2}\right|$ and $\left|\frac{dr}{dt} \frac{d^2 r}{dt^2} \frac{d^2 r}{dt^2}\right|$.

5. Show that, if R = A sin wt + B cos wt, where A, B, w are constants, then $\frac{d^2 R}{dt^2} = -\omega^2 R$ and $R \times \frac{dR}{dt} =$ – wA × B.

6. Find the unit tangent vector at any point on the vurve $x = t^2 + 2$, $y = 4t - 5$, $z = 2t^2 - 6t$, where t is any variable. Also determine the unit tangent vector at the point $t = 2$.

7. A particle moves along a curve $x = e^{-t}$, $y = 2 \cos 3t$, $z = 2 \sin 3t$, where t is the time variable. Determine its velocity and acceleration vectors and also the magnitudes of velocity and acceleration at $t = 0$.

8. The position vector of a particle at time t is $R = \cos(t-1) I + \sinh(t-1) J + at^3 K$. Find the condition imposed on a by requiring that at time $t = 1$, the acceleration is normal to the position vector.

9. If $A = x^2yz\, I - 2xz^3\, J + xz^2\, K$ and $B = 2z\, I + yJ - x^2K$, find $\frac{\partial^2}{\partial x \partial y}(A \times B)$ at (1, 0,–2).

10. If R = (a cos *t*) *I* + (a sin *t*) *J* + (*at* tan a) *K*, find the value of

    (i) $\left|\frac{dR}{dt} \times \frac{d^2 R}{dt^2}\right|$ (ii) $\left|\frac{dR}{dt}, \frac{d^2 R}{dt^2}, \frac{d^3 R}{dt^3}\right|$

    also find the unit tantent vector at any point t of the vurve.

INTEGRATION OF VECTORS

INTRODUCTION

In the previous chapter we have learnt about differentiation of vector valued function and their properties with their application. In this chapter we study the integration of vector function and its application..

Integration of Vector Function

If f and F are two vector functions of the scalar variable t such that

$$\frac{dF}{dt} = f,$$

Then F is called the indefinite integral of w. r .t. and we write

The function f, which is to be integrated is called the integrand.

If c is an arbitrary constant vector independent of t, then

$$\frac{d}{dt}(F+c) = \frac{dF}{dt} = f$$

$$\therefore \qquad \int f dt = F + c$$

Thus, the integral F is indefinite to the extent of an additive arbitrary constant vector c.

If $\frac{dF}{dt} = f$ for all values of t in the interval [a, b], the definite integral between the limits $t = a$ and $t = b$ is written as

$$\int_a^b f dt = [F]_a^b = F(b) - F(a)$$

Theorem: If $f = f_1\, i + f_2\, j + f_3\, k$, where, f_1, f_2, f_3 are scalars functions of the scalar variable t, then

$$\int f dt = i\int f_1 dt + j\int f_2 dt + k\int f_3 dt$$

Thus, to integrate a vector function, we should integrate its components.

Some Standard Results

We are going to dicuss some standard results for integration:

Let f and g be two vector function of scalar variable t.

1. Since $\frac{d}{dt}(f \cdot g) = \frac{df}{dt} \cdot g + f \cdot \frac{df}{dt}$

   Integrating both side, we get

$$\int\left(\frac{df}{dt}.g + f.\frac{df}{dt}\right)dt = f.g + c$$

Where c is the constant of integration and is a scalar because the integrand is a scalar.

2. Since $\frac{d}{dt}(f \cdot f) = \frac{d}{dt}(f \cdot f) = \frac{df}{dt} \cdot f + f \cdot \frac{d}{dt} = 2f \cdot \frac{df}{dt}$

$$\therefore \int 2f \cdot \frac{df}{dt} = f^2 + c$$

Where c is the constant of integration and is a scalar because the integrand is a scalar.

3. Since $\frac{d}{dt}\left(\frac{df}{dt}\right)^2 = 2\frac{df}{dt} \cdot \frac{d^2 f}{dt^2}$

$$\therefore \int\left(2\frac{df}{dt} \cdot \frac{d^2 f}{dt^2} g\right)dt = \left(\frac{df}{dt}\right) + c$$

Where c is the constant of integration and is a scalar because the integrand is a scalar.

4. Since $\frac{d}{dt}(f \times g) == \frac{df}{dt} \times g + f \times \frac{dg}{dt}$

$$\therefore \int\left(\frac{df}{dt} \times g + f \times \frac{dg}{dt}\right)dt = f \times g + c$$

Where c is the constant of integration and is a scalar because the integrand is a scalar

5. Since $\frac{d}{dt}(a \times g) = a \times \frac{df}{dt}$, where a is constant vector.

$$\therefore \int\left(a \times \frac{df}{dt}\right)dt = a \times f + c$$

Where c is the constant of integration.

6. If c is a constant scalar, $\int cfdt = c\int fdt$

7. $\int(f \pm g)dt = \int fdt \pm \int gdt$

Remember: The constant of integration is of the same nature as the integrand

Example: If $r = (7t^2)\, i + 2\, t\, j - 3\, t^2\, k$, find $\int_2^3 r \times \frac{d^2 r}{dt^2} dt$.

Solution: Since, $\frac{d}{dt}\left[r\times\frac{dr}{dt}\right]=r\times\frac{d^2r}{dt^2}+\frac{dr}{dt}\times\frac{dr}{dt}$

$$\Rightarrow \frac{d}{dt}\left[r\times\frac{dr}{dt}\right]=r\times\frac{d^2r}{dt^2}$$

∴ Integrating both sides w. r. t., we have

$$\int\left(r\times\frac{d^2r}{dt^2}\right)dt = r\times\frac{dr}{dt}+c$$

where c is the constant of integration.

$$\therefore \int_2^3\left(r\times\frac{d^2r}{dt^2}\right)dt = \left[r\times\frac{dr}{dt}\right]_2^3$$

Let us find $r\times\frac{dr}{dt}$

Since, $r = (7t^2)\, i + 2\, t\, j - 3\, t^2\, k$

$$\therefore \frac{dr}{dt}=\frac{d}{dt}\{(7t^2)\, i + 2\, t\, j + (-3\, t^2)\, k\}$$

$$\therefore \frac{dr}{dt} = 14\, t\, I + 2\, j - 6\, t\, k$$

$$\therefore \left[r\times\frac{dr}{dt}\right]=\begin{vmatrix} i & j & k \\ 7t^2 & 2t & -3t^2 \\ 14t & 2 & -6t \end{vmatrix}=i\left(-12t^2+6t^2\right)$$

$$+\, j\left(-42t^3+42t^3\right)+k\left(14t^2-28t^2\right)$$

$$\Rightarrow \left[r\times\frac{dr}{dt}\right] = i\left(-6t^2\right)+j(0)+k\left(-14t^2\right)$$

Using (2) in (1), we get

$$\therefore \int_2^3\left(r\times\frac{d^2r}{dt^2}\right)dt=\left[i\left(-6t^2\right)+j(0)+k\left(-14t^2\right)\right]_2^3$$

$$=\left[i\left\{\left(-6.3^2\right)-\left(-6.2^2\right)\right\}+k\left\{\left(-14.3^2\right)-\left(-14.2^2\right)\right\}\right]$$

Hence, $\int_2^3 r\times\frac{d^2r}{dt^2}\,dt \;=\; -30\text{ i} - 70\text{ k}$

EXERCISE

1. Compute the line integral $\int_C (y^2 dx - x^2 dy)$ about the triangle whose vertices are (1, 0), (0, 1) and (–1, 0).
2. If velocity vector is F = yI + 2J + xzK m/sec., show that the flux of water through the parabolic cylinder $y = x^2$, $0 \le x \le 3$, $0 \le z \le 2$ is 69 m^3/sec.
3. Evaluate $\int_S F.dS$ were F = x I + $(z^2 - zx)$ J – xy K and S is the triangular surface with vertices (2, 0, 0), (0, 2, 0) and (0, 0, 4).
4. Evaluate the line integral $\int_C \left[\left(x^2 + xy\right)dx + \left(x^2 + y^2\right)dy\right]$ where C is the square formed by the lines $y = \pm 1$ and $x = \pm 1$.
5. If F = $(5xy - 6x^2)$I + (2y – 4x)J, evaluate $\int_C F.dR$ along the curve C in the xy-plane, $y = x^3$ from the point (1, 1) to (2, 8).
6. A vector field is given by F = (sin y)I + x(1 + cos y)J. Evaluate the line integral over a circular path given by $x^2 + y^2 = a^2$, z = 0.

Chapter 6

Vector Differential Operator

GRADIENT DIVERGENCE AND CURL

In this chapter we learnt about the Gradient, Divergence and Curl with their properties. Also we studied the use of these operator.

VECTOR FIELD

The set of points of the region R together with the vector $f(P)$ is said to be a vector field over R. For example, velocity of a moving fluid at any instant is a vector field.

Note: A function is said to be single valued or uniform function, if it has only one value at each point. If it has more than one value at each point, it is called a multiple valued function.

We have a *vector field* over a space or part of space if we associate with each point in this space, or part of space, a vector. We could write,

$$v = v(x).$$

This equation could consist of an equation for each coordinate like this,

$$V = (v_1, v_2, v_3) = v_1(x)\, i + v_2(x)\, j + v_2(x)\, k.$$

Here each component v_n is a function of x. In the previous example we saw that the gradient is just such a field, with each component being $\frac{\partial f}{\partial x_n}$

Level Surfaces

If is a uniform continuous point function, then through any point $P(x, y, z)$ of the region considered, we can draw a surface such that at each point on it, the function has the same value as at P. Such a surface is called a level surface of the function.

For Example: $f(x, y, z) = c$

where c is a constant, is a level surface.

No two level surfaces corresponding to different values of the function can intersect because otherwise, there would be two different values of the function for each point of intersection, which is contrary to the assumption that the function is uniform.

THE VECTOR DIFFERENTIAL OPERATOR

The vector differential operator is defined by

$$\nabla = i\frac{\partial}{\partial x} + j\frac{\partial}{\partial y} + k\frac{\partial}{\partial z}$$

This vector operator possesses properties like ordinary vectors and is useful in defining three quantities-the gradient, the divergence and the curl, which arise in practical applications.

GRADIENT OF SCALAR FUNCTION

If a scalar function ϕ is defined and differentiable at each point (x, y, z) in a certain region of space (i.e. ϕ defines a differentiable scalar field), then the gradient of ϕ written as grad ϕ or is a vector defined by

$$\nabla f = \left(i\frac{\partial}{\partial x} + j\frac{\partial}{\partial y} + k\frac{\partial}{\partial z} \right)\phi = i\frac{\partial \phi}{\partial x} + j\frac{\partial \phi}{\partial y} + k\frac{\partial \phi}{\partial z}$$

Note that (i) Ñf is a vector whose three components are $\frac{\partial \phi}{\partial x}, \frac{\partial \phi}{\partial y}, \frac{\partial \phi}{\partial z}$

If ϕ is a scalar point function, then $\nabla\phi$ is a vector point function (i.e., $\nabla\phi$ defines a vector field).

PROPERTIES OF GRADIENT

If ϕ and ψ are two scalar point functions , then

1. $\nabla(\phi \pm \psi) = \nabla\phi \pm \nabla\psi$
2. $\nabla(\phi\psi) = \phi\nabla\psi + \psi\nabla\phi$
3. $\nabla\left(\frac{\phi}{\psi}\right) = \frac{\psi\nabla - \phi\nabla\psi}{\psi^2}, \psi \neq 0$
4. $\nabla(c\phi) = c\nabla\phi$, where c is a constant
5. f is a constant if and only if $\nabla f = 0$.

APPLICATION OF DEL (∇) OPERATOR

$\nabla\phi$ is perpendicular to the surface S given by $\phi(x, y, z) = c$, c being constant, at the point $P(x, y, z)$. Let a space curve C lies on the surface S be represented by $r = xi + yj + zk$, where x, y, z are scalar functions of a scalar t, then differentiating ϕ (x, y, z) = c, w. r. t., we have $\frac{d\phi}{dt} = 0$.

$$\Rightarrow \frac{\partial \phi}{\partial x}\frac{dx}{dt} + \frac{\partial \phi}{\partial y}\frac{dy}{dt} + \frac{\partial \phi}{\partial z}\frac{dz}{dt} = 0$$

$$\Rightarrow \left(i\frac{\partial \phi}{\partial x} + j\frac{\partial \phi}{\partial y} + k\frac{\partial \phi}{\partial z}\right)\cdot\left(i\frac{dx}{dt} + j\frac{dy}{dt} + k\frac{dz}{dt}\right) = 0$$

$$\Rightarrow \nabla\phi \cdot \frac{dr}{dt} = 0$$

But $\frac{dr}{dt}$ is the vector tangent to the curve C at the point P.

$\therefore$ $\nabla\phi$ is perpendicular to the curve C at P.

The same reasoning is applicable to every smooth curve on the surface S that passes through P. Therefore, $\nabla\phi$ is perpendicular to every such curve at P. Hence, $\nabla\phi$ is perpendicular to the surface S.

Note: $|\nabla\phi|$ gives the rate of change of ϕ along the normal to the surface at P.

For a curve with position vector $r(t)$, the unit tangent vector $\hat{T}(t)$ is defined by

$$\hat{T}(t) = \frac{\dot{r}}{|\dot{r}|} = \frac{\dot{r}}{\dot{s}} = \frac{dr}{ds}$$

where t is a parameterization variable, s is the arc length, and an over dot denotes a derivative with respect to time t, $\dot{x} = dx/dt$. For a function given parametrically by $(f(t), g(t))$, the tangent vector relative to the point $(f(t), g(t))$ is therefore given by

$$x(t) = \frac{\dot{f}}{\sqrt{\dot{f}^2 + \dot{g}^2}}$$

$$y(t) = \frac{\dot{g}}{\sqrt{\dot{f}^2 + \dot{g}^2}}$$

To actually place the vector tangent to the curve, it must be displaced by $(f(t), g(t))$ It is also true that

$$\frac{d\hat{T}}{ds} = k\hat{N}$$

$$\frac{d\hat{T}}{dt} = k\frac{ds}{dt}\hat{N}$$

$$[\dot{T}, \ddot{T}, \dddot{T}] = k^5 \frac{d}{ds}\left(\frac{\tau}{k}\right),$$

where N is the normal vector, k is the curvature, τ is the torsion, and $[A, B, C]$ is the scalar triple product

TANGENT PLANES TO LEVEL SURFACES

TANGENT PLANES

Let (x_0, y_0) be any point of a surface function $z = f(x, y)$. Then the surface has a non vertical tangent plane at (x_0, y_0) with equation

$$z = f(x_0, y_0) + f_x(x_0, y_0) + (x - x_0) + f_y(x_0, y_0)(y - y_0)$$

Since the equation of a plane requires a point and a normal vector to the plane for finding the equation of a tangent plane to a surface at a given point requires the calculation of a surface normal vector.

Suppose that S is a surface with equations $F(x, y, z) = k$, that is, it is a level surface of a function F of three variables, and let $P(x_0, y_0, z_0)$ be a point on S. Let C be and curve that lies on the surface S and passes through the point P. Recall that the curve C is described by a continuous vector function. $r(t) = \langle x(t), y(t), z(t)\rangle$. Let to be the parameter value corresponding to P; that is, $r(t_0) = \langle x_0, y_0, z_0\rangle$. Since C lies on S, any point $(x(t), y(t), z(t))$ must satisfy the equation of S, that is,

$$F(x(t), y(t), z(t)) = k$$

We can then use the Chain Rule to differentiate both sides of the above equation as follows:

$$\frac{\partial F}{\partial x}\frac{\partial x}{dt} + \frac{\partial F}{\partial y}\frac{dy}{dt} + \frac{\partial F}{\partial z}\frac{\partial z}{\partial t} = 0$$

But since $\nabla F = (F_x, F_y, F_z)$ and $r'(t)(x'(t), y'(t), z'(t))$, the equation above can be written as a dot product as follows:

$$\nabla F \cdot r'(t) = 0$$

In particular, when $t = t_0$ we have $r(t_0) = (x_0, y_0, z_0)$ so

Because the gradient vector at P, $\nabla F\ \langle x_0, y_0, z_0\rangle$, is perpendicular to the tangent vector $r'(t_0)$ to any curve C on S that passes though P. If $\nabla F\ \langle x_0, y_0, z_0\rangle \neq 0$, it is therefore natural to define the tangent plane to the level surface $F(x, y, z) = k$ at $P(x_0, y_0, z_0)$ that passes through P and has the normal vector of $\nabla F\ \langle x_0, y_0, z_0\rangle$ is (in standard equation form):

$$\nabla F_x(x_0, y_0, z_0)(x - x_0) + \nabla F_y(x_0, y_0, z_0)(y - y_0) + \nabla F_z(x_0, y_0, z_0)(z - z_0) = 0$$

or using the symmetric equations:

$$\frac{x - x_0}{\nabla F_x\langle x_0, y_0, z_0\rangle} = \frac{y - y_0}{\nabla F_y\langle x_0, y_0, z_0\rangle} = \frac{z - z_0}{\nabla F_z\langle x_0, y_0, z_0\rangle}$$

Let a be the position vector of a given point A on the surface $\phi\ (x, y, z) = c$, (where c is constant). If r is the position vector of any point P on the tangent plane at A, then

$AP = r - a$ and it lies in the tangent plane .

But grad ϕ is normal to the surface.

$(r - a).\text{grad } \phi = 0$,

which is the equation of the tangent plane.

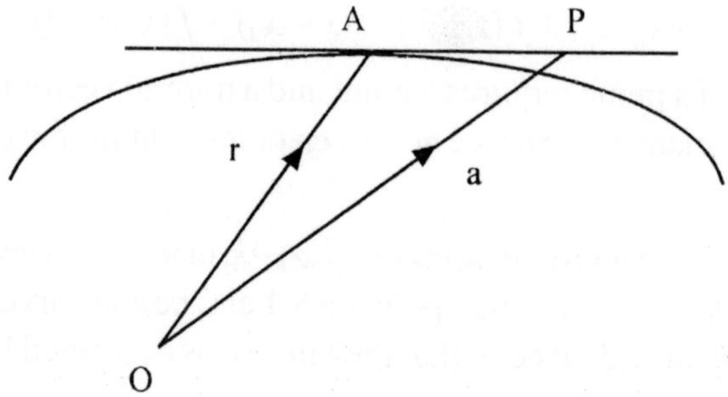

If $a = (x, y, z)$ and $r = (X, Y, Z)$, then the equation of tangent plane can be written as

$$\{(i(X - x) + j(Y - y) + k(Z - z))\} \cdot \left(i\frac{\partial \phi}{\partial x} + j\frac{\partial \phi}{\partial y} + k\frac{\partial \phi}{\partial z} \right) = 0$$

$$\Rightarrow (X - x)\frac{\partial \phi}{\partial x} + (Y - y)\frac{\partial \phi}{\partial y} + (Z - z)\frac{\partial \phi}{\partial z} = 0$$

To begin with, a level surface $U(x,y,z) = k$ is said to be *smooth* if the *gradient* $\nabla U = (U_x, U_y, U_z)$ is continuous and non-zero at each point on the surface. Equivalently, we often write the gradient as

$$\nabla U = U_x e_x + U_y e_y + U_z e_z$$

where $e_x = (1,0,0)$, $e_y = (0,1,0)$, and $e_z = (0,0,1)$.

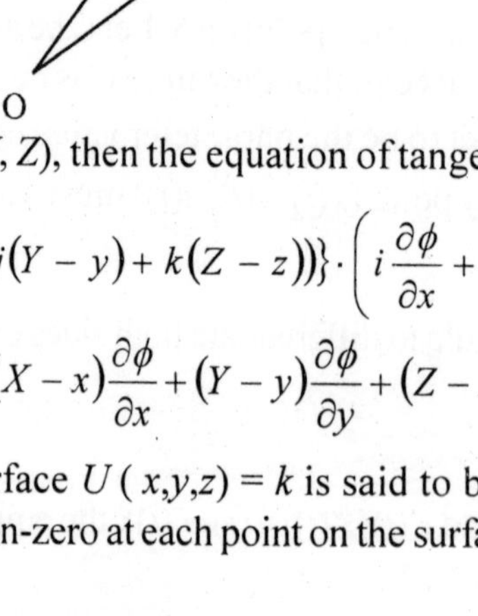

Suppose that $r(t) = (x(t), y(t), z(t))$ is a curve that lies on a smooth surface $U(x,y,z) = k$. Applying the derivative with respect to t to both sides of the equation of the level surface yields

$$\frac{dU}{dt} = \frac{d}{dt}k$$

Since k is a constant, the chain rule implies that

$$\nabla U \cdot v = 0$$

where $v = (x'(t), y'(t), z'(t))$. However, v is tangent to the surface because it is tangent to a curve on the surface, which implies that ∇U is orthogonal to each tangent vector v at a given point on the surface.

That is, $\nabla U(p, q, r)$ at a given point (p, q, r) is *normal* to the tangent plane to the surface $U(x, y, z) = k$ at the point (p, q, r).

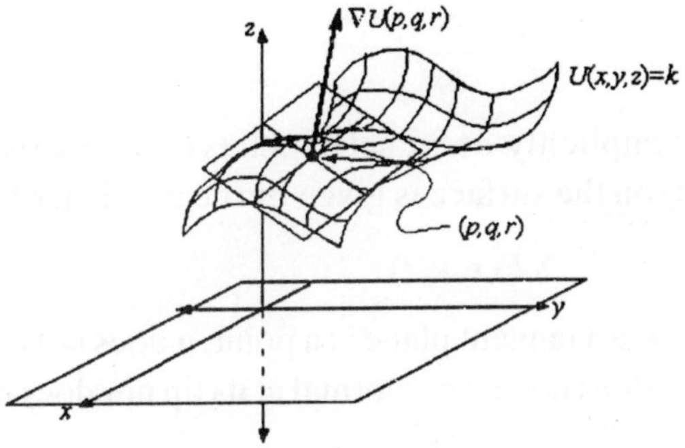

We thus say that the gradient ∇U is *normal* to the surface $U(x, y, z) = k$ at each point on the surface.

NORMAL TO A PLANE

SURFACE NORMAL

A surface normal, or just normal to a flat surface is a three-dimensional vector which is perpendicular to that surface. A normal to a non-flat surface at a point ρ on the surface is a vector which is perpendicular to the tangent plane to that surface at ρ.

NORMAL VECTOR

Here's the plane again with s and t being known vectors which lie in it.

Working out the cross product of s and t will automatically give us a vector perpendicular to the plane in which s and t lie. This is the method for finding a normal vector to this plane.

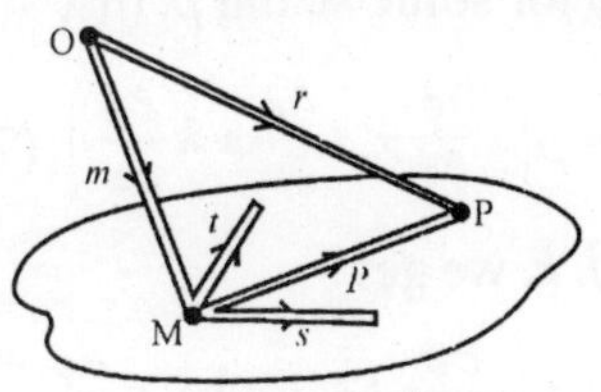

$r = m + as + bt$

Then $s \times t = -2i + j + k = N$, a normal vector to the plane.

It's also possible to find a normal vector by using the dot product. If we call the normal vector N, then N must be perpendicular to both s and t.

Therefore $N.s = 0$ and $N.t = 0$.

CALCULATING A SURFACE NORMAL

For a polygon (such as a triangle), a surface normal can be calculated as the vector cross product of two (non-parallel) edges of the polygon.

For a plane given by the equation $ax + by + cz = d$, the vector (a, b, c) is a normal.

If a (possibly non-flat) surface S is parametrized by a system of curvilinear coordinates $x(s, t)$, with s and t real variables, then a normal is given by the cross product of the partial derivatives

$$\frac{\partial x}{\partial s} \times \frac{\partial x}{\partial t}.$$

If a surface S is given implicitly, as the set of points (x, y, z) satisfying $F(x, y, z) = 0$, then, a normal at a point (x, y, z) on the surface is given by the gradient of $F(x, y, z)$.

$$\nabla F(x, y, z).$$

If a surface does not have a tangent plane at a point, it does not have a normal at that point either. For example, a cone does not have a normal at its tip nor does it have a normal along the edge of its base.

If r is the position vector of any point P on the normal at A having position vector a, then $\overrightarrow{AP} = r - a$ and it lies in the tangent plane.

Also, grad ϕ is along the normal to the surface.

$(r - a) \times \text{grad } \phi = 0$,

which is the equation of the normal to the plane.

If $a = (x, y, z)$ and $r = (X, Y, Z)$,

$$r - a = (i(X - x) + j(Y - y) + k(Z - z))$$

and grad $$\phi = i\frac{\partial \phi}{\partial x} + j\frac{\partial \phi}{\partial y} + k\frac{\partial \phi}{\partial z}$$

$\therefore$ $(r - a)$ is parallel to grad ϕ for some scalar p. $(i(X - x) + j(Y - y) + k(Z - z))$

$$k = p\left(i\frac{\partial \phi}{\partial x} + j\frac{\partial \phi}{\partial y} + k\frac{\partial \phi}{\partial z}\right) \quad (72)$$

Comparing coefficients of i, j, k, we get

$$\Rightarrow (X - x) = p\frac{\partial \phi}{\partial x}, (Y - y) = p\frac{\partial \phi}{\partial y}, (Z - z) = p\frac{\partial \phi}{\partial z}$$

or $$\frac{(X - x)}{\frac{\partial \phi}{\partial x}} = \frac{(Y - y)}{\frac{\partial \phi}{\partial y}} = \frac{(Z - z)}{\frac{\partial \phi}{\partial z}} = p$$

is the required equation of the normal to the plane.

Uniqueness of the Normal

A normal to a surface does not have a unique direction; the vector pointing in the opposite direction of a surface normal is also a surface normal.

EQUATIONS OF PLANE

The direction of perpendicular to a plane is unique to that plane

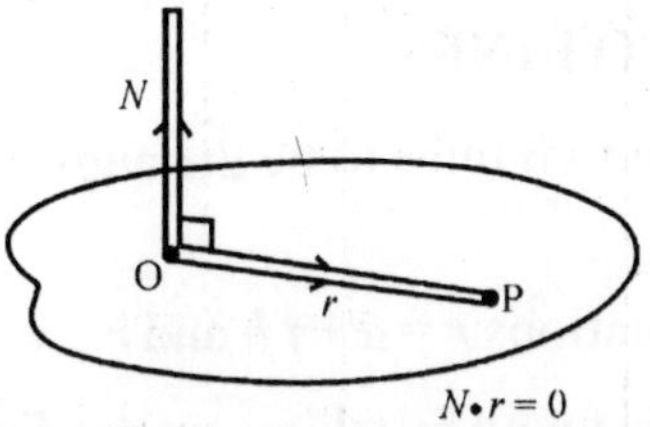

$N \cdot r = 0$

Suppose we know the vector N which is perpendicular to the plane. This means that it must be perpendicular to the position vector r of any point in the plane from the origin.

$$N \cdot r = 0.$$

which is the required equation of plane.

Now we are going to discuss another method to find the equation of a plane which doesn't pass through the origin.

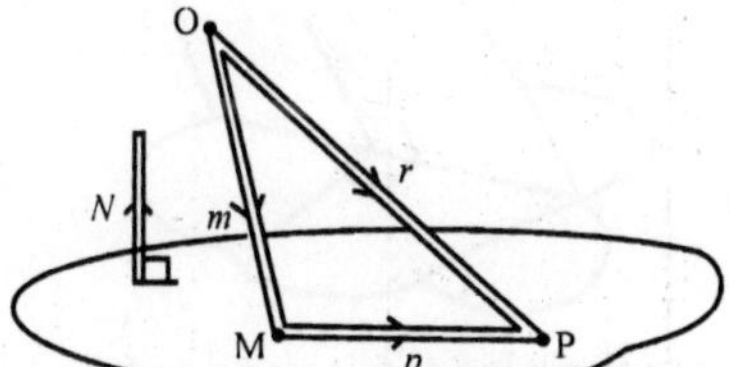

If P is any general point in the plane, so that the vector $MP = p$, then N and p are perpendicular to each other.

Therefore $N.\ p = 0$ but $p = r - m$.

$N.\left(r - m\right) = 0$ or $N.\ r - N.\ m = 0$

Now suppose we are starting with the plane below, with the 2 known vectors s and t lying in this plane.

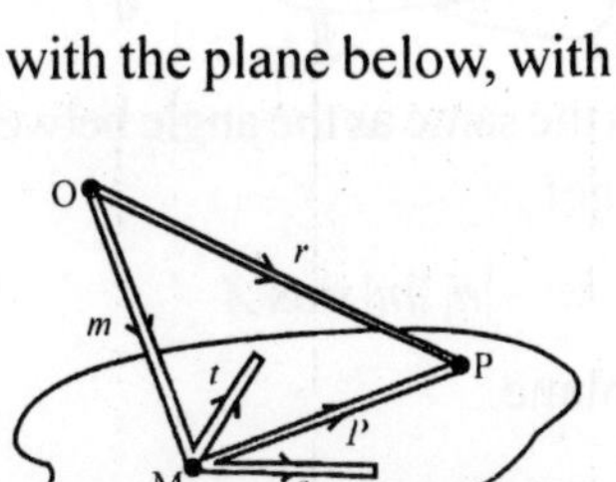

Fig. $r = m + as + bt$

We use s and t to find a vector N which would be perpendicular to this plane $r = m + as + bt$.

ANGLES BETWEEN LINES AND PLANES

Here we are going to discuss how to determine

1. The angle between two lines
2. The angle between two planes
3. The angle between a line and a plane.

THE ANGLE BETWEEN TWO LINES

We define the angle between two lines to be the angle between their direction vectors placed tail to tail.

Lte the two lines have the equations $r = a + t\,b$ and $r = c + s\,d$.

The angle between the lines is found by taking out the dot product of b and d.

We have $b.\ d = |b|\ |d| \cos A$.

THE ANGLE BETWEEN TWO PLANES

It is defined as the angle between two lines, one in each plane, so that they are at right angles to the line of intersection of the two planes as shown in the figure.

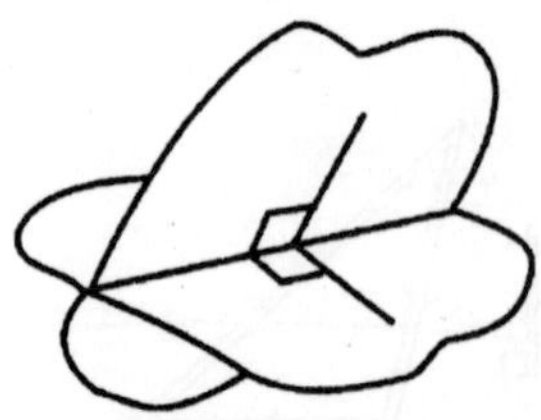

To find the angle say A between the plane, we need to know a normal vector to each of the planes.

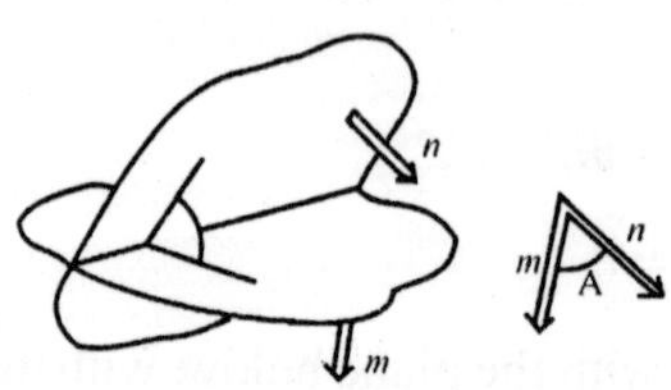

The angle between the planes is the same as the angle between their two normal vectors.Using dot product of normal vectors, we get

$$n.\ m = |n|\ |m| \cos A$$

where A is the angle between the plane..

THE ANGLE BETWEEN A LINE AND A PLANE

Here we use a normal vector to the plane for determining the angle between a line and a plane.

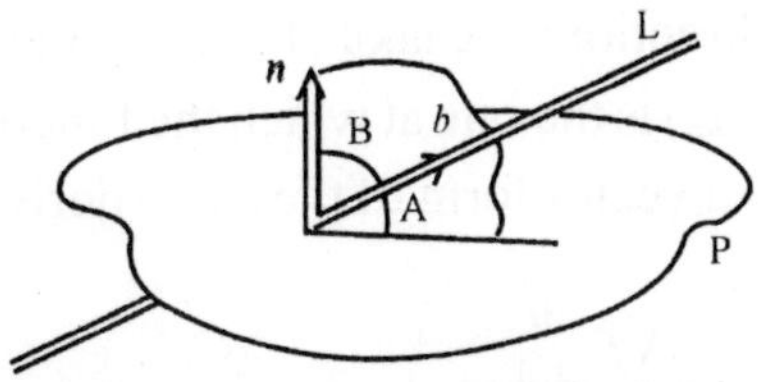

Since A and B together make a right angle, we can find A by using the dot product of n and the direction vector b of line L to first find cos B. Or we can find A even more directly by using the identity

$$\cos B = \cos(90 - A) = \sin A$$

so $n \cdot b = |n|\ |b| \sin A$ so giving A.

DIRECTIONAL DERIVATIVE

Here we are going to discuss a special type of derivative, called a directional derivative, that enables us to find the rate of change of a function of two or more variables in any direction.

Suppose that we wish to find the rate of change of z at (x_0, y_0) in the direction of an arbitrary unit vector $u = \langle a, b \rangle$. To do this we consider the surface S with equation $z = f(x, y)$ and we let $z_0 = f(x_0, y_0)$, then the point $P(x_0, y_0, z_0)$ lies on S. The vertical plane that passes though P in the direction of u intersects S in a curve C. The slope of the tangent line T to C at P is the rate of change of z in the direction of u.

The directional derivative of f at (x_0, y_0) in the direction of a unit vector $u = \langle a, b \rangle$. is

$$D_u f(x_0, y_0) = \lim_{h \to 0} \frac{f(x_0 + ha, y_0 + hb) - f(x_0, y_0)}{h}$$

provided the limit exists.

When we compute the directional derivative of a function defined by a formula, we generally use the following theorem:

If t is a differentiable function of x and y, then t has a directional derivative in the direction of any unit vector $u = (a, b)$ and:

Definition: The directional derivative of a scalar function $f(\vec{x})$ $f(x_1, x_2, ..., x_n)$ along a vector $\vec{v} = (v_1, ..., v_n)$ is the function defined by the limit

$$D_{\vec{v}} f = \lim_{h \to 0} \frac{f(\vec{x} + h\vec{v}) - f(\vec{x})}{h}.$$

If the function is differentiable, it can be written in terms of the gradient $\nabla(f)$ of f by

$$D_{\vec{v}} f = \nabla(f) . \vec{v}$$

At any point ρ the directional derivative of f intuitively represents the rate of change in f

along $\vec{v}$ at the point ρ. Usually directions are taken to be normalized, so $\vec{v}$ is a unit vector. The directional derivative $\nabla_u f(x_0, y_0, z_0)$ is the rate at which the function $f(x, y, z)$ changes at a point (x_0, y_0, z_0) in the direction u. It is a vector form of the usual derivative, and can be defined as

$$\nabla_u f = \nabla f \cdot \frac{u}{|u|}$$

$$= \lim_{h \to 0} \frac{f(x + h\hat{u}) - f(x)}{h},$$

where ∇ is called "nabla" or "del" and $\hat{u}$ denotes a unit vector.

The directional derivative is also often written in the notation

$$\frac{d}{ds} = \hat{s}.\nabla = s_x \frac{\partial}{\partial x} + s_y \frac{\partial}{\partial y} + s_z \frac{\partial}{\partial z},$$

where s denotes a unit vector in any given direction and $\partial f / \partial x = f_x$ denotes a partial derivative.

Let $\hat{u} = (u_x, u_y, u_z)$ be a unit vector in Cartesian coordinates, so

$$|\hat{u}| = \sqrt{u_x^2 + u_y^2 + u_z^2} = 1,$$

then
$$\nabla_{\hat{u}} f = \frac{\partial f}{\partial x} u_x + \frac{\partial f}{\partial y} u_y + \frac{\partial f}{\partial z} u_z.$$

If f were a function of more variables, say $x, y, z, t, \ldots$ we can use exactly the same approach to describe changes, the only difference being that the tangent plane becomes the tangent hyper-plane, and there are partial derivatives in more directions. The conclusions are exactly the same:

1. The gradient vector is in the direction of the projection of the normal to the tangent hyper-plane into the hyper-plane of coordinates.
2. The directional derivative in any direction is given by the dot product of a unit vector in that direction with the gradient vector.
3. The component of the gradient vector in the direction of any axis is the partial derivative of f with respect to the corresponding distance variable in that direction.
4. That partial derivative is the ordinary derivative with respect to that variable assuming all the other variables remain constant.

The upshot of all this is that the gradient vector, whose components can be computed by ordinary one dimensional differentiation for a field in any number of dimensions, is all you need to compute its directional derivative in any direction.

Notice from the above theorem that the directional derivative can be written as the dot product of two vectors:

$$D_u f(x, y) = f_x(x, y)a + f_y(x_0, y_0)b$$

$$D_u f(x,y) = \left(f_x(x,y), f_y(x_0,y_0)\right)\cdot(a,b)$$

$$D_u f(x,y) = \left(f_x(x,y), f_y(x_0,y_0)\right)\cdot u$$

The first vector dot product above is used in more than just this equation so it is given a special name, it is called the gradient of f and has a special representation: ∇f. The definition of ∇f is given below:

$$\nabla f(x,y) = \left(f_x(x,y), f_y(x,y)\right) = \frac{\partial f}{\partial x}i + \frac{\partial f}{\partial y}j$$

Therefore, the directional derivative can be rewritten as the following:

$$D_u f(x,y) = \nabla f(x,y)\cdot u$$

FUNCTIONS OF THREE VARIABLES

Using three variables, we can define the directional derivatives in a similar manner:

The directional derivative of f at (x_0, y_0, z_0) in the direction of a unit vector $u = (a,b,c)$ is

$$D_u f(x_0,y_0,z_0) = \lim_{h\to 0}\frac{f(x_0+ha, y_0+hb, z_0+hc) - f(x_0,y_0,z_0)}{h}$$

if the limit exists.

Also there is this formula for finding the gradient:

$$D_u f(x,y,z) = f_x(x,y,z)a + f_y(x,y,z)b + f_z(x,y,z)c$$

Just like with two variables, there above equation can be seen as the dot product of two vectors, with one of the vectors being the gradient vector $|(\nabla f)$:

$$\nabla f(x,y) = \left(f_x, f_y, f_z\right) = \frac{\partial f}{\partial x}i + \frac{\partial f}{\partial y}j + \frac{\partial f}{\partial z}k$$

Also, just as before, the directional derivative can be written as follows:

$$D_u f(x,y,z) = \nabla f(x,y,z)\cdot u$$

MAXIMIZING THE DIRECTIONAL DERIVATIVE

Suppose we have a function f of two or three variables and we consider all possible directional derivatives of f at a given point. These give the rates of changes of f in all possible directions.

Suppose f is a differentiable function of two or three variables. The maximum value of the directional derivative $D_u f(x,y)$ is $|\nabla f(x)$ and it occurs when u has the same direction as the gradient vector $\nabla f(x)$.

DERIVATIVES OF VECTOR FUNCTIONS; THE DIVERGENCE

The gradient is a vector function of several variables. Such an entity is called a vector field.

Since a vector in three dimensions has three components, and each of these will have partial derivatives in each of three directions, there are actually nine partial derivatives of a vector field in any coordinate system.

Thus in our usual rectangular coordinates we have, with a vector field $v(x, y, z)$, partial derivatives

$$\frac{\partial v_x}{\partial x}, \frac{\partial v_x}{\partial y}, \frac{\partial v_x}{\partial z}, \frac{\partial v_y}{\partial x} \frac{\partial v_y}{\partial y}, \frac{\partial v_y}{\partial z}, \frac{\partial v_z}{\partial x}, \frac{\partial v_z}{\partial y}, \frac{\partial v_z}{\partial z}$$

All of these can be computed by the same rules used for computing partial derivatives of scalar functions (often called scalar fields). The first of these is the divergence, written as div v, or in terms of the differential operator del, ∇ which is the vector operator with components $\left(\frac{\partial}{\partial x}, \frac{\partial}{\partial y}, \frac{\partial}{\partial z}\right)$ as $\nabla \cdot v$.

Explicitly, it is the dot product of this differential operator with the vector v, div $v = \nabla \cdot v = \frac{\partial v_x}{\partial x} + \frac{\partial v_y}{\partial y} + \frac{\partial v_z}{\partial z}$

Being a dot product, it is a number and not a vector.

A vector field which has zero divergence everywhere is called solenoidal.

Definition: Let x, y, z be a system of Cartesian coordinates on a 3-dimensional space, and let i, j, k be the corresponding basis of unit vectors.Given a vector field, $v (x, y, z)$:

$$v(x, y, z) = v_1(x, y, z)i + v_2(x, y, z)j + v_3(x, y, z)k$$

Each of its three components has derivatives with respect to each of the three variables, which gives a total of nine first derivatives;

The Divergence of v, is a combination of these of particular importance; it is a scalar function; it is defined as the dot product of the vector operator $\left(\frac{\partial}{\partial x}, \frac{\partial}{\partial y}, \frac{\partial}{\partial z}\right)$ or equivalently $\frac{\partial}{\partial x}i, \frac{\partial}{\partial y}j, \frac{\partial}{\partial z}k$ with the vector v.The divergence of a vector field $v = iv_1 + jv_2 + kv_3$ in rectangular coordinates is defined as the scalar product of the del operator and the function

$$\text{div } v = \nabla.v = \left[i\frac{\partial}{\partial x} + j\frac{\partial}{\partial y} + k\frac{\partial}{\partial z}\right].(iv_1 + jv_2 + kv_3)$$

$$\nabla.v = \frac{\partial v_1}{\partial x} + \frac{\partial v_2}{\partial y} + \frac{\partial v_3}{\partial z}$$

The divergence is a scalar function of a vector field. The divergence theorem is an important mathematical tool in electricity and magnetism.

Thus we have: div $v = \frac{\partial v_1}{\partial x} + \frac{\partial v_2}{\partial y} + \frac{\partial v_3}{\partial z}$

This vector operator is often written as ∇, and called "del". The divergence is written as either (div v) or ($\nabla \cdot v$) and is also called "del dot v".

The divergence of a continuously differentiable vector field $\mathrm{F} = F_1 i + F_2 j + F_3 k$ is defined to be the scalar-valued function:

$$\operatorname{div} \mathrm{F} = \nabla \cdot F = \frac{\partial F_1}{\partial x} + \frac{\partial F_2}{\partial y} + \frac{\partial F_3}{\partial z}.$$

21Although expressed in terms of coordinates, the result is invariant under orthogonal transformations, as the physical interpretation suggests.

PROPERTIES OF DIVERGENCE

The following properties can all be derived from the ordinary differentiation rules of calculus. Most important of which, the divergence is a linear operator, i.e.

$$div(aF + bG) = adiv(F) + bdiv(G)$$

for all vector fields F and G and all real numbers a and b.

There is a product rule of the following type: if φ is a scalar valued function and F is a vector field, then

$$\operatorname{div}(\varphi F) = \operatorname{grad}(\varphi) \cdot F + \varphi \operatorname{div}(F),$$

or in more suggestive notation

$$\nabla.(\varphi F) = (\nabla \varphi) \cdot F + \varphi(\nabla.F).$$

Another product rule for the cross product of two vector fields F and G in three dimensions involves the curl and reads as follows:

$$\operatorname{div}(F \times G) = \operatorname{curl}(F).G - F.\operatorname{curl}(G),$$

or

$$\nabla.(F \times G) = (\nabla \times F) \cdot G - F \cdot (\nabla \times G).$$

The Laplacian of a scalar field is the divergence of the field's gradient.

GENERALIZATIONS

The divergence of a vector field can be defined in any number of dimensions. If

$$F = (F_1, F_2, ..., F_n),$$

define

$$\text{div}F = \nabla.F = \frac{\partial F_1}{\partial x_1} + \frac{\partial F_2}{\partial x_2} + \ldots + \frac{\partial F_n}{\partial x_n}.$$

For any *n*, the divergence is a linear operator, and it satisfies the "product rule"

$$\nabla.(\varphi F) = (\nabla\varphi).F + \varphi(\nabla.F).$$

for any scalar-valued function φ.

Divergence in Various Coordinatess

Compared to the divergence in rectangular coordinates:

$$\nabla.A = \frac{\partial A_x}{\partial X} + \frac{\partial A_y}{\partial Y} + \frac{\partial A_z}{\partial Z}$$

In cylindrical polar coordinates:

$$\nabla.A = \frac{\partial A_r}{\partial r} + \frac{A_r}{r} + \frac{1}{r}\frac{\partial A_\theta}{\partial \theta} + \frac{\partial A_z}{\partial z}$$

and in spherical polar coordinates:

$$\nabla.A = \frac{1}{r^2}\frac{\partial}{\partial r}(r^2 Ar) + \frac{1}{r\sin\theta}\frac{\partial A_\phi}{\partial \phi} + \frac{1}{r\sin\theta}\frac{\partial}{\partial \theta}(A_\theta \sin\theta)$$

The divergence of a vector field is proportional to the density of point sources of the field. In Gauss' law for the electric field

$$\nabla.E = \frac{\rho}{\varepsilon_0}$$

the divergence gives the density of point charges. In Gauss' law for the magnetic field

$$\nabla.B = 0$$

the zero value for the divergence implies that there are no point sources of magnetic field.

DERIVATIVES OF VECTOR FUNCTIONS; THE CURL

The second important combination of partial derivatives of a vector field *v* is its curl. This is the cross product of the differential operator ∇ with the vector *v*.

The important vector derivative of a vector field *v* is given by its curl, (curl *v*) which is defined to be ($\nabla \times v$). Specifically, it is defined by

$$\text{curl } v = \nabla \times v = \left(\frac{dv_z}{\partial y} - \frac{\partial v_y}{\partial z}\right)i + \left(\frac{\partial v_x}{\partial z} - \frac{\partial v_z}{\partial x}\right)j + \left(\frac{\partial v_y}{\partial x} - \frac{\partial v_x}{\partial y}\right)k.$$

The curl of a vector field, denoted curl(*F*) or $\nabla \times F$ (the notation used in this work), is defined as the vector field having magnitude equal to the maximum "circulation" at each point and to be oriented perpendicularly to this plane of circulation for each point.

If $\nabla \times F = 0$, then the field is said to be an irrotational field. The symbol ∇ is variously known as "nabla" or "del."

In Cartesian coordinates, the curl is defined by

$$\nabla \times F = \left(\frac{\partial F_2}{\partial y} - \frac{\partial F_y}{\partial z}\right)\hat{x} + \left(\frac{\partial F_x}{\partial z} - \frac{\partial F_z}{\partial x}\right)\hat{y} + \left(\frac{\partial F_y}{\partial x} - \frac{\partial F_x}{\partial y}\right)\hat{z}.$$

This provides the motivation behind the adoption of the symbol $\nabla\times$ for the curl, since interpreting ∇ as the gradient operator $\nabla = (\partial/\partial x, \partial/\partial y, \partial/\partial z)$, the "cross product" of the gradient operator with F is given by

$$\nabla \times F = \begin{vmatrix} \hat{x} & \hat{y} & \hat{z} \\ \frac{\partial}{\partial x} & \frac{\partial}{\partial y} & \frac{\partial}{\partial z} \\ F_x & F_y & F_z \end{vmatrix},$$

which is precisely equation (4). A somewhat more elegant formulation of the curl is given by the matrix operator equation

$$\nabla \times F = \begin{vmatrix} 0 & -\frac{\partial}{\partial z} & \frac{\partial}{\partial y} \\ \frac{\partial}{\partial z} & 0 & -\frac{\partial}{\partial x} \\ -\frac{\partial}{\partial y} & \frac{\partial}{\partial x} & 0 \end{vmatrix} F$$

The curl can be similarly defined in arbitrary orthogonal curvilinear coordinates using

$$F = F_1\hat{u}_1 + F_2\hat{u}_2 + F_3\hat{u}_3$$

and defining $$h_i = \left|\frac{\partial r}{\partial u_i}\right|,$$

as $$\nabla \times F = \frac{1}{h_1h_2h_3}\begin{vmatrix} h_1\hat{u}_1 & h_2\hat{u}_2 & h_3\hat{u}_3 \\ \frac{\partial}{\partial u_1} & \frac{\partial}{\partial u_2} & \frac{\partial}{\partial u_3} \\ h_1F_1 & h_2F_2 & h_3F_3 \end{vmatrix}$$

$$= \frac{1}{h_1h_2h_3}\left[\frac{\partial}{\partial u_2}(h_3F_3) - \frac{\partial}{\partial u_3}(h_2F_2)\right]\hat{u}_1 + \frac{1}{h_1h_3}$$

$$\left[\frac{\partial}{\partial u_3}(h_1F_1) - \frac{\partial}{\partial u_1}(h_3F_3)\right]\hat{u}_2 + \frac{1}{h_1h_2}\left[\frac{\partial}{\partial u_1}(h_2F_2) - \frac{\partial}{\partial u_2}(h_1F_1)\right]\hat{u}_3$$

The curl can be generalized from a vector field to a tensor field as

$$(\nabla \times A)^{\alpha} = \epsilon^{\alpha\mu r} A_{r;\mu},$$

where ϵ_{ijk} is the permutation tensor and ";" denotes a covariant derivative

PROPERTIES OF CURL OF VECTOR FUNCTION

1. curl$(cU) = c$ curl U
2. curl$(U + V) =$ curl $U +$ curl V
3. curl$(fU) = f$curl $U + (\nabla f \times U)$
4. curl$(\nabla f +$ curl $U) =$ curl(∇f) + curl(curl U)
5. $\nabla \times (\nabla f) = 0$
6. div(curl$U) = 0$

LAPLACE OPERATOR

Laplace operator or Laplacian is a differential operator, specifically an important case of an elliptic operator, with many applications. In physics, it is used in modeling of wave propagation and heat flow, forming the Helmholtz equation.

Definition: The Laplace operator is a second order differential operator in the *n*-dimensional Euclidean space, defined as the divergence of the gradient:

$$\nabla = \nabla^2 = \nabla.\nabla \cdot$$

Equivalently, the Laplacian is the sum of all the *unmixed* second partial derivatives:

$$\Delta = \sum_{i=1}^{n} \frac{\partial^2}{\partial x_i^2}.$$

Here, it is understood that the x_i are Cartesian coordinates on the space; the equation takes a different form in spherical coordinates and cylindrical coordinates, as shown below.

If we form the scalar ("dot") product of $\vec{\nabla}$ with *itself* we get a scalar *second derivative* operator called the Laplacian:

$$\vec{\nabla} \cdot \vec{\nabla} \equiv \nabla^2 \equiv \frac{\partial^2}{\partial x^2} + \frac{\partial^2}{\partial y^2} + \frac{\partial^2}{\partial z^2}$$

In the three-dimensional space the Laplacian is commonly written as

$$\Delta = \frac{\partial^2}{\partial x^2} + \frac{\partial^2}{\partial y^2} + \frac{\partial^2}{\partial z^2}.$$

ILLUSTRATIVE EXAMPLES

1. *Find the directional derivative of* $f = (x^2 + y^2 + z^2)^{-1/2}$ *at the point* $P(1, 2, 3)$ *in the direction of the vector* $yzi+zxj+xyk$.

Solution: Unit vector in the direction of given vector $\hat{a} = \dfrac{yzi + zxj + xyk}{\sqrt{(y^2z^2 + z^2x + x^2y^2)}}$

Also $\frac{\partial\phi}{\partial x} = \frac{-1}{2}(x^2 + y^2 + z^2)^{-3/2}.\ 2x = -x(x^2 + y^2 + z^2)^{-3/2}$

Similarly $\frac{\partial\phi}{\partial y} = -y\ (x^2 + y^2 + z^2)^{-3/2}$ and $\frac{\partial\phi}{\partial z} = -z\ (x^2+y^2+z^2)^{-3/2}$.

$\therefore$ grad $\phi = i\frac{\partial\phi}{\partial x} + j\frac{\partial\phi}{\partial y} + k\frac{\partial\phi}{\partial z} = -y\ (x^2 + y^2 + z^2)^{-3/2}\ (xi + yj + zk)$.

$\therefore\ \frac{\partial\phi}{\partial s}\ \hat{a}\ \bullet\text{grad}\ \phi = \frac{yzi + zxj + xyk}{\sqrt{(y^2z^2 + z^2x^2 + x^2y^2)}}$

$\bullet\{-(x^2+y^2+z^2)^{-3/2}\ (xi+yj+zk)\} = \frac{-3xyz(x^2 + y^2 + z^2)^{-3/2}}{\sqrt{(y^2z^2 + z^2x^2 + x^2y^2)}}$

Therefor $\frac{\partial\phi}{\partial s}$ at P (1, 2, 3) is $= \frac{-3(6)(9+1+4)^{-3/2}}{\sqrt{(4+36+9)}} = \frac{-18(14)^{-3/2}}{7} = \frac{-18}{7\times 14\sqrt{14}} =$

$\frac{9}{49\sqrt{14}}$ *(neglecting -ive sign).*

2. *Find the equations of the tangent plane and normal to the surface xyz = 4 at the point (1, 2, 2)*

Solution: Tangent plane is $(R\text{-r})\bullet\text{grad}\ f = 0$ *i.e.* $2X + Y + Z = 6$.

Also, the normal to the surface at (1,2,2) is

$$\frac{X-1}{(\partial f/\partial x)} = \frac{X-2}{(\partial f/\partial x)} = \frac{X-2}{(\partial f/\partial x)}\ \ i.e.\ \ \frac{X-1}{2} = \frac{Y-2}{1} = \frac{Z-2}{1}.$$

EXERCISE

1. Find the directional derivative of $f = x^2yz + 4xz^2$ at the point (1, –2, –1) in the direction of the vector $2i - j - 2k$.
2. What is the directional derivative of $f = xy^2 + yz^3$ at the point (2, –1, 1) in the direction of the normal to the surface $x \log z - y^2 = -4$ at (–1, 2, 1)?
3. If $F = (x + y + 1)\ I + J - (x + y)K$, show that F. curl F = 0.
4. In what direction from (3, 1, –2) is the directional derivative of $f = x^2y^2z^4$ maximum ? Find also the magnitude of this maximum.
5. Evaluate (i) div $[3x^2\text{I} + 5xy^2\text{J} + xyz^3\text{K}]$ at the point (1, 2, 3).

   (ii) curl $[e^{xyz}\ (I + J + K)]$.
6. A particle moves on the vurve $x = 2t^2$, $y = t^2 - 4t$, $z = 3t - 5$, where t is the time. Find the components of velocity and acceleration at time $t = 1$ in the direction I – 3J + 2K.

7. (a) Find ∇f, if $f = \log(x^2 + y^2 + z^2)$

   (b) Show that grad $(1/r) = -R/r^3$.

8. Find the value of a if the vector $(ax^2y + yz)I + (xy^2 - xz^2)J + (2xyz - 2x^2y^2)K$ has zero divergence. Find the curl of the above vector which has zero divergence.

3. Find a unit vector normal to the surface $x^3 + y^3 + 3xyz = 3$ at the point $(1, 2, -1)$.

4. Find the constants a and b so that the surface $ax^2 - byz = (a + 2)x$ is orthogonal to the surface $4x2y + z^3 = 4$ at the point $(1, -1, 2)$.

9. If $V = (xI + yJ + zK)/\sqrt{(x^2 + y^2 + z^2)}$ Show that

   $\nabla V = 2/\sqrt{(x^2 + y^2 + z^2)}$ and $\nabla \times V = 0$.

10. If $u = x^2 + y^2 + z^2$ and $V = xI + yJ + zK$, show div (uV) = 5u.

11. If $R = xI + yJ + zK$ and $r \neq 0$, show that

    (a) $\nabla (1/r^2) = -2R/r^4$; $\nabla (R/r^2) = 1/r^2$ (ii) div $(r^nR) = (n + 3)r^n$; curl$(r^nR) = 0$.

    (b) $\nabla\left[\nabla.\dfrac{R}{r}\right] = -\dfrac{2}{r^3}R.$

12. Show that each of following vectors are solenoidal :

    (a) $(x + 3y)I + (y - 3z)J + (x - 2z)K$ (ii) $3y^4z^2I + 4x^3z^2J + 3x^2y^2K$

    (b) $\nabla f \times \nabla y$.

Chapter 7

Functions of Several Variables

FUNCTIONS OF TWO OR THREE VARIABLES

We have discussed vector-valued functions of a single real variable. We will now examine real-valued functions of a point (or vector) in $\mathbb{R}^2$ or $\mathbb{R}^3$. For the most part these functions will be defined on sets of points in $\mathbb{R}^2$, but there will be times when we will use points in $\mathbb{R}^3$, and there will also be times when it will be convenient to think of the points as vectors (or terminal points of vectors).

A real-valued function f defined on a subset D of $\mathbb{R}^2$ is a rule that assigns to each point (x, y) in D a real number $f(x, y)$. The largest possible set D in $\mathbb{R}^2$ on which f is defined is called the domain of f, and the range of f is the set of all real numbers $f(x, y)$ as (x, y) varies over the domain D. A similar definition holds for functions $f(x, y, z)$ defined on points (x, y, z) in $\mathbb{R}^3$.

Example: The domain of the function

$$f(x, y) = xy$$

is all of $\mathbb{R}^2$, and the range of f is all of $\mathbb{R}$.

Example: The domain of the function

$$f(x, y) = \frac{1}{x - y}$$

is all of $\mathbb{R}^2$ except the points (x, y) for which $x = y$. That is, the domain is the set $D = \{(x, y) : x \neq y\}$. The range of f is all real numbers except 0.

Example: The domain of the function

$$f(x, y) = \sqrt{1 - x^2 - y^2}$$

is the set $D = \{(x, y) : x^2 + y^2 \leq 1\}$, since the quantity inside the square root is nonnegative if and only if $1 - (x^2 + y^2) \geq 0$. We see that D consists of all points on and nside the unit circle in $\mathbb{R}^2$ (D is sometimes called the *closed unit disk*). The range of f is the interval $[0, 1]$ in $\mathbb{R}$.

Example: The domain of the function

$$f(x, y, z) = e^{x+y-z}$$

is all of $\mathbb{R}^3$, and the range of f is all positive real numbers.

A function $f(x, y)$ defined in $\mathbb{R}^2$ is often written as $z = f(x, y)$, so that the graph of $f(x, y)$ is the set $\{(x, y, z): z = f(x, y)\}$ in $\mathbb{R}^3$. So we see that this graph is a surface in $\mathbb{R}^3$, since it satisfies an equation of the form $F(x, y, z) = 0$ (namely, $F(x, y, z) = f(x, y) - z$).

The traces of this surface in the planes $z = c$, where c varies over $\mathbb{R}$, are called the level curves of the function. Equivalently, the level curves are the solution sets of the equations $f(x, y) = c$, for c in $\mathbb{R}$. Level curves are often projected onto the xy-plane to give an idea of the various "elevation" levels of the surface (as is done in topography).

Example: The graph of the function

$$f(x,y) = \frac{\sin\sqrt{x^2 + y^2}}{\sqrt{x^2 + y^2}}$$

is shown below. Note that the level curves (shown both on the surface and projected onto the xy-plane) are groups of concentric circles.

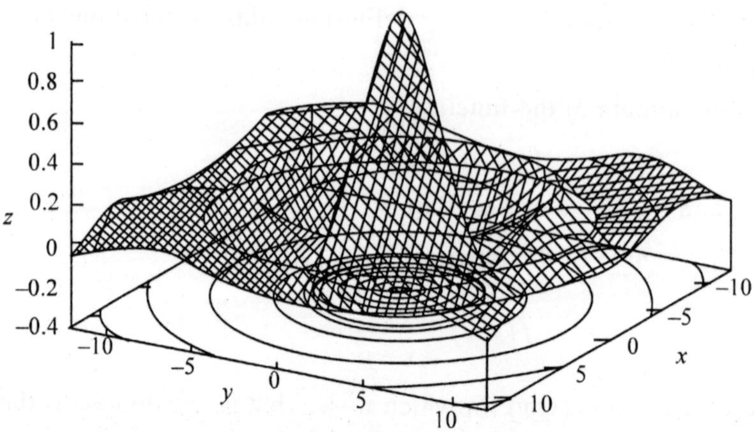

Fig. The function $f(x, y) = \dfrac{\sin\sqrt{x^2 + y^2}}{\sqrt{x^2 + y^2}}$

You may be wondering what happens to the function in Example at the point $(x, y) = (0, 0)$, since both the numerator and denominator are 0 at that point. The function is not defined at $(0, 0)$, but the *limit* of the function exists (and equals 1) as (x, y) *approaches* $(0, 0)$. We will now state explicitly what is meant by the limit of a function of two variables.

Definition: Let (a, b) be a point in $\mathbb{R}^2$, and let $f(x, y)$ be a real-valued function defined on some set containing (a, b) (but not necessarily defined at (a, b) itself). Then we say that the limit of $f(x, y)$ equals L as (x, y) approaches (a, b), written as

$$\lim_{(x,y)\to(a,b)} f(x,y) = L,$$

if given any $\varepsilon > 0$, there exists a $\delta > 0$ such that

$$|f(x, y) - L| < \varepsilon \text{ whenever } 0 < \sqrt{(x-a)^2 + (y-b)^2} < \delta .$$

A similar definition can be made for functions of three variables. The idea behind the above definition is that the values of $f(x, y)$ can get arbitrarily close to L (i.e. within ε of L) if we pick (x, y) sufficiently close to (a, b) (i.e. inside a circle centered at (a, b) with some sufficiently small radius δ).

If you recall the "epsilon-delta" proofs of limits of real-valued functions of a single variable, you may remember how awkward they can be, and how they can usually only be done easily for simple functions.

In general, the multivariable cases are at least equally awkward to go through, so we will not bother with such proofs. Instead, we will simply state that when the function $f(x, y)$ is given by a single formula and is defined at the point (a, b) (e.g. is not some indeterminate form like 0/0) then you can just substitute $(x, y) = (a, b)$ into the formula for $f(x, y)$ to find the limit.

Example: $\lim_{(x,y)\to(1,2)} \dfrac{xy}{x^2+y^2} = \dfrac{(1)(2)}{1^2+2^2} = \dfrac{2}{5}$

since $f(x, y) = \dfrac{xy}{x^2+y^2}$ is properly defined at the point (1, 2).

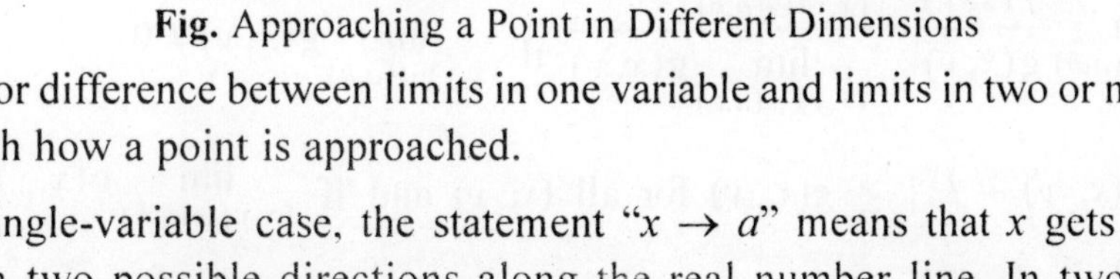

(a) $x \to a$ in R (b) $(x, y) \to (a, b)$ in R^2

Fig. Approaching a Point in Different Dimensions

The major difference between limits in one variable and limits in two or more variables has to do with how a point is approached.

In the single-variable case, the statement "$x \to a$" means that x gets closer to the value a from two possible directions along the real number line. In two dimensions, however, (x, y) can approach a point (a, b) along an infinite number of paths.

Example: $\lim_{(x,y)\to(0,0)} \frac{xy}{x^2+y^2}$ does not exist

Note that we can not simply substitute $(x, y) = (0, 0)$ into the function, since doing so gives an indeterminate form 0/0. To show that the limit does not exist, we will show that the function approaches different values as (x, y) approaches $(0, 0)$ along different paths in $\mathbb{R}^2$. To see this, suppose that $(x, y) \to (0, 0)$ along the positive x-axis, so that $y = 0$ along that path. Then

$$f(x, y) = \frac{xy}{x^2+y^2} = \frac{x0}{x^2+0^2} = 0$$

along that path (since $x > 0$ in the denominator). But if $(x, y) \to (0, 0)$ along the straight line $y = x$ through the origin, for $x > 0$, then we see that

$$f(x, y) = \frac{xy}{x^2+y^2} = \frac{x^2}{x^2+x^2} = \frac{1}{2}$$

which means that $f(x, y)$ approaches different values as $(x, y) \to (0, 0)$ along different paths. Hence the limit does not exist.

Limits of real-valued multivariable functions obey the same algebraic rules as in the single-variable case, as shown in the following theorem, which we state without proof.

Theorem: Suppose that $\lim_{(x,y)\to(a,b)} f(x,y)$ and $\lim_{(x,y)\to(a,b)} g(x,y)$ both exist, and that k is some scalar. Then:

(a) $$\lim_{(x,y)\to(a,b)} [f(x,y) \pm g(x,y)] = \left[\lim_{(x,y)\to(a,b)} f(x,y)\right] \pm \left[\lim_{(x,y)\to(a,b)} g(x,y)\right]$$

(b) $$\lim_{(x,y)\to(a,b)} kf(x,y) = k\left[\lim_{(x,y)\to(a,b)} f(x,y)\right]$$

(c) $$\lim_{(x,y)\to(a,b)} [f(x,y)g(x,y)] = \left[\lim_{(x,y)\to(a,b)} f(x,y)\right]\left[\lim_{(x,y)\to(a,b)} g(x,y)\right]$$

(d) $$\lim_{(x,y)\to(a,b)} \frac{f(x,y)}{g(x,y)} = \frac{\lim_{(x,y)\to(a,b)} f(x,y)}{\lim_{(x,y)\to(a,b)} g(x,y)} \text{ if } \lim_{(x,y)\to(a,b)} g(x,y) \neq 0$$

(e) If $| f(x, y) - L | \leq g(x, y)$ for all (x, y) and if $\lim_{(x,y)\to(a,b)} g(x,y) = 0$, then $\lim_{(x,y)\to(a,b)} f(x, y) = L$.

Note that in part (e), it suffices to have $|f(x, y) - L| \leq g(x, y)$ for all (x, y) "sufficiently close" to (a, b) (but excluding (a, b) itself).

Example: Show that

$$\lim_{(x,y)\to(0,0)} \frac{y^4}{x^2 + y^2} = 0 .$$

Since substituting $(x, y) = (0, 0)$ into the function gives the indeterminate form 0/0, we need an alternate method for evaluating this limit. We will use Theorem. First, notice that $y^4 = \left(\sqrt{y^2}\right)^4$ and so $0 \leq y^4 \leq \left(\sqrt{x^2 + y^2}\right)^4$ for all (x, y). But $\left(\sqrt{x^2 + y^2}\right)^4 = (x^2 + y^2)^2$. Thus, for all $(x, y) \neq (0, 0)$ we have

$$\left|\frac{y^4}{x^2 + y^2}\right| \leq \frac{(x^2 + y^2)^2}{x^2 + y^2} = x^2 + y^2 \to 0 \text{ as } (x, y) \to (0, 0).$$

Therefore $\displaystyle\lim_{(x,y)\to(0,0)} \frac{y^4}{x^2 + y^2} = 0$.

Continuity can be defined similarly as in the single-variable case.

Definition: A real-valued function $f(x, y)$ with domain D in $\mathbb{R}^2$ is continuous at the point (a, b) in D if $\displaystyle\lim_{(x,y)\to(a,b)} f(x, y) = f(a, b)$. We say that $f(x, y)$ is a continuous function if it is continuous at every point in its domain D.

Unless indicated otherwise, you can assume that all the functions we deal with are continuous. In fact, we can modify the function from Example so that it is continuous on all of $\mathbb{R}^2$.

Example: Define a function $f(x, y)$ on all of $\mathbb{R}^2$ as follows:

$$f(x, y) = \begin{cases} 0 & \text{if } (x, y) = (0,0) \\ \dfrac{y^4}{x^2 + y^2} & \text{if } (x, y) \neq (0,0) \end{cases}$$

Then $f(x, y)$ is well-defined for all (x, y) in $\mathbb{R}^2$ (i.e. there are no indeterminate forms for any (x, y)), and we see that

$$\lim_{(x,y)\to(a,b)} f(x, y) = \frac{b^4}{a^2 + b^2} = f(a, b) \text{ for } (a, b) \neq (0, 0).$$

So since $\displaystyle\lim_{(x,y)\to(0,0)} f(x, y) = 0 = f(0, 0)$ by Example,

then $f(x, y)$ is continuous on all of $\mathbb{R}^2$.

PARTIAL DERIVTETATIVES

Now that we have an idea of what functions of several variables are, and what a limit of such a function is, we can start to develop an idea of a derivative of a function of two or more variables. We will start with the notion of a *partial derivative*.

Definition: Let $f(x, y)$ be a real-valued function with domain D in $\mathbb{R}^2$, and let (a, b) be a point in D. Then the partial derivative of f at (a, b) with respect to x, denoted by $\frac{\partial f}{\partial x}$ (a, b) is defined as

$$\frac{\partial f}{\partial x}(a,b) = \lim_{h \to 0} \frac{f(a+h,b) - f(a,b)}{h}$$

and the partial derivative of f at (a, b) with respect to y, denoted by $\frac{\partial f}{\partial y}$ (a, b), is defined as

$$\frac{\partial f}{\partial y}(a,b) = \lim_{h \to 0} \frac{f(a,b+h) - f(a,b)}{h}$$

Note: The symbol " is pronounced "del".

Recall that the derivative of a function $f(x)$ can be interpreted as the rate of change of that function in the (positive) x direction. From the definitions above, we can see that the partial derivative of a function $f(x, y)$ with respect to x or y is the rate of change of $f(x, y)$ in the (positive) x or y direction, respectively. What this means is that the partial derivative of a function $f(x, y)$ with respect to x can be calculated by treating the *y variable* as a *constant*, and then simply differentiating $f(x, y)$ as if it were a function of x alone, using the usual rules from single-variable calculus. Likewise, the partial derivative of $f(x, y)$ with respect to y is obtained by treating the x variable as a constant and then differentiating $f(x, y)$ as if it were a function of y alone.

Example: Find $\frac{\partial f}{\partial x}(x, y)$ and $\frac{\partial f}{\partial y}$ (x, y) for the function $f(x, y) = x^2y + y^3$.

Solution: Treating y as a constant and differentiating $f(x, y)$ with respect to x gives

$$\frac{\partial f}{\partial x}(x,y) = 2xy$$

and treating x as a constant and differentiating $f(x, y)$ with respect to y gives

$$\frac{\partial f}{\partial y}(x,y) = x^2 + 3y^2.$$

We will often simply write $\frac{\partial f}{\partial x}$ and $\frac{\partial f}{\partial y}$ instead of $\frac{\partial f}{\partial x}$ (x, y) and $\frac{\partial f}{\partial y}$ (x, y).

Example: Find $\frac{\partial f}{\partial x}$ and $\frac{\partial f}{\partial y}$ for the function $f(x, y) = \frac{\sin(xy^2)}{x^2+1}$

Solution: Treating y as a constant and differentiating $f(x, y)$ with respect to x gives

$$\frac{\partial f}{\partial x} = \frac{(x^2+1)(y^2\cos(xy^2)) - (2x)\sin(xy^2)}{(x^2+1)^2}$$

and treating x as a constant and differentiating $f(x, y)$ with respect to y gives

$$\frac{\partial f}{\partial y} = \frac{2xy\cos(xy^2)}{x^2+1}$$

Since both $\frac{\partial f}{\partial x}$ and $\frac{\partial f}{\partial y}$ are themselves functions of x and y, we can take *their* partial derivatives with respect to x and y. This yields the *higher-order partial derivatives*:

$$\frac{\partial^2 f}{\partial x^2} = \frac{\partial}{\partial x}\left(\frac{\partial f}{\partial x}\right) \qquad \frac{\partial^2 f}{\partial y^2} = \frac{\partial}{\partial y}\left(\frac{\partial f}{\partial y}\right)$$

$$\frac{\partial^2 f}{\partial y \partial x} = \frac{\partial}{\partial y}\left(\frac{\partial f}{\partial x}\right) \qquad \frac{\partial^2 f}{\partial x \partial y} = \frac{\partial}{\partial x}\left(\frac{\partial f}{\partial y}\right)$$

$$\frac{\partial^3 f}{\partial x^3} = \frac{\partial}{\partial x}\left(\frac{\partial^2 f}{\partial x^2}\right) \qquad \frac{\partial^3 f}{\partial y^3} = \frac{\partial}{\partial y}\left(\frac{\partial^2 f}{\partial y^2}\right)$$

$$\frac{\partial^3 f}{\partial y \partial x^2} = \frac{\partial}{\partial y}\left(\frac{\partial^2 f}{\partial x^2}\right) \qquad \frac{\partial^3 f}{\partial x \partial y^2} = \frac{\partial}{\partial x}\left(\frac{\partial^2 f}{\partial y^2}\right)$$

$$\frac{\partial^3 f}{\partial y^2 \partial x} = \frac{\partial}{\partial y}\left(\frac{\partial^2 f}{\partial y \partial x}\right) \qquad \frac{\partial^3 f}{\partial x^2 \partial y} = \frac{\partial}{\partial x}\left(\frac{\partial^2 f}{\partial x \partial y}\right)$$

$$\frac{\partial^3 f}{\partial x \partial y \partial x} = \frac{\partial}{\partial x}\left(\frac{\partial^2 f}{\partial y \partial x}\right) \qquad \frac{\partial^3 f}{\partial y \partial x \partial y} = \frac{\partial}{\partial y}\left(\frac{\partial^2 f}{\partial x \partial y}\right)$$

Example: Find the partial derivatives $\frac{\partial f}{\partial x}, \frac{\partial f}{\partial y}, \frac{\partial^2 f}{\partial x^2}, \frac{\partial^2 f}{\partial y^2}, \frac{\partial^2 f}{\partial y \partial x}$ and $\frac{\partial^2 f}{\partial x \partial y}$ for the function $f(x, y) = e^{x^2 y} + xy^3$.

Solution: Proceeding as before, we have

$$\frac{\partial f}{\partial x} = 2xy^{x^2 y} + y^3 \qquad \frac{\partial f}{\partial y} = x^2 e^{x^2 y} + 3xy^2$$

$$\frac{\partial^2 f}{\partial x^2} = \frac{\partial}{\partial x}(2xye^{x^2 y} + y^3) \qquad \frac{\partial^2 f}{\partial y^2} = \frac{\partial}{\partial y}(x^2 e^{x^2 y} + 3xy^2)$$

$$= 2ye^{x^2 y} + 4x^2 y^2 e^{x^2 y} \qquad = x^4 e^{x^2 y} + 6xy$$

$$\frac{\partial^2 f}{\partial y \partial x} = \frac{\partial}{\partial y}(2xye^{x^2 y} + y^3) \qquad \frac{\partial^2 f}{\partial x \partial y} = \frac{\partial}{\partial x}(x^2 e^{x^2 y} + 3xy^2)$$

$$= 2xe^{x^2 y} + 2x^3 ye^{x^2 y} + 3y^2 \qquad = 2xe^{x^2 y} + 2x^3 ye^{x^2 y} + 3y^2$$

Higher-order partial derivatives that are taken with respect to different variables, such as $\frac{\partial^2 f}{\partial y \partial x}$ and $\frac{\partial^2 f}{\partial x \partial y}$, are called mixed partial derivatives. Notice in the above example that $\frac{\partial^2 f}{\partial y \partial x} = \frac{\partial^2 f}{\partial x \partial y}$. It turns that this will usually be the case. Specifically, whenever both $\frac{\partial^2 f}{\partial y \partial x}$ and $\frac{\partial^2 f}{\partial x \partial y}$ are continuous at a point (a, b), then they are equal at that point. All the functions we will deal with will have continuous partial derivatives of all orders, so you can assume in the remainder of the text that

$$\frac{\partial^2 f}{\partial y \partial x} = \frac{\partial^2 f}{\partial x \partial y} \text{ for all } (x, y) \text{ in the domain of } f.$$

In other words, it doesn't matter in which order you take partial derivatives. This applies even to mixed partial derivatives of order 3 or higher. The notation for partial derivatives varies. All of the following are equivalent:

$$\frac{\partial f}{\partial x} : f_x(x, y), f_1(x, y), D_x(x, y), D_1(x, y)$$

$$\frac{\partial f}{\partial y} : f_y(x, y), f_2(x, y), D_y(x, y), D_2(x, y)$$

$$\frac{\partial^2 f}{\partial x^2} : f_{xx}(x, y), f_{11}(x, y), D_{xx}(x, y), D_{11}(x, y)$$

$$\frac{\partial^2 f}{\partial y^2} : f_{yy}(x, y), f_{22}(x, y), D_{yy}(x, y), D_{22}(x, y)$$

$$\frac{\partial^2 f}{\partial y \partial x} : f_{xy}(x, y), f_{12}(x, y), D_{xy}(x, y), D_{12}(x, y)$$

$$\frac{\partial^2 f}{\partial x \partial y} : f_{yx}(x, y), f_{21}(x, y), D_{yx}(x, y), D_{21}(x, y)$$

EXERCISES

For Exercises 1–16, find $\frac{\partial f}{\partial x}$ and $\frac{\partial f}{\partial y}$.

1. $f(x, y) = x^2 + y^2$
2. $f(x, y) = \cos(x + y)$
3. $f(x, y) = \sqrt{x^2 + y + 4}$
4. $f(x, y) = \frac{x+1}{y+1}$
5. $f(x, y) = e^{xy} + xy$
6. $f(x, y) = x^2 - y^2 + 6xy + 4x - 8y + 2$
7. $f(x, y) = x^4$
8. $f(x, y) = x + 2y$
9. $f(x, y) = \sqrt{x^2 + y^2}$
10. $f(x, y) = \sin(x + y)$
11. $f(x, y) = \sqrt[3]{x^2 + y + 4}$
12. $f(x, y) = \frac{xy+1}{x+y}$
13. $f(x, y) = e^{-(x^2+y^2)}$
14. $f(x, y) = \ln(xy)$
15. $f(x, y) = \sin(xy)$
16. $f(x, y) = \tan(x + y)$

For Exercises 17-26, find $\frac{\partial^2 f}{\partial x^2}, \frac{\partial^2 f}{\partial y^2}$ and $\frac{\partial^2 f}{\partial y \partial x}$ (use Exercises 1 –8, 14, 15).

17. $f(x, y) = x^2 + y^2$
18. $f(x, y) = \cos(x + y)$

19. $f(x, y) = \sqrt{x^2 + y + 4}$

20. $f(x, y) = \dfrac{x+1}{y+1}$

21. $f(x, y) = e^{xy} + xy$

22. $f(x, y) = x^2 - y^2 + 6xy + 4x - 8y + 2$

23. $f(x, y) = x^4$

24. $f(x, y) = x + 2y$

25. $f(x, y) = \ln(xy)$

26. $f(x, y) = \sin(xy)$

27. Show that the function $f(x, y) = \sin(x + y) + \cos(x - y)$ satisfies the *wave equation*

$$\frac{\partial^2 f}{\partial x^2} - \frac{\partial^2 f}{\partial y^2} = 0$$

The wave equation is an example of a *partial differential equation.*

28. Let u and v be twice-differentiable functions of a single variable, and let $c \neq 0$ be a constant. Show that $f(x, y) = u(x + cy) + v(x - cy)$ is a solution of the *general one-dimensional wave equation*

$$\frac{\partial^2 f}{\partial x^2} - \frac{1}{c^2}\frac{\partial^2 f}{\partial y^2} = 0.$$

Tangent Plane to a Surface

In the previous section we mentioned that the partial derivatives $\frac{\partial f}{\partial x}$ and $\frac{\partial f}{\partial y}$ can be thought of as the rate of change of a function $z = f(x, y)$ in the positive x and y directions, respectively. Recall that the derivative $\frac{dy}{\partial x}$ of a function $y = f(x)$ has a geometric meaning, namely as the slope of the tangent line to the graph of f at the point $(x, f(x))$ in $\mathbb{R}^2$. There is a similar geometric meaning to the partial derivatives $\frac{\partial f}{\partial x}$ and $\frac{\partial f}{\partial y}$ of a function $z = f(x, y)$: given a point (a, b) in the domain D of $f(x, y)$, the trace of the surface described by $z = f(x, y)$ in the plane $y = b$ is a curve in $\mathbb{R}^3$ through the point $(a, b, f(a, b))$, and the slope of the tangent line L^x to that curve at that point is $\frac{\partial f}{\partial x}(a, b)$. Similarly, $\frac{\partial f}{\partial y}(a, b)$ is the slope of the tangent line Ly to the trace of the surface $\dot{z} = f(x, y)$ in the plane $x = a$.

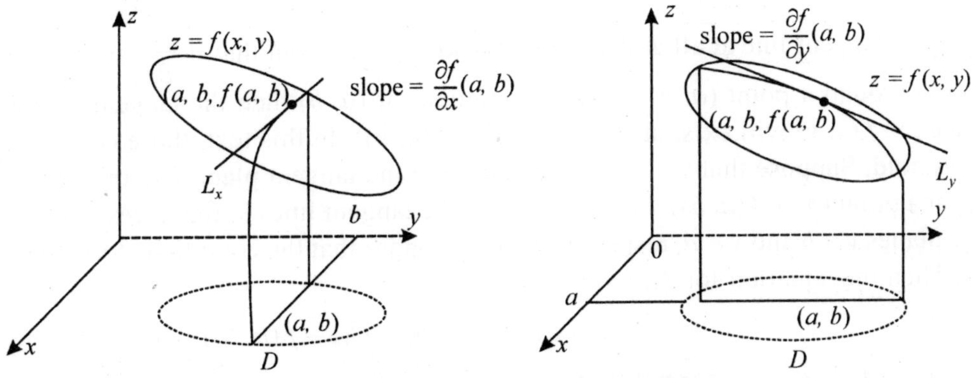

(a) Tangent line L_x in the plane $y = b$ (b) Tangent line L_y in the plane $x = a$

Fig. Partial Derivatives as Slopes

Since the derivative $\dfrac{dy}{\partial x}$ of a function $y = f(x)$ is used to find the tangent line to the graph of f (which is a curve in $\mathbb{R}^2$), you might expect that partial derivatives can be used to define a *tangent plane* to the graph of a surface $z = f(x, y)$. This indeed turns out to be the case. First, we need a definition of a tangent plane. The intuitive idea is that a tangent plane "just touches" a surface at a point. The formal definition mimics the intuitive notion of a tangent line to a curve.

Definition: Let $z = f(x, y)$ be the equation of a surface S in $\mathbb{R}^3$, and let $P = (a, b, c)$ be a point on S. Let T be a plane which contains the point P, and let $Q = (x, y, z)$ represent a generic point on the surface S. If the (acute) angle between the vector $\overrightarrow{PQ}$ and the plane T approaches zero as the point Q approaches P along the surface S, then we call T the tangent plane to S at P. Note that since two lines in $\mathbb{R}^3$ determine a plane, then the two tangent lines to the surface $z = f(x, y)$ in the x and y directions described in Figure are contained in the tangent plane at that point, *if the tangent plane exists at that point*. The existence of those two tangent lines does not by itself guarantee the existence of the tangent plane.

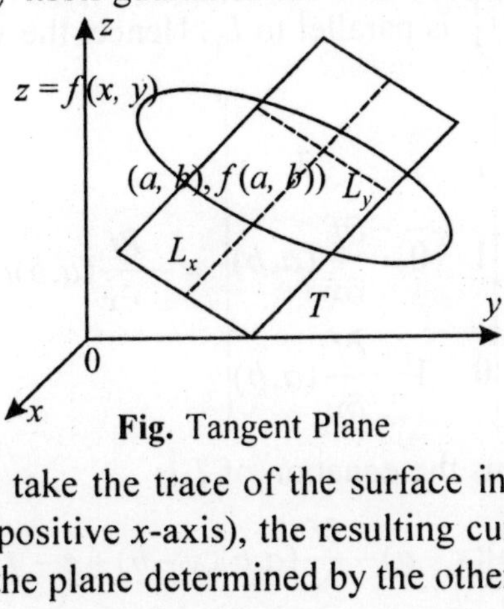

Fig. Tangent Plane

It is possible that if we take the trace of the surface in the plane $x - y = 0$ (which makes a 45° angle with the positive x-axis), the resulting curve in that plane may have a tangent line which is not in the plane determined by the other two tangent lines, or it may not have a tangent line at all at that point. Luckily, it turns out that if $\frac{\partial f}{\partial x}$ and $\frac{\partial f}{\partial y}$ exist in a region around a point (a, b) and are continuous at (a, b) then the tangent plane to the surface $z = f(x, y)$ will exist at the point $(a, b, f(a, b))$. In this text, those conditions will always hold. Suppose that we want an equation of the tangent plane T to the surface $z = f(x, y)$ at a point $(a, b, f(a, b))$. Let L_x and L_y be the tangent lines to the traces of the surface in the planes $y = b$ and $x = a$, respectively, and suppose that the conditions for T to exist do hold. Then the equation for T is

$$A(x - a) + B(y - b) + C(z - f(a, b)) = 0$$

where $n = (A, B, C)$ is a normal vector to the plane T.

Since T contains the lines L_x and L_y, then all we need are vectors vx and vy that are parallel to L_x and L_y, respectively, and then let $n = v_x \times v_y$.

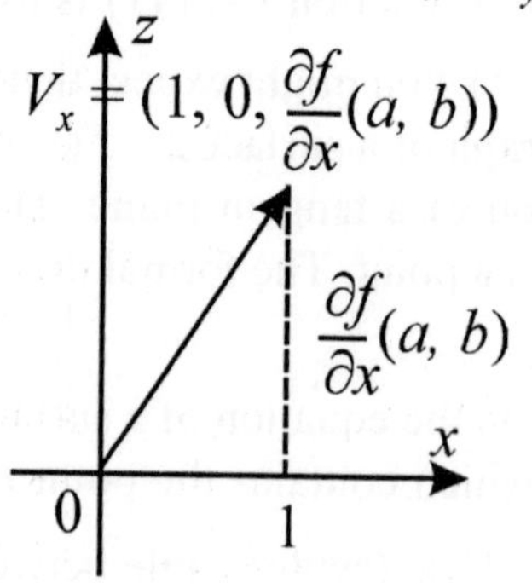

Since the slope of L_x is $\frac{\partial f}{\partial x}(a, b)$, then the vector $vx = \left(1, 0, \frac{\partial f}{\partial x}(a,b)\right)$ is parallel to L_x (since v_x lies in the xz-plane and lies in a line with slope $\frac{\frac{\partial f}{\partial x}(a,b)}{1} = \frac{\partial f}{\partial x}(a,b)$. Similarly, the vector $v_y = \left(0, 1, \frac{\partial f}{\partial y}(a,b)\right)$ is parallel to L_y. Hence, the vector

$$n = v_x \times v_y = \begin{vmatrix} i & j & k \\ 1 & 0 & \frac{\partial f}{\partial x}(a,b) \\ 0 & 1 & \frac{\partial f}{\partial y}(a,b) \end{vmatrix} = -\frac{\partial f}{\partial x}(a,b)i - \frac{\partial f}{\partial y}(a,b)j + k$$

is normal to the plane T. Thus the equation of T is

$$-\frac{\partial f}{\partial x}(a,b)(x - a) - \frac{\partial f}{\partial y}(a,b)(y - b) + z - f(a,b) = 0.$$

Multiplying both sides by –1, we have the following result:

The equation of the tangent plane to the surface $z = f(x, y)$ at the point $(a, b, f(a, b))$ is

$$\frac{\partial f}{\partial x}(a, b)(x - a) + \frac{\partial f}{\partial y}(a, b)(y - b) - z + f(a, b) = 0$$

Example: Find the equation of the tangent plane to the surface $z = x^2 + y^2$ at the point (1, 2, 5).

Solution: For the function $f(x, y) = x^2 + y^2$, we have $\frac{\partial f}{\partial x} = 2x$ and $\frac{\partial f}{\partial y} = 2y$, so the equation of the tangent plane at the point (1, 2, 5) is

$$2(1)(x - 1) + 2(2)(y - 2) - z + 5 = 0, \text{ or}$$

$$2x + 4y - z - 5 = 0.$$

In a similar fashion, it can be shown that if a surface is defined implicitly by an equation of the form $F(x, y, z) = 0$, then the tangent plane to the surface at a point (a, b, c) is given by the equation

$$\frac{\partial F}{\partial x}(a, b, c)(x - a) + \frac{\partial F}{\partial y}(a, b, c)(y - b) + \frac{\partial F}{\partial z}(a, b, c)(z - c) = 0.$$

Note that formula (2.6) is the special case of formula (2.7) where $F(x, y, z) = f(x, y) - z$.

Example: Find the equation of the tangent plane to the surface $x^2 + y^2 + z^2 = 9$ at the point (2, 2, –1).

Solution: For the function $F(x, y, z) = x^2 + y^2 + z^2 - 9$, we have $\frac{\partial F}{\partial x} = 2x$, $\frac{\partial F}{\partial y} = 2y$, and $\frac{\partial F}{\partial z} = 2z$, so the equation of the tangent plane at (2, 2, –1) is

$$2(2)(x - 2) + 2(2)(y - 2) + 2(-1)(z + 1) = 0, \text{ or}$$

$$2x + 2y - z - 9 = 0.$$

EXERCISES

For Exercises 1–6, find the equation of the tangent plane to the surface $z = f(x, y)$ at the point P.

1. $f(x, y) = x^2 + y^3$, $P = (1, 1, 2)$
2. $f(x, y) = xy$, $P = (1, -1, -1)$
3. $f(x, y) = x^2y$, $P = (-1, 1, 1)$

4. $f(x, y) = xe^y$, $P = (1, 0, 1)$
5. $f(x, y) = x + 2y$, $P = (2, 1, 4)$
6. $f(x, y) = \sqrt{x^2 + y^2}$, $P = (3, 4, 5)$

For Exercises 7-10, find the equation of the tangent plane to the given surface at the point P.

7. $\frac{x^2}{4} + \frac{y^2}{9} + \frac{z^2}{16} = 1, P = \left(1, 2, \frac{2\sqrt{11}}{3}\right)$
9. $x^2 + y^2 - z^2 = 0$, $P = (3, 4, 5)$
10. $x^2 + y^2 = 4$, $P = (\sqrt{3}, 1, 0)$

Directional Derivatives and the Gradient

For a function $z = f(x, y)$, we learned that the partial derivatives $\frac{\partial f}{\partial x}$ and $\frac{\partial f}{\partial y}$ represent the (instantaneous) rate of change of f in the positive x and y directions, respectively. What about other directions? It turns out that we can find the rate of change in *any* direction using a more general type of derivative called a *directional derivative*.

Definition: Let $f(x, y)$ be a real-valued function with domain D in $\mathbb{R}^2$, and let (a, b) be a point in D. Let v be a unit vector in $\mathbb{R}^2$. Then the directional derivative of f at (a, b) in the direction of v, denoted by $D_v f(a, b)$, is defined as

$$D_v f(a, b) = \lim_{h \to 0} \frac{f((a,b) + hv) - f(a,b)}{h}$$

Notice in the definition that we seem to be treating the point (a, b) as a vector, since we are adding the vector hv to it. But this is just the usual idea of identifying vectors with their terminal points, which the reader should be used to by now. If we were to write the vector v as $v = (v_1, v_2)$, then

$$D_v f(a,b) = \lim_{h \to 0} \frac{f(a + hv_1, b + hv_2) - f(a,b)}{h}$$

From this we can immediately recognize that the partial derivatives $\frac{\partial f}{\partial x}$ and $\frac{\partial f}{\partial y}$ are special cases of the directional derivative with $v = i = (1, 0)$ and $v = j = (0, 1)$, respectively. That is, $\frac{\partial f}{\partial x} = D_i f$ and $\frac{\partial f}{\partial y} = D_j f$. Since there are many vectors with the same direction, we use a unit vector in the definition, as that represents a "standard" vector for a given direction.

If $f(x, y)$ has continuous partial derivatives $\frac{\partial f}{\partial x}$ and $\frac{\partial f}{\partial y}$ (which will always be the case in this text), then there is a simple formula for the directional derivative:

Theorem: Let $f(x, y)$ be a real-valued function with domain D in $\mathbb{R}^2$ such that the partial derivatives $\frac{\partial f}{\partial x}$ and $\frac{\partial f}{\partial y}$ exist and are continuous in D. Let (a, b) be a point in D, and let $v = (v_1, v_2)$ be a unit vector in $\mathbb{R}^2$. Then

$$D_v f(a, b) = v_1 \frac{\partial f}{\partial x}(a, b) + v_2 \frac{\partial f}{\partial y}(a, b).$$

Proof: Note that if $v = i = (1, 0)$ then the above formula reduces to $D_v f(a, b) = \frac{\partial f}{\partial x}(a, b)$, which we know is true since $D_i f = \frac{\partial f}{\partial x}$, as we noted earlier. Similarly, for $v = j = (0, 1)$ the formula reduces to $D_v f(a, b) = \frac{\partial f}{\partial y}(a, b)$, which is true since $D_j f = \frac{\partial f}{\partial y}$. So since $i = (1, 0)$ and $j = (0, 1)$ are the only unit vectors in $\mathbb{R}^2$ with a zero component, then we need only show the formula holds for unit vectors $v = (v_1, v_2)$ with $v_1 \neq 0$ and $v_2 \neq 0$. So fix such a vector v and fix a number $h \neq 0$. Then

$$\begin{aligned} f(a + hv_1, b + hv_2) - f(a, b) &= f(a + hv_1, b + hv_2) \\ &\quad - f(a + hv_1, b) + f(a + hv_1, b) - f(a, b). \end{aligned}$$

Since $h \neq 0$ and $v_2 \neq 0$, then $hv_2 \neq 0$ and thus any number c between b and $b + hv_2$ can be written as $c = b + \alpha hv_2$ for some number $0 < \alpha < 1$. So since the function $f(a + hv_1, y)$ is a real-valued function of y (since $a + hv_1$ is a fixed number), then the Mean Value Theorem from single-variable calculus can be applied to the function $g(y) = f(a + hv_1, y)$ on the interval $[b, b + hv_2]$ (or $[b + hv_2, b]$ if one of h or v_2 is negative) to find a number $0 < \alpha < 1$ such that

$$\begin{aligned} \frac{\partial f}{\partial y}(a + hv_1, b + \alpha hv_2) = g'(b + \alpha hv_2) &= \frac{g(b + hv_2)}{b + hv_2 - b} \\ &= \frac{f(a + hv_1, b + hv_2) - f(a + hv_1, b)}{hv_2} \end{aligned}$$

and so $\quad f(a + hv_1, b + hv_2) - f(a + hv_1, b) = hv_2 \frac{\partial f}{\partial y}(a + hv_1, b + \alpha hv_2).$

By a similar argument, there exists a number $0 < \beta < 1$ such that

$$f(a + hv_1, b) - f(a, b) = hv_1 \frac{\partial f}{\partial x}(a + \beta hv_1, b).$$

Thus, by equation, we have

$$\frac{f(a+hv_1,b+hv_2)-f(a,b)}{h} = \frac{hv_2\frac{\partial f}{\partial y}(a+hv_1,b+\alpha hv_2)+hv_1\frac{\partial f}{\partial x}(a+\beta hv_1,b)}{h}$$

$$= v_1\frac{\partial f}{\partial y}\ (a+hv_1,\ b+\alpha hv_2)+v_1\frac{\partial f}{\partial x}\ (a+\beta hv_1,\ b)$$

so by formula we have

$$D_v f(a,b) = \lim_{h\to 0}\frac{f(a+hv_1,b+hv_2)-f(a,b)}{h}$$

$$= \lim_{h\to 0}\left[v_2\frac{\partial f}{\partial y}(a+hv_1,b+\alpha hv_2)+v_1\frac{\partial f}{\partial x}(a+\beta hv_1,b)\right]$$

$$= v_2\frac{\partial f}{\partial y}(a,b)+v_1\frac{\partial f}{\partial x}(a,b)$$

by the continuity of $\frac{\partial f}{\partial x}$ and $\frac{\partial f}{\partial y}$, so

$$D_v f(a,b) = v_1\frac{\partial f}{\partial x}(a,b)+v_2\frac{\partial f}{\partial y}(a,b)$$

after reversing the order of summation.

Note that $D_v f(a, b) = v \,.\left(\frac{\partial f}{\partial x}(a,b),\frac{\partial f}{\partial y}(a,b)\right)$. The second vector has a special name:

Definition: For a real-valued function $f(x, y)$, the gradient of f, denoted by ∇f, is the vector

$$\nabla f = \left(\frac{\partial f}{\partial x},\frac{\partial f}{\partial y}\right)$$

in $\mathbb{R}^2$. For a real-valued function $f(x, y, z)$, the gradient is the vector

$$\nabla f = \left(\frac{\partial f}{\partial x},\frac{\partial f}{\partial y},\frac{\partial f}{\partial z}\right)$$

in $\mathbb{R}^3$. The symbol ∇ is pronounced "del".

Corollary: $D_v f = v \,.\, \nabla f$

Example: Find the directional derivative of $f(x, y) = xy^2 + x^3y$ at the point (1, 2) in the direction of $v = \left(\frac{1}{\sqrt{2}},\frac{1}{\sqrt{2}}\right)$.

Solution: We see that $\nabla f = (y^2 + 3x^2y, 2xy + x^3)$, so

$$D_v f(1, 2) = v \,.\, \nabla f(1, 2) = \left(\frac{1}{\sqrt{2}}, \frac{1}{\sqrt{2}}\right) . (2^2 + 3(1)^2(2), 2(1)(2) + 1^3) = \frac{15}{\sqrt{2}}$$

A real-valued function $z = f(x, y)$ whose partial derivatives $\frac{\partial f}{\partial x}$ and $\frac{\partial f}{\partial y}$ exist and are continuous is called *continuously differentiable*. Assume that $f(x, y)$ is such a function and that $\nabla f \neq 0$. Let c be a real number in the range of f and let v be a unit vector in $\mathbb{R}^2$ which is tangent to the level curve $f(x, y) = c$.

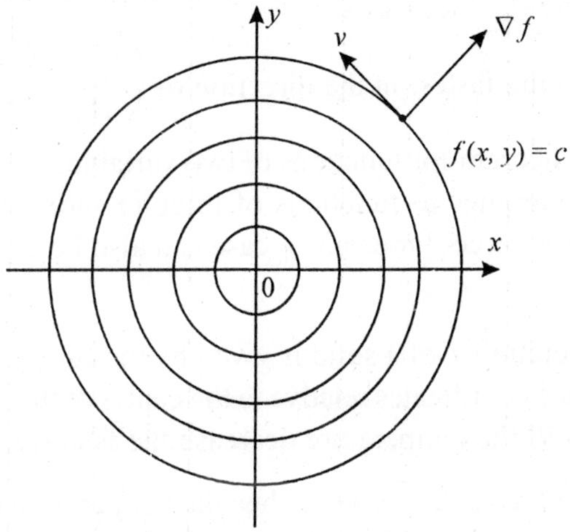

The value of $f(x, y)$ is constant along a level curve, so since v is a tangent vector to this curve, then the rate of change of f in the direction of v is 0, i.e. $D_v f = 0$. But we know that $D_v f = v \,.\, \nabla f = \| v \| \, \| \nabla f \| \cos \theta$, where θ is the angle between v and ∇f.

So since $\| v \| = 1$ then $D_v f = \| \nabla f \| \cos \theta$. So since $\nabla f \neq 0$ then $D_v f = 0 \Rightarrow \cos \theta = 0$ $\Rightarrow \theta = 90°$. In other words, $\nabla f \perp$ v, which means that ∇f is *normal* to the level curve.

In general, for any unit vector v in $\mathbb{R}^2$, we still have $D_v f = \| \nabla f \| \cos \theta$, where θ is the angle between v and ∇f. At a fixed point (x, y) the length $\| \nabla f \|$ is fixed, and the value of $D_v f$ then varies as θ varies. The largest value that $D_v f$ can take is when $\cos \theta = 1$ $(\theta = 0°)$, while the smallest value occurs when $\cos \theta = -1$ $(\theta = 180°)$. In other words, the value of the function f increases the fastest in the direction of ∇f (since $\theta = 0°$ in that case), and the value of f decreases the fastest in the direction of $-\nabla f$ (since $\theta = 180°$ in that case). We have thus proved the following theorem

Theorem: Let $f(x, y)$ be a continuously differentiable real-valued function, with ∇f ∇ 0. Then:

(a) The gradient ∇f is normal to any level curve $f(x, y) = c$.

(b) The value of $f(x, y)$ increases the fastest in the direction of ∇f.

(c) The value of $f(x, y)$ decreases the fastest in the direction of $-\nabla f$.

Example: In which direction does the function $f(x, y) = xy^2 + x^3y$ increase the fastest from the point (1, 2)? In which direction does it decrease the fastest?

Solution: Since $\nabla f = (y^2 + 3x^2y, 2xy + x^3)$, then $\nabla f(1, 2) = (10, 5) \neq 0$. A unit vector in that direction is $v = \dfrac{\nabla f}{\|\nabla f\|} = \left(\dfrac{2}{\sqrt{5}}, \dfrac{1}{\sqrt{5}}\right)$. Thus, f increases the fastest in the direction of $\left(\dfrac{2}{\sqrt{5}}, \dfrac{1}{\sqrt{5}}\right)$ and decreases the fastest in the direction of $\left(\dfrac{-2}{\sqrt{5}}, \dfrac{-1}{\sqrt{5}}\right)$.

Though we proved Theorem for functions of two variables, a similar argument can be used to show that it also applies to functions of three or more variables. Likewise, the directional derivative in the three-dimensional case can also be defined by the formula $D_v f = v \cdot \nabla f$.

Example: The temperature T of a solid is given by the function $T(x, y, z) = e^{-x} + e^{-2y} + e^{4z}$, where x, y, z are space coordinates relative to the center of the solid. In which direction from the point (1, 1, 1) will the temperature decrease the fastest?

Solution: Since $\nabla f = (-e^{-x}, -2e^{-2y}, 4e^{4z})$, then the temperature will decrease the fastest in the direction of $-\nabla f(1, 1, 1) = (e^{-1}, 2e^{-2}, -4e^{4})$.

EXERCISES

For Exercises 1-10, compute the gradient ∇f.

1. $f(x, y) = x^2 + y^2 - 1$
2. $f(x, y) = \dfrac{1}{x^2 + y^2}$
3. $f(x, y) = \sqrt{x^2 + y^2 + 4}$
4. $f(x, y) = x^2e^y$
5. $f(x, y) = \ln(xy)$
6. $f(x, y) = 2x + 5y$
7. $f(x, y, z) = \sin(xyz)$
8. $f(x, y, z) = x^2e^{yz}$
9. $f(x, y, z) = x^2 + y^2 + z^2$

10. $f(x, y, z) = \sqrt{x^2 + y^2 + z^2}$

For Exercises 11-14, find the directional derivative of f at the point P in the direction of $v = \left(\frac{1}{\sqrt{2}}, \frac{1}{\sqrt{2}}\right)$.

11. $f(x, y) = x^2 + y^2 - 1$, $P = (1, 1)$

12. $f(x, y) = \frac{1}{x^2 + y^2}$, $P = (1, 1)$

13. $f(x, y) = \sqrt{x^2 + y^2 + 4}$, $P = (1, 1)$ 14. $f(x, y) = x^2 e^y$, $P = (1, 1)$

For Exercises 15-16, find the directional derivative of f at the point P in the direction of $v = \left(\frac{1}{\sqrt{3}}, \frac{1}{\sqrt{3}}, \frac{1}{\sqrt{3}}\right)$.

15. $f(x, y, z) = \sin(xyz)$, $P = (1, 1, 1)$
16. $f(x, y, z) = x^2 e^{yz}$, $P = (1, 1, 1)$

For Exercises 19-26, let $f(x, y)$ and $g(x, y)$ be continuously differentiable real-valued functions, let c be a constant, and let v be a unit vector in $\mathbb{R}^2$. Show that:

19. $\nabla(cf) = c\,\nabla f$
20. $\nabla(f + g) = \nabla f + \nabla g$
21. $\nabla(fg) = f\,\nabla g + g\,\nabla f$
22. $\nabla(f/g) = \frac{g\nabla f - f\nabla g}{g^2}$ if $g(x, y) \neq 0$
23. $D_{-v} f = -D_v f$
24. $D_v(cf) = c\,D_v f$
25. $D_v(f + g) = D_v f + D_v g$
26. $D_v(fg) = f D_v g + g\,D_v f$
27. The function $r(x, y) = \sqrt{x^2 + y^2}$ is the length of the position vector $r = xi + yj$ for each point (x, y) in $\mathbb{R}^2$. Show that $\nabla r = \frac{1}{r} r$ when $(x, y) \neq (0, 0)$, and that $\nabla(r^2) = 2r$.

Maxima and Minima

The gradient can be used to find *extreme points* of real-valued functions of several variables, that is, points where the function has a *local maximum* or *local minimum*. We will consider only functions of two variables; functions of three or more variables require methods using linear algebra.

Definition: Let $f(x, y)$ be a real-valued function, and let (a, b) be a point in the domain of f. We say that f has a local maximum at (a, b) if $f(x, y) \leq f(a, b)$ for all (x, y) inside some disk of positive radius centered at (a, b), i.e. there is some sufficiently small $r > 0$ such that $f(x, y) \leq f(a, b)$ for all (x, y) for which $(x - a)^2 + (y - b)^2 < r^2$.

Likewise, we say that f has a local minimum at (a, b) if $f(x, y) \geq f(a, b)$ for all (x, y) inside some disk of positive radius centered at (a, b).

If $f(x, y) \leq f(a, b)$ for all (x, y) in the domain of f, then f has a global maximum at (a, b). If $f(x, y) \geq f(a, b)$ for all (x, y) in the domain of f, then f has a global minimum at (a, b).

Suppose that (a, b) is a local maximum point for $f(x, y)$, and that the first-order partial derivatives of f exist at (a, b). We know that $f(a, b)$ is the largest value of $f(x, y)$ as (x, y) goes in all directions from the point (a, b), in some sufficiently small disk centered at (a, b). In particular, $f(a, b)$ is the largest value of f in the x direction (around the point (a, b)), that is, the single-variable function $g(x) = f(x, b)$ has a local maximum at $x = a$. So we know that $g'(a) = 0$. Since $g'(x) = \frac{\partial f}{\partial x}(x, b)$, then $\frac{\partial f}{\partial x}(a, b) = 0$. Similarly, $f(a, b)$ is the largest value of f near (a, b) in the y direction and so $\frac{\partial f}{\partial y}(a, b) = 0$. We thus have the following theorem:

Theorem: Let $f(x, y)$ be a real-valued function such that both $\frac{\partial f}{\partial x}(a, b)$ and $\frac{\partial f}{\partial y}(a, b)$ exist. Then a necessary condition for $f(x, y)$ to have a local maximum or minimum at (a, b) is that $\nabla f(a, b) = 0$.

Note: Theorem can be extended to apply to functions of three or more variables. A point (a, b) where $\nabla f(a, b) = 0$ is called a critical point for the function $f(x, y)$. So given a function $f(x, y)$, to find the critical points of f you have to solve the equations $\frac{\partial f}{\partial x}(x, y) = 0$ and $\frac{\partial f}{\partial y}(x, y) = 0$ simultaneously for (x, y). Similar to the single-variable case, the *necessary* condition that $\nabla f(a, b) = 0$ is not always *sufficient* to guarantee that a critical point is a local maximum or minimum.

Example: The function $f(x, y) = xy$ has a critical point at $(0, 0)$: $\frac{\partial f}{\partial x} = y = 0 \Rightarrow y =$

0, and $\frac{\partial f}{\partial y} = x = 0 \Rightarrow x = 0$, so (0, 0) is the only critical point. But clearly f does not have a local maximum or minimum at (0, 0) since any disk around (0, 0) contains points (x, y) where the values of x and y have the same sign (so that $f(x, y) = xy > 0 = f(0, 0)$) and different signs (so that $f(x, y) = xy < 0 = f(0, 0)$). In fact, along the path $y = x$ in $\mathbb{R}^2$, $f(x, y) = x^2$, which has a local minimum at (0, 0), while along the path $y = -x$ we have $f(x, y) = -x^2$, which has a local maximum at (0, 0). So (0, 0) is an example of a *saddle point*, i.e. it is a local maximum in one direction and a local minimum in another direction. The graph of $f(x, y)$ is shown in Figure, which is a hyperbolic paraboloid.

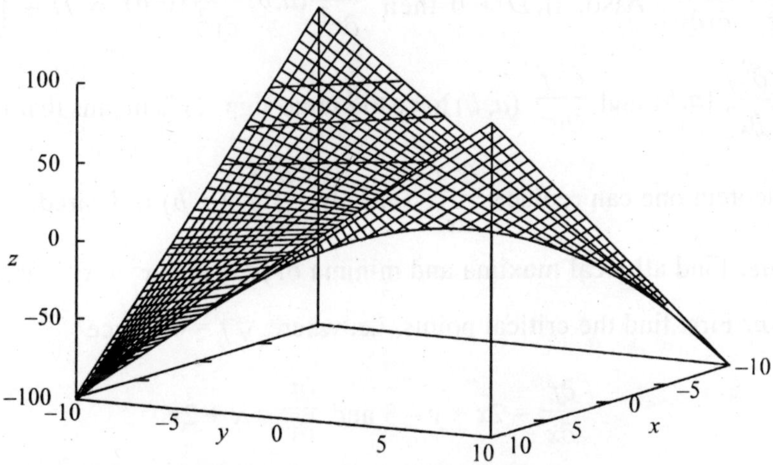

Fig. $f(x, y) = xy$, Saddle Point at (0, 0)

The following theorem gives sufficient conditions for a critical point to be a local maximum or minimum of a *smooth function* (i.e. a function whose partial derivatives of all orders exist and are continuous), which we will not prove here.

Theorem: Let $f(x, y)$ be a smooth real-valued function, with a critical point at (a, b) (i.e. $\nabla f(a, b) = 0$). Define

$$D = \frac{\partial^2 f}{\partial x^2}(a,b)\frac{\partial^2 f}{\partial y^2}(a,b) - \left(\frac{\partial^2 f}{\partial y \partial x}(a,b)\right)^2$$

Then

(a) if $D > 0$ and $\frac{\partial^2 f}{\partial x^2}(a, b) > 0$, then f has a local minimum at (a, b)

(b) if $D > 0$ and $\frac{\partial^2 f}{\partial x^2}(a, b) < 0$, then f has a local maximum at (a, b)

(c) if $D < 0$, then f has neither a local minimum nor a local maximum at (a, b)

(d) if $D = 0$, then the test fails.

If condition (c) holds, then (a, b) is a *saddle point*. Note that the assumption that $f(x, y)$ is smooth means that

$$D = \begin{vmatrix} \frac{\partial^2 f}{\partial x^2}(a,b) & \frac{\partial^2 f}{\partial y \partial x}(a,b) \\ \frac{\partial^2 f}{\partial x \partial y}(a,b) & \frac{\partial^2 f}{\partial y^2}(a,b) \end{vmatrix}$$

since $\frac{\partial^2 f}{\partial y \partial x} = \frac{\partial^2 f}{\partial x \partial y}$. Also, if $D > 0$ then $\frac{\partial^2 f}{\partial x^2}(a,b)\frac{\partial^2 f}{\partial y^2}(a,b) = D + \left(\frac{\partial^2 f}{\partial y \partial x}(a,b)\right)^2$ > 0, and so $\frac{\partial^2 f}{\partial x^2}(a,b)$ and $\frac{\partial^2 f}{\partial y^2}(a, b)$ have the same sign. This means that in parts (a) and (b) of the theorem one can replace $\frac{\partial^2 f}{\partial x^2}(a, b)$ by $\frac{\partial^2 f}{\partial y^2}(a, b)$ if desired.

Example: Find all local maxima and minima of $f(x, y) = x^2 + xy + y^2 - 3x$.

Solution: First find the critical points, i.e. where $\nabla f = 0$. Since

$$\frac{\partial f}{\partial x} = 2x + y - 3 \text{ and } \frac{\partial f}{\partial y} = x + 2y$$

then the critical points (x, y) are the common solutions of the equations

$$2x + y - 3 = 0$$
$$x + 2y = 0$$

which has the unique solution $(x, y) = (2, -1)$. So $(2, -1)$ is the only critical point.

To use Theorem, we need the second-order partial derivatives:

$$\frac{\partial^2 f}{\partial x^2} = 2, \frac{\partial^2 f}{\partial y^2} = 2, \frac{\partial^2 f}{\partial y \partial x} = 1$$

and so

$$D = \frac{\partial^2 f}{\partial x^2}(2,-1)\frac{\partial^2 f}{\partial y^2}(2,-1) - \left(\frac{\partial^2 f}{\partial y \partial x}(2,-1)\right)^2 = (2)(2) - 1^2 = 3 > 0$$

and $\frac{\partial^2 f}{\partial x^2}(2, -1) = 2 > 0$. Thus, $(2, -1)$ is a local minimum.

Example: Find all local maxima and minima of $f(x, y) = xy - x^3 - y^2$.

Solution: First find the critical points, i.e. where $\nabla f = 0$. Since

$$\frac{\partial f}{\partial x} = y - 3x^2 \text{ and } \frac{\partial f}{\partial y} = x - 2y$$

then the critical points (x, y) are the common solutions of the equations

$$y - 3x^2 = 0$$
$$x - 2y = 0$$

The first equation yields $y = 3x^2$, substituting that into the second equation yields $x - 6x^2 = 0$, which has the solutions $x = 0$ and $x = \frac{1}{6}$. So $x = 0 \Rightarrow y = 3(0) = 0$ and $x = \frac{1}{6} \Rightarrow y = 3\left(\frac{1}{6}\right)^2 = \frac{1}{12}$.

So the critical points are $(x, y) = (0, 0)$ and $(x, y) = \left(\frac{1}{6}, \frac{1}{12}\right)$.

To use Theorem, we need the second-order partial derivatives:

$$\frac{\partial^2 f}{\partial x^2} = -6x, \frac{\partial^2 f}{\partial y^2} = -2, \frac{\partial^2 f}{\partial y \partial x} = 1$$

So

$$D = \frac{\partial^2 f}{\partial x^2}(0,0)\frac{\partial^2 f}{\partial y^2}(0,0) - \left(\frac{\partial^2 f}{\partial y \partial x}(0,0)\right)^2$$

$$= (-6(0))(-2) - 1^2 = -1 < 0$$

and thus (0, 0) is a saddle point. Also,

$$D = \frac{\partial^2 f}{\partial x^2}\left(\frac{1}{6}, \frac{1}{12}\right)\frac{\partial^2 f}{\partial y^2}\left(\frac{1}{6}, \frac{1}{12}\right) - \left(\frac{\partial^2 f}{\partial y \partial x}\left(\frac{1}{6}, \frac{1}{12}\right)\right)^2$$

$$= \left(-6\left(\frac{1}{6}\right)\right)(-2) - 1^2 = 1 > 0$$

and $\frac{\partial^2 f}{\partial x^2}\left(\frac{1}{6}, \frac{1}{12}\right) = -1 < 0$. Thus, $\left(\frac{1}{6}, \frac{1}{12}\right)$ is a local maximum.

Example: Find all local maxima and minima of $f(x, y) = (x - 2)^4 + (x - 2y)^2$.

Solution: First find the critical points, i.e. where $\nabla f = 0$. Since

$$\frac{\partial f}{\partial x} = 4(x-2)^3 + 2(x-2y) \text{ and } \frac{\partial f}{\partial y} = 4(x-2y)$$

then the critical points (x, y) are the common solutions of the equations

$$4(x-2)^3 + 2(x-2y) = 0$$

$$-4(x-2y) = 0$$

The second equation yields $x = 2y$, substituting that into the first equation yields $4(2y - 2)^3 = 0$, which has the solution $y = 1$, and so $x = 2(1) = 2$. Thus, (2, 1) is the only critical point.

To use Theorem, we need the second-order partial derivatives:

$$\frac{\partial^2 f}{\partial x^2} = 12(x-2)^2 + 2, \frac{\partial^2 f}{\partial y^2} = 8, \frac{\partial^2 f}{\partial y \partial x} = -4$$

So
$$D = \frac{\partial^2 f}{\partial x^2}(2,1)\frac{\partial^2 f}{\partial y^2}(2,1) - \left(\frac{\partial^2 f}{\partial y \partial x}(2,1)\right)^2 = (2)(8) - (-4)^2 = 0$$

and so the test fails. What can be done in this situation? Sometimes it is possible to examine the function to see directly the nature of a critical point. In our case, we see that $f(x, y) \geq 0$ for all (x, y), since $f(x, y)$ is the sum of fourth and second powers of numbers and hence must be nonnegative. But we also see that $f(2, 1) = 0$. Thus $f(x, y) \geq 0 = f(2, 1)$ for all (x, y), and hence (2, 1) is in fact a *global* minimum for f.

Example: Find all local maxima and minima of $f(x, y) = (x^2 + y^2)\, e^{-(x^2+y^2)}$.

Solution: First find the critical points, i.e. where $\nabla f = 0$. Since

$$\frac{\partial f}{\partial x} = 2x(1 - (x^2 + y^2))\, e^{-(x^2+y^2)}$$

$$\frac{\partial f}{\partial y} = 2y(1 - (x^2 + y^2))\, e^{-(x^2+y^2)}$$

then the critical points are (0, 0) and all points (x, y) on the unit circle $x^2 + y^2 = 1$.

To use Theorem, we need the second-order partial derivatives:

$$\frac{\partial^2 f}{\partial x^2} = 2[1 - (x^2 + y^2) - 2x^2 - 2x^2(1 - (x^2 + y^2))]\, e^{-(x^2+y^2)}$$

$$\frac{\partial^2 f}{\partial y^2} = 2[1 - (x^2 + y^2) - 2y^2 - 2y^2(1 - (x^2 + y^2))]\ e^{-(x^2+y^2)}$$

$$\frac{\partial^2 f}{\partial y \partial x} = -4xy[2 - (x^2 + y^2)]\ e^{-(x^2+y^2)}$$

At (0, 0), we have $D = 4 > 0$ and $\frac{\partial^2 f}{\partial x^2}$ $(0, 0) = 2 > 0$, so (0, 0) is a local minimum. However, for points (x, y) on the unit circle $x^2 + y^2 = 1$, we have

$$D = (-4x^2 e^{-1})(-4y^2 e^{-1}) - (-4xye^{-1})^2 = 0$$

and so the test fails. If we look at the graph of $f(x, y)$, as shown in Figure , it looks like we might have a local maximum for (x, y) on the unit circle $x^2 + y^2 = 1$. If we switch to using polar coordinates (r, θ) instead of (x, y) in $\mathbb{R}^2$, where $r^2 = x^2 + y^2$, then we see that we can write $f(x, y)$ as a function $g(r)$ of the variable r alone: $g(r) = r^2\ e^{-r^2}$. Then $g'(r) = 2r(1 - r^2)$ e^{-r^2}, so it has a critical point at $r = 1$, and we can check that $g''(1) = -4e^{-1} < 0$, so the Second Derivative Test from single-variable calculus says that $r = 1$ is a local maximum. But $r = 1$ corresponds to the unit circle $x^2 + y^2 = 1$. Thus, the points (x, y) on the unit circle $x^2 + y^2 = 1$ are local maximum points for f.

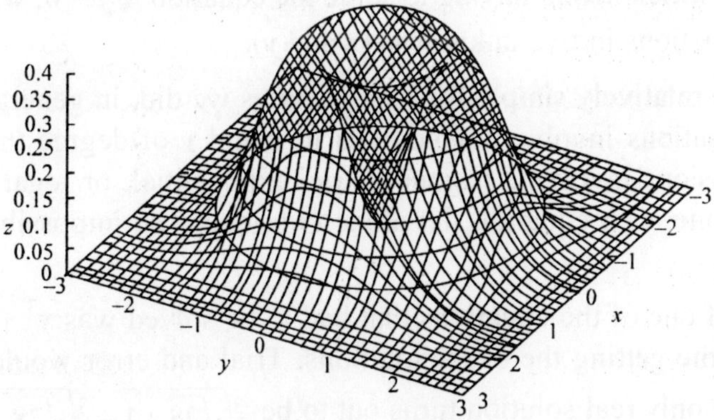

Fig. $f(x, y) = (x^2 + y^2)\ e^{-(x^2+y^2)}$

EXERCISES

For Exercises 1-10, Find all local maxima and minima of the function $f(x, y)$.

1. $f(x, y) = x^3 - 3x + y^2$
2. $f(x, y) = x^3 - 12x + y^2 + 8y$
3. $f(x, y) = x^3 - 3x + y^3 - 3y$
4. $f(x, y) = x^3 + 3x^2 + y^3 - 3y^2$

5. $f(x, y) = 2x^3 + 6xy + 3y^2$
6. $f(x, y) = 2x^3 - 6xy + y^2$
7. $f(x, y) = \sqrt{x^2 + y^2}$
8. $f(x, y) = x + 2y$
9. $f(x, y) = 4x^2 - 4xy + 2y^2 + 10x - 6y$
10. $f(x, y) = -4x^2 + 4xy - 2y^2 + 16x - 12y$
11. For a rectangular solid of volume 1000 cubic meters, find the dimensions that will minimize the surface area.
12. Prove that if (a, b) is a local maximum or local minimum point for a smooth function $f(x, y)$, then the tangent plane to the surface $z = f(x, y)$ at the point $(a, b, f(a, b))$ is parallel to the xy-plane.
13. Find three positive numbers x, y, z whose sum is 10 such that x^2y^2z is a maximum.

Unconstrained Optimization: Numerical Methods

The types of problems that we solved in the previous section were examples of *unconstrained optimization* problems. That is, we tried to find local (and perhaps even global) maximum and minimum points of real-valued functions $f(x, y)$, where the points (x, y) could be any points in the domain of f. The method we used required us to find the critical points of f, which meant having to solve the equation $\nabla f = 0$, which in general is a system of two equations in two unknowns (x and y).

While this was relatively simple for the examples we did, in general this will not be the case. If the equations involve polynomials in x and y of degree three or higher, or complicated expressions involving trigonometric, exponential, or logarithmic functions, then solving even one such equation, let alone two, could be impossible by elementary means.

For example, if one of the equations that had to be solved was $x^3 + 9x - 2 = 0$, you may have a hard time getting the exact solutions. Trial and error would not help much, especially since the only real solution turns out to be $\sqrt[3]{\sqrt{28}+1} - \sqrt[3]{\sqrt{28}-1}$.

In a situation such as this, the only choice may be to find a solution using some numerical method which gives a sequence of numbers which converge to the actual solution. For example, Newton's method for solving equations $f(x) = 0$, which you probably learned in single-variable calculus. In this section we will describe another method of Newton for finding critical points of real-valued functions of two variables.

Let $f(x, y)$ be a smooth real-valued function, and define

$$D(x, y) = \frac{\partial^2 f}{\partial x^2}(x, y)\,\frac{\partial^2 f}{\partial y^2}(x, y) - \left(\frac{\partial^2 f}{\partial y \partial x}(x, y)\right)^2.$$

Newton's algorithm: Pick an initial point (x_0, y_0). For $n = 0, 1, 2, 3, ...,$ define:

$$x_{n+1} = x_n - \frac{\begin{vmatrix} \frac{\partial^2 f}{\partial y^2}(x_n, y_n) & \frac{\partial^2 f}{\partial x \partial y}(x_n, y_n) \\ \frac{\partial f}{\partial y}(x_n, y_n) & \frac{\partial f}{\partial x}(x_n, y_n) \end{vmatrix}}{D(x_n, y_n)},$$

$$y_{n+1} = y_n - \frac{\begin{vmatrix} \frac{\partial^2 f}{\partial x^2}(x_n, y_n) & \frac{\partial^2 f}{\partial x \partial y}(x_n, y_n) \\ \frac{\partial f}{\partial x}(x_n, y_n) & \frac{\partial f}{\partial y}(x_n, y_n) \end{vmatrix}}{D(x_n, y_n)}$$

Then the sequence of points $(x_n, y_n)_{n=1}^{\infty}$ converges to a critical point. If there are several critical points, then you will have to try different initial points to find them.

Example: Find all local maxima and minima of $f(x, y) = x^3 - xy - x + xy^3 - y^4$.

Solution: First calculate the necessary partial derivatives:

$$\frac{\partial f}{\partial x} = 3x^2 - y - 1 + y^3, \quad \frac{\partial f}{\partial y} = -x + 3xy^2 - 4y^3$$

$$\frac{\partial^2 f}{\partial x^2} = 6x, \quad \frac{\partial^2 f}{\partial y^2} = 6xy - 12y^2, \quad \frac{\partial^2 f}{\partial y \partial x} = -1 + 3y^2$$

Notice that solving $\nabla f = 0$ would involve solving two third-degree polynomial equations in x and y, which in this case can not be done easily. We need to pick an initial point (x_0, y_0) for our algorithm. Looking at the graph of $z = f(x, y)$ over a large region may help, though it may be hard to tell where the critical points are.

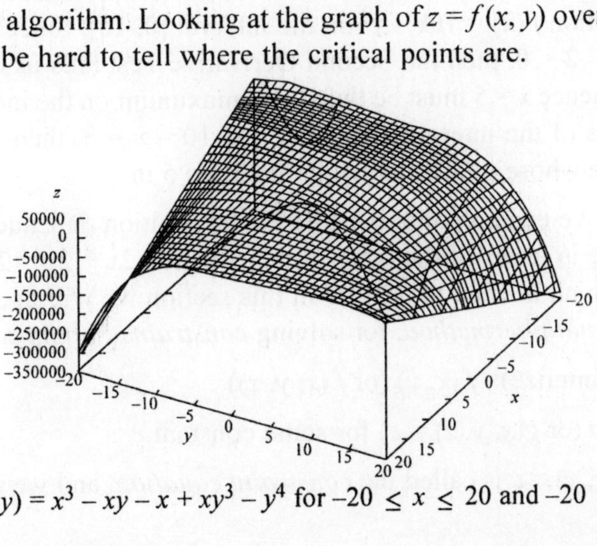

Fig. $f(x, y) = x^3 - xy - x + xy^3 - y^4$ for $-20 \le x \le 20$ and $-20 \le y \le 20$

Notice in the formulas that we divide by D, so we should pick an initial point where D is not zero. And we can see that $D(0, 0) = (0)(0) - (-1)^2 = -1 \neq 0$, so take $(0, 0)$ as our initial point. Since it may take a large number of iterations of Newton's algorithm to be sure that we are close enough to the actual critical point, and since the computations are quite tedious, we will let a computer do the computing. For this, we will write a simple program, using the Java programming language, which will take a given initial point as a parameter and then perform 100 iterations of Newton's algorithm. In each iteration the new point will be printed, so that we can see if there is convergence.

Constrained Optimization: Lagrange Multipliers

Earlier we were concerned with finding maxima and minima of functions without any constraints on the variables (other than being in the domain of the function). What would we do if there were constraints on the variables? The following example illustrates a simple case of this type of problem.

Example: For a rectangle whose perimeter is 20 m, find the dimensions that will maximize the area.

Solution: The area A of a rectangle with width x and height y is $A = xy$. The perimeter P of the rectangle is then given by the formula $P = 2x + 2y$. Since we are given that the perimeter $P = 20$, this problem can be stated as:

Maximize: $f(x, y) = xy$

given: $2x + 2y = 20$

The reader is probably familiar with a simple method, using single-variable calculus, for solving this problem. Since we must have $2x + 2y = 20$, then we can solve for, say, y in terms of x using that equation. This gives $y = 10 - x$, which we then substitute into f to get $f(x, y) = xy = x(10 - x) = 10x - x^2$. This is now a function of x alone, so we now just have to maximize the function $f(x) = 10x - x^2$ on the interval $[0, 10]$. Since $f'(x) = 10 - 2x = 0 \Rightarrow x = 5$ and $f''(5) = -2 < 0$, then the Second Derivative Test tells us that $x = 5$ is a local maximum for f, and hence $x = 5$ must be the global maximum on the interval $[0, 10]$ (since $f = 0$ at the endpoints of the interval). So since $y = 10 - x = 5$, then the maximum area occurs for a rectangle whose width and height both are 5 m.

Notice in the above example that the ease of the solution depended on being able to solve for one variable in terms of the other in the equation $2x + 2y = 20$. But what if that were not possible (which is often the case)? In this section we will use a general method, called the *Lagrangemultipliermethod*, for solving *constrained optimization* problems:

Maximize (or minimize): $f(x, y)$ (or $f(x, y, z)$)

given: $g(x, y) = c$ (or $g(x, y, z) = c$) for some constant c

The equation $g(x, y) = c$ is called the *constraint equation*, and we say that x and y are *constrained* by $g(x, y) = c$. Points (x, y) which are maxima or minima of $f(x, y)$ with the condition that they satisfy the constraint equation $g(x, y) = c$ are called *constrained maximum* or *constrained minimum* points, respectively. Similar definitions hold for functions of three variables.

The Lagrange multiplier method for solving such problems can now be stated:

Theorem: Let $f(x, y)$ and $g(x, y)$ be smooth functions, and suppose that c is a scalar constant such that $\nabla g(x, y) \neq 0$ for all (x, y) that satisfy the equation $g(x, y) = c$. Then to solve the constrained optimization problem

Maximize (or minimize) : $f(x, y)$

given: $g(x, y) = c$,

find the points (x, y) that solve the equation $\nabla f(x, y) = \lambda \nabla g(x, y)$ for some constant λ (the number λ is called the *Lagrange multiplier*). If there is a constrained maximum or minimum, then it must be such a point.

A rigorous proof of the above theorem requires use of the Implicit Function Theorem, which is beyond the scope of this text. Note that the theorem only gives a *necessary* condition for a point to be a constrained maximum or minimum. Whether a point (x, y) that satisfies $\nabla f(x, y) = \lambda \nabla g(x, y)$ for some λ actually *is* a constrained maximum or minimum can *sometimes* be determined by the nature of the problem itself.

For instance, in Example it was clear that there had to be a global maximum. So how can you tell when a point that satisfies the condition in Theorem really is a constrained maximum or minimum? The answer is that it depends on the constraint function $g(x, y)$, together with any implicit constraints. It can be shown that if the constraint equation $g(x, y) = c$ (plus any hidden constraints) describes a *bounded* set B in $\mathbb{R}^2$, then the constrained maximum or minimum of $f(x, y)$ will occur either at a point (x, y) satisfying $\nabla f(x, y) = \lambda \nabla g(x, y)$ or at a "boundary" point of the set B.

In Example the constraint equation $2x + 2y = 20$ describes a line in $\mathbb{R}^2$, which by itself is not bounded. However, there are "hidden" constraints, due to the nature of the problem, namely $0 \leq x, y \leq 10$, which cause that line to be restricted to a *line segment* in $\mathbb{R}^2$ (including the endpoints of that line segment), which *is* bounded.

Example: For a rectangle whose perimeter is 20 m, use the Lagrange multiplier method to find the dimensions that will maximize the area.

Solution: As we saw in Example, with x and y representing the width and height, respectively, of the rectangle, this problem can be stated as:

Maximize: $f(x, y) = xy$

given: $g(x, y) = 2x + 2y = 20$

Then solving the equation $\nabla f(x, y) = \lambda \nabla g(x, y)$ for some λ means solving the equations $\frac{\partial f}{\partial x} = \lambda \frac{\partial g}{\partial x}$ and $\frac{\partial f}{\partial y} = \lambda \frac{\partial g}{\partial y}$, namely:

$$y = 2\lambda,$$
$$x = 2\lambda$$

The general idea is to solve for λ in both equations, then set those expressions equal (since they both equal λ) to solve for x and y. Doing this we get

$$\frac{y}{2} = \lambda = \frac{x}{2} \Rightarrow x = y,$$

so now substitute either of the expressions for x or y into the constraint equation to solve for x and y:

$$20 = g(x, y) = 2x + 2y = 2x + 2x = 4x \Rightarrow x = 5 \Rightarrow y = 5$$

There must be a maximum area, since the minimum area is 0 and $f(5, 5) = 25 > 0$, so the point (5, 5) that we found (called a *constrained critical point*) must be the constrained maximum.

$\therefore$ The maximum area occurs for a rectangle whose width and height both are 5 m.

Example: Find the points on the circle $x^2 + y^2 = 80$ which are closest to and farthest from the point (1, 2).

Solution: The distance d from any point (x, y) to the point (1, 2) is

$$d = \sqrt{(x-1)^2 + (y-2)^2},$$

and minimizing the distance is equivalent to minimizing the square of the distance. Thus the problem can be stated as:

Maximize (and minimize): $f(x, y) = (x - 1)^2 + (y - 2)^2$

given: $g(x, y) = x^2 + y^2 = 80$

Solving $\nabla f(x, y) = \lambda \nabla g(x, y)$ means solving the following equations:

$$2(x - 1) = 2\lambda x,$$
$$2(y - 2) = 2\lambda y$$

Note that $x \neq 0$ since otherwise we would get $-2 = 0$ in the first equation. Similarly, $y \neq 0$. So we can solve both equations for λ as follows:

$$\frac{x-1}{x} = \lambda = \frac{y-2}{y} \Rightarrow xy - y = xy - 2x \Rightarrow y = 2x$$

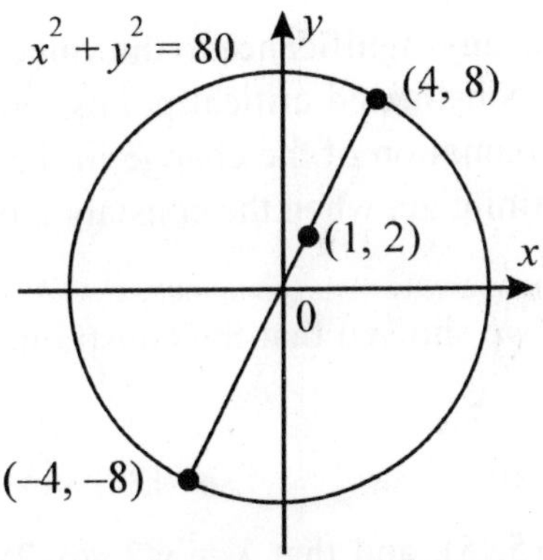

Substituting this into $g(x, y) = x^2 + y^2 = 80$ yields $5x^2 = 80$, so $x = \pm 4$. So the two constrained critical points are (4, 8) and (–4, –8). Since $f(4, 8) = 45$ and $f(-4, -8) = 125$, and since there must be points on the circle closest to and farthest from (1, 2), then it must be the case that (4, 8) is the point on the circle closest to (1, 2) and (–4, –8) is the farthest from (1, 2).

Notice that since the constraint equation $x^2 + y^2 = 80$ describes a circle, which is a bounded set in $\mathbb{R}^2$, then we were guaranteed that the constrained critical points we found were indeed the constrained maximum and minimum. The Lagrange multiplier method can be extended to functions of three variables.

Example: Maximize (and minimize): $f(x, y, z) = x + z$

given: $g(x, y, z) = x^2 + y^2 + z^2 = 1$

Solution: Solve the equation $\nabla f(x, y, z) = \lambda \nabla g(x, y, z)$:

$$1 = 2\lambda x$$
$$0 = 2\lambda y$$
$$1 = 2\lambda z$$

The first equation implies $\lambda \neq 0$ (otherwise we would have $1 = 0$), so we can divide by λ in the second equation to get $y = 0$ and we can divide by λ in the first and third equations to get $x = \dfrac{1}{2\lambda} = z$. Substituting these expressions into the constraint equation $g(x, y, z) = x^2 + y^2 + z^2 = 1$ yields the constrained critical points $\left(\frac{1}{\sqrt{2}}, 0, \frac{1}{\sqrt{2}}\right)$ and $\left(\frac{-1}{\sqrt{2}}, 0, \frac{-1}{\sqrt{2}}\right)$. Since $f\left(\frac{1}{\sqrt{2}}, 0, \frac{1}{\sqrt{2}}\right) > f\left(\frac{-1}{\sqrt{2}}, 0, \frac{-1}{\sqrt{2}}\right)$, and since the constraint equation $x^2 + y^2 + z^2 = 1$ describes a sphere (which is bounded) in $\mathbb{R}^3$, then $\left(\frac{1}{\sqrt{2}}, 0, \frac{1}{\sqrt{2}}\right)$ is the constrained maximum point and $\left(\frac{-1}{\sqrt{2}}, 0, \frac{-1}{\sqrt{2}}\right)$ is the constrained minimum point.

So far we have not attached any significance to the value of the Lagrange multiplier λ. We needed λ only to find the constrained critical points, butmade no use of its value. It turns out that λ gives an approximation of the change in the value of the function $f(x, y)$ that we wish to maximize or minimize, when the constant c in the constraint equation $g(x, y) = c$ is changed by 1.

For example, in Example we showed that the constrained optimization problem

Maximize: $f(x, y) = xy$

given: $g(x, y) = 2x + 2y = 20$

had the solution $(x, y) = (5, 5)$, and that $\lambda = x/2 = y/2$. Thus, $\lambda = 2.5$. In a similar fashion we could show that the constrained optimization problem

Maximize: $f(x, y) = xy$

given: $g(x, y) = 2x + 2y = 21$

has the solution $(x, y) = (5.25, 5.25)$. So we see that the value of $f(x, y)$ at the constrained maximum increased from $f(5, 5) = 25$ to $f(5.25, 5.25) = 27.5625$, i.e. it increased by 2.5625 when we increased the value of c in the constraint equation $g(x, y) = c$ from $c = 20$ to $c = 21$. Notice that $\lambda = 2.5$ is close to 2.5625, that is,

$$\lambda \approx \nabla f = f(\text{new max. pt}) - f(\text{old max. pt}).$$

Finally, note that solving the equation $\nabla f(x, y) = \lambda \nabla g(x, y)$ means having to solve a system of two (possibly nonlinear) equations in three unknowns, which as we have seen before, may not be possible to do. And the 3-variable case can get even more complicated. All of this somewhat restricts the usefulness of Lagrange's method to relatively simple functions. Luckily there are many numerical methods for solving constrained optimization problems, though we will not discuss them here.

EXERCISES

1. Find the constrained maxima and minima of $f(x, y) = 2x + y$ given that $x^2 + y^2 = 4$.
2. Find the constrained maxima and minima of $f(x, y) = xy$ given that $x^2 + 3y^2 = 6$.
3. Find the points on the circle $x^2 + y^2 = 100$ which are closest to and farthest from the point (2, 3).
4. Find the constrained maxima and minima of $f(x, y, z) = x + y^2 + 2z$ given that $4x^2 + 9y^2 - 36z^2 = 36$.
5. Find the volume of the largest rectangular parallelepiped that can be inscribed in the ellipsoid

$$\frac{x^2}{a^2} + \frac{y^2}{b^2} + \frac{z^2}{c^2} = 1.$$

Chapter 8

Multiple Integrals

Having studied single integrals over a straight line (pre-university), and single integrals over a curve in three-dimensional space, we will now look at double integrals over an area, triple integrals over a volume, and double integrals over a surface in three-dimensional space.

There are a couple of extra books that are helpful for these subjects:

1. Calculus by Boyçe and DiPrima (lots of nice pictures);
2. Differential & Integral Calculus by Courant (old fashioned, but precise).

Double Integrals (Area Integrals)

Suppose that we wish to integrate $f: \mathbb{R}^2 \to \mathbb{R}$ over a region $\mathbb{R}$ of the xy-plane, where $\mathbb{R}$ is closed and bounded, where

1. By bounded we mean that there exists a circumscribing circle/rectangle.
2. By closed we mean that the boundary is part of region.

Subdivide R by sectionally smooth curves into N subregions $S_1, S_2, ..., S_N$ with areas $\delta S_1, ..., \delta S_N$.

For a subregion S_j define its greatest diameter $|S_j|$, to be the greatest distance between two points in S_j. In each S_j choose an arbitrary point (ξ_j, η_j).

Definition: By analogy with the Riemann integral, define the double integral

$$\iint_R f(x,y)dS = \lim_{\substack{N\to\infty \\ |S_j|\to 0}} \sum_j f(\xi_j,\eta_j)\delta S_j,$$

If the limit exists as the greatest of the diameters of the subregions tends to zero, and if the limit is both

1. Independent of the subdivision of R, and
2. Independent of the choice of the (ξ_j, η_j).

Properties. Double integrals have the expected properties, e.g. for functions f and g, and real constants λ and μ:

$$\iint_R (\lambda f + \mu g) dS = \lambda \iint_R f dS + \mu \iint_R g dS.$$

Evaluation by Successive Integration

One way to evaluate double integrals is first to keep one variable fixed and to evaluate the inner integral, and then to evaluate the remaining outer integral.

For simplicity assume the region R is convex[a] and is described by

$$a \leq x \leq b, y_1(x) \leq y \leq y_2(x).$$

Then

$$\iint_R f(x,y) dS = \int_{x=a}^{b} \underbrace{\left[\int_{y=y_1(x)}^{y_2(x)} f(x,y) dy \right]}_{\text{only a function of } x} dx.$$

Suppose instead that the region is equivalently specified as

$$c \leq y \leq d, x_1(y) \leq x \leq x_2(y).$$

Then

$$\iint_R f(x,y) dS = \int_{y=c}^{d} \underbrace{\left[\int_{x=x_1(y)}^{x_2(y)} f(x,y) dy \right]}_{\text{only a function of } y} dy.$$

Remarks:

1. For nice functions (e.g. f continuous), give the same result. We will assume that this is true for this course; however it is not always so. Consider

$$f = \begin{cases} +1 & y < x < y+1 \\ -1 & y+1 < x < y+2 \\ 0 & \text{otherwise} \end{cases}$$

   Then

$$\int_{y=0}^{\infty} \left[\int_{x=0}^{\infty} f dx \right] dy = 0,$$

   while using the formula for the area of a triangle

$$\int_{x=0}^{\infty} \left[\int_{y=0}^{\infty} f dy \right] dx = 3\left(\frac{1}{2}\right) - \frac{1}{2} = 1.$$

2. If R is not convex rotate the co-ordinate axes so that lines of either constant x or constant y cut boundaries no more than twice. If that is not possible then split R into regions where the axes can be rotated so that lines of either constant x or constant y cut boundaries no more than twice.
3. NEVER EVER write a double integral as
$$\int_{y=y_1(x)}^{y_2(x)} \left[\int_{x=a}^{b} f(x,y)dx \right] dy.$$
This is MEANINGLESS.
4. An alternative notation is to write the d immediately following the integration sign it refers to, viz.
$$\iint_R f(x,y)dS = \int_a^b dx \int_{y_1(x)}^{y_2(x)} dy f(x,y).$$

Examples:

1. For the region R specified by (4.2a), i.e. $a \leq x \leq b$ and $y_1(x) \leq y \leq y_2(x)$
$$\iint_R \frac{\partial P}{\partial y} dydx = \int_{x=a}^{b} \left[\int_{y=y_1(x)}^{y_2(x)} \frac{\partial P}{\partial y} dy \right] dx$$
$$= \int_{x=a}^{b} [P(x, y_2(x)) - P(x, y_1(x))] dx.$$
While this is not a complete evaluation, the double integral has been reduced to a standard Riemann integral. We will use this result in the proof of Green's theorem.
2. If $f(x, y) = 1$, then
$$\iint_R dS = \text{area of } R.$$
Special Case. For R given by $a \leq x \leq b$ and $0 \leq y \leq y(x)$,
$$\iint_R dS = \int_a^b \left[\int_0^{y(x)} dy \right] dx$$
$$= \int_a^b y(x)dx.$$
Some care is needed with signs if $y(x)$ is negative (since areas are by convention positive).
3. Suppose that $f(x, y) = y$, and that the region is R is specified by
$$1 \leq x \leq 2 \text{ and } 1/x \leq y \leq e^x.$$

Do the y-integration first; then

$$\int_{x=1}^{2} dx \int_{y=1/x}^{e^x} dyy = \int_{x=1}^{2} dx \left[\frac{1}{2} y^2\right]_{1/x}^{e^x}$$

$$= \frac{1}{2}\int_{1}^{2} dx(e^{2x} - 1/x^2)$$

$$= \frac{1}{4}(e^4 - e^2 - 1).$$

Exercise: Do the x-integration first.

Moral: Think before deciding on the order of integration.

1. The volume V beneath the surface $z = f(x, y) > 0$ and above a region R in the xy-plane is

$$V = \iint_R f(x, y) dS.$$

Triple Integrals (Volume Integrals)

Suppose $f : \mathbb{R}^3 \to \mathbb{R}$ and V is a domain in $\mathbb{R}^3$. Subdivide V into N sub-domains $V_1, ..., V_N$, with volumes $\delta\tau_1, ..., \delta\tau_N$, and let $\xi_j \in V_j$.

Definition. Define the triple integral

$$\iiint_v f(x) d\tau = \lim_{N\to\infty} \sum_{N\to\infty} f(\xi_j)\delta\tau_j$$

if the limit exists as the maximum dimension of the largest V_j tends to zero, and the limit is independent both of the subdivision of V and the choice of the ξ_j.

Evaluation

As for double integrals one can often evaluate triple integrals by repeated integration, e.g. for a convex region

$$\iiint_v f(x) d\tau = \int_{x=a}^{b} dx \int_{y=y_1(x)}^{y_2(x)} dy \int_{z=z_1(x,y)}^{x_2(x,y)} dz f(x, y, z).$$

For 'nice' functions the order of integration is unimportant (although it is, of course, important to ensure that the integration ranges are correct).

Example: Let V be the tetrahedron bounded by the four planes

$$x = 0, y = 0, z = 0 \text{ and } x + y + z = a.$$

The vertices are at $O(0, 0, 0)$, $P(a, 0, 0)$, $Q(0, a, 0)$, $R(0, 0, a)$

Suppose that we wish to evaluate

$$J_x = \iiint x^2 d\tau, J_y = \iiint y^2 d\tau, J_z = \iiint z^2 d\tau.$$

First we note that by symmetry $J_x = J_y = J_z$. We will concentrate on J_x.

If we are to evaluate the J_x by repeated integration, which integral should we do first?

1. The z integral? Looks good because the integrand has no explicit z dependence.
2. The y integral? Looks good because the integrand has no explicit y dependence.
3. The x integral? Looks bad because the integrand has explicit x dependence.

We will do 3!

$$\begin{aligned} J_x &= \int_0^a dz \int_0^{a-z} dy \int_0^{a-z-y} dx\, x^2 \\ &= \int_0^a dz \int_0^{a-z} dy \frac{1}{z}(a-z-y)^3 \\ &= \int_0^a dz \frac{1}{12}(a-z)^4 = \frac{1}{60}a^5. \end{aligned}$$

Remark: We can combine four of these shapes to form a pyramid, $\widehat{V}$. The moment of inertia of the pyramid about Oz is

$$I_z = \iiint_{\widehat{V}} (x^2 + y^2) d\tau = \frac{2a^5}{15}.$$

Optional Exercise

1. Calculate the moments of inertia about Ox and Oy for the pyramid.
2. Find the centre of mass of the pyramid.
3. Find the moments of inertia about centre of mass using the Theorem of Parallel Axes.

Jacobians and Change of Variables

The evaluation of some multiple integrals can sometimes be simplified by a change of variables, e.g. $(x, y) \rightarrow (u, v)$.

As well as accounting for the change in the region of integration as a result of the change of variables, we also need to account for any change in the elementary area or volume, cf. change of variable for ordinary integrals where the transformation $x = g(u)$ is accompanied by $dx = g'(u)\, du$ so that

$$\int_{g(a)}^{g(b)} f(x)dx = \int_a^b f(g(u))g'(u)du.$$

Jacobians

Start in 2D and suppose that

$$x(u) : \mathbb{R}^2 \to \mathbb{R}^2$$

is a transformation representing a change of variables. Then in the case of a function of x,

$$f(x) : \mathbb{R}^2 \to \mathbb{R},$$

when viewed as a function of u,

$$f(u) : \mathbb{R}^2 \to \mathbb{R},$$

the partial derivatives are related by the the chain rule; in matrix form this states that

$$\begin{pmatrix} \frac{\partial f}{\partial u} & \frac{\partial f}{\partial v} \end{pmatrix} = \begin{pmatrix} \frac{\partial f}{\partial x} & \frac{\partial f}{\partial y} \end{pmatrix} \begin{pmatrix} \frac{\partial x}{\partial u} & \frac{\partial x}{\partial v} \\ \frac{\partial y}{\partial u} & \frac{\partial y}{\partial v} \end{pmatrix}$$

Definition: The determinant of the Jacobian matrix

$$U = \begin{pmatrix} \frac{\partial x}{\partial u} & \frac{\partial x}{\partial v} \\ \frac{\partial y}{\partial u} & \frac{\partial y}{\partial v} \end{pmatrix},$$

is called the Jacobian, or Jacobian determinant, of the transformation $x(u)$, and is written

$$J = J(x,u) = \frac{\partial(x,y)}{\partial(u,v)} = \begin{vmatrix} \frac{\partial x}{\partial u} & \frac{\partial x}{\partial v} \\ \frac{\partial y}{\partial u} & \frac{\partial y}{\partial v} \end{vmatrix} = \det\left(\frac{\partial x_i}{\partial u_j}\right) = \det(U_{ij}) = \det U.$$

Remarks:

1. There is an important distinction between the Jacobian matrix and the Jacobian (or Jacobian determinant).
2. For a three dimensional transformation $x(u) : \mathbb{R}^3 \to \mathbb{R}^3$

$$J = J(x,u) = \frac{\partial(x,y,z)}{\partial(u,v,w)} = \begin{vmatrix} \frac{\partial x}{\partial u} & \frac{\partial x}{\partial v} & \frac{\partial x}{\partial w} \\ \frac{\partial y}{\partial u} & \frac{\partial y}{\partial v} & \frac{\partial y}{\partial w} \\ \frac{\partial z}{\partial u} & \frac{\partial z}{\partial v} & \frac{\partial z}{\partial w} \end{vmatrix}$$

Co-ordinate Transformations

For a change of variables given by the transformation $x(u)$ to be 'useful', there needs to be a 1-1 correspondence between points in the u co-ordinate system and those in the x co-ordinate system.

In order to appreciate when this is so, start from the definition of a differential applied componentwise to $x : \mathbb{R}^m \to \mathbb{R}^m$:

$$dx_i = \sum_{\alpha=1}^{m} \frac{\partial x_i}{\partial u_\alpha} du_\alpha = \sum_{\alpha=1}^{m} U_{i\alpha} du_\alpha, \text{ i.e., } dx = \mathsf{U}\, du.$$

Short Observation. If there is a 1-1 correspondence between points in the u co-ordinate system and those in the x co-ordinate system, then as well as expressing dx in terms of du it should be possible to express du in terms of dx. If so, then from it follows that

$$du = \mathsf{U}^{-1} dx.$$

This suggests that there will be a 1-1 correspondence between points in the two co-ordinate systems if $\det(\mathsf{U}) \neq 0$.

More Lengthy Comment: If dx and du are interpreted as increments in x and u in their respective co-ordinate systems, equation provides a means for calculating the local change in x corresponding to a given local change in u. Denote by $dx^{(\alpha)}$ for $\alpha = 1, ..., m$ the increment in x corresponding to an increment

$$du^{(\alpha)} = (0, ..., du_\alpha, ..., 0) \neq 0,$$

where the $du^{(\alpha)}$ for $\alpha = 1, ..., m$ can be viewed as a local basis in the u co-ordinate system. Then from

$$dx_i^{(\alpha)} = \sum_{\alpha=1}^{m} U_{i\alpha} du_i^{(\alpha)} = U_{i\alpha} du_\alpha$$

The condition for the transformation $x(u)$ to be 'useful' is that the $dx^{(\alpha)}$ for $\alpha = 1, ..., m$ are a local basis in the x co-ordinate system, i.e. that they are independent, or equivalently that the only solution to

$$\sum_{\alpha=1}^{m} dx^{(\alpha)} \lambda_\alpha = 0, \text{ or } \sum_{\alpha=1}^{m} dx_i^{(\alpha)} \lambda_\alpha = 0 \text{ for } i = 1, ..., m,$$

is $\lambda_\alpha = 0$ for $\alpha = 1, ..., m$. From the results for the solution of linear systems in Algebra and Geometry it is thus necessary that

$$\det\left(dx_i^{(\alpha)}\right) = \det\left(U_{i\alpha} du_\alpha\right) = \det(U_{i\alpha}) du_1 ... du_m \neq 0$$

i.e. that $\det(U) \neq 0$. Moreover, if $\det U \neq 0$ then U^{-1} exists, and it then follows from that the inverse transformation,

$$du = U^{-1}\, dx,$$

is well defined.

It is in fact possible to prove that if $x(u)$ is continuously differentiable and $\det(U) \neq 0$, then the transformation $x(u)$ is non-singular and there exists a 1-1 correspondence between points in the u plane and those in the x plane. This is the generalisation to more than one variable of the Inverse Function Theorem.

Chain Rule for Jacobians

Start in 2D, and suppose that $x = (x, y)$ is a function of $u = (u, v)$, and $u = (u, v)$ is a function of $s = (s, t)$, i.e. $x \equiv x(u)$ and $u \equiv u(s)$. Write

$$x_1 = x,\ u_1 = u,\ s_1 = s$$

$$x_2 = y,\ u_2 = v,\ s_2 = t,$$

and let X_s, X_u and U_s deonte the Jacobian matrices of $x \equiv x(s)$, $x \equiv x(u)$ and $u \equiv u(s)$ respectively. Then from the chain rule for two vector functions.

$$\frac{\partial x_i}{\partial s_k} = \frac{\partial x_i}{\partial u_j}\frac{\partial u_j}{\partial s_k} \text{ i.e. } X_s = X_u\, U_s.$$

Hence from and from the definition of the Jacobian

$$\frac{\partial(x,y)}{\partial(s,t)} = |X_s| = |X_u U_s| = |X_u||U_x| = \frac{\partial(x,y)}{\partial(u,v)}\frac{\partial(u,v)}{\partial(s,t)},$$

i.e.,
$$\frac{\partial(x,y)}{\partial(s,t)} = \frac{\partial(x,y)}{\partial(u,v)}\frac{\partial(u,v)}{\partial(s,t)}.$$

Remark:

1. The extension to three and higher dimensions is straight forward:

$$\frac{\partial(x,y,z)}{\partial(r,s,t)} = \frac{\partial(x,y,z)}{\partial(u,v,w)}\frac{\partial(u,v,w)}{\partial(r,s,t)}.$$

Inverse: Suppose we identify (s, t) with (x, y), then

$$\frac{\partial(x,y)}{\partial(u,v)}\frac{\partial(u,v)}{\partial(x,y)} = \frac{\partial(x,y)}{\partial(x,y)} = \begin{vmatrix} \left(\frac{\partial x}{\partial x}\right)_y & \left(\frac{\partial x}{\partial y}\right)_x \\ \left(\frac{\partial y}{\partial x}\right)_y & \left(\frac{\partial y}{\partial y}\right)_x \end{vmatrix} = 1.$$

Hence

$$\frac{\partial(x,y)}{\partial(u,v)} = \frac{1}{\dfrac{\partial(u,v)}{\partial(x,y)}}.$$

Remark:

1. Calculate the easier of $\dfrac{\partial(x,y)}{\partial(u,v)}$ and $\dfrac{\partial(u,v)}{\partial(x,y)}$ and use if necessary.

Example: *Plane Polar Co-ordinates.*

In plane polar co-ordinates $x = \rho \cos \phi$ and $y = \rho \sin \phi$; so

$$\frac{\partial x}{\partial \rho} = \cos\phi, \quad \frac{\partial x}{\partial \phi} = -\rho \sin\phi,$$

$$\frac{\partial y}{\partial \rho} = \sin\phi, \quad \frac{\partial y}{\partial \phi} = \rho \cos\phi.$$

Hence

$$\frac{\partial(x,y)}{\partial(\rho,\phi)} = \begin{vmatrix} \dfrac{\partial x}{\partial \rho} & \dfrac{\partial x}{\partial \phi} \\ \dfrac{\partial y}{\partial \rho} & \dfrac{\partial y}{\partial \phi} \end{vmatrix} = \rho\cos^2\phi + \rho\sin^2\phi = \rho,$$

and from (4.9)

$$\frac{\partial(\rho,\phi)}{\partial(x,y)} = \frac{1}{\rho}.$$

Exercise: Calculate $\dfrac{\partial(\rho,\phi)}{\partial(x,y)}$ using $\rho = (x^2+y^2)^{\frac{1}{2}}$ and $\phi = \arctan(y/x)$, and remembering that

Change of Variable in Multiple Integrals

Start in 2D. Consider

$$I = \iint_R f(x,y)dS,$$

and suppose, as above, that $x \equiv x(u)$. The question is then how do we express the integral in terms of the new coordinates u and v (in terms of which we hope the integral is easier to evaluate)?

Three Tasks:

1. Find $F(u, v) \equiv f(x(u, v), y(u, v))$.
2. Identify the region of integration $\widehat{R}$ in the uv-plane.

3. Find how the elementary areas dS_{xy} and dS_{uv} map between the different co-ordinate planes.

Three Answers:

1. Usually straightforward.
2. To find the transformed region $\widehat{R}$, consider each component of boundary in turn.

For instance, suppose that one boundary in the *xy*-plane is given by $g_1(x, y) = 0$. Then in the *uv*-plane the boundary is given by

$$G_1(u, v) = g_1(x(u, v), y(u, v)) = 0.$$

Warning:

1. Care is needed in the case of transformation with singular points, i.e. points where $J = 0$. Such a transformation may not be 1-1 because one region may map to two or more regions, e.g. if there is a p fix on one sign!
2. In order to determine how elementary areas transform, consider a rectangular subdivision of the integration region $\widehat{R}$ in the *uv*-plane based on curves defined by u = constant and v = constant. In the *xy*-plane consider a corresponding subdivision also based on curves defined by u = constant and v = constant, i.e. curves defined by $x = V(v)$ and $x = U(u)$ where

$$V(v) = x(u_0, v),\ U(u) = x(u, v_0),$$

for fixed u_0 and v_0.

In particular consider a rectangle with area *uv* in the *uv*-plane and its transformation in the *xy*-plane:

If $|u| \ll 1$ and $|v| \ll 1$ then by Taylor's theorem

$$x(u_0 + \delta u, v_0) - x(u_0, v_0) = \frac{\partial x}{\partial u}(u_0, v_0)\delta u + o(\delta u),$$

$$x(u_0, v_0 + \delta v) - x(u_0, v_0) = \frac{\partial x}{\partial v}(u_0, v_0)\delta u + o(\delta v).$$

Hence as δu, $\delta v \to 0$, the transformed area approaches that of a parallelogram with sides

$$\frac{\partial x}{\partial u}\delta u \text{ and } \frac{\partial x}{\partial v}\delta v,$$

and area

$$\left|\frac{\partial x}{\partial u} \times \frac{\partial x}{\partial v}\right| \delta u\, \delta v = \left|\frac{\partial x}{\partial u}\frac{\partial y}{\partial v} - \frac{\partial x}{\partial v}\frac{\partial y}{\partial u}\right| \delta u\, \delta v$$

$$= abs\left(\begin{vmatrix} \dfrac{\partial x}{\partial u} & \dfrac{\partial x}{\partial v} \\ \dfrac{\partial y}{\partial u} & \dfrac{\partial y}{\partial v} \end{vmatrix}\right) \delta u\, \delta v$$

$$= \left|\frac{\partial(x,y)}{\partial(u,v)}\right| \delta u\, \delta v.$$

We infer that infinitesimal areas in the xy and uv planes are related by

$$dS_{xy} = \left|\frac{\partial(x,y)}{\partial(u,v)}\right| dS_{uv} \quad \left(\left|\frac{\partial(x,y)}{\partial(u,v)}\right| \neq 0\right).$$

Remarks:

1. The absolute value of the Jacobian determinant is appropriate, because areas are positive by convention.
2. The Jacobian determinant is positive if the 'handedness' of the co-ordinate system is preserved.

In particular, if a point in the uv-plane traces out a curve in a counterclockwise direction, then the image point in the xy-plane traces out a curve in the counterclockwise/clockwise direction according as the Jacobian determinant is positive/negative.

Putting the above three components together, we conclude that

$$\iint_R f(x,y)dS_{xy} = \iint_{\hat{R}} F(u,v)\left|\frac{\partial(x,y)}{\partial(u,v)}\right| dS_{uv}.$$

Three Dimensions: Similarly in 3D, except we need the volume of the parallelepiped with generators.

Exercise. Show that this volume is given by

$$\left|\left(\frac{\partial x}{\partial u} \times \frac{\partial x}{\partial v}\right) \cdot \frac{\partial x}{\partial w}\right| \delta u\, \delta v\, \delta w = \left|\frac{\partial(x,y,z)}{\partial(u,v,w)}\right| \delta u\, \delta v\, \delta w.$$

Hence

$$\iiint_V f(x)d\tau_x = \iiint_{\hat{V}} F(u)\left|\frac{\partial(x,y,z)}{\partial(u,v,w)}\right| d\tau_u,$$

where $F(u) = f(x(u))$, and it has been assumed that [small] volumes are positive.

Examples:

1. Evaluate

$$I = \iint_R \frac{1}{x^2} dS,$$

where the region R is bounded by the planes

$$y = 0, y = x, x + y = 1 \text{ and } x + y = 2.$$

Since the boundaries can be expressed as either $y = kx$ or $x + y = c$, this suggests trying the new variables

$$u = x + y \text{ and } v = y/x.$$

Then

$$\frac{\partial(u,v)}{\partial(x,y)} = \begin{vmatrix} u_x & u_y \\ v_x & v_y \end{vmatrix} = \begin{vmatrix} 1 & 1 \\ -\frac{y}{x^2} & \frac{1}{x} \end{vmatrix} = \frac{x+y}{x^2},$$

and from the result for the inverse of Jacobians

$$\frac{\partial(x,y)}{\partial(u,v)} = \frac{x^2}{x+y}.$$

Thus it can be transformed to

$$I = \iint_{\hat{R}} \frac{1}{x^2} \frac{x^2}{x+y} dudv = \iint_{\hat{R}} \frac{1}{u} dudv.$$

From the region of integration in the uv-plane is

$$1 \leq u \leq 2 \text{ and } 0 \leq v \leq 1.$$

Hence

$$I = \int_1^2 du \int_0^1 dv \frac{1}{u} = \ln 2.$$

Comment. If instead we had chosen $u = y/x$ and $v = x + y$ then the Jacobian would have changed sign since the handedness of the system would not have been preserved. As a result a little extra care concerning signs would have been required.

2. Plane Polar Co-ordinates: (ρ, ϕ).

Suppose

$$f(x, y) \equiv F(\rho, \phi)$$

then using (4.13a) and (4.10a)

$$\iint_R f(x,y) dxdy = \iint_R F(\rho,\phi) \rho d\rho d\phi.$$

Example: Evaluate

$$I^2 = \int_0^{\infty} dx \int_0^{\infty} dy e^{-x^2-y^2}.$$

A transformation to plane polar co-ordinates using (4.16) yields

$$I^2 = \int_0^{\infty} d\rho \int_0^{\frac{\pi}{2}} d\phi \rho e^{-\rho^2} \frac{\pi}{2}\left[-\frac{1}{2}e^{-\rho^2}\right]_0^{\infty} = \frac{\pi}{4}.$$

Further we note that

$$I^2 = \int_0^{\infty} dx e^{-x^2} \int_0^{\infty} dy e^{-y^2} \left(\int_0^{\infty} dt e^{-t^2}\right)^2,$$

and thus

$$I = \int_0^{\infty} dt e^{-t^2} = \frac{1}{2}\pi^{\frac{1}{2}}.$$

3. Cylindrical Polar Co-ordinates: (ρ, ϕ, z).

There is potential for notational confusion here. Many use (r, θ, z) for cylindrical polar co-ordinates, while others use (ρ, ϕ, z) — the advantage of the latter is that a notational clash with spherical polar co-ordinates over the meanings of r and θ is then avoided. The relationship to Cartesian co-ordinates is as follows:

	(ρ, ϕ, z)	(r, ϕ, z)
Radial co-ordinate	$\rho^2 = x^2 + y^2 > 0$	$r^2 = x^2 + y^2 > 0$
Polar angle	$0 \le \phi = \tan^{-1}\frac{y}{x} < 2\pi$	$0 \le \theta = \tan^{-1}\frac{y}{x} < 2\pi$
Axial co-ordinate	z	z

From result for plane polar co-ordinates (since the axial co-ordinate z does not depend on ρ or ϕ), or from a re-derivation using $x = \rho \cos \phi$ and $y = \rho \sin \phi$, the Jacobian for the transformation between Cartesian co-ordinates and cylindrical polar co-ordinates is given by

$$\frac{\partial(x, y, z)}{\partial(\rho, \phi, z)} = \rho,$$

and so the infinitesimal volume transforms according to

$$dx\, dy\, dz \to \rho d\rho\, d\phi\, dz.$$

Thus the integral of

$$f(x, y, z) \equiv F(\rho, \phi, z)$$

over a volume V becomes in cylindrical polar co-ordinates

$$\iiint_V f(x,y,z)dxdydz = \iiint_{\hat{V}} F(\rho,\phi,z)\rho d\rho d\phi dz.$$

Example: Suppose that $f = 1$ and V is the cylindrical region $0 \le z \le h$ and $x^2 + y^2 \le a^2$, then

$$\iiint_V dxdydz = \int_0^h dz \int_0^{2\pi} d\phi \int_0^a d\rho\rho = \pi a^2 h.$$

4. Spherical Polar Co-ordinates: (r, θ, ϕ)

	(r, θ, ϕ)
Radial co-ordinate	$r^2 = x^2 + y^2 + z^2 > 0$
Polar/latitudinal angle	$0 \le \theta \le \pi$
Azimuthal / longitudinal angle	$0 \le \phi \le 2\pi$

In terms of Cartesian co-ordinates

$$x = r \sin\theta \cos\phi,\ y = r \sin\theta \sin\phi,\ z = r\cos\theta.$$

Hence the Jacobian for the transformation between Cartesian co-ordinates and spherical polar co-ordinates is given by

$$\frac{\partial(x,y,z)}{\partial(r,\theta,\phi)} = \begin{vmatrix} s_\theta c_\phi & rc_\theta c_\phi & -rs_\theta s_\phi \\ s_\theta c_\phi & rc_\theta s_\phi & rs_\theta c_\phi \\ c_\theta & -rs_\theta & 0 \end{vmatrix}$$

$$= r^2 \sin\theta,$$

and so the infinitesimal volume transforms according to

$$dx\,dy\,dz \to r^2 \sin\theta\, dr\, d\theta\, d\phi.$$

Thus the integral of

$$f(x, y, z) \equiv F(r, \theta, \phi)$$

over a volume V becomes in spherical polar co-ordinates

$$\iiint_V f(x,y,z)dxdydz = \iiint_{\hat{V}} F(r,\theta,\phi)r^2 \sin\theta dr d\theta d\phi.$$

Example: Suppose that $f = 1$, and that V is the interior of the sphere of radius a. Then

$$\iiint_V f dxdydz = \int_0^a dr \int_0^\pi d\theta \int_0^{2\pi} d\phi r^2 \sin\theta$$

$$= \frac{1}{3}a^3.2.2\pi = \frac{4}{3}\pi a^3.$$

Surface Integrals

So far we have evaluated integrals over areas in 2D, and volumes in 3D. We consider now integrals over a general surface in 3D.

Surfaces

For a smooth surface S described by the parametric representation, i.e.

$$x \equiv x(u, v)$$

where u and v are independent variables, we have from that a unit normal is given by

$$n = \frac{x_u \times x_v}{|x_u \times x_v|}.$$

In calculations of surface integrals we need, in addition to the normal, the size of an 'elementary area' in the surface.

By an analogous argument to that in the previous subsection, a rectangle of area $\delta u \delta v$ at (u_0, v_0) in the uv-plane is mapped by the the transformation into a region that, for small $\delta u \delta v$, is roughly an almost planar parallelogram in the surface S. As in the directions of the sides of the parallelogram are given by

$$x(u_0 + \delta u, u_0) - x(u_0, v_0) = \frac{\partial x}{\partial u}(u_0, v_0)\delta u + o(\delta u),$$

$$x(u_0, v_0 + \delta v) - x(u_0, v_0) = \frac{\partial x}{\partial v}(u_0, v_0)\delta v + o(\delta v),$$

Hence as δu, $\delta v \to 0$ the elementary area within the surface S is given by

$$dS = \left|\frac{\partial x}{\partial u} \times \frac{\partial x}{\partial v}\right| dudv,$$

where $du\, dv$ is [sloppyish] notation for dS_{uv}.

Combining we define the vector elementary area to be

$$dS = n\, dS$$

$$= \left(\frac{\partial x}{\partial u} \times \frac{\partial x}{\partial v}\right) dudv,$$

or possibly the negative of this depending on the choice of the positive normal direction.

Examples:

1. Consider a surface specified 'explicitly' by

$$x(x, y) = (x, y, z(x, y)).$$

Then

$$\frac{\partial x}{\partial x} = (1,0,z_x),$$

$$\frac{\partial x}{\partial y} = (0,1,z_y),$$

so that, after identifying x and y with u and v respectively, we have that

$$dS = (-z_x, -z_y, 1)\, dx\, dy,$$

and

$$dS = \left(1+z_x^2+z_y^2\right)^{\frac{1}{2}} dxdy \geq dxdy.$$

2. *Spherical Surface.* Consider a spherical surface of radius a represented parametrically by

$$x(\theta, \phi) = (a \sin \theta \cos \phi, a \sin \theta \sin \phi, a \cos \theta).$$

Then

$$x_\theta = (a \cos \theta \cos \phi, a \cos \theta \sin \phi, -a \sin \theta),$$

$$x_\phi = (-a \sin \theta \sin \phi, a \sin \theta \cos \phi, 0),$$

so that after identifying θ and ϕ with u and v respectively, we have that

$$dS = (\sin \theta \cos \phi, \sin \theta \sin \phi, \cos \theta)\, a^2 \sin \theta\, d\theta\, d\phi$$

$$= \hat{x}\, a^2 \sin \theta\, d\theta\, d\phi,$$

where $\hat{x}$ is the unit vector in the radial direction, and

$$dS = a^2 \sin \theta\, d\theta\, d\phi.$$

3. *Surface of Revolution.* Consider the surface specified with $t = z$)

$$x(\phi, z) = (R(z) \cos \phi, R(z) \sin \phi, z).$$

In terms of cylindrical polar co-ordinates, the surface is given by

$$\rho = (x^2+y^2)^{\frac{1}{2}} = R(z), \;\; 0 \leq \phi < 2\pi.$$

Proceed geometrically. In a plane ϕ = constant, the length of a line element on the surface is given by

$$ds = (d\rho^2 + dz^2)^{\frac{1}{2}} = (R_z^2+1)^{\frac{1}{2}} dz.$$

Hence

$$dS = Rd\phi(R_z^2+1)^{\frac{1}{2}} dz = (1+R_z^2)^{\frac{1}{2}} Rd\phi dz.$$

Exercise: Obtain same result and also show that

$$dS = (\cos\phi, \sin\phi, -R_z)Rd\phi\, dz.$$

Answer: From (4.28)

$$x_\phi = (-R\sin\phi, R\cos\phi, 0),$$
$$x_z = (R_z\cos\phi, R_z\sin\phi, 1).$$

Hence from, and compare with,

$$dS = (R\cos\phi, R\sin\phi, -RR_z)\, d\phi\, dz$$
$$= (\cos\phi, \sin\phi, -R_z)Rd\phi\, dz,$$
$$dS = (1+R_z^2)^{\frac{1}{2}}\, Rd\phi dz.$$

Evaluation of Surface Integrals

The above expressions for dS and dS effectively reduce surface integrals to area integrals. There are at least five types of surface integral encountered in physical applications:

Scalars: $\iint_S f dS, \iint_S u.dS$

Vectors: $\iint_S f dS, \iint_S u.dS, \iint_S u \times dS$

Examples:

1. Calculate

$$I = \iint_S xy\, dS,$$

where S is the portion of the surface $x + y + z = a$ in $x \geq 0, y \geq 0, z \geq 0$.

From with $z = a - x - y$,

$$dS = \left(1 + z_x^2 + z_y^2\right)^{\frac{1}{2}} dxdy = \sqrt{3}dxdy.$$

Hence, if A is the projection of S onto the xy-plane, then

$$I = \sqrt{3}\iint_A xydxdy$$
$$= \sqrt{3}\int_0^a dx \int_0^{a-x} dyxy$$
$$= \sqrt{3}\int_0^a dx \frac{1}{2}x(a-x)^2 = \frac{a^4}{8\sqrt{3}}.$$

2. Integrals such as

$$I = \iint_S u dS$$

can be evaluated component by component.

3. Evaluate

$$I = \iint_S dS$$

where S is a closed surface.

Do this component by component starting with I. $\hat{z}$. Then, noting that the scalar product $\hat{z}.dS$ gives the area of the element of surface when projected onto the xy-plane, we split $I.\ \hat{z}$ into two:

$$I.\hat{z} = \iint_{\hat{z}.dS>0} \hat{z}.dS + \iint_{\hat{z}.dS<0} \hat{z}.dS,$$

where the first/second integral is the projection of the 'upper'/'lower' surface onto (x, y).

For a closed surface the two projections must be equal and opposite, hence

$$I.\hat{z} = 0.$$

Similarly, $I.\hat{x} = 0$ and $I.\hat{y} = 0$. Hence $I = 0$.

4. Evaluate

$$I = \iint_S x.dS$$

where S is the surface of the cone bounded by $(x^2 + y^2)^{\frac{1}{2}} + z = 1$ and $z = 0$, and the positive normal direction is out of the body.

On $z = 0$, including the unit disc S_1 that is the base of the surface,

$$x\ .\ n = -z = 0.$$

Hence

$$I = \iint_{S_1} x.dS + \iint_{S_2} x.dS = \iint_{S_2} x.dS.$$

On S_2, which is a surface of revolution,

$$\rho = (x^2 + y^2)^{\frac{1}{2}} = R(z) \equiv 1 - z.$$

Hence from

$$dS = (\cos\phi,\ \sin\phi,\ -R_z)Rd\phi\ dz$$
$$= (\cos\phi,\ \sin\phi,\ 1)Rd\phi\ dz,$$

where we have ensured that the normal direction is out of the body. Hence on S_2

$$x \,.\, dS = (R\cos^2\phi + R\sin^2\phi + z)Rd\phi\, dz$$
$$= (1 - z)d\phi\, dz,$$

from using. It follows that

$$\iint_{S_1} x.dS = \int_0^1 dz \int_0^{2\pi} d\phi(1-z) = \pi.$$

Exercise: Repeat using

$$x(x, y) = \left(x, y, 1-\left(x^2+y^2\right)^{\frac{1}{2}}\right).$$

Flux

Definition: Let $v(x, t)$ be a vector field. Then the flux, Q, of v through a surface S is defined to be

$$Q = \iint_S v.dS.$$

Example: Suppose that $v(x, t)$ describes the velocity field of the flow of a fluid, e.g. air through or out of a jet engine, water through a hose-pipe, blood flow through arteries/veins, sweat, tears, etc. Suppose we wish to measure how much fluid is coming out of a surface S per unit time.

Fluid flowing parallel to a surface does not cross it, hence it is the component of velocity normal to S that matters, i.e. $v.\ n$. In time δt, the volume of fluid crossing a small area dS is

$$(v.\ n)\ \delta t\ dS = (v. dS)\ \delta t.$$

Therefore in unit time the volume of fluid crossing S is

$$Q = \iint_S v.dS.$$

This is referred to as the volume flux out of S.

Example: Consider the integral

$$I = \iint_D \frac{ky \sin kr_1 \sin kr_2}{r_1 r_2} dxdy,$$

where k is a constant,

$$r_1 = [(x+1)^2 + y^2]^{\frac{1}{2}}, r_2 = [(x-1)^2 + y^2]^{\frac{1}{2}},$$

and D is the half of the ellipse $r_1 + r_2 \leq 4$ which is in $y \geq 0$. By means of the transformation

$$u = \frac{1}{2}(r_1 + r_2), v = \frac{1}{2}(r_1 - r_2),$$

or otherwise, show that the integral is equal to 4 $\sin^3 k \cos k$.

Solution: First consider the $y = 0$ boundary, on which

$$r_1 = | x + 1 | \text{ and } r_2 = | x - 1 |.$$

Hence on $y = 0$,

for $1 \leq x \leq 2 : r_1 = x + 1, r_2 = x - 1$ and $r_1 - r_2 = 2, r_1 + r_2 = 2x$,

for $-1 \leq x \leq 1 : r_1 = x + 1 , r_2 = 1 - x$ and $r_1 - r_2 = 2x, r_1 + r_2 = 2$,

for $-2 \leq x \leq -1 : r_1 = -x - 1, r_2 = 1 - x$ and $r_2 - r_1 = -2, r_1 + r_2 = -2x$.

Thus if, as suggested,

$$u = \frac{1}{2}(r_1 + r_2), v = \frac{1}{2}(r_1 - r_2),$$

it follows that the boundary can be split into four parts specified by:

(a) $v = -1$,

(b) $u = 1$,

(c) $v = 1$,

(d) $u = 2$.

Further, since

$$r_1^2 = (x+1)^2 + y^2, r_2^2 = (x-1)^2 + y^2,$$

it follows that

$$r_1 r_{1x} = x + 1, r_1 r_{1y} = y, r_2 r_{2x} = x - 1 , r_2 r_{2y} = y,$$

and hence that

$$\frac{\partial(r_1, r_2)}{\partial(x, y)} = r_{1x} r_{2y} - r_{1y} r_{2x} = \frac{(x+1)y}{r_1 r_2} - \frac{(x-1)y}{r_1 r_2} = \frac{2y}{r_1 r_2}.$$

Also

$$\frac{\partial(u, v)}{\partial(r_1 r_2)} = u_{r_1} v_{r_2} - u_{r_2} v_{r_1} = \frac{1}{2}.\left(-\frac{1}{2}\right) - \frac{1}{2}.\frac{1}{2} = -\frac{1}{2},$$

and so

$$\frac{\partial(u, v)}{\partial(x, y)} = \frac{\partial(u, v)}{\partial(r_1, r_2)} \frac{\partial(r_1, r_2)}{\partial(x, y)} = -\frac{y}{r_1 r_2}.$$

We note that the Jacobian is negative in D since the handedness is not preserved under the co-ordinate transformation. Further, we also note that for $0 < k \ll 1$

$$I \approx \iint_D k^3 y\,dx\,dy > 0.$$

Hence we conclude that

$$I = \int_1^2 du \int_{-1}^1 dv \left|\frac{\partial(x,y)}{\partial(u,y)}\right| \frac{ky \sin kr_1 \sin kr_2}{r_1 r_2}$$

$$= \frac{1}{2}k \int_1^2 du \int_{-1}^1 dv[\cos k(r_1 - r_2) - \cos k(r_1 + r_2)]$$

$$= \frac{1}{2}k \int_1^2 du \int_{-1}^1 dv[\cos 2kv - \cos 2ku]$$

$$= \frac{1}{2}k \left[\left[\frac{1}{2k}\sin 2kv\right]_{-1}^1 - 2\left[\frac{1}{2k}\sin 2ku\right]_1^2\right]$$

$$= \sin 2k - \frac{1}{2}\sin 4k = \sin 2k(1 - \cos 2k) = 4\sin^3 k \cos k,$$

where we have checked that the sign is correct by noting that $I > 0$ when $0 < k \ll 1$.

EXERCISE

1. Let $f(t)$ be a differentiable curve such that $f(t) \neq 0$ for all t. Show that

$$\frac{d}{dt}\left(\frac{f(t)}{\|f(t)\|}\right) = \frac{f(t) \times (f'(t) \times f(t))}{\|f(t)\|^3}$$

2. Let $f(t)$ be a smooth curve such that $f'(t) \neq 0$ for all t. Then we can define the *unit tangent vector T* by

$$T(t) = \frac{f'(t)}{\|f'(t)\|}$$

Show that

$$T'(t) = \frac{f'(t) \times (f''(t) \times f'(t))}{\|f'(t)\|^3}$$

3. Find T, N, B and κ at each point of the helix $f(t) = (\cos t, \sin t, t)$.
4. Show that the arc length L of a curve whose spherical coordinates are $\rho = \rho(t)$, $\theta = \phi(t)$ and $\phi = \phi(t)$ for t in an interval $[a, b]$ is

$$L = \int_a^b \sqrt{\rho'(t)^2 + (\rho(t)^2 \sin^2 \phi(t))\theta'(t)^2 + \rho(t)^2 \phi'(t)^2}\, dt.$$

Chapter 9

Line and Surface Integrals

Line Integrals

In single-variable calculus you learned how to integrate a real-valued function $f(x)$ over an interval $[a, b]$ in $\mathbb{R}^1$. This integral (usually called a *Riemann integral*) can be thought of as an integral over a *path* in $\mathbb{R}^1$, since an interval (or collection of intervals) is really the only kind of "path" in $\mathbb{R}^1$. You may also recall that if $f(x)$ represented the force applied along the x-axis to an object at position x in $[a, b]$, then the *work* W done in moving that object from position $x = a$ to $x = b$ was defined as the integral:

$$W = \int_a^b f(x)dx$$

In this section, we will see how to define the integral of a function (either realvalued or vector-valued) of two variables over a general path (i.e. a curve) in $\mathbb{R}^2$. This definition will be motivated by the physical notion of work. We will begin with real-valued functions of two variables. In physics, the intuitive idea of work is that Work = Force × Distance.

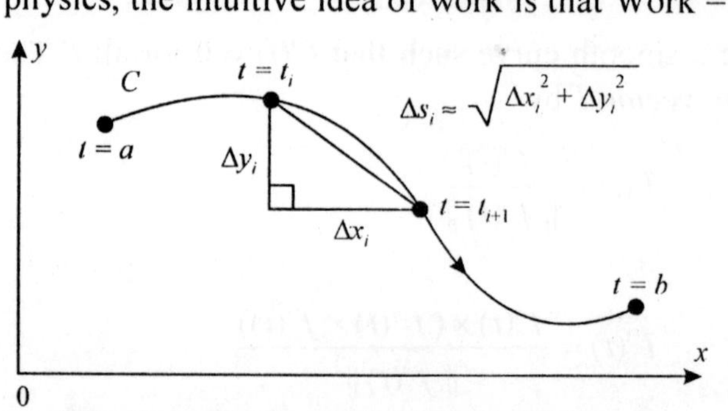

Fig. Curve C : $x = x(t)$, $y = y(t)$ for t in $[a, b]$

Suppose that we want to find the total amount W of work done in moving an object along a curve C in $\mathbb{R}^2$ with a smooth parametrization $x = x(t)$, $y = y(t)$, $a \le t \le b$, with a force $f(x, y)$ which varies with the position (x, y) of the object and is applied in the direction of motion along C.

We will assume for now that the function $f(x, y)$ is continuous and real-valued, so we only consider the magnitude of the force. Partition the interval $[a, b]$ as follows:

$$a = t_0 < t_1 < t_2 < \ldots < t_{n-1} < t_n = b, \text{ for some integer } n \geq 2$$

As we can see from Figure, over a typical subinterval $[t_i, t_{i+1}]$ the distance Δs_i traveled along the curve is approximately $\sqrt{\Delta x_i^2 + \Delta y_i^2}$, by the Pythagorean Theorem. Thus, if the subinterval is small enough than the work done in moving the object along that piece of the curve is approximately

$$\text{Force} \times \text{Distance} \approx f(x_{i*}, y_{i*})\sqrt{\Delta x_i^2 + \Delta y_i^2},$$

where $(x_{i*}, y_{i*}) = (x(t_{i*}), y(t_{i*}))$ for some t_{i*} in $[t_i, t_{i+1}]$, and so

$$W \approx \sum_{i=0}^{n-1} f(x_{i*}, y_{i*})\sqrt{\Delta x_i^2 + \Delta y_i^2}$$

is approximately the total amount of work done over the entire curve. But since

$$\sqrt{\Delta x_i^2 + \Delta y_i^2} = \sqrt{\left(\frac{\Delta x_i}{\Delta t_i}\right)^2 + \left(\frac{\Delta y_i}{\Delta t_i}\right)^2}\,\Delta t_i,$$

where $\Delta t_i = t_{i+1} - t_i$, then

$$W \approx \sum_{i=0}^{n-1} f(x_{i*}, y_{i*})\sqrt{\left(\frac{\Delta x_i}{\Delta t_i}\right)^2 + \left(\frac{\Delta y_i}{\Delta t_i}\right)^2}\,\Delta t_i.$$

Taking the limit of that sum as the length of the largest subinterval goes to 0, the sum over all subintervals becomes the integral from $t = a$ to $t = b$, $\frac{\Delta x_i}{\Delta t_i}$ and $\frac{\Delta y_i}{\Delta t_i}$ become $x'(t)$ and $y'(t)$, respectively, and $f(x_{i*}, y_{i*})$ becomes $f(x(t), y(t))$, so that

$$W = \int_a^b f(x(t), y(t))\sqrt{x'(t)^2 + y'(t)^2}\,dt.$$

The integral on the right side of the above equation gives us our idea of how to define, for *any* real-valued function $f(x, y)$, the integral of $f(x, y)$ along the curve C, called a *line integral*:

Definition: For a real-valued function $f(x, y)$ and a curve C in $\mathbb{R}^2$, parametrized by $x = x(t)$, $y = y(t)$, $a \leq t \leq b$, the line integral of $f(x, y)$ along C with respect to arc length s is

$$\int_C f(x, y)\,ds = \int_a^b f(x(t), y(t))\sqrt{x'(t)^2 + y'(t)^2}\,dt.$$

The symbol ds is the differential of the arc length function

$$s = s(t) = \int_a^t \sqrt{x'(u)^2 + y'(u)^2}\,du,$$

which you may recognize fromSection he length of the curve C over the interval $[a, t]$, for all t in $[a, b]$. That is,

$$ds = s'(t)dt = \sqrt{x'(t)^2 + y'(t)^2}\,dt,$$

by the Fundamental Theorem of Calculus.

For a general real-valued function $f(x, y)$, what does the line integral $\int_C f(x,y)ds$ represent? The preceding discussion of ds gives us a clue. You can think of differentials as infinitesimal lengths. So if you think of $f(x, y)$ as the height of a picket fence along C, then $f(x, y)\,ds$ can be thought of as approximately the area of a section of that fence over some infinitesimally small section of the curve, and thus the line integral $\int_C f(x,y)ds$ is the total area of that picket fence.

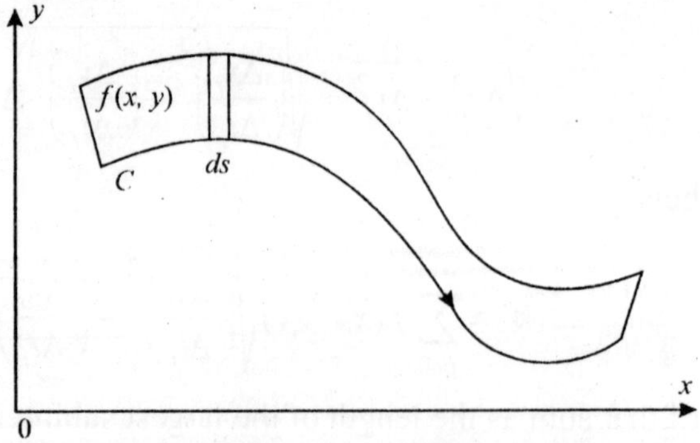

Fig. Area of Shaded Rectangle = Height × Width ≈ $f(x, y)\,ds$

Example: Use a line integral to show that the lateral surface area A of a right circular cylinder of radius r and height h is $2\pi rh$.

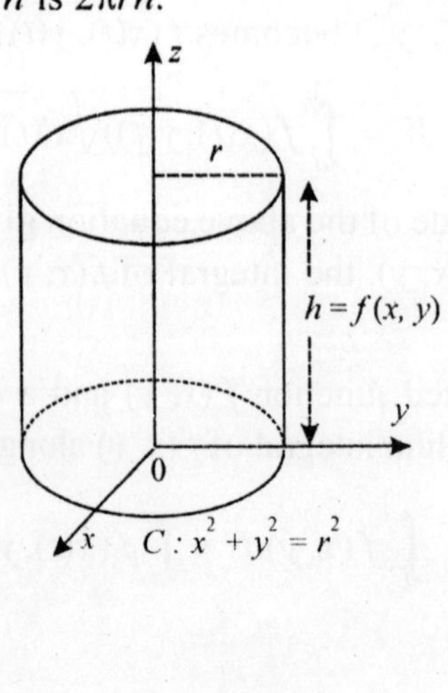

Solution: We will use the right circular cylinder with base circle C given by $x^2 + y^2 = r^2$ and with height h in the positive z direction. Parametrize C as follows:

$$x = x(t) = r\cos t,\ y = y(t) = r\sin t,\ 0 \le t \le 2\pi$$

Let $f(x, y) = h$ for all (x, y). Then

$$A = \int_C f(x,y)ds = \int_a^b f(x(t),y(t))\sqrt{x'(t)^2 + y'(t)^2}\,dt$$

$$= \int_0^{2\pi} h\sqrt{(-r\sin t)^2 + (r\cos t)^2}\,dt$$

$$= h\int_0^{2\pi} r\sqrt{\sin^2 t + \cos^2 t}\,dt$$

$$= rh\int_0^{2\pi} 1dt = 2\pi rh$$

Note in Example that if we had traversed the circle C twice, i.e. let t vary from 0 to 4π, then we would have gotten an area of $4\pi rh$, i.e. twice the desired area, even though the *curve* itself is still the same (namely, a circle of radius r). Also, notice that we traversed the circle in the counter-clockwise direction. If we had gone in the clockwise direction, using the parametrization

$$x = x(t) = r\cos(2\pi - t),\ y = y(t) = r\sin(2\pi - t),\ 0 \le t \le 2\pi,$$

then it is easy to verify that the value of the line integral is unchanged.

In general, it can be shown that reversing the direction in which a curve C is traversed leaves $\int_C f(x,y)ds$ unchanged, for any $f(x, y)$. If a curve C has a parametrization $x = x(t)$, $y = y(t)$, $a \le t \le b$, then denote by $-C$ the same curve as C but traversed in the opposite direction. Then $-C$ is parametrized by

$$x = x(a + b - t),\ y = y(a + b - t),\ a \le t \le b,$$

and we have

$$\int_C f(x,y)ds = \int_{-C} f(x,y)ds.$$

Notice that our definition of the line integral was with respect to the arc length parameter s. We can also define

$$\int_C f(x,y)dx = \int_a^b f(x(t),y(t))x'(t)dt$$

as the *line integral of* $f(x, y)$ *along* C *with respect to* x, and

$$\int_C f(x,y)dy = \int_a^b f(x(t),y(t))y'(t)dt$$

as the *line integral of* $f(x, y)$ *along* C *with respect to* y.

In the derivation of the formula for a line integral, we used the idea of work as force multiplied by distance. However, we know that force is actually a *vector*. So it would be helpful to develop a vector form for a line integral. For this, suppose that we have a function $f(x, y)$ defined on $\mathbb{R}^2$ by

$$f(x, y) = P(x, y)i + Q(x, y)j$$

for some continuous real-valued functions $P(x, y)$ and $Q(x, y)$ on $\mathbb{R}^2$. Such a function f is called a vector field on $\mathbb{R}^2$. It is defined at *points* in $\mathbb{R}^2$, and its values are *vectors* in $\mathbb{R}^2$. For a curve C with a smooth parametrization $x = x(t)$, $y = y(t)$, $a \leq t \leq b$, let

$$r(t) = x(t)\, i + y(t)\, j$$

be the position vector for a point $(x(t), y(t))$ on C. Then $r'(t) = x'(t)\, i + y'(t)\, j$ and so

$$\int_C P(x,y)dx + \int_C Q(x,y)dy = \int_a^b P(x(t),y(t))x'(t)dt + \int_a^b Q(x(t),y(t))y'(t)dt$$

$$= \int_a^b (P(x(t),y(t))x'(t) + Q(x(t),y(t))y'(t))dt$$

$$= \int_a^b f(x(t),y(t)).r'(t)dt$$

by definition of $f(x, y)$. Notice that the function $f(x(t), y(t)) \,.\, r'(t)$ is a *real-valued* function on $[a, b]$, so the last integral on the right looks somewhat similar to our earlier definition of a line integral. This leads us to the following definition:

Definition: For a vector field $f(x, y) = P(x, y)\, i + Q(x, y)\, j$ and a curve C with a smooth parametrization $x = x(t)$, $y = y(t)$, $a \leq t \leq b$, the line integral of f along C is

$$\int_C f.dr = \int_C P(x,y)dx + \int_C Q(x,y)dy$$

$$= \int_a^b f(x(t),y(t).r'(t)dt,$$

where $r(t) = x(t)\, i + y(t)\, j$ is the position vector for points on C.

We use the notation $dr = r'(t)dt = dx\, i + dy\, j$ to denote the differential of the vectorvalued function r. The line integral in Definition is often called a *line integral of a vector field* to distinguish it from the line integral in Definition which is called a *line integral of a scalar field*. For convenience we will often write

$$\int_C P(x,y)dx + \int_C Q(x,y)dy = \int_C P(x,y)dx + Q(x,y)dy,$$

where it is understood that the line integral along C is being applied to both P and Q. The quantity $P(x, y)\, dx + Q(x, y)\, dy$ is known as a differential form. For a realvalued function $F(x, y)$, the differential of F is $dF = \frac{\partial F}{\partial x}dx + \frac{\partial F}{\partial y}dy$. A differential form $P(x, y)\, dx + Q(x, y)\, dy$ is called exact if it equals dF for some function $F(x, y)$.

Recall that if the points on a curve C have position vector $r(t) = x(t)\, i + y(t)\, j$, then $r'(t)$ is a tangent vector to C at the point $(x(t), y(t))$ in the direction of increasing t (which we call the *direction of* C). Since C is a smooth curve, then $r'(t) \neq 0$ on $[a, b]$ and hence

$$T(t) = \frac{r'(t)}{\| r'(t) \|}$$

is the unit tangent vector to C at $(x(t), y(t))$. Putting Definitions together we get the following theorem:

Theorem: For a vector field $f(x, y) = P(x, y)i + Q(x, y)j$ and a curve C with a smooth parametrization $x = x(t)$, $y = y(t)$, $a \leq t \leq b$ and position vector $r(t) = x(t)i + y(t)j$,

$$\int_C f.dr = \int_C f.Tds,$$

where $T(t) = \dfrac{r'(t)}{\| r'(t) \|}$ is the unit tangent vector to C at $(x(t), y(t))$.

If the vector field $f(x, y)$ represents the force moving an object along a curve C, then the work W done by this force is

$$W = \int_C f.Tds = \int_C f.dr.$$

Example: Evaluate $\int_C (x^2 + y^2)dx + 2xydy$, where:

(a) $C : x = t\, , y = 2t\, , 0 \leq t \leq 1$

(b) $C : x = t\, , y = 2t2\, , 0 \leq t \leq 1$

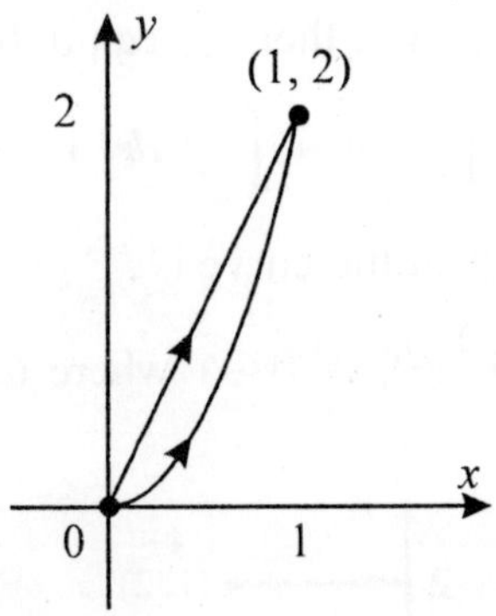

Solution: Figure shows both curves.

(a) Since $x'(t) = 1$ and $y'(t) = 2$, then

$$\int_C (x^2 + y^2)dx + 2xydy = \int_0^1 \Big((x(t)^2 + y(t)^2)x'(t) + 2x(t)y(t)y'(t)\Big)dt$$

$$= \int_0^1 \Big((t^2 + 4t^2)(1) + 2t(2t)(2)\Big)dt$$

$$= \int_0^1 13t^2 dt$$

$$= \left.\frac{13t^3}{3}\right|_0^1 = \frac{13}{3}$$

(b) Since $x'(t) = 1$ and $y'(t) = 4t$, then

$$\int_C (x^2 + y^2)dx + 2xydy = \int_0^1 \left((x(t)^2 + y(t)^2)x'(t) + 2x(t)y(t)'(t)\right)dt$$

$$= \int_0^1 \left((t^2 + 4t^4)(1) + 2t(2t^2)(4t)\right)dt$$

$$= \int_0^1 (t^2 + 20t^4)dt = \left.\frac{t^3}{3} + 4t^5\right|_0^1 = \frac{1}{3} + 4 = \frac{13}{3}$$

So in both cases, if the vector field $f(x, y) = (x^2 + y^2)\, i + 2xy\, j$ represents the force moving an object from (0, 0) to (1, 2) along the given curve C, then the work done is $\frac{13}{3}$. This may lead you to think that work (and more generally, the line integral of a vector field) is independent of the path taken. However, as we will see in the next section, this is not always the case. Although we defined line integrals over a single smooth curve, if C is a *piecewise smooth curve*, that is

$$C = C_1 \cup C_2 \cup .... \cup C_n$$

is the union of smooth curves C_1, ..., C_n, then we can define

$$\int_C f.dr = \int_{C_1} f.dr_1 + \int_{C_n} f.dr_2 + ... \int_{C_n} f.dr_n$$

where each ri is the position vector of the curve C_i.

Example: Evaluate $\int_C (x^2 + y^2)dx + 2xydy$, where C is the polygonal path from (0, 0) to (0, 2) to (1, 2).

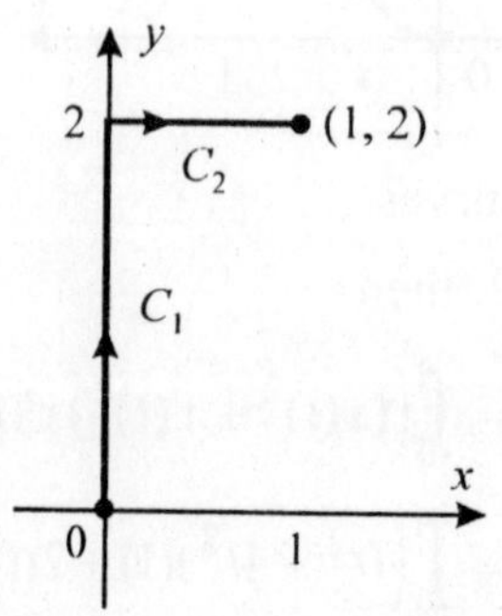

Solution: Write $C = C_1 \cup C_2$, where C_1 is the curve given by $x = 0, y = t, 0 \le t \le 2$ and C_2 is the curve given by $x = t, y = 2, 0 \le t \le 1$. Then

$$\int_C (x^2+y^2)dx + 2xy\,dy = \int_{C_1} (x^2+y^2)dx + 2xydy + \int_C (x^2+y^2)dx + 2xy\,dy$$

$$= \int_0^2 \left((0^2+t^2)(0) + 2(0)t(1)\right)dt + \int_0^1 \left((t^2+4)(1) + 2t(2)(0)\right)dt$$

$$= \int_0^2 0\,dt + \int_0^1 (t^2+4)dt$$

$$= \frac{t^3}{3} + 4t\Big|_0^1 = \frac{1}{3} + 4 = \frac{13}{3}$$

Line integral notation varies quite a bit. For example, in physics it is common to see the notation $\int_a^b f.dl$, where it is understood that the limits of integration a and b are for the underlying parameter t of the curve, and the letter l signifies length. Also, the formulation $\int_C f.Tds$ from Theorem is often preferred in physics since it emphasizes the idea of integrating the tangential component $f.T$ of f in the direction of T (i.e. in the direction of C), which is a useful physical interpretation of line integrals.

EXERCISES

For Exercises 1-4, calculate $\int_C f(x,y)ds$ for the given function $f(x, y)$ and curve C.

1. $f(x, y) = xy$; $C : x = \cos t, y = \sin t, 0 \le t \le \pi/2$
2. $f(x, y) = \dfrac{x}{x^3+1}$; $C : x = t, y = 0, 0 \le t \le 1$
3. $f(x, y) = 2x + y$; C: polygonal path from (0, 0) to (3, 0) to (3, 2)
4. $f(x, y) = x + y^2$; C: path from (2, 0) counterclockwise along the circle $x^2 + y^2 = 4$ to the point (–2, 0) and then back to (2, 0) along the x-axis.
5. Use a line integral to find the lateral surface area of the part of the cylinder $x^2 + y^2 = 4$ below the plane $x + 2y + z = 6$ and above the xy-plane.

For Exercises 6-11, calculate $\int_C f.dr$ for the given vector field $f(x, y)$ and curve C.

6. $f(x, y) = i - j$; $C : x = 3t, y = 2t, 0 \le t \le 1$
7. $f(x, y) = y\,i - x\,j$; $C : x = \cos t, y = \sin t, 0 \le t \le 2\pi$
8. $f(x, y) = x\,i + y\,j$; $C : x = \cos t, y = \sin t, 0 \le t \le 2\pi$
9. $f(x, y) = (x^2 - y)\,i + (x - y^2)\,j$; $C : x = \cos t, y = \sin t, 0 \le t \le 2\pi$

10. $f(x, y) = xy^2 i + xy^3 j$; C : the polygonal path from (0, 0) to (1, 0) to (0, 1) to (0, 0)

11. $f(x, y) = (x^2 + y^2) i$; $C : x = 2 + \cos t, y = \sin t, 0 \leq t \leq 2\pi$

13. Show that if $f \perp r'(t)$ at each point $r(t)$ along a smooth curve C, then $\int_C f.dr = 0$.

14. Show that if f points in the same direction as $r'(t)$ at each point $r(t)$ along a smooth curve C, then $\int_C f.dr = \int_C \| f \| ds$.

16. Let C be a smooth curve with arc length L, and suppose that $f(x, y) = P(x, y) i + Q(x, y) j$ is a vector field such that $\| f(x, y) \| \leq M$ for all (x, y) on C. Show that $\left| \int_C f.dr \right| \leq ML$.

(*Hint: Recall that* $\left| \int_a^b g(x)dx \right| \leq \int_a^b | g(x) | \, dx$ *for Riemann integrals.*)

17. Prove that the Riemann integral $\int_a^b f(x)dx$ is a special case of a line integral.

Properties of Line Integrals

We know from the previous section that for line integrals of real-valued functions (scalar fields), reversing the direction in which the integral is taken along a curve does not change the value of the line integral:

$$\int_C f(x,y)ds = \int_{-C} f(x,y)ds$$

For line integrals of vector fields, however, the value does change. To see this, let $f(x, y) = P(x, y) i + Q(x, y) j$ be a vector field, with P and Q continuously differentiable functions. Let C be a smooth curve parametrized by $x = x(t)$, $y = y(t)$, $a \leq t \leq b$, with position vector $r(t) = x(t) i + y(t) j$ (we will usually abbreviate this by saying that $C : r(t) = x(t) i + y(t) j$ is a smooth curve).

We know that the curve $-C$ traversed in the opposite direction is parametrized by $x = x(a + b - t)$, $y = y(a + b - t)$, $a \leq t \leq b$. Then

$$\int_{-C} P(x,y)dx = \int_a^b P(x(a+b-t), y(a+b-t)) \frac{d}{dt}(x(a+b-t))dt$$

$$= \int_a^b P(x(a+b-t), y(a+b-t))(-x'(a+b-t))dt \quad \text{(by the Chain Rule)}$$

$$= \int_b^a P(x(u), y(u))(-x'(u))(-du) \quad \text{(by letting } u = a + b - t)$$

$$= \int_b^a P(x(u), y(u))x'(u)du$$

$$= -\int_a^b P(x(u), y(u))x'(u)du, \text{ since } \int_b^a = -\int_a^b, \text{ so}$$

$$\int_{-C} P(x, y)dx = -\int_C P(x, y)dx$$

since we are just using a different letter (u) for the line integral along C. A similar argument shows that

$$\int_{-C} Q(x, y)dy = -\int_C Q(x, y)dy,$$

and hence

$$\int_{-C} f.dr = \int_{-C} P(x, y)dx + \int_{-C} Q(x, y)dy$$

$$= -\int_C P(x, y)dx + -\int_C Q(x, y)dy$$

$$= -\left(\int_C P(x, y)dx + \int_C Q(x, y)dy\right)$$

$$\int_{-C} f.dr = -\int_C f.dr.$$

The above formula can be interpreted in terms of the work done by a force $f(x, y)$ (treated as a vector) moving an object along a curve C: the total work performed moving the object along C from its initial point to its terminal point, and then back to the initial point moving backwards along the same path, is zero. This is because when force is considered as a vector, direction is accounted for.

The preceding discussion shows the importance of always taking the *direction* of the curve into account when using line integrals of vector fields. For this reason, the curves in line integrals are sometimes referred to as *directed curves* or *oriented curves*.

Recall that our definition of a line integral required that we have *a* parametrization $x = x(t)$, $y = y(t)$, $a \le t \le b$ for the curve C. But as we know, any curve has infinitely many parametrizations. So could we get a different value for a line integral using some other parametrization of C, say, $x = \tilde{x}(u)$, $y = \tilde{y}(u)$, $c \le u \le d$? If so, this would mean that our definition is not well-defined. Luckily, it turns out that the value of a line integral of a vector field is unchanged as long as the direction of the curve C is preserved by whatever parametrization is chosen:

Theorem: Let $f(x, y) = P(x, y)\, i + Q(x, y)\, j$ be a vector field, and let C be a smooth curve parametrized by $x = x(t)$, $y = y(t)$, $a \le t \le b$. Suppose that $t = \alpha(u)$ for $c \le u \le d$, such that $a = \alpha(c)$, $b = \alpha(d)$, and $\alpha'(u) > 0$ on the open interval (c, d) (i.e. $\alpha(u)$ is strictly increasing on $[c, d]$). Then $\int_C f.dr$ has the same value for the parametrizations $x = x(t)$, $y = y(t)$, $a \le t \le b$ and $x = \tilde{x}(u) = x(\alpha(u))$, $y = \tilde{y}(u) = y(\alpha(u))$, $c \le u \le d$.

Proof: Since $\alpha(u)$ is strictly increasing and maps $[c, d]$ onto $[a, b]$, then we know that $t = \alpha(u)$ has an inverse function $u = \alpha^{-1}(t)$ defined on $[a, b]$ such that $c = \alpha^{-1}(a)$, $d = \alpha^{-1}(b)$, and $\frac{du}{dt} = \frac{1}{\alpha'(u)}$. Also, $dt = \alpha'(u)\,du$, and by the Chain Rule

$$\tilde{x}'(u)\frac{d\tilde{x}}{du} = \frac{d}{du}(x(\alpha(u))) = \frac{dx}{dt}\frac{dt}{du} = x'(t)\alpha'(u)$$

$$\Rightarrow x'(u) = \frac{\tilde{x}'(u)}{\alpha'(u)}$$

so making the susbstitution $t = \alpha(u)$ gives

$$\int_a^b P(x(t), y(t))x'(t)dt = \int_{\alpha^{-1}(a)}^{\alpha^{-1}(b)} P(x(\alpha(u)), Y(\alpha(u)))\frac{\tilde{x}'(u)}{\alpha'(u)}(\alpha'(u)du)$$

$$= \int_c^d P(\tilde{x}(u), \tilde{y}(u))\tilde{x}'(u)du,$$

which shows that $\int_C P(x, y)dx$ has the same value for both parametrizations. A similar argument shows that $\int_C Q(x, y)dy$ has the same value for both parametrizations, and hence $\int_C f.dr$ has the same value.

Notice that the condition $\alpha'(u) > 0$ in Theorem means that the two parametrizations move along C in the same direction. That was *not* the case with the "reverse" parametrization for $-C$: for $u = a + b - t$ we have $t = \alpha(u) = a + b - u \Rightarrow \alpha'(u) = -1 < 0$.

Example: Evaluate the line integral $\int_C (x^2 + y^2)dx + 2xydy$, along the curve $C : x = t$, $y = 2t^2$, $0 \leq t \leq 1$, where $t = \sin u$ for $0 \leq u \leq \pi/2$.

Solution: First, we notice that $0 = \sin 0$, $1 = \sin(\pi/2)$, and $\frac{dt}{du} = \cos u > 0$ on $(0, \pi/2)$. So by Theorem we know that if C is parametrized by

$$x = \sin u,\ y = 2\sin^2 u\ ,\ 0 \leq u \leq \pi/2.$$

then $\int_C (x^2 + y^2)dx + 2xydy$ should have the same value as we found in Example, namely $\frac{13}{3}$. And we can indeed verify this:

$$\int_C (x^2 + y^2)dx + 2xydy = \int_0^{\pi/2} ((\sin^2 u + (2\sin^2 u)^2)\cos u +$$

$$2(\sin u)(2\sin^2 u)4\sin u\cos u)\,du$$

$$= \int_0^{\pi/2} \left(\sin^2 u + 20\sin^4 u\right)\cos u du$$

$$= \frac{\sin^3 u}{3} + 4\sin^5 u \Big|_0^{\pi/2}$$

$$= \frac{1}{3} + 4\frac{13}{3}$$

In other words, the line integral is unchanged whether t or u is the parameter for C.

By a closed curve, we mean a curve C whose initial point and terminal point are the same, i.e. for $C : x = x(t), y = y(t), a \leq t \leq b$, we have $(x(a), y(a)) = (x(b), y(b))$.

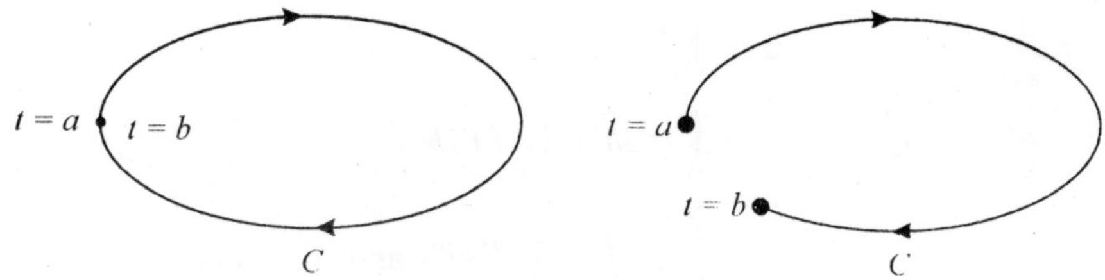

(a) Closed (b) Not closed

Fig. Closed vs Nonclosed Curves

A simple closed curve is a closed curve which does not intersect itself. Note that any closed curve can be regarded as a union of simple closed curves. We use the special notation

$$\oint_C f(x, y)ds \text{ and } \oint_C f.dr$$

to denote line integrals of scalar and vector fields, respectively, along closed curves.

In some older texts you may see the notation $\oint$ or $\oint$ to indicate a line integral traversing a closed curve in a counterclockwise or clockwise direction, respectively.

So far, the examples we have seen of line integrals have had the same value for different curves joining the initial point to the terminal point. That is, the line integral has been independent of the path joining the two points. As we mentioned before, this is not always the case. The following theorem gives a necessary and sufficient condition for this *path independence*:

Theorem: In a region R, the line integral $\oint_C f.dr$ is independent of the path between any two points in R if and only if $\oint_C f.dr = 0$ for every closed curve C which is contained in R.

Proof: Suppose that $\oint_C f.dr = 0$ for every closed curve C which is contained in R. Let P_1 and P_2 be two distinct points in R. Let C_1 be a curve in R going from P_1 to P_2, and let C_2 be another curve in R going from P_1 to P_2.

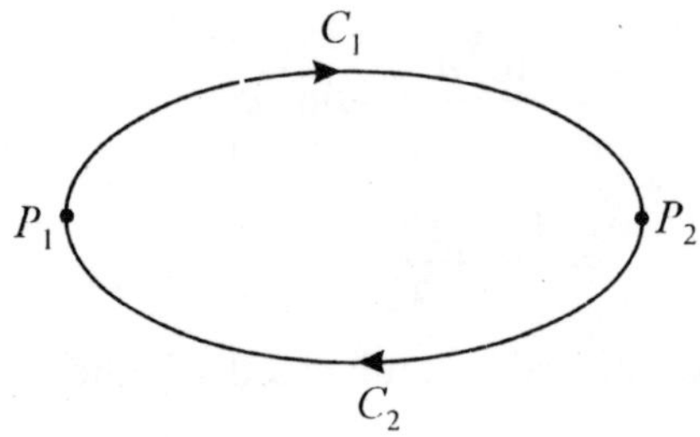

Then $C = C_1 \cup - C_2$ is a closed curve in R (from P_1 to P_1), and so $\oint_C f.dr = 0$. Thus,

$$0 = \oint_C f.dr$$

$$= \oint_C f.dr + \int_{-C_2} f.dr$$

$$= \oint_{C_1} f.dr - \int_{C_2} f.dr, \text{ and so}$$

$\oint_{C_1} f.dr = \int_{C_2} f.dr$. This proves path independence.

Conversely, suppose that the line integral $\oint_C f.dr$ is independent of the path between any two points in R. Let C be a closed curve contained in R. Let P_1 and P_2 be two distinct points on C. Let C_1 be a part of the curve C that goes from P_1 to P_2, and let C_2 be the remaining part of C that goes from P_1 to P_2. Then by path independence we have

$$\oint_{C_1} f.dr = \int_{C_2} f.dr$$

$$\oint_{C_1} f.dr - \int_{C_2} f.dr = 0$$

$$\oint_{C_1} f.dr + \int_{-C_2} f.dr = 0, \text{ so}$$

$$\oint_C f.dr = 0$$

since $C = C_1 \cup - C_2$.

Clearly, the above theorem does not give a practical way to determine path independence, since it is impossible to check the line integrals around all possible closed curves in a region. What it mostly does is give an idea of the way in which line integrals behave, and how seemingly unrelated line integrals can be related (in this case, a specific line integral between two points and *all* line integrals around closed curves).

For a more practical method for determining path independence, we first need a version of the Chain Rule for multivariable functions:

Theorem Chain Rule: If $z = f(x, y)$ is a continuously differentiable function of x and y, and both $x = x(t)$ and $y = y(t)$ are differentiable functions of t, then z is a differentiable function of t, and

$$\frac{dz}{dt} = \frac{\partial z}{\partial x}\frac{dx}{dt} + \frac{\partial z}{\partial y}\frac{dy}{dt}$$

at all points where the derivatives on the right are defined.

The proof is virtually identical to the proof of Theorem from Section (which uses the Mean Value Theorem), so we omit it. We will now use this Chain Rule to prove the following *sufficient* condition for path independence of line integrals:

Theorem: Let $f(x, y) = P(x, y)\, i + Q(x, y)\, j$ be a vector field in some region R, with P and Q continuously differentiable functions on R.

Let C be a smooth curve in R parametrized by $x = x(t)$, $y = y(t)$, $a \le t \le b$. Suppose that there is a real-valued function $F(x, y)$ such that $\nabla F = f$ on R. Then

$$\oint_C f.dr = F(B) - F(A),$$

where $A = (x(a), y(a))$ and $B = (x(b), y(b))$ are the endpoints of C. Thus, the line integral is independent of the path between its endpoints, since it depends only on the values of F at those endpoints.

Proof: By definition of $\oint_C f.dr$, we have

$$\oint_C f.dr = \int_a^b (P(x(t), y(t))x'(t) + Q(x(t), y(t))y'(t))dt$$

$$= \int_a^b \left(\frac{\partial F}{\partial x}\frac{dx}{dt} + \frac{\partial F}{\partial y}\frac{dy}{dt}\right) dt$$

(since $\nabla F = f \Rightarrow \dfrac{\partial F}{\partial x} = P$ and $\dfrac{\partial F}{\partial y} = Q$)

$$= \int_a^b F'(x(t), y(t))dt \text{ (by the Chain Rule in Theorem)}$$

$$= f(x(t), y(t))\Big|_a^b = F(B) - F(A)$$

by the Fundamental Theorem of Calculus.

Theorem can be thought of as the line integral version of the Fundamental Theorem of Calculus. A real-valued function $F(x, y)$ such that $\nabla F(x, y) = f(x, y)$ is called a potential for f. A conservative vector field is one which has a potential.

Example: Recall from Examples that the line integral $\int_C (x^2 + y^2)dx + 2xydy$ was found to have the value $\frac{13}{3}$ for three different curves C going from the point (0, 0) to the point (1, 2). Use Theorem to show that this line integral is indeed path independent.

Solution: We need to find a real-valued function $F(x, y)$ such that

$$\frac{\partial F}{\partial x} = x^2 + y^2 \text{ and } \frac{\partial F}{\partial y} = 2xy$$

Suppose that $\frac{\partial F}{\partial x} = x^2 + y^2$, Then we must have $F(x, y) = \frac{1}{3}x^3 + xy^2 + g(y)$ for some function $g(y)$. So $\frac{\partial F}{\partial y} = 2xy + g'(y)$ satisfies the condition $\frac{\partial F}{\partial y} = 2xy$ if $g'(y) = 0$, i.e. $g(y) = K$, where K is a constant. Since any choice for K will do (why?), we pick $K = 0$. Thus, a potential $F(x, y)$ for $f(x, y) = (x^2 + y^2)\, i + 2xy\, j$ exists, namely

$$F(x, y) = \frac{1}{3}x^3 + xy^2.$$

Hence the line integral $\int_C (x^2 + y^2)dx + 2xydy$ is path independent.

Note that we can also verify that the value of the line integral of f along any curve C going from (0, 0) to (1, 2) will always be $\frac{13}{3}$, since by Theorem

$$\int_C f.dr = F(1,2) - F(0,0) = \frac{1}{3}(1)^3 + (1)(2)^2 - (0+0) = \frac{1}{3} + 4 = \frac{13}{3}.$$

A consequence of Theorem in the special case where C is a closed curve, so that the endpoints A and B are the same point, is the following important corollary:

Corollary: If a vector field f has a potential in a region R, then $\oint_C f.dr = 0$ for any closed curve C in R (i.e. $\oint_C \nabla F.dr = 0$ for any real-valued function $F(x, y)$).

Example: Evaluate $\oint_C xdx + ydy$ for $C : x = 2\cos t,\ y = 3\sin t,\ 0 \le t \le 2\pi$.

Solution: The vector field $f(x, y) = x\, i + y\, j$ has a potential $F(x, y)$:

$$\frac{\partial F}{\partial x} = x \Rightarrow F(x, y) = \frac{1}{2}x^2 + g(y), \text{ so}$$

$$\frac{\partial F}{\partial y} = y \Rightarrow g'(y) = y \Rightarrow g(y) = \frac{1}{2}y^2 + K$$

for any constant K, so $F(x, y) = \frac{1}{2}x^2 + \frac{1}{2}y^2$ is a potential for $f(x, y)$. Thus,

$$\oint_C x\,dx + y\,dy = \oint_C f.dr = 0$$

by Corollary since the curve C is closed (it is the ellipse $\frac{x^2}{4} + \frac{y^2}{9} = 1$.

EXERCISES

1. Evaluate $\oint_C (x^2 + y^2)dx + 2xy$ for $C : x = \cos t, y = \sin t, 0 \leq t \leq 2\pi$.
2. Evaluate $\int_C (x^2 + y^2)dx + 2xy\,dy$ for $C : x = \cos t, y = \sin t, 0 \leq t \leq \pi$.
3. Is there a potential $F(x, y)$ for $f(x, y) = y\,i - x\,j$? If so, find one.
4. Is there a potential $F(x, y)$ for $f(x, y) = x\,i - y\,j$? If so, find one.
5. Is there a potential $F(x, y)$ for $f(x, y) = xy^2\,i + x^3y\,j$? If so, find one.
6. Let $f(x, y)$ and $g(x, y)$ be vector fields, let a and b be constants, and let C be a curve in R^2. Show that

$$\int_C (af \pm bg).dr = a\int_C f.dr \pm b\int_C g.dr.$$

7. Let C be a curve whose arc length is L. Show that $\int_C 1ds = L$.
8. Let $f(x, y)$ and $g(x, y)$ be continuously differentiable real-valued functions in a region R. Show that

$$\oint_C f\nabla g.dr = -\oint_C g\nabla f.dr$$

   for any closed curve C in R.

9. Let $f(x, y) = \frac{-y}{x^2 + y^2}i + \frac{x}{x^2 + y^2}j$ for all $(x, y) \neq (0, 0)$, and $C : x = \cos t, y = \sin t,$ $0 \leq t \leq 2\pi$.

   (a) Show that $f = \nabla F$, for $F(x, y) = \tan^{-1}(y / x)$.

   (b) Show that $\oint_C f.dr = 2\pi$.

10. Let $g(x)$ and $h(y)$ be differentiable functions, and let $f(x, y) = h(y)\,i + g(x)\,j$. Can f have a potential $F(x, y)$?

    If so, find it. You may assume that F would be smooth. (*Hint: Consider the mixed partial derivatives of F.*)

GREEN'S THEOREM

We will now see a way of evaluating the line integral of a *smooth* vector field around a simple closed curve. A vector field $f(x, y) = P(x, y)\, i + Q(x, y)\, j$ is smooth if its component functions $P(x, y)$ and $Q(x, y)$ are smooth. We will use *Green's Theorem*(sometimes called *Green's Theorem in the plane*) to relate the *line* integral around a closed curve with a *double* integral over the region inside the curve:

Theorem: Let R be a region in $\mathbb{R}^2$ whose boundary is a simple closed curve C which is piecewise smooth. Let $f(x, y) = P(x, y)\, i + Q(x, y)\, j$ be a smooth vector field defined on both R and C. Then

$$\oint_C f.dr = \iint_R \left(\frac{\partial Q}{\partial x} - \frac{\partial P}{\partial y}\right) dA.$$

where C is traversed so that R is always on the left side of C.

Proof: We will prove the theorem in the case for a *simple* region R, that is, where the boundary curve C can be written as $C = C_1 \cup C_2$ in two distinct ways:

C_1 = the curve $y = y_1(x)$ from the point X_1 to the point X_2

C_2 = the curve $y = y_2(x)$ from the point X_2 to the point X_1,

where X_1 and X_2 are the points on C farthest to the left and right, respectively; and

C_1 = the curve $x = x_1(y)$ from the point Y_2 to the point Y_1

C_2 = the curve $x = x_2(y)$ from the point Y_1 to the point Y_2,

where Y_1 and Y_2 are the lowest and highest points, respectively, on C.

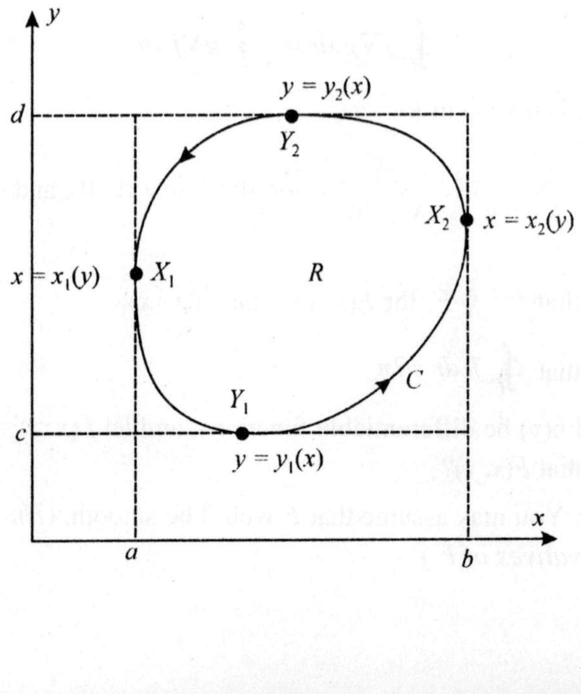

Integrate $P(x, y)$ around C using the representation $C = C_1 \cup C_2$. Since $y = y_1(x)$ along C_1 (as x goes from a to b) and $y = y_2(x)$ along C_2 (as x goes from b to a), then we have

$$\begin{aligned}\oint_C P(x,y)dx &= \int_{C_1} P(x,y)dx + \int_{C_2} P(x,y)dx \\ &= \int_a^b P(x,y_1(x))dx + \int_a^b P(x,y_2(x))dx \\ &= \int_a^b P(x,y_1(x))dx - \int_a^b P(x,y_2(x))dx \\ &= -\int_a^b (P(x,y_2(x))) - (P(x,y_1(x)))dx \\ &= -\int_a^b \left(P(x,y)\Big|_{y=y_1(x)}^{y=y_2(x)}\right)dx \\ &= -\int_a^b \int_{y_1(x)}^{y_2(x)} \frac{\partial P(x,y)}{\partial y} dydx\end{aligned}$$

(by the Fundamental Theorem of Calculus)

$$= -\iint_R \frac{\partial P}{\partial y} dA.$$

Likewise, integrate $Q(x, y)$ around C using the representation $C = C_1 \cup C_2$. Since $x = x_1(y)$ along C_1 (as y goes from d to c) and $x = x_2(y)$ along C_2 (as y goes from c to d), then we have

$$\begin{aligned}\oint_C Q(x,y)dy &= \int_{C_1} Q(x,y)dy + \int_{C_2} Q(x,y)dy \\ &= \int_d^c Q(x_1(y),y)dy + \int_c^d Q(x_2(y),y)dy \\ &= -\int_c^d Q(x_1(y),y)dy + \int_c^d Q(x_2(y),y)dy \\ &= \int_c^d (Q(x_2(y),y) - Q(x_1(y),y)), dy \\ &= \int_c^d \left(Q(x,y)\Big|_{x=x_1(y)}^{x=x_2(y)}\right)dy \\ &= \int_c^d \int_{x_1(y)}^{x_2(y)} \frac{\partial Q(x,y)}{\partial x} dxdy\end{aligned}$$

(by the Fundamental Theorem of Calculus)

$$= \iint_R \frac{\partial Q}{\partial x} dA \text{, and so}$$

$$= \oint_C f.dr = \oint_C P(x,y)dx + \oint_C Q(x,y)dy$$

$$= -\iint_R \frac{\partial P}{\partial y} dA + \iint_R \frac{\partial Q}{\partial x} dA$$

$$= \iint_R \left(\frac{\partial P}{\partial x} - \frac{\partial P}{\partial y} \right) dA.$$

Though we proved Green's Theorem only for a simple region R, the theorem can also be proved for more general regions.

Example: Evaluate $\oint_C (x^2 + y^2)dx + 2xydy$, where C is the boundary (traversed counterclockwise) of the region $R = \{(x, y) : 0 \leq x \leq 1, 2x^2 \leq y \leq 2x\}$.

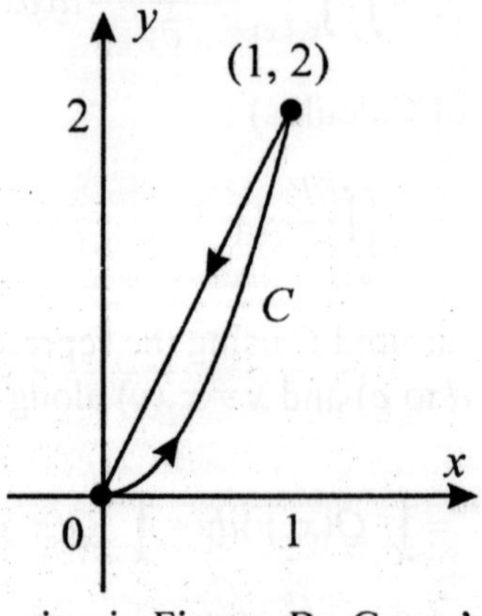

Solution: R is the shaded region in Figure. By Green's Theorem, for $P(x, y) = x^2 + y^2$ and $Q(x, y) = 2xy$, we have

$$\oint_C (x^2 + y^2)dx + 2xydy = \iint_R \left(\frac{\partial Q}{\partial x} - \frac{\partial P}{\partial y} \right) dA$$

$$= \iint_R (2y - 2y)dA = \iint_R 0dA = 0.$$

We actually already knew that the answer was zero. Recall from Example that the vector field $f(x, y) = (x^2 + y^2)\, i + 2xy\, j$ has a potential function $F(x, y) = \frac{1}{3}x^3 + xy^2$, and so $\oint_C f.dr = 0$.

Example: Let $f(x, y) = P(x, y)\, i + Q(x, y)\, j$, where

$$P(x, y) = \frac{-y}{x^2 + y^2} \text{ and } Q(x, y) = \frac{x}{x^2 + y^2},$$

and let $R = \{(x, y) : 0 < x^2 + y^2 \leq 1\}$. For the boundary curve $C : x^2 + y^2 = 1$, traversed counterclockwise, that $\oint_C f.dr = 2\pi$. But

$$\frac{\partial Q}{\partial x} = \frac{y^2 - x^2}{(x^2 + y^2)^2} = \frac{\partial P}{\partial y}$$

$$\Rightarrow \iint_R \left(\frac{\partial Q}{\partial x} - \frac{\partial P}{\partial y}\right) dA = \iint_R 0\, dA = 0.$$

This would seem to contradict Green's Theorem. However, note that R is not the *entire* region enclosed by C, since the point (0, 0) is not contained in R. That is, R has a "hole" at the origin, so Green's Theorem does not apply.

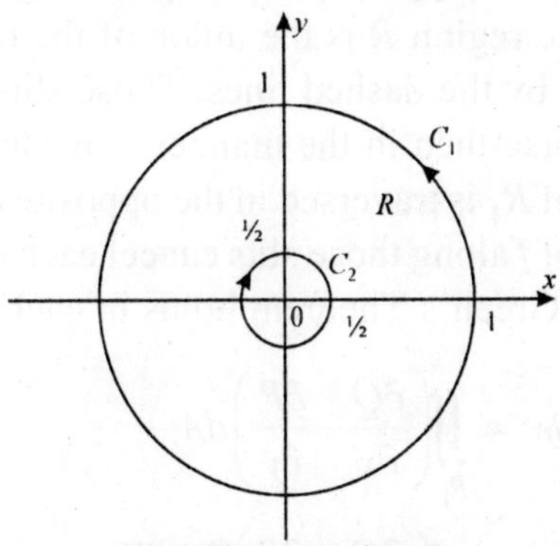

Fig. The Annulus R

If we modify the region R to be the *annulus* $R = \{ (x, y) : 1/4 \leq x^2 + y^2 \leq 1\}$, and take the "boundary" C of R to be $C = C_1 \cup C_2$, where C_1 is the unit circle $x^2 + y^2 = 1$ traversed counterclockwise and C_2 is the circle $x^2 + y^2 = 1/4$ traversed *clockwise*, then it can be shown that

$$\oint_C f.dr = 0$$

We would still have $\iint_R \left(\frac{\partial Q}{\partial x} - \frac{\partial P}{\partial y}\right) dA = 0$, so for this R we would have

$$\oint_C f.dr = \iint_R \left(\frac{\partial Q}{\partial x} - \frac{\partial P}{\partial y}\right) dA,$$

which shows that Green's Theorem holds for the annular region R.

It turns out that Green's Theorem can be extended to *multiply connected* regions, that is, regions like the annulus in Example, which have one or more *regions* cut out from the interior, as opposed to discrete *points* being cut out. For such regions, the "outer" boundary and the "inner" boundaries are traversed so that R is always on the left side.

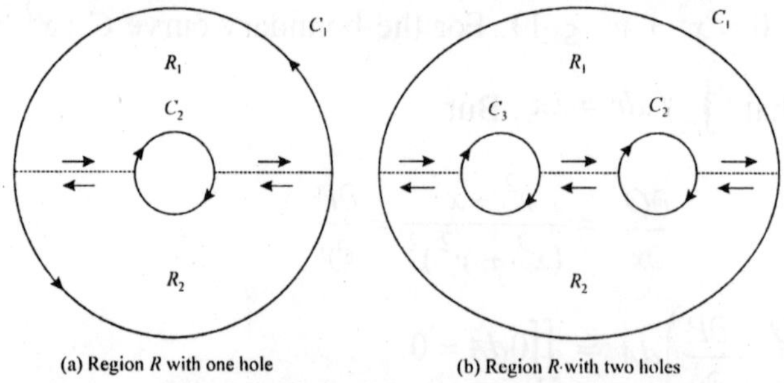

(a) Region R with one hole (b) Region R with two holes

Fig. Multiply Connected Regions

The intuitive idea for why Green's Theorem holds for multiply connected regions is shown in Figure above. The idea is to cut "slits" between the boundaries of a multiply connected region R so that R is divided into subregions which do *not* have any "holes". For example, in Figure (a) the region R is the union of the regions R_1 and R_2, which are divided by the slits indicated by the dashed lines. Those slits are part of the boundary of both R_1 and R_2, and we traverse then in the manner indicated by the arrows. Notice that along each slit the boundary of R_1 is traversed in the opposite direction as that of R_2, which means that the line integrals of f along those slits cancel each other out. Since R_1 and R_2 do not have holes in them, then Green's Theorem holds in each subregion, so that

$$\oint_{\substack{bdy\\ of\, R_1}} f.dr = \iint_{R_1}\left(\frac{\partial Q}{\partial x}-\frac{\partial P}{\partial y}\right)dA$$

and $$\oint_{\substack{bdy\\ of\, R_2}} f.dr = \iint_{R_2}\left(\frac{\partial Q}{\partial x}-\frac{\partial P}{\partial y}\right)dA.$$

But since the line integrals along the slits cancel out, we have

$$\oint_{C_1\cup C_2} f.dr = \oint_{\substack{bdy\\ of\, R_1}} f.dr + \oint_{\substack{bdy\\ of\, R_2}} f.dr,$$

and so $$\oint_{C_1\cup C_2} f.dr = \iint_{R_1}\left(\frac{\partial Q}{\partial x}-\frac{\partial P}{\partial y}\right)dA + \iint_{R_2}\left(\frac{\partial Q}{\partial x}-\frac{\partial P}{\partial y}\right)dA$$

$$= \iint_{R}\left(\frac{\partial Q}{\partial x}-\frac{\partial P}{\partial y}\right)dA,$$

which shows that Green's Theorem holds in the region R. A similar argument shows that the theorem holds in the region with two holes shown in Figure.

We know from Corollary that when a smooth vector field $f(x, y) = P(x, y)\, i + Q(x, y)\, j$ on a region R (whose boundary is a piecewise smooth, simple closed curve C) has a potential in R, then $\oint_C f.dr = 0$. And if the potential $F(x, y)$ is smooth in R, then $\frac{\partial F}{\partial x} = P$ and $\frac{\partial F}{\partial y} = Q$, and so we know that

$$\frac{\partial^2 F}{\partial y \partial x} = \frac{\partial^2 F}{\partial x \partial y} \Rightarrow \frac{\partial P}{\partial y} = \frac{\partial Q}{\partial x} \text{ in } R.$$

Conversely, if $\frac{\partial P}{\partial y} = \frac{\partial Q}{\partial x}$ in R then

$$\oint_C f.dr = \iint_R \left(\frac{\partial Q}{\partial x} - \frac{\partial P}{\partial y}\right) dA = \iint_R 0\, dA = 0.$$

For a simply connected region R (i.e. a region with no holes), the following can be shown: The following statements are equivalent for a simply connected region R in $\mathbb{R}^2$:

(a) $f(x, y) = P(x, y)\, i + Q(x, y)\, j$ has a smooth potential $F(x, y)$ in R

(b) $\int_C f.dr$ is independent of the path for any curve C in R

(c) $\oint_C f.dr = 0$ for every simple closed curve C in R

(d) $\frac{\partial P}{\partial y} = \frac{\partial Q}{\partial x}$ in R (in this case, the differential form $Pdx + Qdy$ is exact)

EXERCISES

For Exercises 1-4, use Green's Theorem to evaluate the given line integral around the curve C, traversed counterclockwise.

1. $\oint_C (x^2 - y^2)\, dx + 2xy\, dy$; C is the boundary of $R = \{(x, y) : 0 \le x \le 1, 2x^2 \le y \le 2x\}$
2. $\oint_C x^2 y dx + 2xy\, dy$; C is the boundary of $R = \{(x, y) : 0 \le x \le 1, x^2 \le y \le x\}$
3. $\oint_C 2y dx - 3x dy$; C is the circle $x^2 + y^2 = 1$
4. $\oint_C \left(e^{x^2} + y^2\right) dx + \left(e^{y^2} + x^2\right) dy$; C is the boundary of the triangle with vertices (0, 0), (4, 0) and (0, 4)
5. Is there a potential $F(x, y)$ for $f(x, y) = (y^2 + 3x^2)\, i + 2xy\, j$? If so, find one.
6. Is there a potential $F(x, y)$ for $f(x, y) = (x^3 \cos(xy) + 2x \sin(xy))i + x^2 y \cos(xy)j$? If so, find one.

7. Is there a potential $F(x, y)$ for $f(x, y) = (8xy + 3)i + 4(x^2 + y)j$? If so, find one.

8. Show that for any constants a, b and any closed simple curve C, $\oint_C adx + bdy = 0$.

10. Evaluate $\oint_C e^x \sin y dx + (y^3 + ex \cos y)dy$, where C is the boundary of the rectangle with vertices $(1, -1)$, $(1, 1)$, $(-1, 1)$ and $(-1, -1)$, traversed counter clockwise.

11. For a region R bounded by a simple closed curve C, show that the area A of R is

$$A = -\oint_C ydx = \oint_C xdy = \frac{1}{2}\oint_C xdy - ydx,$$

where C is traversed so that R is always on the left. (*Hint: Use Green's Theorem and the fact that* $A = \iint_R 1dA$.)

Surface Integrals and the Divergence Theorem

We have learned how to integrate along a curve. We will now learn how to perform integration over a *surface* in $\mathbb{R}^3$, such as a sphere or a paraboloid. Recall how we identified points (x, y, z) on a curve C in $\mathbb{R}^3$, parametrized by $x = x(t)$, $y = y(t)$, $z = z(t)$, $a \leq t \leq b$, with the terminal points of the position vector

$$r(t) = x(t)i + y(t)j + z(t)k \text{ for } t \text{ in } [a, b].$$

The idea behind a parametrization of a curve is that it "transforms" a subset of $\mathbb{R}^1$ (normally an interval $[a, b]$) into a curve in $\mathbb{R}^2$ or $\mathbb{R}^3$.

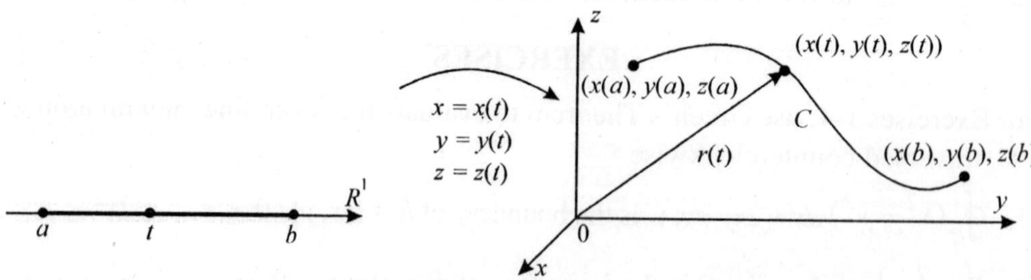

Fig. Parametrization of a Curve C in $\mathbb{R}^3$

Similar to how we used a parametrization of a curve to define the line integral along the curve, we will use a parametrization of a surface to define a *surface integral*. We will use *two* variables, u and v, to parametrize a surface Σ in $\mathbb{R}^3$: $x = x(u, v)$, $y = y(u, v)$, $z = z(u, v)$, for (u, v) in some region R in $\mathbb{R}^2$.

In this case, the position vector of a point on the surface Σ is given by the vectorvalued function

$$r(u, v) = x(u, v)i + y(u, v)j + z(u, v)k \text{ for } (u, v) \text{ in } R.$$

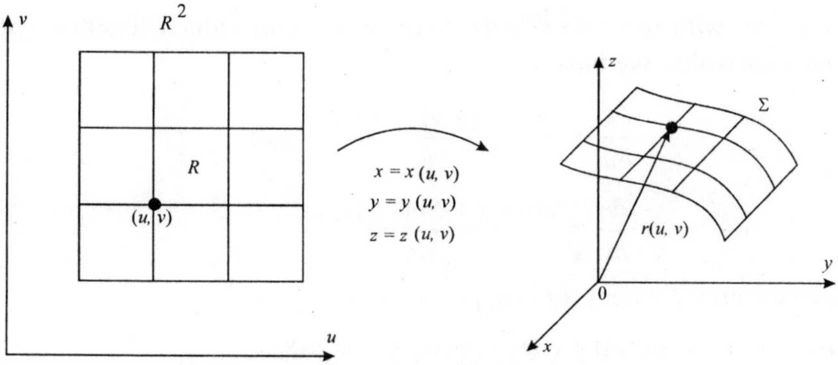

Fig. Parametrization of a Surface Σ in $\mathbb{R}^3$

Since $r(u, v)$ is a function of two variables, define the partial derivatives $\frac{\partial r}{\partial u}$ and $\frac{\partial r}{\partial v}$ for (u, v) in R by

$$\frac{\partial r}{\partial u}(u,v) = \frac{\partial x}{\partial u}(u,v)i + \frac{\partial y}{\partial u}(u,v)j + \frac{\partial z}{\partial u}(u,v)k, \text{ and}$$

$$\frac{\partial r}{\partial v}(u,v) = \frac{\partial x}{\partial v}(u,v)i + \frac{\partial y}{\partial v}(u,v)j + \frac{\partial z}{\partial v}(u,v)k .$$

The parametrization of Σ can be thought of as "transforming" a region in $\mathbb{R}^2$ (in the uv-plane) into a 2-dimensional surface in $\mathbb{R}^3$. This parametrization of the surface is sometimes called a *patch*, based on the idea of "patching" the region R onto Σ in the grid-like manner.

In fact, those gridlines in R lead us to how we will define a surface integral over Σ. Along the vertical gridlines in R, the variable u is constant. So those lines get mapped to curves on Σ, and the variable u is constant along the position vector r(u, v). Thus, the tangent vector to those curves at a point (u, v) is $\frac{\partial r}{\partial v}$. Similarly, the horizontal gridlines in R get mapped to curves on Σ whose tangent vectors are $\frac{\partial r}{\partial u}$.

Now take a point (u, v) in R as, say, the lower left corner of one of the rectangular grid sections in R. Suppose that this rectangle has a small width and height of Δu and Δv, respectively. The corner points of that rectangle are (u, v), $(u + \Delta u, v)$, $(u + \Delta u, v + \Delta v)$ and $(u, v + \Delta v)$. So the area of that rectangle is $A = \Delta u\, \Delta v$.

Then that rectangle gets mapped by the parametrization onto some section of the surface Σ which, for Δu and Δv small enough, will have a surface area (call it $d\sigma$) that is very close to the area of the parallelogram which has adjacent sides $r(u + \Delta u, v) - r(u, v)$ (corresponding to the line segment from (u, v) to $(u + \Delta u, v)$ in R) and $r(u, v + \Delta v) - r(u, v)$ (corresponding to the line segment from (u, v) to $(u, v + \Delta v)$ in R). But by combining our usual notion of a partial derivative with that of the derivative of a vector-valued function applied to a function of two variables, we have

$$\frac{\partial r}{\partial u} \approx \frac{r(u+\Delta u, v) - r(u,v)}{\Delta u}, \text{ and}$$

$$\frac{\partial r}{\partial v} \approx \frac{r(u+v+\Delta v) - r(u,v)}{\Delta v},$$

and so the surface area element $d\sigma$ is approximately

$$\|(r(u + \Delta u, v) - r(u, v)) \times (r(u, v + \Delta v) - r(u, v))\|$$

$$\approx \left\| \left(\Delta u \frac{\Delta r}{\Delta u}\right) \times \left(\Delta v \frac{\Delta r}{\Delta v}\right) \right\| = \left\| \frac{\Delta r}{\Delta u} \times \frac{\Delta r}{\Delta v} \right\| \Delta u \, \Delta v$$

by Theorem. Thus, the total surface area S of Σ is approximately the sum of all the quantities $\left\| \frac{\Delta r}{\Delta u} \times \frac{\Delta r}{\Delta v} \right\| \Delta u \, \Delta v$, summed over the rectangles in R. Taking the limit of that sum as the diagonal of the largest rectangle goes to 0 gives

$$S = \iint_R \left\| \frac{\partial r}{\partial u} \times \frac{\partial r}{\partial v} \right\| du \, dv.$$

We will write the double integral on the right using the special notation

$$\iint_\Sigma d\sigma = \iint_R \left\| \frac{\partial r}{\partial u} \times \frac{\partial r}{\partial v} \right\| du \, dv.$$

This is a special case of a *surface integral* over the surface Σ, where the surface area element $d\sigma$ can be thought of as 1 $d\sigma$. Replacing 1 by a general real-valued function $f(x, y, z)$ defined in $\mathbb{R}^3$, we have the following:

Definition: Let Σ be a surface in $\mathbb{R}^3$ parametrized by $x = x(u, v)$, $y = y(u, v)$, $z = z(u, v)$, for (u, v) in some region R in $\mathbb{R}^2$. Let $r(u, v) = x(u, v)i + y(u, v)j + z(u, v)k$ be the position vector for any point on Σ, and let $f(x, y, z)$ be a real-valued function defined on some subset of $\mathbb{R}^3$ that contains Σ. The surface integral of $f(x, y, z)$ over Σ is

$$\iint_\Sigma f(x,y,z) d\sigma = \iint_R f(x(u,v), y(u,v), z(u,v)) \left\| \frac{\partial r}{\partial u} \times \frac{\partial r}{\partial v} \right\| du \, dv.$$

In particular, the surface area S of Σ is

$$S = \iint_\Sigma 1 d\sigma.$$

Example: A *torus* T is a surface obtained by revolving a circle of radius a in the yz-

plane around the z-axis, where the circle's center is at a distance b from the z-axis ($0 < a < b$). Find the surface area of T.

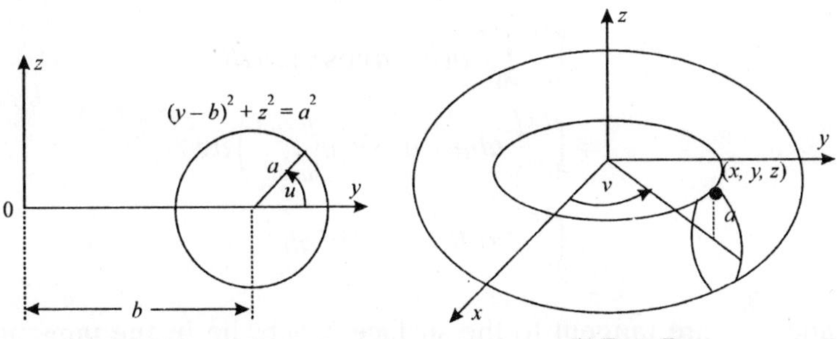

(a) Circle in the yz-plane (b) Torus T

Solution: For any point on the circle, the line segment from the center of the circle to that point makes an angle u with the y-axis in the positive y direction. And as the circle revolves around the z-axis, the line segment from the origin to the center of that circle sweeps out an angle v with the positive x-axis. Thus, the torus can be parametrized as:

$$x = (b + a \cos u) \cos v,\ y = (b + a \cos u) \sin v,\ z = a \sin u,\ 0 \le u \le 2\pi,\ 0 \le v \le 2\pi$$

So for the position vector

$$\begin{aligned} r(u, v) &= x(u, v)i + y(u, v)j + z(u, v)k \\ &= (b + a \cos u) \cos vi + (b + a \cos u) \sin vj + a \sin uk \end{aligned}$$

we see that

$$\frac{\partial r}{\partial u} = -a \sin u \cos vi - a \sin u \sin vj + a \cos uk$$

$$\frac{\partial r}{\partial v} = -(b + a \cos u) \sin vi + (b + a \cos u) \cos vj + 0k,$$

and so computing the cross product gives

$$\begin{aligned} \frac{\partial r}{\partial u} \times \frac{\partial r}{\partial v} &= -a(b + a \cos u) \cos v \cos ui \\ &\quad - a(b + a \cos u) \sin v \cos uj - a(b + a \cos u) \sin uk, \end{aligned}$$

which has magnitude

$$\left\| \frac{\partial r}{\partial u} \times \frac{\partial r}{\partial v} \right\| = a(b + a \cos u).$$

Thus, the surface area of T is

$$S = \iint_{\Sigma} 1 d\sigma$$

$$= \int_0^{2\pi}\int_0^{2\pi}\left\|\frac{\partial r}{\partial u}\times\frac{\partial r}{\partial v}\right\|du\,dv$$

$$= \int_0^{2\pi}\int_0^{2\pi} a(b+a\cos u)\,du\,dv$$

$$= \int_0^{2\pi}\left(abu + a^2 \sin u\Big|_{u=0}^{u=2\pi}\right)dv$$

$$= \int_0^{2\pi} 2\pi ab\,dv = 4\pi^2 ab$$

Since $\frac{\partial r}{\partial u}$ and $\frac{\partial r}{\partial v}$ are tangent to the surface Σ (i.e. lie in the tangent plane to Σ at each point on Σ), then their cross product $\frac{\partial r}{\partial u}\times\frac{\partial r}{\partial v}$ is perpendicular to the tangent plane to the surface at each point of Σ. Thus,

$$\iint_\Sigma f(x,y,z)d\sigma \iint_R f(x(u,v),y(u,v),z(u,v))\,\|\,n\,\|d\sigma,$$

where $n = \frac{\partial r}{\partial u}\times\frac{\partial r}{\partial v}$. We say that n is a normal vector to Σ.

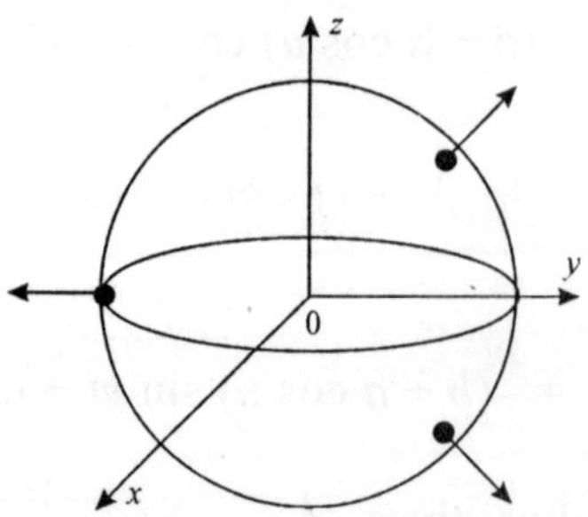

Recall that normal vectors to a plane can point in two opposite directions. By an outward unit normal vector to a surface Σ, we will mean the unit vector that is normal to and points away from the "top" (or "outer" part) of the surface. This is a hazy definition, but the picture in Figure gives a better idea of what outward normal vectors look like, in the case of a sphere. With this idea in mind, we make the following definition of a surface integral of a 3-dimensional *vector field* over a surface:

Definition: Let Σ be a surface in $\mathbb{R}^3$ and let $f(x, y, z) = f_1(x, y, z)i + f_2(x, y, z)j + f_3(x, y, z)k$ be a vector field defined on some subset of $\mathbb{R}^3$ that contains Σ. The surface integral of f over Σ is

$$\iint_\Sigma f.d\sigma = \iint_\Sigma f.nd\sigma,$$

where, at any point on Σ, n is the outward unit normal vector to Σ.

Note in the above definition that the dot product inside the integral on the right is a real-valued function, and hence we can use Definition to evaluate the integral.

Example: Evaluate the surface integral $\iint_{\Sigma} f.d\sigma$, where $f(x, y, z) = yzi + xzj + xyk$ and Σ is the part of the plane $x + y + z = 1$ with $x \geq 0, y \geq 0$, and $z \geq 0$, with the outward unit normal n pointing in the positive z direction.

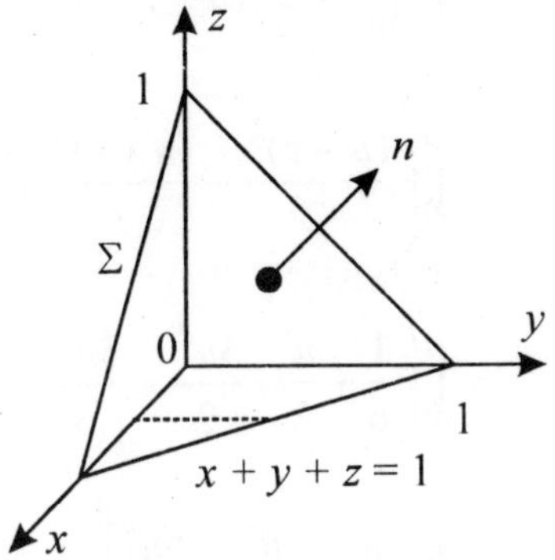

Solution: Since the vector $v = (1, 1, 1)$ is normal to the plane $x + y + z = 1$ (why?), then dividing v by its length yields the outward unit normal vector $n = \left(\frac{1}{\sqrt{3}}, \frac{1}{\sqrt{3}}, \frac{1}{\sqrt{3}}\right)$. We now need to parametrize Σ. As we can see from Figure, projecting Σ onto the xy-plane yields a triangular region $R = \{(x, y) : 0 \leq x \leq 1, 0 \leq y \leq 1 - x\}$. Thus, using (u, v) instead of (x, y), we see that

$$x = u, y = v, z = 1 - (u + v), \text{ for } 0 \leq u \leq 1, 0 \leq v \leq 1 - u$$

is a parametrization of Σ over R (since $z = 1 - (x + y)$ on Σ). So on Σ,

$$f \,.\, n = (yz, xz, xy) \,.\left(\frac{1}{\sqrt{3}}, \frac{1}{\sqrt{3}}, \frac{1}{\sqrt{3}}\right) = \frac{1}{\sqrt{3}}(yz + xz + xy)$$

$$= \frac{1}{\sqrt{3}}((x + y)z + xy) = \frac{1}{\sqrt{3}}((u + v)(1 - (u + v)) + uv)$$

$$= \frac{1}{\sqrt{3}}((u + v) - (u + v)^2 + uv)$$

for (u, v) in R, and for $r(u, v) = x(u, v)i + y(u, v)j + z(u, v)k = ui + vj + (1 - (u + v))k$ we have

$$\frac{\partial r}{\partial u} \times \frac{\partial r}{\partial v} = (1, 0, -1) \times (0, 1, -1) = (1, 1, 1)$$

$$\Rightarrow \left\| \frac{\partial r}{\partial u} \times \frac{\partial r}{\partial v} \right\| = \sqrt{3}.$$

Thus, integrating over R using vertical slices gives

$$\iint_{\Sigma} f.d\sigma = \iint_{\Sigma} f.nd\sigma$$

$$= \iint_{R} (f(x(u,v), y(u,v), z(u,v)).n) \left\| \frac{\partial r}{\partial u} \times \frac{\partial r}{\partial v} \right\| dv\, du$$

$$= \int_0^1 \int_0^{1-u} \frac{1}{\sqrt{3}} ((u+v) - (u+v)^2 + uv)\sqrt{3} dv\, du$$

$$= \int_0^1 \left(\frac{(u+v)^2}{2} - \frac{(u+v)^3}{3} + \frac{uv^2}{2} \Bigg|_{v=0}^{v=1-u} \right) du$$

$$= \int_0^1 \left(\frac{1}{6} + \frac{u}{2} - \frac{3u^2}{2} + \frac{5u^3}{6} \right) du$$

$$= \frac{u}{6} + \frac{u^2}{4} - \frac{u^3}{2} + \frac{5u^4}{24} \Bigg|_0^1 = \frac{1}{8}.$$

Computing surface integrals can often be tedious, especially when the formula for the outward unit normal vector at each point of Σ changes.

The following theorem provides an easier way in the case when Σ is a closed surface, that is, when Σ encloses a bounded solid in $\mathbb{R}^3$.

For example, spheres, cubes, and ellipsoids are closed surfaces, but planes and paraboloids are not.

Theorem: (Divergence Theorem) Let Σ be a closed surface in $\mathbb{R}^3$ which bounds a solid S, and let $f(x, y, z) = f_1(x, y, z)i + f_2(x, y, z)j + f_3(x, y, z)k$ be a vector field defined on some subset of $\mathbb{R}^3$ that contains Σ. Then

$$\iint_{\Sigma} f.d\sigma = \iiint_{S} \text{div}\, f dV,$$

where

$$\text{div}\, f = \frac{\partial f_1}{\partial x} + \frac{\partial f_2}{\partial y} + \frac{\partial f_3}{\partial z}$$

is called the divergence of f.

The proof of the Divergence Theoremis very similar to the proof of Green's Theorem, i.e. it is first proved for the simple case when the solid S is bounded above by one surface, bounded below by another surface, and bounded laterally by one or more surfaces. The proof can then be extended to more general solids.

Example: Evaluate $\iint_{\Sigma} f.d\sigma$, where $f(x, y, z) = xi + yj + zk$ and is the unit sphere $x^2 + y^2 + z^2 = 1$.

Solution: We see that div $f = 1 + 1 + 1 = 3$, so

$$\iint_{\Sigma} f.d\sigma = \iiint_{S} \text{div } f dV = \iiint_{S} 3dV$$

$$= 3\iiint_{S} 1\,dV = 3vol(S) = 3.\frac{4\pi(1)^3}{3} = 4\pi.$$

In physical applications, the surface integral $\iint_{\Sigma} f.d\sigma$ is often referred to as the flux of f through the surface Σ. For example, if f represents the velocity field of a fluid, then the flux is the net quantity of fluid to flow through the surface Σ per unit time. A positive flux means there is a net flow *out* of the surface (i.e. in the direction of the outward unit normal vector n), while a negative flux indicates a net flow inward (in the direction of $-n$).

The term divergence comes from interpreting div f as a measure of how much a vector field "diverges" from a point. This is best seen by using another definition of div f which is equivalent to the definition given by formula. Namely, for a point (x, y, z) in $\mathbb{R}^3$,

$$\text{div } f(x, y, z) = \lim_{V \to 0} \frac{1}{V} \iint_{\Sigma} f.d\sigma,$$

where V is the volume enclosed by a closed surface Σ around the point (x, y, z). In the limit, $V \to 0$ means that we take smaller and smaller closed surfaces around (x, y, z), which means that the volumes they enclose are going to zero. It can be shown that this limit is independent of the shapes of those surfaces. Notice that the limit being taken is of the ratio of the flux through a surface to the volume enclosed by that surface, which gives a rough measure of the flow "leaving" a point, as we mentioned. Vector fields which have zero divergence are often called *solenoidal* fields. The following theorem is a simple consequence of formula.

Theorem: If the flux of a vector field f is zero through every closed surface containing a given point, then div $f = 0$ at that point.

Proof: By formula, at the given point (x, y, z) we have

$$\text{div } f(x, y, z) = \lim_{V \to 0} \frac{1}{V} \iint_{\Sigma} f.d\sigma \text{ for closed surfaces } \Sigma \text{ containing } (x, y, z), \text{ so}$$

$$= \lim_{V \to 0} \frac{1}{V} (0) \text{ by our assumption that the flux through each } \Sigma \text{ is zero, so}$$

$$= \lim_{V \to 0} 0 = 0.$$

Lastly, we note that sometimes the notation

$$\oiint_{\Sigma} f(x, y, z)d\sigma \quad \text{and} \quad \oiint_{\Sigma} f.d\sigma$$

is used to denote surface integrals of scalar and vector fields, respectively, over closed surfaces. Especially in physics texts, it is common to see simply $\oint_{\Sigma}$ instead of $\oiint_{\Sigma}$.

EXERCISES

For Exercises 1-4, use the Divergence Theorem to evaluate the surface integral $\oint_{\Sigma} f.d\sigma$ of the given vector field f(x, y, z) over the surface Σ.

1. $f(x, y, z) = xi + 2yj + 3zk,\ \Sigma : x^2 + y^2 + z^2 = 9$
2. $f(x, y, z) = xi + yj + zk,\ \Sigma$: boundary of the solid cube $S = \{(x, y, z) : 0 \leq x, y, z \leq 1\}$
3. $f(x, y, z) = x^3 i + y^3 j + z^3 k,\ \Sigma : x^2 + y^2 + z^2 = 1$
4. $f(x, y, z) = 2i + 3j + 5k,\ \Sigma : x^2 + y^2 + z^2 = 1$
5. Show that the flux of any constant vector field through any closed surface is zero.
7. Evaluate the surface integral $\iint_{\Sigma} f.d\sigma$, where $f(x, y, z) = x^2 i + xyj + zk$ and Σ is the part of the plane $6x + 3y + 2z = 6$ with $x \geq 0$, $y \geq 0$, and $z \geq 0$, with the outward unit normal n pointing in the positive z direction.
8. Use a surface integral to show that the surface area of a sphere of radius r is $4\pi r^2$. (*Hint: Use spherical coordinates to parametrize the sphere.*)
9. Use a surface integral to show that the surface area of a right circular cone of radius R and height h is $\Sigma R\sqrt{h^2 + R^2}$. (*Hint: Use the parametrization* $x = r\cos\theta$, $y = r\sin\theta$, $z = \frac{h}{R}r$, *for* $0 \leq r \leq R$ *and* $0 \leq \theta \leq 2\pi$.)
10. The ellipsoid $\frac{x^2}{a^2} + \frac{y^2}{b^2} + \frac{z^2}{c^2} = 1$ can be parametrized using *ellipsoidal coordinates*

$$x = a\sin\phi\cos\theta,\ y = b\sin\phi\sin\theta,\ z = c\cos\phi,$$

$$\text{for } 0 \leq \theta \leq 2\pi \text{ and } 0 \leq \phi \leq \pi.$$

Show that the surface area S of the ellipsoid is

$$S = \int_0^{\pi}\int_0^{2\pi} \sin\phi\sqrt{a^2b^2\cos^2\phi + c^2(a^2\sin^2\theta + b^2\cos^2\theta)\sin^2\phi}\,d\theta d\phi.$$

Stokes' Theorem

So far the only types of line integrals which we have discussed are those along curves in $\mathbb{R}^2$. But the definitions and properties which were covered can easily be extended to include functions of three variables, so that we can now discuss line integrals along curves in $\mathbb{R}^3$.

Definition: For a real-valued function $f(x, y, z)$ and a curve C in $\mathbb{R}^3$, parametrized by $x = x(t)$, $y = y(t)$, $z = z(t)$, $a \leq t \leq b$, the line integral of $f(x, y, z)$ along C with respect to arc length s is

$$\int_C f(x,y,z)ds = \int_a^b f(x(t),y(t),z(t))\sqrt{x'(t)^2 + y'(t)^2 + z'(t)^2}\,dt.$$

The line integral of $f(x, y, z)$ along C with respect to x is

$$\int_C f(x,y,z)dx = \int_a^b f(x(t),y(t),z(t))x'(t)dt.$$

The line integral of $f(x, y, z)$ along C with respect to y is

$$\int_C f(x,y,z)dy = \int_a^b f(x(t),y(t),z(t))y'(t)dt.$$

The line integral of $f(x, y, z)$ along C with respect to z is

$$\int_C f(x,y,z)dz = \int_a^b f(x(t),y(t),z(t))z'(t)dt.$$

Similar to the two-variable case, if $f(x, y, z) \geq 0$ then the line integral $\int_C f(x,y,z)ds$ can be thought of as the total area of the "picket fence" of height $f(x, y, z)$ at each point along the curve C in $\mathbb{R}^3$. Vector fields in $\mathbb{R}^3$ are defined in a similar fashion to those in $\mathbb{R}^2$, which allows us to define the line integral of a vector field along a curve in $\mathbb{R}^3$.

Definition: For a vector field $f(x, y, z) = P(x, y, z)\, i + Q(x, y, z)\, j + R(x, y, z)\, k$ and a curve C in $\mathbb{R}^3$ with a smooth parametrization $x = x(t)$, $y = y(t)$, $z = z(t)$, $a \leq t \leq b$, the line integral of f along C is

$$\int_C f.dr = \int_C P(x,y,z)dx + \int_C Q(x,y,z)dy + \int_C R(x,y,z)dz$$

$$= \int_a^b f(x(t),y(t),z(t)).r'(t)dt,$$

where $r(t) = x(t)\, i + y(t)\, j + z(t)\, k$ is the position vector for points on C.

Similar to the two-variable case, if $f(x, y, z)$ represents the force applied to an object at a point (x, y, z) then the line integral $\int_C f.dr$ represents the work done by that force in moving the object along the curve C in $\mathbb{R}^3$.

Some of the most important results we will need for line integrals in $\mathbb{R}^3$ are stated below without proof (the proofs are similar to their two-variable equivalents).

Theorem: For a vector field $f(x, y, z) = P(x, y, z)\, i + Q(x, y, z)\, j + R(x, y, z)k$ and a curve C with a smooth parametrization $x = x(t)$, $y = y(t)$, $z = z(t)$, $a \leq t \leq b$ and position vector $r(t) = x(t)i + y(t)j + z(t)k$,

$$\int_C f.dr = \int_C f.Tds,$$

where $T(t) = \dfrac{r'(t)}{\| r'(t) \|}$ is the unit tangent vector to C at $(x(t), y(t), z(t))$.

Theorem: (Chain Rule) If $w = f(x, y, z)$ is a continuously differentiable function of x, y, and z, and $x = x(t)$, $y = y(t)$ and $z = z(t)$ are differentiable functions of t, then w is a differentiable function of t, and

$$\frac{dw}{dt} = \frac{\partial w}{\partial x}\frac{dx}{dt} + \frac{\partial w}{\partial y}\frac{dy}{dt} + \frac{\partial w}{\partial z}\frac{dz}{dt}.$$

Also, if $x = x(t_1, t_2)$, $y = y(t_1, t_2)$ and $z = z(t_1, t_2)$ are continuously differentiable function of (t_1, t_2), then

$$\frac{\partial w}{\partial t_1} = \frac{\partial w}{\partial x}\frac{\partial x}{\partial t_1} + \frac{\partial w}{\partial y}\frac{\partial y}{\partial t_1} + \frac{\partial w}{\partial z}\frac{\partial z}{\partial t_1}$$

and

$$\frac{\partial w}{\partial t_2} = \frac{\partial w}{\partial x}\frac{\partial x}{\partial t_2} + \frac{\partial w}{\partial y}\frac{\partial y}{\partial t_2} + \frac{\partial w}{\partial z}\frac{\partial z}{\partial t_2}.$$

Theorem: Let $f(x, y, z) = P(x, y, z)i + Q(x, y, z)j + R(x, y, z)k$ be a vector field in some solid S, with P, Q and R continuously differentiable functions on S. Let C be a smooth curve in S parametrized by $x = x(t)$, $y = y(t)$, $z = z(t)$, $a \leq t \leq b$. Suppose that there is a real-valued function $F(x, y, z)$ such that $\nabla F = f$ on S. Then

$$\int_C f.dr = F(B) - F(A),$$

where $A = (x(a), y(a), z(a))$ and $B = (x(b), y(b), z(b))$ are the endpoints of C.

Corollary: If a vector field f has a potential in a solid S, then $\int_C f.dr \mid = 0$ for any closed curve C in S (i.e. $\int_C \nabla F.dr = 0$ for any real-valued function $F(x, y, z)$).

Example: Let $f(x, y, z) = z$ and let C be the curve in $\mathbb{R}^3$ parametrized by

$$x = t \sin t,\ y = t \cos t,\ z = t,\ 0 \leq t \leq 8\pi.$$

Evaluate $\int_C f(x, y, z)ds$.

Solution: Since $x'(t) = \sin t + t\cos t$, $y'(t) = \cos t - t\sin t$, and $z'(t) = 1$, we have

$$\begin{aligned} x'(t)^2 + y'(t)^2 + z'(t)^2 &= (\sin^2 t + 2t\sin t\cos t + t^2\cos^2 t) \\ &\quad + (\cos^2 t - 2t\sin t\cos t + t^2\sin^2 t) + 1 \\ &= t^2(\sin^2 t + \cos^2 t) + \sin^2 t + \cos^2 t + 1 \\ &= t^2 + 2, \end{aligned}$$

so since $f(x(t), y(t), z(t)) = z(t) = t$ along the curve C, then

$$\begin{aligned} \int_C f(x,y,z)ds. &= \int_0^{8\pi} f(x(t),y(t),z(t))\sqrt{x'(t)^2 + y'(t)^2 + z'(t)^2}\,dt \\ &= \int_0^{8\pi} t\sqrt{t^2 + 2}dt \\ &= \left(\frac{1}{3}(t^2+2)^{3/2}\right)\Bigg|_0^{8\pi} = \frac{1}{3}\left((64\pi^2+2)^{3/2} - 2\sqrt{2}\right). \end{aligned}$$

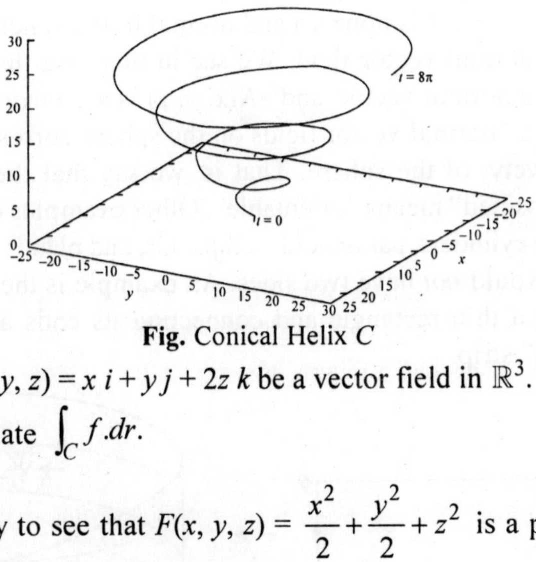

Fig. Conical Helix C

Example: Let $f(x, y, z) = x\,i + y\,j + 2z\,k$ be a vector field in $\mathbb{R}^3$. Using the same curve C from Example, evaluate $\int_C f.dr$.

Solution: It is easy to see that $F(x, y, z) = \frac{x^2}{2} + \frac{y^2}{2} + z^2$ is a potential for $f(x, y, z)$ (i.e. $\nabla F = f$). So by Theorem we know that

$$\begin{aligned} \int_C f.dr &= F(B) - F(A), \text{ where } A = (x(0), y(0), z(0)) \text{ and } B = (x(8\pi), y(8\pi), z(8\pi)), \text{ so} \\ &= F(8\pi\sin 8\pi, 8\pi\cos 8\pi, 8\pi) - F(0\sin 0, 0\cos 0, 0) \\ &= F(0, 8\pi, 8\pi) - F(0, 0, 0) \\ &= 0 + \frac{(8\pi)^2}{2} + (8\pi)^2 - (0 + 0 + 0) = 96\pi^2. \end{aligned}$$

We will now discuss a generalization of Green's Theorem in $\mathbb{R}^2$ to *orientable* surfaces in $\mathbb{R}^3$, called *Stokes' Theorem.* A surface Σ in $\mathbb{R}^3$ is orientable if there is a continuous vector field N in $\mathbb{R}^3$ such that N is nonzero and normal to Σ (i.e. perpendicular to the tangent plane) at each point of Σ. We say that such an N is a *normal vector field.*

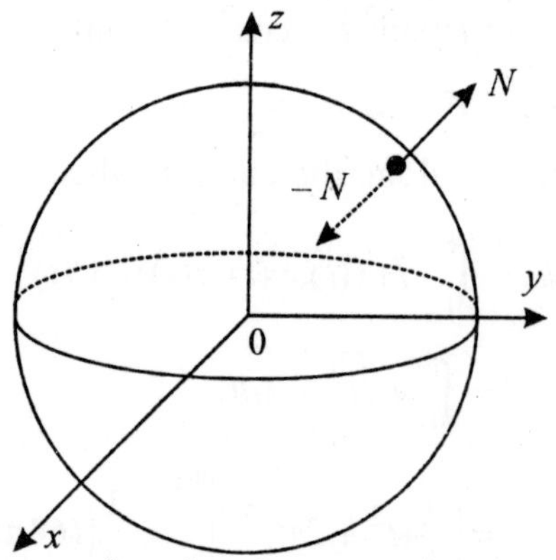

For example, the unit sphere $x^2 + y^2 + z^2 = 1$ is orientable, since the continuous vector field $N(x, y, z) = x\,i + y\,j + z\,k$ is nonzero and normal to the sphere at each point. In fact, $-N(x, y, z)$ is another normal vector field. We see in this case that $N(x, y, z)$ is what we have called an outward normal vector, and $-N(x, y, z)$ is an *inward* normal vector. These "outward" and "inward" normal vector fields on the sphere correspond to an "outer" and "inner" side, respectively, of the sphere. That is, we say that the sphere is a *two-sided* surface. Roughly, "twosided" means "orientable". Other examples of two-sided, and hence orientable, surfaces are cylinders, paraboloids, ellipsoids, and planes. You may be wondering what kind of surface would *not* have two sides. An example is the Möbius strip, which is constructed by taking a thin rectangle and connecting its ends at the opposite corners, resulting in a "twisted" strip.

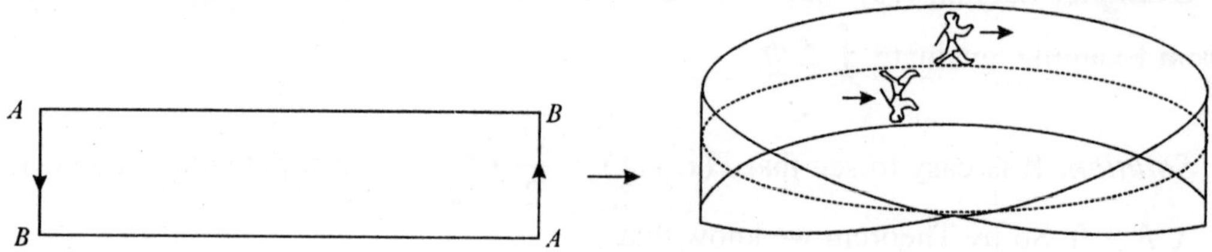

(a) Connect A to A and B to B along the ends (b) Not orientable

Fig. Möbius Strip

If you imagine walking along a line down the center of theMöbius strip, then you arrive back at the same place from which you started but upside down! That is, your *orientation* changed even though your motion was continuous along that center line. Informally, thinking of your vertical direction as a normal vector field along the strip, there is a discontinuity at your starting point (and, in fact, at every point) since your vertical direction takes two different values there. The Möbius strip has only *one side*, and hence is nonorientable.

For an orientable surface Σ which has a boundary curve C, pick a unit normal vector n such that if you walked along C with your head pointing in the direction of n, then the surface would be on your left. We say in this situation that n is a *positive unit normal vector* and that C is traversed *n-positively*. We can now state Stokes' Theorem:

Theorem: (Stokes' Theorem) Let Σ be an orientable surface in $\mathbb{R}^3$ whose boundary is a simple closed curve C, and let $f(x, y, z) = P(x, y, z)i + Q(x, y, z)j + R(x, y, z)k$ be a smooth vector field defined on some subset of $\mathbb{R}^3$ that contains Σ. Then

$$\int_C f.dr = \iint_{\Sigma} (\text{curl} f).n\, d\sigma,$$

where

$$\text{curl} f = \left(\frac{\partial R}{\partial y} - \frac{\partial Q}{\partial z}\right) i + \left(\frac{\partial P}{\partial z} - \frac{\partial R}{\partial x}\right) j + \left(\frac{\partial Q}{\partial x} - \frac{\partial P}{\partial y}\right) k,$$

n is a positive unit normal vector over Σ, and C is traversed n-positively.

Proof: As the general case is beyond the scope of this text, we will prove the theorem only for the special case where Σ is the graph of $z = z(x, y)$ for some smooth real-valued function $z(x, y)$, with (x, y) varying over a region D in $\mathbb{R}^2$.

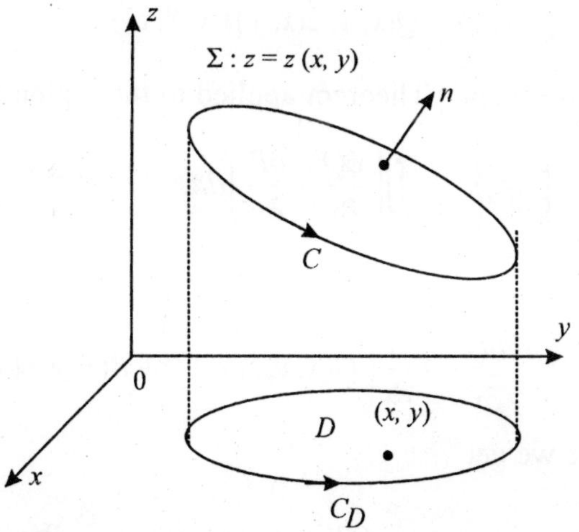

Projecting Σ onto the xy-plane, we see that the closed curve C (the boundary curve of Σ) projects onto a closed curve C_D which is the boundary curve of D. Assuming that C has a smooth parametrization, its projection C_D in the xy-plane also has a smooth parametrization, say

$$CD : x = x(t),\ y = y(t),\ a \le t \le b,$$

and so C can be parametrized (in $\mathbb{R}^3$) as

$$C : x = x(t),\ y = y(t),\ z = z(x(t), y(t)),\ a \le t \le b,$$

since the curve C is part of the surface $z = z(x, y)$. Now, by the Chain Rule, for $z = z(x(t), y(t))$ as a function of t, we know that

$$z'(t) = \frac{\partial z}{\partial x}x'(t) + \frac{\partial z}{\partial y}y'(t),$$

and so

$$\int_C f.dr = \int_C P(x,y,z)dx + Q(x,y,z)dy + R(x,y,z)dz$$

$$= \int_a^b \left(Px'(t) + Qy'(t) + R\left(\frac{\partial z}{\partial x}x'(t) + \frac{\partial z}{\partial y}y'(t)\right)\right) dt$$

$$= \int_a^b \left(\left(P + R\frac{\partial z}{\partial x}\right)x'(t) + \left(Q + R\frac{\partial z}{\partial y}\right)y'(t)\right) dt$$

$$= \int_{C_D} P(x,y)dx + Q(x,y)dy,$$

where $\tilde{P}(x, y) = P(x, y, z(x, y)) + R(x, y, z(x, y)) \dfrac{\partial z}{\partial x}(x, y)$, and

$$\tilde{Q}(x, y) = Q(x, y, z(x, y)) + R(x, y, z(x, y))\frac{\partial z}{\partial y}(x, y)$$

for (x, y) in D. Thus, by Green's Theorem applied to the region D, we have

$$\oint_C f.dr = \iint_D \left(\frac{\partial \tilde{Q}}{\partial x} - \frac{\partial \tilde{P}}{\partial y}\right) dA.$$

Thus,

$$\frac{\partial \tilde{Q}}{\partial x} = \frac{\partial}{\partial x}\left(Q(x,y,z(x,y)) + R(x,y,z(x,y))\frac{\partial z}{\partial y}(x,y)\right),$$

so by the Product Rule we get

$$= \frac{\partial}{\partial x}(Q(x,y,z(x,y))) + \left(\frac{\partial}{\partial x}R(x,y,z(x,y))\right)\frac{\partial z}{\partial y}(x,y)$$

$$+ R(x,y,z(x,y))) + \frac{\partial}{\partial x}\left(\frac{\partial z}{\partial y}(x,y)\right).$$

Now, by formula in Theorem, we have

$$\frac{\partial}{\partial x}(Q(x,y,z(x,y))) = \frac{\partial Q}{\partial x}\frac{\partial x}{\partial x} + \frac{\partial Q}{\partial y}\frac{\partial y}{\partial x} + \frac{\partial Q}{\partial z}\frac{\partial z}{\partial x}$$

$$= \frac{\partial Q}{\partial x}.1 + \frac{\partial Q}{\partial y}.0 + \frac{\partial Q}{\partial z}\frac{\partial z}{\partial x}$$

$$= \frac{\partial Q}{\partial x} + \frac{\partial Q}{\partial z} + \frac{\partial z}{\partial x}.$$

Similarly,

$$\frac{\partial}{\partial x}(R(x, y, z(x, y))) = \frac{\partial R}{\partial x} + \frac{\partial R}{\partial z}\frac{\partial z}{\partial x}.$$

Thus,

$$\frac{\partial \tilde{Q}}{\partial x} = \frac{\partial Q}{\partial x} + \frac{\partial Q}{\partial z}\frac{\partial z}{\partial x} + \left(\frac{\partial R}{\partial x} + \frac{\partial R}{\partial z}\frac{\partial z}{\partial x}\right)\frac{\partial z}{\partial y} + R(x, y, z(x, y))\frac{\partial^2 z}{\partial x \partial y}$$

$$= \frac{\partial Q}{\partial x} + \frac{\partial Q}{\partial z}\frac{\partial z}{\partial x} + \frac{\partial R}{\partial x}\frac{\partial z}{\partial y} + \frac{\partial R}{\partial z}\frac{\partial z}{\partial x}\frac{\partial z}{\partial y} + R\frac{\partial^2 z}{\partial x \partial y}.$$

In a similar fashion, we can calculate

$$\frac{\partial \tilde{P}}{\partial y} = \frac{\partial P}{\partial y} + \frac{\partial P}{\partial z} + \frac{\partial z}{\partial y} + \frac{\partial R}{\partial y}\frac{\partial z}{\partial x} + \frac{\partial R}{\partial z}\frac{\partial z}{\partial y}\frac{\partial z}{\partial x} + R\frac{\partial^2 z}{\partial y \partial x}.$$

So subtracting gives

$$\frac{\partial \tilde{Q}}{\partial x} - \frac{\partial \tilde{P}}{\partial y} = \left(\frac{\partial Q}{\partial z} - \frac{\partial R}{\partial y}\right)\frac{\partial z}{\partial x} - \left(\frac{\partial P}{\partial z} - \frac{\partial R}{\partial x}\right)\frac{\partial z}{\partial y} + \left(\frac{\partial Q}{\partial x} - \frac{\partial P}{\partial y}\right)$$

since $\frac{\partial^2 z}{\partial x \partial y} = \frac{\partial^2 z}{\partial y \partial x}$ by the smoothness of $z = z(x, y)$. Hence, by equation,

$$\oint_C f dr. = \iint_D \left(\left(\frac{\partial R}{\partial y} - \frac{\partial Q}{\partial z}\right)\frac{\partial z}{\partial x} - \left(\frac{\partial P}{\partial z} - \frac{\partial R}{\partial x}\right)\frac{\partial z}{\partial y} + \left(\frac{\partial Q}{\partial x} - \frac{\partial P}{\partial y}\right)\right) dA$$

after factoring out a -1 from the terms in the first two products in equation.

Now, recall that the vector $N = -\frac{\partial z}{\partial x}i - \frac{\partial z}{\partial y}j + k$ is normal to the tangent plane to the surface $z = z(x, y)$ at each point of Σ. Thus,

$$n = \frac{N}{\| N \|} = \frac{-\frac{\partial z}{\partial x}i - \frac{\partial z}{\partial y}j + k}{\sqrt{1 + \left(\frac{\partial z}{\partial x}\right)^2 + \left(\frac{\partial z}{\partial y}\right)^2}}$$

is in fact a positive unit normal vector to Σ. Hence, using the parametrization $r(x, y) = x\,i + y\,j + z(x, y)\,k$, for (x, y) in D, of the surface Σ, we have $\frac{\partial r}{\partial x} = i + \frac{\partial z}{\partial x}k$ and $\frac{\partial r}{\partial y} = j + \frac{\partial z}{\partial y}k$, and so $\left\|\frac{\partial r}{\partial x} \times \frac{\partial r}{\partial y}\right\| = \sqrt{1 + \left(\frac{\partial z}{\partial x}\right)^2 + \left(\frac{\partial z}{\partial y}\right)^2}$. So we see that using formula for curl f, we have

$$\iint_\Sigma (\text{curl}\, f).n\,d\sigma = \iint_D (\text{durl}\, f).n\left\|\frac{\partial r}{\partial x} \times \frac{\partial r}{\partial y}\right\| dA$$

$$\iint_D \left(\left(\frac{\partial R}{\partial y} - \frac{\partial Q}{\partial z}\right)i + \left(\frac{\partial P}{\partial z} - \frac{\partial R}{\partial x}\right)j + \left(\frac{\partial Q}{\partial x} - \frac{\partial P}{\partial y}\right)k\right).\left(-\frac{\partial z}{\partial x}i - \frac{\partial z}{\partial y}j + k\right) dA$$

$$\iint_D \left(-\left(\frac{\partial R}{\partial y} - \frac{\partial Q}{\partial z}\right)\frac{\partial z}{\partial x} - \left(\frac{\partial P}{\partial z} - \frac{\partial R}{\partial x}\right)\frac{\partial z}{\partial y} + \left(\frac{\partial Q}{\partial x} - \frac{\partial P}{\partial y}\right)\right) dA,$$

which, upon comparing to equation, proves the Theorem.

Note: The condition in Stokes' Theorem that the surface Σ have a (continuously varying) positive unit normal vector n and a boundary curve C traversed n-positively can be expressed more precisely as follows: if $r(t)$ is the position vector for C and $T(t) = r'(t)/\|r'(t)\|$ is the unit tangent vector to C, then the vectors T, n, $T \times n$ form a right-handed system.

Also, it should be noted that Stokes' Theorem holds even when the boundary curve C is piecewise smooth.

Example: Verify Stokes' Theorem for $f(x, y, z) = z\,i + x\,j + yk$ when Σ is the paraboloid $z = x^2 + y^2$ such that $z \leq 1$.

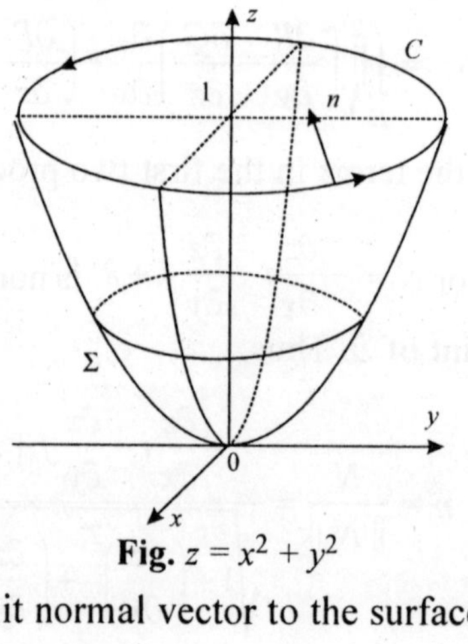

Fig. $z = x^2 + y^2$

Solution: The positive unit normal vector to the surface $z = z(x, y) = x^2 + y^2$ is

$$n = \frac{-\frac{\partial z}{\partial x}i - \frac{\partial z}{\partial y}j + k}{\sqrt{1+\left(\frac{\partial z}{\partial x}\right)^2 + \left(\frac{\partial z}{\partial y}\right)^2}} = \frac{-2xi + 2yj + k}{\sqrt{1+4x^2+4y^2}},$$

and curl $f = (1-0)i + (1-0)j + (1-0)k = i + j + k$, so

$$(\text{curl } f) \,.\, n = (-2x - 2y + 1)/\sqrt{1+4x^2+4y^2}$$

Since can be parametrized as $r(x, y) = xi + yj + (x^2 + y^2)k$ for (x, y) in the region $D = \{(x, y) : x^2 + y^2 \leq 1\}$, then

$$\iint_{\Sigma}(\text{curl } f).n d\sigma = \iint_{D}(\text{curl } f).n\left\|\frac{\partial r}{\partial x}\times\frac{\partial r}{\partial y}\right\|dA$$

$$= \iint_{D}\frac{-2x-3y+1}{\sqrt{1+4x^2+4y^2}}\sqrt{1+4x^2+4y^2}\,dA$$

$$= \iint_{D}(-2x-2y+1)dA,$$

so switching to polar coordinates gives

$$= \int_0^{2\pi}\int_0^1(-2r\cos\theta - 2r\sin\theta + 1)r dr d\theta$$

$$= \int_0^{2\pi}\int_0^1(-2r^2\cos\theta - 2r^2\sin\theta + r)dr d\theta$$

$$= \int_0^{2\pi}\left(-\frac{2r^3}{3}\cos\theta - \frac{2r^3}{3}\sin\theta + \frac{r^2}{2}\Bigg|_{r=0}^{r=1}\right)d\theta$$

$$= \int_0^{2\pi}\left(-\frac{2}{3}\cos\theta - \frac{2}{3}\sin\theta + \frac{1}{2}\right)d\theta$$

$$= -\frac{2}{3}\sin\theta + \frac{2}{3}\cos\theta + \frac{1}{2}\theta\Bigg|_0^{2\pi} = \pi.$$

The boundary curve C is the unit circle $x^2 + y^2 = 1$ laying in the plane $z = 1$, which can be parametrized as $x = \cos t$, $y = \sin t$, $z = 1$ for $0 \leq t \leq 2\pi$. So

$$\oint_C f.dr \text{ s} = \int_0^{2\pi}((1)(-\sin t) + (\cos t)(\cos t) + (\sin t)(0))dt$$

$$= \int_0^{2\pi} \left(-\sin t + \frac{1+\cos 2t}{2} \right) dt \left(\text{here we used } \cos^2 t = \frac{1+\cos 2t}{2} \right)$$

$$= \cos t + \frac{t}{2} + \frac{\sin 2t}{4} \Bigg|_0^{2\pi} = \pi.$$

So we see that $\oint_C f.dr = \iint_\Sigma (\text{curl} f).n d\sigma$, as predicted by Stokes' Theorem.

The line integral in the preceding example was far simpler to calculate than the surface integral, but this will not always be the case.

Example: Let Σ be the elliptic paraboloid $z = \frac{x^2}{4} + \frac{y^2}{9}$ for $z \le 1$, and let C be its boundary curve. Calculate $\oint_C f.dr$ for $f(x, y, z) = (9xz + 2y)i + (2x + y^2)j + (-2y^2 + 2z)k$, where C is traversed counterclockwise.

Solution: The surface is similar to the one in Example except now the boundary curve C is the ellipse $\frac{x^2}{4} + \frac{y^2}{9} = 1$ laying in the plane $z = 1$. In this case, using Stokes' Theorem is easier than computing the line integral directly. As in Example, at each point $(x, y, z(x, y))$ on the surface $z = z(x, y) = \frac{x^2}{4} + \frac{y^2}{9}$ the vector

$$n = \frac{-\frac{\partial z}{\partial x} i - \frac{\partial z}{\partial y} j + k}{\sqrt{1 + \left(\frac{\partial z}{\partial x}\right)^2 + \left(\frac{\partial z}{\partial y}\right)^2}} = \frac{-\frac{x}{2} i + \frac{2y}{9} j + k}{\sqrt{1 + \frac{x^2}{4} + \frac{4y^2}{9}}},$$

is a positive unit normal vector to Σ And calculating the curl of f gives

$$\text{curl} f = (-4y - 0)i + (9x - 0)j + (2 - 2)k = -4y\, i + 9x\, j + 0k,$$

so

$$(\text{curl} f).n = \frac{(-4y)\left(-\frac{x}{2}\right) + (9x)\left(-\frac{2y}{9}\right) + (0)(1)}{\sqrt{1 + \frac{x^2}{4} + \frac{4y^2}{9}}} = \frac{2xy - 2xy + 0}{\sqrt{1 + \frac{x^2}{4} + \frac{4y^2}{9}}} = 0$$

and so by Stokes' Theorem

$$\oint_C f.dr = \iint_\Sigma (\text{curl} f).n d\sigma = \iint_\Sigma 0 d\sigma = 0.$$

In physical applications, for a simple closed curve C the line integral $\oint_C f.dr$ is often called the circulation of f around C. For example, if E represents the electrostatic field due to a point charge, then it turns out that curl $E = 0$, which means that the circulation $\oint_C E.dr = 0$ by Stokes' Theorem. Vector fields which have zero curl are often called *irrotational* fields.

In fact, the term curl was created by the 19th century Scottish physicist James Clerk Maxwell in his study of electromagnetism, where it is used extensively. In physics, the curl is interpreted as a measure of *circulation density*. This is best seen by using another definition of curl f which is equivalent to the definition given by formula. Namely, for a point (x, y, z) in $\mathbb{R}^3$,

$$n \,.\, (\text{curl}\, f)(x, y, z) = \lim_{S \to 0} \frac{1}{S} \oint_C f.dr,$$

where S is the surface area of a surface containing the point (x, y, z) and with a simple closed boundary curve C and positive unit normal vector n at (x, y, z). In the limit, think of the curve C shrinking to the point (x, y, z), which causes Σ, the surface it bounds, to have smaller and smaller surface area. That ratio of circulation to surface area in the limit is what makes the curl a rough measure of circulation density (i.e. circulation per unit area).

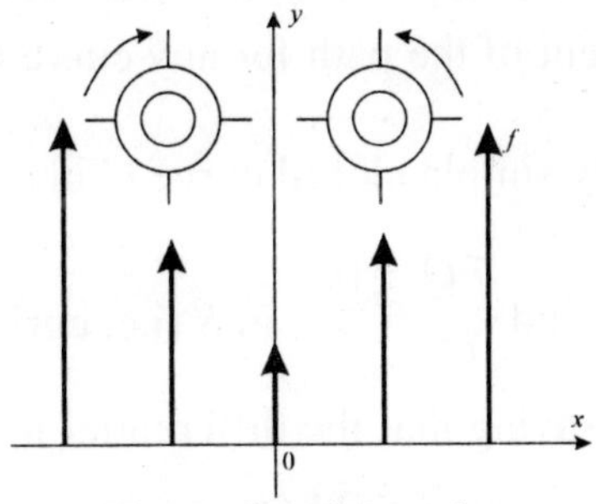

Fig. Curl and Rotation

An idea of how the curl of a vector field is related to rotation is shown in Figure. Suppose we have a vector field $f(x, y, z)$ which is always parallel to the xy-plane at each point (x, y, z) and that the vectors grow larger the further the point (x, y, z) is from the y-axis. For example, $f(x, y, z) = (1 + x^2)\, j$.

Think of the vector field as representing the flow of water, and imagine dropping two wheels with paddles into that water flow. Since the flow is stronger (i.e. the magnitude of f is larger) as you move away from the y-axis, then such a wheel would rotate counterclockwise if it were dropped to the right of the y-axis, and it would rotate clockwise if it were dropped to the left of the y-axis.

In both cases the curl would be nonzero (curl $f(x, y, z) = 2x\, k$ in our example) and would obey the right-hand rule, that is, curl $f(x, y, z)$ points in the direction of your thumb as you cup your right hand in the direction of the rotation of the wheel. So the curl points outward (in the positive z-direction) if $x > 0$ and points inward (in the negative z-direction) if $x < 0$.

Notice that if all the vectors had the same direction *and* the same magnitude, then the wheels would not rotate and hence there would be no curl (which is why such fields are called irrotational, meaning no rotation).

Finally, by Stokes' Theorem, we know that if C is a simple closed curve in some solid region S in $\mathbb{R}^3$ and if $f(x, y, z)$ is a smooth vector field such that curl $f = 0$ in S, then

$$\oint_C f.dr = \iint_\Sigma (\text{curl } f).n d\sigma = \iint_\Sigma 0.n d\sigma = \iint_\Sigma 0 d\sigma = 0,$$

where Σ is any orientable surface inside S whose boundary is C (such a surface is sometimes called a *capping surface* for C). So similar to the two-variable case, we have a three-dimensional version, for solid regions in $\mathbb{R}^3$ which are simply connected (i.e. regions having no holes):

The following statements are equivalent for a simply connected solid region S in $\mathbb{R}^3$:

1. $f(x, y, z) = P(x, y, z)\, i + Q(x, y, z)\, j + R(x, y, z)\, k$ has a smooth potential $F(x, y, z)$ in S
2. $\oint_C f.dr$ is independent of the path for any curve C in S
3. $\oint_C f.dr = 0$ for every simple closed curve C in S
4. $\dfrac{\partial R}{\partial y} = \dfrac{\partial Q}{\partial z}, \dfrac{\partial P}{\partial z} = \dfrac{\partial R}{\partial x}$, and $\dfrac{\partial Q}{\partial x} = \dfrac{\partial P}{\partial y}$ in S (i.e. curl $f = 0$ in S)

Part (d) is also a way of saying that the differential form $Pdx + Qdy + Rdz$ is exact.

Example: Determine if the vector field $f(x, y, z) = xyz\, i + xz\, j + xyk$ has a potential in $\mathbb{R}^3$.

Solution: Since $\mathbb{R}^3$ is simply connected, we just need to check whether curl $f = 0$ throughout $\mathbb{R}^3$, that is,

$$\frac{\partial R}{\partial y} = \frac{\partial Q}{\partial z}, \frac{\partial P}{\partial z} = \frac{\partial R}{\partial x}; \text{ and } \frac{\partial Q}{\partial x} = \frac{\partial P}{\partial y}$$

throughout $\mathbb{R}^3$, where $P(x, y, z) = xyz$, $Q(x, y, z) = xz$, and $R(x, y, z) = xy$. But we see that

$$\frac{\partial P}{\partial z} = xy, \frac{\partial R}{\partial x} = y \Rightarrow \frac{\partial P}{\partial z} \neq \frac{\partial R}{\partial x} \text{ for some } (x, y, z) \text{ in } \mathbb{R}^3.$$

Thus, $f(x, y, z)$ does not have a potential in $\mathbb{R}^3$.

EXERCISES

For Exercises 1-3, calculate $\oint_C f(x,y,z)ds$ for the given function $f(x, y, z)$ and curve C.

1. $f(x, y, z) = z$; $C : x = \cos t, y = \sin t, z = t, 0 \le t \le 2\pi$.
2. $f(x, y, z) = \dfrac{x}{y} + y + 2yz$; $C : x = t^2, y = t, z = 1, 1 \le t \le 2$.
3. $f(x, y, z) = z^2$; $C : x = t \sin t, y = t \cos t, z = \dfrac{2\sqrt{2}}{3}t^{3/2}, 0 \le t \le 1$.

For Exercises 4-9, calculate $\oint_C f dr$ for the given vector field $f(x, y, z)$ and curve C.

4. $f(x, y, z) = i - j + k$; $C : x = 3t, y = 2t, z = t, 0 \le t \le 1$.
5. $f(x, y, z) = y\, i - x\, j + z\, k$; $C : x = \cos t, y = \sin t, z = t, 0 \le t \le 2\pi$.
6. $f(x, y, z) = x\, i + y\, j + z\, k$; $C : x = \cos t, y = \sin t, z = 2, 0 \le t \le 2\pi$.
7. $f(x, y, z) = (y - 2z)\, i + xy\, j + (2xz + y)\, k$; $C : x = t, y = 2t, z = t^2 - 1, 0 \le t \le 1$.
8. $f(x, y, z) = yz\, i + xz\, j + xyk$; C : the polygonal path from (0, 0, 0) to (1, 0, 0) to (1, 2, 0).
9. $f(x, y, z) = xy\, i + (z - x)\, j + 2yz\, k$; C : the polygonal path from (0, 0, 0) to (1, 0, 0) to (1, 2, 0) to (1, 2, –2).

For Exercises 10-13, state whether or not the vector field $f(x, y, z)$ has a potential in $\mathbb{R}^3$ (you do not need to find the potential itself).

10. $f(x, y, z) = y\, i - x\, j + z\, k$.
11. $f(x, y, z) = a\, i + b\, j + ck$ (a, b, c constant).
12. $f(x, y, z) = (x + y)\, i + x\, j + z^2\, k$.
13. $f(x, y, z) = xy\, i - (x - yz^2)\, j + y^2z\, k$.

For Exercises 14-15, verify Stokes' Theorem for the given vector field $f(x, y, z)$ and surface Σ.

14. $f(x, y, z) = 2y\, i - x\, j + z\, k$; $\Sigma : x^2 + y^2 + z^2 = 1, z \ge 0$.
15. $f(x, y, z) = xy\, i + xz\, j + yz\, k$; $\Sigma : z = x2 + y2, z \le 1$.
16. Construct a Möbius strip from a piece of paper, then draw a line down its center. Cut the Möbius strip along that center line completely around the strip. How many surfaces does this result in? How would you describe them? Are they orientable?

17. Use Gnuplot to plot the Möbius strip parametrized as:

$$r(u, v) = \cos u\left(1+v\cos\frac{u}{2}\right)i + \sin u\left(1+v\cos\frac{u}{2}\right)j + v$$

$$\sin\frac{u}{2}k,\ 0 \le u \le 2\pi,\ -\frac{1}{2}\le v\le\frac{1}{2}$$

18. Let Σ be a closed surface and $f(x, y, z)$ a smooth vector field. Show that

$$\iint_{\Sigma}(\text{curl}\, f).n\,d\sigma = 0$$

. (*Hint: Split Σ in half.*)

19. Show that Green's Theorem is a special case of Stokes' Theorem.

Gradient, Divergence, Curl and Laplacian

In this final section we will establish some relationships between the gradient, divergence and curl, and we will also introduce a new quantity called the *Laplacian*. We will then show how to write these quantities in cylindrical and spherical coordinates.

For a real-valued function $f(x, y, z)$ on $\mathbb{R}^3$, the gradient $\nabla f(x, y, z)$ is a vector-valued function on $\mathbb{R}^3$, that is, its value at a point (x, y, z) is the vector

$$\nabla f(x,y,z) = \left(\frac{\partial f}{\partial x},\frac{\partial f}{\partial y},\frac{\partial f}{\partial z}\right) = \frac{\partial f}{\partial x}i+\frac{\partial f}{\partial y}j+\frac{\partial f}{\partial z}k$$

in $\mathbb{R}^3$, where each of the partial derivatives is evaluated at the point (x, y, z). So in this way, you can think of the *symbol* ∇ as being "applied" to a real-valued function f to produce a vector ∇f.

It turns out that the divergence and curl can also be expressed in terms of the symbol ∇. This is done by thinking of ∇ as a *vector* in $\mathbb{R}^3$, namely

$$\nabla = \frac{\partial}{\partial x}i+\frac{\partial}{\partial y}j+\frac{\partial}{\partial z}k.$$

Here, the symbols $\frac{\partial}{\partial x},\frac{\partial}{\partial y}$ and $\frac{\partial}{\partial z}$ are to be thought of as "partial derivative operators" that will get "applied" to a real-valued function, say $f(x, y, z)$, to produce the partial derivatives $\frac{\partial}{\partial x},\frac{\partial}{\partial y}$ and $\frac{\partial}{\partial z}$. For instance, $\frac{\partial}{\partial x}$ "applied" to $f(x, y, z)$ produces $\frac{\partial f}{\partial x}$.

Is ∇ *really* a vector? Strictly speaking, no, since $\frac{\partial}{\partial x},\frac{\partial}{\partial y}$ and $\frac{\partial}{\partial z}$ are not actual numbers. But it helps to *think* of ∇ as a vector, especially with the divergence and curl, as we will soon see. The process of "applying" $\frac{\partial}{\partial x}, \frac{\partial}{\partial y}, \frac{\partial}{\partial z}$ to a real-valued function $f(x, y, z)$ is normally thought of as *multiplying* the quantities:

$$\left(\frac{\partial}{\partial x}\right)(f) = \frac{\partial f}{\partial x}, \left(\frac{\partial}{\partial y}\right)(f) = \frac{\partial f}{\partial y}, \left(\frac{\partial}{\partial z}\right)(f) = \frac{\partial f}{\partial z}$$

For this reason, ∇ is often referred to as the "del operator", since it "operates" on functions.

For example, it is often convenient to write the divergence div f as $\nabla . f$, since for a vector field $f(x, y, z) = f_1(x, y, z)i + f_2(x, y, z)j + f_3(x, y, z)k$, the dot product of f with ∇ (thought of as a vector) makes sense:

$$\begin{aligned}\nabla . f &= \left(\frac{\partial}{\partial x}i + \frac{\partial}{\partial y}j + \frac{\partial}{\partial z}k\right).(f_1(x, y, z)i + f_2(x, y, z)j + f_3(x, y, z)k)\\ &= \left(\frac{\partial}{\partial x}\right)(f_1) + \left(\frac{\partial}{\partial y}\right)(f_2) + \left(\frac{\partial}{\partial z}\right)(f_3)\\ &= \frac{\partial f_1}{\partial x} + \frac{\partial f_2}{\partial y} + \frac{\partial f_3}{\partial z}\\ &= \text{div } f.\end{aligned}$$

We can also write curl f in terms of ∇, namely as $\nabla \times f$, since for a vector field $f(x, y, z) = P(x, y, z)i + Q(x, y, z)j + R(x, y, z)k$, we have:

$$\begin{aligned}\nabla \times f &= \begin{vmatrix} i & j & k \\ \frac{\partial}{\partial x} & \frac{\partial}{\partial y} & \frac{\partial}{\partial z} \\ P(x,y,z) & Q(x,y,z) & R(x,y,z) \end{vmatrix}\\ &= \left(\frac{\partial R}{\partial y} - \frac{\partial Q}{\partial z}\right)i - \left(\frac{\partial R}{\partial x} - \frac{\partial P}{\partial z}\right)j + \left(\frac{\partial Q}{\partial x} - \frac{\partial P}{\partial y}\right)k\\ &= \left(\frac{\partial R}{\partial y} - \frac{\partial Q}{\partial z}\right)i + \left(\frac{\partial P}{\partial z} - \frac{\partial R}{\partial x}\right)j + \left(\frac{\partial Q}{\partial x} - \frac{\partial P}{\partial y}\right)k\\ &= \text{curl } f.\end{aligned}$$

For a real-valued function $f(x, y, z)$, the gradient $\nabla f(x, y, z) = \frac{\partial f}{\partial x}i + \frac{\partial f}{\partial y}j + \frac{\partial f}{\partial z}k$ is a vector field, so we can take its divergence:

$$\text{div } \nabla f = \nabla \cdot \nabla f$$

$$= \left(\frac{\partial}{\partial x} i + \frac{\partial}{\partial y} j + \frac{\partial}{\partial z} k\right) . \left(\frac{\partial f}{\partial x} i + \frac{\partial f}{\partial y} j + \frac{\partial f}{\partial z} k\right)$$

$$= \frac{\partial}{\partial x}\left(\frac{\partial f}{\partial x}\right) + \frac{\partial}{\partial y}\left(\frac{\partial f}{\partial y}\right) + \frac{\partial}{\partial z}\left(\frac{\partial f}{\partial z}\right)$$

$$= \frac{\partial^2 f}{\partial x^2} + \frac{\partial^2 f}{\partial y^2} + \frac{\partial^2 f}{\partial z^2}$$

Note that this is a real-valued function, to which we will give a special name:

Definition: For a real-valued function $f(x, y, z)$, the Laplacian of f, denoted by Δf, is given by

$$\Delta f(x, y, z) = \nabla . \nabla f = \frac{\partial^2 f}{\partial x^2} + \frac{\partial^2 f}{\partial y^2} + \frac{\partial^2 f}{\partial z^2}.$$

Often the notation $\nabla^2 f$ is used for the Laplacian instead of Δf, using the convention $\nabla^2 = \nabla . \nabla$

Example: Let $r(x, y, z) = x\, i + y\, j + z\, k$ be the position vector field on $\mathbb{R}^3$. Then $\|r(x, y, z)\|^2 = r . r = x^2 + y^2 + z^2$ is a real-valued function. Find

(a) The gradient of $\| r \|^2$

(b) The divergence of r

(c) The curl of r

(d) The Laplacian of $\| r \|^2$

Solution: (a) $\nabla \| r \|^2 = 2x\, i + 2y\, j + 2z\, k = 2r$

(b) ∇ " $. r = = \frac{\partial}{\partial x}(x) + \frac{\partial}{\partial y}(y) + \frac{\partial}{\partial z}(z) = 1 + 1 + 1 = 3$

(c) $\nabla \times r = \begin{vmatrix} i & j & k \\ \frac{\partial}{\partial x} & \frac{\partial}{\partial y} & \frac{\partial}{\partial z} \\ x & y & z \end{vmatrix} = (0 - 0)i - (0 - 0)j + (0 - 0)k = 0$

(d) $\Delta \| r \|^2 = \frac{\partial^2}{\partial x^2}(x^2 + y^2 + z^2) + \frac{\partial^2}{\partial y^2}(x^2 + y^2 + z^2) + \frac{\partial^2}{\partial z^2}(x^2 + y^2 + z^2) = 2 + 2 + 2 = 6$

Note that we could have calculated $\Delta\| r \|^2$ another way, using the ∇ notation along with parts (a) and (b):

$$\Delta \| r \|^2 = \nabla \cdot \nabla \| r \|^2 = \nabla \times 2r = 2\nabla \cdot r = 2(3) = 6$$

Notice that in Example if we take the curl of the gradient of $\Delta\| r \|^2$ we get

$$\nabla \times (\nabla \| r \|^2) = \nabla \times 2r = 2\nabla \times r = 2\,0 = 0.$$

The following theorem shows that this will be the case in general:

Theorem: For any smooth real-valued function $f(x, y, z)$, $\nabla \times (\nabla f) = 0$.

Proof: We see by the smoothness of f that

$$\nabla \times (\nabla f) = \begin{vmatrix} i & j & k \\ \dfrac{\partial}{\partial x} & \dfrac{\partial}{\partial y} & \dfrac{\partial}{\partial z} \\ \dfrac{\partial f}{\partial x} & \dfrac{\partial f}{\partial x} & \dfrac{\partial f}{\partial z} \end{vmatrix}$$

$$= \left(\frac{\partial^2 f}{\partial y \partial z} - \frac{\partial^2 f}{\partial z \partial y}\right) i - \left(\frac{\partial^2 f}{\partial x \partial z} - \frac{\partial^2 f}{\partial z \partial x}\right) j + \left(\frac{\partial^2 f}{\partial x \partial y} - \frac{\partial^2 f}{\partial y \partial x}\right) k = 0,$$

since the mixed partial derivatives in each component are equal.

Corollary: If a vector field $f(x, y, z)$ has a potential, then curl $f = 0$.

Another way of stating Theorem is that gradients are irrotational. Also, notice that if we take the divergence of the curl of r we trivially get

$$\nabla \cdot (\nabla \times r) = \nabla \cdot 0 = 0.$$

The following theorem shows that this will be the case in general:

Theorem: For any smooth vector field $f(x, y, z)$, $\nabla \cdot (\nabla \times f) = 0$.

The proof is straightforward and left as an exercise for the reader.

Corollary: The flux of the curl of a smooth vector field $f(x, y, z)$ through any closed surface is zero.

Proof: Let Σ be a closed surface which bounds a solid S. The flux of $\nabla \times f$ through Σ is

$$\iint_{\Sigma} (\nabla \times f) . d\sigma = \iiint_{S} \nabla . (\nabla \times f) dV \quad \text{(by the Divergence Theorem)}$$

$$= \iiint_{S} 0\, dV \quad \text{(by Theorem)}$$

$$= 0.$$

There is another method for proving Theorem which can be useful, and is often used in physics. Namely, if the surface integral $\iint_{\Sigma} f(x,y,z)d\sigma = 0$ for *all* surfaces Σ in some solid region (usually all of $\mathbb{R}^3$), then we must have $f(x, y, z) = 0$ throughout that region. The proof is not trivial, and physicists do not usually bother to prove it. But the result is true, and can also be applied to double and triple integrals.

For instance, to prove Theorem, assume that $f(x, y, z)$ is a smooth real-valued function on $\mathbb{R}^3$. Let C be a simple closed curve in $\mathbb{R}^3$ and let Σ be any capping surface for C (i.e. Σ is orientable and its boundary is C). Since ∇f is a vector field, then

$$\iint_{\Sigma} (\nabla \times (\nabla f)).n d\sigma = \oint_C \nabla f.dr \text{ by Stokes' Theorem, so}$$

$$= 0 \text{ by Corollary.}$$

Since the choice of Σ was arbitrary, then we must have $(\nabla \times (\nabla f)).n = 0$ throughout $\mathbb{R}^3$, where n is any unit vector. Using *i, j* and *k* in place of *n*, we see that we must have $\nabla \times (\nabla f) = 0$ in $\mathbb{R}^3$, which completes the proof.

Example: A system of electric charges has a *charge density* $\rho(x, y, z)$ and produces an electrostatic field $E(x, y, z)$ at points (x, y, z) in space. *Gauss' Law* states that

$$\iint_{\Sigma} E.d\sigma = 4\pi \iiint_S \rho dV$$

for any closed surface Σ which encloses the charges, with S being the solid region enclosed by Σ. Show that $\nabla . \mathrm{E} = 4\pi\rho$. This is one of *Maxwell's Equations*.10

Solution: By the Divergence Theorem, we have

$$\iiint_S \nabla.E dV = \iint_{\Sigma} E.d\sigma$$

$$4\pi \iiint_S \rho dV$$

by Gauss' Law, so combining the integrals gives

$$\iiint_S (\nabla.E - 4\pi\rho)dV = 0, \text{ so}$$

$$\nabla . \mathrm{E} - 4\pi\rho = 0 \text{ since and hence } S \text{ was arbitrary, so}$$

$$\nabla . \mathrm{E} = 4\pi\rho.$$

Often (especially in physics) it is convenient to use other coordinate systems when dealing with quantities such as the gradient, divergence, curl and Laplacian. We will present the formulas for these in cylindrical and spherical coordinates.

Recall that a point (x, y, z) can be represented in cylindrical coordinates (r, θ, z), where $x = r\cos\theta$, $y = r\sin\theta$, $z = z$. At each point (r, θ, z), let e_r, e_θ, e_z be unit vectors in the direction of increasing r, θ, z, respectively. Then e_r, e_θ, e_z form an orthonormal set of vectors. Note, by the right-hand rule, that $e_z \times e_r = e_\theta$.

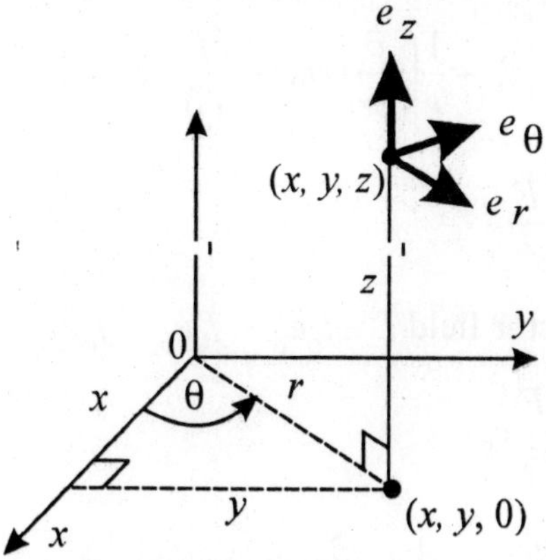

Fig. Orthonormal Vectors e_r, e_θ, e_z in Cylindrical Coordinates

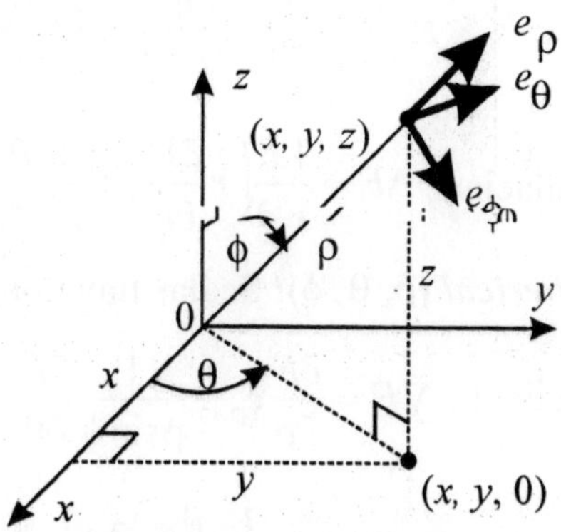

Fig. Orthonormal Vectors e_r, e_θ, e_z in Spherical Coordinates

Similarly, a point (x, y, z) can be represented in spherical coordinates (ρ, θ, ϕ), where $x = \rho\sin\phi\cos\theta$, $y = \rho\sin\phi\sin\theta$, $z = \rho\cos\phi$.

At each point (ρ, θ, ϕ), let e_ρ, e_θ, e_ϕ be unit vectors in the direction of increasing ρ, θ, ϕ, respectively. Then the vectors e_ρ, e_θ, e_ϕ are orthonormal. By the right-hand rule, we see that $e_\theta \times e_{\rho}, = e_\phi$.

We can now summarize the expressions for the gradient, divergence, curl and Laplacian in Cartesian, cylindrical and spherical coordinates in the following tables:

***Cartesian* (*x*, *y*, *z*):** Scalar function F; Vector field $f = f_1\, i + f_2\, j + f_3\, k$

gradient: $\nabla F = \dfrac{\partial F}{\partial x} i + \dfrac{\partial F}{\partial y} j + \dfrac{\partial F}{\partial z} k$

divergence: $\nabla . F = \dfrac{\partial f_1}{\partial x} + \dfrac{\partial f_2}{\partial y} + \dfrac{\partial f_3}{\partial z}$

curl: $\nabla \times f = \left(\dfrac{\partial f_3}{\partial y} - \dfrac{\partial f_2}{\partial z}\right) i + \left(\dfrac{\partial f_1}{\partial z} - \dfrac{\partial f_3}{\partial x}\right) j + \left(\dfrac{\partial f_2}{\partial x} - \dfrac{\partial f_1}{\partial y}\right) k$

Laplacian: $\Delta F = \dfrac{\partial^2 F}{\partial y} + \dfrac{\partial^2 F}{\partial y^2} + \dfrac{\partial^2 F}{\partial z^2}$

***Cylindrical* (r, *θ*, *z*):** Scalar function *F*; Vector field $f = f_r\, e_r + f_\theta\, e_\theta + f_z e_z$

gradient: $\Delta F = \frac{\partial F}{\partial r} e_r + \frac{1}{r}\frac{\partial F}{\partial \theta} e_\theta + \frac{\partial F}{\partial z} e_z$

divergence: $\Delta . f = \frac{1}{r}\frac{\partial}{\partial r}(r f_r) + \frac{1}{r}\frac{\partial f_\theta}{\partial \theta} + \frac{\partial f_z}{\partial z}$

curl: $\nabla \times f = \left(\frac{1}{r}\frac{\partial f_z}{\partial \theta} - \frac{\partial f_\theta}{\partial z}\right) e_r + \left(\frac{\partial f_r}{\partial z} - \frac{\partial f_z}{\partial r}\right) e_\theta + \frac{1}{r}\left(\frac{\partial}{\partial r}(r f_\theta) - \frac{\partial f_r}{\partial \theta}\right) e_z$

Laplacian: $\Delta F = \frac{1}{r}\frac{\partial}{\partial}\left(r \frac{\partial F}{\partial r}\right) + \frac{1}{r^2}\frac{\partial^2 F}{\partial \theta^2} + \frac{\partial^2 F}{\partial z^2}$

Spherical **(ρ, θ, ϕ):** Scalar function *F*; Vector field $f = f_\rho e_\rho + f_\theta e_\theta + f_\phi e_\phi$

gradient : $\nabla F = \frac{\partial F}{\partial \rho} e_\rho + \frac{1}{\rho \sin\phi}\frac{\partial F}{\partial \theta} e_\theta + \frac{1}{\rho}\frac{\partial F}{\partial \phi} e_\phi$

divergence : $\nabla . f = \frac{1}{\rho^2}\frac{\partial}{\partial \rho}(\rho^2 f_\rho) + \frac{1}{\rho \sin\phi}\frac{\partial f_\theta}{\partial \theta} + \frac{1}{\rho \sin\phi}\frac{\partial}{\partial \phi}(\sin\phi f_\theta)$

curl : $\nabla \times f = \frac{1}{\rho \sin\phi}\left(\frac{\partial}{\partial \phi}(\sin\phi f_\theta) - \frac{\partial f_\phi}{\partial \theta}\right) e_\rho + \frac{1}{\rho}\left(\frac{\partial}{\partial \rho}(\rho f_\phi) - \frac{\partial f_\rho}{\partial \phi}\right) e_\theta$

$$+\left(\frac{1}{\rho \sin\phi}\frac{\partial f_\rho}{\partial \theta} - \frac{1}{\rho}\frac{\partial}{\partial \rho}(\rho f_\theta)\right) e_\phi$$

Laplacian: $\Delta F = \frac{1}{\rho^2}\frac{\partial}{\partial \rho}\left(\rho^2 \frac{\partial F}{\partial \rho}\right) + \frac{1}{\rho^2 \sin^2\phi}\frac{\partial^2 F}{\partial \theta^2} + \frac{1}{\rho^2 \sin\phi}\frac{\partial}{\partial \phi}\left(\sin\phi \frac{\partial F}{\partial \phi}\right)$

The derivation of the above formulas for cylindrical and spherical coordinates is straightforward but extremely tedious. The basic idea is to take the Cartesian equivalent of the quantity in question and to substitute into that formula using the appropriate coordinate transformation. As an example, we will derive the formula for the gradient in spherical coordinates.

Goal: Show that the gradient of a real-valued function $F(\rho, \theta, \phi)$ in spherical coordinates is:

$$\nabla F = \frac{\partial F}{\partial \rho} e_\rho + \frac{1}{\rho \sin\phi}\frac{\partial F}{\partial \theta} e_\theta + \frac{1}{\rho}\frac{\partial F}{\partial \phi} e_\phi$$

Idea: In the Cartesian gradient formula $\nabla F(x, y, z) = \frac{\partial F}{\partial x} i + \frac{\partial F}{\partial y} j + \frac{\partial F}{\partial z} k$, put the

Cartesian basis vectors *i, j, k* in terms of the spherical coordinate basis vectors e_ρ, e_θ, e_ϕ and functions of ρ, θ and ϕ. Then put the partial derivatives $\frac{\partial F}{\partial x}, \frac{\partial F}{\partial y}, \frac{\partial F}{\partial z}$ in terms of $\frac{\partial F}{\partial \rho}, \frac{\partial F}{\partial \theta}, \frac{\partial F}{\partial \phi}$ and functions of ρ, θ and ϕ.

Step 1: Get formulas for e_ρ, e_θ, e_ϕ in terms of *i, j, k*.

We can see from Figure that the unit vector e_θ in the ρ direction at a general point (ρ, θ, ϕ) is $e_\rho = r\frac{r}{\|r\|}$, where $r = xi + yj + zk$ is the position vector of the point in Cartesian coordinates. Thus,

$$e_\rho = \frac{r}{\|r\|} = \frac{xi + yj + kz}{\sqrt{x^2 + y^2 + z^2}},$$

so using $x = \rho \sin \phi \cos \theta$, $y = \rho \sin \phi \sin \theta$, $z = \rho \cos \phi$, and $\rho = \sqrt{x^2 + y^2 + z^2}$, we get:

$$e_\rho = \sin \phi \cos \theta i + \sin \phi \sin \theta j + \cos \phi k$$

Now, since the angle θ is measured in the *xy*-plane, then the unit vector e_θ in the θ direction must be parallel to the *xy*-plane. That is, e_θ is of the form $ai + bj + 0k$. To figure out what *a* and *b* are, note that since $e_\theta \perp e_\rho$, then in particular $e_\theta \perp e_\rho$ when e_ρ is in the *xy*-plane. That occurs when the angle ϕ is π/2. Putting ϕ = π/2 into the formula for e_ρ gives $e_\rho = \cos \theta i + \sin \theta j + 0k$, and we see that a vector perpendicular to that is $- \sin \theta i + \cos \theta j + 0k$. Since this vector is also a unit vector and points in the (positive) θ direction, it must be e_θ:

$$e_\theta = - \sin \theta i + \cos \theta j + 0k$$

Lastly, since $e\phi = e\theta \times e_\rho$, we get:

$$e_\phi = \cos \phi \cos \theta i + \cos \phi \sin \theta j - \sin \phi k$$

Step 2: Use the three formulas from Step 1 to solve for *i, j, k* in terms of e_ρ, e_θ, e_ϕ.

This comes down to solving a system of three equations in three unknowns. There are many ways of doing this, but we will do it by combining the formulas for e_ρ and $e\phi$ to eliminate *k*, which will give us an equation involving just *i* and *j*.

This, with the formula for e_θ, will then leave us with a system of two equations in two unknowns (*i* and *j*), which we will use to solve first for *j* then for *i*. Lastly, we will solve for *k*.

First, note that

$$\sin \theta \, e_\rho + \cos \theta \, e_\phi = \cos \theta i + \sin \theta j$$

so that

$$\sin\theta(\sin\phi e_\rho + \cos\phi e_\phi) + \cos\theta e_\theta = (\sin^2\theta + \cos^2\theta)j = j,$$

and so:

$$j = \sin\phi\sin\theta e_\rho + \cos\theta e_\theta + \cos\phi\sin\theta e_\phi$$

Likewise, we see that

$$\cos\theta(\sin\phi e_\rho + \cos\phi e_\phi) - \sin\theta e_\theta = (\cos^2\theta + \sin^2\theta)i = i,$$

and so:

$$i = \sin\phi\cos\theta e_\rho - \sin\theta e_\theta + \cos\phi\cos\theta e_\phi$$

Lastly, we see that:

$$k = \cos\phi e_\rho - \sin\phi e_\phi$$

Step 3: Get formulas for $\frac{\partial F}{\partial \rho}, \frac{\partial F}{\partial \theta}, \frac{\partial F}{\partial \phi}$ in terms of $\frac{\partial F}{\partial x}, \frac{\partial F}{\partial y}, \frac{\partial F}{\partial z}$.

By the Chain Rule, we have

$$\frac{\partial F}{\partial \rho} = \frac{\partial F}{\partial x}\frac{\partial x}{\partial \rho} + \frac{\partial F}{\partial y}\frac{\partial y}{\partial \rho} + \frac{\partial F}{\partial z}\frac{\partial z}{\partial \rho},$$

$$\frac{\partial F}{\partial \theta} = \frac{\partial F}{\partial x}\frac{\partial x}{\partial \theta} + \frac{\partial F}{\partial y}\frac{\partial y}{\partial \theta} + \frac{\partial F}{\partial z}\frac{\partial z}{\partial \theta},$$

$$\frac{\partial F}{\partial \phi} = \frac{\partial F}{\partial x}\frac{\partial x}{\partial \phi} + \frac{\partial F}{\partial y}\frac{\partial y}{\partial \phi} + \frac{\partial F}{\partial z}\frac{\partial z}{\partial \phi},$$

which yields:

$$\frac{\partial F}{\partial \rho} = \sin\phi\cos\theta\frac{\partial F}{\partial x} + \sin\phi\sin\theta\frac{\partial F}{\partial y} + \cos\phi\frac{\partial F}{\partial z}$$

$$\frac{\partial F}{\partial \theta} = -\rho\sin\phi\sin\theta\frac{\partial F}{\partial x} + \rho\sin\phi\cos\theta\frac{\partial F}{\partial y}$$

$$\frac{\partial F}{\partial \phi} = \rho\sin\phi\cos\theta\frac{\partial F}{\partial x} + \rho\cos\phi\sin\theta\frac{\partial F}{\partial y} + -\rho\sin\phi\frac{\partial F}{\partial z}$$

Step 4: Use the three formulas from Step 3 to solve for $\frac{\partial F}{\partial x}, \frac{\partial F}{\partial y}, \frac{\partial F}{\partial z}$ in terms of $\frac{\partial F}{\partial \rho}, \frac{\partial F}{\partial \theta}, \frac{\partial F}{\partial \phi}$

Again, this involves solving a system of three equations in three unknowns. Using a similar process of elimination as in Step 2, we get:

$$\frac{\partial F}{\partial x} = \frac{1}{\rho\sin\phi}\left(\rho\sin^2\phi\cos\theta\frac{\partial F}{\partial\rho} - \sin\theta\frac{\partial F}{\partial\theta} + \sin\phi\cos\phi\cos\theta\frac{\partial F}{\partial\phi}\right)$$

$$\frac{\partial F}{\partial y} = \frac{1}{\rho\sin\phi}\left(\rho\sin^2\phi\cos\theta\frac{\partial F}{\partial\rho} + \cos\theta\frac{\partial F}{\partial\theta} + \sin\phi\cos\phi\cos\theta\frac{\partial F}{\partial\phi}\right)$$

$$\frac{\partial F}{\partial z} = \frac{1}{\rho}\left(\rho\sin\phi\frac{\partial F}{\partial\rho} - \sin\phi\frac{\partial F}{\partial\phi}\right)$$

Step 5: Substitute the formulas for *i, j, k* from Step 2 and the formulas for $\frac{\partial F}{\partial x}, \frac{\partial F}{\partial y}, \frac{\partial F}{\partial z}$ from Step 4 into the Cartesian gradient formula $\nabla(x, y, z) = \frac{\partial F}{\partial x}i + \frac{\partial F}{\partial y}j + \frac{\partial F}{\partial z}k.$

Doing this last step is perhaps the most tedious, since it involves simplifying 3 × 3 + 3 × 3 + 2 × 2 = 22 terms! Namely,

$$\nabla F = \frac{1}{\rho\sin\phi}\left(\rho\sin^2\phi\cos\theta\frac{\partial F}{\partial\rho} - \sin\theta\frac{\partial F}{\partial\theta} + \sin\phi\cos\phi\cos\theta\frac{\partial F}{\partial\phi}\right)$$

$$(\sin\phi\cos\theta e_\rho - \sin\theta e_\theta + \cos\phi\cos\theta e_\phi)$$

$$+\frac{1}{\rho\sin\phi}\left(\rho\sin^2\phi\cos\theta\frac{\partial F}{\partial\rho} + \cos\theta\frac{\partial F}{\partial\theta} + \sin\phi\cos\phi\sin\theta\frac{\partial F}{\partial\phi}\right)$$

$$(\sin\phi\sin\theta e_\rho + \cos\theta e_\theta + \cos\phi\sin\theta e_\phi)$$

$$+\frac{1}{\rho}\left(\rho\cos\phi\frac{\partial F}{\partial\rho} - \sin\phi\frac{\partial F}{\partial\phi}\right)(\cos\phi e_\rho - \sin\phi e_\phi),$$

which we see has 8 terms involving e_ρ, 6 terms involving e_θ, and 8 terms involving e_ϕ. But the algebra is straightforward and yields the desired result:

$$\nabla F = \frac{\partial F}{\partial\rho}e_\rho + \frac{1}{\rho\sin\phi}\frac{\partial F}{\partial\theta}e_\theta + \frac{1}{\rho}\frac{\partial F}{\partial\phi}e_\phi$$

Example: In Example we showed that $\nabla\| r \|^2 = 2r$ and $\Delta\| r \|^2 = 6$, where $r(x, y, z) = xi + yj + zk$ in Cartesian coordinates. Verify that we get the same answers if we switch to spherical coordinates.

Solution: Since $\| r \|^2 = x^2 + y^2 + z^2 = \rho^2$ in spherical coordinates, let $F(\rho, \theta, \phi) = \rho^2$ (so that $F(\rho, \theta, \phi) = \| r \|^2$). The gradient of F in spherical coordinates is

$$\nabla F = \frac{\partial F}{\partial \rho} e_\rho + \frac{1}{\rho \sin\phi} \frac{\partial F}{\partial \theta} e_\theta + \frac{1}{\rho}\frac{\partial F}{\partial \phi} e_\phi$$

$$= 2\rho e_\rho + \frac{1}{\rho \sin\phi}(0)e_\theta + \frac{1}{\rho}(0)e_\phi$$

$$= 2\rho e_\rho = 2\rho \frac{r}{\|r\|}, \text{ as we showed earlier, so}$$

$$= 2\rho \frac{r}{\rho} = 2r, \text{ as expected. And the Laplacian is}$$

$$\nabla F = \frac{1}{\rho^2}\frac{\partial}{\partial \rho}\left(\rho^2 \frac{\partial F}{\partial \rho}\right) + \frac{1}{\rho^2 \sin^2\phi}\frac{\partial^2 F}{\partial \theta^2} + \frac{1}{\rho^2 \sin\phi}\frac{\partial}{\partial \phi}\left(\sin\phi \frac{\partial F}{\partial \phi}\right)$$

$$= \frac{1}{\rho^2}\frac{\partial}{\partial \rho}\left(\rho^2 2\rho\right) + \frac{1}{\rho^2 \sin\phi}(0) + \frac{1}{\rho^2 \sin\phi}\frac{\partial}{\partial \phi}(\sin\phi(0))$$

$$= \frac{1}{\rho^2}\frac{\partial}{\partial \rho}(2\rho^3) + 0 + 0$$

$$= \frac{1}{\rho^2}(6\rho^2) = 6, \text{, as expected.}$$

EXERCISES

For Exercises 1-6, find the Laplacian of the function $f(x, y, z)$ in Cartesian coordinates.

1. $f(x, y, z) = x + y + z$
2. $f(x, y, z) = x^5$
3. $f(x, y, z) = (x^2 + y^2 + z^2)^{3/2}$
4. $f(x, y, z) = e^{x+y+z}$
5. $f(x, y, z) = x^3 + y^3 + z^3$
6. $f(x, y, z) = e^{-x^2-y^2-z^2}$
7. Let $f(x, y, z) = \dfrac{z}{x^2 + y^2}$ in Cartesian coordinates. Find ∇f in cylindrical coordinates.
8. For $f(r, \theta, z) = r\, e_r + z \sin\theta\, e_\theta + r_z e_z$ in cylindrical coordinates, find div f and curl f.
9. For $f(\rho, \theta, \phi) = e_\rho + \theta \cos\theta\, e_\theta + \theta\, e_\theta$ in spherical coordinates, find div f and curl f.

For Exercises 10-21, prove the given formula ($r = \| r \|$ is the length of the position vector field $r(x, y, z) = xi + yj + zk$).

10. $\nabla(1/r) = -r/r^3$
11. $\Delta(1/r) = 0$
12. $\nabla.(r/r^3) = 0$
13. $\nabla(\ln r) = r/r^2$
14. div $(F + G)$ = div F + div G
15. curl $(F + G)$ = curl F + curl G
16. div $(fF) = f$ div $F + F + \nabla f$
17. div $(F \times G) = G$. curl $F - F$. curl G
18. div $(\nabla f \times \nabla g) = 0$
19. curl $(fF) = f$ curl $F + (\nabla f) \times F$
20. curl (curl F) = ∇(div F) $- \Delta F$
21. $\Delta(fg) = f\Delta g + g\Delta f + 2(\nabla f . \nabla g)$
22. Derive the gradient formula in cylindrical coordinates: $\nabla F = \frac{\partial F}{\partial r}e_r + \frac{1}{r}\frac{\partial F}{\partial \theta}e_\theta + \frac{\partial F}{\partial z}e_z$
23. Use $f = u\nabla v$ in the Divergence Theorem to prove:

    (a) *Green's first identity:* $\iiint_S (u\Delta v + (\nabla u).(\nabla v))dV = \iint_\Sigma (u\nabla v).d\sigma$

    (b) *Green's second identity:* $\iiint_S (u\Delta v - v\Delta u)dV = \iint_\Sigma (u\nabla v - v\nabla u).d\sigma$

24. Suppose that $\Delta u = 0$ (i.e. u is *harmonic*) over $\mathbb{R}^3$. Define the *normal derivative* $\frac{\partial u}{\partial v}$ of u over a closed surface Σ with outward unit normal vector n by $\frac{\partial u}{\partial n} = D_n u = n.\nabla u$. Show that $\iint_\Sigma \frac{\partial u}{\partial n}d\sigma = 0$. (*Hint: Use Green's second identity.*)

Chapter 10

Applications of Vector Calculus

The divergence and Stokes' theorems (and their related results) supply fundamental tools which can be used to derive equations which can be used to model a number of physical situations. Essentially, these theorems provide a mathematical language with which to express physical laws such as conservation of mass, momentum and energy. The resulting equations are some of the most fundamental and useful in engineering and applied science.

In the following sections the derivation of some of these equations will be outlined. The goal is to show how vector calculus is used in applications. Generally speaking, the equations are derived by first using a conservation law in integral form, and then converting the integral form to a diûerential equation form using the divergence theorem, Stokes' theorem, and vector identities. The diûerential equation forms tend to be easier to work with, particularly if one is interested in solving such equations, either analytically or numerically.

DYNAMICS OF IDEAL FLUIDS

Most of our knowledge regarding the motion of fluids comes from laboratory and terrestrial flows. We will use this opportunity to review elementary fluid dynamics with an emphasis on application to everyday flows.

THE EQUATIONS OF MOTION

First, we will discuss abstract fluid motion and then consider some special cases. To describe the motion of a fluid, we must give its properties at every point. For example, at different places, the fluid is moving with different velocities. To specify the character of the flow, therefore, we must give the three components of velocity at every point and for any time. If we can find the equations that determine the velocity, then we would know how the liquid moves at all times. The velocity, however, is not the only property that the fluid has which varies from point to point. We have just discussed the variation of the pressure from point to point. And there are still other variables.

There may also be a variation of density from point to point. In addition, the fluid may carry an electric current whose density varies from point to point in magnitude and direction. There may be a temperature which varies from point to point, or a magnetic field, and so on. So the number of fields needed to describe the complete situation will depend on how complicated the problem is. There are interesting phenomena when currents and magnetism play a dominant part in determining the behavior of the fluid; this subject is called magnetohydrodynamics.

We will take the situation where there is no magnetic field and no conductivity, and we will not worry about the temperature because we will suppose that the density and pressure determine in a unique manner the temperature at any point. We will reduce the complexity by making the assumption that the density is a constant--we imagine that the fluid is essentially incompressible. Putting it another way, we are supposing that the variations of pressure are so small that the changes in density produced thereby are negligible.

If that is not the case, we would encounter phenomena additional to the ones we will be discussing here, for example, the propagation of sound or of shock waves. It is easy to determine when the approximation of constant ρ is a good one if we can say that if the velocities of flow are much less than the speed of sound. In the general theory of fluids one must begin with an equation of state for the fluid which connects the pressure to the density. In our approximation this equation of state is simply ρ = const.

This is the first relation for our variables. The escape that fluids makes in our attempts at understanding it is not related to the approximation of constant density.

THE EQUATION OF CONTINUITY

The equation of continuity expresses the conservation of matter--if matter flows away from a point, there must be a decrease in the quantity remaining. If the fluid velocity is *u*, then the mass which flows in a unit time across a unit area of surface is the component of ρu normal to the surface. There is a similar relation for electric charge and current. We know from the divergence of the current gives the rate of decrease of the density per unit time. If instead of charge we consider the change of mass in a volume *V*, this is balanced by the flow out of that volume

$$\frac{d}{dt}\int_v \rho\, dV = -\int_s \rho u . n dA,$$

where the volume *V* is fixed in space, and the minus sign is necessary because when *u* and the unit vector *n* points in the same direction, i.e., if *u* is outward and $u . n > 0$, the mass within *V* decreases.

Since the volume is fixed in space, the limits of integration time independent so the time differentiation may be included in the integrand on the LHS, and by Gauss's divergence theorem the second term can be writtten as a volume integral:

$$\int_v \frac{\partial \rho}{\partial t} dV + \int_s \nabla.(\rho u) dV = 0$$

Since the volume is arbitrary, the continuity equation,

$$\frac{\partial \rho}{\partial t} + \nabla.(\rho u) = 0$$

Must hold everywhere. If ρ = const. then the equation of continuity simplifies to

$$\nabla.u = 0$$

The velocity field of an incompressible fluid, like the magnetic field B, has zero divergence. The hydrodynamic equations are often closely analogous to the electrodynamics equations, but electrodynamics is much easier than hydrodynamics

THE EQUATION OF MOTION

We will get our next equation from Newton's law which tells us how the velocity changes because of the forces. The mass of an element of volume of the fluid times its acceleration, a, must be equal to the force on the element. Taking an element of unit volume, and writing the force per unit volume as f, we have

$$\rho a = f$$

We will write the force density as the sum of three terms. We have already considered the pressure force per unit volume, $-\nabla p$. Then there are the body forces which act at a distance, e.g., gravity or electricity. When they are conservative forces with a potential per unit mass, ϕ, they give a force density

$$-\rho\nabla\phi.$$

If the external forces are not conservative, we would have to write f_{ext} for the external force per unit volume. Then there are the surface forces per unit volume, which is due to the fact that in a flowing fluid there can also be a shearing stress. This is called the viscous force, which we will write f_{visc}. Our equation of motion is

$$\rho a = -\nabla p - \rho\nabla\phi + f_{visc}.$$

For now we set $f_{visc} = 0$ and by ignoring viscosity we will be making an approximation which describes an ideal fluid. The mathematician and computer scientist John von Neumann was well aware of the profound impact of this approximation, and he was also aware that, during most of the development of hydrodynamics until about 1900, almost the main interest was in solving beautiful mathematical problems with this approximation which had almost nothing to do with real fluids.

He characterized the theorist who worked on this problem as scientists who studied "dry water." Such analyses leave out an essential property of the fluid.

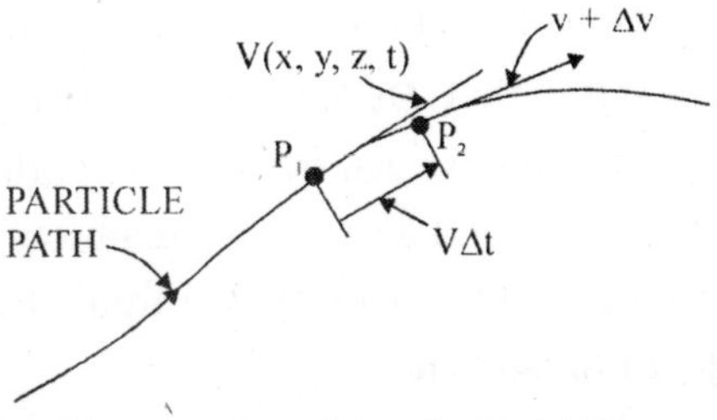

Fig. The Acceleration of a Fluid Element.

If we leave out f_{visc}, we have everything we need except an expression for the acceleration. Naïvely, you might think that the formula for the acceleration of a fluid particle would be very simple, for it seems obvious that if u is the velocity of a fluid particle at some place in the fluid, the acceleration would just be $\partial u / \partial t$; it is not.

The derivative $\partial u / \partial t$, is the rate at which the velocity $u(x, t)$ changes at a fixed point in space. What we need is how fast the velocity changes for a given fluid element. Imagine that we mark one of the drops of fluid with ink so we can follow it. In a small interval of time Δt, this drop will move from P_1 to a different location P_2.

$$u_j(x_i + \Delta x_i, t + \Delta t) = u_j(x_i, t) + \frac{\partial u_j}{\partial t}\Delta t + \frac{\partial u_j}{\partial x_i}\Delta x_i,$$

where repeated indices implies summation. Using $\Delta x_i = u_i \Delta t$

$$u_j(x_i + \Delta x_i, t + \Delta t) = u_j(x_i, t) + \left(\frac{\partial u_j}{\partial t} + u_j \frac{\partial u_j}{\partial x_i}\right)\Delta t$$

From the definition of the derivative

$$a_j \lim_{\Delta t \to 0}\left(\frac{\Delta u_j}{\Delta t}\right) = \frac{\partial u_j}{\partial t} + u_i \frac{\partial u_j}{\partial x_i}$$

and so the acceleration of a fluid element is

$$\boxed{a = \frac{\partial u}{\partial t} + (u.\nabla)u.}$$

We define the differential operator

$$\frac{D}{Dt} \equiv \frac{\partial}{\partial t} + (u.\nabla),$$

which is known as the material derivative. The notation D/Dt is used to emphasize that the material time derivative includes both space and time partial derivatives. The material derivative is a scalar operator so that the material derivative of a scalar variable, such as density, is a scalar quantity. The material derivative of a vector is a vector.

Thus, there can be acceleration of fluid even when the velocity at a given point is constant, i.e., even if $\partial u / \partial t = 0$. It is possible that $Du / Dt \neq 0$ because $(u.\nabla)u \neq 0$. For steady flow, ($\partial u / \partial t = 0$), u is a static vector field and we can speak of streamlines, which are the trajectories followed by fluid elements. To understand describe a streamline mathematically consider two points on a streamline r and $r + \delta r$). The velocity of a fluid element on such a streamline is perpendicular to r and parallel to δr so that

$$u \times \delta r = 0 \cdot$$

The cross product can be written as

$$\begin{vmatrix} i & j & k \\ u_x & u_y & u_z \\ dx & dy & dz \end{vmatrix} = 0$$

Thus, the differential equation describing a streamline is

$$\frac{dx}{u_x} = \frac{dy}{u_y} = \frac{dz}{u_z}.$$

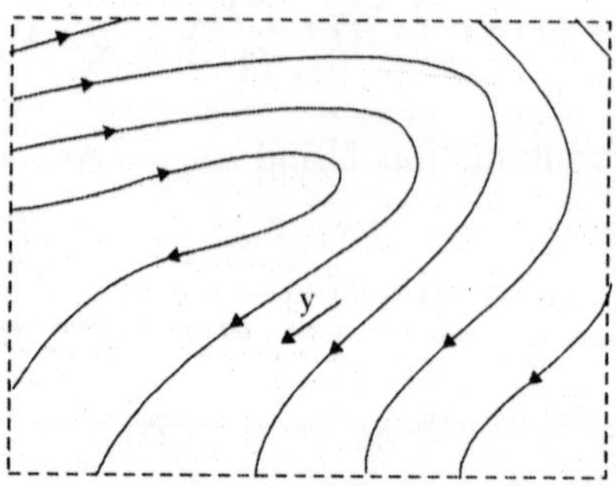

Fig. Streamlines in a Steady Fluid flow Track the path of Fluid Elements.

The term $(u.\nabla)u$ represents the change of u in the direction of u, and vanishes only when there is no variation of u in the direction of u. If all the fluid elements follow straight streamlines then this acceleration term is zero. Acceleration occurs where streamlines change direction. As an example, water flowing through a converging nozzle can be in steady state so that at any location the velocity remains constant, however the water accelerates as the cross-section of the nozzle decreases and the streamlines converge. Of course, a fluid element can also accelerate by changing its speed along a straight streamline. The physical origin of this term becomes clearer in terms of curvilinear coordinates attached to a streamline. Let s be the coordinate measured along the streamline, $V(s)$ the speed $| u |$, and $ds = V\,dt$.

$$(u.\nabla)u = V\frac{dV}{ds}\hat{e}_t - \frac{V^2}{R}\hat{e}_n,$$

where R is the radius of curvature of the streamline, and $\hat{e}_t$ and $\hat{e}_n$ are unit vectors tangential and normal to the streamline. The first part of this expression is the rate of change of speed $dV/dt = V\, dV/ds$, while the second is the centripetal acceleration, which is directed towards the center of curvature of the streamline and is associated with the change in direction of the velocity of a particle.

Now we have all we need for the equation of motion, and the rest is just math,

$$\rho\left[\frac{\partial u}{\partial t}+(u.\nabla)u\right]=-\nabla p-\rho\nabla\phi$$

or

$$\frac{\partial u}{\partial t}+(u.\nabla)u=-\frac{1}{\rho}-\nabla p-\nabla\phi$$

where viscosity has been omitted. It is often useful to use the vector identity

$$(u.\nabla)u\equiv(\nabla\times u)\times u+\frac{1}{2}\nabla|u|^2$$

and identify a new vector field $w=\nabla\times u$ so that

$$(u.\nabla)u\equiv w\times u+\frac{1}{2}\nabla|u|^2$$

Giving the final form for the equation of motion

$$\frac{\partial u}{\partial t}+w\times u+\frac{1}{2}\nabla|u|^2=-\frac{1}{\rho}\nabla p-\nabla\phi,$$

which is Euler equation.

The vector field w is called the vorticity. If the vorticity is zero everywhere, we say that the flow is irrotational. The circulation, Γ around any closed loop, C, in a fluid is the line integral of the fluid velocity, at a given instant of time, around that loop,

$$\Gamma=\oint_c u.dl.$$

The circulation per unit area is, by Stokes' theorem, equal to $\nabla\times u$ in a simply connected region. So the vorticity is the circulation round a unit area in the direction perpendicular to the surface. The vorticity is a measure of local rotation. It has nothing to do with global rotation. It follows that if you put a speck of dirt, i.e., not an infinitesimal point, at any place in the liquid it will spin with the angular velocity $w/2$. A simple example of vorticity and circulation is a shear flow described by

$$v_x=\beta_y$$

In this case, from the definition of the curl,

$$w_x=-\frac{\partial v_x}{\partial y}=-\beta$$

We can also use the vorticity to calculate that angular velocity of a fluid element. Since w is the circulation per unit area

$$w = \frac{\oint u.dl}{\pi r^2} = \frac{2\pi r v}{\pi r^2}$$

which means that

$$w = \frac{2v}{r}$$

If the angular velocity, $\Omega = v/r$, then the angular velocity of a fluid element is,

$$\Omega = \frac{w}{2}$$

POTENTIAL FLOW AND ELECTROSTATICS

If we are interested only in the velocity field, we can eliminate the pressure from our equations. Taking the curl of both sides of the Euler equation, remembering that the curl of any gradient is zero and that ρ is a constant so that $\nabla . u = 0$, we get

$$\frac{\partial w}{\partial t} + \nabla \times (w \times u) = 0$$

This equation, together with

$$w = \nabla \times u$$

and

$$\nabla . u = 0$$

completely describe the velocity field. Mathematically speaking, if we know w at some time, then we know the curl of the velocity vector, and we also know that its divergence is zero, so given the physical situation we have all we need to determine u everywhere. It is just like the situation in magnetism where we had

$$\nabla . B = 0, \quad \nabla \times B = \frac{4\pi}{c} j$$

Thus, a given w determines u just as a given j determines B. We can therefore hijack the Biot-Savart law from electromagnetic theory to invert the relationship $w = \nabla . u$. That is to say, in infinite domains,

$$u = \frac{1}{4\pi} \int \frac{w(x') \times r}{r^3} d^3 x', \; r = x - x'.$$

Note that, like B, the vorticity field is solenoidal, $\nabla . w = 0$, since it is the curl of another vector. Consequently, we may invoke the idea of vortex tubes, which are analogous to magnetic flux tubes or streamtubes. This procedure gives the velocity field only; we have lost all information about the pressure. The absence of information regarding the pressure is not surprising since pressure acts normally and cannot change the angular momentum of a fluid element.

We point out one special consequence of our equation. If $w = 0$ everywhere at any time t, $\frac{\partial w}{\partial t}$ also vanishes, so that w is still zero everywhere at $t + \Delta t$. We have a solution to the equation; the flow is permanently irrotational. If a flow was started with zero rotation, it would always have zero rotation. The equations to be solved then are

$$\Delta . u = 0, \quad \nabla \times u = 0.$$

The problem is mathematically equivalent to magnetostatics or electrostatics, and can be reduced to finding the solution of Laplace's equation,

$$\nabla^2 \phi = 0$$

where the velocity is related to a scalar potential

$$u = \nabla \phi$$

Hence, incompressible, irrotational flow is called potential flow. Steady flow Consider Euler's equation of motion in the limit of steady flow in which the velocity picture remains unchanged and u is a static vector field, i.e.,

$$\frac{\partial u}{\partial t} = 0$$

The fluid velocity is constant at all locations; the fluid is not static, but is always replaced by new fluid moving in exactly the same way. In analogy field lines in magnetostatics, we can draw lines, called streamlines, which are always tangent to the fluid velocity. For steady flow, they represent the paths of fluid elements. In unsteady flow the streamline pattern changes in time, and the streamlines do not represent the path of a fluid element at any instant.

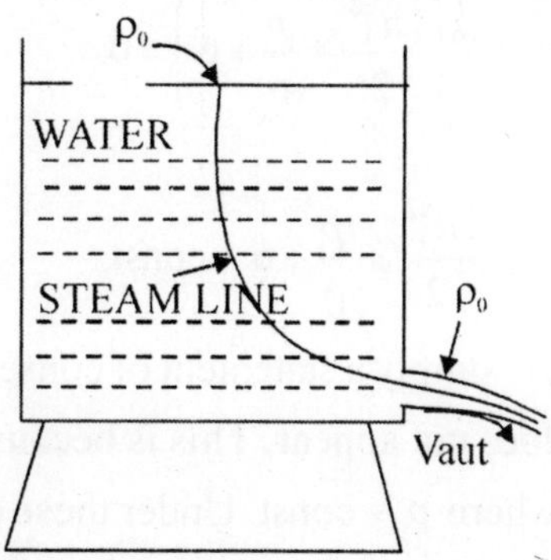

Fig. Flow from a tank. The Stream Contracts to 62% of the area of the Aperture.

Starting with the Euler equation,

$$\frac{\partial u}{\partial t} + w \times u + \frac{1}{2} \Delta |u|^2 = -\frac{1}{\rho} \nabla_p - \nabla \phi.$$

and setting time derivative to zero for steady flow,

$$w \times u + \frac{1}{2}\Delta |u|^2 = -\frac{1}{\rho}\nabla_p - \nabla\phi.$$

Now take the dot product with u, and since

$$u \,.\, (w \times u) = 0$$

we are left with

$$u.\nabla\left(\frac{|u|^2}{2} + \frac{p}{\rho} + \phi\right) = 0$$

for incompressible flow. For steady flow we have that $\frac{\partial u}{\partial t}$, thus the material derivative D/Dt becomes $(u.\nabla)$, meaning that the quantity in parenthesis does not change for a material fluid element as is passes through the flow. Another way of thinking about this is to note that the quantity in parenthesis does not change for a small displacement in the direction of the fluid velocity. In other words, the quantity

$$\boxed{\frac{|u|^2}{2} + \frac{p}{\rho} + \phi = \text{const.}}$$

along a stream line. This is Bernoulli's theorem (pronounced Ber-noo-yee). Furthermore, for an irrotational flow, $w = 0$, Euler's equation of motion gives us for incompressible flow,

$$\frac{1}{2}\nabla |u|^2 = -\frac{1}{\rho}\nabla_p - \nabla\phi.$$

or

$$\nabla\left(\frac{|u|^2}{2} + \frac{p}{\rho} + \phi\right) = 0,$$

that is

$$\frac{|u|^2}{2} + \frac{p}{\rho} + \phi = \text{const.}$$

everywhere. Bernoulli's theorem is simply a statement of conservation of energy. Note that the fluid's internal energy $\in = c_v T$ does not appear. This is because we have derived Bernoulli's theorem for the restrictive case where ρ = const. Under these conditions no pdV work can be done on the fluid and, since the flow is also isentropic (no heat flow and no energy dissipation by friction) the internal energy remains constant.

Bernoulli's theorem can be demonstrated by holding two pieces of paper close together and trying to blow them apart? Try it! They come together. The reason, of course, is that the air has a higher speed going through the constricted space between the sheets than it does when it gets outside. The pressure between the sheets is lower than atmospheric pressure, so they come together rather than separating.

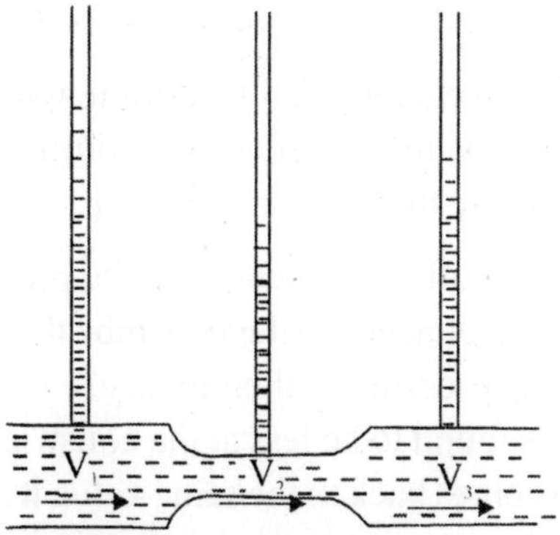

Fig. According to Bernoulli's Theorem the Pressure is Lowest where the Velocity is Highest.

There is an important class of problems in astrophysics associated with accretion and mass loss known as the Bondi problem that uses Bernoulli's theorem. But first consider some examples where the Bernoulli integral gives us a description of the flow. Suppose we have water flowing out of a hole near the bottom of a tank. We take a situation in which the flow speed u_{out} at the hole is much larger than the flow speed near the top of the tank; in other words, we imagine that the diameter of the tank is so large that we can neglect the drop in the liquid level.

At the top of the tank the pressure is p_0, the atmospheric pressure, and the pressure at the sides of the jet is also p_0. Now we write our Bernoulli equation for a streamline, such as the one shown in the figure. At the top of the tank, we take $u = 0$ equal to zero and we also take the gravity potential $\phi = 0$. At the speed u_{out}, and $\phi = -gh$, so that

$$p_0 = p_0 + \frac{1}{2}\rho u^2{}_{out} - pgh.$$

Hence,

$$u_{out} = \sqrt{2gh}.$$

This velocity is just what we would get for something which falls the distance h. It is not too surprising, since the water at the exit gains kinetic energy at the expense of the potential energy of the water at the top.

You cannot figure out the rate that the fluid flows out of the tank by multiplying this velocity by the area of the hole. The fluid velocities as the jet leaves the hole are not all parallel to each other but have components inward toward the center of the stream; the jet is converging. After the jet has gone a little way, the contraction stops and the velocities do become parallel. So the total flow is the velocity times the area at that point. In fact, if we have a discharge opening which is just a round hole with a sharp edge, the jet contracts to 62% of the area of the hole. The reduced effective area of the discharge varies for different shapes of discharge tubes, and experimental contractions are available as tables of efflux coefficients. If the discharge tube is re-entrant, it is possible to prove that the efflux coefficient is exactly 50%.

Consider a horizontal pipe with changing cross section, with water flowing in one end and out the other. Conservation of energy, in the form of Bernoulli's formula, says that the pressure is lower in the constricted area where the velocity is higher.

We can easily demonstrate this effect by measuring the pressure at different cross sections with small vertical columns of water attached to the flow tube through holes small enough so that they do not disturb the flow. The pressure is then measured by the height of water in these vertical columns. The pressure is found to be less at the constriction than it is on either side. If the area beyond the constriction comes back to the same value it had before the constriction, the pressure rises again.

Bernoulli's formula predicts that the pressure downstream of the constriction should be the same as it was upstream, but actually it is noticeably less. The reason that our prediction is wrong is that we have neglected the frictional, viscous forces which cause a pressure drop along the tube. Despite this pressure drop the pressure is definitely lower at the constriction (because of the increased speed) than it is on either side of it-as predicted by Bernoulli.

The speed u_2 must certainly exceed u_1 to get the same amount of water through the narrower tube. So the water accelerates in going from the wide to the narrow part. The force that gives this acceleration comes from the drop in pressure. We can check our results with another simple demonstration. Suppose we have on a tank a discharge tube which throws a jet of water upward.

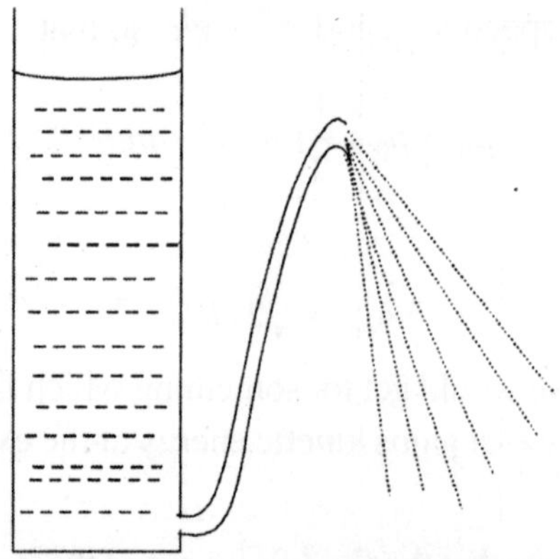

Fig. The jet does not Reach the level of the free Liquid surface Hence the jet Speed is less than $\sqrt{2gh}$ because of Friction due to Viscosity.

If the efflux velocity were exactly $\sqrt{2gh}$, the discharge water should rise to a level even with the surface of the water in the tank. Experimentally, it falls somewhat short. Our prediction is roughly right, but again viscous friction which has not been included in our energy conservation formula has resulted in a loss of energy.

CIRCULATION

The flow of an incompressible fluid with no vorticity satisfies:

$$\nabla.u = 0 \quad \nabla \times u = 0$$

which are identical to the equations of electrostatics or magnetostatics in a vacuum. The divergence of the E is zero when there are no charges, and the curl of the E always zero. The curl of B is zero if there are no currents, and the divergence of B is always zero. Therefore, incompressible, irrotational flow has the same solutions as the equations for B in electrostatics or for B in magnetostatics. For example, the electrostatic analog of flow past a sphere is a uniform electric field plus a dipole. The dipole field is so adjusted that the flow velocity normal to the surface of the sphere is zero. The same problem for the flow past a cylinder can be worked out in a similar way by using a suitable line dipole plus a uniform flow field. This solution holds for a situation in which the fluid velocity at large distances is constant, both in magnitude and direction.

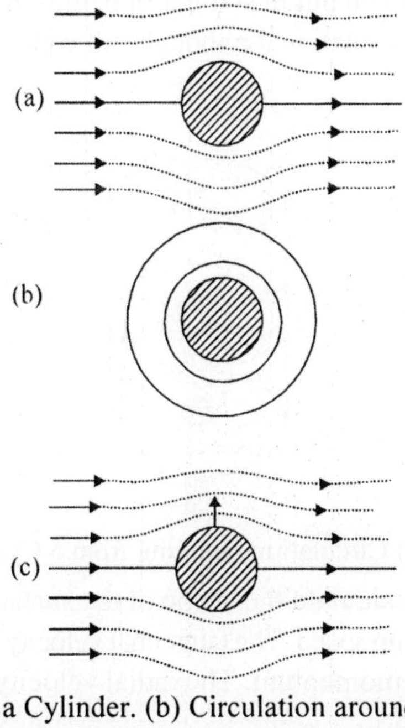

Fig. Ideal flow past a Cylinder. (b) Circulation around a Cylinder ($w = 0$). (c) The Superposition of (a) and (b).

There is a solution for the flow around a cylinder when the fluid at large distances moves in circles around the cylinder. The flow is, then, circular everywhere. Such a flow has a circulation around the cylinder, although $\nabla \times u = 0$ in the fluid. How can there be circulation without curl? We have a circulation around the cylinder because the line integral of u around any loop enclosing the cylinder is not zero. At the same time, the line integral of u around any closed path which does not include the cylinder is zero. The circumstances are the same for the magnetic field around a wire. The curl of B is zero outside of the wire, although a line integral of B around a path which encloses the wire did not vanish. The velocity field in an irrotational circulation around a cylinder is identical to B around a wire carrying a current. For a circular path with its center at the center of the cylinder, the line integral of the velocity is

$$\oint_c u.dl = 2\pi r u_\theta$$

For irrotational flow the integral must be independent of r. Call this constant value Γ, then we have that

$$u_\theta = \frac{\Gamma}{2\pi r}$$

where u_θ is the tangential velocity, and r is the distance from the axis.

There is a nice demonstration of a fluid circulating around a hole. Take a large clear plastic Coke bottle and cut off the base. Fill this funnel with water, stir up some circulation with a stick, and remove the cap. Although you put in some w at beginning, it soon dies down because of viscosity and the flow becomes irrotational-although still with some circulation around the hole.

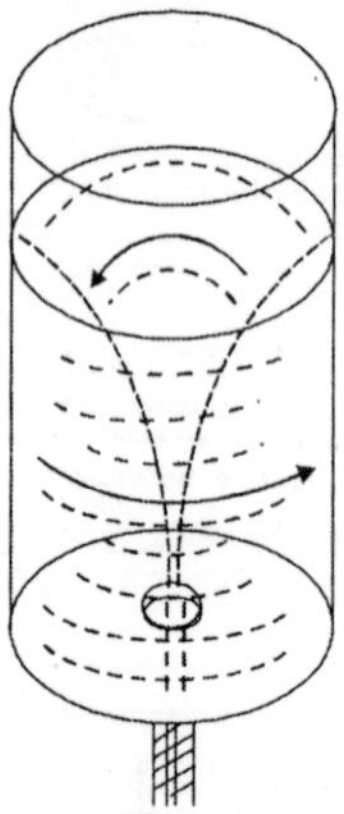

Fig. Water with Circulation Draining from a Cylindrical Funnel.

From the theory, we can calculate the shape of the surface of the water. As a particle of the water moves inward it picks up speed. The tangential velocity goes as $1/r$ is just a manifestation of the conservation of angular momentum. The radial velocity also goes as $1/r$. In cylindrical coordinates, $\nabla.u$ is written as

$$\frac{1}{r}\frac{\partial}{\partial r}(ru_r) + \frac{1}{r}\frac{\partial}{\partial \theta}u_\theta + \frac{\partial}{\partial z}u_z = 0$$

If we assume azimuthal symmetry, $(\partial u_\theta / \partial \theta) = 0$, and that the hole in the funnel is small compared to the diamter of the vessel so that $u_z \simeq 0$, we have water going radially inward toward a hole it follows that the radial velocity is proportional to $1/r$. So the total velocity also increases as $1/r$, and the water goes in along Archimedean spirals. The air-water surface is at atmospheric pressure, so from Bernoulli's theorem it must satisfy

$$gz + \frac{1}{2}u^2 = \text{const.}$$

But u is proportional to $1/r$, so the shape of the surface is

$$(z - z_0) = \frac{k}{r^2}.$$

An interesting point-which is for incompressible, irrotational flow-is that if we have one solution and a second solution, then the sum is also a solution. This is true because the incompressible, irrotational fluid equations, like those of electrostatics or magnetostatics are linear. The complete equations of hydrodynamics are highly non-linear. For the irrotational flow about the cylinder, however, we can superpose the flows and get the new flow pattern shown. This flow is of special interest. The flow velocity is higher on the upper side of the cylinder than on the lower side. The pressures are therefore lower on the upper side than on the lower side. So when we have a combination of a circulation around a cylinder and a net horizontal flow, there is a net vertical force on the cylinder called a lift force. Of course, if there is no circulation, there is no net force on any body according to our theory of "dry" water.

VORTEX LINES

We have already written down the general equations for the flow of an incompressible fluid when there may be vorticity. They are

I. $\nabla . u = 0,$

II. $w = \nabla \times u,$

III. $\frac{\partial w}{\partial t} + \nabla \times (w \times u) = 0.$

(a)

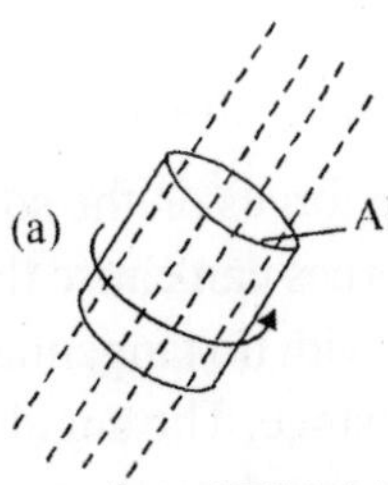

(b)

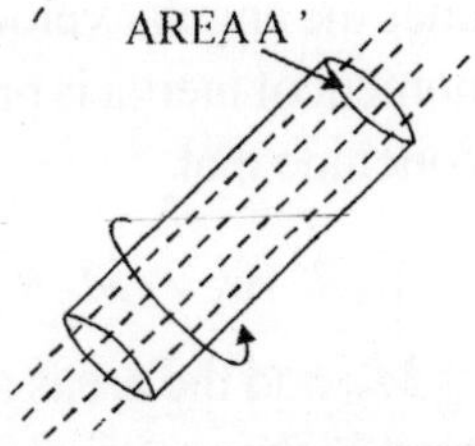

Fig. (a) A group of vortex lines at t; (b) the same group of lines later at t'.

The physical content of these equations has been described in words by Helmholtz in terms of three theorems. First, imagine that in the fluid we were to draw vortex lines rather than streamlines. By vortex lines we mean field lines that have the direction of w and have a density in any region proportional to the magnitude of w. From II the divergence of w is always zero (the divergence of a curl is always zero). Vortex lines are like lines of B, they never start or stop, and tend to go in closed loops. Helmholtz described III in words by the following statement: vortex lines move with the fluid. This means that if you mark the fluid particles along some vortex lines then as the fluid moves and carries those particles along, they will always mark the new positions of the vortex lines. In whatever way the atoms of the liquid move, the vortex lines move with them.

That is one description of the laws that also suggests a method for solving any problem. Given the initial flow pattern, say u everywhere, then you can calculate w. From the u you can also tell where the vortex lines are going to be a little later, they move with the speed u. With the new w you can use I and II to find the new u. That's just like the problem of finding B, given the currents. If we are given the flow pattern at one instant we can in principle calculate it for all subsequent times. We have the general solution for nonviscous flow.

We would like to show how Helmholtz's statement III can be partly understood. It is just the law of conservation of angular momentum applied to the fluid. Imagine a small cylindrical fluid element whose axis is parallel to the vortex lines. At some time later, this same piece of fluid will be somewhere else. Generally it will occupy a cylinder with a different diameter and be in a different place. It may also have a different orientation.

If the diameter has changed, however, the length will have increased to keep the volume constant (since we are assuming an incompressible fluid). Also, since the vortex lines are stuck with the material, their density will go up as the cross-sectional area goes down. The product of the vorticity w and area A of the cylinder will remain constant, so according to Helmholtz, we should have

$$w_1 A_2 = w_1 A_1 .$$

Notice that with zero viscosity all the forces on the surface of the cylinder (or any volume) are normal to the surface. The pressure forces can cause the volume to be moved from place to place, or can cause it to change shape; but with no tangential forces the magnitude of the angular momentum of the material inside cannot change. The angular momentum of the liquid in the little cylinder is its moment of inertia I times the angular velocity of the liquid, which is proportional to the vorticity w. For a cylinder, the moment of inertia is proportional to mr^2. From the conservation of angular momentum, we would conclude that

$$(M_1 R_1^2) w_1 = (M_2 R_2^2) w_2$$

But the mass is the same, $M_1 = M_2$, and the areas are proportional to R^2, so this reduces to $w_2 A_2 = w_1 A_1$. Helmholtz's statement which is equivalent to III is just a consequence of the fact that in the absence of viscosity the angular momentum of an element of the fluid cannot change.

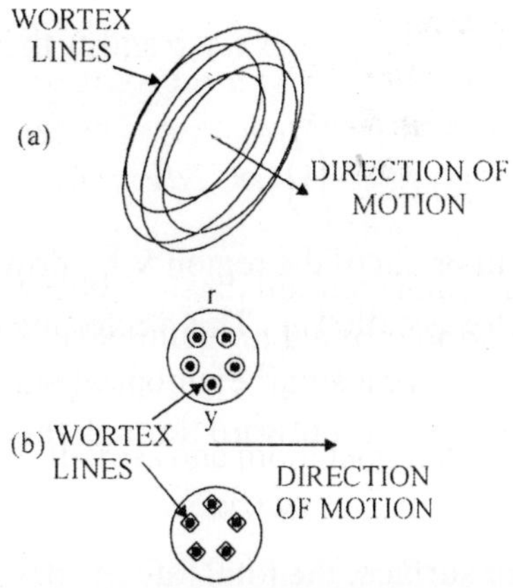

Fig. A Moving Vortex ring. (a) The Vortex lines; (b) a Cross-section of the Ring.

A smoke ring is a torus-shaped bundle of vortex lines (a). Since $w = \nabla \times u$, these vortex lines represent also a circulation of u as shown in part (b) of the figure. We can understand the forward motion of a vortex ring in the following way: The circulating velocity around the bottom of the ring extends up to the top of the ring, having there a forward motion. Since the lines of *w* move with the fluid, they also move ahead with the velocity *u*. Of course, the circulation of *u* around the top part of the ring is responsible for the forward motion of the vortex lines at the bottom.

We must now mention a serious difficulty. We have already noted that III says that, if *w* is initially zero, it will always be zero. This result is a great failure of the theory of ``dry" water, because it means that once *w* is zero it is always zero -- it is impossible to produce any vorticity under any circumstance. Yet, in our simple demonstration with the drum, we can generate a vortex ring starting with air which was initially at rest. Also, we all know that we can start some vorticity in a lake with a paddle. Clearly, we must go to a theory of "wet" water to get a complete understanding of the behavior of a fluid.

Another feature of the "dry" water theory which is incorrect is the supposition we make regarding the flow at the boundary between it and the surface of a solid. When we discussed the flow past a cylinder we permitted the fluid to slide along the surface of the solid. The velocity at a solid surface could have any value depending on how it got started, and we did not consider any friction between the fluid and the solid. It is an experimental fact, however, that the velocity of a real fluid always goes to zero at the surface of a solid object. Therefore, our solution for the cylinder, with or without circulation, is wrong, as is our result regarding the generation of vorticity.

THE HEAT EQUATION

Consider a solid material occupying a region of space V. The region has a boundary surface, which we shall designate as S. Suppose the solid has a density ñ and a heat capacity c. If the temperature of the solid at any point in V is T (r, t), where

$$\vec{r} = x\hat{i} + y\hat{j} + z\hat{k}$$

is the position vector (so that T depends upon x, y, z and t), then the total heat energy contained in the solid is

$$\iiint_V \rho CTdV$$

Heat energy can only get in or out of the region V by flowing across the boundary S. Let's suppose that the heat flux vector is called $\vec{q}$. This vector measures rate of energy flow past a point per unit area. Thus, if we have a small element of surface area d S with outward unit normal vector n, the rate of energy flow outward through this element of surface area is

$$\vec{q} \cdot \hat{n} dS$$

Integrating over the entire surface, the total rate of energy flow (i.e., the flux) out of the region is therefore

$$\oiint_S \vec{q} \cdot \hat{n} dS$$

Conservation of energy then requires that the rate at which the energy within the region V and the rate at which energy crosses the boundary of region must balance, i.e.,

$$\frac{d}{dt}\left(\iiint_V \rho CTdV\right) + \oiint_S \vec{q} \cdot \hat{n} dS = 0$$

The signs of the terms must appear in this manner, because if the flux of energy out of the region is positive, then the total energy inside the region must be decreasing, hence its derivative will be negative.

To simplify this, we take the time derivative inside the volume integral.

$$\frac{d}{dt}\iiint_V \rho CTVdV = \iiint_V \frac{\partial}{\partial t}(\rho CT)dV$$

We can do this because the region V over which the integral is being taken is independent of time; V is the same region for all time. Thus, at any fixed time, integration over V just gives a single number, i.e., the energy inside V at that time. The energy will change with time because T (and possibly ñ and c) changes with time.

Note that the time derivative must be turned into a partial derivative when it is moved inside the volume integral, however, because the functions being integrated depend upon both position and time. Similarly, we apply the divergence theorem to the flux integral

$$\oiint_S \vec{q} \cdot \vec{n} dS = \iiint_V \vec{\nabla} \cdot \vec{q} dV$$

The result is

$$\iiint_V \frac{\partial}{\partial t}(\rho CT)dV + \iiint_V \vec{\nabla} \cdot \vec{q} dV$$

Or equivalently,

$$\iiint_V \left[\frac{\partial}{\partial t}(\rho CT) + \vec{\nabla}\cdot\vec{q}\right] dV = 0$$

It would be nice to say at this point that the quantity in brackets is zero; this would allow us to get rid of the integral. This would be true, but the right argument is needed, because an integral can be zero either because the integrand is zero or because it is equally positive and negative, and when integrated there is perfect cancellation.

In the present case, however, there is one more fact: the integral must be zero for arbitrary volumes V. This forces the integrand to be zero everywhere. The proof is by contradiction. First, assume that the integrand is not zero everywhere. Pick a point at which it is positive, and then choose a region V that is entirely within the region where it's positive. Since we are integrating a quantity that is entirely positive, the resulting integral must be zero. This can't be true, however, since we know the integral must be zero. Therefore, we must have

$$\frac{\partial}{\partial t}(\rho CT) + \vec{\nabla}\cdot\vec{q} = 0$$

This is the partial diûerential equation form of the fundamental principle of energy conservation for heat transfer. More can be done if one has an explicit result for the heat flux q in terms of the temperature. In many cases, observations show that the heat flux is given by Fick's law,

$$\vec{q} = -k\vec{\nabla}T$$

i.e., the heat transfer is proportional to the negative of the temperature gradient. Thus, heat tends to flow from hotter regions to colder regions. While this is true in most cases, it is not always true; in particular, if phase transitions are possible it can be violated.

If we use Fick's law in the equation for conservation of heat energy, it becomes

$$\frac{\partial}{\partial t}(\rho CT) = \vec{\nabla}\cdot\left(k\vec{\nabla}T\right)$$

Finally if r, *C* and *k* are all constants, this equation simplifies to the heat equation.

$$\frac{\partial T}{\partial t} = k\nabla^2 T$$

where

k = *k*/r*C*

is the thermal diûusivity.

A special case of this equation results if the object is in thermal equilibrium, i.e., the temperature doesn't change with time:

$$\nabla^2 T = 0$$

This is known as Laplace's equation.

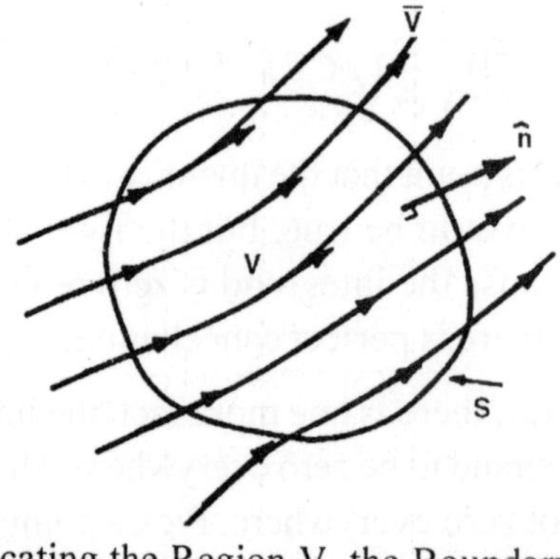

Fig. Schematic Diagram Indicating the Region V, the Boundary Surface S, the Normal to the Surface ^n, the Fluid Velocity Vector Field *v*, and the Particle paths (Dashed Lines).

FLUID MECHANICS

Conservation of Mass

Conservation laws are also the basis for the equations of fluid mechanics. Let ñ be the fluid density and $\vec{v}$ be the fluid velocity. Then the mass in a region V is given by

$$\iiint_V \rho dV$$

and the rate at which mass leaves the region through its boundary S is

$$\oiint_S \rho \vec{v} \cdot \vec{n} dS$$

Here ñ *v* is the mass flux vector; this has units of kg/m^3 × m/sec = kg/m^{2}sec, i.e., mass per unit area per second. The rate of change of the mass in V must be balanced by the rate at which mass leaves through S,

$$\frac{d}{dt} \iiint_V \rho dV + \oiint_S \rho \vec{v} \cdot \vec{n} dS = 0$$

Pulling the time derivative inside the V integral, and converting the surface integral into a volume integral using the divergence theorem, we have

$$\iiint_V \left[\frac{\partial \rho}{\partial t} + \vec{\nabla} \cdot (\rho \vec{v}) \right] dV = 0$$

As before, because the region V is arbitrary, we must have the terms between the brackets be identically zero, which gives

$$\frac{\partial \rho}{\partial t} + \vec{\nabla} \cdot (\rho \vec{v}) = 0$$

This is the partial diûerential equation form of conservation of mass.

Equation (20) can be simplified somewhat by using the vector identity

$$\vec{\nabla}\cdot(\rho\vec{v}) = \vec{\nabla}\rho\cdot\vec{v} + \rho\vec{\nabla}\cdot\vec{v} = \vec{v}\cdot\vec{\nabla}\rho + \rho\vec{\nabla}\cdot\vec{v}$$

Which gives.

$$\frac{\partial\rho}{\partial t} + \vec{v}\cdot\vec{\nabla}\rho + \rho\vec{\nabla}\cdot\vec{v} = 0$$

The combination $\frac{\partial}{\partial t} + \vec{v}\cdot\vec{\nabla}$ occurs so much in fluid mechanics that it is given a special notation,

$$\frac{\partial}{\partial t} + \vec{v}\cdot\vec{\nabla} \equiv \frac{D}{Dt}$$

This is called the material or substantial derivative, and physically it represents taking the time derivative 'following the fluid', i.e., taking the time derivative of something transported along by the fluid. One can see this by considering ñ(x, t) not at a fixed position *x*, but instead ñ(x(t), t), where d*x*/d*t* = *v*; because the position x moves at the fluid velocity *v*, this is the density of a specific fluid element. Taking the time derivative using the chain rule,

$$\rho(\vec{x},t) = \vec{\nabla}\rho\cdot\frac{d\vec{x}}{dt} + \frac{\partial\rho}{\partial t} = \frac{\partial\rho}{\partial t} + \frac{d\vec{x}}{dt}\cdot\vec{\nabla}\rho = \frac{\partial\rho}{\partial t} + \vec{v}\cdot\vec{\nabla}\rho = \frac{D\rho}{Dt}$$

In any event, with this notation we have

$$\frac{D\rho}{Dt} + \rho\vec{\nabla}\cdot\vec{v} = 0$$

This has a nice physical interpretation: if $\vec{\nabla}.\vec{v} > 0$, so that the fluid velocity is locally expanding, then Dñ/Dt < 0, i.e., the local density is decreasing. Similarly, if $\vec{\nabla}.\vec{v} < 0$, so the fluid velocity is locally compressing, then Dñ/Dt > 0 and the density is increasing.

We can also consider the case of an incompressible fluid : if ñ is constant, then Dñ/Dt = 0 and therefore $\vec{\nabla}\cdot\vec{v} = 0$

Water is an example of a fluid that is almost incompressible.

CONSERVATION OF MOMENTUM

The next main conservation law is momentum. Actually, of course, the law is that the time rate of change of momentum is equal to force, d(momentum)/dt = force. To apply it to a fluid, we need to identify all of the sources of momentum and all of the forces involved.

First, the momentum density (per unit volume) of the fluid is ñv. Therefore, the momentum in a volume V of fluid is

$$\iiint_V \rho\vec{v}dV$$

Of course, what we really need is the time rate of change of this momentum,

$$\frac{d}{dt}\iiint_V \rho\vec{v}dV$$

Another way that the momentum in V can change is if the fluid transports momentum out of the region across the boundary S .

$$\oiint_S \rho\vec{v}(\vec{v}\cdot\vec{n})dS$$

This is similar to the previous case of conservation of mass, where we had mass density times the normal component of the velocity, giving mass flux. In this case, we have momentum density times the normal component of the velocity, giving momentum flux.

We also need the forces acting on V. First of all, we have the net pressure force, which is

$$-\oiint_S p\vec{n}dS$$

since the pressure acts normally to each element of the surface (with an inward force when the pressure is positive, hence the minus sign). Similarly, if we have any body forces, such as gravity, we have to include them. It's traditional to use $\vec{f}$ as the body force per unit mass, so that

$$\iiint_V \vec{f}\,\vec{v}dV$$

is the total body force.

Putting everything together, we have conservation of momentum:

$$\frac{d}{dt}\iiint_V \rho\vec{v}dV + \oiint_S \rho\vec{v}(\vec{v}\cdot\vec{n})dS = -\oiint_S p\vec{n}dS + \iiint_V \vec{f}\,\vec{v}dV$$

Note that other forces have been neglected, such as forces due to viscosity, where a bit of fluid exerts a drag force on nearby fluid elements.

The next step should be to convert the surface integrals to volume integrals using the divergence theorem. Unfortunately, the form of the surface integrals is not right for the divergence theorem, because the integrands of the surface integrals are vectors, not scalars. Thus, we need to somehow modify the divergence theorem. Let's look first at the force due to pressure. Here, we are missing a vector. We can fix this by taking the dot product of the integral (neglecting the " sign for the moment) with a constant vector a:

$$\vec{a}\cdot\oiint_S p\vec{n}dS = \oiint_S p\vec{a}\cdot\vec{n}dS = \oiint_S (p\vec{a})\cdot\vec{n}dS$$

Now we can apply the divergence theorem

$$\oiint_S (p\vec{a})\cdot\vec{n}dS = \iiint_V \vec{\nabla}\cdot(p\vec{a})dV$$

which give the desired result.

EXERCISE

1. A pump has a 30 cm diameter suction pipe and 25 cm diameter delivery pipe. When 220 L/s of water was being pumped, the pressure on the suction side of the pump was 4 m of vacuum and on the delivery side the pressure was 100 kPa. Assuming an efficiency of 50% for the pump-motor set, estimate the electrical power consumed.

2. The velocity in a flow is found to vary as $\upsilon_x = V_0 / \left(1-\frac{x}{L}\right)^2$. Determine the acceleration at a point $x = 0.5$ m when $L = 1.5$ m and $V_0 = 3.0$ m/s.

3. A mercury-water defferential manometer connected to the 15 cm diameter suction pipe and 12 cm delivery pipe of a pump shows a deflection of 40 cm. The centrelines of the suction and delivery pipes are at the same level. If the pump is discharging 70 L/s of water, estimate the head developed by the pump.

4. Calculate the pressure and density of air at an elevation of 4500 m above sea level, if the atmospheric pressure and density at sea level are 101.35 kPa and 1.2255 kg/m^3. Assume isothermal process.

5. A hydrometer is to be built that the mark corresponding to relative density of 1.0 and 1.5 are 8 cm apart on a 5 mm diameter steam. How far from the mark of 1.0 will the relative density mark of 1.25 be located?

6. A 40 cm diameter pipe is reduced uniformly to 20 cm diameter in a length of 0.5 m. If the steady discharge of water through this pipe is 100 L/s, determine the acceleration at 10 cm from the 20 cm diameter and (i) when the flow is steady and (ii) when the flow is increasing at a rate of 600 L/s per minute.

7. A bourdon gauge connected to a closed tank indicates 3f kPa at a place where the atmospheric pressure is 95.480 kPa. If the tank is now carried to ahigher elevation where the atmospheric pressure is 70.11 kPa, what reading will be indicated by the gauge?

8. A 15 cm diameter jet of water with a velocity of 15 m/s strikes a plane normally. If the plate is moving with a velocity of 6 m/s in the direction of the jet calculate the work done per second on the plate and plate and the efficiency (η) of energy transfer.

9. A hydrometer has a 6 mm diameter stem. The distance between markings of relative density 1.0 and 0.90 is 10 cm. determine the weight of the hydrometer.

10. A cylinder of 0.8 m diameter is made of 5 mm thick plates. If the maximum permissible stress in the plate is 100 MPa, calculate the maximum permissible pressure inside the cylinder.

11. A solid cube of side 0.5 m is made of a material of relative density 0.5. The cube floats in aliquid of relative density 0.95 with two of its faces horizontal. Examine its stability.

Encyclopaedia
of
VECTOR CALCULUS

Encyclopaedia
of
VECTOR CALCULUS

Volume 2

Dr. Rakesh Kumar Pandey

ANMOL PUBLICATIONS PVT. LTD.
NEW DELHI - 110 002 (INDIA)

ANMOL PUBLICATIONS PVT. LTD.

Regd. Office: 4360/4, Ansari Road, Daryaganj,
New Delhi-110 002 (India)
Ph.: 23278000, 23261597

Branch Office: No. 1015, Ist Main Road, BSK IIIrd Stage
IIIrd Phase, IIIrd Block,
Bangalore-560 085 (India)
Tel.: 080-41723429
Visit us at: www.anmolpublications.com

Encyclopaedia of Vector Calculus

First Edition, 2009

ISBN 978-81-261-4110-4 (Set)

PRINTED IN INDIA

Printed at Mehra Offset Press, Delhi.

Contents

Preface

Encyclopaedia of Vector Calculus is the foundation stone on which a vast amount of Applied Mathematics is based. Topics such as Fluid Dynamics, Solid Mechanics and Electromagnetism depend heavily on the Calculus of Vector Quantities in Three Dimensions. This Encyclopaedia covers the material in a comprehensive but concise manner, combining mathematical rigour with Physical Insight. There are many diagrams to illustrate the physical meaning of the Mathematical concepts, which is essential for a full understanding of the subject. There are worked examples that cover all of the material. The a complete chapter is devoted to most important applications of Vector Calculus including *Heat, Mechanics and Electromagnetism*.

Author

List of Symbols

$\pm$	$\not\subset$	$\subset$	$\supset$	$\subseteq$	$\in$	$\notin$	∂
$\neq$	$\equiv$	$\approx$	$\angle$	∇	$\sum$	Π	Ω
ϕ	α	β	χ	ε	γ	η	λ
μ	ν	$\perp$	π	θ	ρ	σ	τ
$\Re$	ω	ξ	ψ	ζ	Υ	$\exists$	$\int$
$\iint$	$\iiint$	$\oint$	$\times$	$\sqrt{}$	$\therefore$	$\because$	Δ
$\geq$	$\leq$	$\Rightarrow$,	$\Downarrow$	$\Leftrightarrow$	∞	$\propto$	$\mathbb{R}$
$\mathbb{Z}$	$\mathbb{N}$	$\mathbb{C}$	$\mathbb{Q}$				

Chapter 1

Vectors in Space

A vector is a directed line segment. The vector represented by the arrow from a point P to a point Q is denoted by $\overrightarrow{PQ}$. Two vectors are regarded as equal when they have the same length and direction. $\overrightarrow{P_1Q_1} = \overrightarrow{P_2Q_2}$ means that the directed line segments from P_1 to Q_1 and P_2 to Q_2 represent the same vector.

A vector represents a parallel translation of the space:

$$\vec{a} = \overrightarrow{PQ}.$$

corresponds to the translation of the point P into the point Q, pictured by an arrow.

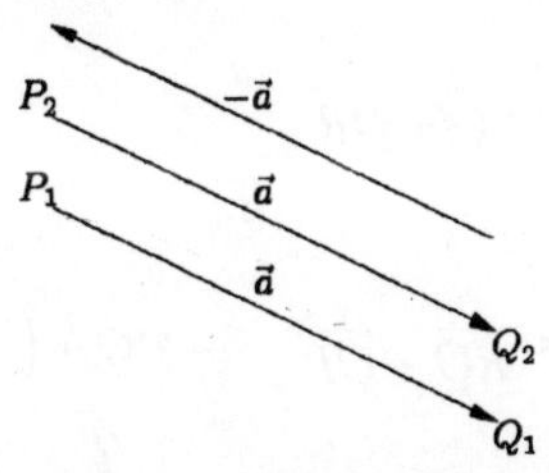

As the figure shows, arrows with the same direction and length represent the same vector. The special representation with respect to the origin of the coordinate sytem is called position vector and defines the vector's coordinates:

$$\overrightarrow{OA} = \begin{pmatrix} a_1 \\ a_2 \\ a_3 \end{pmatrix}.$$

The coordinates of $\vec{a}$ can also be obtained by calculating the difference of the recpective coordinates of the points P and Q. Finally, the zero vector is denoted by $\overrightarrow{OO} = \overrightarrow{PP} = \vec{0}$.

ADDITION OF VECTORS

The sum of two vectors corresponds to the composition of two translations:

$$\overrightarrow{PR} = \overrightarrow{PQ} + \overrightarrow{QR}.$$

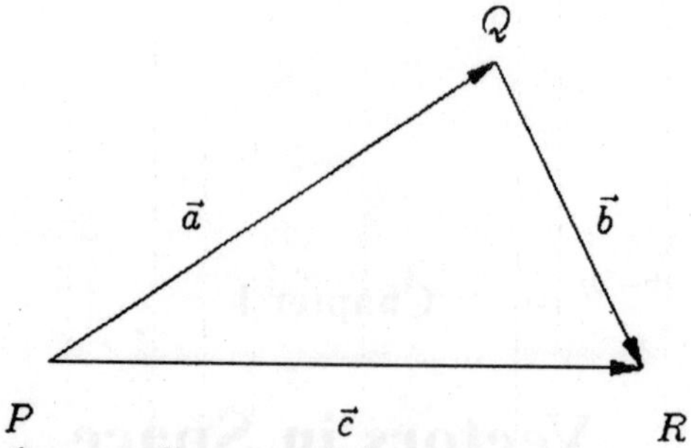

Correspondingly for the coordinates, we have

$$\vec{c} = \vec{a} + \vec{b} = \begin{pmatrix} a_1 \\ a_2 \\ a_3 \end{pmatrix} + \begin{pmatrix} b_1 \\ b_2 \\ b_3 \end{pmatrix} = \begin{pmatrix} a_1 + b_1 \\ a_2 + b_2 \\ a_3 + b_3 \end{pmatrix}.$$

The translation opposite to $\overrightarrow{PQ}$ is denoted by

$$\overrightarrow{PQ} = - \overrightarrow{PQ}.$$

For the points

$$P = (1, 1, 1), \quad Q = (4, 2,7), \quad R = (1, 4, 3)$$

we compute the vectors

$$\overrightarrow{PR} = \overrightarrow{PQ} + \overrightarrow{QR}$$

and

$$\overrightarrow{RP} = \overrightarrow{RQ} + \overrightarrow{QP} = -\overrightarrow{PQ} + \left(-\overrightarrow{QR}\right).$$

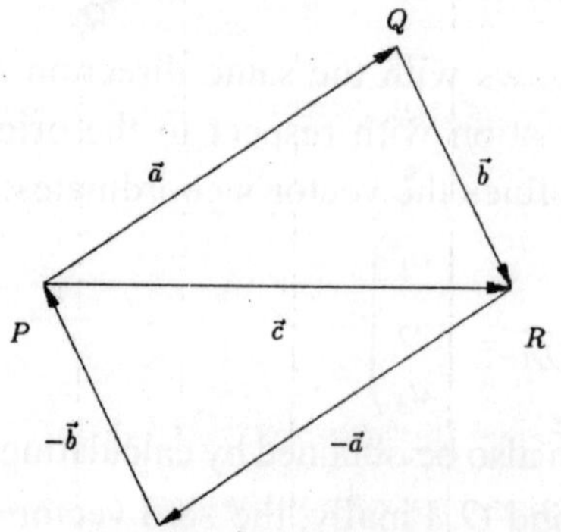

As illustrated in the figure, we have for the coordinates

$$\vec{c} = \vec{a} + \vec{b} = \begin{pmatrix} 4-1 \\ 2-1 \\ 7-1 \end{pmatrix} + \begin{pmatrix} 1-4 \\ 4-2 \\ 3-7 \end{pmatrix} = \begin{pmatrix} 3 \\ 1 \\ 6 \end{pmatrix} + \begin{pmatrix} -3 \\ 2 \\ -4 \end{pmatrix} = \begin{pmatrix} 0 \\ 3 \\ 2 \end{pmatrix}$$

$$-\vec{b} + (-\vec{a}) = \begin{pmatrix} 4-1 \\ 2-4 \\ 7-3 \end{pmatrix} + \begin{pmatrix} 1-4 \\ 1-2 \\ 1-7 \end{pmatrix} = \begin{pmatrix} -3 \\ 2 \\ -4 \end{pmatrix} + \begin{pmatrix} -3 \\ -1 \\ -6 \end{pmatrix} = \begin{pmatrix} 0 \\ -3 \\ -2 \end{pmatrix} = -\vec{c}$$

By an inclined plane we understand a plane that is tilted with respect to the horizontal. As illustrated in the figure, the weight force can be resolved into two mutually perpendicular components (resolving of a force by means of the parallelogram of forces):

$$\vec{f}_G = \vec{f}_N + \vec{f}_H.$$

The normal force $\vec{f}_N$ acts orthogonally to the inclined plane. Whereas the slope deviation $\vec{f}_H$ force is parallel to the inclined plane and acts downwards.

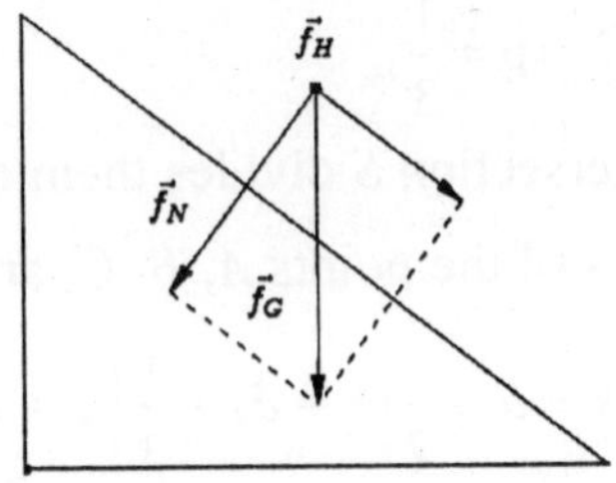

SCALAR MULTIPLICATION

The vector $s\vec{a}$ corresponds to an ' s-fold' translation, that is

$$s\begin{pmatrix} a_1 \\ a_2 \\ a_3 \end{pmatrix} = \begin{pmatrix} sa_1 \\ sa_2 \\ sa_3 \end{pmatrix}.$$

Particularly, we have $0\vec{a} = \vec{0}$.

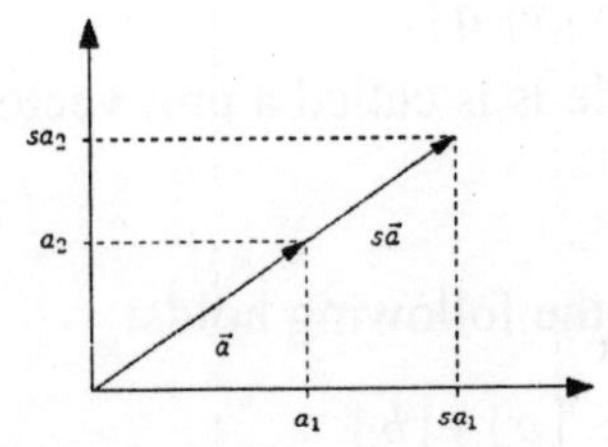

POINT OF INTERSECTION OF THE MEDIANS

One can reach the point along f different ways:

$$\begin{aligned} \overrightarrow{AS} &= \frac{1}{2}\overrightarrow{AB} + \lambda\overrightarrow{M_cC} \\ &= \frac{1}{2}\overrightarrow{AB} + \lambda\left(\overrightarrow{AC} - \frac{1}{2}\overrightarrow{AB}\right) \end{aligned}$$

and

$$\overrightarrow{AS} = \frac{1}{2}\overrightarrow{AB} + \mu\overrightarrow{M_cC}$$

$$= \frac{1}{2}\overrightarrow{AB} + \mu\left(\overrightarrow{AC} - \frac{1}{2}\overrightarrow{AC}\right)$$

Equating we obtain

$$\left(\frac{1}{2} - \frac{1}{2}\lambda - \mu\right)\overrightarrow{AB} = \left(\frac{1}{2} - \lambda - \frac{1}{2}\mu\right)\overrightarrow{AC}.$$

Since the vectors $\overrightarrow{AB}$ and $\overrightarrow{AC}$ have different directions the coefficients must be 0. From this we conclude

$$\lambda = \mu = \frac{1}{3},$$

that is, the medians' point of intersection S divides them at a ratio of 1:2.

Denoting the postion vectors of the points A, B, C, and S by $\vec{a}, \vec{b}, \vec{c}$ and, we obtain

$$\vec{s} = \vec{a} + \frac{1}{2}(\vec{b} - \vec{a}) + \frac{1}{3}\left(\vec{c} - \vec{a} - \frac{1}{2}(\vec{b} - \vec{a})\right)$$

$$= \frac{1}{3}\left(\vec{a} + \vec{b} + \vec{a}\right).$$

ABSOLUTE VALUE

The magnitude $\vec{a} = \overline{PQ}$ of is the length of the arrow pointing from P to Q or from Q to A, resp., that is,

$$|\vec{a}| = \sqrt{a_1^2 + a_2^2 + a_3^2}\,.$$

In particular we have $|s\vec{a}| = |s||\vec{a}|$.

Any vector whose magnitude is is called a unit vector.

TRIANGLE INEQUALITY

For the sum of two vectors the following holds:

$$|\vec{a} + \vec{b}| \le |\vec{a}| + |\vec{b}|$$

We have equality if $\vec{a}$ and $\vec{b}$ only if and are parallel, that is $\vec{a} = s\vec{b}$ with $s \ge 0$.

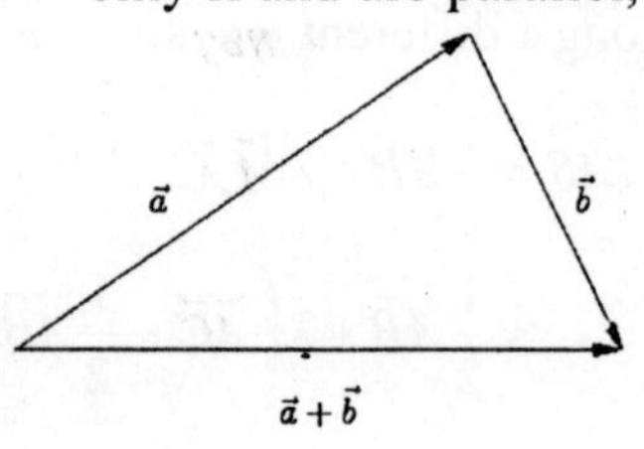

CALCULATION RULES

The following calculation rules apply to vectors:

$$\vec{a} + \vec{0} = \vec{a},$$

$$\vec{a} + (-\vec{a}) = \vec{0},$$

$$\vec{a} + \vec{b} = \vec{b} + \vec{a} \qquad \text{(commutative law)},$$

$$\vec{a} + (\vec{b} + \vec{c}) = (\vec{a} + \vec{b})\vec{c} \qquad \text{(associative law)}$$

$$s(\vec{a} + \vec{b}) = s\vec{a} + s\vec{b} \qquad \text{(distributive law)}$$

DRIFT OF AN AIRPLANE

An airplane travels with speed km/h due east while the wind velocity is km/h from WSW.

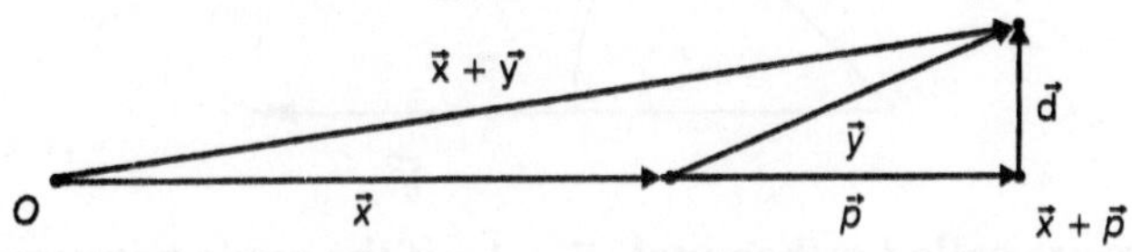

In the picture east is the direction of the horizontal axis and north is the direction of the vertical one. The velocity vector of the air plane itself is therefore

$$\vec{x} = \begin{pmatrix} 800 \\ 0 \end{pmatrix}$$

and the vector of the wind velocity is given by

$$\vec{y} = \begin{pmatrix} 50\cos(\pi/8) \\ 50\sin(\pi/8) \end{pmatrix}.$$

Starting at O the coordinates of the air plane are after one hour (superposition of the motions)

$$\vec{x} + \vec{y} = \begin{pmatrix} 800 + 50\cos(\pi/8) \\ 50\sin(\pi/8) \end{pmatrix} \approx \begin{pmatrix} 846.19 \\ 19.13 \end{pmatrix}$$

If $\vec{p}$ denotes the orthogonal projection of $\vec{y}$ on the line with direction $\vec{x}$, then the velocity of the air plane relative to the earth is the sum of the component due east

$$\vec{x} + \vec{p} = \begin{pmatrix} 800 + 50\cos(\pi/8) \\ 0 \end{pmatrix} \approx \begin{pmatrix} 846.19 \\ 0 \end{pmatrix}$$

and the drift

$$\vec{d} = \vec{x} + \vec{y} - (\vec{x} + \vec{p}) = \begin{pmatrix} 0 \\ 50\cos(\pi/8) \end{pmatrix} \approx \begin{pmatrix} 0 \\ 19.13 \end{pmatrix}$$

towards north.

SCALAR PRODUCT

ANGLE

Given two vectors $\vec{a}, \vec{b} \neq \vec{0}$. The associated arrows, attached with their tails to a given point (vertex), form two angels of which the smaller one is denoted by $\sphericalangle(\vec{a}, \vec{b}) \in [0, \pi]$.

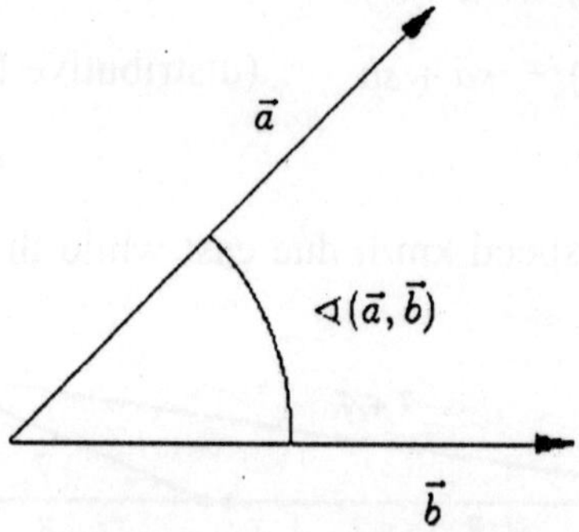

The two vectors are called orthogonal, $\vec{a} \perp \vec{b}$, if the angle between them equals $\pi/2$. By convention, the zero vector is orthogonal to every vector.

The following figure illustrates some typical cases.

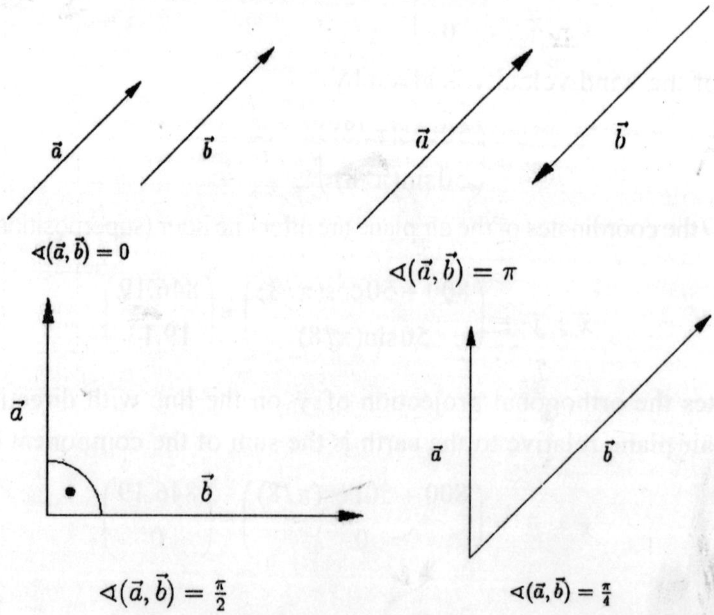

LAW OF COSINES

n a triangle, for the angle γ opposite to side $\overline{AB}$ the following holds true:

$$c^2 = a^2 + b^2 - 2ab\cos\gamma.$$

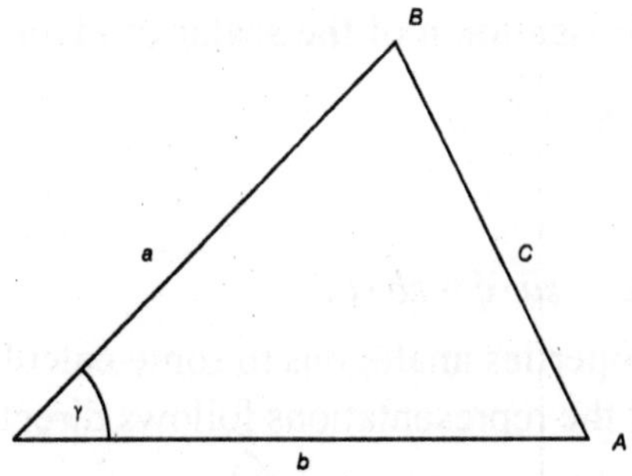

If $\gamma = \pi/2$, then we obtain the special case of the Pythagorean Theorem: $c^2 = a^2 + b^2$.

In the proof we will use the Pythagorean Theorem.

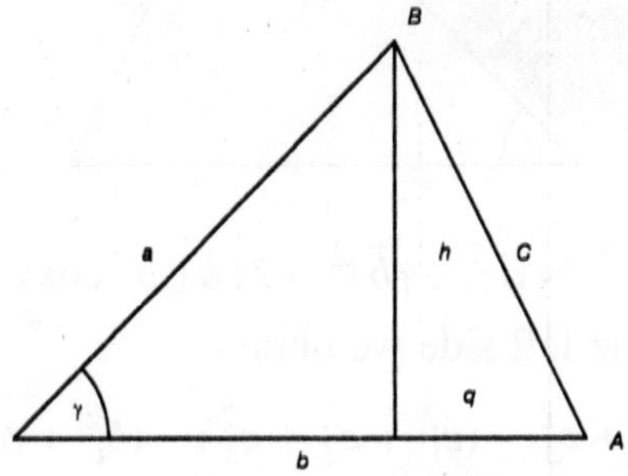

We have

$$c^2 = h^2 + q^2, \quad h^2 = a^2 + p^2$$

and

$$q = b - p, \qquad p = a\cos\gamma.$$

Consequently we obtain

$$\begin{aligned} c^2 &= (a^2 - p^2) + (b - p)^2 \\ &= a^2 - b^2 + b^2 - 2b(a\cos\gamma) + p^2 \\ &= a^2 + b^2 - 2ab\cos\gamma, \end{aligned}$$

as desired.

SCALAR PRODUCT, DOT PRODUCT

The scalar product of two vectors is defined by

$$\begin{aligned} \vec{a} \cdot \vec{b} &= |\vec{a}||\vec{b}|\cos\sphericalangle(\vec{a}, \vec{b}) \\ &= a_1b_1 + a_2b_2 + a_3b_3. \end{aligned}$$

In particular, we have

$$\vec{a} \cdot \vec{b} = |\vec{a}|^2$$

and $\vec{a}\cdot\vec{b}=0 \quad \Leftrightarrow \vec{a}\perp\vec{b}.$

From the coordinate representation of the scalar product it follows

$$\vec{a}\cdot\vec{b} = \vec{b}\cdot\vec{a}$$

and

$$(s\vec{a}+r\vec{b})\cdot\vec{c} = s\vec{a}\cdot\vec{c} + r\vec{b}\cdot\vec{c},$$

i.e., the scalar product has properties analogous to some calculation rules of the product of real numbers. Equivalence of the representations follows directly from the Law of Cosines:

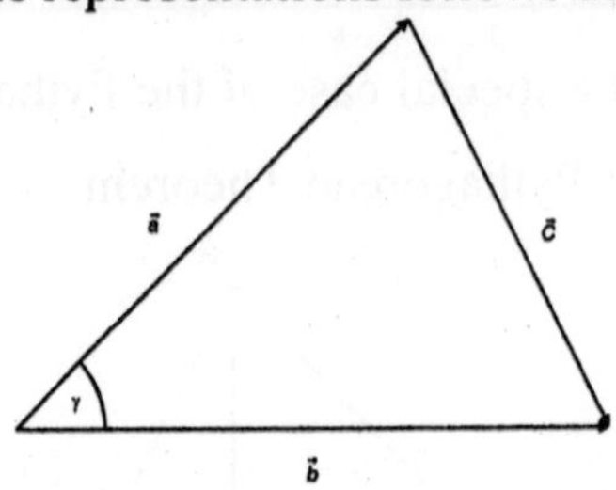

$$|\vec{c}|^2 = |\vec{a}|^2 + |\vec{b}|^2 - 2|\vec{a}||\vec{b}|\cos\gamma.$$

Bringing all squares to the left side we obtain

$$c_1^2 + c_2^2 + c_3^2 - (c_1^2 + c_2^2 + c_3^2) - (b_1^2 + b_2^2 + b_3^2)$$

After substituting $c_i^2 = (b_i - a_i)^2$ all squares nullify each other. The remaining mixed terms coincide with the scalar product multiplied by -2.

The depicted triangle has the vertices

$$A = (6, 0), \qquad B = (4, 4), \qquad C = (0,0)$$

and we have

$$\vec{a} = \overrightarrow{CB} = \begin{pmatrix}4\\4\end{pmatrix}, \quad \vec{b} = \overrightarrow{CA} = \begin{pmatrix}6\\0\end{pmatrix}, \quad \vec{c} = \overrightarrow{BA} = \begin{pmatrix}2\\-4\end{pmatrix}$$

The angle γ can be computed by means of the scalar product:

$$\cos\gamma = \frac{\begin{pmatrix}4\\4\end{pmatrix}\cdot\begin{pmatrix}6\\0\end{pmatrix}}{\left|\begin{pmatrix}4\\4\end{pmatrix}\right|\left|\begin{pmatrix}6\\0\end{pmatrix}\right|} = \frac{24}{6\sqrt{32}} = \frac{1}{\sqrt{2}}.$$

We obtain $\gamma = 45°$, as it can be seen directly from the coordinates.

By means of our example we can also verify the Law of Cosines. We have

$$|\vec{c}|^2 - |\vec{a}|^2 - |\vec{b}|^2 = 20 - 32 - 36 = -48,$$

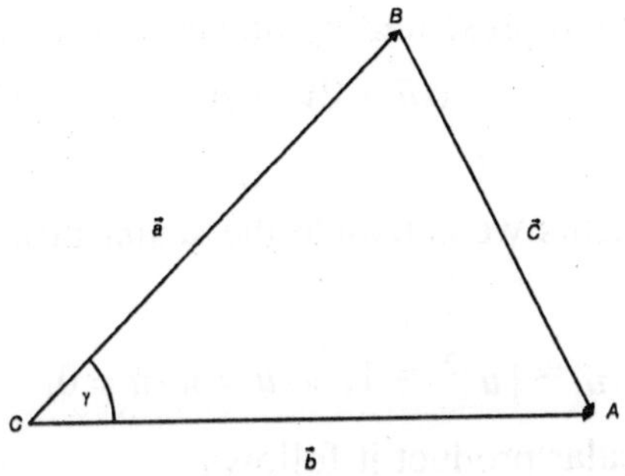

which is in accordance with

$$-2\,|\,\vec{a}\,|-|\,\vec{b}\,|\cos\gamma = -2(4\sqrt{2})\,6/\sqrt{2}.$$

ORTHOGONAL BASIS

An orthonormal basis consists of mutually orthogonal unit vectors $\vec{u}, \vec{v}, \vec{w}$.

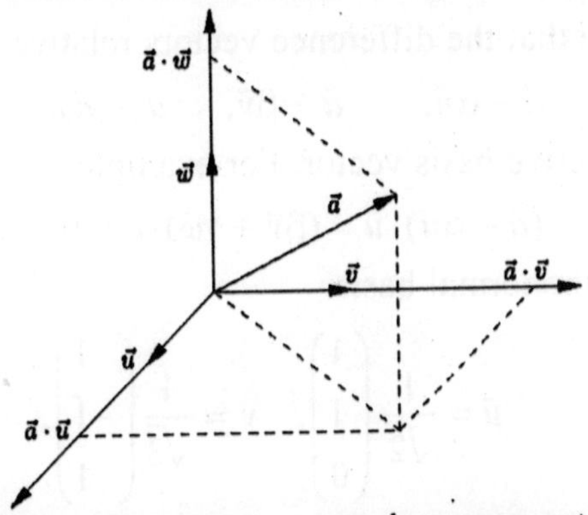

As illustrated in the figure, any vector $\vec{a}$ can be represented by a linear combination

$$\vec{a} = (\vec{a}\cdot\vec{u})\vec{u} + (\vec{a}\cdot\vec{v}) + (\vec{a}\cdot\vec{w})\vec{w}.$$

The addends are the projections of $\vec{a}$ onto the axes generated by the vectors of the ONB. For the coefficients we have

$$|\,\vec{a}\cdot\vec{u}\,|^2 + |\,\vec{a}\cdot\vec{v}\,|^2 + |\,\vec{a}\cdot\vec{w}\,|^2 = |\,\vec{a}\,|^2$$

In particular, we have

$$\begin{pmatrix} a_1 \\ a_2 \\ a_3 \end{pmatrix} = a_1\vec{e}_x + a_2\vec{e}_y + a_3\vec{e}_z$$

for the canonical orthonormal basis of the Cartesian coordinate system

$$\vec{e}_x = \begin{pmatrix} 1 \\ 0 \\ 0 \end{pmatrix},\quad \vec{e}_y \begin{pmatrix} 0 \\ 1 \\ 0 \end{pmatrix},\quad \vec{e}_z = \begin{pmatrix} 0 \\ 0 \\ 1 \end{pmatrix}.$$

That any vector $\vec{a}$ can be represented by linear combination

$$\alpha\vec{u} + \beta\vec{v} + \gamma\vec{w}$$

is geometrically evident.

To compute the coefficients we calculate the scalar products with the basis vectors. Since

$$\vec{u}\cdot\vec{u} = |\vec{u}|^2 = 1,\ \vec{v}\cdot\vec{u} = \vec{w}\cdot\vec{u} = 0,$$

and by the linearity of the scalar product it follows

$$\vec{a}\cdot\vec{u} = (\alpha\vec{u} + \beta\vec{v} + \gamma\vec{\omega})\cdot\vec{u} = \alpha.$$

Similarly we may affirm the formulas for β und γ. We can verify the formula for the sum of squares of the coefficients by computing

$$|\vec{a}|^2 = (\alpha\vec{u} + \beta\vec{v} + \gamma\vec{\omega})\cdot(\alpha\vec{u} + \beta\vec{v} + \gamma\vec{\omega}).$$

We can also easily see that the difference vectors relative to the projections,

$$\vec{a} - \alpha\vec{u},\qquad \vec{a} - \beta\vec{v},\qquad \vec{a} - \gamma\vec{\omega},$$

are orthogonal to the respective basis vector. For example:

$$(\vec{a} - \alpha\vec{u})\cdot\vec{u} = (\beta\vec{v} + \gamma\vec{\omega})\cdot\vec{u} = 0.$$

With respect to the orthonormal basis

$$\vec{u} = \frac{1}{\sqrt{2}}\begin{pmatrix}1\\1\\0\end{pmatrix},\quad \vec{v} = \frac{1}{\sqrt{3}}\begin{pmatrix}1\\-1\\1\end{pmatrix},\quad \vec{w} = \frac{1}{\sqrt{6}}\begin{pmatrix}-1\\1\\2\end{pmatrix}$$

the vector

$$\vec{a} = \begin{pmatrix}1\\2\\4\end{pmatrix} = \vec{e}_x + 2\vec{e}_y + 4\vec{e}_z$$

has the coefficients

$$\vec{a}\cdot\vec{u} = \frac{1}{\sqrt{2}}\begin{pmatrix}1\\1\\0\end{pmatrix}\cdot\begin{pmatrix}1\\2\\4\end{pmatrix}\frac{3}{2}\sqrt{2}$$

$$\vec{a}\cdot\vec{v} = \frac{1}{\sqrt{3}}\begin{pmatrix}1\\-1\\1\end{pmatrix}\cdot\begin{pmatrix}1\\2\\4\end{pmatrix} = \sqrt{3}$$

$$\vec{a}\cdot\vec{w} = \frac{1}{\sqrt{6}}\begin{pmatrix}-1\\1\\2\end{pmatrix}\cdot\begin{pmatrix}1\\2\\4\end{pmatrix}\frac{3}{2}\sqrt{6}.$$

Thus we obtain the representation

$$\vec{a} = \frac{3}{2}\sqrt{2}\vec{u} \quad \sqrt{3}\vec{v} + \frac{3}{2}\sqrt{6}\vec{w}$$

$$= \frac{3}{2}\sqrt{2}\cdot\frac{1}{\sqrt{2}}\begin{pmatrix}1\\1\\0\end{pmatrix} + \sqrt{3}\cdot\frac{1}{\sqrt{3}}\begin{pmatrix}1\\-1\\1\end{pmatrix} + \frac{3}{2}\sqrt{6}\cdot\frac{1}{\sqrt{6}}\begin{pmatrix}-1\\1\\2\end{pmatrix}$$

$$= \begin{pmatrix}1\\2\\4\end{pmatrix}$$

VECTOR PRODUCT

VECTOR PRODUCT, CROSS PRODUCT

The vector

$$\vec{c} = \vec{a} \times \vec{b}$$

is orthogonal to $\vec{a}$ and $\vec{b}$. It is oriented by the„Right-Hand-Rule" and has the length

$$|\vec{c}| = |\vec{c}||\vec{b}|\sin(\sphericalangle(\vec{a}, \vec{b})).$$

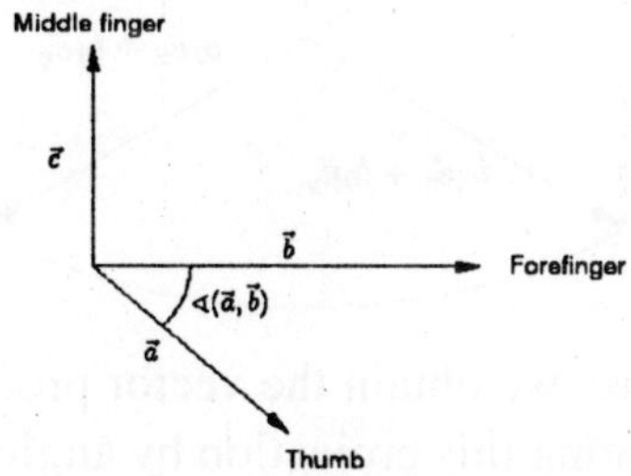

Alternatively, we have

$$\vec{a} \times \vec{b} = \begin{pmatrix} a_2b_3 - a_3b_2 \\ a_3b_2 - a_1b_3 \\ a_1b_2 - a_2b_1 \end{pmatrix}$$

In order to prove the equivalence of the two definitions, first, we show that both definitions are linear in both components, i.e. for the first component

$$\vec{a} \times (\vec{b} + \vec{c}) = \vec{a} \times \vec{b} + \vec{a} \times \vec{c}.$$

and for an arbitrary $\alpha \in \mathbb{R}$

$$\alpha(\vec{a} \times \vec{b}) = \vec{a} \times (\alpha\vec{b}).$$

Similarly for the second component

$$\vec{a} \times (\vec{b} + \vec{c}) = \vec{a} \times \vec{b} + \vec{a} \times \vec{c}$$

and for an arbitrary $\alpha \in \mathbb{R}$

$$\alpha \times (\vec{c} \times \vec{b}) = \vec{a} \times (a\vec{b}).$$

For the definition of the vector product by coordinates this is a routine calculation. For the geometric definition only additivity is not evident.

To show this additivity we show

$$\vec{e}_z \times \vec{b} = \begin{pmatrix} -b_2 \\ b_1 \\ 0 \end{pmatrix}.$$

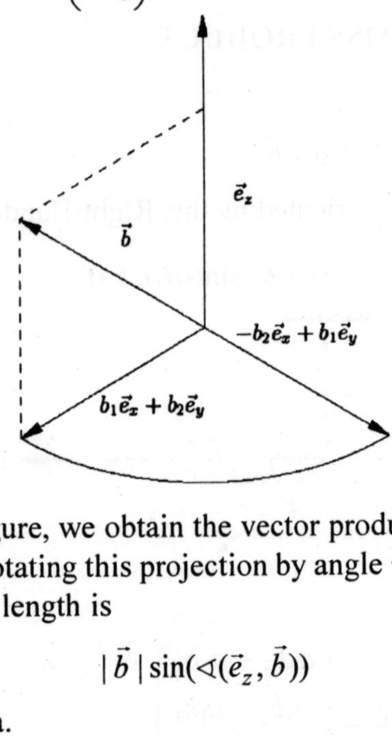

As we can see in the figure, we obtain the vector product by projecting the vector $\vec{b}$ onto the *xy*-plane and then rotating this projection by angle $\pi/2$(this follows the right hand rule). Since the projection's length is

$$|\vec{b}|\sin(\sphericalangle(\vec{e}_z, \vec{b}))$$

we obtain the above formula.

Linearity in the second component follows now easily with this formula, because we may w. l. o. g. choose the coordinate system in such a way that $\vec{e}_z$ coincides with the first vector. Similarly one shows the additivity of the geometric definition in the first component.

Now, it is sufficient to verify the equivalence of the definitions for all combinations of the canonical basis vectors $\vec{e}_x$, $\vec{e}_y$, $\vec{e}_z$.

For the standard basis (canonical basis) we have

$$\vec{e}_x \times \vec{e}_y = \vec{e}_z, \quad \vec{e}_y \times \vec{e}_z = \vec{e}_x, \quad \vec{e}_z \times \vec{e}_x = \vec{e}_y$$

$$\vec{e}_x \times \vec{e}_x = \vec{0}, \quad \vec{e}_y \times \vec{e}_y = \vec{0}, \quad \vec{e}_z \times \vec{e}_z = \vec{0}.$$

Respective formulas apply to any orthonormal basis oriented according to the„Right-Hand Rule“.

LORENTZ FORCE

If an electron with charge $-e$ passes through a magnetic field $\vec{b}$ at velocity $\vec{v}$, then a force, called Lorentz force, acts on it.

$$\vec{f} = e\vec{b} \times \vec{v}.$$

In the test arrangement as shown this force causes a move of the current carrying wire in direction $\vec{f}$.

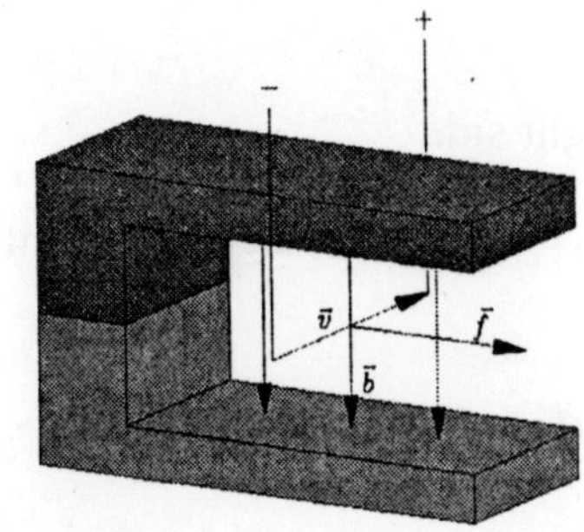

RULES FOR VECTOR PRODUCT

The vector product has the following properties:

1. Antisymmetry

$$\vec{a} \times \vec{b} = (\vec{b} \times \vec{a})$$

2. Linearity

$$(\alpha_1\vec{a}_1 + \alpha_2\vec{a}_2) \times (\beta_1\vec{b}_1 + \beta_2\vec{b}_2) = \alpha_1\vec{b}_1(\vec{a}_1 \times \vec{b}_1) + \alpha_1\beta_2(\vec{a}_1 \times \vec{b}_2) + \alpha_2\beta_1(\vec{a}_1 \times \vec{b}_1) + \alpha_2\beta_2(\vec{a}_2 \times \vec{b}_2)$$

3. The area of the parallelogram generated by the vectors and coincides with $|a \times b|$.

$$(\vec{a} \times \vec{b}) \times \vec{c} = (\vec{a} \cdot \vec{c})\vec{b} - (\vec{b} . \vec{a})\vec{a}$$

4. Grassmann Identity

$$(\vec{a} \times \vec{b}) \cdot (\vec{c} \times \vec{d}) = (\vec{a} \times \vec{c})(\vec{b} \cdot \vec{d}) - (\vec{a} \cdot \vec{d})(\vec{b} \cdot \vec{c})$$

5. Lagrange Identity

Antisymmetry and linearity follow directly from the coordinate representation of the vector product.

In order to prove the identities it is no loss of generality to assume $\vec{a} = \vec{e}_x$. Thus, because of linearity in $\vec{b}$, we have to consider only the cases $\vec{b} = \vec{e}_y$ and $\vec{b} = \vec{e}_z$. For $\vec{a} = \vec{b}$ both sides of the identities equal zero obviously.

$$(\vec{e}_z \times \vec{e}_y) \times \vec{c} = \begin{pmatrix} 0 \\ 0 \\ 1 \end{pmatrix} \times \begin{pmatrix} c_1 \\ c_2 \\ c_3 \end{pmatrix} = \begin{pmatrix} -c_2 \\ c_1 \\ 0 \end{pmatrix}$$

In the first case we obtain for the left side of the Grassmann-Identity

$$c_1\vec{e}_y - c_2\vec{c}_x.$$

This coincides with the right side

In this special case $(\vec{a} = \vec{e}_x, \vec{b} = \vec{e}_y)$ the Lagrange-Identity has the form $(\vec{a} = \vec{e}_x, \vec{b} = \vec{e}_y)$

$$\begin{pmatrix} 0 \\ 0 \\ 1 \end{pmatrix} \cdot \begin{pmatrix} \cdots \\ \cdots \\ c_1d_2 - c_2d_1 \end{pmatrix} = c_1d_2 - c_2d_1 .$$

This is obviously correct.

The other case, $\vec{b} = \vec{e}_z$, may be proved similarly.

1. Antisymmetry

$$\begin{pmatrix} -2 \\ 4 \\ 3 \end{pmatrix} \times \begin{pmatrix} 5 \\ -1 \\ 2 \end{pmatrix} = \begin{pmatrix} 11 \\ 19 \\ -18 \end{pmatrix} = -\left(\begin{pmatrix} 5 \\ -1 \\ 2 \end{pmatrix} \times \begin{pmatrix} -2 \\ 4 \\ 3 \end{pmatrix}\right)$$

2. Linearity

$$\left(2\begin{pmatrix} 1 \\ 0 \\ 0 \end{pmatrix} + 3\begin{pmatrix} 0 \\ 1 \\ 0 \end{pmatrix}\right) \times \left(5\begin{pmatrix} 0 \\ 1 \\ 0 \end{pmatrix} + 7\begin{pmatrix} 0 \\ 0 \\ 1 \end{pmatrix}\right)$$

$$= 10\begin{pmatrix} 0 \\ 0 \\ 1 \end{pmatrix} + 14\begin{pmatrix} 0 \\ -1 \\ 0 \end{pmatrix} + 15\begin{pmatrix} 0 \\ 0 \\ 0 \end{pmatrix} + 21\begin{pmatrix} 1 \\ 0 \\ 0 \end{pmatrix} = \begin{pmatrix} 21 \\ -14 \\ 10 \end{pmatrix}$$

3. Grassmann Identity

$$\left(\begin{pmatrix}1\\2\\3\end{pmatrix}\times\begin{pmatrix}2\\3\\4\end{pmatrix}\right)\times\begin{pmatrix}-2\\1\\0\end{pmatrix}=0.\begin{pmatrix}2\\3\\4\end{pmatrix}-(-1).\begin{pmatrix}1\\2\\3\end{pmatrix}=\begin{pmatrix}1\\2\\3\end{pmatrix}$$

4. Lagrange Identity

$$\left(\begin{pmatrix}0\\1\\2\end{pmatrix}\times\begin{pmatrix}1\\2\\3\end{pmatrix}\right)\cdot\left(\begin{pmatrix}0\\3\\-2\end{pmatrix}\times\begin{pmatrix}-2\\1\\0\end{pmatrix}\right)=3\cdot0-1\cdot0=0$$

can be concluded without any calculation, since the second vector is orthogonal to both

EXERCISE

1. Show that the line of intersection of the planes $x + 2y - z - 3 = 0$, $3x - y + 2z - 1 = 0$.
2. Find the area of the triangle whose vertices are the points (1, 2, 3), (–2, 1, –4), (3, 4, –2).
3. Find the volume of the tetrahedron formed by the planes $2x + 3y + z = 6$, $2x + 3y = 0$, $3y + z = 0$, $2x + z = 0$.
4. Show that the equation $2x^2 - 2y^2 + 4z^2 + 2yz + 6zx + 3xy = 0$.
5. Find the equations of the planes through the intersection of the planes $x + 3y + 6 = 0$, $3x - y - 4z = 0$, whose perpendicular distance from the origin is unity.
6. Find the equation of the bisector of the obtuse angle between the planes $3x + 4y - 5z + 1 = 0$, $5x + 12y - 13z = 0$.
7. Find the volume of the tetrahedron formed by the points (1, 1, 3), (4, 3, 2), (5, 2, 7) and (6, 4, 8).
8. Show that the volume of the tetrahedron formed by the planes

   $my + nz = 0$, $nz + lx = 0$, $lx + my = 0$ and $lx + my + nz = p$ is $\dfrac{2p^3}{3lmn}$.
9. If θ is the angle between the planes $a_1x + b_1y + c_1z + d_1 = 0$ and $a_2x + b_2y + c_2z = 0$, show that

$$\tan\theta=\frac{\sqrt{(b_1c_2-b_2c_1)^2+(c_1a_2-c_2a_1)^2+(a_1b_2-a_2b_1)^2}}{a_1a_2+b_1b_2+c_1c_2}.$$

10. Find the bisector of that angle between the planes $3x - 6y + 2z + 5 = 0$ and $4x - 12y + 3z - 3 = 0$ which contains the origin.

11. Find the equation of the plane passing through the line of intersection of the planes $2x - y = 0$, $3z - y = 0$ and perpendicular to the plane $4x + 5y - 3z = 8$.

12. Find the equations of the planes bisecting the angles between the planes $x + 2y + 2z = 19$, $4x - 3y + 12z + 3 = 0$ and point out which bisects the acute angle.

13. The plane $lx + my = 0$ is rotated about its line of intersection with the plane $z = 0$ through an angle α. Prove that the equation of the plane in its new position is $lx + my = z\sqrt{(l^2 + m^2)} \tan \alpha = 0$.

14. Find the equation of the plane passing through the intersection of the planes $x + 2y + 3z + 4 = 0$, $4x + 3y + 2z + 1 = 0$ and the point (1, 2, 3).

15. A variable plane makes with the co-ordinate planes a tetrahedron of constant volume $64k^3$. Find the locus of the foot of the perpendicular from the origin to the plane.

16. Find equation of the plane through the point (2, –3, 1) and perpendicular to the line joining the points (3, 4, –1), (3, –1, 5).

17. A point P moves on the plane $\frac{x}{a} + \frac{y}{b} + \frac{z}{c} = 1$.

18. If p is the point (2, 3, –1). Find equation of the plane through p and at right angles to OP, where O is the origin.

19. Find equation of the plane through the point (2, –3, 4) and parallel to the plane $2x - 5y - 7z = 6$.

20. Find equation of the plane through (–2, 3, 4) and perpendicular to the planes $2x + 3y + 4z = 6$, $3x + 2y + 2z = 8$.

21. Find the direction cosines of the perpendicular drawn from the origin to the plane $2x + 3y - z + 1 = 0$.

Chapter 2

Linear Coordinates of Curve

Any linear coordinates which are not Cartesian coordinates; frequently used curvilinear coordinates are polar coordinates and cylindrical coordinates.

CURVILINEAR COORDINATES

Curvilinear coordinates are a coordinate system based on some transformation of the standard coordinate system. We need the same number of coordinates. If we consider the 2D case, then instead of Cartesian coordinates x and y we use e.g. p and q; the level curves of p and q in the xy-plane, as well as those of x and y in the pq-plane are in general curved. Required is that the transformation is locally invertible at each point.

This means that we can convert a point given in one coordinate system to its curvilinear coordinates and back. Depending on the application, a curvilinear coordinate system may be simpler than the Cartesian coordinate system. This also has consequences that we can express many of the concepts in vector calculus which are given in Cartesian or spherical coordinates or any other arbitrary coordinate system, also in curvilinear coordinates.

Definition: Let x, y and z be functions of some variables u_1, u_2 and u_3:

$$x = x(u_1, u_2, u_3)$$

$$y = y(u_1, u_2, u_3)$$

$$z = z(u_1, u_2, u_3)$$

or, in vector notation

$$r = r\ (u_1, u_2, u_3)$$

u_1, u_2 and u_3 are called *curvilinear coordinates.*

In R^3, for example, we have some transformation: $x_i = x_i(x'_1, x'_2, x'_3); i = 1,2,3$ giving *curvilinear coordinates* x_1', x_2', x_3', for x_1, x_2, x_3. If this transformation is locally invertible everywhere, the Jacobian determinant

$$\frac{\partial(x_1, x_2, x_3)}{\partial(x'_1, x'_2\, x'_3)}$$

is nonzero, and for this to happen, the vectors

$$\frac{\partial x}{\partial x'_i}$$

must form a basis for R^3.

From these basis vectors, we define *scale factors*

$$hx'_1 = h = \left|\frac{\partial x}{\partial x'_1}\right| = \sqrt{\sum_{k=0}^{3}\left(\frac{\partial x_k}{\partial x'_i}\right)^2}$$

and thus arrive at the unit basis vectors for the curvilinear coordinates to be

$$ex'_i = 1/h_i \frac{\partial x}{\partial x'_1}$$

Note that the coordinate system we choose need not be orthogonal, but for the purposes of this article, they are treated as being so. The system is orthogonal iff

$$e_{x'_i} . e_{x'_j} = \delta_{ij}$$

where δ_{ij} is the Kronecker delta.

COORDINATE CURVES

Let us held u_2 and u_3 constant, and change u_1. The vector **r** will change along a curve. This curve is called *coordinate curve* for u_1. Analogously we can define coordinate curves for u_2 and u_3. Examples:

I. Cylindrical coordinates:

1. Only ρ changes--lines parallel to *xy* plane and containing the point (0, 0, *z*)
2. Only ϕ changes--circles parallel to *xy* plane with the center in the point (0, 0, *z*)
3. Only *z* changes–lines parallel to *z* axis

II. Spherical coordinatess:

1. Only *r* changes–lines containing origin
2. Only θ changes–circles with the center in origin an normal to *xy* plane
3. Only ϕ changes–circles parallel to *xy* plane with the center in the point (0, 0, z)

COORDINATE SURFACES

If we take u_1 constant and allow u_2 and u_3 change, we will have *coordinate surface*. Analogously we can define other two surfaces. Examples:

I. Cylindrical coordinates:

1. ρ = *con st*: cylinders with the axis z (hence the name!)
2. ϕ = *con st*: planes containing the axis z
3. z = *con st*: planes normal to the axis z

II. Spherical coordinates:

1. r = *con st*: spheres with the center at the origin (hence the name!)
2. θ = *con st*: cones with the center at origin and the axis z
3. ϕ = *con st*: planes containing the axis

Cartesian coordinates x_1, x_2, x_3 which have the scalar product, are called Euclidean coordinates. It is often convenient to associate the points of Euclidean space with vectors, for example, with each point P we associate the vector (or arrow) with its tail at the origin of coordinates and its tip at P. This vector is called the radius vector with components (x_1, x_2, x_3). At any point P of Euclidean space we can construct the small line element

$$dx = (dx_1, dx_2, dx_3)$$

which is vector too. Two vectors $h = (x_1, x_2, x_3)$ and $f = (y_1, y_2, y_3)$ from the same origin can be added and result is the vector with coordinates $(x_1 + y_1, x_2 + y_2, x_3 + y_3)$. A vector can also be multiplied by any real number. The Euclidean scalar product of two (real) vectors is the number

$$\langle f, h\rangle = \sum_{i=1}^{n} x_i y_i$$

The scalar product of the vector with itself give the square of the vector length. The square of the length of a line element in space with scalar product is called the metric of the space. The metric of Euclidean space is

$$< \mathrm{dx}, \mathrm{dx} > = dx_1^2 + dx_2^2 + dx_3^2.$$

The same Euclidean metric in curvilinear coordinates is

$$< dx, dx > = \sum_{k=1}^{3} \frac{\partial x_k \partial x_k}{\partial x'_i \partial x'_j} dx'_i \, dx'_j$$

The symmetric tensor

$$g_{i,j}(x'_i, x'_j) = \sum_{k=1}^{3} \frac{\partial x_k \partial x_k}{\partial x'_i \partial x'_j}$$

are called the fundamental (or metric) tensor of the Euclidean space in curvilinear coordinates. Connection between fundamental tensor and Lamé coefficients is

$$g_{i,i}(x'_i, x'_j) = h_i^2$$

Example: If we consider polar coordinates for R^2, note that

$$(x, y) = (r\cos\theta, r\sin\theta)$$

(r, θ) are the curvilinear coordinates, and the Jacobian determinant of the transformation $(r, \theta) \to (r\cos\theta, r\sin\theta)$ is r.

The basis vectors are $b_r = (\cos\theta, \sin\theta)$, $b_\theta = (-r\sin\theta, r\cos\theta)$, with unit basis vectors $e_r = (\cos\theta, \sin\theta)$, $e_\theta = (-\sin\theta, \cos\theta)$ with scale factors $h_r = 1$ and $h_\theta = r$. The fundamental tensor is $g_{1,1} = 1$, $g_{2,2} = r^2$, $g_{1,2} = g_{2,1} = 0$.

EXAMPLES OF CURVILINEAR COORDINATES

SCALING COEFFICIENTS AND LOCAL BASIS

Let e_i be the tangents to coordinate curves. Vectors e_i are proportional to $\partial r / \partial u_i$:

$$e_i = \frac{1}{h_i}\frac{\partial r}{\partial u_i}, \quad h_i = \left|\frac{\partial r}{\partial u_i}\right|$$

Vectors e_i are local basis, numbers h_i are *scaling coefficients*

Taylor expansion:

$$dr = \frac{\partial r}{\partial u_1}du_1 + \frac{\partial r}{\partial u_2}du_2 + \frac{\partial r}{\partial u_3}du_3 = e_1 h_1 du_1 + e_2 h_2 du_2 + e_3 h_1 du_3$$

Cylindrical coordinates:

$$r = \begin{pmatrix} \rho\cos\phi \\ \rho\sin\phi \\ z \end{pmatrix}$$

Differentiating:

Along ρ : $\dfrac{\partial r}{\partial \rho} = \begin{pmatrix} \cos\phi \\ \sin\phi \\ 0 \end{pmatrix}$

$$h_\rho = 1, e_\rho = \begin{pmatrix} \cos\phi \\ \sin\phi \\ 0 \end{pmatrix}$$

Along ϕ:
$$\frac{\partial r}{\partial \phi} = \begin{pmatrix} -\rho \sin\phi \\ \rho\cos\phi \\ 0 \end{pmatrix}$$

$$h_\phi = \rho,\ e_\phi = \begin{pmatrix} -\sin\phi \\ \cos\phi \\ 0 \end{pmatrix},$$

Along z:
$$\frac{\partial r}{\partial z} = \begin{pmatrix} 0 \\ 0 \\ 1 \end{pmatrix}$$

$$h_z = 1,\ e_z = \begin{pmatrix} 0 \\ 0 \\ 1 \end{pmatrix}$$

Spherical coordinates:
$$r = \begin{pmatrix} r\sin\theta\cos\phi \\ r\sin\theta\sin\phi \\ r\cos\theta \end{pmatrix}$$

Differentiating:

Along r:
$$\frac{\partial r}{\partial r} = \begin{pmatrix} \sin\theta\cos\phi \\ \sin\theta\sin\phi \\ \cos\theta \end{pmatrix}$$

$$h_r = 1,\ e_r = \begin{pmatrix} \sin\theta\cos\phi \\ \sin\theta\sin\phi \\ \cos\theta \end{pmatrix}$$

Along θ:
$$\frac{\partial r}{\partial \theta} = \begin{pmatrix} r\cos\theta\cos\phi \\ r\cos\theta\sin\phi \\ -r\sin\theta \end{pmatrix}$$

$$h_\theta = e_\theta = \begin{pmatrix} \cos\theta\cos\phi \\ \cos\theta\sin\phi \\ -\sin\theta \end{pmatrix}$$

Along ϕ : $\dfrac{\partial r}{\partial \phi} = \begin{pmatrix} -r\sin\theta\sin\phi \\ r\sin\theta\cos\phi \\ 0 \end{pmatrix}$

$$h_\phi = r\sin\theta,\ e_\phi = \begin{pmatrix} -\sin\phi \\ \cos\phi \\ 0 \end{pmatrix}$$

Circular Cylindrical Coordinates:

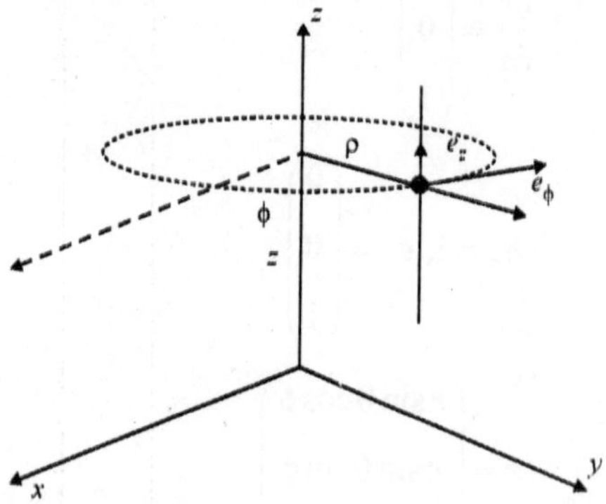

$x = \rho\cos\phi,\ y = \rho\sin\phi,\ z = z$

Spherical Coordinates

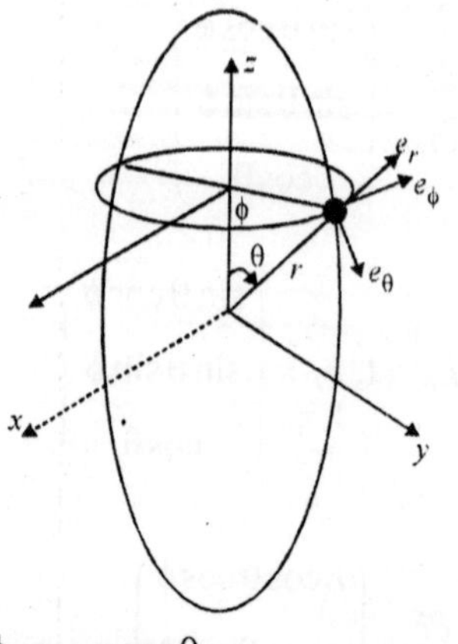

$x = r\sin\theta\cos\phi,\ y = r\sin\theta\sin\phi,\ z = r\cos\theta$

PHYSICAL COMPONENTS OF VECTORS AND TENSORS

Vectors: Suppose we have a vector a. At any point we can expand it in the local basis:

$$a = a_1 e_1 + a_2 e_2 + a_3 e_3$$

The numbers a_1, a_2 and a_3 are called *physical components* of the vector a.

Example: Flow and sphere. Suppose we have a flow with the constant velocity $v = ve_z$. What is the flow on the surface of the sphere r = const??

Solution: In Cartesian coordinates

$$v = \begin{pmatrix} 0 \\ 0 \\ v \end{pmatrix}$$

We can write down:

$$v = v_r e_r + v_\theta e_\theta + v_\phi e_\phi$$

Multiplying this equation by e_r, e_θ and e_ϕ, we obtain:

$$v_r = e_r.v = v\cos\theta,\ v_\theta = e_\theta.v = -v\sin\theta,$$

$$v_\phi = e_\phi.v = 0$$

Physical components of vectors: Once again, use local basis:

$$\hat{U} = u_{ij} e_i \otimes e_j$$

LENGTH, SURFACE AND VOLUME

Element of length: Suppose we have a curve $r = r(u_1, u_2, u_3)$. Its element of length:

$$dl^2 = dr^2$$

Use Taylor expansion . Since the system is orthogonal,

$$dl^2 = h_1^2 du_1^2 + h_2^2 du_2^2 + h_3^2 du_3^2$$

Examples:

1. Cylindrical coordinates:

$$dl^2 = d\rho^2 + r^2 d\phi^2 + dz^2$$

2. Spherical coordinates:

$$dl^2 = dr^2 + r^2 d\theta^2 + r^2 \sin^2\theta d\phi^2$$

Meaning of scaling coefficients: they measure the real length of du_i.

Element of surface: The surface S is a vector with the components:

$$dS = e_1 h_2 h_3 du_2 du_3 + e_2 h_1 h_3 du_1 du_3 + e_3\ h_2 h_1 du_2 du_1$$

Element of volume: Scalar

$$dV = h_1 h_2 h_3 du_1 du_2 du_3$$

LINE AND SURFACE INTEGRALS

Since we use curvilinear coordinates to aid in the calculation in vector calculus, there are adjustments we need to make in the calculation of line, surface and volume integrals.

Line integrals: Normally in the calculation of line integrals we are interested in calculating

$$\int_C f ds = \int_a^b f(x(t)) \left| \frac{\partial x}{\partial t} \right| dt$$

where x(t) parametrizes C in Cartesian coordinates. In curvilinear coordinates, the term

$$\left| \frac{\partial x}{\partial t} \right| = \left| \sum \frac{\partial x}{\partial x'_i} \frac{\partial x'_i}{\partial t} \right|$$

by the chain rule. But from the definition of the curvilinear coordinates,

$$\frac{\partial x}{\partial x'_i} = h_i ex'_i$$

and thus

$$\left| \frac{\partial x}{\partial t} \right| = \sqrt{\sum h_i ex'_i \frac{\partial x'_i}{\partial t}}$$

and we can proceed normally.

Surface integrals: Likewise, if we are interested in a surface integral, the relevant calculation, with the parameterization of the surface in Cartesian coordinates is:

$$\int_S f ds = \iint_T f(x(s,t)) \left| \frac{\partial x}{\partial s} \times \frac{\partial x}{\partial t} \right| dxdt$$

Again, in curvilinear coordinates, the term

$$\left| \frac{\partial x}{\partial s} \times \frac{\partial x}{\partial t} \right| = \left| \frac{\partial x}{\partial x'_i} \frac{\partial x'_i}{\partial s} \times \frac{\partial x}{\partial x'_i} \frac{\partial x'_i}{\partial t} \right|$$

and we make use of the definition of curvilinear coordinates again to yield

$$\frac{\partial x}{\partial x'_i} \frac{\partial x'_i}{\partial s} = \sum \frac{\partial x'_i}{\partial s} hx'_i e_{x'_i}$$

and

$$\frac{\partial x}{\partial x'_i} \frac{\partial x'_i}{\partial t} = \sum \frac{\partial x'_i}{\partial t} hx'_i e_{x'_i}$$

where the cross product, in terms of curvilinear coordinates, will be:

$$\begin{vmatrix} e_{x'_1} & e_{x'_2} & e_{x'_3} \\ h_1 \frac{\partial x'_1}{\partial s} & h_2 \frac{\partial x'_2}{\partial s} & h_3 \frac{\partial x'_3}{\partial s} \\ h_1 \frac{\partial x'_1}{\partial t} & h_2 \frac{\partial x'_2}{\partial t} & h_3 \frac{\partial x'_3}{\partial t} \end{vmatrix}$$

GRAD, CURL, DIV

Gradient

Gradient of a function f is such a vector that grad $f = g = g_1e_1 + g_2e_2 + g_3e_3$ that $g.dr = df$

But, using expansion (1) and chain rule

$$g.dr = g_1h_1du_1 + g_2h_2du_2 + g_3h_3du_3$$

$$df = \frac{\partial f}{\partial u_1}du_1 + \frac{\partial f}{\partial u_2}du_2 + \frac{\partial f}{\partial u_3}du_3$$

We have:

$$gi = \frac{1}{h_i}\frac{\partial f}{\partial u_i}$$

or

$$\nabla = e_1\frac{1}{h_1}\frac{\partial}{\partial u_1} + e_2\frac{1}{h_2}\frac{\partial}{\partial u_2} + e_3\frac{1}{h_3}\frac{\partial}{\partial u_3}$$

Examples:

1. Cylindrical coordinates:

$$\text{grad } f = e_\rho\frac{\partial f}{\partial \rho} + e_\phi\frac{1}{\rho}\frac{\partial f}{\partial \phi} + e_z\frac{\partial f}{\partial z}$$

2. Spherical coordinates:

$$\text{grad } f = e_r\frac{\partial f}{\partial r} + e_\theta\frac{1}{r}\frac{\partial f}{\partial \theta} + e_\phi\frac{1}{\sin\theta}\frac{\partial f}{\partial \phi}$$

Divergence

Suppose we have a vector a. In the local basis

$$a = a_1e_1 + a_2e_2 + a_3e_3$$

What is a?

Simple idea

We can multiply ∇ from equation (2) by a:

$$\text{div}\, a = e_1\frac{1}{h_1}\frac{\partial}{\partial u_1}(a_1e_1) + \ldots\ldots.$$

Problem: Both a_i and e_i depend on u!. So

$$e_1\frac{1}{h_1}\frac{\partial}{\partial u_1}(a_1e_1) = \frac{1}{h_1}\frac{\partial a_1}{\partial u_1} + e_1a_1\frac{1}{h_1}\frac{\partial e_1}{\partial u_1}$$

We do not know what to do with the second term!

A useful theorem: There are two simple relations:

$$\text{curl}\frac{e_1}{h_1} = 0, \ \text{div}\frac{e_1}{h_2 h_3} = 0$$

Proof:

1. First, let us take grad u_1. From the general formula

$$\nabla u_1 = e_1 \frac{1}{h_1}\frac{\partial u_1}{\partial u_1} = \frac{e_1}{h_1}$$

Since e_1/h_1 is a gradient, its curlis zero.

2. We have

$$\frac{e_1}{h_2 h_3} = \frac{e_2 \times e_3}{h_2 h_3} = \frac{e_2}{h_2} = \frac{e_3}{h_3}$$

This is a vector product of two gradients. Each of them has zero curl. From the formula

$$\nabla.(a \times b) = b = \text{b. curla a} - \text{a . curl b}$$

we conclude that the divergence of this product is zero.

Calculation of div a: We can write down:

$$a_1 e_1 = a_1 h_2 h_3 \frac{e_1}{h_2 h_3}$$

but
$$e_1 \frac{1}{h_1}\frac{\partial}{\partial u_1}\left(\frac{e_1}{h_2 h_3}\right) = \text{div}\frac{e_1}{h_2 h_3} = 0$$

and therefore

$$e_1 \frac{1}{h_1}\frac{\partial}{\partial u_1}\left(a_1 h_2 h_3 \frac{e_1}{h_2 h_3}\right) = \frac{1}{h_1 h_2 h_3}\frac{\partial}{\partial u_1}(h_2 h_3 a_1)$$

We obtained
$$\text{div } a = \frac{1}{h_1 h_2 h_3}\left[\frac{\partial}{\partial u_1}(h_2 h_3 a_1) + \frac{\partial}{\partial u_3}(h_1 h_3 a_2) + \frac{\partial}{\partial u_3}(h_1 h_2 a_3)\right]$$

Examples:

1. Cylindrical coordinates:

$$\text{div } a = \frac{1}{\rho}\left[\frac{\partial}{\partial \rho}(\rho a_\rho) + \frac{\partial}{\partial \phi}(a\phi) + \frac{\partial}{\partial_z}(\rho a_z)\right] = \frac{1}{\rho}\frac{\partial}{\partial \rho}(\rho a_\rho) + \frac{1}{\rho}\frac{\partial a_\phi}{\partial \phi} + \frac{\partial a_z}{\partial_z}$$

2. Spherical coordinates:

$$\text{div } a = \frac{1}{r^2 \sin\theta}\left[\frac{\partial}{\partial r}(r^2 \sin\theta a_r) + \frac{\partial}{\partial \theta}(r \sin\theta a_\theta) + \frac{\partial}{\partial \phi}(r a_\phi)\right]$$

$$= \frac{1}{r^2}\frac{\partial}{\partial r}(r^2 a_r) + \frac{1}{r}\frac{\partial}{\partial \theta}(\sin\theta a_\theta)$$

LAPLACIAN

Question: What happens if the vector a is ∇?

Answer: Laplacian:

$$\nabla = \frac{1}{h_1h_2h_3}\left[\frac{\partial}{\partial u_1}\left(\frac{h_2h_3}{h_1}\frac{\partial}{\partial u_1}\right)+\frac{\partial}{\partial u_2}\left(\frac{h_1h_2}{h_2}\frac{\partial}{\partial u_2}\right)+\frac{\partial}{\partial u_3}\left(\frac{h_1h_2}{h_3}\frac{\partial}{\partial u_3}\right)\right]$$

Examples:

1. Cylindrical coordinates:

$$\Delta = \frac{1}{\rho}\frac{\partial}{\partial_\rho}\left(\rho\frac{\partial}{\partial_\rho}\right)+\frac{1}{\rho^2}\frac{\partial^2}{\partial\phi^2}+\frac{\partial^2}{\partial z^2}$$

2. Spherical coordinates:

$$\Delta = \frac{1}{r^2}\frac{\partial}{\partial r}\left(r^2\frac{\partial}{\partial r}\right)+\frac{1}{r^2\sin\theta}\frac{\partial^2}{\partial\theta}\sin\theta\frac{\partial}{\partial\theta}+\frac{1}{r^2\sin^2\theta}\frac{\partial^2}{\partial\phi^2}$$

CURL *a*

Let us calculate a. The same idea as for gradient: we will factor out the term with zero curl!

Result:

$$\text{curl } a = \frac{1}{h_1h_2h_3}\begin{vmatrix} h_1e_1 & h_2e_2 & h_2e_2 \\ \frac{\partial}{\partial u_1} & \frac{\partial}{\partial u_2} & \frac{\partial}{\partial u_3} \\ h_1a_1 & h_2a_2 & h_3a_3 \end{vmatrix}$$

Examples:

1. Cylindrical coordinates:

$$\text{curl } a = \frac{1}{\rho}\begin{vmatrix} e_\rho & \rho e_\phi & e_z \\ \frac{\partial}{\partial\rho} & \frac{\partial}{\partial\phi} & \frac{\partial}{\partial z} \\ a_\rho & r\,a_\phi & a_z \end{vmatrix}$$

2. Spherical coordinates:

$$\text{curl } a = \frac{1}{r^2\sin\theta}\begin{vmatrix} e_r & re_\theta & r\sin\theta e_\phi \\ \frac{\partial}{\partial r} & \frac{\partial}{\partial\theta} & \frac{\partial}{\partial\phi} \\ a_r & r\,a_\theta & r\sin\theta a_\phi \end{vmatrix}$$

In orthogonal curvilinear coordinates, one can express the gradient, curl and divergence of a function or vector field as follows:

$$\nabla f = \sum \frac{1}{h}\frac{\partial f}{\partial x_i}\hat{e}_{x_i}$$

$$\nabla \times \vec{v} = \frac{1}{\Omega h_i}\begin{pmatrix} h_1\partial_1 \\ \vdots \\ h_n\partial_n \end{pmatrix} \times \vec{v}$$

$$\nabla . \vec{v} = \sum \frac{1}{\Omega}\frac{\partial}{\partial_{x_i}}\left(\frac{\Omega v_i}{h_i}\right),$$

where Ω is the product of all h_i.

VECTOR CALCULUS

Vector calculus (also called *vector analysis*) is a field of mathematics concerned with multivariate real analysis of vectors in two or more dimensions. It consists of a suite of formulas and problem solving techniques very useful for engineering and physics. Vector analysis has its origin in quaternion analysis, and was formulated by the American scientist, J. Willard Gibbs [1].

Vector calculus is concerned with scalar fields, which associate a scalar to every point in space, and vector fields, which associate a vector to every point in space. For example, the temperature of a swimming pool is a scalar field: to each point we associate a scalar value of temperature. The water flow in the same pool is a vector field: to each point we associate a velocity vector.

Three operations are important in vector calculus:

Gradient: Measures the rate and direction of change in a scalar field; the gradient of a scalar field is a vector field.

Curl: Measures a vector field's tendency to rotate about a point; the curl of a vector field is another vector field.

Divergence: Measures a vector field's tendency to originate from or converge upon a given point.

A fourth operation, the Laplacian, is combination of the divergence and gradient operations.

Likewise, there are three important theorems related to these operators:

Gradient theorem
Stokes' theorem
Divergence theorem

Most of the analytic results are easily understood, in a more general form, using the machinery of differential geometry, of which vector calculus forms a subset.

DIFFUSION AROUND A SPHERE

Let us discuss a sphere with radius *R* with the concentration of some reagent c_0 on the surface. The diffusion coefficient is *D*. What is the steady flow of reagent?

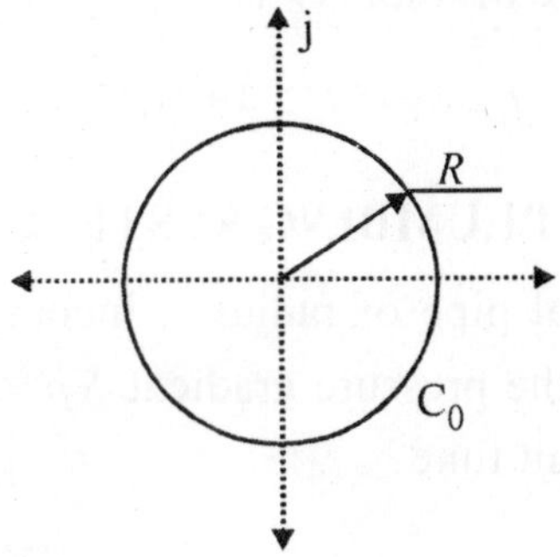

Solution: the diffusion equation

$$\frac{\partial c}{\partial t} = D\Delta c$$

Steady state: no time dependence,

$$\Delta c = 0$$

Let us use spherical coordinates. Obviously *c* depends only on *r*, and diffusion equation is

$$\frac{1}{r^2}\frac{\partial}{\partial r}\left(r^2\frac{\partial c}{\partial r}\right) = 0$$

Its solution should satisfy:

$$c(R) = c_0,\ c(\infty) = 0$$

Let us seek the solution in the form

$$c = \frac{A}{r}$$

1. Equation 3 is satisfied
2. Obviously $c(\infty) = 0$
3. $c(R) = c_0$. if $A = c_0 R$

So the distribution of reagent is:

$$c(r) = c_0\frac{R}{r}$$

The flow:

$$j = -D \text{ grad } c$$

We have:

$$j_r = D\frac{R}{r^2}e_r,\ j_\phi = j_\phi = 0$$

The total flow over a sphere of radius r is

$$J = 4\pi r^2 j_r = 4\pi D R$$

PLUMBING SYSTEM

Let us discuss a cylindrical pipe of radius r. Incompressible liquid with viscosity coefficient η is pumped with the pressure gradient $\nabla p = -G = const$. How much liquid flows through the pipe in the unit time?

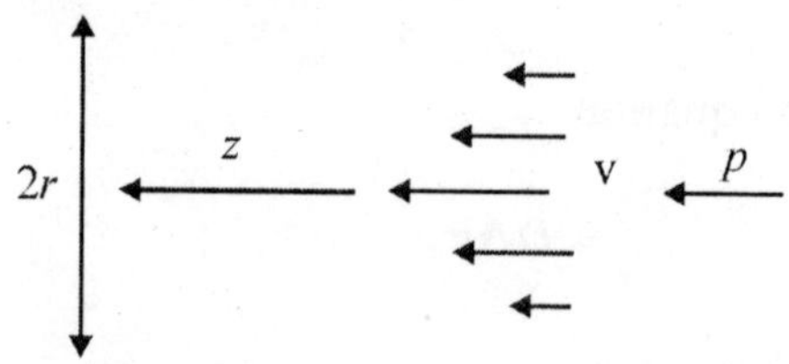

Let us introduce cylindrical coordinates with the axis z along the pipe axis. The velocity v is along the z axis and does not depend on ϕ or z:

$$v = v(\rho)e_z$$

The amount of liquid flowing through the tube in the init time is the integral through the cross-section of the tube:

$$Q = \iint_s v\,dS$$

Since

$$dS = \rho\,d\rho\,d\phi$$

we have
$$Q = \int_0^r d\rho \int_0^{2\pi} d\phi\,\rho v(\rho) = 2\pi \int_0^r \rho v(\rho)\,d\rho$$

Now let us calculate $v(\rho)$. Use the Navier-Stokes equation and the following considerations:

(i) Steady flow, so $\partial v/\partial t = 0$

(ii) In this particular geometry $(v.\nabla)v = 0$

Therefore

$$-\operatorname{grad} p + \eta \Delta v = 0$$

This equation has only z component:

$$G + \frac{\eta}{\rho}\frac{\partial}{\partial \rho}\left(\rho \frac{\partial v}{\partial \rho}\right) = 0$$

The velocity v must be zero near the wall:

$$v\ (r) = 0$$

Let us seek the solution in the form:

$$v(\rho) = a - b\rho^2$$

Then

$$\frac{\eta}{\rho}\frac{\partial}{\partial \rho}\left(\rho \frac{\partial v}{\partial \rho}\right) = -4b\eta$$

and therefore

$$b = \frac{G}{4\eta}$$

From the boundary condition

$$a = br^2 = \frac{G}{4\eta} r^2$$

We obtained

$$Q = 2\pi \int_0^r \rho v(\rho) d\rho = \frac{\pi G}{8\eta} r^4$$

The total amount of liquid

$$Q = 2\pi \int_0^r \rho v(\rho) d\rho = \frac{\pi G}{8\eta} r^4$$

The answer is proportional to r^4. It means that enlarging pipe diameter two times means increasing the capacity $2^4 = 16$ times.

EXRECISE

1. Express $\nabla^2 \phi$ in orthogonal curvilinear coordinates.
2. Compute gradf in orthogonal curvilinear coordinates system

   $x = u - v,\ y = u + v,\ z = w^2$ for $\phi(u, v, w) = uv + w^2$

3. Prove that V = $\frac{\cos\theta}{r^3}\left(T_r/\sin\theta - T_\theta/\cos\theta + r^4 T_\phi\right)$ is solenoidal.

4. Show that r^aR is an irrotational vector for any value of a, but is solenoidal only if a + 3 = 0, where R = xI + yJ + zK and r is the magnitude of R.

5. If f is a solution of the Laplace equation, prove that $\nabla\phi$ is both solenoidal and irrotational.

6. Show that $xi + yj + zk$ is a irrotational vector.

7. Show that $\left(z^2 + 2x + 3y\right)i + \left(3x + 2y + z\right)j + \left(y + 2zx\right)k$ is a irrotational vector.

8. Consider the transformation $x = u^2 - v^2,\ y = 2uv,\ z = w$

   (i) Show that $\left(u, v, w\right)$ from orthogonal curvilinear coordinates.

   (ii) Express $\nabla^2 g\left(u, v, w\right)$.

9. Find the divergence and curl of the vector field

   $F = w\hat{e}_1 + u\hat{e}_2 + v\hat{e}_3$

10. If A and B are irrotational, prove that A × B is solenoidal.

11. Evaluate $\nabla\phi$ = xyz in cylindrical coordinates.

12. Show that $\nabla\left(r/\sin\theta\right)\nabla\theta = \nabla\phi$.

13. If u, v, w are orthogonal curvilinear coordinates with h_1, h_2, h_3 as scale factors, prove that

$$\left[\frac{\partial R}{\partial u}, \frac{\partial R}{\partial v}, \frac{\partial R}{\partial w}\right] = \frac{1}{\left[\nabla v, \nabla \mathbf{v}, \nabla w\right]} = h_1 h_2 h_3.$$

14. Express the following vectors in cylindrical coordinates

    (i) $2yI - zJ + 3xK$. (ii) $2xI - 3y^2J + zxK$.

15. Express the following vectors in spherical polar coordinates

    (i) $xi + 2yJ + yzK$ (ii) $xyI + yzJ + zxK$

Chapter 3

Curves in Vector Field

LINE INTEGRALS

Suppose that a wire has as density $f(x, y, z)$ at the point (x, y, z) on the wire. Then the line integral will equal the total mass of the wire. Below is the definition in symbols.

DEFINITION OF THE LINE INTEGRAL

Let f be a function defined on a curve C of finite length. Then the *line integral* of f along C is

$$\int_C f(x, y)ds = \lim_{n\to\infty} \sum_{i=1}^{n} f(x_i, y_i)\Delta s_i \text{ (for two dimension)}$$

$$\int_C f(x, y, z)ds = \lim_{n\to\infty} \sum_{i=1}^{n} f(x_i, y_i, z_i)\Delta s_i \text{ (for three dimension)}$$

EVALUATING LINE INTEGRALS

Let, $r(t) \quad = x(t)i + y(t)j$

be a differentiable vector valued function. Then

$$ds = \|r'(t)\|dt = \sqrt{x'(t)^2 + y'(t)^2}\,dt$$

We are now ready to state the theorem that shows us how to compute a line integral.

LINE INTEGRALS OVER VECTOR VALUED FUNCTIONS

Theorem: Let $r(t) = x(t)i + y(t)j \quad a \leq t \leq b$

be a differentiable vector valued function that defines a smooth curve C. Then

$$\int_C f(x, y)ds = \int_a^b f(x(t), y(t))\,\sqrt{x'(t)^2 + y'(t)^2}\,dt$$

and for three dimensions, if

$$r(t) = x(t)i + y(t)j + z(t)k \qquad a \leq t \leq b$$

then

$$\int_C f(x, y, z)ds = \int_a^b f(x(t), y(t), z(t)) \sqrt{x'(t)^2 + y'(t)^2 + z'(t)^2}\,dt$$

Line Integrals in Differential Form

We can rewrite r′(t) dt as

$$\frac{dr}{dt}dt = \left(\frac{dx}{dt}i + \frac{dy}{dt}j + \frac{dz}{dt}k\right)dt$$

$$\Rightarrow dr = dxi + dyj + dzk$$

So that if

$$F = Mi + Nj + Pk$$

then

$$F \cdot r'(t)dt = Mdx + Ndy + Pdz$$

This is called the differential form.

Let $r = xi + yj + zk$ be the position vector of appoint $P(x, y, z)$ on a curve C, where x, y, z are functions of scalar parameter t, $a \leq t \leq b$.

Let $t = a$ at A and $t = b$ at B.

Let $f = f_1 i + f_2 j + f_3 k$ be a vector point function defined and continuous along curve C.

If s denotes the arc length of the curve C, then $\frac{dr}{dt} = t$ is a unit vector along the tangent to the curve C at P. The component of the vector f along the tangent at P is

$$f_t = f \cdot t = f \cdot \frac{dr}{ds}$$

For points on C, f_t is a unction of s and $\int_A^B f_t ds$ is known as the tangent line integral or just line integral of f along C.

$$\text{But } f_t = f \cdot t = f \cdot \frac{dr}{ds} \text{ ds} = \text{f .dr}$$

$$\therefore \int_A^B f_t ds = \int_A^B f \cdot tds = \int_A^B f \cdot dr \text{ and is also written as } \int_C f \cdot dr$$

Note that $\int_A^B f \cdot dr = -\int_B^A f \cdot dr$

The differential displacement vector dr along the curve C is defined by

$$dr = dxi + dyj + dzk$$

$$\begin{aligned} f \cdot dr &= (f_1 i + f_2 j + f_3 k) \cdot (dxi + dyj + dzk) \\ &= f_1 dx + f_2 dy + f_3 dz \end{aligned}$$

Therefore , in component form the above line can be written as

$$\begin{aligned} \int_C f \cdot dr &= \int_C (f_1 i + f_2 j + f_3 k) \cdot (dxi + dyj + dzk) \\ &= \int_C (f_1 dx + f_2 dy + f_3 dz) \end{aligned}$$

If the given curve is given by the parametric equations:

$x = x(t), y = y(t), z = z(t)$, then can be write

$$\int_C f \cdot dr = \int_a^b \left(f_1 \frac{dx}{dt} + f_2 \frac{dy}{dt} + f_3 \frac{dz}{dt} \right) dt,$$

Where *a*, *b* are the values of the parameter *t* at *A* and *B*.

Circulation

If *C* is a simple closed curve, then the tangent line integral of f around *C* is called the circulation of about *C* and is often denoted by

$$\oint_C f \cdot dr = \oint_C (f_1 dx + f_2 dy + f_3 dz)$$

Applications of line integral

The main application of line integrals is finding the work done on an object in a force field. The work done W along each piece will be approximately equal to

$$dW = F \cdot T\, ds$$

Now recall that

$$T = \frac{r'(t)}{\|r'(t)\|}$$

and that

$$ds = \|r'(t)\| dt$$

Hence,

$$dW = F \cdot r'(t) dt$$

As usual, we add up all the small pieces of work and take the limit as the pieces get small to end up with an integral.

Definition: Let F be a vector field and C be a curve defined by the vector valued function r. Then the work done by F on an object moving along C is given by

$$\text{Work done} = \int_C F \cdot dr = \int_a^b F(x(t), y(t), z(t)) \cdot r'(t)\,dt$$

SURFACE INTEGRAL

Let S be a surface in three-dimensional space R^3 and let f be a function defined on this surface. Then the *integral of f with respect to the surface area* is usually notated as

$$\int_S f(u,v)d^2A$$

To compute above integral, one may use the following formula to convert it to a double integral:

$$\int_S f(u,v)d^2A = \int f(u,v)\sqrt{\left(\frac{\partial(x,y)}{\partial(u,v)}\right)^2 + \left(\frac{\partial(y,z)}{\partial(u,v)}\right)^2 + \left(\frac{\partial(z,x)}{\partial(u,v)}\right)^2}\,dudv$$

If the surface is described as the graph of a function g, then we may also use the following formula:

$$\int_S f(x,y)d^2A = \int f(x,y)\sqrt{1 + \left(\frac{\partial g}{\partial x}\right)^2 + \left(\frac{\partial g}{\partial y}\right)^2}\,dxdy$$

EVALUATING SURFACE INTEGRAL

For a scalar function f over a surface parameterized by u and v, the surface integral is given by

$$\phi = \int_S f d\alpha = \int_S f(u,v)\left|T_u \times T_v\right| dudv$$

where T_u and T_v are tangent vectors and $a \times b$ is the cross product.

Let P be any point on the surface S and let n be the unit vector at P in the direction of the outward drawn normal to the surface S at P. Then, $f \cdot n$ is the normal component of f at P. The integral of $f \cdot n$ over S is called the (normal) surface integral of f over S and is denoted by

$$\iint_S f \cdot ndS$$

It is also called the flux of f over S. If we associate with the differential of surface area dS a vector dS whose magnitude is dS and whose direction is that of n, then

$$dS = ndS$$

where $f_n = f \cdot n$ denotes the component of f in the direction of the outward normal at P.

If the outward drawn normal to the surface S at P makes angles 2, 3, 4 with the positive direction of x, y, z axes respectively, and l, m, n are the direction cosines of this outward normal, then

$$l = \cos\alpha, m = \cos\beta, n = \cos\gamma$$

Also, $\hat{n} = \cos\alpha\, i + \cos\beta\, j + \cos\gamma\, k = li + mj + nk$

If $$f = f_1 i + f_2 j + f_3 k, \text{ then } f \cdot \hat{n}$$

$$= (f_1 i + f_2 j + f_3 k)\cdot(li + mj + nk)$$

$$= f_1 l + f_2 m + f_3 n$$

$$= f_1 \cos\alpha + f_2\cos\beta + f_3\cos\gamma$$

$$\iint_S f\cdot\hat{n}\, dS = \iint_S (f_1\cos\alpha + f_2\cos\beta + f_3\cos\gamma) dS$$

Now, $dS \cos\alpha$ = orthogonal projection of the area of the dS on the yz – plane = $dy\,dz$

$dS\cos\beta$ = orthogonal projection of the area of the dS on the yz – plane = $dz\,dx$: and

$dS\cos\gamma$ = orthogonal projection of the area of the dS on the yz –plane = $dx\,dy$

$$\therefore \iint_S f\cdot\hat{n}\, dS = \iint_S (f_1 dydz + f_2 dzdx + f_3 dxdy)$$

Also, we can evaluate the surface integral using the method given below.

$$\phi = \int_S F\cdot da = \int_S (F\cdot\hat{n})\, d\alpha$$

$$= \int_S (f_x dy\,dz + f_y dz\,dx + f_z dx\,dy)$$

where $a\cdot b$ is a dot product and $\hat{n}$ is a unit normal vector.

If $z = f(x, y)$, then $d\,a$ is given explicitly by

$$da = \pm\left(-\frac{\partial z}{\partial x}\hat{x} - \frac{\partial z}{\partial y}\hat{y} + \hat{z}\right) dxdy$$

If the surface is surface parameterized using u and v, then

$$\phi = \int_S F\cdot(T_u \times T_v)\, dudv$$

Let S be a surface such that any line perpendicular to xy – plane meets S in not more than one point. Then, the equation of the surface S can be written as $z = h(x, y)$.

Let the projection of S on xy– plane = $dS\cos\gamma$,

where γ is the acute angle which the normal n at P to the surface S makes with z- axis.

$$\therefore\ dx = \cos\gamma = dxdy$$

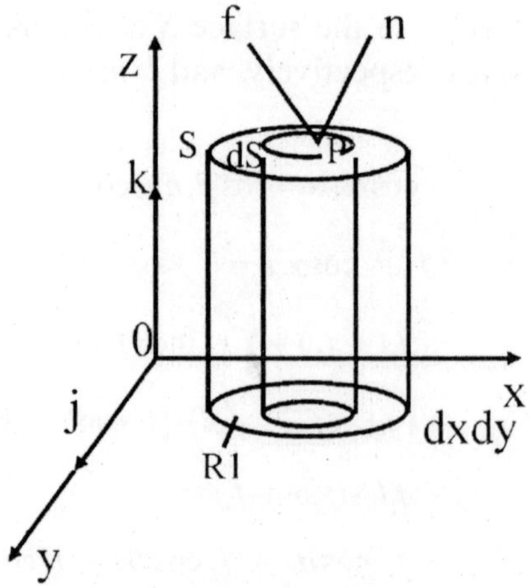

Fig. Surface Integral

Also, $\cos\gamma = |n \cdot k|$, where k is unit vector along z-axis

$$\therefore \quad dS = \frac{dxdy}{|n \cdot k|}$$

Hence, $$\iint_S f \cdot ndS = \iint_{R_1} f \cdot n\frac{dxdy}{|n \cdot k|}.$$

Similarly,

$$\iint_S f \cdot ndS = \iint_{R_2} f \cdot n\frac{dydz}{|n \cdot i|} = \iint_{R_2} f \cdot n\frac{dzdx}{|n \cdot j|}$$

VOLUME INTEGRAL

Let be a scalar point function defined throughout a given region of volume V. Divide the given region into n sub-regions of volumes $\delta V_1, \delta V_2, \ldots, \delta V_n$. Let $P(x_i, y_i, z_i)$ be any point inside or on the boundary of the sub-regions of volume δV_i. From the sum $\sum_{i=1}^{n} \phi(P_i)\delta V_i$ where $\phi(P_i) = \phi(x_i, y_i, z_i)$. If as $n \to \infty$ in such a way that each δV_i tends to zero, the sum tends to a finite limit and is independent of the mode of sub-division, then this limit is called the volume integral of ϕ over the volume V and is denoted by

$$\iiint_V \phi\, dV$$

If we sub-divide the region of volume V into small a cuboids by drawing lines parallel to the coordinate axes, then $dV = dx\, dy\, dz$ and

$$\iiint_V \phi\, dxdydz$$

Likewise if f is a vector point function defined in the given region of volume V, we define and denote the vector volume integral of f over V by

$$\iiint_V f\, dV .$$

A triple integral over three coordinates giving the volume within some region G,

$$V = \iiint_G \phi\, dxdydz$$

VOLUME ELEMENT

A volume element is the differential element dV whose volume integral over some range in a given coordinate system gives the volume of a solid,

$$V = \iiint_G \phi\, dxdydz$$

In $\mathfrak{R}^n$, the volume of the infinitesimal n-hypercube bounded by $dx_1, dx_2, dx_3, \ldots, dx_n$ has volume given by the wedge product

$$dV = dx_1 \wedge dx_2 \wedge dx_3 \wedge \ldots \wedge dx_n$$

The use of the antisymmetric wedge product instead of the symmetric product $dx_1, dx_2, dx_3, \ldots, dx_n$ is a technical refinement often omitted in informal usage. Dropping the wedges, the volume element for curvilinear coordinates in $\mathfrak{R}^3$ is given by

$$dV = \left|(h_1 u_1 du_1)\cdot(h_2 u_2 du_2)\times(h_3 u_3 du_3)\right|$$

$$= h_1 h_2 h_3\ du_1 du_2 du_3$$

$$= \left|\frac{\partial r}{\partial u_1}\cdot\frac{\partial r}{\partial u_2}\times\frac{\partial r}{\partial u_3}\right| du_1 du_2 du_3$$

$$= \begin{vmatrix} \frac{\partial x}{\partial u_1} & \frac{\partial x}{\partial u_2} & \frac{\partial x}{\partial u_3} \\ \frac{\partial y}{\partial u_1} & \frac{\partial y}{\partial u_2} & \frac{\partial y}{\partial u_3} \\ \frac{\partial z}{\partial u_1} & \frac{\partial z}{\partial u_2} & \frac{\partial z}{\partial u_3} \end{vmatrix} du_1 du_2 du_3$$

$$= \left|\frac{\partial(x,y,z)}{\partial(u_1,u_2,u_3)}\right| du_1 du_2 du_3$$

where the latter is the Jacobian and the h_i are scale factors.

REDUCTION OF VOLUME TO SURFACE INTEGRALS

One of the most important theorems in vector analysis is known as the Divergence

Theorem, which is also sometimes called Gauss' Theorem. This is essentially just an application of the fundamental theorem of calculus

$$\int_a^b \frac{df(x)}{dx}dx = f(b) - f(a)$$

This enables us to express the integral of the quantity $\frac{df}{dx}$ along an interval in terms of the values of f itself at the endpoints of that interval.

Now suppose we are given a scalar function f(*x*, *y*, *z*) throughout a region *V* enclosed by a surface *S*, and we want to evaluate the integral of the quantity $\frac{\partial f}{\partial x}$ over this entire region. This can be written as

$$\int_V \frac{\partial f}{\partial x}dV = \iiint_{xyz} \frac{\partial f}{\partial x}\, dxdydz$$

where the three integrals are evaluated over suitable ranges to cover the entire region *V*. For any fixed values of $y = y_0$ and $z = z_0$ the function $f(x, y_0, z_0)$ is totally dependent on *x*, so we can evaluate the integral along any line parallel to the *x* axis through the region *V* for any particular y_0, z_0 using the fundamental theorem of calculus

$$\int_a^b \frac{df(x, y_0, z_0)}{dx}dx = f(b, y_0, z_0) - f(a, y_0, z_0)$$

where a,b are the *x* values at which the line intersects the surface *S*. Hence, we can cover all the contributions by integrating the value of the function *f*(*x*, *y*, *z*) over the entire surface *S*.

Consequently, we can write our original integral as

$$\int_V \frac{\partial f}{\partial x}dV = \int_S f\cos(n \cdot i)\, dS$$

where cos (*n*, *i*) denotes the cosine of the angle between a unit vector normal to the surface and a basis vector in the positive x direction. Naturally, since the coordinate axes are symmetrical, we could arrive at analogous results with x replaced by either y or z, and *i* replaced by *j* or *k* respectively.

Now suppose we are given a vector field *F*(*x*, *y*, *z*) with the components $F_x(x, y, z)$, $F_y(x, y, z)$, and $F_z(x, y, z)$. Using the above results and simple additivity we have

$$\int_V \left(\frac{\partial F_x}{\partial x} + \frac{\partial F_y}{\partial y} + \frac{\partial F_z}{\partial z}\right) dV = \int_S [F_x \cos(n \cdot i) + F_y \cos(n \cdot j) \; + F_z \cos(n.k)] dS$$

The integrand on the left side is ∇×*F*, i.e. the divergence of *F*. Also, notice that $\cos(n \cdot i)$, $\cos(n \cdot j)$, and $\cos(n \cdot k)$ are the components of the normal unit vector *n*, so the integrand on the right side is simply $F \times n$, i.e., the dot product of F and the unit normal to the surface. Hence we can express the Divergence Theorem in its familiar form

$$\int_V \nabla \cdot F dV = \int_S F \cdot n \, dS$$

GAUSS'S DIVERGENCE THEOREM

Statement: The surface integral of the normal component of a vector f taken over a closed surface is equal to the integral of the divergence of f taken over the volume enclosed by the surface.

Symbolically: If f is a continuously differentiable vector point function and S is closed surface enclosing the volume V with an outward pointing unit normal, n, the Divergence Theorem says:

$$\iint_S F \cdot n \, dS = \iiint_V div \text{ F dV}$$

$$\text{Or } \iint_S F \cdot n \, d\sigma = \iiint_V div \text{ F dV}$$

We take $F = Mi + Nj + Pk$

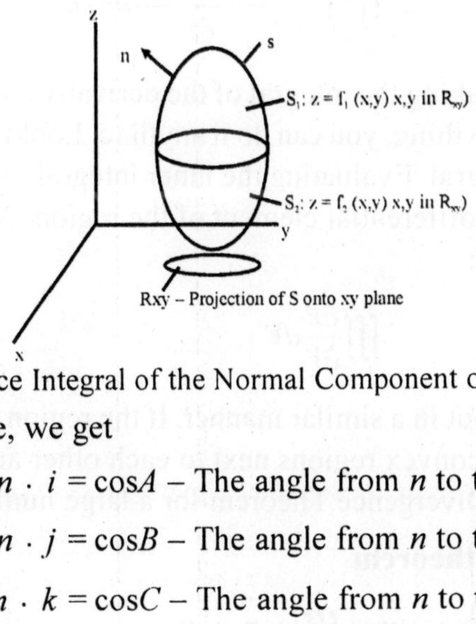

Fig. surface Integral of the Normal Component of a Vector

Using the above figure, we get

$n \cdot i = \cos A$ – The angle from n to the x-axis

$n \cdot j = \cos B$ – The angle from n to the y-axis

$n \cdot k = \cos C$ – The angle from n to the z-axis

$n = \cos Ai + \cos Bj + \cos Ck$ since the dot product of n with each basis vector is that component of n.

Now we restate the Divergence Theorem using the components of the vector field F.

$$\iint_S [M \cos A + N \cos B + P \cos C] \, d\sigma = \iiint_V \left(\frac{\partial M}{\partial x} + \frac{\partial N}{\partial y} + \frac{\partial P}{\partial z} \right) dV$$

Taking just the z terms

$$\iint_S P\cos C\, d\sigma = \iiint_V \frac{\partial P}{\partial z} dV$$

Splitting into integrals over the two surfaces S_1 and S_2

$$\iint_{S_1} P\left(x, y, f_1(x,y)\right)\cos C\left(x, y, f_1(x,y)\right) dxdy$$

$$+\iint_{S_2} P\left(x, y, f_2(x,y)\right)\cos C\left(x, y, f_2(x,y)\right) dxdy$$

Project onto $R_{xy}\, dS = \frac{dA}{\|n\cdot k\|} = \frac{dA}{|\cos C|}$

On top(S_1) $\cos C \geq 0 \Rightarrow |\cos C| = \cos C$

On bottom (S_2) $\cos C \leq 0 \Rightarrow |\cos C| = -\cos C$

$$\iint_{R_{xy}} P\left(x, y, f_1(x,y)\right)\cos C \frac{dA}{\cos C} - \iint_{S_2} P\left(x, y, f_2(x,y)\right)\cos C \frac{dA}{\cos C}$$

$$= \iint_{R_{xy}} \left\{ \int_{f_2(x,y)}^{f_1(x,y)} \frac{\partial P(x,y,z)}{\partial z} dz \right\} dA$$

The last step involves taking the integral of the derivative of P. This is like multiplying by one. It doesn't change anything, you can do it anytime. Looking at the last term, what we really have is a volume integral. Evaluating the inner integral, we have $[f_1(x, y) - f_2(x, y)] \times dA$. This is the volume of a differential element of the region. Adding all of them up gives us the volume of the region:

$$\iiint_V \frac{\partial P}{\partial V} dV$$

The other terms work out in a similar manner. If the region is not a convex region, we can construct it by placing convex regions next to each other and summing up the results. Thus, we have proved the Divergence Theorem for a large number of regions

Deduction from Gauss's theorem

1. $\iiint_V f\cdot\nabla\phi\, dV = \iint_S \phi f\cdot n dS - \iiint_V \phi\, \text{div} f\, dV$
2. $\iint_S f\times n\, dS = -\iiint_V curlf\, dV$
3. $\iint_S \phi\, n\, dS = \iiint_V grad\phi\, dV$

GREEN'S THEOREM

Let S be a sufficiently nice region in the plane, and let δS be its boundary; then we have

$$\iint_S \left(\frac{\partial v_2}{\partial x} - \frac{\partial v_1}{\partial y} \right) dxdy = \oint_{\delta S} (v_1 dx + v_2 dy)$$

where the boundary, δS is traversed counterclockwise on its outside cycle, (and clockwise on any internal cycles).

Statement: If ϕ and ψ are two scalar point functions having continuous second order derivatives in a region V bounded by a closed surface S, then

$$\iiint_V (\phi\nabla^2\psi - \psi\nabla^2\phi) dv = \iint_S (\phi\nabla\psi - \psi\nabla\phi) \cdot n \, dS$$

Proof: By Gauss divergence theorem, we have

$$\iiint_V \nabla \cdot f dv = \iint_S f \cdot n \, dS$$

Putting $f = \phi\nabla\psi$, we get

$$\nabla \cdot f = \nabla \cdot (\phi\nabla\psi) = \phi(\nabla \cdot \nabla\psi) + (\nabla\phi)\cdot(\nabla\psi)$$
$$= \phi(\nabla^2\psi) + (\nabla\phi)\cdot(\nabla\psi)$$

Also, $f \cdot n = (\phi\nabla\psi)\cdot n$

$\therefore$ (68), becomes

$$\iiint_V [\phi\nabla^2\psi + (\nabla\phi)\cdot(\nabla\psi)] dv = \iint_S (\phi\nabla\psi)\cdot n \, dS$$

This is called the Green's first theorem (or identity)

Interchanging ϕ and ψ, we get

$$\iiint_V [\psi\nabla^2\phi + (\nabla\psi)\cdot(\nabla\phi)] dv = \iint_S (\psi\nabla\phi)\cdot n \, dS$$

we get,

$$\iiint_V (\phi\nabla^2\psi - \psi\nabla^2\phi) dv = \iint_S (\phi\nabla\psi - \psi\nabla\phi) \cdot n \, dS$$

This is called the Green's second theorem (or identity)

Hence, the Green's theorem can also be stated as

$$\iiint_V (\phi\nabla^2\psi - \psi\nabla^2\phi) dv = \iint_S \left(\phi\frac{\partial\psi}{\partial n} - \psi\frac{\partial\phi}{\partial n} \right) dS$$

Where $\frac{\partial\psi}{\partial n}$ and $\frac{\partial\phi}{\partial n}$ denote the directional derivative of ϕ and ψ respectively along the outward drawn normal at any point of S.

OTHER FORMS OF GREEN'S THEOREM

Green's theorem is a vector identity which is equivalent to the curl theorem in the plane. Over a region D in the plane with boundary ∂D, Green's theorem states

$$\int_{\partial D}\left[f(x,y)dx + g(x,y)dy\right] = \iint_D\left(\frac{\partial g}{\partial x} - \frac{\partial f}{\partial y}\right) dxdy$$

where the left side is a line integral and the right side is a surface integral. This can also be written compactly in vector form as

$$\int_{\partial D} F \cdot dS = \iint_D (\nabla \times F) \cdot \text{da}$$

If the region D is on the left when traveling around ∂D, then area of D can be computed using the elegant formula

$$\text{A} = \frac{1}{2}\int_{\partial D} xdy - ydx$$

giving a surprising connection between the area of a region and the line integral around its boundary. For a plane curve specified parametrically as $(x(t), y(t))$, equation becomes

$$A = \frac{1}{2}\int\left(x\frac{dy}{dt} - y\frac{dx}{dt}\right)dt$$

$$A = \frac{1}{2}\int(xy' - yx')\,dt$$

which gives the *signed* area enclosed by the curve.

Green's Theorem for Area in Plane Relating to its Bounding Curve

The analog of the "*Fundamental Theorem of Differential and Integral Calculus*" for a region $\Re$ bounded in a plane with normal k that is bounded by a curve $\partial\Re$ is:

$$\iint_R\left(\nabla \times \vec{F}\right)\cdot \hat{k}\; dxdy = \oint_{\partial R} \vec{F}\cdot d\vec{r}$$

Within the domain, the rotational flow $(\nabla \times F)$ from one cell moves into its neighbors; however, at the edges the local rotation is a net loss or gain. The local net loss or gain is $\vec{F}\cdot(dx, dy)$. The generalization of this idea to a surface $\partial\hat{A}$ bounding a domain $\hat{A}$ results in Stokes' theorem.

REDUCTION OF SURFACE INTEGRAL TO LINE INTEGRAL

Let the region R on the surface be break up into tiny pieces each one looking planar, apply Green's theorem on each and add up the mess to get the same theorem on all of R. The result is called *Stokes' Theorem*.

Theorem: With R an arbitrary reasonably smooth and nice region on a surface that is locally planar, and with δ R its boundary, we have

$$\iint_R \overline{\nabla \times v dS} = \oint_{\delta R} \nabla \cdot d\bar{l}$$

We can use it as a tool for rewriting integrals of one kind as integrals of another and for proving peculiar facts about irrotational fields (those for which curl = 0).

It can be stated in the following form:The circulation integral of v over a bounding curve is the flux integral of the curl of *v* over any surface it bounds.

Notice that this theorem immediately implies that if R is simply connected and curl *v* = 0 throughout it, then the potential *f* (with $v = \nabla f$) defined as path integral is path independent.

STOKES' THEOREM

The divergence theorem is used to find a surface integral over a closed surface and Green's theorem is use to find a line integral that encloses a surface (region) in the *xy*-plane. This theorem relates the surface integral to a line integral.

Let S be a oriented surface with unit normal vector *N* and let *C* be the boundary of *S*. Then *C* is *positively oriented* if its orientation follows the right hand rule, that is if you right hand curls around *N* in the direction of C's orientation, then your thumb will be pointing in the direction of *N*.

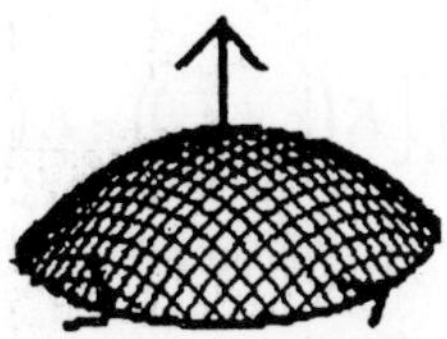

Fig. Enclosed Region in the *xy*-plane

Statement: Let S be an oriented surface with unit normal vector *n* and *C* be the positively oriented boundary of *S*. If *F* is a vector field with continuous first order partial derivatives then

$$\int_C F \cdot dr = \iint_S (Curl F) \cdot n dS$$

In other words: The line integral of the tangential component of a vector f taken around a simple closed curve *C* is equal to the surface integral of the normal component of the curl of f taken over any surface *S* having *C* as its boundary.

CARTESIAN FORM OF STOKES' THEOREM FOR PLANE

Let us take a system of Cartesian rectangular coordinate axes such that the plane of the given surface *S* is the *xy*-plane and *z*-axis lies along the direction of the normal vector *n*.

Here, the normal vector is constant

$$\text{Let } F = F_i i + F_2 j + F_3 k$$

$$\therefore F \cdot dr = (F_i i + F_2 j + F_3 k) \cdot (dxi + dyj + dzk)$$

$$\Rightarrow F \cdot dr = F_i dx + F_2 dy + F_3 dz$$

Now,

$$\iint_S (Curl F)\cdot n dS = \iint_S (Curl F)\cdot k dS$$

$$= \iint_S \left(\frac{\partial F_2}{\partial x} - \frac{\partial F_1}{\partial y}\right) dxdy$$

∴ Stokes' Theorem for plane becomes

$$\oint_C (F_i dx + F_2 dy) = \iint_S \left(\frac{\partial F_2}{\partial x} - \frac{\partial F_1}{\partial y}\right) dxdy$$

Let us, now prove it. Let the region *S* be such that any line parallel to either axis meet the boundary *C* in at the most two points.

Let the region be $x = a$, $y = \phi_1(x)$, $y = \phi_2(x)$, where $\phi_2(x) \geq \phi_1(x)$.

The boundary *C* is split into two arc C_1 and C_2 as shown.

$$\therefore \iint_S \frac{\partial F_1}{\partial y}\, dxdy = \int_a^b \left[\int_{\phi_1}^{\phi_2} \frac{\partial F_1}{\partial y} dy\right] dx$$

$$= \int_a^b \left[F_1(x, \phi_2(x)) - F_1(x, \phi_1(x))\right] dx$$

$$= -\iint_{C_2} F_1(x, y) dx - \iint_{C_1} F_1(x, y) dx$$

$$= -\oint_C F_1(x, y)\, dx$$

Similarly, $\therefore \iint_S \frac{\partial F_2}{\partial y}\, dxdy = -\oint_C F_2(x, y)\, dx$

∴ From (82) and (83), we get

$$\oint_C (F_i dx + F_2 dy) = \iint_S \left(\frac{\partial F_2}{\partial x} - \frac{\partial F_1}{\partial y}\right) dxdy$$

Hence the result.

If the region *S* is such that a line parallel to either axes meets the boundary *C* in more than tow points, we divide *S* into a finite number of sub-regions such that the boundary of each is met in at the most two points by any line parallel to either axis. By applying the above result to each sub-region and then adding the results, we find that the theorem is true for *S*. This because the line integrals along the boundary curves will cancel in pairs.

CARTESIAN FORM OF STOKES' THEOREM IN SPACE

Let $f = f_1 i + f_2 j + f_3 k$. Let the outward drawn normal vector *n* makes angles α, β, γ with the positive direction of *x*, *y*, *z* axes respectively so that

$$n = \cos\alpha i + \cos\beta j + \cos\gamma k$$

Then, $$\nabla \times f = \begin{vmatrix} i & j & k \\ \frac{\partial}{\partial x} & \frac{\partial}{\partial y} & \frac{\partial}{\partial z} \\ f_1 & f_2 & f_3 \end{vmatrix}$$

$$= \left[i\left(\frac{\partial f_3}{\partial x} - \frac{\partial f_2}{\partial z}\right) + j\left(\frac{\partial f_1}{\partial z} - \frac{\partial f_3}{\partial x}\right) + k\left(\frac{\partial f_2}{\partial x} - \frac{\partial f_1}{\partial y}\right)\right]$$

$$(\nabla \times f)\cdot n = \left[i\left(\frac{\partial f_3}{\partial x} - \frac{\partial f_2}{\partial z}\right) + j\left(\frac{\partial f_1}{\partial z} - \frac{\partial f_3}{\partial x}\right) + k\left(\frac{\partial f_2}{\partial x} - \frac{\partial f_1}{\partial y}\right)\right] \cdot (i\cos\alpha + j\cos\beta + k\cos\gamma)$$

$$= \left[\left(\frac{\partial f_3}{\partial x} - \frac{\partial f_2}{\partial z}\right)\cos\alpha + \left(\frac{\partial f_1}{\partial z} - \frac{\partial f_3}{\partial x}\right)\cos\beta + \left(\frac{\partial f_2}{\partial x} - \frac{\partial f_1}{\partial y}\right)\cos\gamma\right]$$

Also, $$f\cdot dr = (f_1 i + f_2 j + f_3 k)\cdot(dxi + dyj + dzk)$$

$$= f_1 dx + f_2 dy + f_3 dz$$

$\therefore$ Stokes' Theorem becomes

$$\int_C (f_1 dx + f_2 dy + f_3 dz) = \iint \left[\left(\frac{\partial f_3}{\partial x} - \frac{\partial f_2}{\partial z}\right)\cos\alpha + \left(\frac{\partial f_1}{\partial z} - \frac{\partial f_3}{\partial x}\right)\cos\beta + \left(\frac{\partial f_2}{\partial x} - \frac{\partial f_1}{\partial y}\right)\cos\gamma\right] dS$$

ILLUSTRATIVE EXAMPLES

1. *Prove that* $\int\int\int_V (u\nabla^2 w - w\nabla^2 u)dv = 3 \int\int\int_S (u\nabla w - w\nabla u)\bullet ds.$

Solution: By divergence theorem, we know that $\int\int\int_V \nabla \bullet F\, dv = \int\int_S F\bullet \hat{n}\, ds$

Now let $F = u\nabla w. \therefore \int\int\int_V \nabla \bullet \{u\nabla w\} dv \int\int_S (u\nabla w)\bullet \hat{n}\, ds.$

But we already know that $\nabla\bullet(u\nabla w) = u(\nabla\bullet\nabla w) + (\nabla u)\bullet(\nabla w) = u\nabla^2 w + (\nabla u)\bullet(\nabla w)$.

Putting this value of $\nabla\bullet(u\nabla w)$, we get

$$*\int\int\int_V \{u\nabla^2 w + (\nabla u)\bullet(\nabla w)\} dv = \int\int_S (u73\tilde{o}w)\bullet ds.$$

This Result is known as Green's first identity or theorem.

Now interchanging u and w we get $\int\int\int_V \{w\nabla^2 u + (\nabla w)\bullet(\nabla u)\} dv = \int\int_S (w\nabla u)\bullet ds.$

Subtracting we get $\int\int\int_V \{u\nabla^2 w - w\nabla^2 u\} dv = \int\int_S \{u\nabla w - w\nabla u\} \bullet ds.$

This is called Green's second identity or symmetrical theorem.

2. *Evaluate* $\iint_s F \bullet n\, ds$ *where S is the surface of the sphere* $x^2+y^2+z^2=16$ *and F* $z3xi + 4yj + 5zk$

Solution: By Gauss's Divergence theorem, we have $\iint_s F \bullet n\,ds = \iiint_V \nabla \bullet F\, dv =$

$$\iiint_V \left[\frac{\partial}{\partial x}(3x) + \frac{\partial}{\partial z}(4y) + \frac{\partial}{\partial z}(5z)\right] dV = \iiint_V (3+4+5) dV = 12\iiint_V dv$$ $= 12$ V, where V is the voloume of the given sphere. $= 12.\frac{4}{3}\pi$. Ex.2. *Prove that*

3. *Verify Gauss's Theorem and show that* $\int_S \left[(x^3 - yz)i - 2x^2yj + 2k\right] \bullet \hat{n}\, ds = \frac{1}{3}a^5,$ *where S denotes the surface of the cube bounded by the planes* $x = 0$, $x = a$, $y = 0$ $y = a$, $z = 0$, $z = a$.

Solution: Let $F = (x^3 - yz)i - 2x^2yj + 2k$.

$$\therefore \text{div } A = \nabla \bullet F = \left(i\frac{\partial}{\partial x} + j\frac{\partial}{\partial y} + k\frac{\partial}{\partial z}\right) \bullet \{(x^3 - yz)i - 2x^{3\,\prime} yj + 2k\}$$

$$= \frac{\partial}{\partial x}(x^3 - yz) + \frac{\partial}{\partial y}(-2x^2y) + \frac{\partial}{\partial z}(2) = 3x^2 - 2x^2 = x^2.$$

Now applying Gauss's divergence Theorem, we have

$$\int F \bullet \hat{n}\, ds = \int \textbf{div}\, F dv = \iiint x^2 dx\, dy\, dz, \qquad \{\text{using (A)}\}$$

$$= \left[\int_a^0 x^2 dx\right]\left[\int_a^0 dy\right]\left[\int_a^0 dz\right] = \frac{1}{3}a^2.a.a = \frac{1}{3}a^5$$

Verification.On the faces *PQRS*

$$\int F \bullet \hat{n}\, ds = \iint \{(x^3 - yz)i - 2x^2yj + 2k\} \bullet i dy\, dz \qquad (\because \hat{n} = i)$$

$$= \int_0^a \int_0^a (x^3 - yz\, dy\, dz) = \int_{z=0}^a \int_{y=0}^a (a^3 - yz) dy\, dz, \qquad (\because i \cdot k = 0 \text{ etc.})(\because x = 0)$$

$$= a^3 \int_0^a dy \int_0^a dz - \int_0^a y dy \int_0^a z dz = a^3 \cdot a \cdot a - \frac{1}{2}a^2 \frac{1}{2}a^2 = a^2 - \frac{1}{4}a^4$$

On the face $OQ'R'R \int F \bullet \hat{n} ds = \int\int (x^3 - yz)i - 2x^2 j + 2k) \bullet (-idy\,dz)73 \quad (\because \hat{n} = -i)$

$$= \int_{y=0}^{a} \int_{z=0}^{a} (x^3 - yz) dy\,dz = -\int_0^a \int_0^a yz\,dy\,dz, \qquad (\because x = o) = \frac{1}{2}a^2.a^2 = \frac{1}{4}a^4.$$

On the face $QQ'R'R'$

$$\int F \bullet \hat{n}\, ds = \int\int\int \{x^3 - yz)i - 2x^2 yj = 2k\} \bullet jdxdz \qquad (\because \hat{n} = j)$$

$$= -\int_{y=0}^{a} \int_{z=0}^{a} 2x^2 ydx\,dz = -2a \int_0^a x^2 dx \int_0^a dz \qquad (\because j \cdot k = 0 = i \cdot j; \because y = a)$$

$$= -2a\left(\frac{1}{2}a^3\right)(a) = -\frac{2}{3}a^5.$$

$$\int F \bullet \hat{n} ds = \int\int\int \{x^3 - yz)i - 2x^2 yj + 2k\} \bullet (-jdx\,dz) \text{OS'SP} \qquad (\because \hat{n} = -j)$$

$$= \int_0^a \int_0^a 2x^2 ydx\,dz = 0, (\because y = 0) \qquad (\because \hat{n} = k)$$

$$\int_{RR'S'S} F \bullet \hat{n}\, ds = \int\int \{(x^3 - yz)2 - 2x^2\, yj - 2k\{\bullet k\, dx\, dy = \int_{x=0}^{a} \int_{y=0}^{a} 2dx\,dy = 2a(a) = 2a^2$$

$$\int_{OPQQ'} F \bullet \hat{n}\, ds = \int\int \{(x^3 - yz)i - 2x^2\, yj + 2k\} \bullet (-kdx\,dy) \qquad (\because \hat{n} = -k)$$

$$= -\int_o^a \int_0^a 2dx\,dy = -2(a)(a) = -2a^2$$

get $\int_s F \bullet \hat{n} ds = a^5 - \frac{1}{4}a^4 + \frac{1}{4}a^4 - \frac{2}{3}a^5 + 0 + 2a^2 = \frac{1}{3}a^5$.

So the theorem is verified by the ddirect integration method.

4. *Prove that* $\int\int\int_V \frac{1}{r^2} dV = \int\int_S \frac{r \bullet \hat{n}}{r^2} ds$ where the symbols have their usual meanings.

Solution: By Gauss's Divergence theorem $\int\int\int_V \nabla \bullet FdV = \int\int_S F \bullet \hat{n} ds$

Here$\boldsymbol{F} = \frac{r}{r^2}$ But $\nabla \bullet F = \nabla \bullet \frac{r}{r^2} = div\left(\frac{r}{r^2}\right) = \frac{1}{r^2} div\, r + grad\left(\frac{r}{r^2}\right) \bullet r = \frac{3}{r^2} + \left(\frac{-2}{r^4} r\right) \bullet r$

$$\left[\because grad\, r^n = nr^{n-2} r \because grad r^{-2} = \frac{-2}{r^4} r\right] = \frac{3}{r^2} - \frac{2}{r^4} r^2 = \frac{1}{r^2} = \frac{3}{r^2} - \frac{2}{r^4}(r \bullet r)$$

Then (1) implies $\int\int\int_V \frac{1}{2} dv = \int\int_S \frac{r}{r^2} \bullet \hat{n} ds.$

EXERCISE

1. Evaluate $\int_S F.Nds$ where F = 6z I – 4 J + y K and S is the portion of the plane 2x + 3y + 6z = 12 in the first octant.
2. Verify Green's theorem for $\int_C \left[\left(3x-8y^2\right)dx+\left(4y-6xy\right)dy\right]$ where C is the boundary of the region bounded by x = 0, y = 0 and x + y = 1.
3. Applying Green's theorem, evaluate $\int_C \left[\left(y-\sin x\right)dx+\cos x dy\right]$ where C is the plane triangle enclosed by the lines y = 0, x = p/2 and y = $\frac{2}{\pi}x$.
4. Apply Green's theorem to prove that the area enclosed by a plane curve is $\frac{1}{2}\int_C (xdy-ydx)$. Hence find the area of an ellipse whese semi-major and semi-minor axes are of lengths a and b.
5. Verify Stoke's theorem for the vector field F = $(s^2 - y^2)$I + $2xy$J
   (i) Integrated round the rectangle in the plane $z = 0$ and bounded by the lines $x = 0$, $y = 0$, x = a and y = b.
   (ii) Over the box bounded by the planes x = 0, x = a; y = 0, y = b; z = 0, z = c; if the face z = o is cut.
6. Verify Stoke's theorem for a vector field defined by F = $-y^3$I + x^3J, in the region $x^2 + y^2 \le 1$, z = 0.
7. Verify Stoke's theorem for the vector field F = $(2x - y)$I – yz^2J –y^2zK over the upper half surface of $x^2 + y^2 + z^2 = 1$, bounded by its projection on the xy-plane.
8. Verify divergence theorem for F = $4xz$I – y^2J + yzK, taken over the cube bounded by $x = 0$, $x = 1$; $y = 0$, $y = 1$; $z = 0$, $z = 1$.
9. Verify Gauss divergence theorem for the function F = yI + xJ + z^2K over the cylindrical region bounded by $x^2 + y^2 = 0$, $z = 0$ and $z = 2$.
10. Using divergence theorem, prove that

    (i) $\int_S R.dS = 3V.$ (ii) $\int_S \nabla r^2.dS = 6V.$

    where S is any closed surface enclosing a volume V and $r^2 = x^2 + y^2 + z^2$.
11. Using divergence theorem, evaluate $\int_S R.dS$ where S is the surface of the sphere $x^2 + y^2 + z^2 = 0$.
12. Use divergence theorem to evaluate $\int_S F.dS$, where F = x^3I + + y^3J + y^3K, and s is the surface of the sphere $x^2 + y^2 + z^2 = a^2$.

Chapter 4

Quadratic Curves

Linear relationships are those that produce a straight line when plotted on a graph. They are used to describe an enormous number of situations, from the slope of a road to the expansion of the universe. However straight lines are not enough to describe fully the complicated world in which we live.

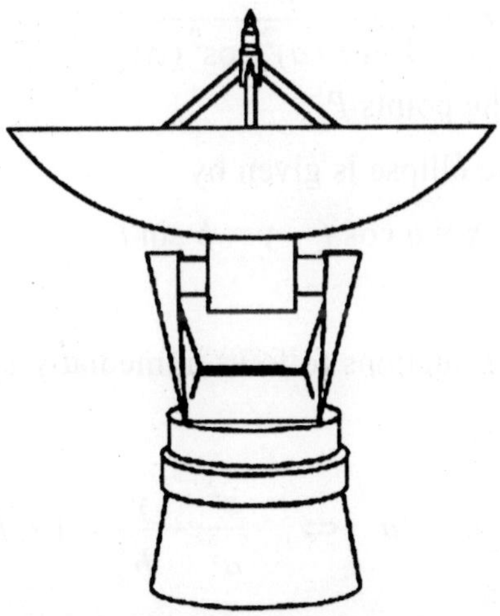

Fig. The Satellite Dish is a Curve

For example, the smooth shape of the satellite dish is a curve described by quadratic equation. In this unit you'll explore the properties of quadratic equations and learn how to plot them.

ELLIPSE

For the points $P(x, y)$ on an ellipse the sum of the distances from two fixed points $F_{\pm}$ (the foci) is a given positive constant:

$$|\overrightarrow{PF_-}| + |\overrightarrow{PF_+}| = 2a$$

with $2a > |\overrightarrow{F_-F_+}|$.

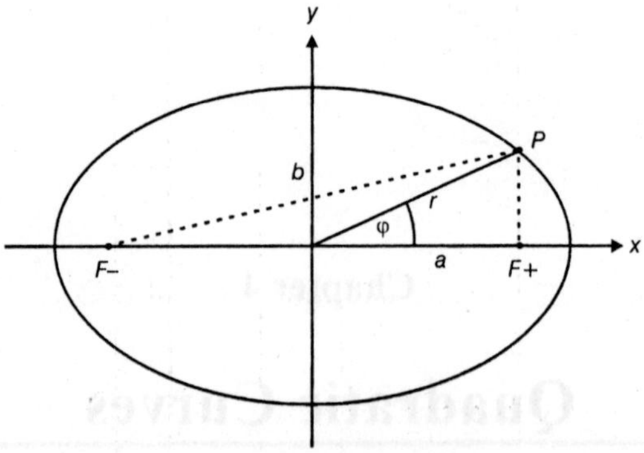

If $F_{\pm} = (\pm f, 0)$, then we have for the coordinates

$$\frac{x^2}{a^2} + \frac{y^2}{b^2} = 1, \qquad b^2 = a^2 - f^2,$$

and

$$r^2 = \frac{b^2}{1-(f/a)^2 \cos^2(\varphi)}$$

for the polar coordinates of the points P.

A parameterisation of the ellipse is given by

$$x = a \cos t, \quad y = b \sin t$$

with $t \in [0, 2\pi)$.

Equivalence of the representations follows immediatly by calculation.

To proof

$$|\overrightarrow{PF_-}| + |\overrightarrow{PF_+}| = 2a \quad \Leftrightarrow \quad \frac{x^2}{a^2} + \frac{y^2}{b^2} = 1, \quad b^2 = a^2 - f^2$$

by squaring of

$$\underbrace{2a - \sqrt{(x+f)^2 + y^2}}_{>0} = \sqrt{(x-f)^2 + y^2}$$

we obtain the equivalent equation

$$\underbrace{4a^2 + 4xf}_{>0} = 4a\sqrt{(x-f)^2 + y^2}\,.$$

Squaring again and dividing by $4a$ yield

$$a^2 + 2xf + \frac{f^2}{a^2}x^2 = x^2 + 2xf + f^2 + y^2.$$

Substituting $f^2 = a^2 - b^2$ and transforming we obtain the coordinate form.

To proof

$$r^2 = \frac{b^2}{1-(f/a)^2 \cos^2(\varphi)}$$

we multiply by the denominator and take into account that

$$x^2 = (x^2 + y^2) \cos^2(\varphi)$$

This implies

$$x^2 + y^2 - \frac{a^2 - b^2}{a^2} x^2 = b^2$$

and division by b^2 yields the coordinate form.

FOCUS RAYS

A ray of light originating at a focus will pass through the other focus after a simple reflection.

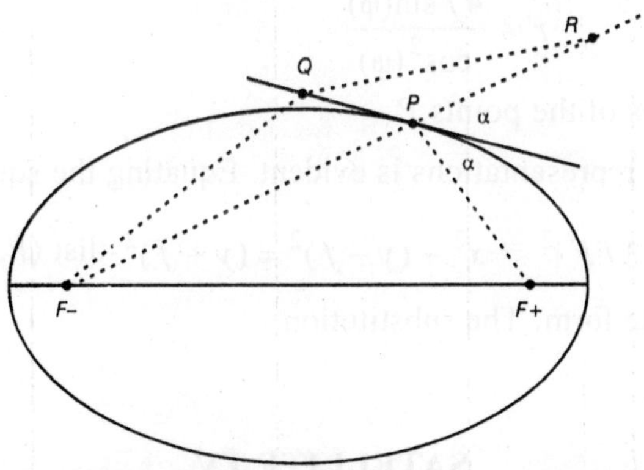

To prove this, select the following auxiliary points: the mirror image R of focus F_+ with respect to tangent and an arbitrary $Q \neq P$ point on g. Since Q lies outside of the ellipse we have

$$2a = |\overrightarrow{PF_-}| + |\overrightarrow{PF_+}| < |\overrightarrow{QF_-}| + |\overrightarrow{PF_+}|.$$

Sustituting the line segments $\overrightarrow{PF_+}$ and $\overrightarrow{QF_+}$ by its mirror images we obtain

$$|\overrightarrow{PF_-}| + |\overrightarrow{PR}| < |\overrightarrow{QF_-}| + |\overrightarrow{QR}|,$$

and hence the points F_-, P, R lie on a line. Thus we have $\sphericalangle(F_-PQ) = \alpha$.

This reflection property is used to break up kidney stones (lithotripsy). The rays of a radial source located in focus F_- can be focused in F_+ by an elliptic reflector (spheroid).

PARABOLA

The points $P = (x, y)$ of a parabola have the same distance from a given point (focus) and a given line g (directrix).

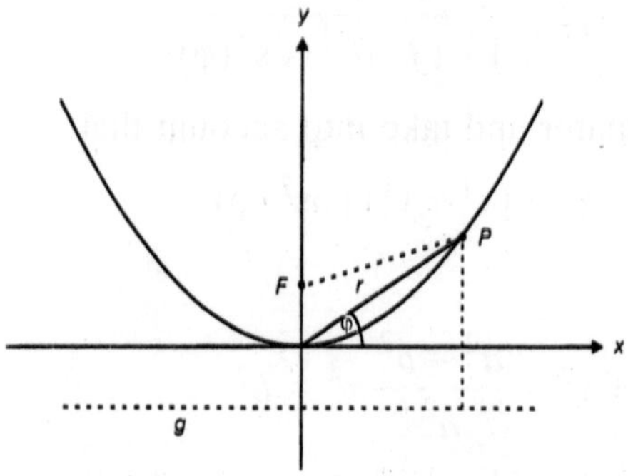

If $F = (0, f)$ and $g : y = -f$, then we have for the coordinates

$$4fy = x^2$$

and

$$r = \frac{4f\sin(\varphi)}{\cos^2(\varphi)}$$

for the polar coordinates of the points P.

Equivalence of the representations is evident. Equating the squared distances,

$$|\overrightarrow{PF}|^2 = x^2 + (y-f)^2 = (y+f)^2 \text{ dist } (P, g),$$

we obtain the coordinate form. The substitution

yields the polar form.

SATELLITE TV

Parallel rays perpendicular to the directrix are bundled in the focus. This property is used to amplify incoming signals in parabol aerials ('satellite dishes').

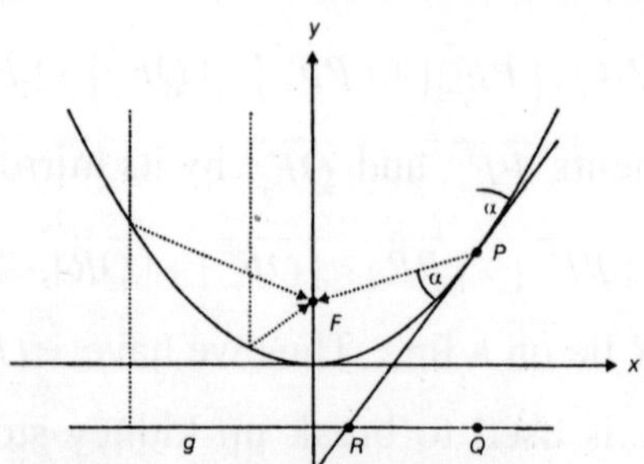

To prove this property, first of all, we see that

$$\overrightarrow{FP} + \overrightarrow{QP} = \left(\begin{pmatrix} x \\ y \end{pmatrix} - \begin{pmatrix} 0 \\ f \end{pmatrix}\right) + \left(\begin{pmatrix} x \\ y \end{pmatrix} - \begin{pmatrix} x \\ -f \end{pmatrix}\right) = \begin{pmatrix} x \\ 2y \end{pmatrix}$$

is parallel to the direction of the tangent in point P,

$$\begin{pmatrix} 1 \\ x/(2f) \end{pmatrix}.$$

Since $$|\overrightarrow{FP}| = |\overrightarrow{FQ}|,$$

it follows equality of the angles $\sphericalangle(F, P, R)$ and $\sphericalangle(R, P, Q)$, where R is the intersection point of the tangent and g.

HYPERBOLA

For the points $P = (x, y)$ on a hyperbola the difference of the distances from two given points (foci) $F_\pm$ is constant:

$$|\overrightarrow{PF_-}| - |\overrightarrow{PF_+}| = \pm 2a$$

with $2a < |\overrightarrow{F_-F_+}|$.'

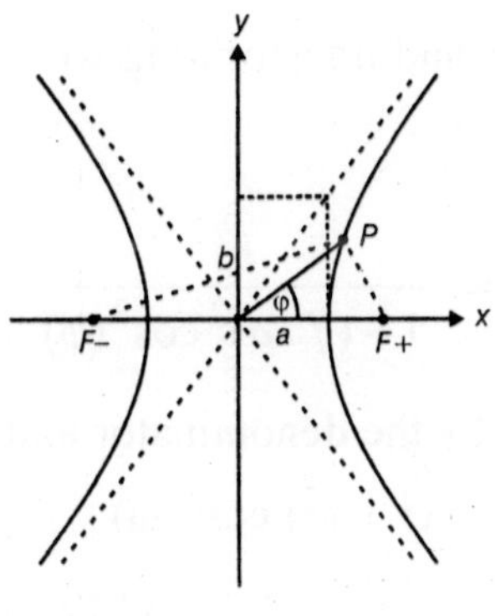

If $F_\pm = (\pm f, 0)$, then we have for the coordinates

$$\frac{x^2}{a^2} - \frac{y^2}{b^2} = 1, \quad b^2 = f^2 - a^2,$$

and $$r^2 = -\frac{b^2}{1 - (f/a)^2 \cos^2 \theta}$$

for the polar coordinates of the points P. The asymptotes have the slope $\pm b/a$.

A parametrisation of the hyperbola branches is given by

$$x = \pm a \cosh t, \quad y = b \sinh t$$

with $r \in \mathbb{R}$.

Equivalence of the representations follows immediatly by calculation.

To proof

$$|\overrightarrow{PF_-}| - |\overrightarrow{PF_+}| = \pm\ 2a \quad \Leftrightarrow \quad \frac{x^2}{a^2} - \frac{y^2}{b^2} = 1, \quad b^2 = a^2 - f^2$$

squaring of

$$\underbrace{\sqrt{(x+f)^2 + y^2} - 2a}_{>0} = \sqrt{(x-f)^2 + y^2}$$

leads to the equivalent equation

$$\underbrace{4a^2 + 4xf}_{>0} = 4a\sqrt{(x+f)^2 + y^2}\,.$$

Squaring again and dividing by $4a$ yield

$$a^2 + 2xf + \frac{f^2}{a^2}x^2 = x^2 + 2xf + f^2 + y^2.$$

By Substituting $f^2 = a^2 + b^2$ and transforming we obtain the coordinate form.

To proof

$$r^2 = -\frac{b^2}{1 - (f/a)^2 \cos^2(\varphi)}$$

we multiply the polar form by the denominator and observe that

$$x^2 = (x^2 + y^2)\cos^2(\varphi).$$

Thus follows

$$x^2 + y^2 - \frac{a^2 - b^2}{a^2}x^2 = -b^2,$$

and division by $-b^2$ yields the coordinate form.

NAVIGATION

To determine the position P of a ship, the time differences $2a_{j,k}$, while receiving synchronous radio signals of different transmitting stations F_i, are compared.

Hence, the point P can be determined as point of intersection of hyperbolas with foci F_j, F_k.

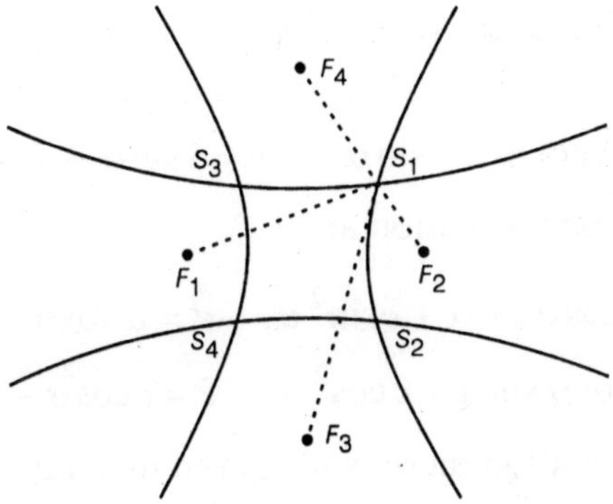

As a concrete example put

$$F_1 = (-2, 0), \quad F_2 = (0, 2), \quad F_3 = (0, -3), \quad F_4 = (3, 0)$$

and. $a_{1,2} = a_{3,4} = 1$ The corresponding equations of the hyperbolas are

$$x^2 - \frac{y^2}{3} = 1$$

$$x^2 - \frac{y^2}{8} = 1,$$

and we obtain

$$x_2 = \frac{32}{23}, y^2 = \frac{27}{23}$$

for the coordinates of the possible points of intersection S_i.

ROTATION OF CONIC SECTIONS

If the conic section with the equation

$$ax^2 + bxy + cy^2 + dx + ey + f = 0$$

is transformed by a rotation about the origin through the angle α, then the equation in the new coordinates $\tilde{x}$ and $\tilde{y}$ has the form

$$\tilde{a}\tilde{x}^2 + \tilde{c}\tilde{y}^2 + \tilde{d}\tilde{x} + \tilde{e}\tilde{y} + \tilde{f} = 0,$$

provided

$$\tan 2\alpha = \frac{b}{a-c} \quad (a \neq c)$$

or

$$\alpha = \frac{\pi}{4} \quad (a = c)$$

The coordinate transformation which expresses the old coordinates in terms of the new ones is given by

$$x = \tilde{x}\cos\alpha - \tilde{y}\sin\alpha, \quad y = \tilde{x}\sin\alpha + \tilde{y}\cos\alpha.$$

The parameters of the new equation are

$$\tilde{a} = a\cos^2\alpha + b\cos\alpha\sin\alpha + c\sin^2\alpha, \quad \tilde{d} = d\cos\alpha + e\sin\alpha, \quad \tilde{b} = 0,$$

$$\tilde{c} = a\sin^2\alpha - b\cos\alpha\sin\alpha + c\cos^2\alpha, \quad \tilde{e} = e\cos\alpha - d\sin\alpha, \quad \tilde{f} = f.$$

Note that in the new equation there is no mixed quadratic term, i.e. the coeffcient $\tilde{b}$ of $\tilde{x}\tilde{y}$ is zero. Thus this equation may be easily further transformed into a normal form of the conic section by completing squares.

Consider the conic section

$$x^2 - 2\sqrt{3}xy + 3y^2 - \sqrt{3x} - y + f = 0.$$

The condition on the angle α of a suitable rotation of the coordinate axes is

$$\tan 2\alpha = \sqrt{3}.$$

Thus with $a = \frac{\pi}{6}$

the coordinate transformation

$$x = \sqrt{3}\tilde{x}/2 - \tilde{y}/2, \quad y = \tilde{x}/2 + \sqrt{3}\tilde{y}/2$$

yields

$$\tilde{a} = 1\cdot 3/4 + (-2\sqrt{3})\cdot\sqrt{3}/4 + 3\cdot 1/4 = 0,$$

$$\tilde{c} = 1\cdot 1/4 - (-2\sqrt{3})\cdot\sqrt{3}/4 + 3\cdot 3/4 = 4,$$

$$\tilde{d} = -\sqrt{3}\cdot\sqrt{3}/2 + (-1)\cdot 1/2 = -2,$$

$$\tilde{e} = -1\cdot\sqrt{3}/2 - (-\sqrt{3})\cdot 1/2 = 0,$$

and the equation gets the form

$$4\tilde{y}^2 - 2\tilde{x} + f = 0.$$

Consequently the given conic section is a parabola (independent from *f*).

CURVES AND SURFACES

Every graphics system has some form of primitive to draw lines. Using these primitives we can draw many complex shapes. However, as these shapes get ever more complex and finely detailed, so does the data needed to describe them accurately.

The worst case scenario is the curve. A curve can be described by a finite number of short straight segments. However, on close inspection this is only an approximation. To get a better approximation we can use more segments per unit length. This increases the amount of data required to store the curve and makes it difficult to manipulate. We clearly need a way of representing these curves in a more mathematical fashion.

The types of curve will be discussed fall into two broad categories: interpolating or approximation curves. Interpolating curves will pass through the points used to describe it, whereas an approximating curve will get near to the points, though exactly what is called near will be discussed later. The points through which the curve passes are known as knots; the curve described by the equation is often referred to as a spline. This term originated in manual design, where a spline is a thin strip. This strip was held in place by weights to create a curve which could then be traced. In the same way we now use knots to describe a curve.

Curves

A curve may be represented parametrically by

$$x = \psi(t),$$

where x is a point on the curve, and t is a parameter that varies along the curve.

Remarks:

1. Pure mathematicians define a 'curve' (or 'path' if the curve is 'smooth' enough) to be a mapping, while the image of the mapping is a 'set of points of the curve'. Applied mathematicians tend to identify the set of points with the curve.
2. The parametrisation of a given curve is not unique.
3. We will concentrate on $\mathbb{R}^3$, but there are straightforward extensions to $\mathbb{R}^m$.
4. One of our aims is to identify what characterises a curve.

Examples:

1. $x = \psi(t) = (at, bt^2, 0)$.

   This is the parabola $a^2y = bx^2$ in the xy-plane.

2. $x = \psi(t) = (e^t \cos t, e^t \sin t, 0)$.

   This is a spiral.

3. $x = \psi(t) = (\cos t, \sin t,)$.

   This is a helix on the cylinder of radius one, i.e. the cylinder $\rho^2 = x^2 + y^2 = 1$.

4. Suppose $x = (\rho(t) \cos \phi(t), \rho(t) \sin \phi(t), z(t))$, where

   $$\rho = \gamma + \cos \alpha t, \phi = \beta t, z = \sin \alpha t,$$

   and α, β and $\gamma > 1$ are constants. This curve lies on the torus $(\rho - \gamma)^2 + z^2 = 1$.

Unlectured paragraph discussing when the curve is closed. For the curve defined by $0 \leq t \leq t_c$, to be closed it is necessary that the curve

1. Returns to $x = \gamma + 1$ at t_c, which implies that
$$(\gamma + \cos \alpha t_c) \cos \beta t_c = \gamma + 1,$$
i.e. that $\alpha t_c = p\pi$ and $\beta t_c = q\pi$, where p and q are integers;
2. Returns to $y = 0$ at t_c, which is ensured if
$$\beta\, t_c = q\pi \text{ since } \sin q\pi = 0;$$
3. Returns to $z = 0$ at t_c, which is ensured if
$$\alpha t_c = \pi \text{ since } \sin p\pi = 0.$$

It follows from the above three results that for the curve to be closed it is necessary that

$$\frac{\alpha}{\beta} = \frac{p}{q},$$

i.e. that α / β is rational.

Reading Exercise. What happens if α / β is irrational?

Definitions:

1. A simple curve is one that does not intersect or touch itself. simple intersects touches
2. A curve is smooth if the derivative $\psi'(t)$ exists. It is piecewise smooth if the curve can be divided into a finite number of smooth curves.

The Length of a Simple Curve

Let C be a simple curve defined by

$$x = \psi(t) \text{ for } t_a \leq t \leq t_b.$$

Form a dissection D by splitting $[t_a, t_b]$ into N intervals

$$t_a = t_0 < t_1 < \ldots < t_N = t_b.$$

The points $\{\psi(t_0), \psi(t_1), \ldots \psi(t_N)\}$ can be viewed as the vertices of an inscribed polygonal path. The length of this path is

$$\ell_D = \sum_{j=1}^{N} |\psi(t_j)\,\psi(t_j - 1)|,.$$

Define

$$\ell(C) = \sup_D \ell_D.$$

Then if $\ell(C) < \infty$, C is said to be rectifiable and $\ell(C)$ is the length of the curve.

Remarks:

1. If C is not simple, but can be split into a finite number of simple curves C_j of length $\ell(C_j)$, then
$$\ell(C) = \sum_j \ell(C_j).$$
2. If is continuously dierentiable (i.e. if the vector derivative $\psi'(t) = (\psi_1'(t), \psi_2'(t), \psi_3'(t))$ is continuous), then
$$\ell = \int_{t_a}^{t_b} |\psi'(t)|\, dt.$$

This result can be derived using the definition of a Riemann integral and the idea of uniform continuity.

The Arc Length

Definition: The arc length [function] of a smooth curve C parameterised by $\psi(t)$ is defined to be
$$s(t) = \int_{t_a}^{t_b} |\psi'(\tau)|\, d\tau.$$

Remarks:

1. At one end of the curve the arc length is zero, while at the other end the arc length equals the length of the curve,
$$s(t_a) = 0, \text{ and } s(t_b) = \ell.$$
2. The arc length is an increasing function since
$$\frac{ds}{dt} = |\psi'(t)| \geq 0.$$
3. Using $x = \psi(t)$, the infinitesimal arc length can be re-expressed (admittedly slightly cavalierly) as
$$\frac{ds}{dt} = |\psi'(t)|\, dt = \left[\left(\frac{dx}{dt}\right)^2 + \left(\frac{dy}{dt}\right)^2 + \left(\frac{dz}{dt}\right)^2\right]^{1/2}$$
$$dt = (dx^2 + dy^2 + dz^2)^{\frac{1}{2}}.$$

This is as expected on geometrical grounds. Further, on geometrical grounds we would expect that

$$\ell = \int (dx^2 + dy^2 + dz^2)^{1/2}.$$

$$= \int_{t_a}^{t_b} |\psi'(t)|\, dt.$$

4. If $\dfrac{ds}{dt} = |\psi'(t)| > 0$ then (by the inverse function theorem this term's Analysis course) s(t) is invertible, and there is an inverse function such that

$$t = \lambda(s).$$

We can then parametrise the curve in terms of s:

$$x = \psi(t) = \psi(\lambda(s)) = r(s).$$

Further, $\dfrac{d\lambda}{ds} = \dfrac{dt}{ds} = \dfrac{1}{ds/dt} = \dfrac{1}{|\psi'(t)|} = \dfrac{1}{|\psi'(\lambda(s))|}$, it follows from using the chain rule that

$$r'(s) = \psi'(\lambda(s))\frac{d\lambda}{ds} = \frac{\psi'(\lambda(s))}{|\psi'(\lambda(s))|},$$

and hence that (as required for consistency)

$$|r'(s)| = 1.$$

The Tangent

Definition: Let L be the straight line through ψ (t) and ψ $(t + h)$. The tangent at ψ (t) is the limiting position of L as $h \to 0$.

Remarks:

1. L has the direction of the vector

$$\frac{1}{h}(\psi(t+h) - \psi(t)).$$

Hence if

$$\psi'(t) = \lim_{h\to 0} \frac{(\psi(t+h) - (\psi(t))}{h}$$

is non-zero, $\psi'(t)$ has the direction of the tangent at ψ (t).

2. For a smooth curve C parameterised using arc length, that $r'(s) \neq 0$; and moreover that

$$r' \equiv \frac{dr}{ds}$$

is the unit tangent vector.

The Osculating Plane

If $\psi''(t) \neq 0$, then three neighbouring points, say t_1, t_2 and t_3, on a curve determine a plane. The limiting plane as t_2, t_3 tend to t_1 is called the osculating plane at the point t_1.

The Normal of the Osculating Plane. A normal to the plane defined by t_1, t_2 and t_3 is given by

$$n = (\psi(t_2) - \psi(t_1)) \times (\psi(t_3) - (\psi(t_1).$$

Let $t_2 = t_1 + \delta t_2$ and $t_3 = t_1 + \delta t_3$, then using Taylor's theorem for vector functions, or alternatively applied componentwise,

$$\psi(t_2) - \psi(t_1) = \psi(t_1 + \delta t_2) - \psi(t_1)$$

$$= \delta t_2 \psi'(t_1) + \frac{1}{2}\delta t_2^2 + \psi''(t_1) + ...,$$

and so

$$n = (\delta t_2 \psi'(t_1) + \frac{1}{2}\delta t_2^2 \psi''(t_1) + ...) \times (\delta t_3 \psi'(t_1) + \frac{1}{2}\delta t_3^2 + \psi''(t_1) + ...)$$

$$= \frac{1}{2}\delta t_2\, \delta t_3(\delta t_3 - \delta t_2)(\psi'(t_1) \times \psi''(t_1)) + ...$$

Hence the unit normal of the osculating plane at t_1 is given, up to a sign, by

$$\hat{n} = \frac{\psi'(t_1) \times \psi''(t_1)}{|\psi'(t_1) \times \psi''(t_1)|}$$

and the plane itself has the equation

$$(x - (\psi(t_1)).\ \hat{n} = 0.$$

Remark: We should check under what conditions the normal is well defined, i.e. we should determine under what conditions $\psi'(t) \times \psi''(t) \neq 0$.

We have already seen that there is an advantage in parameterising the curve using arc length, since we then know that $r'(s) \neq 0$. Moreover, if we define

$$u(s) = \frac{dr}{ds}(s),$$

then because $u.u = 1$, it follows from using our earlier result for the derivative of a scalar product, that

$$0 = \frac{d}{ds}(u, u) = \frac{d}{ds}(u_j u_j) = 2u_j \frac{du_j}{ds} = 2u\frac{du}{ds},$$

and that u is perpendicular to

$$\frac{du}{ds}(s) = \frac{d^2r}{ds^2}(s).$$

Hence, except when $r''(s) = 0$, we conclude that $r'(s) \times r''(s) \neq 0$ and that the normal of the osculating plane is well defined.

The Serret-Frenet Equations

The first equation is

$$\frac{d^2r}{ds^2} = \frac{du}{ds} = \kappa p,$$

where κ is defined by

$$k(s) = |r''(s)| \geq 0,$$

and p is a unit vector.

Definitions: *Curvature.* $\kappa(s)$ is referred to as the curvature, while $\rho(s) = 1/k(s)$ is referred to as the radius of curvature.

Principle Normal: If $k > 0$, i.e. if $r''(s) \neq 0$, then p is defined to be the unit principle normal.

Bi-normal: Let

$$b = u \times p = \frac{r'(s) \times r''(s)}{\kappa}.$$

Then b is defined to be the unit bi-normal vector. It follows that b is, up to a sign, the unit normal of the osculating plane.

Moving Trihedral: Since u is perpendicular to p, it follows from that u, p and b form a right handed co-ordinate system — the moving trihedral.

Remark: Knowledge of the the curvature is necessary when calculating certain physical forces such as surface tension or bending moments.

Torsion: Since $b.b = 1$ it follows as above that

$$b\frac{db}{ds} = 0.$$

Further, since u, p and b are mutually orthogonal, and in particular $u.b = 0$ and $p.b = 0$, it follows using definition of p that

$$0 = \frac{d}{ds}(u.b) = u.\frac{db}{ds} + \frac{du}{ds}.b = u,\frac{db}{ds} + kp.b = u.\frac{db}{ds}.$$

Thus $\frac{db}{ds}$ is perpendicular to both b and u, i.e. $\frac{db}{ds}$ is parallel to p. We write

$$\frac{db}{ds} = -\tau p,$$

where τ is defined to be the torsion; this is the second Serret-Frenet equation.

The Third Equation. Further,

$$b \times u = (u \times p) \times u = (u \cdot u)\, p - (p \cdot u)\, u = p.$$

Hence

$$\frac{dp}{ds} = \frac{db}{ds} \times u + b \times \frac{du}{ds}$$

$$= -\tau p \times u + \kappa b \times p.$$

$$\frac{dp}{ds} = \tau b - \kappa u.$$

This is the third Serret-Frenet equation.

Summary: To summarise the three Serret-Frenet equations are

$$\frac{du}{ds} = \kappa p,$$

$$\frac{db}{ds} = -\tau p,$$

$$\frac{dp}{ds} = \tau b - \kappa u,$$

Remarks:

1. Since $p = b \times u$ we can write the above vector system of equations as

$$\frac{du}{ds} = \kappa(b \times u), \frac{db}{ds} = \tau(u \times b).$$

This system can be viewed as six scalar equations in six scalar unknowns. Hence if κ(s), τ(s), $u(0)$ and $b(0)$ are known, it is possible to solve for $u(s)$ and $b(s)$, and subsequently for r(s).

1. $\kappa(s)$ and $\tau(s)$ contain all the information about the geometry of a curve up to translation and orientation.

Surfaces

A curve is the locus of a point with 1 degree of freedom. A surface is the locus of a point with ∈ degrees of freedom.

Representations

We will concentrate on $\mathbb{R}^3$.

Explicit Representation:

$$z = f(x, y),$$

e.g. a hemisphere of radius a

$$z = (a^2 - x^2 - y^2)^{1/2}.$$

Implicit Representation:

$$F(x) = F(x, y, z) = 0,$$

e.g.

$$z - f(x, y) = 0;$$

a sphere of radius a: $x^2 + y^2 + z^2 - a^2 = 0$.

Parametric Representation:

$$x \equiv x(u) = (x(u, v), y(u, v), z(u, v)),$$

e.g. a sphere of radius a

$$x = (a \sin u \cos v, a \sin u \sin v, a \cos u).$$

Remark: Implicit and parametric representations are not unique.

Normals

Parametric Representation: Suppose a surface has the parametric representation

$$x \equiv x(u, v).$$

Then the functions $x(u, v_0)$ of u and $x(u_0, v)$ of v, for fixed v_0 and u_0 respectively, describe curves in the surface.

Further, if we assume that both $x(u, v_0)$ and $x(u_0, v)$ are dierentiable, then from the definition of a tangent, it follows that

a tangent at (u_0, v_0) to the line $v = v_0$ is $\frac{\partial x}{\partial u}(u_0, v_0)$,

a tangent at (u_0, v_0) to the line $u = u_0$ is $\frac{\partial x}{\partial v}(u_0, v_0)$,

Hence a unit normal to the surface at (u_0, v_0) is given by

$$n = \frac{x_u \times x_v}{|x_u \times x_v|}.$$

Implicit Representation: For an implicit representation, let $x = \psi(t)$ describe a curve in the surface.

Then, if the surface is described by $F(x) = 0$,

$$F(\psi(t)) = 0.$$

From the chain rule (1.30c) we conclude that

$$0 = \frac{d}{dt}F(\psi(t)) = \frac{\partial}{\partial x_j} F(\psi(t))\frac{d\psi_j}{dt} = \nabla F(\psi(t)).\frac{d\psi}{dt}.$$

Hence if $\nabla F \neq 0$, ∇F is perpendicular to the tangent of the curve $\psi(t)$. But $\psi(t)$ can be any curve on the surface, hence ∇F must be normal to the surface, i.e. up to a sign

$$n = \frac{\nabla F}{|\nabla F|}.$$

Remarks:

1. Since ∇F is in the direction of the maximum rate of change of F, the normal to the surface is in the direction of the maximum rate of change of F.
2. If n is a normal, then so is $(-n)$.
3. For surfaces it is conventional to denote the unit normal by n, rather than by, say, $\hat{n}$.
4. The plane normal to n touching the surface is known as the tangent plane.

Examples:

1. Consider a surface of revolution defined parametrically by

$$x(\phi, t) = (R(t) \cos \phi, R(t) \sin \phi, t)$$

In cylindrical polar co-ordinates (ρ, ϕ, z), the surface is given by

$$z = t,$$

$$\rho = (x^2 + y^2)^{\frac{1}{2}} = R(z).$$

Further,

$$x(\phi, t) = (-R \sin \phi, R \cos \phi, 0),$$

$$x_t(\phi, t) = (R' \cos \phi, R' \sin \phi, 1),$$

and hence from, after identifying ϕ and t with u and v respectively, a normal to the surface is given by

$$n = \frac{x_\phi \times x_t}{|x_\phi \times x_t|} = \frac{(R\cos\phi,\ R\sin\phi, -RR')}{|R(1+R'^2)^{\frac{1}{2}}|}$$

$$= \frac{(\cos\phi, \sin\phi, -R')}{(1+R'^2)^{\frac{1}{2}}}$$

Remember that $-n$ is also a normal.

2. Consider the surface with the implicit representation

$$F(x) = a \,.\, x - c = 0,$$

where $a \in \mathbb{R}^3$ and $c \in \mathbb{R}$. Since

$$(\nabla F)_i = \frac{\partial}{\partial x_i}(a_j x_j - c) = a_j \delta_{ij} = a_i, \text{ i.e. } \nabla F = a,$$

the unit normal is given by

$$n = \frac{a}{|a|}.$$

This is as expected is the equation of a plane with normal a.

3. A surface is given by the implicit representation

$$F(x, y, z) = x^2 - y^2 - z^2 - c^2 = 0$$

where c is a real constant. Since

$$x^2 - \sigma^2 = c^2 \text{ where } \sigma^2 = y^2 + z^2,$$

this is a surface of revolution about the x-axis. Further, since

$$\nabla F = 2(x, -y, -z),$$

the unit normal (up to a sign) is given by

$$n = \frac{1}{r}(x, -y, -z).$$

Unlectured remark. $\nabla F = 0$ at $x = 0$; hence if $c = 0$ so that the surface is given by

$$x^2 = y^2 + z^2,$$

the normal at (0, 0, 0) is not defined (but then this is the apex of a cone).

Definitions: *Smooth Surfaces.* A smooth surface is a surface with a unique (upto a sign) normal whose direction depends continuously on the points of the surface. A piecewise smooth surface is one that can be divided into finitely many portions each of which is smooth. smooth piecewise smooth

Remark:

1. An alternative definition of. a smooth surface for a given parametric representation, e.g. $x \equiv x(u, v)$ is that x_u and x_v are continuous and that $| x_u \times x_v | \neq 0$. However, with this definition the same surface can fail to be smooth at dierent points, depending on the parameterisation.

Orientable Surfaces. Let S be a smooth surface. Choose a unit normal n of S at a point P, i.e. fix on one of the two possible normals at point P. The direction of n is then said to be the positive normal direction.

The surface S is said to be orientable if the positive normal direction once given at an arbitrary point P can be continued in a unique and continuous way to the entire surface. sphere mobius strip klein bottle orientable non-orientable non-orientable

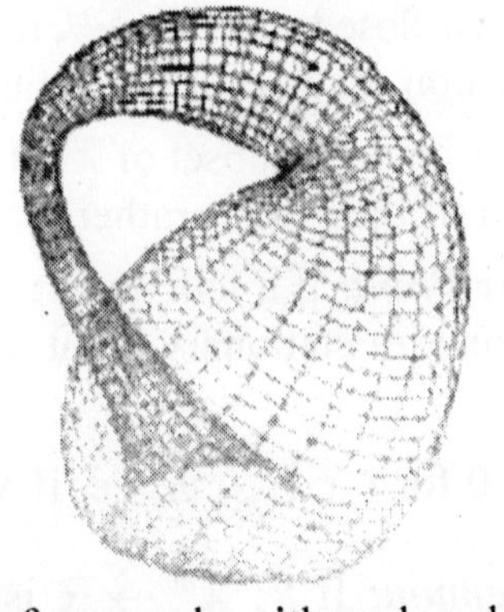

Open/Closed Surfaces. A surface can be either closed, if 'any' straight line intersects the surface an even number of times (where tangents count twice or not at all), e.g. a sphere, or open, if the surface is bounded by a curve, e.g. a mobius strip or a hemisphere.

A closed surface bounds a convex region if 'any' straight line intersects the surface a maximum of twice (where tangents count twice or not at all), e.g. a sphere. The Orientation of a Bounding Curve: If an open orientable surface is bounded by a simple closed curve C, then an orientation of C can be identified by the 'right-hand-rule'.

A piecewise smooth surface is orientable if the orientations of common boundaries can be arranged to be opposite.

Critical Points or Stationary Points

In this subsection we extend the ideas of maxima and minima for functions of one variable, i.e. $f : \mathbb{R} \rightarrow \mathbb{R}$, to functions of several variables, i.e. $f : \mathbb{R}^m \rightarrow \mathbb{R}$.

Definitions: *Local Maximum or Minimum.* A function $f : \mathbb{R}^m \to \mathbb{R}$ defined on a domain D is said to have a local maximum or minimum at $a \in D$ if for all $x \in D$ that are suciently close to a

$$f(x) < f(\text{a}) \text{ or } f(x) > f(\text{a})$$

respectively if $x \neq a$.

Weak Local Maximum or Minimum: If

$$f(x) \leq f(a) \text{ or } f(x) \geq f(a),$$

then f is said to have a weak local maximum or weak local minimum respectively.

Extremum: An extremum is equivalent to a maximum or minimum, while a weak extremum is equivalent to a weak maximum or minimum.

Remarks:

1. The above are local definitions. That a local maximum or minimum has been found does not imply that a global maximum or minimum has been found.
2. For a function defined on a closed subset of $\mathbb{R}^m$, remember that the global maximum or minimum of the function may occur on a boundary.
3. For a function defined on an open subset of $\mathbb{R}^m$ it may be more appropriate to seek local suprema or infima (if they exist) rather than local maxima or minima.

Critical Point or Stationary Point: If a function $f : \mathbb{R}^m \to \mathbb{R}$ is dierentiable suciently close to a, then a is a critical point, or stationary point, if

$$\frac{\partial f}{\partial x_j}(a) = 0 \text{ for } j = 1,\ldots, m, \text{ i.e. if } \nabla f(a) = 0.$$

Property of a Weak Extremum: If $f : \mathbb{R}^m \to \mathbb{R}$ is dierentiable, and f has a weak extremum at a point a in the interior of its domain of definition, then a is a critical point of f, i.e. $\nabla f(\text{a}) = 0$.

To see this suppose, without loss of generality, that f has a weak local minimum. Then for $|h|$ suciently small,

$$f(a + he_j) \geq f(\underline{a}).$$

Hence

$$\lim_{h \to 0+} \frac{f(a + he_j) - f(a)}{h} \geq 0,$$

and

$$\lim_{h \to 0-} \frac{f(a + he_j) - f(a)}{h} \leq 0,$$

But f is dierentiable, which implies that the above two limits must both be equal to $\frac{\partial f}{\partial x_j}$ (a); hence $\frac{\partial f}{\partial x_j}$ (a) = 0.

Remark: A weak extremum at a point a need not be a critical point if a lies on the boundary of the domain of f.

Saddle Point. A critical point that is not a weak extremum is called a saddle point.

Level Set. The set $L(c)$ defined by $L(c) = \{x \mid f(x) = c\}$ is called a level set of f. In $\mathbb{R}^2$ it is called a level curve or contour. In $\mathbb{R}^3$ it is called a level surface.

Functions of Two Variables

For a function of two variables, $f: \mathbb{R}^2 \to \mathbb{R}$, the function can be visualised by thinking of it as the surface with the explicit representation

$$z = f(x, y);$$

z is constant on the level curves of f. Further, from the equivalent implicit representation

$$F(x, y, z) = z - f(x, y) = 0 .$$

Figure of $z = \in - x^2 - y^2$, $z = x^2 + y^2$, $z = xy$ from Apostol.

Figure of $z = x^3 = 3xy^2$, $z = x^2y^2$, $z = 1 - x^2$ from Apostol.

we note that

$$\nabla F = (-f_x, -f_y, 1).$$

Hence from the following remarks, the tangent plane is 'horizontal' at a critical point of f (e.g. the tangent plane is 'horizontal' at the top of a 'hill').

Examples:

1. For the function

$$f = \in - x^2 - y^2, \text{ we have that } f_x = -2x\ , f_y = -2y.$$

Hence there is a critical point at $x = y = 0$. Further, since

$$f(x, y) - f(0, 0) = f(x, y) - \in = - x^2 - y^2 \leq 0,$$

the critical point is a maximum.

2. For the function

$$f = x^3 - 3xy^2,$$

we have that

$$f_x = 3x^2 - 3y^2, f_y = -6xy.$$

Hence, as in the previous example, there is a critical point at $x = y = 0$. However in this case we note that

$$f(x, 0) - f(0, 0) = x^3 \begin{cases} > 0 & if\ x > 0 \\ < 0 & if\ x > 0 \end{cases}$$

and thus the critical point is a saddle point.

Maximum, Minimum or Saddle?

How do we determine if an interior critical point is a maximum, a minimum or a saddle point? First a definition.

Definition. Suppose for a function $f : \mathbb{R}^m \to \mathbb{R}$ that all the second-order partial derivatives exist, then define the Hessian matrix $H(a)$ to have components

$$H_{ij}(a) = \frac{\partial^2 f}{\partial x_i \partial x_j}(a)$$

Suppose now that a function $f: \mathbb{R}^m \to \mathbb{R}$ is continuously dierentiable up to and including second order, and that it has a critical point at a. We examine the variation in f near a in order to determine the nature of the critical point.

From Taylor's Theorem with $N = 2$, we have with the help of definition of a critical point, i.e. $\frac{\partial f}{\partial x_i}$ (a) = 0, (a) = 0, that near a

$$f(a+h) = f(a) + 0 + \frac{1}{2}\left(\sum_{i=1}^{m} h_i \frac{\partial}{\partial x_i}\right)\left(\sum_{j=i}^{m} h_j \frac{\partial}{\partial x_j}\right) f(a) + o(|h|^2)$$

$$= f(a) + \frac{1}{2}\sum_{i,J=1}^{m} h_i \left(\frac{\partial^2}{\partial y_i \partial y_j}\right) f(a) h_j + o(|h|^2)$$

from using

$$f(a+h) - f(a) = \frac{1}{2}\sum_{i,J=1}^{m} h_i H_{ij}\ (a) h_j + o(|h|^2) = \frac{1}{2} h^t H h + o(|h|^2),$$

where expressions of the general type h^tHh are known as quadratic forms. We note that

1. For small $|h|$, we expect the sign of $(f(a+h) - f(a))$ to be same as that of the quadratic form;
2. Since f has been assumed to have continuous second order partial derivatives, the mixed second order partial derivatives commute, and hence, in this case, the Hessian matrix H is symmetric.

We can exploit the symmetry of the Hessian matrix in order to simplify the quadratic form. From Algebra and Geometry we know that (or at least in principle we know that):

1. The eigenvalues of H are real, say $\lambda_1, ..., \lambda_m$, although not necessarily distinct;
2. It is possible to find an orthonormal set of eigenvectors, say $e^{(j)}$, which span $\mathbb{R}^m$ (even if the λ_j are not distinct), where

$$He^{(j)} = \lambda_j e^{(j)} \text{ for } j = 1, ..., m \text{ (no s.c.).}$$

Thus, because the $e^{(j)}$ span $\mathbb{R}^m$ and so are a basis, we can write

$$h = \sum_{j=1}^{m} h_j e^{(j)},$$

where, from the fact that the $e^{(j)}$ are orthonormal (i.e. $e^{(i)}.\, e^{(j)} = \delta_{ij}$),

$$h_i = h \cdot e^{(i)}.$$

Hence from

$$Hh = \sum_{j=1}^{m} h_j \lambda_j e^{(j)},$$

$$h' Hh = \sum_{j=1}^{m} h_j e^{(j)}, \sum_{j=1}^{m} h_j \lambda_j e^{(j)},$$

$$= \sum_{i,j} h_i h_j \lambda_j \delta_{ij} \text{ since } e^{(i)}.\, e^{(j)} = \delta_{ij}$$

$$= \sum_{j=1}^{m} \lambda_j h_j^2$$

It follows from that close to a critical point

$$f(a+h) - f(a) = \frac{1}{2} = \sum_{j=1}^{m} \lambda_j h_j^2 + o(|h|^2)$$

The Result:

1. If all the eigenvalues of $H(a)$ are strictly positive, f has a local minimum at a.
2. If all the eigenvalues of $H(a)$ are strictly negative, f has a local maximum at a.
3. If $H(a)$ has at least one strictly positive eigenvalue and at least one strictly negative eigenvalue, then f has a saddle point at a.
4. Otherwise, it is not possible to determine the nature of the critical point from the eigenvalues of $H(a)$.

Outline Proof.

1. Suppose that all eigenvalues are positive, then

$$\lambda_{\min} = \min_j \lambda_j > 0.$$

Hence

$$\sum_{j=1}^{m} \lambda_j h_j^2 \geq \sum_{j=1}^{m} \lambda_{\min} h_j^2 = \lambda_{\min} |h|^2,$$

$$f(a+h) - f(a) \geq \frac{1}{2}\lambda_{\min} |h|^2 + o(|h|^2)$$

It follows for $|h|$ suciently small, i.e. for $a + h$ suciently close to a, that

$$f(a+h) - f(a) > 0,$$

i.e. that a is a local minimum.

2. Apply the above argument to the function $-f$, since the transformation $f \to -f$ changes the sign of the eigenvalues.
3. Without loss of generality suppose that $\lambda_1 > 0$ and that $\lambda_2 < 0$. For $h = (h_1, 0, ..., 0)$ it follows from that if h_1 is suciently small

$$f(a+h) - f(a) = \frac{1}{2}\lambda_1 h_1^2 + o(h_1^2) > 0.$$

Similarly if $h = (0, h_2, 0, ..., 0)$, then if h_2 is suffciently small

$$f(a+h) - f(a) = \frac{1}{2}\lambda_2 h_2^2 + o(h_2^2) > 0$$

Hence a is not an extremum of f, which means that it's a saddle point.

Examples:

1. For the earlier example, i.e. $f = \in - x^2 - y^2$, we have seen that there is a critical point at (0, 0). Further

$$H(x,y) = \begin{pmatrix} f_{xx} & f_{xy} \\ f_{yx} & f_{yy} \end{pmatrix} = \begin{pmatrix} -2 & 0 \\ 0 & -2 \end{pmatrix}.$$

Hence the eigenvalues of the Hessian are –2 and –2, i.e. both are strictly negative, which implies from above that the critical point is a maximum.

2. For the earlier example, i.e. $f = x^3 - 3xy^2$, we have seen that there is again a critical point at (0, 0). Further, since $f_x = 3x^2 - 3y^2$ and $f_y = -6xy$,

$$H(x,y) = \begin{pmatrix} f_{xx} & f_{xy} \\ f_{yx} & f_{yy} \end{pmatrix} = \begin{pmatrix} 6x & -6y \\ -6y & -6x \end{pmatrix},$$

and

$$H(0,0) = \begin{pmatrix} 0 & 0 \\ 0 & 0 \end{pmatrix}$$

Hence both eigenvalues of the Hessian are 0; therefore we can deduce nothing from the eigenvalues of the Hessian (actually, as we saw before, the critical point is a saddle).

Unlectured remark. For the case when $f : \mathbb{R}^2 \to \mathbb{R}$, the calculation of the eigenvalues of the Hessian matrix is overkill since there is a simpler 'recipe'. In particular if $\Delta = \det H(a)$ then if $\Delta > 0$ and $f_{xx}(a) > 0$, f has a local minimum at a; if $\Delta > 0$ and $f_{xx}(a) < 0$, f has a local maximum at a; if $\Delta < 0$, then f has a saddle point at a; if $\Delta = 0$, the test is inconclusive.

EXERCISE

1. If θ is the angle between the planes $a_1x + b_1y + c_1z + d_1 = 0$ and $a_2x + b_2y + c_2z = 0$, show that

$$\tan\theta = \frac{\sqrt{(b_1c_2 - b_2c_1)^2 + (c_1a_2 - c_2a_1)^2 + (a_1b_2 - a_2b_1)^2}.}{a_1a_2 + b_1b_2 + c_1c_2}$$

2. Find the bisector of that angle between the planes $3x - 6y + 2z + 5 = 0$ and $4x - 12y + 3z - 3 = 0$ which contains the origin.
3. Find the equation of the plane passing through the line of intersection of the planes $2x - y = 0$, $3z - y = 0$ and perpendicular to the plane $4x + 5y - 3z = 8$.
4. Find the equations of the planes bisecting the angles between the planes $x + 2y + 2z = 19$, $4x - 3y + 12z + 3 = 0$ and point out which bisects the acute angle.
5. The plane $lx + my = 0$ is rotated about its line of intersection with the plane $z = 0$ through an angle α. Prove that the equation of the plane in its new position is $lx + my = z\sqrt{(l^2 + m^2)} \tan\alpha = 0$.
6. Show that the line of intersection of the planes $x + 2y - z - 3 = 0$, $3x - y + 2z - 1 = 0$.
7. Find the area of the triangle whose vertices are the points (1, 2, 3), (–2, 1, –4), (3, 4, –2).
8. Find the volume of the tetrahedron formed by the planes $2x + 3y + z = 6$, 2x s+ 3y = 0, $3y + z = 0$, $2x + z = 0$.
9. Show that the equation $2x^2 - 2y^2 + 4z^2 + 2yz + 6zx + 3xy = 0$.

10. Find the equations of the planes through the intersection of the planes x + 3y + 6 = 0, 3x – y – 4z = 0, whose perpendicular distance from the origin is unity.

11. Find the equation of the bisector of the obtuse angle between the planes 3x + 4y – 5z + 1 = 0, 5x + 12y – 13z = 0.

12. Find the volume of the tetrahedron formed by the points (1, 1, 3), (4, 3, 2), (5, 2, 7) and (6, 4, 8).

13. Show that the volume of the tetrahedron formed by the planes

$my + nz = 0$, $nz + lx = 0$, $lx + my = 0$ and $lx + my + nz = p$ is $\frac{2p^3}{3lmn}$.

14. Find the equation of the plane passing through the intersection of the planes x + 2y + 3z + 4 = 0, 4x + 3y + 2z + 1 = 0 and the point (1, 2, 3).

15. A variable plane makes with the co-ordinate planes a tetrahedron of constant volume $64k^3$. Find the locus of the foot of the perpendicular from the origin to the plane.

16. Find equation of the plane through the point (2, –3, 1) and perpendicular to the line joining the points (3, 4, –1), (3, –1, 5).

17. A point *P* moves on the plane $\frac{x}{a}+\frac{y}{b}+\frac{z}{c}=1$.

18. Find the direction cosines of the perpendicular drawn from the origin to the plane 2x + 3y – z + 1 = 0.

19. If *p* is the point (2, 3, –1). Find equation of the plane through *p* and at right angles to *OP*, where *O* is the origin.

20. Find equation of the plane through the point (2, –3, 4) and parallel to the plane 2x – 5y – 7z = 6.

21. Find equation of the plane through (–2, 3, 4) and perpendicular to the planes 2x + 3y + 4z = 6, 3x + 2y + 2z = 8.

Chapter 5

Line Integrals of Vector Fields

LINE INTEGRALS WITH RESPECT TO ARC LENGTH

Consider the following problem: a piece of string, corresponding to a curve C, lies in the xy-plane. The mass per unit length of the string is f(x,y). What is the total mass of the string?

The formula for the mass is

$$Mass = \int_C f(x,y)\,ds \cdot$$

The integral above is called a line integral of f(x,y) along C. It is also called a line integral with respect to arc length.

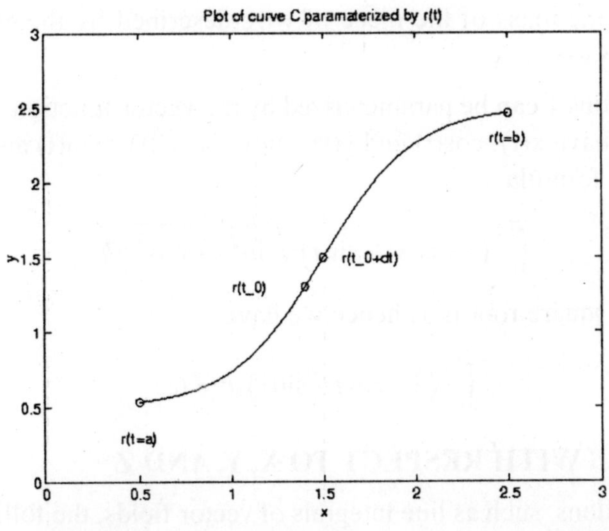

Question: how do we actually evaluate the above integral? The strategy is: (0) parameterize the curve C, (1) cut up the curve C into infinitesimal pieces, (2) determine the mass of each infinitesimal piece, (3) integrate to determine the total mass. It is assumed that C is piecewise smooth. That is, it is a union of finite number of smooth curves.

Suppose that we can describe the curve by the vector function r(t) = <x(t), y(t)> where $a \leq t \leq b$. Consider a portion of the curve corresponding to the infinitesimal interval $t_0 \leq t \leq t_0+dt$.

The arc length of the curve, ds, on this interval is

$$ds = \sqrt{\left[x'(t)\right]^2 + \left[y'(t)\right]^2}\,dt$$

Hence, the mass of the piece (density times length) is

$$f(x(t), y(t))\sqrt{\left[x'(t)\right]^2 + \left[y'(t)\right]^2}\,dt$$

The total mass of the string is the sum of all the masses of all infinitesimal pieces

$$Mass \int_C f(x,y)\,ds = \int_a^b f(x(t), y(t))\sqrt{\left[x'(t)\right]^2 + \left[y'(t)\right]^2}\,dt.$$

The integral on the right is an integral of one variable.

Line integrals are not restricted to curves in the xy plane. If C is a curve in three dimensions parameterized by r(t)=<x(t),y(t),z(t)> with a £ t £ b, then

$$\int_C f(x,y,z)\,ds = \int_a^b f(x(t), y(t), z(t))\sqrt{\left[x'(t)\right]^2 + \left[y'(t)\right]^2 + \left[z'(t)\right]^2}\,dt.$$

Example: Find the mass of the piece of wire described by the curve $x^2+y^2=1$ with density function f(x,y)=3+x+y.

The circle of radius 1 can be parameterized by the vector function r(t)=<cos(t),sin(t)> with $0 \leq t \leq 2\pi$. We have x(t)=cos(t) and y(t)=sin(t), so x'(t)=-sin(t) and y'(t)=cos(t). The mass is given by the formula

$$\int_0^{2\pi} (3 + \cos t + \sin t)\sqrt{\sin^2 t + \cos^2 t}\,dt$$

The term in the square root is 1, hence we have

$$\int_0^{2\pi} (3 + \cos t + \sin t)\,dt = 6\pi$$

LINE INTEGRALS WITH RESPECT TO X, Y, AND Z

In some applications, such as line integrals of vector fields, the following line integral with respect to x arises:

$$\int_C f(x,y,z)\,dx$$

This is an integral over some curve C in xyz space. It can be converted to integral in one variable. Suppose that C can be parameterized by r(t)=<x(t),y(t),z(t)> with a ≤ t ≤ b. Then,

$$\int_C f(x,y,z)\,dx = \int_a^b f(x(t),y(t),z(t))\,x'(t)\,dt$$

There are analogous formulas for integrals with respect to y and z.

In some applications, integrals with respect to x, y, and z occur in a sum:

$$\int_C \left(P(x,y,z)\,dx + Q(x,y,z)\,dy + R(x,y,z)\,dz\right)$$

If C is a curve in the xy plane and R=0, it might be possible to evaluate the line integral using Green's theorem.

Using the standard parameterization for C, this last integral becomes

$$\int_a^b \left(P(x(t),y(t),z(t))\,x'(t) + Q(x(t),y(t),z(t)),y'(t) + R(x(t),y(t),z(t))\,z'(t)\right)dt$$

Example: Evaluate the line integral

$$\int_C (3+x+y)\,dy,$$

where C is the circle in the figure above.

Recall that the circle is parameterized by r(t)=<cos(t),sin(t)> with 0 ≤ t 2π. Hence, x(t)=cos(t), y(t)=sin(t), and y'(t)=cos(t). According to the discussion above,

$$\int_C (3+x+y)\,dy = \int_0^{2\pi} (3+\cos t+\sin t)\cos t\,dt = \pi$$

LINE INTEGRALS OF VECTOR FIELDS

WORK

If you have taken a physics class, you have probably encountered the notion of work in mechanics.

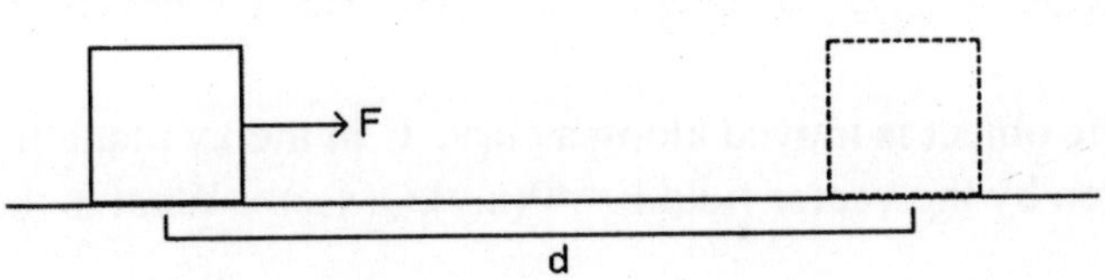

If a constant force of F (in the direction of motion) is applied to move an object a distance d in a straight line, then the work exerted is

$$\text{Work} = \text{Force} \cdot \text{distance}$$

The unit for force is N (newton) and the unit for distance is m (meter). The unit for work is joule=(newton)(meter).

Now suppose that the there is an angle θ between direction in which the constant force is applied and the direction of motion.

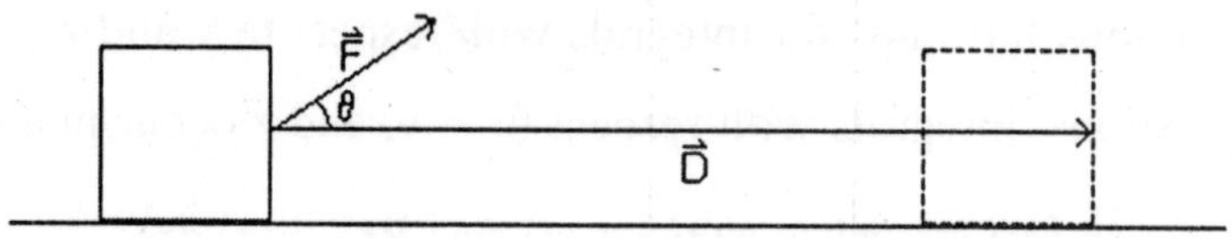

In this case the work is given by:

$$W = |F||d|\cos\theta$$

Here |F| is the magnitude of the force and |d| is the distance. Let us treat force as a vector and let D be the vector from the starting point to the ending point. We can express the above formula in terms of vectors, we have

$$W = F \cdot D$$

where the dot means the vector dot product. This formula is valid in 2 or three dimensions.

LINE INTEGRAL OF A VECTOR FIELD

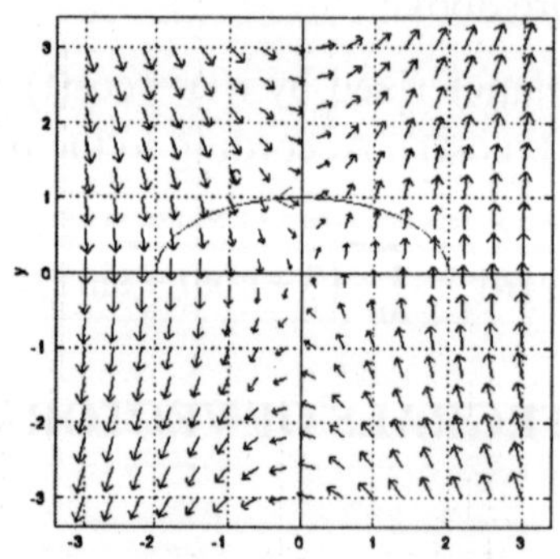

The above formulas for work assume that the object moves in a straight line and that the force and the angle between the direction of the force and the direction of motion are constant.

Suppose that the object is moved along a curve C in the xy plane In addition, suppose that the force is given by the vector field F=<P(x,y),Q(x,y)>. What is the work required to move the object?

The figure above shows the curve C and the vector field. The object is moved in the direction indicated by the arrow on the curve. To compute the work, parameterize the curve C by the vector function r(t)=<x(t),y(t)> with a ≤ t ≤ b, where r(a) is the initial point and r(b) is the final point. Let us consider the work required to move the object on an infinitesimal piece of the curve from position r(t) to r(t+dt).

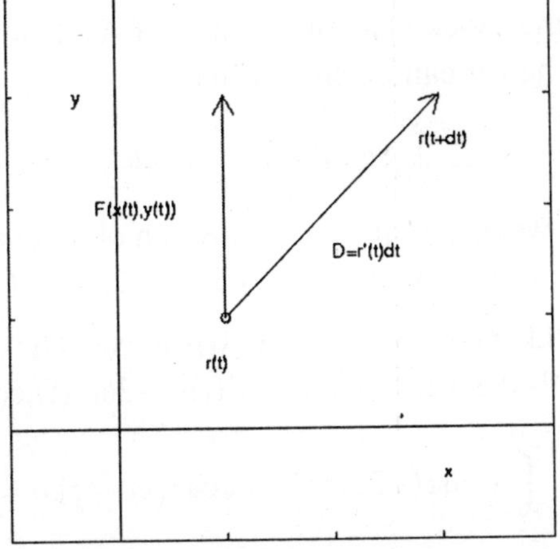

This piece of the curve is a straight line, since dt is infinitesimal. The force on the object and the angle between the force and the direction of motion is constant along the infinitesimal piece. We will assume that the force is the value at t, F(x(t),y(t)). It can be shown

$$D = r(t+dt) - r(t) = r'(t)dt = \langle x'(t), y'(t) \rangle dt$$

Hence, the work to move the object on this part of the curve is

$$F \cdot D = F(x(t), y(t)) \cdot r'(t)dt$$

The total work to move the object along the entire curve is

$$W = \int_a^b F(x(t), y(t)) \cdot r'(t)dt = \int_C F \cdot dr$$

The last integral above is the notation for the line integral of a vector field along a curve C.

Notice that

$$F(x(t), y(t)) \cdot r'(t) + P(x(t), y(t))x'(t) + Q(x(t), y(t))y'(t)$$

Hence, from previous work on line integrals, we have

$$\int_C F \cdot dr = \int_C P(x, y)dx + Q(x, y)dy$$

Line integrals of vector fields extend to three dimensions. If F = < P(x, y, z), Q(x, y, z), R(x, y, z) >, then

$$\int_C F \cdot dr = \int_C P(x,y,z)dx + Q(x,y,z)dy + R(x,y,z)dz$$

In the figure above it is shown that C is traversed in the counter clockwise direction. What if we traverse C in the clockwise direction? Let -C denote the line integral on C in the clockwise direction. Then it can be shown that

$$\int_{-C} F \cdot dr = -\int_C F \cdot dr$$

Example: How much work is required to move an object in the vector field F=<y,3x> along the upper part of the ellipse $x^2/4 + y^2 = 1$ from (2,0) to (-2,0)?

A plot of the vector field and the curve is shown above. The curve can be parameterized as r(t)=<2cos(t),sin(t)> with $0 \le t \le \pi$. Hence, r'(t)=<-2sin(t),cos(t)>. The total work is

$$W = \int_0^\pi \left(\sin t(-2\sin t) + 6\cos t(\cos t)\right)dt = 2\pi$$

If one traverses the curve starting at (-2,0) and ending at (2,0), then the value of the line integral is -2π.

DEPENDENCE OF WORK ON THE PATH

In general, the amount of work required to move the object from (2,0) to (-2,0) depends on the path taken! Suppose that we move the object in the above vector field on the real axis instead of on the ellipse. In this case, r(t)=<2-t,0>, where t goes from 0 to 4. In this case r'(t)=<-1,0>. The line integral is

$$W = \int_0^4 \left((-1)(0) + 3(2-t)(0)\right)dt = 0$$

No work is required! This can be seen in the plot above. The vector field F is orthogonal to the horizontal direction if y=0. (Note we have F=<0,3x> for y=0.) Hence, if one moves an object along the x-axis, no work is expended.

If the vector field is conservative, then the line integral from point A to point B is independent of path.

FUNDAMENTAL THEOREM FOR LINE INTEGRALS

The following result for line integrals is analogous to the Fundamental Theorem of Calculus.

Let C be a curve in the xyz space parameterized by the vector function r(t)=<x(t),y(t),z(t)> for $a \le t \le b$. Suppose that f(x,y,z) is a differentiable function whose gradient grad f=<f_x,f_y,f_z> is continuous on C. Then

$$\int_C \text{grad} f \cdot dr = f(x(b),y(b),z(b)) - f(x(a),y(a),z(a))$$

The above result states that the line integral of a vector field derived from a gradient depends only on the function f(x,y,z) and on the initial point (x(a),y(a),z(a)) and final point (x(b),y(b),z(b)) and not on the particular curve C. Hence, the integral is path independent. (Compare this with an example of a path dependent line integral.)

We have given the result for a function f(x,y,z) of three variables and a curve in 3 dimensional space. The above result is valid for functions of any number of variables.

Example: Verify the Fundamental Theorem for line integrals for the case that C is the top half of the circle $x^2 + y^2=1$ traversed in the counter clockwise direction and

$$f(x,y) = xy + x.$$

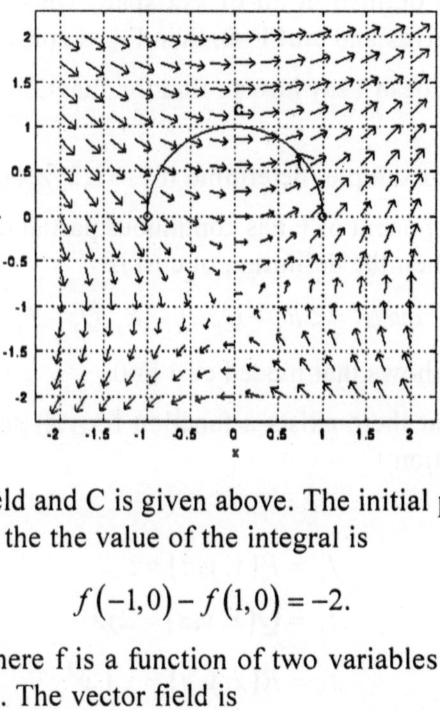

A plot of the vector field and C is given above. The initial point is (1,0) and the final point (-1,0). It follows that the the value of the integral is

$$f(-1,0) - f(1,0) = -2.$$

This is an example where f is a function of two variables so we are dealing with a vector field in the xy plane. The vector field is

$$Gradf =< f_x, y_y >=< y+1, x >$$

We can parameterize the curve C by vector function r(t)=<cos(t),sin(t)> for $0 \le t \le \pi$. On this curve the vector field is <sin(t)+1,cos(t)> and r'(t)=<-sin(t),cos(t)>. It follows that

$$\int_C gradf \cdot dr = \int_0^{\pi} < \sin t + 1, \cos > \cdot < -\sin t, \cos t > dt$$

Cleaning this up, we have

$$\int_0^{\pi} \left(-\sin t - \sin^2 t + \cos^2 t\right) dt = \int_0^{\pi} \left(-\sin t + \cos(2t)\right) dt = -2.$$

Here we have used the identity $\cos^2(t) - \sin^2(t) = \cos(2t)$.

In this example the gradient function <y+1,x> is continuous at all points in the xy plane. The line integral of <y+1,x> from (1,0) to (-1,0) is equal to -2 for *any* curve joining these two points.

CONSERVATIVE VECTOR FIELDS

Recall that a vector field F is conservative if there is a function f such that F=grad f. If we know that a vector field is conservative, then we can apply the Fundamental Theorem. The following result gives a test for determining if a vector field is conservative.

If F is a vector field defined in all of xyz space whose component functions have continuous partial derivatives and curl F=0, then F is a conservative vector field.

The above criterion doesn't say how to find the function f. It just says that a function f exists.

Example: Here is an example: determine if $F=<z,2yz,x+y^2>$ is conservative.

Notice that each component of F has continuous partial derivatives with respect to x, y, and z. If F=<P,Q,R>, then the definition of curl is

$$curlF =< R_y - Q_z, P_z - R_x, Q_x - P_y >$$

A quick calculation shows that indeed curl F=0.

The above shows that there exists a function f(x,y,z) such that grad f=F=<P,Q,R>. How do we find the function f?

We know that

$$f_x = P(x,y,z) = z$$
$$f_y = Q(x,y,z) = 2yz$$
$$f_x = R(x,y,z) = x + y^2$$

We integrate to determine f. By the Fundamental Theorem of Calculus

$$f(x.y.z) = \int f_x(x,y,z)dx + G(y,z)$$

In the above integral y and z are treated as constants. We must add a function of G(y,z) because $G_x(y,z)=0$. Doing this calculation, we find

$$f(x,y,z) = \int zdx + G(y,z) = xz + G(y,z)$$

To determine G(y,z) we use the information that $f_y=2yz$ and $f_z=x+y^2$. We have

$$f_y = 2yz = \frac{\partial}{\partial y}(xz + G(y.z)) = G_y(y,z)$$

This last equation implies

$$G_y(y,z) = 2yz \cdot$$

To determine G(y,z) we integrate with respect to y and add a constant function of z:

$$G(y,z) = \int G_y(y,z)\,dy + H(z) = \int 2yzdy + H(z) = H^2 z + H(z)$$

This means

$$f(x,y,z) = xz + y^2 z + H(z)$$

To find H(z) we differentiate with respect to z:

$$f_x = x + y^2 = \frac{\partial}{\partial z}\left(xz + y^2 z + H(z)\right) = x + y^2 + H'(z)$$

This last equation implies H'(z)=0. Hence, H(z)=constant. The final result is

$f(x,y,z) = xy + y^2 z +$ constant

GREEN'S THEOREM

Green's theorem relates the value of a line integral to that of a double integral.

Let C be a positively oriented, piecewise smooth, simple closed curve that bounds the region R in the xy plane. Let us define the terms in this last sentence. Positively oriented means that C is traversed in a counter-clockwise manner. Simple means that C does not intersect itself between its endpoints. If we parameterize C by the vector function r(t) with a ≤ t ≤ b, then closed means r(a)=r(b). See the figure below:

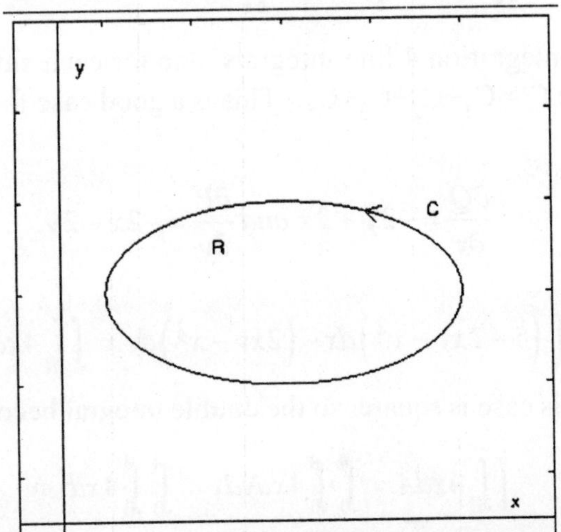

Green's Theorem states that

$$\int_C P(x,y)\,dx + Q(x,y)\,dy = \iint_R \left(\frac{\partial Q}{\partial x}(x,y) - \frac{\partial P}{\partial y}(x,y) \right) dA$$

Here it is assumed that P and Q have continuous partial derivatives on an open region containing R.

Example: Evaluate the line integral

$$\int_C \left(5 - 2xy - y^2\right) dx - \left(2xy - x^2\right) dy,$$

where C is the boundary of the square R with vertices (0,0), (1,0), (1,1), (0,1) traversed in the counter-clockwise direction.

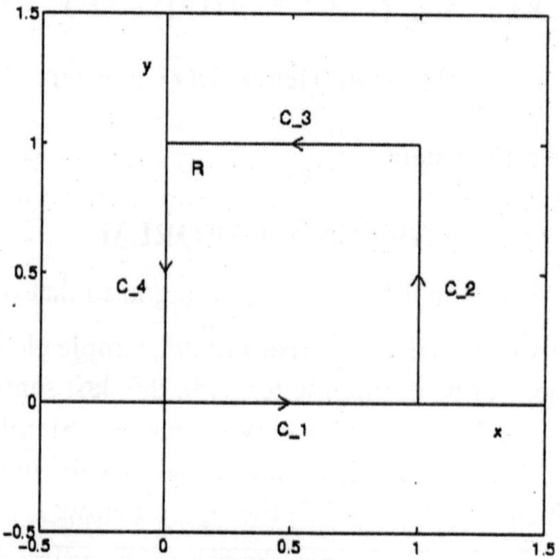

To do the above integration 4 line integrals, one for each side of the square, must be evaluated. (In this case $C = C_1 + C_2 + C_3 + C_4$.) This is a good case for using Green's theorem. We have

$$\frac{\partial Q}{\partial x} = -2y + 2x \text{ and } \frac{\partial P}{\partial y} = -2x - 2y$$

Hence,

$$\int_C \left(5 - 2xy - y^2\right) dx - \left(2xy - x^2\right) dy = \iint_R 4x\,dA.$$

The region R in this case is square, so the double integral becomes the iterated integral

$$\iint_R 4x\,dA = \int_0^1 \int_0^1 4x\,dy\,dx = \int_0^1 \int_0^1 4x\,dx\,dy$$

The iterated integral can be evaluated in a straightforward manner, yielding the result 2.

PARAMETRIC SURFACES

As we have seen previously, z=f(x,y) describes a surface in xyz space. Alternatively, a surface can be described in parametric form:

$$r(u,v) =< x(u,v), y(u,v), z(u,v) >,$$

where the points (u,v) lie in some region R of the uv plane.

Let us compare and contrast the parameterization of a surface with that of a space curve. A space curve is described by the vector function:

$$r(t) =< x(t), y(t), z(t) >$$

where $a \le t \le b$. A space curve is a one-dimensional object, similar to a piece of string. Each coordinate x,y and z depends only on one parameter, t. Each point on the curve corresponds to a different value of t.

A surface is a two dimensional object. One needs TWO pieces of information to uniquely define a point on a surface. These two pieces of information are the parameters u and v. These notions will hopefully become clearer as you go through the examples below.

Examples: A natural example is a sphere. It takes two pieces of information to describe a point on a sphere: the latitude and longitude. Let u, with $0 \le u \le 2\pi$ be the longitude. Let v, with $0 \le v \le \pi$ be the latitude. Here u and v correspond, respectively, to the the spherical coordinates θ and f. Using the formulas for spherical coordinates we have

$$x = a \sin v \cos u, \quad y = a \sin v \sin u, \quad z = a \cos v$$

Here a is a constant, not a variable.

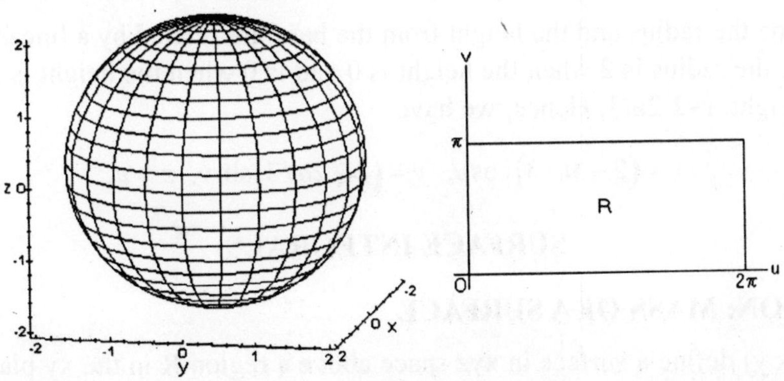

This is an example of surface that CANNOT be described by a single function z=f(x,y). A sphere of radius a centered at the origin can be defined by the relationship

$$x^2 + y^2 + z^2 = a^2$$

The top half of the sphere is defined by the surface

$$z = \sqrt{a^2 - x^2 - y^2},$$

and the bottom half is the defined by the surface

$$z = -\sqrt{a^2 - x^2 - y^2},$$

A second example is a cone, as shown in the figure. The height is 3, the base radius is 2, and the cone is centered at the origin.

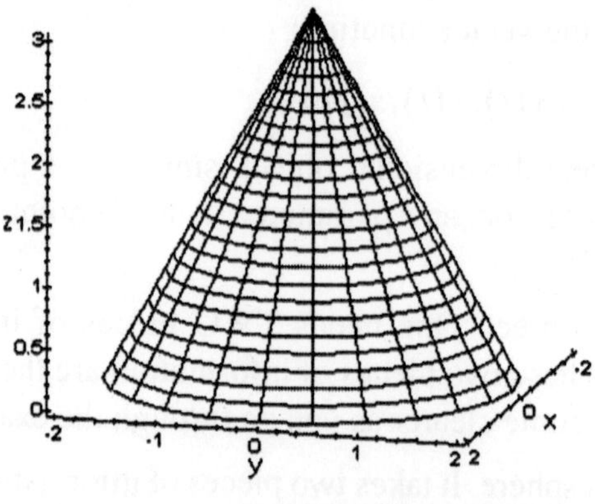

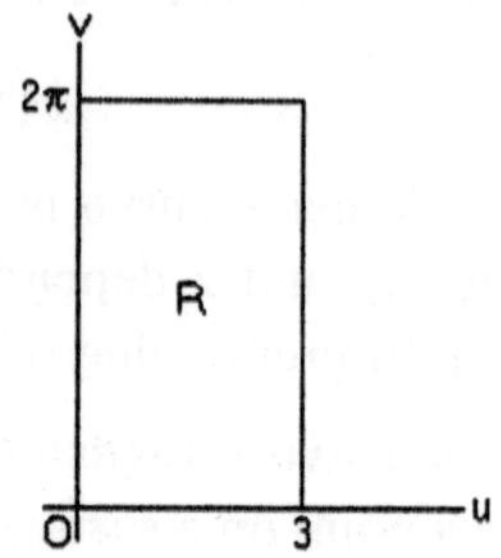

Two parameters are required to define a point on the surface. We choose them to be u, the height from the base, and v, the angle with respect to the x-axis. v is the same as the polar angle θ. We can describe any point on the surface by:

$$x = r\cos u,\ \ y = r\sin v,\ z = u,$$

$0 \le u \le 3$, and $0 \le v \le 2\pi$. Here r is the radius, the same as the variable in polar coordinates. The variable r must be eliminated since it is a third parameter.

For a cone the radius and the height from the base are related by a linear function. In this example, the radius is 2 when the height is 0 and is 0 when the height is 3. If we let u denote the height, r=2-2u/3. Hence, we have

$$x = (2 - 3u/3)\cos u,\ \ y = (2 - 2u/3)\sin v,\ \ z = u,$$

SURFACE INTEGRALS

MOTIVATION: MASS OF A SURFACE

Let z=f(x,y) define a surface in xyz space above a region R in the xy plane. Suppose that the density per unit area of the surface is given by the function P(x,y,z). What is the mass of the surface?

Consider a patch of the surface above an infinitesimal region in the xy plane around the point (x,y) with area dA. It can be shown that the surface area of this patch of the surface is

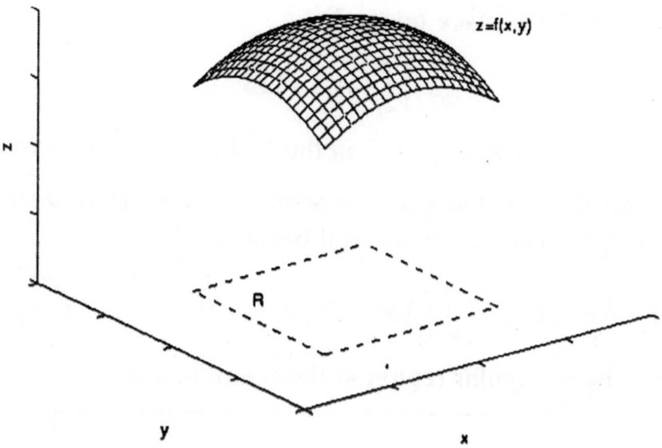

$$\sqrt{1+\left[f_x(x,y)\right]^2+\left[f_y(x,y)\right]^2}\,dA.$$

Hence, the mass of the surface is (density times area) is

$$P(x,y,f(x,y))\sqrt{1+\left[f_x(x,y)\right]^2+\left[f_y(x,y)\right]^2}\,dA.$$

In this last formula we have used the facts that z=f(x,y) and that the density P(x,y,z) is essentially constant on the small patch of surface. The total mass is the sum of the masses of the patches of surface above all infinitesimal regions in R:

$$Mass=\iint_R P(x,y,f(x,y))\sqrt{1+\left[f_x(x,y)\right]^2+\left[f_y(x,y)\right]^2}\,dA.$$

This is a double integral.

The notation for a surface integral of a function P(x,y,z) on a surface S is

$$\iint_S P(x,y,z)\,dS$$

Note that if P(x,y,z)=1, then the above surface integral is equal to the surface area of S.

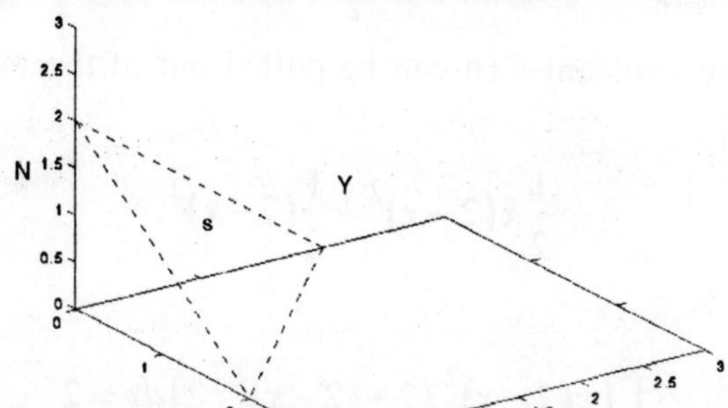

Example: Compute the surface integral

$$\iint_S (xy+z)\,dS$$

where S is that part of the plane x+y+z=2 in the first octant. A plot of S is given below.

Rewriting the equation of the plane in terms of z, we have z=f(x,y)=2-x-y. In this case, we have f_x=-1, f_y=-1 Hence the integral becomes:

$$\iint_S P(x,y,z)\,dS = \iint_R (xy+2-x-y)\sqrt{1+(-1)^2+(-1)^2}\,dA$$

The region R is the triangular region in the figure below:

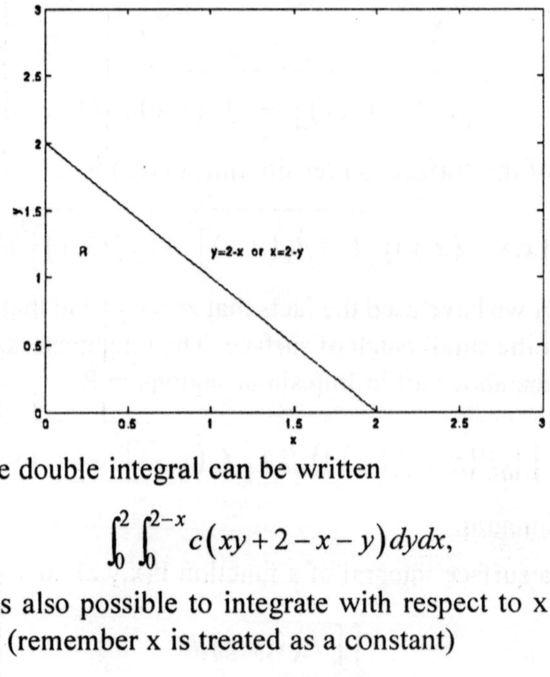

It follows that the double integral can be written

$$\int_0^2\int_0^{2-x} c(xy+2-x-y)\,dydx,$$

where c=sqrt(3). (It is also possible to integrate with respect to x first.) Computing the inner integral, we get (remember x is treated as a constant)

$$\int_0^2\int_0^{2-x}(xy+2-x-y)\,dy = \frac{1}{2}xy^2+2y-xy-\frac{1}{2}y^2\Big|_0^{y=2-x}$$

Here we ignore the constant c (it can be pulled out of the integral). Evaluating this last expression, we get

$$\frac{1}{2}x(2-x)^2+\frac{1}{2}(2-x)^2$$

The outer integral is

$$\int_0^2\left(x(2-x)^2/2+(2-x)^2/2\right)dx = 2$$

The final answer is 2*c=2*sqrt(3).

SURFACE INTEGRALS OF SURFACES DEFINED IN PARAMETRIC FORM

Suppose that the surface S is defined in the parametric form

$$r(u.v) = < x(u,v), y(u,v), z(u,v) >,$$

where (u,v) lies in a region R in the uv plane. In this case the surface integral is given by

$$\iint_S P(x,y,z)\,dS = \iint_R P(x(u,v), y(u,v), z(u,v))\,|r_u \times r_u|\,dA.$$

Here

$$r_u = < \frac{\partial x}{\partial u}, \frac{\partial y}{\partial u}, \frac{\partial z}{\partial u} > \ and\ r_u = < \frac{\partial x}{\partial v}, \frac{\partial y}{\partial v}, \frac{\partial z}{\partial v} >\cdot$$

The x means cross product. A derivation of this formula can be found in textbooks.

Example: Find the mass of the cone below (centered at the origin with base radius 2 and height 3), if the density satisfies

$$P(x,y,z) = \sqrt{x^2 + y^2}$$

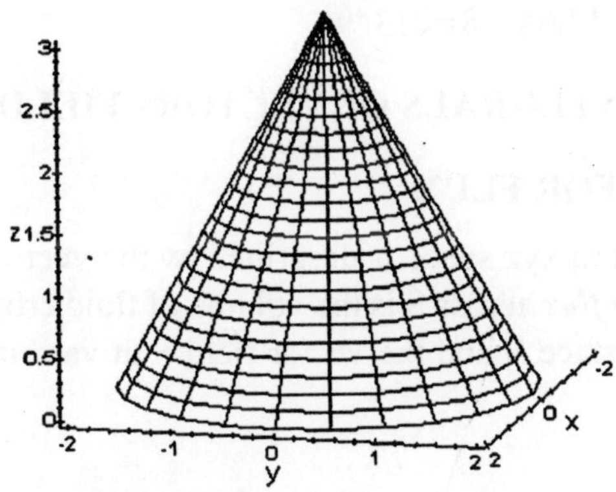

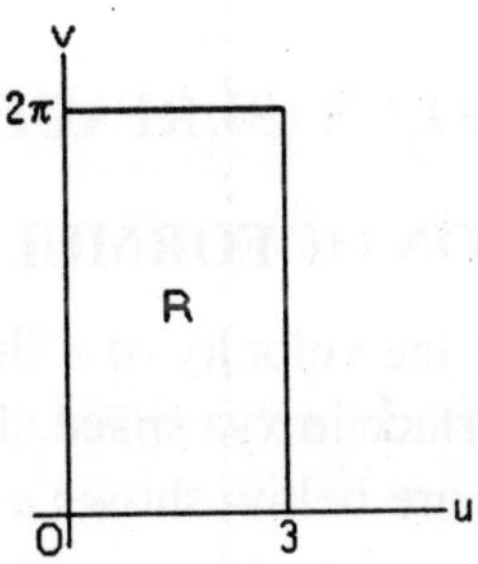

A parameterization of the cone is

$$r(u,v) = < (2 - 2u/3)\cos v, (2 - 2u/3)\sin u, v >,$$

where (u,v) lies in the rectangle in the uv plane shown above. In this case, r_u (the partial derivative of the vector with respect to u) is

$$r_u = < -(2/3)\cos u, (-2/3)\sin v, 1 >$$

and the partial derivative wrt v is

$$r_u = < -(2 - 2u/3)\cos v, -(2 - 2u/3)\cos v, 0 >$$

Taking the cross product, we have

$$r_u \times r_v = < -(2 - 2u/3)\cos v, -(2 - 2u/3)\sin v, -(2/3)(2 - 2u/3) > \cdot$$

The norm of this vector is

$$|r_u \times r_u| = \sqrt{13/9}\,(2-2u/3)$$

The function P(x,y,z) can be expressed in terms of u and v as well by substituting for x and y. We have

$$P(x,y,z) = \sqrt{(2-2u/3)^2\cos^2 v + (2-2u/3)^2\sin^2 v} = (2-2u/3)$$

Hence, the mass is given by

$$\text{Mass} = \int_0^3\int_0^{2\pi} c(2-2u/3)^2\,dA$$

where C=sqrt(13/9). The integral can be expressed as an iterated integral in u and v:

$$\text{Mass} = \int_0^3\int_0^{2\pi} c(2-2u/3)^2\,dv\,du = \int_0^{2\pi}\int_0^3 c(2-2u/3)^2\,du\,dv$$

The integral is independent of the order of integration. Evaluating the integral, we have

$$\text{Mass} = 8\pi\sqrt{13/9}$$

FLUX (SURFACE INTEGRALS OF VECTORS FIELDS)

DERIVATION OF FORMULA FOR FLUX

Suppose the velocity of a fluid in xyz space is described by the vector field F(x,y,z). Let S be a surface in xyz space. The *flux* across S is the volume of fluid crossing S per unit time. The figure below shows a surface S and the vector field F at various points on the surface.

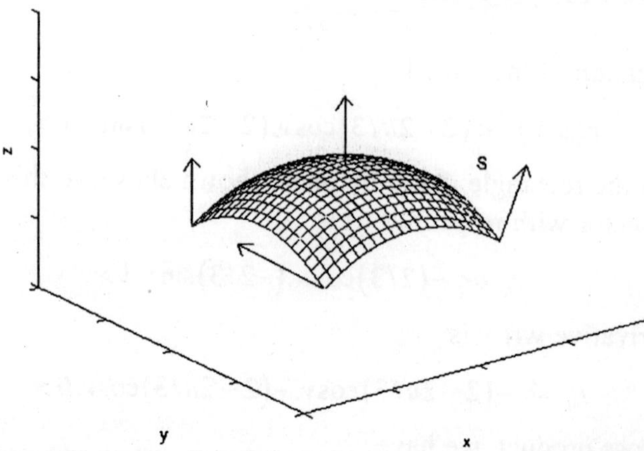

What is the formula for the flux? Let us consider the flux through an infinitesimal piece of the surface with area dS above the point (x_0,y_0) in the xy plane. (The surface is denoted by the dotted region.) Let n denote the unit normal vector to the surface. Let us suppose that the velocity vector $F(x_0,y_0,z_0)$ of the fluid makes an angle θ with the unit normal. In one unit of time a blob of fluid of length $|F|=|F(x_0,y_0,z_0)|$ will pass through the surface. (The blob is in the shape of a *parallelepiped.)*

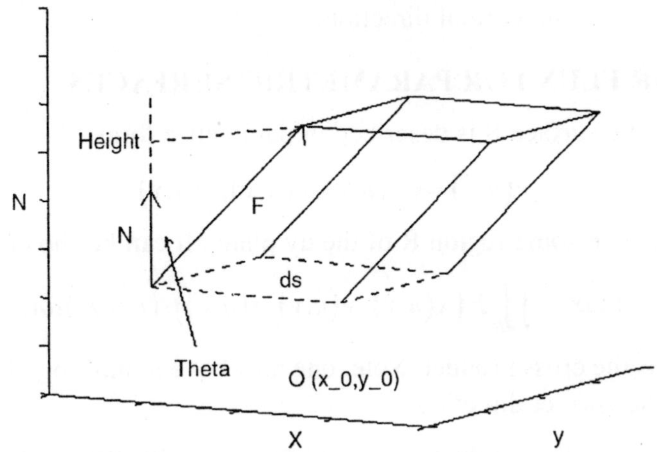

The volume of this fluid is

Volume = height × base area

The base area is dS. The height is equal to $|F|\cos\theta$. Hence, we have

$$\text{Volume} = \left(|F|\cos\theta\right)dS = \left(|F||n|\cos\theta\right)dS = (F\cdot n)\,dS$$

Here we have used the fact that $|n|=1$ and the property of the dot product. The total flux through the surface is

$$\text{Flux} = \iint_S (F\cdot n)\,dS$$

This is a surface integral.

We can write the above integral as an iterated double integral. Suppose that the surface S is described by the function $z=g(x,y)$, where (x,y) lies in a region R of the xy plane. The unit normal vector on the surface above (x_0,y_0) (pointing in the positive z direction) is

$$n = \frac{<-\partial x(x_0,y_0),\partial_y(x_0 y_0),1>}{\sqrt{1+\left[\partial_x(x_0,y_0)\right]^2+\left[\partial_y(x_0,y_0)\right]^2}}$$

The surface area of an infinitesimal piece of the surface above a infinitesimal region in the xy plane with area dA containing (x_0,y_0) is given by

$$dS = \sqrt{1+\left[\partial_x(x_0,y_0)\right]^2+\left[\partial_y(x_0,y_0)\right]^2}\,dA$$

Hence, it follows that the total flux is

$$\text{Flux} = \iint_R F(x,y,z) \cdot < -\partial_x(x,y), -\partial_y(x,y), 1 > dA$$

If we are asked for the flux in the negative z direction, then we use the vector $<g_x(x,y), g_y(x,y), -1>$ for the normal direction.

FORMULA FOR FLUX FOR PARAMETRIC SURFACES

Suppose that the surface S is described in parametric form:

$$r(u,v) = < x(u,v), y(u,v), z(u,v) >$$

where (u,v) lies in some region R of the uv plane. It can be shown that

$$\text{Flux} = \iint_R F(x(u,v), y(u,v), z(u,v)) \cdot (r_u \times r_v)\, dA,$$

Here, x means the cross product. Note, one may have to multiply the normal vector r_u x r_v by -1 to get the correct direction.

Example: Find the flux of the vector field <y,x,z> in the negative z direction through the part of the surface $z=g(x,y)=16-x^2 - y^2$ that lies above the xy plane.

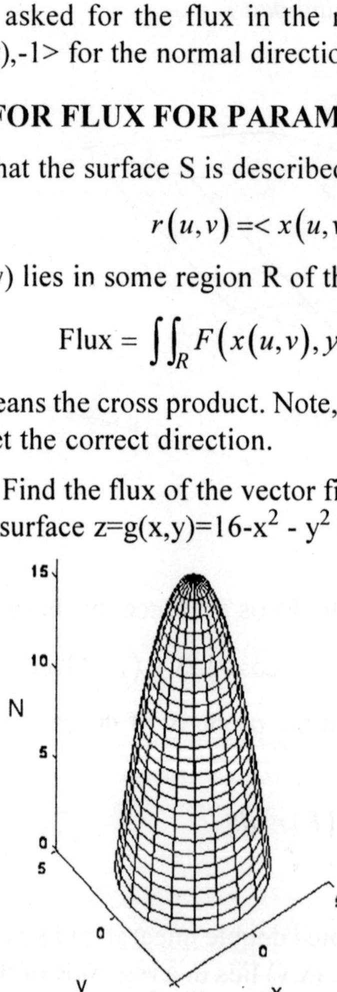

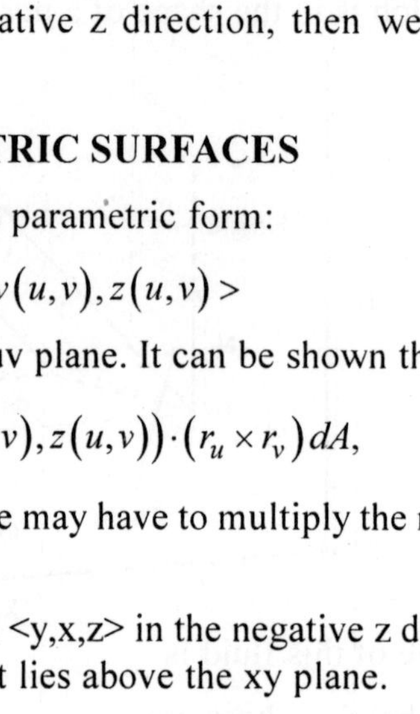

For this problem:

$$gx = -2x \;\; and \;\; g_y = -2y$$

It follows that the normal vector is <-2x,-2y,-1>. Computing Fo<-2x,-2y,-1>, we have

$$< y, x, 16 - x^2 - y^2 > . < -2x, -2y, -1 > = -4xy - 16 + x^2 + y^2$$

Here we use the fact that $z=16-x^2 - y^2$. Hence, the integral becomes

$$\text{Flux} = \iint_R \left(-4xy - 16 + x^2 + y^2\right) dA$$

The region R is the disk of radius 4 centered at the origin shown below. It is convenient to convert to polar coordinates. The region of integration is $0 \le r \le 4$ and $0 \le \theta \le 2\pi$. Substituting,

$$x = \tau\cos\theta \ \ and \ \ y = \tau\sin\theta,$$

we have

$$\text{Flux} = \int_0^{2\pi}\int_0^{4}\left(-4\tau^2\cos\theta\sin\theta - 16 + \tau^2\right)\tau\, d\tau\, d\theta$$

Here we have used the fact that $x^2 + y^2 = r^2$. (We can also integrate with respect to θ first, then r.) The inner integral with respect to r yields (remember θ is held constant)

$$\int_0^{4}\left(-4\tau^3\cos\theta\sin\theta - 16\tau + \tau^3\right)d\tau = -\tau^4\cos\theta\sin\theta - 8\tau^2 +^2/4\Big|_{\tau=0,}^{\tau=4}$$

which is equal to

$$-256\cos\theta\sin\theta - 64$$

The inner integral is

$$\int_0^{2\pi}(-256\cos\theta\sin\theta - 64)d\theta = -128\pi$$

Hence, the flux through the surface in the downward z direction is -128π cubic units per unit time. This means a negative amount of fluid crosses the surface in the downward direction. In other words, there is a flux of 128π cubic units per unit time in the upward z direction.

STOKES' THEOREM

Stokes' Theorem relates line integrals of vector fields to surface integrals of vector fields.

Consider the surface S described by the parabaloid $z=16-x^2 - y^2$ for $z \ge 0$, as shown in the figure below.

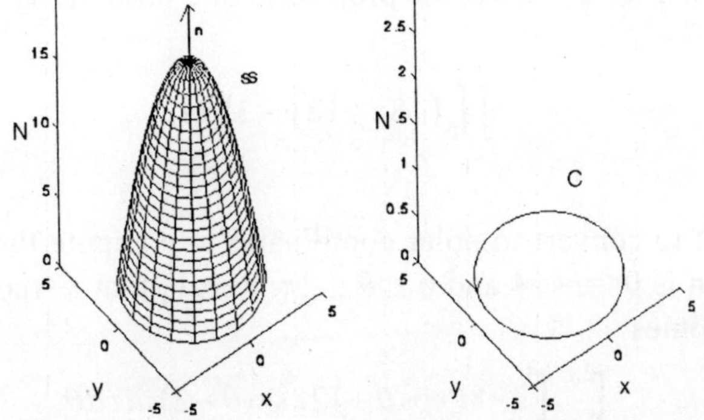

Let n denote the unit normal vector to S with positive z component. The intersection of S with the z plane is the circle $x^2 + y^2 = 16$. Let C denote this circle traversed in the counter clockwise direction. If F is a vector field with continuous partial derivatives in some region containing S, then Stokes' Theorem states

$$\int_C F \cdot dr = \iint_S (\text{curl F.n})\, dS$$

In general C is the boundary of S and is assumed to be piecewise smooth.

For the above equality to hold the direction of the normal vector n and the direction in which C is traversed must be consistent. Suppose that n points in some direction and consider a person walking on the curve C with their head pointing in the same direction as n. For consistency C must be traversed in such a way so that the surface is always on the left.

Example: Verify Stokes' Theorem for the surface S described above and the vector field F=<3y,4z,-6x>.

Let us first compute the line integral. The curve C can be parameterized by the vector function r(t)=<4cos(t),4sin(t),0> and r'(t)=<-4sin(t),4cos(t),0> for $0 \le t \le 2\pi$. Recall that

$$\int_C F \cdot dr = \int_0^{2\pi} F(x(t).y(t)) \cdot \tau'(t)\, dt$$

In this case, F=<12sin(t),0,-24cos(t)>. The line integral is given by

$$\int_0^{2\pi} -48\sin^2 t dt = -48\pi.$$

Now, let us compute the surface integral. Recall that for F=<P,Q,R>,

$$\text{curl F} = < R_y - Q_z, P_z - R_x, Q_x - P_y > \cdot$$

For our vector field, the curl is <-4,6,-3>. For $z=g(x,y)=16-x^2-y^2$ the normal vector to S pointing in the positive z direction is given by $<-g_x, -g_y, 1^3 <2x,2y,1>$. Using the formula for the surface integral of a vector field, we have

$$\iint_S \text{curl F} \cdot n dS = \iint_R < -4, 6, -3 > \cdot < 2x.2y.1 > dA$$

where R is the disk $0 \le x^2 + y^2 \le 16$, the projection of S onto the xy plane. The expression on the right is

$$\iint_R (-8x + 12y - 3)\, dA,$$

a double integral.

It is convenient to convert to polar coordinates to compute the double integral. The region of integration is 0<=r<=4 and $0 \le \theta \le 2\pi$. Recall, that x=rcos θ and y=rsin θ. The double integral becomes

$$\int_0^{2\pi} \int_0^4 (-8\tau\cos\theta + 12\tau\sin\theta - 3)\tau\, d\tau\, d\theta.$$

The inner integral is

$$\int_0^4\left(-8\tau^2\cos\theta+12\tau^2\sin\theta-3\tau\right)=-\left(8/3\right)\tau^2\cos\theta+4\tau^3\sin\theta-3\tau^2/2\Big|_{\tau=0}^{\tau=4}$$

The final result for the inner integral is

$$-\left(512/3\right)\cos\theta+256\sin\theta-24$$

The outer integral is

$$\int_0^{2\pi}\left(-\left(512/3\right)\cos\theta+256\sin\theta-24\right)d\theta=-48\pi$$

This result is same as that for the line integral!

THE DIVERGENCE THEOREM

The Divergence Theorem relates relates volume integrals to surface integrals of vector fields. Let R be a region in xyz space with surface S. Let n denote the unit normal vector to S pointing in the outward direction. Let F(x,y,z) = <P(x,y,z),Q(x,y,z), R(x,y,z)> be a vector field whose components P, Q, and R have continuous partial derivatives. The Divergence Theorem states:

$$\iint_S (F.n)\,dS=\iiint_R (div\ F)\,dV$$

Here div F is the divergence of F. There are various technical restrictions on the region R and the surface S; see the references for the details. The theorem is valid for regions bounded by ellipsoids, spheres, and rectangular boxes, for example.

Example: Verify the Divergence Theorem in the case that R is the region satisfying $0 \le z \le 16 - x^2 - y^2$ and F = < y, x, z >.

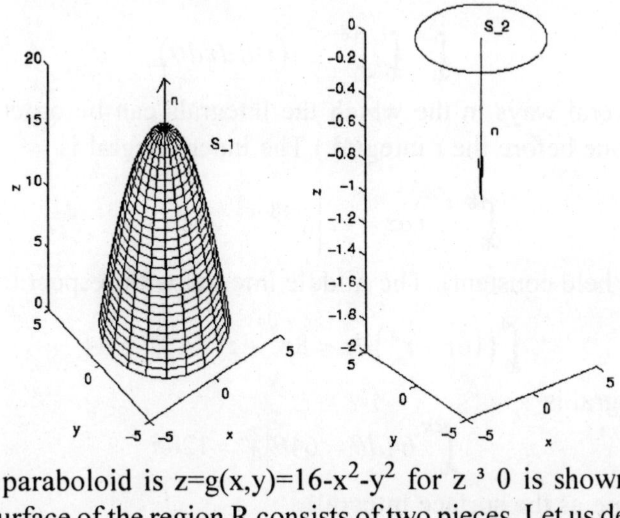

A plot of the paraboloid is $z=g(x,y)=16-x^2-y^2$ for z ³ 0 is shown on the left in the figure above. The surface of the region R consists of two pieces. Let us denote the paraboloid by S_1. The intersection of the parabaloid with the z plane is the circle $x^2+y^2=16$. It follows that that the bottom of R, which we denote by S_2, is the disk $x^2+y^2 \leq 16$. The surface integral of the vector field is

$$\iint_S (F.n)\,dS = \iint_{S_1} (F.n)\,dS + \iint_{S_2} (F.n)\,dS,$$

where the unit normal vector n points away from the region R.

Let us first consider the surface S_1. The outward normal vector points in the positive z direction. Our previous computation in the flux web page shows that if the normal vector points in the negative direction, then the surface integral equals -128π. For a normal vector pointing in the positive z direction the surface integral equals 128π. On the surface S_2, the normal vector points in the negative z direction. In fact, since S_2 lies in the plane z=0, the unit normal vector has x component and y component equal to 0. Hence, <0,0,-1>. On S_2, F=<y,x,z>=<y,x,0>, since S_2 is in the xy plane. It follows that

$$\text{F.n} =< 0,0,-1 > \cdot < y,x,0 >= 0.$$

Hence, the surface integral on S_2 is 0. The surface integral of F on the entire surface is 128π. Now, let us compute the volume integral. The divergence of F=<P,Q,R >=<y,x,z> is

$$\text{div F}=\frac{\partial P}{\partial x}+\frac{\partial Q}{\partial y}+\frac{\partial R}{\partial z}=1$$

It follows that the relevant volume integral is

$$\iiint_R 1dV$$

The circular symmetry of the region R suggests we convert to cylindrical coordinates. In cartesian coordinates the region R is described by the inequalities $0 \leq z \leq 16\text{-}x^2\text{-}y^2$ and $0 \leq x^2+y^2\ \ 16$. In cylindrical coordinates, we have $0 \leq z\ \ 16\text{-}r^2$, $0 \leq r \leq 4$, and $0 \leq \theta \leq 2\pi$. The above volume integral becomes:

$$\int_0^{2\pi}\int_0^4\int_0^{18-\tau^2} 1(\tau dz d\tau d\theta)$$

(There are several ways in the which the integrals can be ordered. However, the z integral must be done before the r integral.) The inner integral is

$$\int_0^{18-\tau^2} \tau dz = z\tau\Big|_0^{z=18-\tau^3} = 16\tau - \tau^3$$

(Note that r is held constant). The middle integral with respect to r is

$$\int_0^4 \left(16\tau - \tau^3\right)d\tau = 8\tau^2 - \tau^4/4\Big|_0^{2\pi} = 64$$

The outer integral is

$$\int_0^{2\pi} 64d\theta = 64\theta\Big|_0^{2\pi} = 128\pi$$

This is the same as the surface integral!

Properties of Line Integrals

We know from the previous section that for line integrals of real-valued functions (scalar fields), reversing the direction in which the integral is taken along a curve does not change the value of the line integral:

$$\int_C f(x,y)ds = \int_{-C} f(x,y)ds$$

For line integrals of vector fields, however, the value does change. To see this, let $f(x, y) = P(x, y)\, i + Q(x, y)\, j$ be a vector field, with P and Q continuously differentiable functions. Let C be a smooth curve parametrized by $x = x(t)$, $y = y(t)$, $a \le t \le b$, with position vector $r(t) = x(t)\, i + y(t)\, j$ (we will usually abbreviate this by saying that $C : r(t) = x(t)\, i + y(t)\, j$ is a smooth curve).

We know that the curve $-C$ traversed in the opposite direction is parametrized by $x = x(a + b - t)$, $y = y(a + b - t)$, $a \le t \le b$. Then

$$\int_{-C} P(x,y)dx = \int_a^b P(x(a+b-t), y(a+b-t))\frac{d}{dt}(x(a+b-t))dt$$

$$= \int_a^b P(x(a+b-t), y(a+b-t))(-x'(a+b-t))dt \text{ (by the Chain Rule)}$$

$$= \int_b^a P(x(u), y(u))(-x'(u))(-du) \text{ (by letting } u = a + b - t)$$

$$= \int_b^a P(x(u), y(u))x'(u)du$$

$$= -\int_a^b P(x(u), y(u))x'(u)du, \text{ since } \int_b^a = -\int_a^b, \text{ so}$$

$$\int_{-C} P(x,y)dx = -\int_C P(x,y)dx$$

since we are just using a different letter (u) for the line integral along C. A similar argument shows that

$$\int_{-C} Q(x,y)dy = -\int_C Q(x,y)dy,$$

and hence

$$\int_{-C} f.dr = \int_{-C} P(x,y)dx + \int_{-C} Q(x,y)dy$$

$$= -\int_C P(x,y)dx + -\int_C Q(x,y)dy = -\left(\int_C P(x,y)dx + \int_C Q(x,y)dy\right)$$

$$\int_{-C} f.dr = -\int_C f.dr.$$

The above formula can be interpreted in terms of the work done by a force $f(x, y)$ (treated as a vector) moving an object along a curve C: the total work performed moving the object along C from its initial point to its terminal point, and then back to the initial point moving backwards along the same path, is zero. This is because when force is considered as a vector, direction is accounted for.

The preceding discussion shows the importance of always taking the *direction* of the curve into account when using line integrals of vector fields. For this reason, the curves in line integrals are sometimes referred to as *directed curves* or *oriented curves*.

Recall that our definition of a line integral required that we have *a* parametrization $x = x(t)$, $y = y(t)$, $a \le t \le b$ for the curve C. But as we know, any curve has infinitely many parametrizations. So could we get a different value for a line integral using some other parametrization of C, say, $x = \tilde{x}(u)$, $y = \tilde{y}(u)$, $c \le u \le d$? If so, this would mean that our definition is not well-defined. Luckily, it turns out that the value of a line integral of a vector field is unchanged as long as the direction of the curve C is preserved by whatever parametrization is chosen:

Theorem: Let $f(x, y) = P(x, y)\, i + Q(x, y)\, j$ be a vector field, and let C be a smooth curve parametrized by $x = x(t)$, $y = y(t)$, $a \le t \le b$. Suppose that $t = \alpha(u)$ for $c \le u \le d$, such that $a = \alpha(c)$, $b = \alpha(d)$, and $\alpha'(u) > 0$ on the open interval (c, d) (i.e. $\alpha(u)$ is strictly increasing on $[c, d]$). Then $\int_C f.dr$ has the same value for the parametrizations $x = x(t)$, $y = y(t)$, $a \le t \le b$ and $x = \tilde{x}(u) = x(\alpha(u))$, $y = \tilde{y}(u) = y(\alpha(u))$, $c \le u \le d$.

Proof: Since $\alpha(u)$ is strictly increasing and maps $[c, d]$ onto $[a, b]$, then we know that $t = \alpha(u)$ has an inverse function $u = \alpha^{-1}(t)$ defined on $[a, b]$ such that $c = \alpha^{-1}(a)$, $d = \alpha^{-1}(b)$, and $\frac{du}{dt} = \frac{1}{\alpha'(u)}$. Also, $dt = \alpha'(u)\, du$, and by the Chain Rule

$$\tilde{x}'(u)\frac{d\tilde{x}}{du} = \frac{d}{du}(x(\alpha(u))) = \frac{dx}{dt}\frac{dt}{du} = x'(t)\alpha'(u)$$

$$\Rightarrow x'(u) = \frac{\tilde{x}'(u)}{\alpha'(u)}$$

so making the susbstitution $t = \alpha(u)$ gives

$$\int_a^b P(x(t), y(t))x'(t)dt = \int_{\alpha^{-1}(a)}^{\alpha^{-1}(b)} P(x(\alpha(u)), Y(\alpha(u)))\frac{\tilde{x}'(u)}{\alpha'(u)}(\alpha'(u)du)$$

$$= \int_c^d P(\tilde{x}(u), \tilde{y}(u))\tilde{x}'(u)du,$$

which shows that $\int_C P(x, y)dx$ has the same value for both parametrizations. A similar argument shows that $\int_C Q(x,y)dy$ has the same value for both parametrizations, and hence $\int_C f.dr$ has the same value.

Notice that the condition $\alpha'(u) > 0$ in Theorem means that the two parametrizations move along C in the same direction. That was *not* the case with the "reverse" parametrization for $-C$: for $u = a + b - t$ we have $t = \alpha(u) = a + b - u \Rightarrow \alpha'(u) = -1 < 0$.

Example: Evaluate the line integral $\int_C (x^2 + y^2)dx + 2xydy$, along the curve $C : x = t$, $y = 2t^2$, $0 \le t \le 1$, where $t = \sin u$ for $0 \le u \le \pi/2$.

Solution: First, we notice that $0 = \sin 0$, $1 = \sin(\pi/2)$, and $\dfrac{dt}{du} = \cos u > 0$ on $(0, \pi/2)$. So by Theorem we know that if C is parametrized by

$$x = \sin u,\ y = 2\sin^2 u,\ 0 \le u \le \pi/2.$$

then $\int_C (x^2 + y^2)dx + 2xydy$ should have the same value as we found in Example, namely $\dfrac{13}{3}$. And we can indeed verify this:

$$\begin{aligned}
\int_C (x^2 + y^2)dx + 2xydy &= \int_0^{\pi/2} ((\sin^2 u + (2\sin^2 u)^2)\cos u + \\
&\qquad 2(\sin u)(2\sin^2 u)4\sin u\cos u)\,du \\
&= \int_0^{\pi/2} \left(\sin^2 u + 20\sin^4 u\right)\cos u\,du \\
&= \left.\frac{\sin^3 u}{3} + 4\sin^5 u\right|_0^{\pi/2} \\
&= \frac{1}{3} + 4\frac{13}{3}
\end{aligned}$$

In other words, the line integral is unchanged whether t or u is the parameter for C.

By a closed curve, we mean a curve C whose initial point and terminal point are the same, i.e. for $C : x = x(t)$, $y = y(t)$, $a \le t \le b$, we have $(x(a), y(a)) = (x(b), y(b))$.

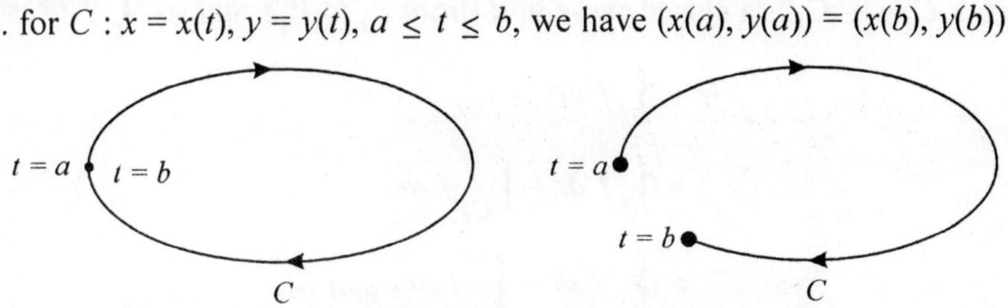

(a) Closed (b) Not closed

Fig. Closed vs Nonclosed Curves

A simple closed curve is a closed curve which does not intersect itself. Note that any closed curve can be regarded as a union of simple closed curves. We use the special notation

$$\oint_C f(x,y)ds \text{ and } \oint_C f.dr$$

to denote line integrals of scalar and vector fields, respectively, along closed curves.

In some older texts you may see the notation $\oint$ or $\oint$ to indicate a line integral traversing a closed curve in a counterclockwise or clockwise direction, respectively.

So far, the examples we have seen of line integrals have had the same value for different curves joining the initial point to the terminal point. That is, the line integral has been independent of the path joining the two points. As we mentioned before, this is not always the case. The following theorem gives a necessary and sufficient condition for this *path independence*:

Theorem: In a region R, the line integral $\oint_C f.dr$ is independent of the path between any two points in R if and only if $\oint_C f.dr = 0$ for every closed curve C which is contained in R.

Proof: Suppose that $\oint_C f.dr = 0$ for every closed curve C which is contained in R. Let P_1 and P_2 be two distinct points in R. Let C_1 be a curve in R going from P_1 to P_2, and let C_2 be another curve in R going from P_1 to P_2.

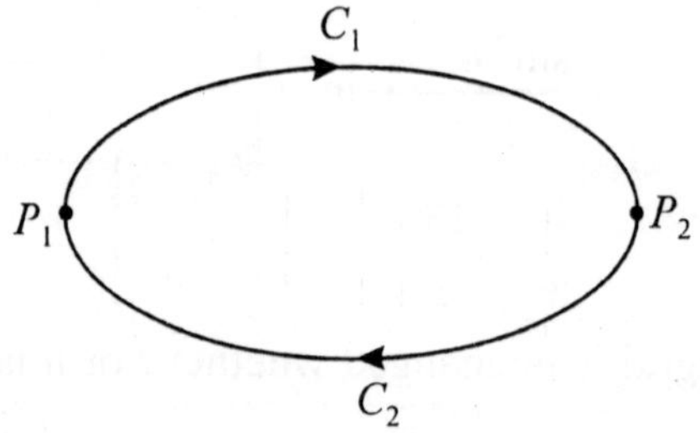

Then $C = C_1 \cup - C_2$ is a closed curve in R (from P_1 to P_1), and so $\oint_C f.dr = 0$. Thus,

$$\begin{aligned} 0 &= \oint_C f.dr \\ &= \oint_C f.dr + \int_{-C_2} f.dr \\ &= \oint_{C_1} f.dr - \int_{C_2} f.dr, \text{ and so} \end{aligned}$$

$\oint_{C_1} f.dr = \int_{C_2} f.dr$. This proves path independence.

Conversely, suppose that the line integral $\oint_C f.dr$ is independent of the path between any two points in R. Let C be a closed curve contained in R. Let P_1 and P_2 be two distinct points on C. Let C_1 be a part of the curve C that goes from P_1 to P_2, and let C_2 be the remaining part of C that goes from P_1 to P_2. Then by path independence we have

$$\oint_{C_1} f.dr = \int_{C_2} f.dr$$

$$\oint_{C_1} f.dr - \int_{C_2} f.dr = 0$$

$$\oint_{C_1} f.dr + \int_{-C_2} f.dr = 0, \text{ so}$$

$$\oint_{C} f.dr = 0$$

since $C = C_1 \cup - C_2$.

Clearly, the above theorem does not give a practical way to determine path independence, since it is impossible to check the line integrals around all possible closed curves in a region. What it mostly does is give an idea of the way in which line integrals behave, and how seemingly unrelated line integrals can be related (in this case, a specific line integral between two points and *all* line integrals around closed curves).

For a more practical method for determining path independence, we first need a version of the Chain Rule for multivariable functions:

Theorem Chain Rule: If $z = f(x, y)$ is a continuously differentiable function of x and y, and both $x = x(t)$ and $y = y(t)$ are differentiable functions of t, then z is a differentiable function of t, and

$$\frac{dz}{dt} = \frac{\partial z}{\partial x}\frac{dx}{dt} + \frac{\partial z}{\partial y}\frac{dy}{dt}$$

at all points where the derivatives on the right are defined.

The proof is virtually identical to the proof of Theorem from Section (which uses the Mean Value Theorem), so we omit it. We will now use this Chain Rule to prove the following *sufficient* condition for path independence of line integrals:

Theorem: Let $f(x, y) = P(x, y)\, i + Q(x, y)\, j$ be a vector field in some region R, with P and Q continuously differentiable functions on R.

Let C be a smooth curve in R parametrized by $x = x(t)$, $y = y(t)$, $a \le t \le b$. Suppose that there is a real-valued function $F(x, y)$ such that $\nabla F = f$ on R. Then

$$\oint_C f.dr = F(B) - F(A),$$

where $A = (x(a), y(a))$ and $B = (x(b), y(b))$ are the endpoints of C. Thus, the line integral is independent of the path between its endpoints, since it depends only on the values of F at those endpoints.

Proof: By definition of $\oint_C f.dr$, we have

$$\oint_C f.dr = \int_a^b (P(x(t), y(t))x'(t) + Q(x(t), y(t))y'(t))dt$$

$$= \int_a^b \left(\frac{\partial F}{\partial x}\frac{dx}{dt} + \frac{\partial F}{\partial y}\frac{dy}{dt} \right) dt$$

(since $\nabla F = f \Rightarrow \frac{\partial F}{\partial x} = P$ and $\frac{\partial F}{\partial y} = Q$)

$$= \int_a^b F'(x(t), y(t))dt \text{ (by the Chain Rule in Theorem)}$$

$$= f(x(t), y(t))\Big|_a^b = F(B) - F(A)$$

by the Fundamental Theorem of Calculus.

Theorem can be thought of as the line integral version of the Fundamental Theorem of Calculus. A real-valued function $F(x, y)$ such that $\nabla F(x, y) = f(x, y)$ is called a potential for f. A conservative vector field is one which has a potential.

Example: Recall from Examples that the line integral $\int_C (x^2 + y^2)dx + 2xydy$ was found to have the value $\frac{13}{3}$ for three different curves C going from the point (0, 0) to the point (1, 2). Use Theorem to show that this line integral is indeed path independent.

Solution: We need to find a real-valued function $F(x, y)$ such that

$$\frac{\partial F}{\partial x} = x^2 + y^2 \text{ and } \frac{\partial F}{\partial y} = 2xy$$

Suppose that $\frac{\partial F}{\partial x} = x^2 + y^2$, Then we must have $F(x, y) = \frac{1}{3} x^3 + xy^2 + g(y)$ for some function $g(y)$. So $\frac{\partial F}{\partial y} = 2xy + g'(y)$ satisfies the condition $\frac{\partial F}{\partial y} = 2xy$ if $g'(y) = 0$, i.e. $g(y) = K$, where K is a constant. Since any choice for K will do (why?), we pick $K = 0$. Thus, a potential $F(x, y)$ for $f(x, y) = (x^2 + y^2)\, i + 2xy\, j$ exists, namely

$$F(x, y) = \frac{1}{3}x^3 + xy^2.$$

Hence the line integral $\int_C (x^2 + y^2)dx + 2xydy$ is path independent.

Note that we can also verify that the value of the line integral of f along any curve C going from (0, 0) to (1, 2) will always be $\frac{13}{3}$, since by Theorem

$$\int_C f.dr = F(1,2) - F(0,0) = \frac{1}{3}(1)^3 + (1)(2)^2 - (0+0) = \frac{1}{3} + 4 = \frac{13}{3}.$$

A consequence of Theorem in the special case where C is a closed curve, so that the endpoints A and B are the same point, is the following important corollary:

Corollary: If a vector field f has a potential in a region R, then $\oint_C f.dr = 0$ for any closed curve C in R (i.e. $\oint_C \nabla F.dr = 0$ for any real-valued function $F(x, y)$).

Example: Evaluate $\oint_C xdx + ydy$ for $C : x = 2\cos t, y = 3\sin t, 0 \le t \le 2\pi$.

Solution: The vector field $f(x, y) = x\,i + y\,j$ has a potential $F(x, y)$:

$$\frac{\partial F}{\partial x} = x \Rightarrow F(x,y) = \frac{1}{2}x^2 + g(y), \text{ so}$$

$$\frac{\partial F}{\partial y} = y \Rightarrow g'(y) = y \Rightarrow g(y) = \frac{1}{2}y^2 + K$$

for any constant K, so $F(x, y) = \frac{1}{2}x^2 + \frac{1}{2}y^2$ is a potential for $f(x, y)$. Thus,

$$\oint_C xdx + ydy = \oint_C f.dr = 0$$

by Corollary 4.6, since the curve C is closed (it is the ellipse $\frac{x^2}{4} + \frac{y^2}{9} = 1$.

EXERCISE

1. Let be the three a $= 2i - j + k$, b $= i + 2j - k$ and c $= i + j - 2k$ vectors . Find the vector in the plane of b and c whose projection on a is of magnitude $\sqrt{\frac{2}{3}}$.
2. If the vectors $ai + j + k$, $i + bj + k$ and $i + j + ck$ $(a \ne b \ne c \ne 1)$ are coplanar vectors> find the value of $\frac{1}{1-a} + \frac{1}{1-b} + \frac{1}{1-c}$.
3. If a $= i + j - k$, b $= i - j + k$ and c are unit vectors perpendicular to the vector a and coplanar with a nd b. Find the unit vector d perpendicular to both a and c.
4. If $d = \lambda(a \times b) + \mu(b \times c) + \nu(c \times a)$ and $[a\ b\ c] = \frac{1}{8}$. Find the value of $\lambda + \mu + \nu$.

Chapter 6

Double and Triple Integrals

Double Integrals

In single-variable calculus, differentiation and integration are thought of as inverse operations. For instance, to integrate a function $f(x)$ it is necessary to find the antiderivative of f, that is, another function $F(x)$ whose derivative is $f(x)$. Is there a similar way of defining integration of real-valued functions of two or more variables? The answer is yes, as we will see shortly. Recall also that the definite integral of a nonnegative function $f(x) \geq 0$ represented the area "under" the curve $y = f(x)$. As we will now see, the *double integral* of a nonnegative real-valued function $f(x, y) \geq 0$ represents the *volume* "under" the surface $z = f(x, y)$.

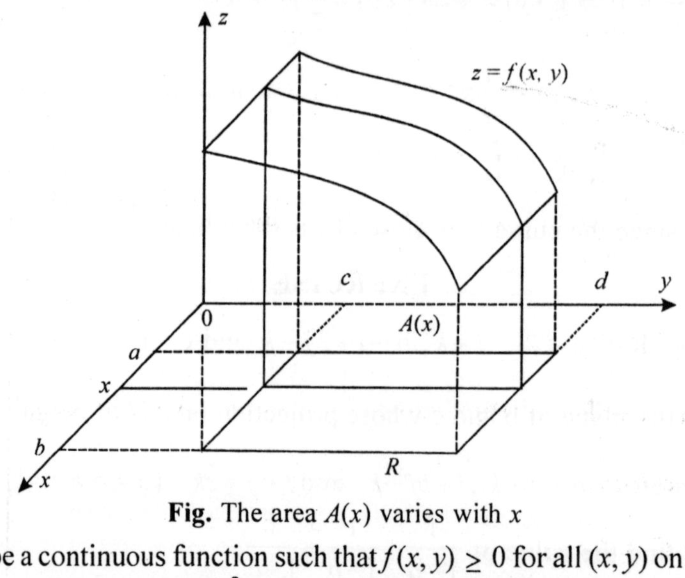

Fig. The area $A(x)$ varies with x

Let $f(x, y)$ be a continuous function such that $f(x, y) \geq 0$ for all (x, y) on the rectangle $R = \{(x, y) : a \leq x \leq b, c \leq y \leq d\}$ in $\mathbb{R}^2$. We will often write this as $R = [a, b] \times [c, d]$. For any number x^* in the interval $[a, b]$, slice the surface $z = f(x, y)$ with the plane $x = x^*$ parallel to the yz-plane. Then the trace of the surface in that plane is the *curve* $f(x^*, y)$, where x^* is fixed and only y varies. The area A under that curve (i.e. the area of the region between the curve and the

xy-plane) as y varies over the interval $[c, d]$ then depends only on the value of x^*. So using the variable x instead of x^*, let $A(x)$ be that area.

Then $A(x) = \int_c^d f(x, y)dy$ since we are treating x as fixed, and only y varies. This makes sense since for a fixed x the function $f(x, y)$ is a continuous function of y over the interval $[c, d]$, so we know that the area under the curve is the definite integral.

The area $A(x)$ is a function of x, so by the "slice" or cross-section method from singlevariable calculus we know that the volume V of the *solid* under the surface $z = f(x, y)$ but above the xy-plane over the rectangle R is the integral over $[a, b]$ of that crosssectional area $A(x)$:

$$V = \int_c^d A(x)dy = \int_a^b \left[\int_c^d f(x, y) \right] dx$$

We will always refer to this volume as "the volume under the surface". The above expression uses what are called iterated integrals. First the function $f(x, y)$ is integrated as a function of y, treating the variable x as a constant (this is called *integrating with respect to y*). That is what occurs in the "inner" integral between the square brackets in equation. This is the first iterated integral. Once that integration is performed, the result is then an expression involving only x, which can then be *integrated with respect to x*. That is what occurs in the "outer" integral above (the second iterated integral). The final result is then a number (the volume). This process of going through two iterations of integrals is called *double integration*, and the last expression in equation is called a double integral.

Notice that integrating $f(x, y)$ with respect to y is the inverse operation of taking the partial derivative of $f(x, y)$ with respect to y. Also, we could just as easily have taken the area of cross-sections under the surface which were parallel to the xz-plane, which would then depend only on the variable y, so that the volume V would be

$$V = \int_c^d \int_a^b \left[\int_a^b f(x, y)\,dx \right] dy$$

It turns out that in general the order of the iterated integrals does not matter. Also, we will usually discard the brackets and simply write

$$V = \int_c^d \int_a^b f(x, y)dxdy,$$

where it is understood that the fact that dx is written before dy means that the function $f(x, y)$ is first integrated with respect to x using the "inner" limits of integration a and b, and then the resulting function is integrated with respect to y using the "outer" limits of integration c and d. This order of integration can be changed if it is more convenient.

Example: Find the volume V under the plane $z = 8x + 6y$ over the rectangle $R = [0, 1] \times [0,$

Example: Find the volume V under the plane $z = 8x + 6y$ over the rectangle $R = [0, 1] \times [0, 2]$.

Solution: We see that $f(x, y) = 8x + 6y \geq 0$ for $0 \leq x \leq 1$ and $0 \leq y \leq 2$, so:

$$V = \int_0^2 \int_0^1 (8x + 6y)dxdy$$

$$= \int_0^2 \left(4x^2 + 6xy \Big|_{x=0}^{x=1} \right) dy$$

$$= \int_0^2 (4 + 6y)dy$$

$$= 4y + 3y^2 \Big|_0^2 = 20$$

Suppose we had switched the order of integration. We can verify that we still get the same answer:

$$V = \int_0^2 \int_0^2 (8x + 6y)dydx$$

$$= \int_0^2 \left(8xy + 3y^2 \Big|_{y=0}^{y=2} \right) dx$$

$$= \int_0^1 (16x + 12)dx$$

$$= 8x^2 + 12x \Big|_0^1 = 20$$

Example: Find the volume V under the surface $z = e^{x+y}$ over the rectangle $R = [2, 3] \times [1, 2]$.

Solution: We know that $f(x, y) = e^{x+y} > 0$ for all (x, y), so

$$V = \int_1^2 \int_2^3 e^{x+y} dxdy$$

$$= \int_1^2 \left(e^{x+y} \Big|_{x=2}^{x=3} \right) dy$$

$$= \int_1^2 (e^{y+3} - e^{y+2})dy$$

$$= e^{y+3} - e^{y+2} \Big|_1^2 = e^5 - e^4 - (e^4 - e^3) = e^5 - 2e^4 + e^3.$$

Recall that for a general function $f(x)$, the integral $\int_a^b f(x)dx$ represents the difference of the area below the curve $y = f(x)$ but above the x-axis when $f(x) \geq 0$, and the area above the curve but below the x-axis when $f(x) \leq 0$. Similarly, the double integral of any continuous function $f(x, y)$ represents the difference of the volume below the surface $z = f(x, y)$ but above the xy-plane when $f(x, y) \geq 0$, and the volume above the surface but below the xy-plane when $f(x, y) \leq 0$. Thus, our method of double integration by means of iterated integrals can be used to evaluate the double integral of *any* continuous function over a rectangle, regardless of whether $f(x, y) \geq 0$ or not.

Example: Evaluate $\int_a^{2\pi} \int_0^{\pi} \sin(x + y)dxdy.$

Solution: Note that $f(x, y) = \sin(x + y)$ is both positive and negative over the rectangle $[0, \pi] \times [0, 2\pi]$. We can still evaluate the double integral:

$$\int_0^{2\pi} \int_0^{\pi} \sin(x + y)dxdy = \int_0^{2\pi}\left(-\cos(x + y)\Big|_{x=0}^{x=\pi}\right)dy$$

$$= \int_0^{2\pi} (-\cos(y + \pi) + \cos y)dy$$

$$= - \ -\sin(y + \pi)\sin y\Big|_0^{2\pi}$$

$$= -\sin 3\pi + \sin 2\pi - (-\sin \pi + \sin 0) = 0$$

EXERCISES

For Exercises 1-4, find the volume under the surface $z = f(x, y)$ over the rectangle R.

1. $f(x, y) = 4xy$, $R = [0, 1] \times [0, 1]$
2. $f(x, y) = e^{x+y}$, $R = [0, 1] \times [-1, 1]$
3. $f(x, y) = x^3 + y^2$, $R = [0, 1] \times [0, 1]$
4. $f(x, y) = x^4 + xy + y^3$, $R = [1, 2] \times [0, 2]$

For Exercises 5-12, evaluate the given double integral.

5. $\int_0^1 \int_1^2 (1 - y)x^2 dxdy$
6. $\int_0^1 \int_0^2 x(x + y)dxdy$
7. $\int_0^2 \int_0^1 (x + 2)dxdy$
8. $\int_{-1}^2 \int_{-1}^1 x(xy + \sin x)dxdy$

9. $\int_0^{\pi/2}\int_0^1 xy\cos(x^2y)\,dxdy$

10. $\int_0^{\pi}\int_0^{\pi/2} \sin x\cos(y-\pi)\,dxdy$

11. $\int_0^2\int_1^4 xy\,dxdy$

12. $\int_{-1}^1\int_{-1}^2 1\,dxdy$

13. Let M be a constant. Show that $\int_c^d\int_a^b M\,dxdy = M(d-c)(b-a)$.

Double Integrals Over a General Region

In the previous section we got an idea of what a double integral over a rectangle represents. We can now define the double integral of a real-valued function $f(x, y)$ over more general regions in $\mathbb{R}^2$.

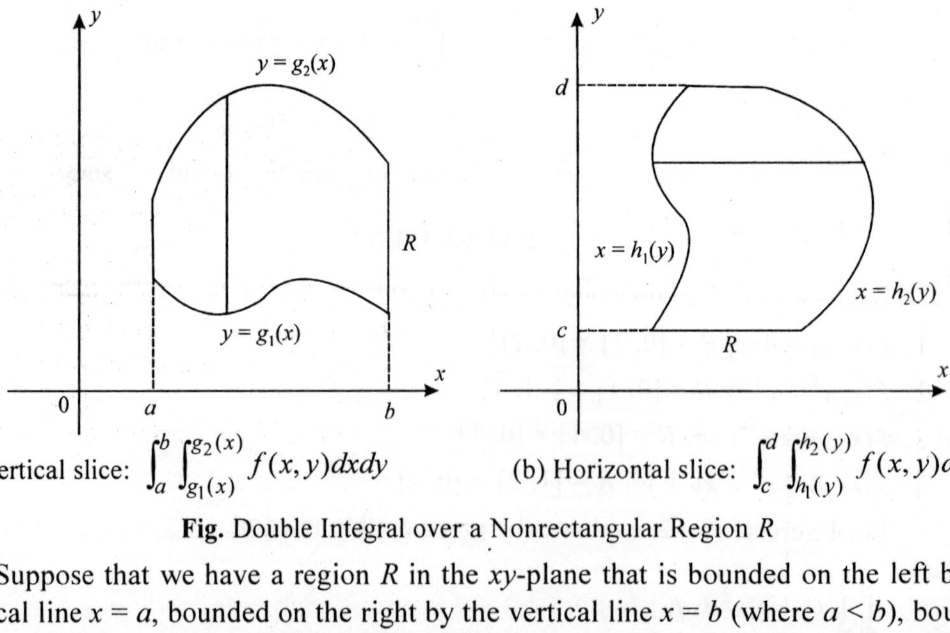

(a) Vertical slice: $\int_a^b\int_{g_1(x)}^{g_2(x)} f(x,y)\,dxdy$ (b) Horizontal slice: $\int_c^d\int_{h_1(y)}^{h_2(y)} f(x,y)\,dxdy$

Fig. Double Integral over a Nonrectangular Region R

Suppose that we have a region R in the xy-plane that is bounded on the left by the vertical line $x = a$, bounded on the right by the vertical line $x = b$ (where $a < b$), bounded below by a curve $y = g_1(x)$, and bounded above by a curve $y = g_2(x)$. We will assume that $g_1(x)$ and $g_2(x)$ do not intersect on the open interval (a, b) (they could intersect at the endpoints $x = a$ and $x = b$, though).

Then using the slice method from the previous section, the double integral of a real-valued function $f(x, y)$ over the region R, denoted by" $\iint\limits_R f(x,y)\,dA$, is given by

$$\iint\limits_R f(x,y)dA = \int_a^b \left[\int_{g_1(x)}^{g_2(x)} f(x,y)dy \right] dx$$

This means that we take vertical slices in the region R between the curves $y = g_1(x)$ and $y = g_2(x)$. The symbol dA is sometimes called an *area element* or *infinitesimal*, with the A signifying area. Note that $f(x, y)$ is first integrated with respect to y, with functions of x as the limits of integration. This makes sense since the result of the first iterated integral will have to be a function of x alone, which then allows us to take the second iterated integral with respect to x.

Similarly, if we have a region R in the xy-plane that is bounded on the left by a curve $x = h_1(y)$, bounded on the right by a curve $x = h_2(y)$, bounded below by the horizontal line $y = c$, and bounded above by the horizontal line $y = d$ (where $c < d$) (assuming that $h_1(y)$ and $h_2(y)$ do not intersect on the open interval (c, d)), then taking horizontal slices gives

$$\iint\limits_{R.} f(x,y)dA = \int_c^d \left[\int_{h_1(x)}^{h_2(x)} f(x,y)dy \right] dy$$

Notice that these definitions include the case when the region R is a rectangle. Also, if $f(x, y) \geq 0$ for all (x, y) in the region R, then $\iint\limits_R f(x,y)dA$ is the volume under the surface $z = f(x, y)$ over the region R.

Example: Find the volume V under the plane $z = 8x + 6y$ over the region $R = \{(x, y) : 0 \leq x \leq 1, 0 \leq y \leq 2x^2\}$.

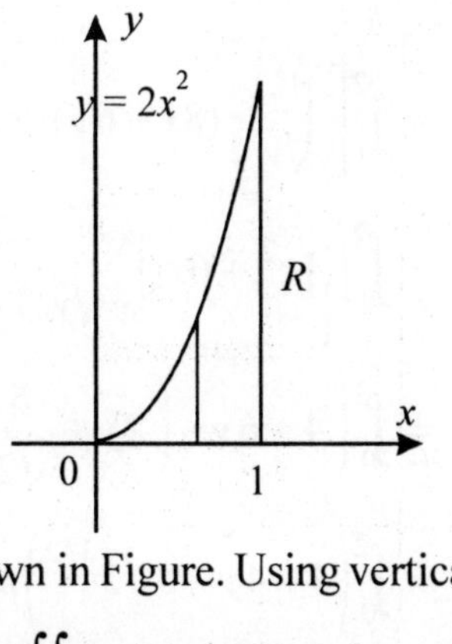

Solution: The region R is shown in Figure. Using vertical slices we get:

$$V = \iint\limits_R (8x + 6y)dA$$

$$= \int_0^1 \left[\int_0^{2x^2} (8x + 5y)dy \right] dx$$

$$= \int_0^1 \left(8xy + 5y^2 \Big|_{y=0}^{y=2x^2} \right) dx$$

$$= \int_0^1 (16x^3 + 12x^4) dx$$

$$= 4x^4 + \frac{12}{5}x^5 \Big|_0^1 = 4 + \frac{12}{5} = \frac{32}{5} = 6.4.$$

We get the same answer using horizontal slices:

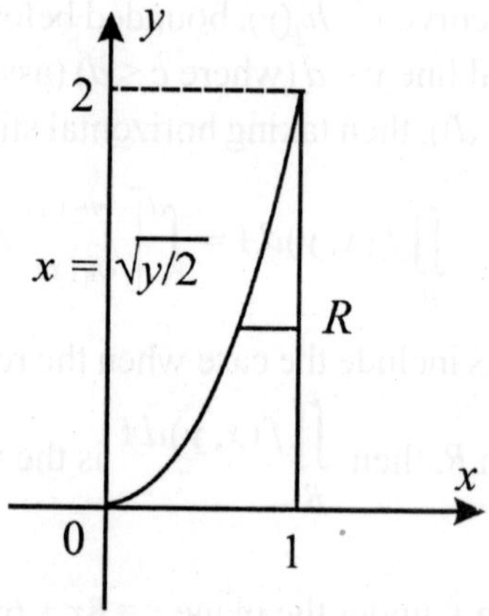

$$V = \iint_R (8x + 6y) dA$$

$$= \int_0^2 \left[\int_{\sqrt{y/2}}^1 (8x + 6y) dx \right] dy$$

$$= \int_0^2 \left(4x^2 + 6xy \Big|_{x=\sqrt{y/2}}^{x=1} \right) dy$$

$$= \int_0^2 \left(4 + 6y - \left(2y + \frac{6}{\sqrt{2}} y\sqrt{y} \right) \right) dy$$

$$= \int_0^2 \left(4 + 4y - 3\sqrt{2}^{3/2} \right) dy$$

$$= 4y + 2y^2 - \frac{6\sqrt{2}}{5} y^{5/2} \Big|_0^2$$

$$= 8 + 8 - \frac{6\sqrt{2}\sqrt{32}}{5} = 16 - \frac{48}{5} = \frac{32}{5} = 6.4$$

Example: Find the volume V of the solid bounded by the three coordinate planes and the plane $2x + y + 4z = 4$.

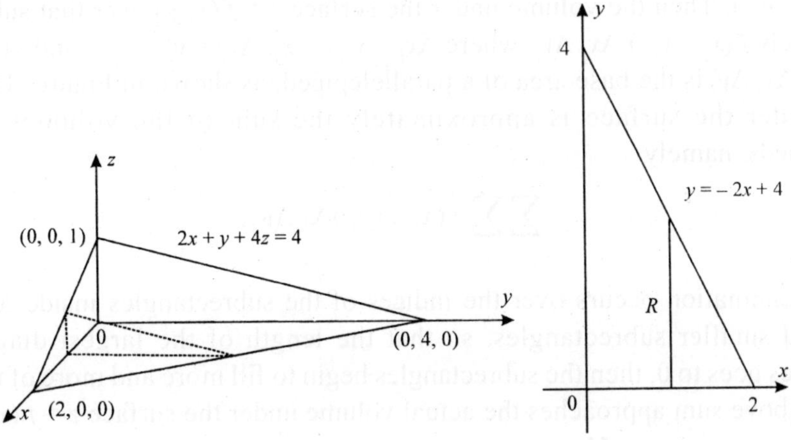

(a) (b)

Solution: The solid is shown in Figure with a typical vertical slice. The volume V is given by $\iint_R f(x, y)dA$, where $f(x, y) = z = \frac{1}{4}(4 - 2x - y)$ and the region R, shown in Figure, is $R = \{(x, y) : 0 \le x \le 2, 0 \le y \le -2x + 4\}$. Using vertical slices in R gives

$$V = \iint_R \frac{1}{4}(4 - 2x - y)dA$$

$$= \int_0^2 \left[\int_0^{-2x+4} \frac{1}{4}(4 - 2x - y)dy\right]dx$$

$$= \int_0^2 \left(-\frac{1}{8}(4 - 2x - y)^2 \Big|_{y=0}^{y=-2x+4}\right)dx$$

$$= \int_0^2 \frac{1}{8}(4 - 2x)^2\, dx$$

$$= -\frac{1}{48}(4 - 2x)^3 \Big|_0^2 = \frac{64}{48} = \frac{4}{3}$$

For a general region R, which may not be one of the types of regions we have considered so far, the double integral $\iint_R f(x, y)dA$ is defined as follows. Assume that $f(x, y)$ is a nonnegative real-valued function and that R is a bounded region in $\mathbb{R}^2$, so it can be enclosed in some rectangle $[a, b] \times [c, d]$. Then divide that rectangle into a grid of subrectangles. Only consider the

Only consider the subrectangles that are enclosed completely within the region R, as shown by the shaded subrectangles in Figure. In any such subrectangle $[x_i, x_{i+1}] \times [y_j, y_{j+1}]$, pick a point (x_{i+}, y_{j+}). Then the volume under the surface $z = f(x, y)$ over that subrectangle is approximately $f(x_{i+}, y_{j+})\, \Delta x_i\, \Delta y_j$, where $\Delta x_i = x_{i+1} - x_i$, $\Delta y_j = y_{j+1} - y_j$, and $f(x_{i+}, y_j)$ is the height and $\Delta x_i\, \Delta y_j$ is the base area of a parallelepiped, as shown in Figure. Then the total volume under the surface is approximately the sum of the volumes of all such parallelepipeds, namely

$$\sum_j \sum_i f(x_{i+}, y_{j+}) \Delta x_i \Delta y_j,$$

where the summation occurs over the indices of the subrectangles inside R. If we take smaller and smaller subrectangles, so that the length of the largest diagonal of the subrectangles goes to 0, then the subrectangles begin to fill more and more of the region R, and so the above sum approaches the actual volume under the surface $z = f(x, y)$ over the region R. We then *define* $\iint\limits_R f(x, y)\,dA$ as the limit of that double summation (the limit is taken over all subdivisions of the rectangle $[a, b] \times [c, d]$ as the largest diagonal of the subrectangles goes to 0).

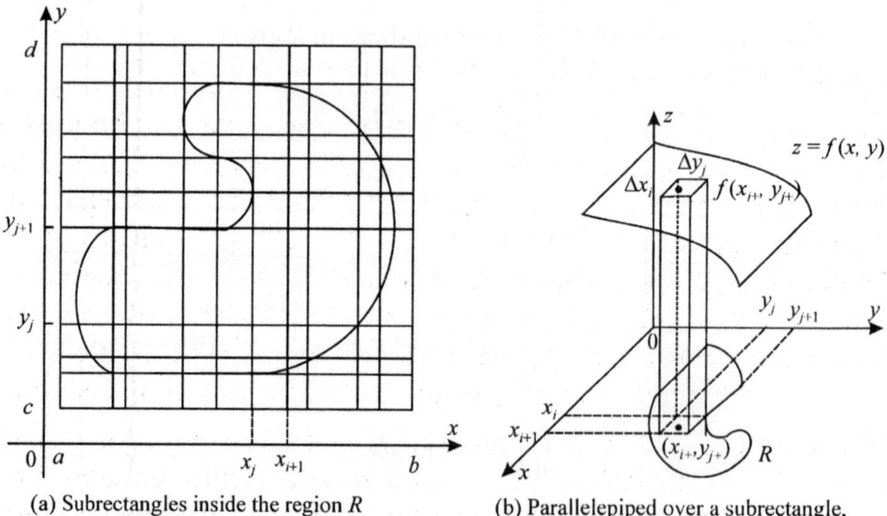

(a) Subrectangles inside the region R

(b) Parallelepiped over a subrectangle, with volume $f(x_{i+}, y_{j+})\, \Delta x_i\, \Delta y_j$

Fig. Double Integral Over a General Region R

A similar definition can be made for a function $f(x, y)$ that is not necessarily always nonnegative: just replace each mention of volume by the negative volume in the description above when $f(x, y) < 0$. In the case of a region of the type shown in Figure, using the definition of the Riemann integral from single-variable calculus, our definition of $\iint\limits_R f(x, y)\,dA$ reduces to a sequence of two iterated integrals.

integrals (i.e. over an unbounded region, or over a region which contains points where the function $f(x, y)$ is not defined) as a sequence of iterated improper single-variable integrals.

Example: Evaluate $\int_1^{\infty}\int_0^{1/x^2} 2y\,dy\,dx$.

Solution: $\int_1^{\infty}\int_0^{1/x^2} 2y\,dy\,dx = \int_1^{\infty}\left(y^2\Big|_{y=0}^{y=1/x^2}\right)dx$

$$= \int_1^{\infty} x^{-4}dx = -\frac{1}{3}x^{-3}\Big|_1^{\infty} = 0-\left(-\frac{1}{3}\right) = \frac{1}{3}$$

EXERCISES

For Exercises 1-6, evaluate the given double integral.

1. $\int_0^1\int_{\sqrt{x}}^1 24x^2y\,dy\,dx$
2. $\int_0^{\pi}\int_0^y \sin x\,dx\,dy$
3. $\int_1^2\int_0^{\ln x} 4x\,dy\,dx$
4. $\int_0^2\int_0^{2y} e^{y^2}\,dx\,dy$
5. $\int_0^{\pi/2}\int_0^y \cos x \sin y\,dx\,dy$
6. $\int_0^{\infty}\int_0^{\infty} xye^{-(x^2+y^2)}\,dx\,dy$
7. $\int_0^2\int_0^y 1\,dx\,dy$
8. $\int_0^1\int_0^{x^2} 2\,dy\,dx$
9. Find the volume V of the solid bounded by the three coordinate planes and the plane $x + y + z = 1$.
10. Find the volume V of the solid bounded by the three coordinate planes and the plane $3x + 2y + 5z = 6$.

12. Prove that the volume of a tetrahedron with mutually perpendicular adjacent sides of lengths a, b, and c, as in Figure, is $\frac{abc}{6}$.

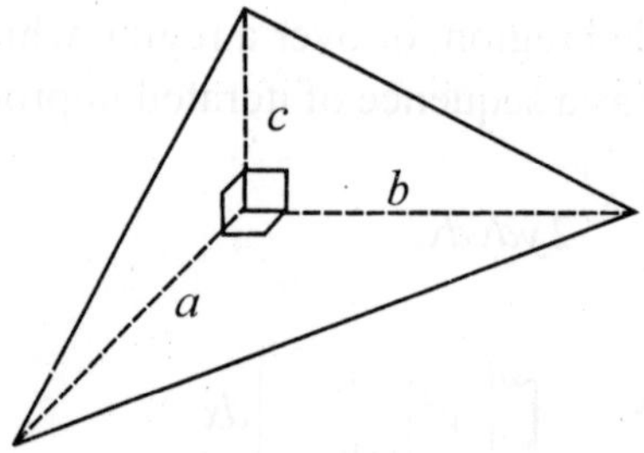

Triple Integrals

Our definition of a double integral of a real-valued function $f(x, y)$ over a region R in $\mathbb{R}^2$ can be extended to define a *triple integral* of a real-valued function $f(x, y, z)$ over a *solid* S in $\mathbb{R}^3$. We simply proceed as before: the solid S can be enclosed in some rectangular parallelepiped, which is then divided into subparallelepipeds. In each subparallelepiped inside S, with sides of lengths Δx, Δy and Δz, pick a point (x_+, y_+, z_+) Then define the triple integral of $f(x, y, z)$ over S, denoted by $\iiint\limits_S f(x,y,z)dV$, by

$$\iiint\limits_S f(x,y,z)dV = \lim \sum\sum\sum f(x_+, y_+, z_+)\,\Delta x\,\Delta y\Delta z,$$

where the limit is over all divisions of the rectangular parallelepiped enclosing S into subparallelepipeds whose largest diagonal is going to 0, and the triple summation is over all the subparallelepipeds inside S. It can be shown that this limit does not depend on the choice of the rectangular parallelepiped enclosing S. The symbol dV is often called the *volume element*.

Physically, what does the triple integral represent? We saw that a double integral could be thought of as the volume under a two-dimensional surface. It turns out that the triple integral simply generalizes this idea: it can be thought of as representing the *hypervolume* under a three-dimensional *hypersurface* $w = f(x, y, z)$ whose graph lies in $\mathbb{R}^4$. In general, the word "volume" is often used as a general term to signify the same concept for any n-dimensional object (e.g. length in $\mathbb{R}^1$, area in $\mathbb{R}^2$). It may be hard to get a grasp on the concept of the "volume" of a four-dimensional object, but at least we now know how to calculate that volume!

In the case where S is a rectangular parallelepiped $[x_1, x_2] \times [y_1, y_2] \times [z_1, z_2]$, that is, $S = \{(x, y, z) : x_1 \le x \le x_2, y_1 \le y \le y_2, z_1 \le z \le z_2\}$, the triple integral is a sequence of three iterated integrals, namely

$$\iiint\limits_S f(x,y,z)dV = \int_{z_1}^{z_2}\int_{y_1}^{y_2}\int_{x_1}^{x_2} f(x,y,z)dzdydz,$$

where the order of integration does not matter. This is the simplest case.

A more complicated case is where S is a solid which is bounded below by a surface $z = g_1(x, y)$, bounded above by a surface $z = g_2(x, y)$, y is bounded between two curves $h_1(x)$ and $h_2(x)$, and x varies between a and b. Then

$$\iiint_S f(x,y,z)dV = \int_a^b \int_{h_1(x)}^{h_2(x)} \int_{g_1(x,y)}^{g_2(x,y)} f(x,y,z)dzdydx.$$

Notice in this case that the first iterated integral will result in a function of x and y (since its limits of integration are functions of x and y). There are, of course, many variations on this case (for example, changing the roles of the variables x, y, z), so as you can probably tell, triple integrals can be quite tricky. At this point, just learning how to evaluate a triple integral, regardless of what it represents, is the most important thing. We will see some other ways in which triple integrals are used later in the text.

Example: Evaluate $\int_0^3 \int_0^2 \int_0^1 (xy+z)dx\,dy\,dz$.

Solution:

$$\begin{aligned}
\int_0^3 \int_0^2 \int_0^1 (xy+z)dx\,dy\,dz &= \int_0^3 \int_0^2 \left(\frac{1}{2}x^2y + xz \Big|_{x=0}^{x=1} \right) dydz \\
&= \int_0^3 \int_0^2 \left(\frac{1}{2}y + z \right) dydz \\
&= \int_0^3 \left(\frac{1}{4}y^2 + yz \Big|_{y=0}^{y=2} \right) dz \\
&= \int_0^3 (1+2z)dz \\
&= z + z^2 \Big|_0^3 = 12
\end{aligned}$$

Example: Evaluate $\int_0^1 \int_0^{1-x} \int_0^{y-x-y} (x+y+z)dzdydx$.

Solution:

$$\begin{aligned}
\int_0^1 \int_0^{1-x} \int_0^{y-x-y} (x+y+z)dzdydx &= \int_0^1 \int_0^{1-x} \left((x+y)z + \frac{1}{2}z^2 \Big|_{z=0}^{z=2-x-y} \right) dydx \\
&= \int_0^1 \int_0^{1-x} \left((x+y)(2-x-y) + \frac{1}{2}(2-x-y)^2 \right) dydx
\end{aligned}$$

$$= \int_0^1 \int_0^{1-x} \left(2 - \frac{1}{2}x^2 - xy - \frac{1}{2}y^2\right) dydx$$

$$= \int_0^1 \left(2y - \frac{1}{2}x^2 y - xy - \frac{1}{2}xy^2 - \frac{1}{6}y^3 \Big|_{y=0}^{y=1-x} \right) dx$$

$$= \int_0^1 \left(\frac{11}{6} - 2x + \frac{1}{6}x^3\right) dx$$

$$= \frac{11}{6}x - x^2 + \frac{1}{24}x^4 \Big|_0^1 = \frac{7}{8}$$

Note that the volume V of a solid in $\mathbb{R}^3$ is given by

$$V = \iiint_S 1 dV.$$

Since the function being integrated is the constant 1, then the above triple integral reduces to a double integral of the types that we considered in the previous section if the solid is bounded above by some surface $z = f(x, y)$ and bounded below by the xy-plane $z = 0$. There are many other possibilities. For example, the solid could be bounded below and above by surfaces $z = g_1(x, y)$ and $z = g_2(x, y)$, respectively, with y bounded between two curves $h_1(x)$ and $h_2(x)$, and x varies between a and b. Then

$$V = \iiint_S 1 dV = \int_a^b \int_{h_1(x)}^{h_2(x)} \int_{g_1(x,y)}^{g_2(x,y)} 1 dzdydx = \int_a^b \int_{h_1(x)}^{h_2(x)} (g_2(x,y) - g_1(x,y))dydx$$

just like in equation.

EXERCISES

For Exercises 1-8, evaluate the given triple integral.

1. $\int_0^3 \int_0^2 \int_0^1 xyz\, dx\, dy\, dz$

2. $\int_0^1 \int_0^x \int_0^y xyz\, dz\, dy\, dx$

3. $\int_0^\pi \int_0^x \int_0^{xy} x^2 \sin z\, dz\, dy\, dx$

4. $\int_0^1 \int_0^z \int_0^y ze^{y^2}\, dx\, dy\, dz$

5. $\int_1^e \int_0^y \int_0^{1/y} x^2 z\, dx\, dz\, dy$

6. $\int_1^2 \int_0^{y^2} \int_0^{z^2} yz\,dx\,dz\,dy$

7. $\int_1^2 \int_2^4 \int_0^3 1\,dx\,dy\,dz$

8. $\int_0^1 \int_0^{1-x} \int_0^{1-x-y} 1\,dz\,dy\,dx$

9. Let M be a constant. Show that $\int_{z_1}^{z_2} \int_{y_1}^{y_2} \int_{x_1}^{x_2} M\,dx\,dy\,dz = M(z_2 - z_1)(y_2 - y_1)(x_2 - x_1)$

10. Find the volume V of the solid S bounded by the three coordinate planes, bounded above by the plane $x + y + z = 2$, and bounded below by the plane $z = x + y$.

11. Show that $\int_a^b \int_a^z \int_a^y f(x)\,dx\,dy\,dz = \int_a^b \frac{(b-x)^2}{2} f(x)dx$.

    (***Hint:*** Think of how changing the order of integration in the triple integral changes the limits of integration.)

Numerical Approximation of Multiple Integrals

As you have seen, calculating multiple integrals is tricky even for simple functions and regions. For complicated functions, it may not be possible to evaluate one of the iterated integrals in a simple closed form. Luckily there are numerical methods for approximating the value of a multiple integral. The method we will discuss is called the *Monte Carlo method*. The idea behind it is based on the concept of the *average value* of a function, which you learned in single-variable calculus. Recall that for a continuous function $f(x)$, the average value $\bar{f}$ of f over an interval $[a, b]$ is defined as

$$\bar{f} = \frac{1}{b-a} \int_a^b f(x)dx.$$

The quantity $b - a$ is the length of the interval $[a, b]$, which can be thought of as the "volume" of the interval. Applying the same reasoning to functions of two or three variables, we define the **average value** of $f(x, y)$ over a region R to be

$$\bar{f} = \frac{1}{A(R)} \iint_R f(x,y)dA,$$

where $A(R)$ is the area of the region R, and we define the **average value** of $f(x, y, z)$ over a solid S to be

$$\bar{f} = \frac{1}{V(S)} \iiint_S f(x,y,z)dV,$$

where $V(S)$ is the volume of the solid S. Thus, for example, we have

$$\iint_R f(x,y)\,dA = A(R)\bar{f}.$$

The average value of $f(x, y)$ over R can be thought of as representing the sum of all the values of f divided by the number of points in R. Unfortunately there are an infinite number (in fact, *uncountably* many) points in any region, i.e. they can not be listed in a discrete sequence. But what if we took a very large number N of *random* points in the region R (which can be generated by a computer) and then took the average of the values of f for those points, and used that average as the value of $\bar{f}$? This is exactly what the Monte Carlo method does. So in formula the approximation we get is

$$\iint_R f(x,y)\,dA \approx A(R)\bar{f} \pm A(R)\sqrt{\frac{\overline{f^2}-(\bar{f})^2}{N}},$$

where

$$\bar{f} = \frac{\sum_{i=1}^{N} f(x_i, y_i)}{N} \text{ and } \overline{f^2} = \frac{\sum_{i=1}^{N} (f(x_i, y_i))^2}{N},$$

with the sums taken over the N random points $(x_1, y_1), \ldots, (x_N, y_N)$. The ± "error term" in formula does not really provide hard bounds on the approximation. It represents a single *standard deviation* from the *expected* value of the integral. That is, it provides a *likely* bound on the error. Due to its use of random points, the Monte Carlo method is an example of a *probabilistic* method (as opposed to *deterministic* methods such as Newton's method, which use a specific formula for generating points).

For example, we can use formula to approximate the volume V under the plane $z = 8x + 6y$ over the rectangle $R = [0, 1] \times [0, 2]$. Below is a code listing (montecarlo.java) for a Java program that calculates the volume, using a number of points N that is passed on the command line as a parameter.

EXERCISES

1. Write a program that uses the Monte Carlo method to approximate the double integral $\iint_R e^{xy}\,dA$, where $R = [0, 1] \times [0, 1]$. Show the program output for N = 10, 100, 1000, 10000, 100000 and 1000000 random points.
2. Write a program that uses the Monte Carlo method to approximate the triple integral $\iint_S e^{xyz}\,dV$, where $S = [0, 1] \times [0, 1] \times [0, 1]$. Show the program output for N = 10, 100, 1000, 10000, 100000 and 1000000 random points.

5. Use the Monte Carlo method to approximate the volume of a sphere of radius 1.
6. Use theMonte Carlo method to approximate the volume of the ellipsoid $\frac{x^2}{9}+\frac{y^2}{4}+\frac{z^2}{1}=1.$

Change of Variables in Multiple Integrals

Given the difficulty of evaluating multiple integrals, the reader may be wondering if it is possible to simplify those integrals using a suitable substitution for the variables. The answer is yes, though it is a bit more complicated than the substitution method which you learned in single-variable calculus.

Recall that if you are given, for example, the definite integral

$$\int_1^2 x^3\sqrt{x^2-1}dx,$$

then you would make the substitution

$$u = x^2 - 1 \Rightarrow x^2 = u + 1$$

$$du = 2x\, dx$$

which changes the limits of integration

$$x = 1 \Rightarrow u = 0$$

$$x = 2 \Rightarrow u = 3$$

so that we get

$$\begin{aligned}\int_1^2 x^3\sqrt{x^2-1}dx &= \int_1^2 \frac{1}{2}x^2.2x\sqrt{x^2-1}dx\\ &= \int_0^3 \frac{1}{2}(u+1)\sqrt{u}\,du\\ &= \frac{1}{2}\int_0^3 (u^{3/2}+u^{1/2})\,du,\end{aligned}$$

which can be easily integrated to give

$$= \frac{14\sqrt{3}}{5}.$$

Let us take a different look at what happened when we did that substitution, which will give some motivation for how substitution works in multiple integrals. First, we let $u = x^2 - 1$. On the interval of integration [1, 2], the function $x \mapsto x^2 - 1$ is strictly increasing (and maps [1, 2] onto [0, 3]) and hence has an inverse function (defined on the interval [0, 3]). That is, on [0, 3] we can define x as a function of u, namely

$$x = g(u) = \sqrt{u+1}.$$

Then substituting that expression for x into the function $f(x) = x^3 \sqrt{x^2-1}$ gives

$$f(x) = f(g(u)) = (u+1)^{3/2}\sqrt{u},$$

and we see that

$$\frac{dx}{du} = g'(u) \Rightarrow dx = g'(u)du$$

$$dx = \frac{1}{2}(u+1)^{-1/2}du,$$

so since

$$g(0) = 1 \Rightarrow 0 = g^{-1}(1)$$

$$g(3) = 2 \Rightarrow 3 = g^{-1}(2)$$

then performing the substitution as we did earlier gives

$$\int_1^2 f(x)dx = \int_1^2 x^3\sqrt{x^2-1}dx$$

$$= \int_0^2 \frac{1}{2}(u+1)\sqrt{u}du, \text{ which can be written as}$$

$$= \int_0^3 (u+1)^{3/2}\sqrt{u}.\frac{1}{2}(u+1)^{-1/2}du, \text{ which means}$$

$$= \int_1^2 f(x)dx = \int_{g^{-1}(1)}^{g^{-1}(2)} f(g(u))g'(u)du.$$

In general, if $x = g(u)$ is a one-to-one, differentiable function from an interval $[c, d]$ (which you can think of as being on the "u-axis") onto an interval $[a, b]$ (on the x-axis), which means that $g'(u) \neq 0$ on the interval (c, d), so that $a = g(c)$ and $b = g(d)$, then $c = g^{-1}(a)$ and $d = g^{-1}(b)$, and

$$\int_a^b f(x)dx = \int_{g^{-1}(a)}^{g^{-1}(b)} f(g(u))g'(u)du.$$

This is called the *change of variable* formula for integrals of single-variable functions, and it is what you were implicitly using when doing integration by substitution. This formula turns out to be a special case of a more general formula which can be used to evaluate multiple integrals. We will state the formulas for double and triple integrals involving real-valued functions of two and three variables, respectively. We will assume that all the functions involved are continuously differentiable and that the regions and solids involved all have "reasonable" boundaries. The proof of the following theorem is beyond the scope of the text.

Theorem: Change of Variables Formula for Multiple Integrals.

Let $x = x(u, v)$ and $y = y(u, v)$ define a one-to-one mapping of a region R' in the uv-plane onto a region R in the xy-plane such that the determinant

$$J(u,v) = \begin{vmatrix} \dfrac{\partial x}{\partial u} & \dfrac{\partial x}{\partial y} \\ \dfrac{\partial y}{\partial u} & \dfrac{\partial y}{\partial v} \end{vmatrix}$$

is never 0 in R'. Then

$$\iint_R f(x,y)\,dA(x,y) = \iint_{R'} f(x(u,v),y(u,v))\,|\,J(u,v)\,|\,dA(u,v).$$

We use the notation $dA(x, y)$ and $dA(u, v)$ to denote the area element in the (x, y) and (u, v) coordinates, respectively.

Similarly, if $x = x(u, v, w)$, $y = y(u, v, w)$ and $z = z(u, v, w)$ define a one-to-one mapping of a solid S' in uvw-space onto a solid S in xyz-space such that the determinant

$$J(u,v,w) = \begin{vmatrix} \dfrac{\partial x}{\partial u} & \dfrac{\partial x}{\partial v} & \dfrac{\partial x}{\partial w} \\ \dfrac{\partial y}{\partial u} & \dfrac{\partial y}{\partial v} & \dfrac{\partial y}{\partial w} \\ \dfrac{\partial z}{\partial u} & \dfrac{\partial z}{\partial v} & \dfrac{\partial z}{\partial w} \end{vmatrix}$$

is never 0 in S', then

$$\iiint_S f(x,y,z)\,dV(x,y,z) = \iiint_{S'} f(x(u,v,w),y(u,v,w),z(u,v,w))\,|\,J(u,v,w)\,|\,dV(u,v,w).$$

The determinant $J(u, v)$ in formula is called the **Jacobian** of x and y with respect to u and v, and is sometimes written as

$$J(u, v) = \frac{\partial(x,y)}{\partial(u,v)}.$$

Similarly, the Jacobian $J(u, v, w)$ of three variables is sometimes written as

$$J(u,v,w) = \frac{\partial(x,y,z)}{\partial(u,v,w)}.$$

Notice that formula (3.19) is saying that $dA(x, y) = |\,J(u, v)|\,dA(u, v)$, which you can think of as a two-variable version of the relation $dx = g'(u)\,du$ in the single-variable case.

The following example shows how the change of variables formula is used.

Example: Evaluate $\iint_R e^{\frac{x-y}{x+y}} dA$, where $R = \{(x, y) : x \geq 0, y \geq 0, x + y \leq 1\}$.

Solution: First, note that evaluating this double integral *without* using substitution is probably impossible, at least in a closed form. By looking at the numerator and denominator of the exponent of e, we will try the substitution $u = x - y$ and $v = x + y$.

To use the change of variables formula, we need to write both x and y in terms of u and v. So solving for x and y gives $x = \frac{1}{2}(u + v)$ and $y = \frac{1}{2}(v - u)$. In Figure below, we see how the mapping $x = x(u, v) = \frac{1}{2}(u + v)$, $y = y(u, v) = \frac{1}{2}(v - u)$ maps the region R' onto R in a one-to-one manner.

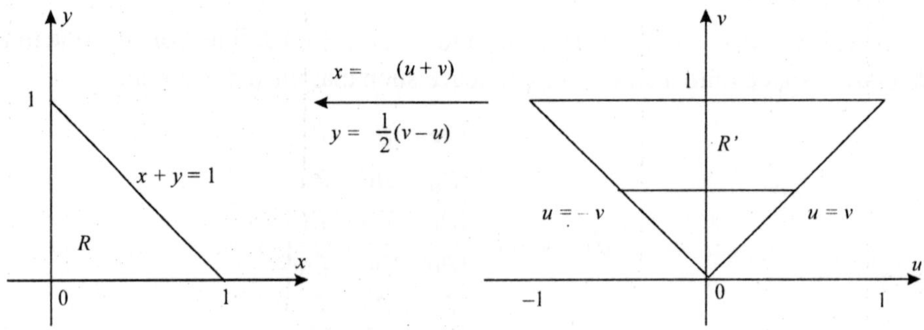

Fig. The Regions R and R'

Now we see that

$$J(u,v) = \begin{vmatrix} \frac{\partial x}{\partial u} & \frac{\partial x}{\partial v} \\ \frac{\partial y}{\partial u} & \frac{\partial y}{\partial v} \end{vmatrix} = \begin{vmatrix} \frac{1}{2} & \frac{1}{2} \\ -\frac{1}{2} & \frac{1}{2} \end{vmatrix} = -\frac{1}{2}$$

$$\Rightarrow |J(u,v)| = \left|\frac{1}{2}\right| = \frac{1}{2},$$

so using horizontal slices in R', we have

$$\iint_R e^{\frac{x-y}{x+y}} dA = \iint_{A'} f(x(u,v), y(u,v)) \, | J(u,v) | \, dA$$

$$= \int_0^1 \int_{-v}^{v} e^{\frac{u}{v}} \frac{1}{2} du dv$$

$$= \int_0^1 \left(\frac{v}{2} e^{\frac{u}{v}} \Bigg|_{u=-v}^{u=v} \right) dv$$

$$= \int_0^1 \frac{v}{2}(e - e^{-1}) dv$$

$$= \frac{v^2}{4}(e - e^{-1}) \Bigg|_0^1 = \frac{1}{4}\left(e - \frac{1}{e}\right) = \frac{e^2 - 1}{4e}$$

The change of variables formula can be used to evaluate double integrals in polar coordinates. Letting

$$x = x(r, \theta) = r \cos \theta \text{ and } y = y(r, \theta) = r \sin \theta,$$

we have

$$J(u, v) = \begin{vmatrix} \frac{\partial x}{\partial r} & \frac{\partial x}{\partial \theta} \\ \frac{\partial y}{\partial r} & \frac{\partial y}{\partial \theta} \end{vmatrix} = \begin{vmatrix} \cos\theta & -r\sin\theta \\ \sin\theta & r\cos\theta \end{vmatrix}$$

$$= r \cos^2 \theta + r \sin^2 \theta = r \Rightarrow |J(u, v)| = |r| = r,$$

so we have the following formula:

Double Integral in Polar Coordinates

$$\iint_R f(x, y) dx dy = \iint_{R'} f(r\cos\theta, r\sin\theta) r dr d\theta,$$

where the mapping $x = r \cos \theta$, $y = r \sin \theta$ maps the region R' in the $r\theta$-plane onto the region R in the xy-plane in a one-to-one manner.

Example: Find the volume V inside the paraboloid $z = x^2 + y^2$ for $0 \le z \le 1$.

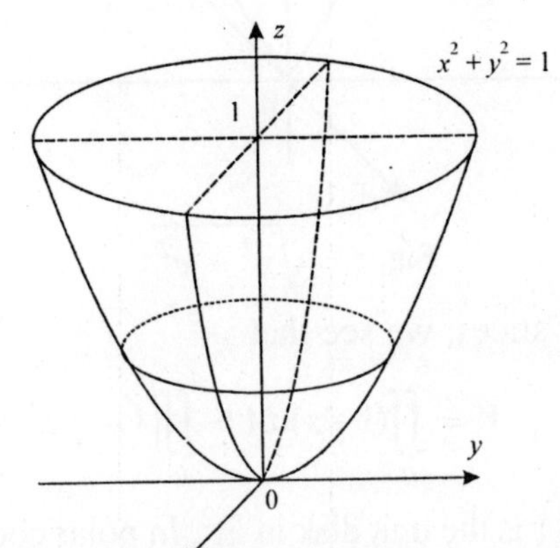

Fig. $z = x^2 + y^2$

Solution: Using vertical slices, we see that

$$V = \iint_R (1 - z)\,dA = \iint_R (1 - (x^2 + y^2))\,dA,$$

where $R = \{(x, y) : x^2 + y^2 \leq 1\}$ is the unit disk in $\mathbb{R}^2$. In polar coordinates (r, θ) we know that $x^2 + y^2 = r^2$ and that the unit disk R is the set $R' = \{(r, \theta) : 0 \leq r \leq 1, 0 \leq \theta \leq 2\pi\}$. Thus,

$$V = \int_0^{2\pi} \int_0^1 (1 - r^2) r\,dr\,d\theta$$

$$= \int_0^{2\pi} \int_0^1 (r - r^3)\,dr\,d\theta$$

$$= \int_0^{2\pi} \left(\frac{r^2}{2} - \frac{r^4}{4} \Bigg|_{r=0}^{r=1} \right) d\theta$$

$$= \int_0^{2\pi} \frac{1}{4}\,d\theta = \frac{\pi}{2}$$

Example: Find the volume V inside the cone $z = \sqrt{x^2 + y^2}$ for $0 \leq z \leq 1$.

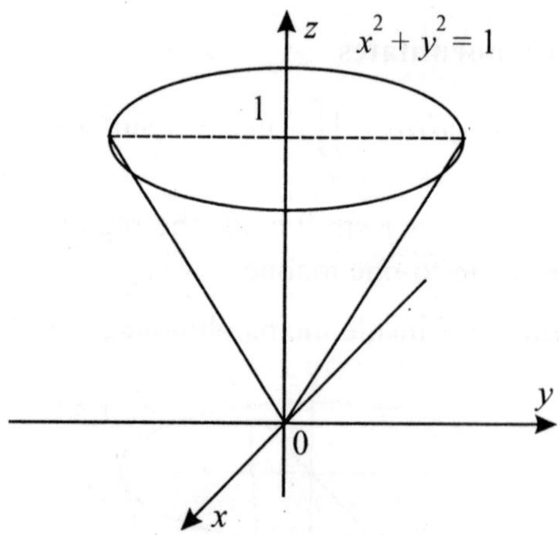

Fig. $z = \sqrt{x^2 + y^2}$

Solution: Using vertical slices, we see that

$$V = \iint_R (1 - z)\,dA = \iint_R \left(1 - \sqrt{x^2 + y^2}\right) dA,$$

where $R = \{(x, y) : x^2 + y^2 \leq 1\}$ is the unit disk in $\mathbb{R}^2$. In polar coordinates (r, θ) we know that $\sqrt{x^2 + y^2} = r$ and that the unit disk R is the set $R' = \{(r, \theta) : 0 \leq r \leq 1, 0 \leq \theta \leq 2\pi\}$. Thus,

$$= \int_0^{2\pi} \int_0^1 (r - r^2)\, dr d\theta$$

$$= \int_0^{2\pi} \left(\frac{r^2}{2} - \frac{r^3}{3} \Bigg|_{r=0}^{r=1} \right) d\theta$$

$$= \int_0^{2\pi} \frac{1}{6} d\theta$$

$$= \frac{\pi}{3}$$

In a similar fashion, it can be shown that triple integrals in cylindrical and spherical coordinates take the following forms:

Triple Integral in Cylindrical Coordinates

$$\iiint_S f(x, y, z)\, dx dy dz = \iiint_{S'} f(r\cos\theta, r\sin\theta, z)\, r\, dr\, d\theta\, dz,$$

where the mapping $x = r \cos \theta$, $y = r \sin \theta$, $z = z$ maps the solid S' in $r\theta z$-space onto the solid S in xyz-space in a one-to-one manner.

Triple Integral in Spherical Coordinates

$$\iiint_S f(x, y, z)\, dx\, dy\, dz = \iiint_{S'} f(\rho\sin\phi\cos\theta, \rho\sin\phi\sin\theta, \rho\cos\phi)\rho^2 \sin\phi\, d\rho\, d\phi\, d\theta,$$

where the mapping $x = \rho \sin \phi \cos \theta$, $y = \rho \sin \phi \sin \theta$, $z = \rho \cos \phi$ maps the solid S' in $\rho\phi\theta$-space onto the solid S in xyz-space in a one-to-one manner.

Example: For $a > 0$, find the volume V inside the sphere $S = x^2 + y^2 + z^2 = a^2$.

Solution: We see that S is the set $\rho = a$ in spherical coordinates, so

$$V = \iiint_S 1 dV = \int_0^{2\pi} \int_0^{\pi} \int_0^{a} 1\rho^2 \sin\phi\, d\rho\, d\phi\, d\theta$$

$$= \int_0^{2\pi} \int_0^{\pi} \left(\frac{\rho^3}{3} \Bigg|_{\rho=0}^{\rho=a} \right) \sin\phi\, d\phi\, d\theta = \int_0^{2\pi} \int_0^{\pi} \frac{a^3}{3} \sin\phi\, d\phi\, d\theta$$

$$= \int_0^{2\pi} \left(-\frac{a^3}{3} \cos\phi \Bigg|_{\phi=0}^{\phi=\pi} \right) d\theta = \int_0^{2\pi} \frac{2a^3}{3} d\theta = \frac{4\pi a^3}{3}.$$

EXERCISES

EXERCISES

1. Find the volume V inside the paraboloid $z = x^2 + y^2$ for $0 \le z \le 4$.
2. Find the volume V inside the cone $z = \sqrt{x^2 + y^2}$ for $0 \le z \le 3$.
3. Find the volume V of the solid inside both $x^2 + y^2 + z^2 = 4$ and $x^2 + y^2 = 1$.
4. Find the volume V inside both the sphere $x^2 + y^2 + z^2 = 1$ and the cone $z = \sqrt{x^2 + y^2}$.
7. Evaluate $\iint_R \sin\left(\frac{x+y}{2}\right)\cos\left(\frac{x-y}{2}\right) dA$, where R is the triangle with vertices (0, 0), (2, 0) and (1, 1). (Hint: Use the change of variables $u = (x + y) / 2$, $v = (x - y) / 2$.)
8. Find the volume of the solid bounded by $z = x^2 + y^2$ and $z^2 = 4(x^2 + y^2)$.
9. Find the volume inside the elliptic cylinder $\frac{x^2}{a^2} + \frac{y^2}{b^2} = 1$ for $0 \le z \le 2$.
10. Show that the volume inside the ellipsoid $\frac{x^2}{a^2} + \frac{y^2}{b^2} + \frac{z^2}{c^2} = 1$ is $\frac{4\pi abc}{3}$.
11. Show that the *Beta function*, defined by

$$B(x, y) = \int_0^1 t^{x-1}(1-t)^{y-1}\,dt, \text{ for } x > 0, y > 0,$$

satisfies the relation $B(y, x) = B(x, y)$ for $x > 0$, $y > 0$.

12. Using the substitution $t = u / (u + 1)$, show that the Beta function can be written as

$$B(x, y) = \int_0^\infty \frac{u^{x-1}}{(u+1)^{x+y}}\,du, \text{ for } x > 0, y > 0.$$

Application: Center of Mass

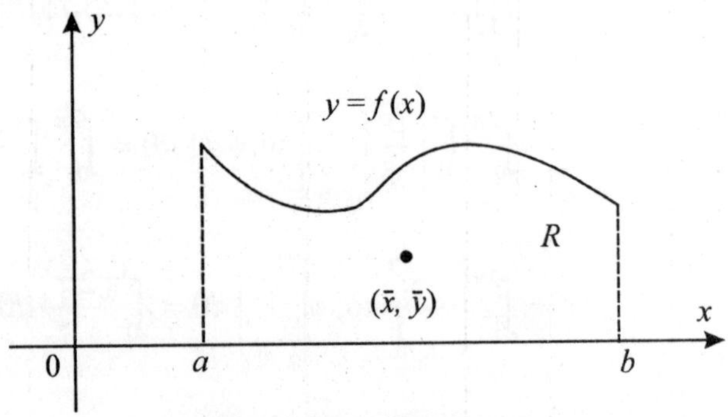

Fig. Center of mass of R

of mass of R has coordinates $(\overline{x}, \overline{y})$ given by

$$\overline{x} = \frac{M_y}{M} \text{ and } \overline{y} = \frac{M_x}{M},$$

where $$M_x = \int_a^b \frac{(f(x))^2}{2} dx, M_y = \int_a^b xf(x)dx,\ M = \int_a^b f(x)dx,$$

assuming that R has *uniform density*, i.e the *mass* of R is uniformly distributed over the region. In this case the area M of the region is considered the mass of R (the density is constant, and taken as 1 for simplicity).

In the general case where the density of a region (or *lamina*) R is a continuous function $\delta = \delta(x, y)$ of the coordinates (x, y) of points inside R (where R can be *any* region in $\mathbb{R}^2$) the coordinates $(\overline{x}, \overline{y})$ of the center of mass of R are given by

$$\overline{x} = \frac{M_y}{M} \text{ and } \overline{y} = \frac{M_x}{M},$$

where $$M_y = \iint_R x\delta(x, y)dA, M_x = \iint_R y\delta(x, y)dA, M = \iint_R \delta(x, y)dA,$$

The quantities M_x and M_y are called the *moments* (or *first moments*) of the region R about the x-axis and y-axis, respectively. The quantity M is the mass of the region R. To see this, think of taking a small rectangle inside R with dimensions Δx and Δy close to 0. The mass of that rectangle is approximately $\delta(x_*, y_*)\Delta x\ \Delta y$, for some point (x_*, y_*) in that rectangle. Then the mass of R is the limit of the sums of the masses of all such rectangles inside R as the diagonals of the rectangles approach 0, which is the double integral $\iint_R \delta(x, y)dA.$

Note that the formulas in represent a special case when $\delta(x, y) = 1$ throughout R in the formulas.

Example: Find the center of mass of the region $R = \{(x, y) : 0 \leq x \leq 1, 0 \leq y \leq 2x^2\}$, if the density function at (x, y) is $\delta(x, y) = x + y$.

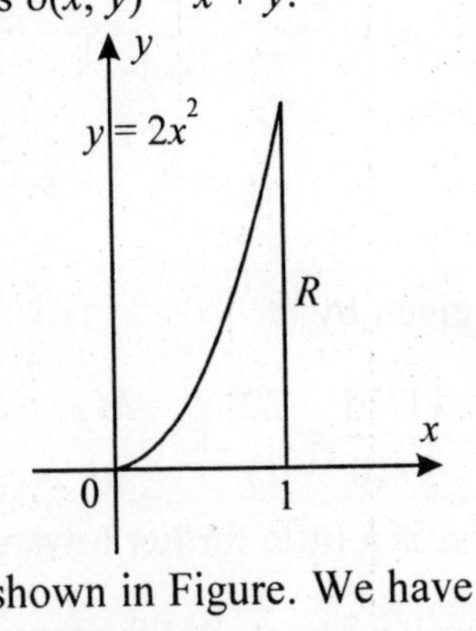

Solution: The region R is shown in Figure. We have

Solution: The region R is shown in Figure. We have

$$M = \iint_R \delta(x, y)dA$$

$$= \int_0^1 \int_0^{2x^2} (x + y)dydx$$

$$= \int_0^1 \left(xy + \frac{y^2}{2} \Big|_{y=0}^{y=2x^2} \right) dx$$

$$= \int_0^1 (2x^3 + 2x^4)dx$$

$$= \frac{x^2}{2} + \frac{2x^5}{5} \Big|_0^1 = \frac{9}{10}$$

and

$$M_x = \iint_R y\delta(x, y)dA \qquad M_y = \iint_R x\delta(x, y)dA$$

$$= \int_0^1 \int_0^{2x^2} y(x + y)dydx \qquad = \int_0^1 \int_0^{2x^2} x(x + y)dydx$$

$$= \int_0^1 \left(\frac{xy^2}{2} + \frac{y^3}{3} \Big|_{y=0}^{y=2x^2} \right) dx \qquad = \int_0^1 \left(x^2 y + \frac{xy^3}{2} \Big|_{y=0}^{y=2x^2} \right) dx$$

$$= \int_0^1 \left(2x^5 + \frac{8x^6}{3} \right) dx \qquad = \int_0^1 \left(2x^4 + 2x^5 \right) dx$$

$$= \frac{x^6}{3} + \frac{8x^7}{21} \Big|_0^1 = \frac{5}{7} \qquad = \frac{2x^5}{5} + \frac{x^6}{3} \Big|_0^1 = \frac{11}{15},$$

so the center of mass $(\overline{x}, \overline{y})$ is given by

$$\overline{x} = \frac{M_y}{M} = \frac{11/15}{9/10} = \frac{22}{27}, \overline{y} = \frac{M_x}{M} = \frac{5/7}{9/10} = \frac{50}{63}.$$

Note how this center of mass is a little further towards the upper corner of the region R than when the density is uniform(it is easy to use the formulas to show that $(\overline{x}, \overline{y}) = \left(\frac{3}{4}, \frac{3}{5}\right)$

that upper corner, where there is quite a bit of area.

In the special case where the density function $\delta(x, y)$ is a constant function on the region R, the center of mass $(\bar{x}, \bar{y})$ is called the *centroid* of R.

The formulas for the center of mass of a region in $\mathbb{R}^2$ can be generalized to a solid S in $\mathbb{R}^3$. Let S be a solid with a continuous mass density function $\delta(x, y, z)$ at any point (x, y, z) in S $\mathbb{R}$ Then the center of mass of S has coordinates $(\bar{x}, \bar{y}, \bar{z})$, where

$$\bar{x} = \frac{M_y}{M} =, \bar{y} = \frac{M_x}{M}, \bar{z} = \frac{M_{xy}}{M},$$

where

$$M_{yz} = \iiint_S x\delta(x,y,z)dV, M_{xz} = \iiint_S y\delta(x,y,z)dV, M_{xy} = \iiint_S z\delta(x,y,z)dV,$$

$$M = \iiint_S \delta(x,y,z)dV.$$

In this case, M_{yz}, M_{xz} and M_{xy} are called the *moments* (or *first moments*) of S around the yz-plane, xz-plane and xy-plane, respectively. Also, M is the mass of S.

Example: Find the center of mass of the solid $S = \{(x, y, z) : z \geq 0, x^2 + y^2 + z^2 \leq a^2\}$, if the density function at (x, y, z) is $\delta(x, y, z) = 1$.

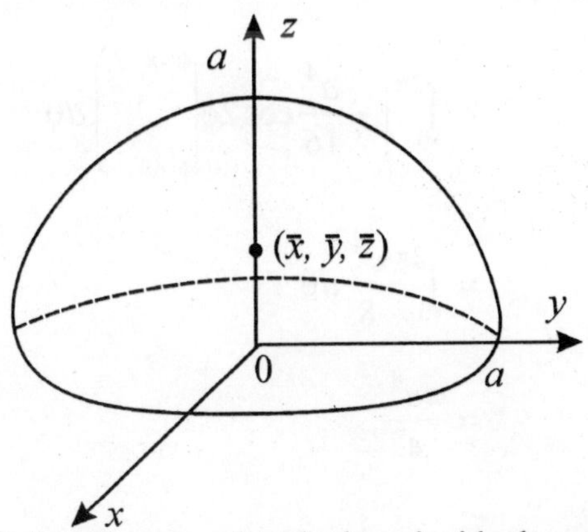

Solution: The solid S is just the upper hemisphere inside the sphere of radius a centered at the origin. So since the density function is a constant and S is symmetric about the z-axis, then it is clear that $\bar{x} = 0$ and $\bar{y} = 0$, so we need only find $\bar{z}$. We have

$$M = \iiint_S \delta(x,y,z)dV = \iiint_S 1dV = Volume(S).$$

But since the volume of S is half the volume of the sphere of radius a, which we know by Example is $\frac{4\pi a^3}{3}$, then $M = \frac{2\pi a^3}{3}$. And

$$M_{xy} = \iiint_S z\delta(x,y,z)dV$$

$$= \iiint_S z dV \text{, which in spherical coordinates is}$$

$$= \int_0^{2\pi}\int_0^{\pi/2}\int_0^a (\rho\cos\phi)\rho^2 d\phi\, d\rho\, d\phi\, d\theta$$

$$= \int_0^{2\pi}\int_0^{\pi/2} \sin\phi\cos\phi\left(\int_0^a \rho^3 d\rho\right) d\phi\, d\theta$$

$$= \int_0^{2\pi}\int_0^{\pi/2} \frac{a^4}{4}\sin\phi\cos\phi d\phi\, d\theta$$

$$M_{xy} = \int_0^{2\pi}\int_0^{\pi/2} \frac{a^4}{4}\sin 2\phi d\phi\, d\theta$$

(since $\sin 2\phi \doteq 2 \sin\phi \cos\phi$)

$$= \int_0^{2\pi}\left(-\frac{a^4}{16}\cos 2\phi\Big|_{\phi=0}^{\phi=\pi/2}\right) d\theta$$

$$= \int_0^{2\pi} \frac{a^4}{8} d\theta$$

$$= \frac{\pi a^4}{4},$$

so

$$\bar{z} = \frac{M_{xy}}{M} = \frac{\frac{\pi a^4}{4}}{\frac{2\pi a^3}{3}} = \frac{3a}{8}.$$

Thus, the center of mass of S is $(\bar{x}, \bar{y}, \bar{z})\bar{z} = \left(0, 0, \frac{3a}{8}\right)$..

EXERCISES

For Exercises 1-5, find the center of mass of the region R with the given density function $\delta(x, y)$.

1. $R = \{(x, y) : 0 \le x \le 2, 0 \le y \le 4\}$, $\delta(x, y) = 2y$
2. $R = \{(x, y) : 0 \le x \le 1, 0 \le y \le x^2\}$, $\delta(x, y) = x + y$
3. $R = \{(x, y) : y \ge 0, x^2 + y^2 \le a^2\}$, $\delta(x, y) = 1$
4. $R = \{(x, y) : y \ge 0, x \ge 0, 1 \le x^2 + y^2 \le 4\}$, $\delta(x, y) = \sqrt{x^2 + y^2}$
5. $R = \{(x, y) : y \ge 0, x^2 + y^2 \le 1\}$, $\delta(x, y) = y$

For Exercises 6-10, find the center of mass of the solid S with the given density function $\delta(x, y, z)$.

6. $S = \{(x, y, z) : 0 \le x \le 1, 0 \le y \le 1, 0 \le z \le 1\}$, $\delta(x, y, z) = xyz$
7. $S = \{(x, y, z) : z \ge 0, x^2 + y^2 + z^2 \le a^2\}$, $\delta(x, y, z) = x^2 + y^2 + z^2$
8. $S = \{(x, y, z) : x \ge 0, \ge 0, z \ge 0, x^2 + y^2 + z^2 \le a^2\}$, $\delta(x, y, z) = 1$
9. $S = \{(x, y, z) : 0 \le x \le 1, 0 \le y \le 1, 0 \le z \le 1\}$, $\delta(x, y, z) = x^2 + y^2 + z^2$
10. $S = \{(x, y, z) : 0 \le x \le 1, 0 \le y \le 1, 0 \le z \le 1 - x - y\}$, $\delta(x, y, z) = 1$

Application: Probability and Expected Value

In this section we will briefly discuss some applications of multiple integrals in the field of probability theory. In particular we will see ways in which multiple integrals can be used to calculate *probabilities* and *expected values*.

Probability

Suppose that you have a standard six-sided (fair) die, and you let a variable X represent the value rolled. Then the *probability* of rolling a 3, written as $P(X = 3)$, is $\frac{1}{6}$, since there are six sides on the die and each one is equally likely to be rolled, and hence in particular the 3 has a one out of six chance of being rolled. Likewise the probability of rolling *at most* a 3, written as $P(X \le 3)$, is $\frac{3}{6} = \frac{1}{2}$, since of the six numbers on the die, there are three equally likely numbers (1, 2, and 3) that are less than or equal to 3. Note that $P(X \le 3) = P(X = 1) + P(X = 2) + P(X = 3)$. We call X a *discrete random variable* on the *sample space* (or *probability space*) Ω consisting of all possible outcomes. In our case, $\Omega = \{1, 2, 3, 4, 5, 6\}$. An *event* A is a subset of the sample space. For example, in the case of the die, the event $X \le 3$ is the set $\{1, 2, 3\}$.

Now let X be a variable representing a random real number in the interval (0, 1). Note that the set of all real numbers between 0 and 1 is *not* a discrete (or *countable*) set of values, i.e. it can not be put into a one-to-one correspondence with the set of positive integers. In this case, for any real number x in (0, 1), it makes no sense to consider $P(X = x)$ since it *must* be 0 (why?). Instead, we consider the probability $P(X \leq x)$, which is given by $P(X \leq x) = x$. The reasoning is this: the interval (0, 1) has length 1, and for x in (0, 1) the interval (0, x) has length x. So since X represents a *random* number in (0, 1), and hence is *uniformly distributed* over (0, 1), then

$$P(X \leq x) = \frac{\text{length of } (0, x)}{\text{length of } (0,1)} \frac{x}{1} = x.$$

We call X a *continuous random variable* on the *sample space* $\Omega = (0, 1)$. An *event* A is a subset of the sample space. For example, in our case the event $X \leq x$ is the set (0, x). In the case of a discrete random variable, we saw how the probability of an event was the *sum* of the probabilities of the individual outcomes comprising that event (e.g. $P(X \leq 3) = P(X = 1) + P(X = 2) + P(X = 3)$ in the die example). For a continuous random variable, the probability of an event will instead be the *integral* of a function, which we will now describe.

Let X be a continuous real-valued random variable on a sample space Ω in $\mathbb{R}$. For simplicity, let $= (a, b)$. Define the *distribution function* F of X as

$$F(x) = P(X \leq x)\text{, for } -\infty < x < \infty$$

$$= \begin{cases} 1, & \text{for } x \geq b \\ P(X \leq x) & \text{for } a < x < b \\ 0, & \text{for } x \leq a \end{cases}$$

Suppose that there is a nonnegative, continuous real-valued function f on $\mathbb{R}$ such that

$$F(x) = \int_{-\infty}^{x} f(y)dy, \text{ for } -\infty < x < \infty,$$

and

$$\int_{-\infty}^{\infty} f(x)dx = 1.$$

Then we call f the *probability density function* (or *p.d.f.* for short) for X. We thus have

$$P(X \leq x) = \int_{a}^{x} f(y)dy, \text{ for } a < x < b.$$

Also, by the Fundamental Theorem of Calculus, we have

$$F'(x) = f(x), \text{ for } -\infty < x < \infty.$$

Example: Let X represent a randomly selected real number in the interval (0, 1). We say that X has the *uniform distribution* on (0, 1), with distribution function

$$F(x) = P(X \le x) = \begin{cases} 1, & \text{for } x \ge 1 \\ x, & \text{for } 0 < x < 1 \\ 0, & \text{for } x \le 0 \end{cases}$$

and probability density function

$$f(x) = F'(x) = \begin{cases} 1, & 0 < x < 1 \\ 0, & \text{elsewhere.} \end{cases}$$

In general, if X represents a randomly selected real number in an interval (a, b), then X has the uniform distribution function

$$F(x) = P(X \le x) = \begin{cases} 1, & \text{for } x \ge b \\ \dfrac{x}{b-a}, & \text{for } a < x < b \\ 0, & \text{for } x \le a \end{cases}$$

and probability density function

$$f(x) = F'(x) = \begin{cases} \dfrac{1}{b-a} & \text{for } a < x < b \\ 0, & \text{elsewhere.} \end{cases}$$

Example: A famous distribution function is given by the *standard normal disribution*, whose probability density function f is

$$f(x) = \frac{1}{\sqrt{2\pi}} e^{-x^2/2}, \text{ for } -\infty < x < \infty.$$

This is often called a "bell curve", and is used widely in statistics. Since we are claiming that f is a p.d.f., we *should* have

$$\int_{-\infty}^{\infty} \frac{1}{\sqrt{2\pi}} e^{-x^2/2} dx = 1$$

by formula, which is equivalent to

$$\int_{-\infty}^{\infty} e^{-x^2/2} dx = \sqrt{2\pi}.$$

We can use a double integral in polar coordinates to verify this integral. First,

$$\int_{-\infty}^{\infty}\int_{-\infty}^{\infty} e^{-(x^2+y^2)} dx dy = \int_{-\infty}^{\infty} e^{-y^2/2}\left(\int_{-\infty}^{\infty} e^{-x^2/2} dx\right) dy$$

$$= \left(\int_{-\infty}^{\infty} e^{-x^2/2} dx\right)\left(\int_{-\infty}^{\infty} e^{-y^2/2} dy\right) = \left(\int_{-\infty}^{\infty} e^{-x^2/2} dy\right)^2$$

since the same function is being integrated twice in the middle equation, just with different variables. But using polar coordinates, we see that

$$\int_{-\infty}^{\infty}\int_{-\infty}^{\infty} e^{-(x^2+y^2)/2}dxdy = \int_0^{2\pi} e^{-r^2/2} rdrd\theta$$

$$= \int_0^{2\pi}\left(e^{-r^2/2}\Big|_{r=0}^{r=\infty}\right) d\theta$$

$$= \int_0^{2\pi} (0-(-e^0))d\theta = \int_0^{2\pi} 1d\theta = 2\pi,$$

and so

$$\left(\int_{-\infty}^{\infty} e^{-x^2/2}dx\right)^2 = 2\pi, \text{ and hence}$$

$$\int_{-\infty}^{\infty} e^{-x^2/2}dx. = \sqrt{2\pi}$$

In addition to individual random variables, we can consider *jointly distributed* random variables. For this, we will let X, Y and Z be three real-valued continuous random variables defined on the same sample space Ω in $\mathbb{R}$ (the discussion for two random variables is similar).

Then the *joint distribution function* F of X, Y and Z is given by

$$F(x, y, z) = P(X \leq x, Y \leq y, Z \leq z), \text{ for } -\infty < x, y, z < \infty.$$

If there is a nonnegative, continuous real-valued function f on $\mathbb{R}^3$ such that

$$F(x, y, z) = \int_{-\infty}^{z}\int_{-\infty}^{y}\int_{-\infty}^{x} f(u,v,w)dudvdw, \text{ for } -\infty < x, y, z < \infty$$

and

$$\int_{-\infty}^{\infty}\int_{-\infty}^{\infty}\int_{-\infty}^{\infty} f(x,y,z)dxdydz = 1,$$

then we call f the *joint probability density function* (or *joint p.d.f.* for short) for X, Y and Z. In general, for $a_1 < b_1$, $a_2 < b_2$, $a_3 < b_3$, we have

$$P(a_1 < X \leq b_1, a_2 < Y \leq b_2, a_3 < Z \leq b_3) = \int_{a_3}^{b_3}\int_{b_3}^{b_2}\int_{a_1}^{b_1} f(x,y,z)dxdydz,$$

with the $\leq$ and $<$ symbols interchangeable in any combination. A triple integral, then, can be thought of as representing a probability (for a function f which is a p.d.f.).

Example: Let a, b, and c be real numbers selected randomly from the interval (0, 1). What is the probability that the equation $ax^2 + bx + c = 0$ has at least one real solution x?

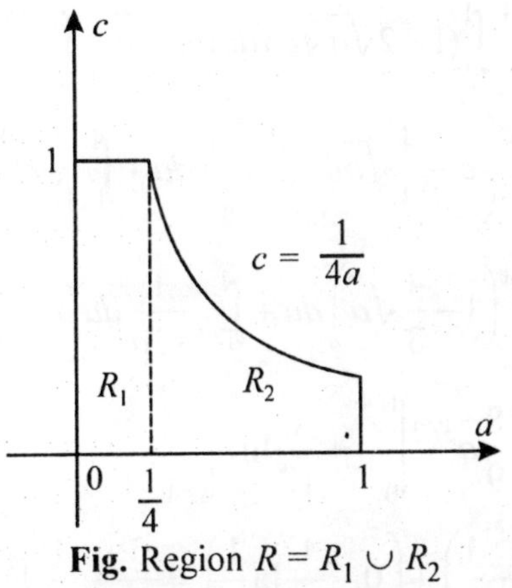

Fig. Region $R = R_1 \cup R_2$

Solution: We know by the quadratic formula that there is at least one real solution if $b^2 - 4ac \geq 0$. So we need to calculate $P(b^2 - 4ac \geq 0)$. We will use three jointly distributed random variables to do this. First, since $0 < a, b, c < 1$, we have

$$b^2 - 4ac \geq 0 \Leftrightarrow 0 < 4ac \leq b^2 < 1 \Leftrightarrow 0 < 2\sqrt{a}\sqrt{c} \leq b < 1,$$

where the last relation holds for all $0 < a, c < 1$ such that

$$0 < 4ac < 1 \Leftrightarrow 0 < c < \frac{1}{4a}.$$

Considering a, b and c as real variables, the region R in the ac-plane where the above relation holds is given by $R = \{(a, c) : 0 < a < 1, 0 < c < 1, 0 < c < \frac{1}{4a}\}$, which we can see is a union of two regions R_1 and R_2.

Now let X, Y and Z be continuous random variables, each representing a randomly selected real number from the interval (0, 1) (think of X, Y and Z representing a, b and c, respectively). Then, similar to how we showed that $f(x) = 1$ is the p.d.f. of the uniform distribution on (0, 1), it can be shown that $f(x, y, z) = 1$ for x, y, z in (0, 1) (0 elsewhere) is the joint p.d.f. of X, Y and Z. Now,

$$P(b^2 - 4ac \geq 0) = P((a, c) \in R, 2\sqrt{a}\sqrt{c} \leq b < 1),$$

so this probability is the triple integral of $f(a, b, c) = 1$ as b varies from $2\sqrt{a}\sqrt{c}$ to 1 and as (a, c) varies over the region R. Since R can be divided into two regions R_1 and R_2, then the required triple integral can be split into a sum of two triple integrals, using vertical slices in R:

$$P(b^2 - 4ac \geq 0) = \underbrace{\int_0^{1/4} \int_0^1}_{R_1} \int_{0\sqrt{a}\sqrt{c}}^1 1\,db\,dc\,da + \underbrace{\int_{1/4}^1 \int_0^{1/4a}}_{R_2} \int_{0\sqrt{a}\sqrt{c}}^1 1\,db\,dc\,da$$

$$= \int_0^{1/4}\int_0^1 (1-2\sqrt{a}\sqrt{c})\,dc\,da + \int_{1/4}^1\int_0^{1/4a}(1-2\sqrt{a}\sqrt{c})\,dc\,da$$

$$= \int_0^{1/4}\left(c - \frac{4}{3}\sqrt{a}c^{3/2}\Big|_{c=0}^{c=1}\right) da + \int_{1/4}^1 \left(c - \frac{4}{3}\sqrt{a}c^{3/2}\Big|_{c=0}^{c=1/4a}\right) da$$

$$= \int_0^{1/4}\left(1 - \frac{4}{3}\sqrt{a}\right) da + \int_{1/4}^1 \frac{1}{12a}\,da$$

$$= a - \frac{8}{9}a^{3/2}\Big|_0^{1/4} + \frac{1}{12}\ln a\Big|_{1/4}^1$$

$$= \left(\frac{1}{4} - \frac{1}{9}\right) + \left(0 - \frac{1}{12}\ln\frac{1}{4}\right) = \frac{5}{36} + \frac{1}{12}\ln 4$$

$$P(b^2 - 4ac \geq 0) = \frac{5 + 3\ln 4}{36} \approx 0.2544$$

In other words, the equation $ax^2 + bx + c = 0$ has about a 25% chance of being solved!

Expected Value

The *expected value* EX of a random variable X can be thought of as the "average" value of X as it varies over its sample space. If X is a discrete random variable, then

$$EX = \sum_x xP(X = x),$$

with the sum being taken over all elements x of the sample space. For example, if X represents the number rolled on a six-sided die, then

$$EX = \sum_{x=1}^{6} xP(X = x) = \sum_{x=1}^{6} x\frac{1}{6} = 3.5$$

is the expected value of X, which is the average of the integers 1 – 6.

If X is a real-valued continuous random variable with p.d.f. f, then

$$EX = \int_{-\infty}^{\infty} xf(x)dx.$$

For example, if X has the uniform distribution on the interval (0, 1), then its p.d.f. is

$$EX = \begin{cases} 1, & \text{for } 0 < x < 1 \\ 0, & \text{elsewhere,} \end{cases}$$

and so $$EX = \int_{-\infty}^{\infty} xf(x)dx = \int_0^1 x\,dx = \frac{1}{2}.$$

For a pair of jointly distributed, real-valued continuous random variables X and Y with joint p.d.f. $f(x, y)$, the expected values of X and Y are given by

$$EX = \int_{-\infty}^{\infty}\int_{-\infty}^{\infty} xf(x,y)\,dx\,dy \quad \text{and} \quad EY = \int_{-\infty}^{\infty}\int_{-\infty}^{\infty} yf(x,y)\,dx\,dy,$$

respectively.

Example: If you were to pick $n > 2$ random real numbers from the interval (0, 1), what are the expected values for the smallest and largest of those numbers?

Solution: Let $U_1, \ldots, U_n$ be n continuous random variables, each representing a randomly selected real number from (0, 1), i.e. each has the uniform distribution on (0, 1). Define random variables X and Y by

$$X = \min(U_1, \ldots, U_n) \text{ and } Y = \max(U_1, \ldots, U_n).$$

Then it can be shown that the joint p.d.f. of X and Y is

$$f(x,y) = \begin{cases} n(n-1)(y-x)^{n-2}, & \text{for } 0 \le x \le y \le 1 \\ 0, & \text{elsewhere.} \end{cases}$$

Thus, the expected value of X is

$$\begin{aligned} EX &= \int_0^1\int_x^1 n(n-1)x(y-x)^{n-2}\,dydx \\ &= \int_x^1\left(nx(y-x)^{n-2}\Big|_{y=x}^{y=1}\right)dx \\ &= \int_x^1 nx(y-x)^{n-2}\,dx \text{ so integration by parts yields} \\ &= -x(1-x)^n - \frac{1}{n+1}(1-x)^{n+1}\Big|_0^1 \end{aligned}$$

$$EX = \frac{1}{n+1},$$

and similarly it can be shown that

$$EY = \int_0^1\int_0^y n(n-1)y(y-x)^{n-2}\,dxdy = \frac{n}{n+1}.$$

So, for example, if you were to repeatedly take samples of $n = 3$ random real numbers from (0, 1), and each time store the minimum and maximum values in the sample, then the average of the minimums would approach $\frac{1}{4}$ and the average of the maximums would approach $\frac{3}{4}$ as the number of samples grows. It would be relatively simple to write a computer program to test this.

EXERCISES

1. Evaluate the integral $\int_{-\infty}^{\infty} e^{-x^2} dx$ using anything you have learned so far.

2. For $\sigma > 0$ and $\mu > 0$, evaluate $\int_{-\infty}^{\infty} \frac{1}{\sigma\sqrt{2\pi}} e^{-(x-\mu)^2/2\sigma^2} dx$.

3. For continuous random variables X, Y with joint p.d.f. $f(x, y)$, define the *second moments* $E(X_2)$ and $E(Y_2)$ by

$$E(X^2) = \int_{-\infty}^{\infty}\int_{-\infty}^{\infty} x^2 f(x,y)\,dx\,dy \text{ and } E(Y^2) = \int_{-\infty}^{\infty}\int_{-\infty}^{\infty} y^2 f(x,y)\,dx\,dy$$

and the *variances* Var(X) and Var(Y) by

$$\text{Var}(X) = E(X^2) - (EX)^2 \text{ and } \text{Var}(Y) = E(Y^2) - (EY)^2.$$

Chapter 7

Big Integral Theorems

These results are analogous to the second fundamental theorem of calculus in that they undo a derivative by intergrating. They have many important applicants in electromagnetism, fluid dynamics, etc.

Divergence Theorem (Gauss' Theorem)

Divergence Theorem. Let S be a bounded, piecewise smooth, orientated, non-interescting surface enclosing a volume V in $\mathbb{R}^3$, with a normal n that points outwards from V. Let u be a vector field with continuous first – order partial derivatives throughout V. Then

$$\iiint_v \nabla.u dr = \iint_S u.dS.$$

The divergence theorem thus startes that $\nabla.u.$ Integrated over a volume V is equal to the flux of u across the closed surface S surrouinding the vloume.

Outeline Proof. Suppose that S is a surface enclosing a volume V such that Cartesian axes can be chosen so thaty any line parallel to any one of the axes meets S in just one or two points (e.g. a convex surface). We observe that

$$\iiint_v \nabla.u dr = \iiint_v \left(\frac{\partial u_1}{\partial x} + \frac{\partial u_2}{\partial y} + \frac{\partial u_3}{\partial z}\right) dr,$$

comprises of three trerms; we initially concentrate on the $\iiint_v \frac{\partial u_3}{\partial z} d\tau,$ term.

Let region A be the projection of S onto the xy–plane. As when we first introduced tripple intergrals, let the lower / upper surfaces, S_1 / S_2 respectively, be parametrised by

$$S_1 : x = (x, y, z_1(x, y))$$
$$S_1 : x = (x, y, z_2(x, y)).$$

Then using the second foundametnal theorem of calculus

$$\iiint_V \frac{\partial u_3}{\partial z} dxdydz = \iint_A \left[\int_{z=z_1}^{z_2} \frac{\partial u_3}{\partial z} dz\right] dxdy \quad \text{s}$$

$$= \iint_A (u_3(x, y, z_2(x, y)) - u_3(x, y, z_1(x, y)) dxdy.$$

From for $x = (x, y, z(x, y))$

$$ds = (-z_x, -z_y, 1)dxdy,$$

and hence (this is a bit of a diry trick)

$$\hat{z} ds = dxdy \text{ on } S_2,$$

$$\hat{z} ds = dxdy \text{ on } S_2,$$

Where the minus sign in the latter result is because the outward normal points in the opposite direction. Thus can be rewritten as

$$\iiint_v \frac{\partial u_2}{\partial y} d\tau = \iint_{s_2} u_3 \hat{z}.dS + \iint_{s_1} u_3 \hat{z} dS = \iint_s u_3 \hat{x}.dS.$$

Similarly $\iiint_v \frac{\partial u_2}{\partial y} d\tau = \iint_{s_2} u_2 \hat{y}.dS, \iiint_v \frac{\partial u_1}{\partial x} d\tau == \iint_s u_1 \hat{x}.dS.$

Adding the above results we obtain

$$\iiint_v \nabla.u d\tau = \iint_s u.dS.$$

Remarks:

1. If the volume is not of the specified type, then divided the volume into parrts that are, and add the results for each sub-volume – the 'common' surface intergrals cancel.
2. The divergence theorem relates a triple intergral to a double integral (cf. the second fundamental theorem of calculus which related a single integral to a function).

Examples:

1. We confirm the divergence theorem for $u = x$, and V a sphere of a radius a. First,m using,

$$\iiint_V \nabla.u d\tau = \iiint_V \nabla.x d\tau = 3\iiint_V d\tau = 3.\frac{4}{3}\pi a^3 = 4\pi a^3;$$

Second , since $n = \hat{x}$ on the surface of a sphere,

$$\iint_S u.dS = \iint_S x.ndS = \iint_S rdS = a\iint_S dS = 4\pi a^3.$$

2. Show that if $\nabla . u = 0$ in $\mathbb{R}^3$, then

$$I = \iint_S u.dS$$

is the same for every open surface spanning a curve C.

Answer: Let S_1 and S_2 be two such surfaes, and let $\hat{s}$ be any other such surface that does not interesect S_1 and S_2. Let V_j be the volume enclosed by $s_j + \hat{s}$. Then

$$0 = \iiint_{V_j} \nabla.u = \iint_{S_1} u.dS + \iint_{\hat{S}} u.dS$$

Thus

$$\iint_{S_1} u.dS = \iint_{\hat{S}} u.dS + \iint_{S} u.dS$$

3. Unlectrured Example, Let $u = (x, y, 0)$ and let S be the hemispherical surface $(x^2 + y^2 + z^2)$

Evaluate

$$I = \iint_S u.dS.$$

Answer: Close the volume by adding A, the disc of radius a in the $z = 0$ plane, Then

$$\iiint_V \nabla.u d\tau = \iint_S u.dS + \iint_A u.dS.$$

On A, rthe surface elements $dS = (0, 0, -1)$ dxdy, and hence $u.\ dS = 0$. Further, since $\nabla . u = 2$,

$$\iiint_V \nabla.u d\tau = 2.\frac{2}{3}\pi a^3 = \frac{4}{3}\pi a^3.$$

Hence

$$I = \frac{4}{3}\pi a^3.$$

Exercise: Calculate I directly.

Extensions

A boundary with moe than one component. Suppose the boundary S has more than one component, e.g. suppose that V lies between closed surfaces S_1 ands S_2. Then

$$\iiint_v \nabla.udr = \iint_{S_1} u.ds + \iint_{S_2} u.ds.$$

where for each surface the normal pointsd away from V.

This result follows from subvdividng the volume into surfaces with only one component, and using the fact that 'common' surface integrals so introduced cancel.

The two dimensional divergence therorem,. Suppose that $u = (u_1(x, y), u_2\ (x, y,),(0)$, and consider a cylinder of volume V, height h, cross section A with bounding curve C, ends S_E and sides S_s. Then from the z – independence

$$\iiint_V \nabla.ud\tau = h\iint_A \nabla.udS.$$

Also at the endsd $u\ .\ n = u.\hat{z} = 0, \hat{z} = 0,$ and thus

$$\iint_{S_E} u.dS = 0$$

while on the side

$$dS = n\ ds\ d\,z \text{ where } ds^2 = dx^2 + dx^2 + dy^2$$

and hence

$$\iint_{S_E} u.dS = \int_0^h dz \oint_C u.nds.$$

Combining the above results and using (6.1) yields the two dimensional divergence theorem

$$\iint_{S_E} \nabla u.dS = \oint_C u.nds.$$

Unlectured Remark. Now suppose that $u = (u(x), 0, 0)$ and let the area A be a rectangle. Then

$$\iint_A \nabla u.dS = \int_c^d dy \frac{du}{dx},$$

and

$$\oint_c u.nds = \int_c^d dyu(b) + \int_d^c dyu(a)),$$

Hence after factoring $(d - c) \neq 0$,

$$\int_a^b \frac{du}{dx} dx = (u(b) - u(a)),$$

which is just the second fundamental theorem of calculus.

The generalisation for a scalar field. For a scalar field $f(x)$ with continuous first – order partial derivatives in V

$$\iiint_V \nabla f \, d\tau = \iint_S fdS.$$

To see this let a be an arbitrary constant vector, and apply the divergence therorem to $u = f\,a$, to obtrain

$$\iiint_V \nabla.(fa)d\tau = \iint_S fa.dS.$$

Now

$$\nabla.(fa) = \frac{\partial}{\partial x_j}\left(fa_j\right) = a_j \frac{\partial f}{\partial x_j} a.(\nabla f),$$

and because a is a constant

$$a.\iiint_v \nabla fs\tau = a \iint_S fdS.$$

Symbolically we have that $a.(p - q) = 0$ for arvitirary a. Choose $a = (p - q)$ to conclude tha $|\, p - q \,| = 0$, and thus that

$$\iiint_V \nabla fd\tau = \iint_S fdS.$$

alternatively we may proceed from by taking $a = i$ to conclude that $\iiint_v \frac{\partial f}{\partial x} d\tau = \iint_S fn_1 dS.$ etc.

Exercise: Show by setting $u = a \times A$, where a is an arbitrary vector and A is a vector field, that

$$\iiint_V \nabla \times Ad\tau = \iint_S n \times AdS.$$

A further 'generalisation'. Equation can be written in component form as

$$\iiint_V \frac{\partial f}{\partial x_j} d\tau = \iint_S f n_j dS \text{ and } \iiint_V \xi_{ijk} \frac{\partial A_k}{\partial x_j} d\tau = \iint_S \varepsilon_{ijk} n_j A_k dS.$$

This suggests the following expression for rthe general form of the divrgence therorem

$$\iiint_V \frac{\partial \bullet}{\partial x_j} d\tau = \iint_S \bullet n_j dS,$$

Where $\bullet$ is any function of x,e.g., $f, u_j, \varepsilon_{ijk} A_k$. This is in fact no more than a statement in suffix notation withi u_j replaced withn $\bullet$ in each of the formulae.

An example: For a closed surface S evaluate, i.e.

$$I = \iint_S dS.$$

with $f = 1$

$$\iint_S dS = \iiint_V \nabla f d\tau = 0,$$

as was done component by comnponent.

Co–ordiante Independent Definition of div

Let a volumes V be enclosed by a surface S, and consider a limit process in which the greatest diameter of V tends to zero whiklekeeping trhe pointg x_0 inside V. If the vector field u is twice differentiable at x_0, then from Taylor's theorem with $x = x_0 + \delta x$,

$$\iiint_v \nabla.u(x) d\tau = \iiint_v (\nabla.u(x0) + o(|\nabla.u|)) d\tau = \nabla.u(x_0)|V| + o(|\nabla.u||V|),$$

Where $|V|$ is the volume of V. Thus using the divergence theorem, a consistent co–ordinate free definition of div is

$$\nabla.u - \lim_{|V|\to 0} \frac{1}{|V|} \iint_S u.dS,$$

where S is any 'nice'small closed surface enclosing a volume V.

Green's Theorem in the plane

Let C be a simple, piecewise smooth, two – dimensioanal, closed curve (descred in the anit – clockwise sence) bounding a closed region A. Suppose $P(x, y)$ and $Q(x, y)$ have continuous first–order partial deriavatives on A. Then

$$\iint_A \left(\frac{\partial Q}{\partial x} - \frac{\partial P}{\partial y} \right) dxdy = \oint_C (Pdx + Qdy).$$

Outline Proof: Let C be a curve such that Cartesian axes can be chosen so that by line parallel to one of the axes meets C in just one or two pointsd (e.g. suppose C is convex).

Divide C into C_1 and C_2 where C_1 and C_2 are the lower and upper curves respectively, so that a line of constant x intersect C_1 at a smaller value of y than C_2. Parameterise C_j by

$$x = (x, y_j(x), 0).$$

Then

$$\iint_A \frac{\partial P}{\partial y} dxdy = \int_{y_1}^{b} dx \int_{y_1}^{y_2} dy \frac{\partial P}{\partial y}$$

$$= \int_a^b dx\left[P(x, y_2(x)) - P(x, y_1(x, y_1)\right]$$

$$= -\int_a^b dxP(x, y_2) - \int_a^b dxP(x, y_1)$$

$$= -\oint_C p(x, y)dx.$$

Similarly for $\widehat{c_j}$ parameterised by $x = (x_j(y), y, 0)$

$$\iint_A \frac{\partial Q}{\partial x} = dxdy = \oint_C Qdy.$$

Subtract the previous result from this to obtain

$$\iint_A \left(\frac{\partial Q}{\partial x} - \frac{\partial P}{\partial y}\right) dxdy = \oint_C (Pdx + Qdy).$$

Remark: This result can be extended to more general curves by surgery.

Unilectured Altrèrnative Outline Poof. From the two–dimensiaonal divergence theorem

$$\iint_A \nabla.udS = \oint_C u.nds.$$

Suppose that the curve C is paramerisėd by arclength so that $x \equiv x\ (s)$. Then sinc $\frac{dx}{ds}$ is tangent to C, the outward normal n to the curve is given by

$$n = \frac{dx}{ds} \times \hat{z} = \left(\frac{dy}{ds}, \frac{dx}{ds}, 0\right).$$

Hence, if we let $u \equiv (Q, -P, 0)$ so that

$$\nabla . u = \frac{\partial Q}{\partial x} - \frac{\partial P}{\partial y},$$

then becomes

$$\iint_A \left(\frac{\partial Q}{\partial x} - \frac{\partial P}{\partial y} \right) dS = \oint_c (Q, P, 0) - (dy, -dx, 0) = \oint_c (Pdx + Qdy).$$

Example: Express

$$I = \oint_c (x^2 ydx + xy^2 dy).$$

for some curve C bounding a surface A. as a surface integral.

Identify

$$P = x^2 y \text{ and } Q = xy^2$$

Then from Green's theorem

$$I = \iint_A \left(\frac{\partial Q}{\partial x} - \frac{\partial P}{\partial y} \right) dxdy$$

$$= \iint_C (y^2 - x^2) dxdy.$$

Remark. The equivalence of these two expression for I when C is one particular curve is question on Example sheet.

Exact Differentials

Suppose $Pdx + Qdy$ is exct. From we have that

$$\frac{\partial P}{\partial y} = \frac{\partial Q}{\partial x}.$$

Green's theorem then gives

$$\oint_c (Pdx + Qdy) = \iint_A \left(\frac{\partial Q}{\partial x} - \frac{\partial P}{\partial y} \right) dxdy = 0.$$

Remark: This is rederivation of result that the line integral of an exact differential around a closed path in a simply connected region is zero.

Stokes's Theorem

This is a generalisation of green's theorem to a three dimnensional surface.

Stokes' Theorem

Let S be piecewise smooth, open orientated, non- intersecting surface bounded by a simple , piecewise smooth, closed curve. C Let $u(x)$be a vector field with continuous first-order partial derivatives on S. Then

$$\iint_S \nabla \times u.dS = \oint_C u.dx,$$

where the line integral is taken in the positive direction of C (right hand rule).

Stoke's theorem thus states that the flux of $\nabla \times u$ across an open surface S is equal to the circulation of u round the bounding curve C.

Remark: If S is in the xy -plane , then (6.13) becomes

$$\iint_S \left(\frac{\partial u_2}{\partial x} - \frac{\partial u_1}{\partial y}\right) dxdy = \oint_C (u_1 dx + u_2 dy),$$

i.e. Green's theorem after writing $u = (P, Q, 0)$.

Heuristic Argument. Divide up S by sectionally smooth curves into N subregions S_1 ....S_N. For small enough subregions, each S_j can be approximate by a plane surface A_j, bounded by curves C_1 ...., C_N. Apply Green's theorem in vector form to each individul subregion A_j, then on summing over the subregions

$$\sum_j \iint_{A_j} (\nabla \times u).dS = \sum_j \oint_{C_j} u.dx.$$

Furher on letting $N \to \infty$

$$\sum_j \iint_{A_j} (\nabla \times u).dS \to \iint_S (\nabla \times u).dS,$$

while

$$\sum_j \oint_{C_j} u.dx \to \oint_C u.dx,$$

Since, the integrals over curves that are 'common' to the plane approximation to two adjacent subregions cancel in the limit. Stokes ' theorme follows.

Outline proof. Let the open surface S be parameterised by $p, q, x \equiv x\,(p, q)$. Then the local vector elementry area is given by ,

$$dS = \left(\frac{\partial x}{\partial p} \times \frac{\partial x}{\partial y}\right) dpdq,$$

where the parameters p and q are assumed to be ordered so that dS in the chosen positive normal direction.

Further, from using the definition of curl, it follows that

$$\nabla \times u.dS = \left(\varepsilon_{ijk}\frac{\partial u_k}{\partial x_j}\right)\left(\varepsilon_{i\ell m}\frac{\partial x_\ell}{\partial p}\frac{\partial x_m}{\partial q}\right)dpdq$$

$$= (\delta_{j\ell}\delta_{km} - \delta_{jm}\delta_{k\ell})\frac{\partial u_k}{\partial x_j}\frac{\partial x_\ell}{\partial p}\frac{\partial x_m}{\partial q}dpdq$$

$$= \left(\frac{\partial u_k}{\partial x_j}\frac{\partial x_j}{\partial p}\frac{\partial x_k}{\partial q} - \frac{\partial u_k}{\partial x_j}\frac{\partial x_j}{\partial q}\frac{\partial x_k}{\partial p}\right)dpdq$$

$$= \left(\frac{\partial x_k}{\partial p}\frac{\partial x_k}{\partial q} - \frac{\partial u_k}{\partial q}\frac{\partial x_k}{\partial p}\right)dpdq$$

using the chain rule

$$= \left[\frac{\partial}{\partial p}\left(u_k\frac{\partial x_k}{\partial q}\right) - \frac{\partial}{\partial q}\left(u_k\frac{\partial x_k}{\partial p}\right)\right]dpdq$$

assuming that $\frac{\partial^2 x_k}{\partial p \partial q} = \frac{\partial^2 x_k}{\partial q \partial p}$.

hence

$$\iint_S (\nabla \times u).dS = \iint_A \left[\frac{\partial}{\partial p}\left(u_k\frac{\partial x_k}{\partial q}\right) - \frac{\partial}{\partial q}\left(u_k\frac{\partial x_k}{\partial p}\right)\right]dpdq,$$

where A is the region in the pq-plane over which S is parmetrised.

Next apply Green's theorem to the right-hand-side by identifying p and q with x and y respectively, to obtain

$$\iint_S (\nabla \times u).dS = \oint_C \left(u_k\frac{\partial x_k}{\partial p}dp + u_k\frac{\partial x_k}{\partial q}dp\right) = \oint_C u_k\left(\frac{\partial x_k}{\partial p}dp + \frac{\partial x_k}{\partial q}dq\right),$$

where $\hat{C}$ is the bounding curve for A and the integral round $\hat{C}$ is in an anti-clockwise direction. Moreover, since

$$dx_k = \frac{\partial x_k}{\partial p}dp + \frac{\partial x_k}{\partial q}dp =,$$

the right -hand-side can be transformed to aclosed line integral around C. Thus

$$\iint_S (\nabla \times u).dS = \oint_C u_k dx_k = \oint_C u.dx,$$

since we claim that as $\hat{C}$ is traced out in the pq-plane in an anti clockwise sense, the curve C is traced out in its oriented direction.

For Our Choice of Positive Normal Anti -Clockwise in $\widehat{C}$ $\Rightarrow$ Oriented in *C*

Suppose tha $p = P(s) \equiv (P(s), Q(s))$ parameterises $\widehat{C}$ in an anti-clockwise sense using arc length. Then from (2.8) the unit tangent , $\hat{u}$ to $\widehat{C}$ is

$$\hat{u} = \frac{dp}{dx} = \left(\frac{dp}{ds}, \frac{dQ}{ds}\right),$$

while the equation of a normal pointing into a from a point P (s), and parameterised by t, is given by

$$p = P(s) + t\hat{a}, \text{ where } \hat{a} = \left(-\frac{dQ}{ds}, \frac{dp}{ds}\right).$$

Hence a [non -unit] tangent, u, to C, corresponding to being traced out in an anti-closckwise sense, is

$$u = \frac{d}{ds}(x(p(s))) = -\frac{\partial x}{\partial p}\frac{dP}{ds} + \frac{\partial x}{\partial q}\frac{dQ}{ds},$$

while a [non- unit]normal, a to C that is also tangent to the the surface S is arrange

$$a = \frac{d}{dt}(x(p(s) + t\hat{a})) = -\frac{\partial x}{\partial p}\frac{dQ}{ds} + \frac{\partial x}{\partial p}\frac{dP}{ds}.$$

It follows that a [non- unit]normal to S, arranged so that is C is traced out in its oriented direction as $\hat{c}$ is traced out out in an anti-clockwise sense, is

$$u \times a = +\left(\frac{\partial x}{\partial p} \times \frac{dx}{dq}\right) \cdot \left(\left(\frac{dP}{ds}\right)^2 + \left(\frac{dQ}{ds}\right)^2\right).$$

We note that this is in the same direction as $n = dS/\, dS$.

Examples:

1. For $u = (x^2, z^2, -y^2)$ evaluate

$$I = \iint_S (\nabla \times u).dS,$$

where S is the surface $z = 4 - 4 - x^2 - y^2, z \geq 0$.

Since the bounding curve C is the circle $x^2 + y^2 = 4$ in the plane $z = 0$, it follows form Stokes's theorem that

$$\iint_S (\nabla \times u).dS = \oint_C u.dx, = \oint_C (x^2 dx + x^2 dy) \text{ but } z = 0 \text{ on } C = \left[\frac{1}{3}x^3\right]_2^2 = 0.$$

Exercise: Evaluate directly without using Stokes's theorme.

1. *Unlectured Example.* Evaluate $I = \int_S (\nabla \times u).dS$ where S is closed surface.

   (a) First divide S into two parts S_1 and S_2 seprated by a curve C such that

   S_1 has oriented boundary curve C

   S_2 has oriented boundary curve -C

   Then from Stokes's theorem

$$= \iint_S (\nabla \times u).dS = \iint_{S_1} (\nabla \times u).dS + \iint_{S_2} (\nabla \times u).dS$$

$$= \oint_C u.dx + \oint_{-C} u.dx$$

$$= 0.$$

   (b) Alternatively from the divergence theorem

$$I = \iiint_v \nabla.(\nabla \times u) d\tau = 0,$$

   because div(curl) is always zero.

Extension

1. Suppose S is bounded by two (or more) curve, say C_1 and C_2. Then

$$\iint_S (\nabla \times u).dS = \oint_{C_1} u.dx + \oint_{C_2} u.dx,$$

   where the oriented direction of each C_j is fixed by the right-hand rule.

   Outline Proof: By surgery /cross- cuts and cancellation

2. For a scalar field f with with continuous first -order partial derivatives in S

$$\iint_S \nabla f \times dS = -\oint_C f dx.$$

   To see this let $u = af$ where a is an arbitrary vector. Then from Stokes' theorem

$$\iint_S \nabla f \times (af).dS = \oint_C fa.dx.$$

But $\nabla \times (af).dS = n.\nabla \times (af) dS$

$$= n_i \varepsilon_{ijk} \frac{\partial(a_k f)}{\partial x_j} dS$$

$$= -a_k \varepsilon_{kji} \frac{\partial f}{\partial x_j} \eta_i dS$$

$$= -a.(\nabla f \times n) dS,$$

and hence

$$-a. \iint_S \nabla f \times dS = a. \oint_C f dx.$$

it follows that since a is arbitary (*cf.* the argument below equation)

$$\iint_S \nabla f \times dS = -\oint_C f dx.$$

Co-ordinate independent Definition of Curl

Let an open smooth surface S be bounded by a curve C. Consider a limit process in which the point x_0 remains on S, the greatest diameter of S tends to zero, and the normals at all points on the surface tend to a specifec direction (i.e the value of n at x_0).

If vector field u is twice differentiable at x_0, then from Taylor's theorem with $x = x_0 + x\delta$,

$$\iint_S (\nabla \times u(x)).dS = \iint_S (\nabla \times u(x_0) + o(|\nabla \times u|)).dS$$

$$= \nabla \times u(x_0) n |S| + o(|\nabla \times u||S||),$$

where $|S|$ is the area of S. Thus using Stokes's theorem, we concluede that a consistrent co-ordinate free defintion of curl has the form for a rectangle)

$$n.(\nabla \times u) = \lim_{S \to 0} \oint_C u.dx,$$

where S in any ' niece ' small open surface with a bounding curve C. All components of $\nabla \times u$ can be recovered by considering limiting surfaces with normals in different direction

Irrotational Fields

Suppose that F has conntinuous first -order partial derivatives in simply connected region V. Then the following are equivalent.

(a) $\nabla \times F = 0$ throughout V i.e, F is irrotational throughout V.

(b) $\oint_C F.dx = 0$ for all closed curves C in V.

(c) F is consevative i.e $\int_{x_a}^{x_b} F.dx$ is independent of path.

(d) There exists a single valued function f called the scalar potential, such that $F = \nabla f$.

(e) $F.dx$ is exact differential

Outline proof:

(a) $\Rightarrow$ (b). If $\nabla \times F = 0$, then by stokes ' theorem.

$$\oint_C F.dx = \iint_S (\nabla \times F).dS = 0.$$

(b) $\Rightarrow$ (c). Let C_1 and C_2 be any two paths from x_a to x_b. Then

$$\int_{C_1} F.dx - \int_{C_2} F.dx = \oint_{C_1 - C_2} F.dx = 0.$$

Hence

$$\int_{C_1} F.dx - \int_{C_2} F.dx.$$

(c) $\Rightarrow$ (d). This was proved in particular it was shown that if F. is counservative, then for the scalar potential definde by

$$f(x) = \int_a^x F.dx,$$

it follows that

$F = \nabla f$

(c) $\Rightarrow$ (d). This was also proved in x 3, 4, 1 particular, if $F = \nabla f$ then

$$F.dx = \nabla f, dx = df.$$

ie. $F.dx$ in an exact differential.

(e) $\Rightarrow$ (a) We prove this in directly by showing '(e) $\Rightarrow$ (d)' andthen that'(d) $\Rightarrow$ (a)'.

(e) $\Rightarrow$ (d)'. If $F.dx$ is an exact, differential, say equal to df, then from the relatioship between a differential and a gradient

$$F.dx = df = \nabla f.dx$$

Next we observe that this result can only hold for all dx if

$$F = \nabla f.$$

(d) $\Rightarrow$ (a)'. If $F = \nabla f$.

$$\nabla \times F = \nabla \times (\nabla f) = 0.$$

Remarks: All for one and one for all. We conclude tht one result implies all the others.

If F is irrotational the $F.\,dx$ is exact. The result '(a) $\Rightarrow$ (e)' proves as promised earlier, that condition , ie. $\nabla \times F = 0$, is a sufficient condition for $F.\,dx$ to be an exact differential in $3D$. Further , in the special case when F is a two dimensionsl vector field, say $F = (P, Q, 0)$ we recover equation, i.e

$$\frac{\partial P}{\partial y} = \frac{\partial Q}{\partial x},$$

as a sufficient condition for $Pdx + Qdy$ to be exact,

Poincare's Lemma for irrotational field. The result '(a) $\Rightarrow$ (d)' proves Poincare's lemma if '(a) $\Rightarrow$ (e)' in a simply connected region then there exists a scalar potential f such that

$$F = \nabla f.$$

Orthogonal Curvilinear Co-ordinates

So far we have concentrated on Cartesian co- ordinates. What do grad, div and curl look like in cylindrical or spherical polar co-ordinates etc?

The key difficulty aries because the basis vectors for cylindrical and spherical polar co- ordinates are function of position (hence the use word curvilinear).

BASIS VECTORS

Denote the orthogonal curvillinear co-ordinates by q_1, q_2, q_3.

	Cartesian Co- ordinates	Cylindrical polar Co –ordinates	Spherical polar Co ordinate
q_1	x	$\rho = (x^2 + y^2)^{1/2}$	$r(x^2 + y^2 + z^2)^{1/2}$
q_2	y	$\phi = \tan^{-1}(y/x)$	$q = \tan^{-2}((x^2 + y^2)^{1/2} / z)$
q_3	z	z	$\phi = \tan^{-1}(y/x)$

The position vector x is a function of $q = (q_1, q_2, q_3)$, i.e $x \equiv x(q)$. Define

$$h_j(q) = \left|\frac{\partial x(q)}{\partial q_j}\right|,$$

and let $e_j(q)$ be the unit vector in the direction of $\frac{\partial x}{\partial q_j}$; then (no s.c)

$$e_\alpha(q) = \frac{1}{h_\alpha(q)}\frac{\partial x(q)}{\partial q_\alpha} \text{ or equivalently} \frac{\partial x(q)}{\partial q_\alpha} = h_\alpha(q)e_\alpha(q).$$

For an orthogonal curvillinear co-ordinate system we demands that for all q

$$e_i(q).\ e_j\ (q) = d_{ij}.$$

Also by convention we required the system to be right-handed i.e (s.c)

$$\varepsilon_{ijk} ei = e_j \times e_k,\ \text{e.g. } e_1 = e_2 \times e_3.$$

Unlectured Remark. With the right -handed covention the Jacobian [determinant] of the co-ordinate transformation $x \equiv x\ (q)$ is positive (as might be expected), using it is given (s.c)

$$\begin{aligned} J &= \varepsilon_{ijk} h_1 [e_1] i h_2 [e_2] h_3 [e_3]_k \\ &= h_1\ h_2\ h_3\ e_1 .e_2 \times e_3 \\ &= h_1\ h_2\ h_3 \end{aligned}$$

Cartesian Co- ordinates

In this case

$$x = (x,\ y,\ z).$$

hence

$$\frac{\partial x}{\partial q_1} = \frac{\partial x}{\partial x} = i;$$

thus

$$h_1 = \left| \frac{\partial x}{\partial q_1} \right| = 1, e_1 = i.$$

Similarly

$e_j.e_j = d_{ij}$ and $e_{ijk}\ e_i = e_j \times e_k$, i.e.m Cartesian co-ordinates are a right handed orthogonal co-ordinate system !

Cylindrical Polar Co -ordinates

In this case

$$x = (\rho \cos\phi, \rho \sin\phi, z)$$

Hence,

$$\frac{\partial x}{\partial q_1} = \frac{\partial x}{\partial \rho} = (\cos\phi, \sin\phi, 0),$$

$$\frac{\partial x}{\partial q_2} = \frac{\partial x}{\partial \phi} = (-\rho \cos\phi, \rho \sin\phi, 0),$$

$$\frac{\partial x}{\partial q_3} = \frac{\partial x}{\partial z} = (0,0,1).$$

It follows that

$$h_1 = \left|\frac{\partial x}{\partial q_1}\right| = 1, \ = e_1 = e_\rho = (\cos\phi, \sin\phi, 0),$$

$$h_2 = \left|\frac{\partial x}{\partial q_2}\right| = \rho, \ e_2 = e_\phi = (-\sin\phi, \cos\phi, 0),$$

$$h_3 = \left|\frac{\partial x}{\partial q_3}\right| = 1, \ = e_3 = e_z = (0, 0, 1).$$

Remarks:

1. $e_i.e_j = d_{ij}$ and $e_{ijk}\ e_i = e_j \times e_k$, ie. cylindrical polar co-ordinates are a right-handed orthogonal curvilinear co-ordinate system.
2. e_r and e_f are function of position.

Spherical Polar Co-ordinates

In this case

$$x = (r \sin\theta \cos\phi, r \sin\theta \sin\phi, r \sin\theta).$$

Hence

$$\frac{\partial x}{\partial q_1} = \frac{\partial x}{\partial r} = (\sin\theta\cos\phi, \sin\theta\sin\phi, \cos\theta),$$

$$\frac{\partial x}{\partial q_2} = \frac{\partial x}{\partial \theta} = (r\cos\theta\cos\phi, r\cos\theta\sin\phi, -r\sin\theta),$$

$$\frac{\partial x}{\partial q_3} = \frac{\partial x}{\partial \phi} = (-r\sin\theta\sin\phi, r\sin\theta\cos\phi, 0).$$

it follows that

$$h_1 = 1, e_1 = e_r(\sin\theta\cos\phi, \sin\theta\sin\phi, \cos\theta),$$

$$h_2 = r, e_2 = e_\theta(\cos\theta\cos\phi, \cos\theta\sin\phi, -\sin\theta),$$

$$h_3 = r\sin\theta, = e_3 = e_\phi(-\sin\phi, \cos\phi, 0).$$

Remarks:

1. $e_{i,}e_j = \delta_{ij}$ *and* $\varepsilon jke_i = e_j \times e_k, i.e.$ spherical polar co-ordinates are a right-handed orthogonal curvilinear co-ordinate system.
2. $e_r, e\theta$ and $e\phi$ are functions of position.

Incremental Change in position or Length.

Consider an incremental change in positon. Then from

$$dx = \sum_j \frac{\partial x}{\partial q_j} dq_j = \sum_j h_j e_j dq_j.$$

Hence from the orthogonality requirement, an incremental change in [squad] length is given by

$$ds^2 = dx^2 \sum_j h_j e_j dq_{j.} = \sum_j h_k e_k dq_k = \sum_j h_j e_j dq_j.$$

Example: For spherical polar co–ordinates from (7.3b),(7,3b) and (7,3c)

$$ds^2 = dr^2 + r^2 d\theta^2 + r^2 \sin^2 \theta d\phi^2.$$

Gradient

We need a co-ordinate independent definiton of the gradient for the case when the basis vectors are functions of position.

We recall from the definition of a differential, that for a fixed basis system in we concluded that

$$ds^2 = dr^2 + r^2 d\theta + r^2 \sin^2 \theta d\phi^2.$$

Definition: For curvilinear4 orthogonal co–ordintes(for which the basis vectors are in general a function of positions), we define ∇f to be the vector such that for all dx

$$df = \nabla f . dx.$$

This is a consistent co-ordinte free definition of grad.

In order to determine the components of ∇f wrtite.

$$\nabla f = \sum_i e_1 \alpha_1,$$

$$df = \sum_i e_1 \alpha_1 . \sum_j h_j e_j dq_j = \sum_j \alpha_1 h_j dq_j.$$

But from the definition of a differential for $f \equiv f(q_1, q_2, q_3)$,

$$df = \sum_i \frac{\partial f}{\partial q_1} dq_1,$$

and hence, must hold for all dq_i

$$\alpha_1 = \frac{1}{h_1}\frac{\partial f}{\partial q_1} \quad (no\ s.c.).$$

$$\nabla f = \sum_i \frac{e_i}{h_i}\frac{\partial f}{\partial q_i} = \left(\frac{1}{h_i}\frac{\partial f}{\partial q_1}, \frac{1}{h_2}\frac{\partial}{p\partial\phi} + e_z\frac{\partial}{\partial z}\right).$$

Remark: Each term has dimensions '*f*/ length'

Examples: Cylindrical Polar Co-ordintes. In culindrical polar co-ordintes, the gradient is given to be

$$\nabla = e_p\frac{\partial}{\partial p} + e_\phi\frac{1}{p}\frac{\partial}{r\partial\theta} + e_z\frac{\partial}{\partial z}.$$

Divergence

From the co-ordiantes independent definition of divergence , for any 'nice' surface *S* enclosing a volume *V*.

$$\nabla.u = \lim_{|v|\to 0}\frac{1}{|v|}\iint_S u.dS.$$

In order to derive an expression for the divergence in orthogonal curvilinear co-ordiantes we consider the flux out of a slightly deformedd cuboid with one vetex at *X*(*p*), where $P = (p_1, p_2, p_3)$ and sides of length $h_1\,(p)\,\delta q_2$ and $h_3(p)\delta q_3$.

On a surface where q_1 = constant, it follows from the definition of a vector elementary are and, that

$$dS = \left(\frac{\partial x}{\partial q_2}\times\frac{\partial x}{\partial q_3}\right)dq_2 dq_3$$

$$= (h_2 e_2 \times h_3 e)dq_2 dq_3$$

$$= e_1\, h_2\, h_3\, dq_3.$$

Hence on the surface $q_1 = p_1 + q_1 = p_1 + \delta q_1$

$$\iint_{Sp_1+\delta q_1} u.dS = \int_{p_2}^{p_2+\delta q_2} dq_2 \int_{p_2}^{p_3+\delta q_2} dq_3 u_1(p_1+\delta q_1, q_2, q_3)$$

$$h_2(p_1+\delta q_1, q_2, q_3)h_3(p_1+\delta q_1, q_2, q_3).$$

Adding the contributions from the surfaces $q_1 = P_1 + \delta q_1$ and $q_1 = p_1$

$$\iint\limits_{S_{p_1}+\delta_{q_1}} u.dS + \iint\limits_{S_{p_1}} u.dS = \int_{p_2}^{p_2+\delta_{q_2}} dq_2 \int_{p_3}^{p_3+\delta q_3} dq_3$$

$$[(u_1h_2h_3)(p_1+\delta q_1, q_2, q_3) - (u_1h_2h_3)(p_1, p_2, p_3)]$$

$$= \int_0^1 d\beta \int_0^1 d\gamma \left[\frac{\partial(u_1u_2u_3)}{\partial q_1}(p_1, p_2, \beta\delta q_2, p_3+\gamma\delta q_3) + 0(\delta q_1)\right]\delta q_2 \delta q_3$$

from Taylor's theorem and writing $q_2 = p_2 + \beta\delta q_2, q_3 = p_3 + \gamma\delta q_3$

$$= \underbrace{\int_0^1 d\beta\, d\gamma}_{1}\left[\partial\frac{(u_1u_2u_3)}{\delta q_1}\right](p_1p_2p_3)\delta q_1\delta q_2\delta q_3 + 0(\delta q_1\delta q_2\delta q_3)$$

from another applicant of Taylor's theorem.

Similar results can be deduced for the surfaces q_2 = constant and q_3 = constnat here

$$\iint\limits_{S} u.dS = \left(\frac{\partial(u_1h_2h_3)}{\partial q_1} + \frac{\partial(u_2h_3h_1)}{\partial q_2} + \frac{\partial(h_3h_1h_2)}{\partial q_3}\right)$$

$$\delta q_1\delta q_2\delta q_3 + h_2e_2dq_2 + h_3e_3dq_3$$

Further from the equation for an incremental change in position

$$dx = h_1e_1dq_1 + h_3e_3dq_3$$

and thus

$$|V| = h_1h_2h_3\delta q_1\delta q_2\delta q_3 + 0(\delta q_1\delta q_2\delta q_3)$$

It follows from the co-ordinate independent definition of devergence

$$\nabla.u = \frac{1}{h_1h_2h_3}\left(\frac{\partial(u_1h_2h3)}{\delta q_1} + \frac{\partial(u_2h_3h_1)}{\partial q_1} + \frac{\partial(u_3h_1h_2)}{\partial q_3}\right).$$

Remark: Each term has dimensions 'u / length'.

Examples: Cylindrical Polar Co-ordinates

$$\nabla.u = \frac{1}{\rho}\frac{\partial}{\partial\rho}(\rho u_\rho) + \frac{1}{\rho}\frac{\partial u_\phi}{\partial_\phi} + \frac{\partial u_z}{\partial_z}.$$

Spherical Polar Co– ordinates.

$$\nabla.u = \frac{1}{r^2}\frac{\partial}{\partial r}(r^2u_r) + \frac{1}{r\sin\theta}\frac{\partial}{\partial\theta}(\sin\theta u_\theta) + \frac{1}{r\sin\theta}\frac{\partial u_\phi}{\partial_\phi}.$$

Curl

From the co-ordinate independent defintion of curl, for any 'nice' open surface S bounded by a Curve C

$$n.\,\mathrm{curl}\, u = \lim_{S\to 0}\frac{1}{|S|}\oint_c u.dx$$

Without loss of generality take $n = e_3$, and consider the circulation round a slightly deformed recrtanfgle C with a vertex at $x(P)$, where $P = (p_1, p_2, p_3)$, and sides of lenth $h_1(P)\,\delta q_2$

On C_1(where $q_2 = p_2$ and $q_3 = p_3$), $dx = h_1 e_1 dq_1$. Hence

$$\int_{C_1} u.dx = \int_{p_1}^{p_1+\delta q_1} dq_1 h_1[(q_1, p_2, p_3)u_1(q_1, p_2, p_3)]$$

Adding the contributions from C_1 and C_3:

$$\int_{C_1} u.dx + \int_{C_3} u.dx = \int_{p_1}^{p_1+\delta q_1} dq_1[(u_1, h_1)(q_1, p_2, p_3) - (u_1 h_1)(q_1, p_2 + \delta q_2, p_3)]$$

$$= -\int_0^1 d\alpha \left[\frac{(u_1, h_1)}{\delta q_2}(p_1 + \delta q_1, p_2, p_3)\delta q_2 + o(\delta q_2)\right]\delta q_1$$

From Taylor's theorem and writing $q_1 = p_1 + \alpha\delta q_1$

$$= -\frac{\partial(u_1 h_1)}{\partial q_2}(p_1, p_2, p_3)\delta q_1 \delta q_2 + o(\delta q_1 \delta q_2)$$

Similarly

$$\int_{C_2} u.dx + \int_{C_2} u.dx = \frac{\delta(u_2 h_2)}{\delta q_1}\delta q_1 \delta q_2 + o(\delta q_1 \delta q_2).$$

Further

$$|s| = h_1 h_2 \delta q_1 + o(\delta q_1 \delta q_2).$$

Hence from

$$e_3.\mathrm{curl}\, u = \frac{1}{h_1 h_2 h_3}\begin{vmatrix} e_{1e_1} & h_2 e_2 & h_3 e_3 \\ \partial q_1 & \partial q_2 & \partial q_3 \\ h_1 u_1 & h_2 u_2 & h_3 u_3 \end{vmatrix}.$$

Remark: Each term has dimensions 'u/length'.

Examples: Cylindrical Polar Co-ordiantes.

$$\text{curl}\, u = \frac{1}{\rho}\begin{vmatrix} e_\rho & \rho e_\phi & e_z \\ \partial_\rho & \partial_\phi & \delta_3 \\ h_\rho & \rho u_\phi & h_z \end{vmatrix}$$

$$= \left(\frac{1}{\rho}\frac{\partial u_z}{\partial_\phi} - \frac{\partial u_\phi}{\partial z}, \frac{\partial u_\rho}{\partial z} - \frac{\partial u_z}{\partial \rho}, \frac{1}{\rho}\frac{\partial(\rho u_\phi)}{\partial \rho} - \frac{1}{\rho}\frac{\partial u_\rho}{\partial_\phi}\right).$$

Spherical Polar Co-ordiantes.

$$\text{Curl } u = \frac{1}{r^2 \sin\theta}\begin{vmatrix} e_r & re_\theta & r\sin\theta e_\phi \\ \partial_r & \partial\theta & \partial\phi \\ u_r & ru_\theta & r\sin\theta u_\phi \end{vmatrix}$$

$$= \left(\frac{1}{r\sin\theta}\left(\frac{\partial(\sin\theta u_\phi)}{\partial\theta} \frac{\partial u_\theta}{\partial\phi}\right) - \frac{1}{r\sin\theta\partial\phi} \frac{1}{r}\frac{1\partial(ru_\theta)}{\partial r} - \frac{1}{r}\frac{\partial u_r}{\partial\theta}\right).$$

Laplacian

$$\nabla^2 f = \nabla.(\nabla f) = \frac{1}{h_1h_2h_3}\left(\frac{\partial}{\partial q_1}\left(\frac{h_2h_3}{h_1}\frac{\partial f}{\partial q_1}\right) + \frac{\partial}{\partial q_2}\left(\frac{h_3h_1}{h_2}\frac{\partial f}{\partial q_2}\right) + \frac{\partial}{\partial q_3}\left(\frac{h_1h_2}{h_3}\frac{\partial f}{\partial q_3}\right)\right).$$

Cylindrical Polar Co-ordinates

$$\nabla^2 f = \frac{1}{\rho}\frac{\partial}{\partial\rho}\left(\rho\frac{\partial f}{\partial\rho}\right) + \frac{1}{\rho^2}\frac{\partial^2 f}{\partial\phi^2} + \frac{\partial^2 f}{\partial z^2},$$

where the two-dimensional version of this with $\frac{\partial}{\partial z} \equiv 0$ has previously been derived in

Spherical Polar Co-ordinates. From equation

$$\nabla^2 f = \frac{1}{r^2} = \frac{\partial}{\partial r}\left(r^2\frac{\partial f}{\partial r}\right) + \frac{1}{r^2\sin\theta}\frac{\partial}{\partial\theta}\left(\sin\theta\frac{\partial f}{\partial\theta}\right) + \frac{1}{r^2\sin^2\theta}\frac{\partial^2 f}{\partial\phi^2}.$$

Remark: To find $\nabla^2 A$, for vector A, i.e.

$$\nabla^2 A = \nabla(\nabla.A) - \nabla\times(\nabla\times A).$$

LINE INTEGRALS AND EXACT DIFFERENTIALS

The Riemann Integral

First some revision of integration via a clarification of definitions.

Dissection. A dissection, partition or subdivision D of the interval $[t_a, t_b]$ is a finite set of points $t_0, ..., t_N$ such that

$$t_a = t_0 < t_1 < ... < t_N = t_b.$$

Modulus. Define the modulus, gauge or norm of a dissection D, written $| D |$, to be the length of the longest subinterval $(t_j - t_j - 1)$ of D, i.e.

$$| D | = \max_{1 \le j \le N} | t_j - t_{j-1} |.$$

Riemann Sum. A Riemann sum, $\sigma(D, \zeta)$, for a bounded function $f: \mathbb{R} \to \mathbb{R}$ is any sum

$$\sigma(D, \zeta) = \sum_{j=1}^{N} f(\zeta_j)(t_j - t_{j-1}) \text{ where } \zeta_j \in [t_{j-1}, t_j].$$

Note that if the subintervals all have the same length so that $t_j = (a + jh)$ and $t_j - t_{j-1} =$ h where $h = (t_N - t_0)/N$, and if we take $\zeta_j = t_j$, then

$$\rho(D, \zeta) = \sum_{j=1}^{N} f(a + jh)h.$$

Integrability. A bounded function $f: \mathbb{R} \to \mathbb{R}$ is integrable if there exists $I \in f: \mathbb{R}$ such that

$$\lim_{|D| \to 0} \sigma(D, \zeta) = I,$$

where the limit of the Riemann sum must exist independent of the dissection (subject to the condition that $| D | \to 0$) and independent of the choice of ζ for a given dissection D[6]. Note that this is more restrictive than saying that the sum converges as $h \to 0$.

Definite Integral. For an integrable function f the Riemann definite integral of f over the interval $[t_a, t_b]$ is defined to be the limiting value of the Riemann sum, i.e.

$$\int_{t_a}^{t_b} f(t)dt = I$$

An integral should be thought of as the limiting value of a sum, NOT as the area under a curve.

First Fundamental Theorem of Calculus. This states that the derivative of the integral of f is f, i.e. if f is suitably 'nice' (e.g. f is continuous) then

$$\frac{d}{dx}\left(\int_{t_a}^{x} f(t)dt \right) = I(x).$$

Second Fundamental Theorem of Calculus. This essentially states that the integral of the derivative of f is f, e.g. if f is dierentiable then

$$\int_{t_a}^{t_b} \frac{df}{dt} dt = f(t_b) - f(t_a).$$

Substitution or Change of Variables. If $f(t)$ and $\frac{d\lambda}{ds}(s)$ are continuous functions of t and s respectively, then by setting $t = \lambda(s)$ and by using the chain rule

$$\int f(t)dt = \int f(\lambda(s))\frac{d\lambda}{ds}ds.$$

Finally the statement above stands repeating

An integral should be thought of as the limiting value of a sum, NOT as the area under a curve.

Indeed, even repeating again. Life will be simpler if you stop thinking, if you ever did, of an integral as the area under a curve.

Of course this is not to say that integrals are not a handy way of calculating the areas under curves.

Line Integrals

Let C be a smooth curve in $\mathbb{R}^3$ parameterised by arc length s:

$$x = r(s).$$

Line Integrals of Scalar Functions

We could define a line integral by dissecting C and then taking a limit of a Riemann sum. However, this is unnecessary.

Definition: Suppose

$$f(x): \mathbb{R}^3 \to \mathbb{R}.$$

Then the line integral of the scalar function f along the curve C is defined to be

$$\int_C f(x)ds = \int_{s_a}^{s_b} f(r(s))ds,$$

where the right hand side is a Riemann integral.

Remarks:

1. Suppose instead that the curve is parameterised by other than the arc length, e.g. by t where $x = \psi(t)$ for $t_a \le t \le t_b$. From the definition of arc length we have that

$$\frac{ds}{dt} = |\psi'(t)|,$$

and thus from the rules governing the change of variable in a Riemann integral

$$\int_C f(x)ds = \int_{s_a}^{s_b} f(r(s))ds = \int_{t_a}^{t_b} f(\psi(t))\,|\psi'(t)|\,dt.$$

Alternatively if we employ the more concise notation $x \circ x(t)$,

$$\int_C f(x)ds = \int_{t_a}^{t_b} f(x(t)) \left[\left(\frac{dx}{dt}\right)^2 + \left(\frac{dy}{dt}\right)^2 + \left(\frac{dz}{dt}\right)^2 \right]^{\frac{1}{2}} dt.$$

- Recall again that the definition of a line integral is best thought of a the limiting value of a Riemann sum. However, if you insist on thinking of integrals as areas, suppose that $x = \psi(t) = (\psi_1(t), \psi_2(t), 0)$ specifies the base position of a curvy wall, and $f(x)$ is the height of the wall at x; then is the area of the wall.

Properties: For functions f and g, and constants λ and μ:

(i) $\int_C fds = -\int_C fds,$

(ii) $\int_C (\lambda f + \mu g)ds = \lambda \int_C fds + \mu \int_C gds,$

(iii) $\int_C fds = \int_{C_1} fds + \int_{C_2} fds.$

Example: Suppose that $f(x, y)$ is a scalar function defined by

$$f = 2xy.$$

We evaluate the line integral $\int fds$ for two different paths.

1. Consider first the path

   $$C : x = (a\cos t, a\sin t), 0 \le t \le \pi/2,$$

   where $a > 0$. Then

   $$\frac{dx}{dt} = (-a\sin t, a\cos t),$$

   and hence from the definition of arc length

   $$\frac{dx}{dt} = \left|\frac{dx}{dt}\right| = a.$$

   $$\int_C fds = \int_0^{\frac{\pi}{2}} 2a^2 \cos t \sin t dt = a^3.$$

2. Next consider the path $\widehat{C} = \widehat{C}_1 + \widehat{C}_2$, where

   $$\widehat{C}_1 : x = (x, 0), a \ge x \ge 0,$$

   $$\widehat{C}_2 : x = (0, y), a \le y \le a.$$

Then $f = 0$ on both $\widehat{C}_1$ and $\widehat{C}_2$, and hence

$$\int_{\widehat{C}} f ds = 0.$$

Remark:

1. In general, if different paths C and $\widehat{C}$ join the same end points, then the line integrals along these different paths are not equal, i.e. in general

$$\int_C f ds \neq \int_{\widehat{C}} f ds.$$

For instance if $f = 1$, then

$$\int_C ds = \text{length of } C,$$

$$\int_{\widehat{C}} ds = \text{length of } \widehat{C},$$

which for most choices of two distinct paths C and $\widehat{C}$ are not equal.

Line Integrals of Vector Functions

For a vector function $F(x)$: $\mathbb{R}^3 \rightarrow \mathbb{R}^3$, there are at least two types of line integral that can be defined along a path C.

1. The first is a vector, and can be evaluated component by component using definition of a line integral of a scalar function:

$$I = \int_C F ds = \int_C \sum_{j=1}^{3} (F_j e_j) ds = \sum_{j=1}^{3} e_j \int_C F_j ds = \sum_{j=1}^{3} e_j \int_{s_a}^{s_b} F_j(r(s)) ds.$$

2. The second integral is a scalar and is defined for a parameterisation t by

$$I = \int_C F.dx = \int_{t_a}^{t_b} F(x(t)).\frac{dx}{dt} dt,$$

where

$$dx = \sum_{i=1}^{3} e_i dx_i = (dx, dy, dz).$$

Hence

$$\int_C F.dx = \int_C (F_1 dx + F_2 dy + F_3 dz) = \int_C F.ds,$$

where in the final expression we have used the alternative notation $ds = dx$; note that then $ds = | ds | = | dx |$.

Examples:

1. *Work.* Suppose that we identify F with the force acting on a particle transversing C. The work done in moving the particle an infinitesimal distance dx is known to be $F.dx$; this is the generalisation to 3D of the 1D formula

   work done = force × distance.

Hence the total work done in transversing the curve C is

$$W = \int_C F.dx.$$

We can parameterise the curve by the time at which the particle is at a point. Then from and the use of Newton's second law of motion (i.e. $F = m\ddot{x}$, assuming no gravity, etc.)

$$\begin{aligned} W &= \int_{t_a}^{t_b} F.\frac{dx}{dt}dt \\ &= \int_{t_a}^{t_b}\left(m\frac{d^2x}{dt^2}\right)\frac{dx}{dt}dt \\ &= \int_{t_a}^{t_b}\frac{d}{dt}\left(\frac{1}{2}m\left(\frac{dx}{dt}\right)^2\right)dt \\ &= \left[\frac{1}{2}m\left(\frac{dx}{dt}\right)^2\right]_{t_a}^{t_b} \end{aligned}$$

= gain in kinetic energy.

2. Suppose that F is a vector function defined by

   $F = (y, x, 0)$.

We evaluate the line integral $\int F.dx$ for the family of paths C parameterised by $x = t$, $y = t^a$ where $a > 0$ and $0 \leq t \leq 1$. Then

$$\int_C F.dx = \int_0^1\left(y\frac{dx}{dt} + x\frac{dy}{dt}\right)dt = \int_0^1 (t^a + tat^{a-1})dt = \int_0^1 (a+1)t^a dt = 1.$$

We note that the answer is independent of a, i.e. the same result is obtained for a number of different paths; this is not a co-incidence.

However, in general, if different paths join the same end points, then the line integrals of vector functions along these different paths are not equal.

Unlectured check. As a confirmation of the above result consider the $a \to \infty$ limiting path, i.e. the piecewise smooth curve $\widehat{C} = \widehat{C}_1 + \widehat{C}_2$ defined by

$$\widehat{C}_1 : x = t_1, y = 0 \text{ for } 0 \le t_1 \le 1,$$
$$\widehat{C}_2 : x = 1, y = t_2 \text{ for } 0 \le t_2 \le 1.$$

On $\widehat{C}_1$

$$F.dx = F.\frac{dx}{dt_1}dt_1 = (0.1 + t_1.0)dt_1 = 0,$$

while on $\widehat{C}_2$

$$F.dx = F.\frac{dx}{dt_1}dt_2 = (t_2.0 + 1.1)dt_2 = dt_2.$$

Hence

$$\int_{\widehat{C}} F.dx = \int_0^1 dt_2 = 1.$$

3. For a scalar differentiable function f consider

$$\int_C \nabla f.dx = \int_{t_a}^{t_b} f(x(t)).\frac{dx}{dt}dt \quad \text{from definition}$$
$$= \int_{t_a}^{t_b} \frac{\partial f}{\partial x_j}\frac{dx_j}{dt}dt$$
$$= \int_{t_a}^{t_b} \frac{d}{dt}(f(x(t)))dt \quad \text{from the chain rule}$$
$$= f(x_b) - f(x_a),$$

where the final line follows from the second fundamental theorem of calculus. Hence, again, the value of the line integral is independent of the path.

Result might be viewed as a [line] integral result for gradients, which is equivalent to the second fundamental theorem of calculus for derivatives (i.e. the line integral 'undoes' the gradient).

Exercise: Find the connection between the preceding two examples.

Differentials

Definition: Suppose the scalar function $f : \mathbb{R}^m \to \mathbb{R}$ is differentiable at a, then the differential df of $f(x_1, ..., x_m)$ at a is defined to be

$$df(a) = \sum_{j=1}^{m} \frac{\partial f}{\partial x_j}(a)dx_j.$$

Alternatively, using the summation convention

$$_{\mathrm{S}}df = \frac{\partial f}{\partial x_j} dx_j.$$

Hence for $m = 1$, i.e. $f \equiv f(x)$

$$df = \frac{\partial f}{\partial x} dx;$$

while for $m = 2$, i.e. $f \equiv f(x, y)$

$$df = \frac{\partial f}{\partial x} dx + \frac{\partial f}{\partial y} dy; \text{ etc.}$$

Remarks:

1. Definition is consistent for $f = x_i'$,

$$df = \frac{\partial x_i}{\partial x_j} dx_j = \delta_{ij} dx_j = dx_i.$$

2. Definition is sometimes referred to as the chain rule for differentials.

Unlectured example: Rederivation of the chain rule. Suppose that $f(x)$ and $x(u)$ are differentiable. If $f(x) = F(u)$, $u = (u, v)$ and $x = (x(u, v), y(u, v))$, then from we have for the change in f due to changes in x and y

$$df = f_x\, dx + f_y\, dy.$$

After a notation change we also have from for changes in x and y due to changes in u and v,

$$dx = x_u\, du + x_v\, dv,\ dy = y_u\, du + y_v\, dv.$$

Hence from substituting (3.16b) into (3.16a)

$$\begin{aligned} df &= f_x(x_u\, du + x_v\, dv) + f_y(y_u\, du + y_v\, dv) \\ &= (f_x x_u + f_y y_u)du + (f_x x_v + f_y y_v)dv. \end{aligned}$$

But it also follows from that the change in $f(x(u, v), y(u, v)) = F(u, v)$ due to changes in u and v is given by

$$df = dF = F_u\, du + F_v\, dv.$$

Hence from comparing the above two results and using the fact that du and dv can be varied independently,

$$F_u = f_x x_u + f_y y_u \text{ and } F_v = f_x x_v + f_y yv.$$

Similarly if $f : \mathbb{R}^m \to \mathbb{R}$ and $x : \mathbb{R}^\ell \to \mathbb{R}^m$

$$df = \frac{\partial f}{\partial x_j} dx_j,\ dx_j = \frac{\partial x_j}{\partial u_k} du_k \text{ and } dF = \frac{\partial F}{\partial u_k} du_k.$$

Hence

$$\frac{\partial F}{\partial u_k} du_k = \frac{\partial f}{\partial x_j} \frac{\partial x_j}{\partial u_k} du_k,$$

or, since the du_k can be varied independently,

$$\frac{\partial F}{\partial u_k} = \frac{\partial f}{\partial x_j} \frac{\partial x_j}{\partial u_k}.$$

Interpretations

Local Linear Approximation. Suppose that f is differentiable, and also suppose that we interpret dx as the increment in x from a, i.e. suppose we let

$$dx = (x - a).$$

Then it follows from the definition of a differentiable function, that df can be interpreted as the linear local approximation to the increment in f corresponding to an increment dx in x; i.e. for $m = 1$

$$f(a + dx) \approx f(a) + df(a) = f(a) + \frac{df}{dx}(a)\, dx.$$

Relationship to Gradient. For a given fixed orthonormal basis, e_j ($j = 1, \ldots, m$), the gradient of a scalar function f is defined as

$$\nabla f = \sum_{j=1}^{m} e_j \frac{\partial f}{\partial x_j}.$$

Hence, using $\mathbb{R}^m$

$$dx = \sum_{i=1}^{m} e_i dx_i,$$

and the property of an orthonormal basis that $e_i \,.\, e_j = \delta_{ij}$, it follows that

$$\nabla f . dx = \sum_{i,j=1}^{m} e_j \frac{\partial f}{\partial x_j} . e_i dx_i = \sum_{j}^{m} \frac{\partial f}{\partial x_j} dx_j = df.$$

Tangent Plane. If we again interpret dx as the increment in x from a, i.e. $dx = (x - a)$, then the equation

$$df(a) = 0$$

can be rewritten using as

$$df(a) = \nabla f(a) \,.\, dx = \nabla f(a) \,.\, (x - a) = 0.$$

This is the equation of the plane with normal $\nabla f(a)$ passing through $x = a$. Moreover from equation for the normal to the surface $F(x) = 0$, we conclude that is the equation of the tangent plane at a to the surface

$$F(x) = f(x) - f(a) = 0.$$

Unlectured Example. For $f: \mathbb{R}^2 \to \mathbb{R}$ and $g : \mathbb{R} \to \mathbb{R}$ suppose that

$$f(x, y) = y - g(x).$$

Then the tangent 'plane'/line at $x = a = (a, b)$ to the 'surface'/curve $f(x) = f(a)$ is given by

$$0 = df(a) = -\frac{dg}{dx}(a)dx + dy = -\frac{dg}{dx}(a)(x-a) + (y-b).$$

In other words the tangent to the curve

$$y = b + g(x) - g(a)$$

is
$$y = b + (x - a)\frac{dg}{dx}(a).$$

Properties: For $f, g : \mathbb{R}^m \to \mathbb{R}$ and , $\lambda, \mu \in \mathbb{R}$, it follows in a straightforward manner from the definition of a differential that

$$d(\lambda f + \mu g) = df + \mu dg,$$

$$d(fg) = f\,dg + g\,df.$$

Unlectured Example. From with $f = g = r$ it follows that $dr^2 = 2rdr$. Hence if

$$r^2 = x^2 + y^2 + z^2,$$

then, using in addition (3.20a), it follows that

$$2r\,dr = 2x\,dx + 2y\,dy + 2z\,dz,$$

i.e.
$$dr = \frac{x}{r}dx + \frac{y}{r}dy + \frac{z}{r}dz.$$

Exact Differentials

Definitions: Differential Form. For vector functions $F : \mathbb{R}^m \to \mathbb{R}^m$, the forms

$$\omega(x) = F(x).dx = \sum_{j=1}^{m} F_j(x)dx_j,$$

that appear in line integrals such as (3.7a) are called differential [one-] forms. For instance if $m = 3$:

$$\omega(x) = F_1(x, y, z)\,dx + F_2(x, y, z)\,dy + F_3(x, y, z)\,dz.$$

Exact Differential (2D): Suppose $P(x, y)$ and $Q(x, y)$ are defined in some domain $\mathbb{R}$ of $\mathbb{R}^2$, and that they have continuous first order partial derivatives. Then if for all $(x, y) \in \mathbb{R}$ there exists a single valued function $f : \mathbb{R}^2 \to \mathbb{R}$ such that

$$\omega(x, y) = P(x, y)\, dx + Q(x, y)\, dy = df(x, y),$$

then ω is an exact differential [form] on $\mathbb{R}$.

Simply Connected: A simply connected domain in $\mathbb{R}^2$ is one where any closed curve in the domain can be shrunk to a point in the domain without leaving the domain.

In $\mathbb{R}^3$ the definition is more complicated. First we define path connected: a domain $\mathbb{R}$ is said to be path-connected if for any two points a and b in $\mathbb{R}$ there exists a continuous function from the unit interval [0, 1] to R defining a curve with $\psi(0) = a$ and $\psi(1) = b$. A domain is simply connected if it is path-connected and every path between two points can be continuously transformed into every other.

A Necessary Condition for ω to be an Exact Differential (2D). If ω is an exact differential, i.e. if there exists a single valued function f such that $\omega = df$,

$$Pdx + Qdy = \frac{\partial f}{\partial x}dx + \frac{\partial f}{\partial y}dy.$$

Since this condition must hold for arbitrary dx, it follows from considering, say, $dx = (dx, 0)$ and $dx = (0, dy)$, that

$$P = \frac{\partial f}{\partial x} \text{ and } Q = \frac{\partial f}{\partial y}$$

respectively. Moreover,

$$\frac{\partial P}{\partial y} = \frac{\partial^2 f}{\partial y \partial x} = \frac{\partial^2 f}{\partial x \partial y} = \frac{\partial Q}{\partial x},$$

since P and Q are assumed to have continuous first order partial derivatives. We conclude that a necessary condition for ω to be exact is that

$$\frac{\partial P}{\partial y} = \frac{\partial Q}{\partial x}.$$

A Sufficient Condition for ω to be an Exact Differential (2D).

If the domain $\mathbb{R}$ is simply connected then is a sufficient condition for ω to be an exact differential.

Unlectured Outline Proof: This outline proof is for those cannot wait for an idea of why this result is true. If you are prepared to wait then, once Stokes' theorem has been discussed, we will obtain a more general result of which this is a special case.

In what follows we will assume for simplicity that $\mathbb{R} = \mathbb{R}^2$. We can then be certain that all the line integrals of P and Q below remain within the domains of P and Q.

If ω is to be an exact differential then we require that

$$f_x(x, y) = P(x, y).$$

This suggests that we consider the function

$$f(x,y) = \int_{x_0}^{x} P(x,y)dx + Y(y),$$

where x_0 is a constant and Y is to be determined. By differentiating with respect to y and using condition it follows that for this function

$$f_y(x,y) = \int_{x_0}^{x} P_y(x,y)dx + Y'(y)$$

$$= \int_{x_0}^{x} Q_x(x,y)dx + Y'(y) = Q_x(x, y) - Q_x(x_0, y) + Y'(y).$$

Now if ω is to be an exact differential then we also require that $f_y = Q_x$. This is achieved if

$$Y = \int_{y_0}^{y} Q_x(x_0, y)dy,$$

where y_0 is a constant. Hence, for the function f defined by

$$f(x,y) = \int_{x_0}^{x} P(x,y)d + \int_{y_0}^{y} Q_x(x_0,y)dy$$

$$= \int_{C_1+C_2} F.dx,$$

where $F = (P, Q)$, we conclude that

$$df = f_x dx + f_y dy = Pdx + Qdy = \omega.$$

Thus, by the construction of an appropriate function, we have shown that if condition is satisfied, then ω is an exact differential.

Exact Differentials (3D). In an analogous way to above, the differential form

$$\omega = F_1 dx + F_2 dy + F_3 dz$$

can be shown to be exact in a simply connected domain $\mathbb{R}$ of $\mathbb{R}^3$ if and only if

$$\frac{\partial F_1}{\partial y} = \frac{\partial F_2}{\partial x}, \frac{\partial F_2}{\partial z} = \frac{\partial F_3}{\partial y}, \frac{\partial F_3}{\partial x} = \frac{\partial F_1}{\partial z}.$$

Unlectured Proof of Necessity: If ω is exact then for some function f

$$F_1dx + F_2dy + F_3dz = f_xdx + f_ydy + f_zdz.$$

Hence from independently varying *dx*, *dy* and *dz*,

$$F_1 = f_x,\ F_2 = f_y \text{ and } F_3 = f_z.$$

Condition then follows from the requirements that

$$f_{xy} = f_{yx}, f_{yz} = \text{fzy and } f_{zx} = f_{xz}.$$

Solution of Differential Equations

The differential equation

$$\frac{dy}{dz} = -\frac{P(x,y)}{Q(x,y)}$$

can alternatively be written as

$$\omega = P(x, y)dx + Q(x, y)dy = 0.$$

The case when ω is an exact differential. If ω is an exact differential and equal to *df*, then since *df* = 0 the solution to the differential equation is

$$f(x, y) = \kappa,$$

where κ is a constant.

Unlectured Example. Suppose that

$$\frac{dy}{dx} = -\frac{\sin y}{2y + x\cos y},$$

Hence solutions are given by

$$y^2 + x \sin y = \kappa \text{ i.e. } x = \frac{\kappa - y^2}{\sin y}.$$

The case when ω is not an exact differential. If the [initial] rewrite of does not yield an exact differential, then the aim is to find an integrating factor, μ(*x*, *y*), such that

$$\mu\omega = \mu(x, y)P(x, y)\, dx + \mu(x, y)Q(x, y)\, dy$$

is exact. It follows from the necessary and sufficient condition that we then require

$$\frac{\partial}{\partial y}(\mu P) = \frac{\partial}{\partial x}(\mu Q),$$

i.e. that $$\mu\left(\frac{\partial P}{\partial y} - \frac{\partial Q}{\partial x}\right) + P\frac{\partial \mu}{\partial y} - Q\frac{\partial \mu}{\partial x} = 0.$$

This is a partial differential equation for μ, which it is not necessarily straightforward to solve. However if either

$$\frac{1}{Q}\left(\frac{\partial P}{\partial y}-\frac{\partial Q}{\partial x}\right)=X(x) \text{ or } \frac{1}{P}\left(\frac{\partial P}{\partial y}-\frac{\partial Q}{\partial x}\right)=Y(y),$$

where X and Y are functions only of x and y respectively, then an integrating factor exists that is only a function of x or y respectively, i.e. either

$$\mu \equiv \mu(x) \text{ and } \frac{1}{\mu}\frac{d\mu}{dx}=X,$$

or,

$$\mu \equiv \mu(y) \text{ and } \frac{1}{\mu}\frac{d\mu}{dy}=-Y,$$

respectively.

Unlectured Example. Note that while this example illustrates integrating factors, the underlying differential equation is actually best solved using separation of variables.

Suppose that

$$\omega = y\,dx - x\,dy, \text{ i.e. } P = y,\ Q = -x.$$

Then

$$\frac{1}{Q}\left(\frac{\partial P}{\partial y}-\frac{\partial Q}{\partial x}\right)=-\frac{2}{x}$$

and

$$\frac{1}{P}\left(\frac{\partial P}{\partial y}-\frac{\partial Q}{\partial x}\right)=\frac{2}{y}.$$

Hence in this case we can choose either $\mu \equiv \mu(x)$ or $\mu \equiv \mu(y)$; specifically either

$$\frac{1}{\mu}\frac{d\mu}{dx}=-\frac{2}{3}\Rightarrow \mu=\frac{1}{x^2}$$

or

$$\frac{1}{\mu}\frac{d\mu}{dy}=-\frac{2}{y}\Rightarrow \mu=\frac{1}{y^2}.$$

As a check note that we then have either

$$\frac{ydx}{x^2}-\frac{dy}{x}=-d\left(\frac{y}{x}\right)$$

or

$$\frac{dx}{y}-\frac{xdy}{y^2}=d\left(\frac{x}{y}\right).$$

Line Integrals of Exact Differentials

Suppose that

$$\omega = P\,dx + Q dy$$

is an exact differential. Then from definition there exists a single valued function $f(x)$ such that

$$df = P\,dx + Q dy.$$

Hence using (i) the relationship between the differential of a scalar function and the gradient of the function, and (ii) example, it follows that

$$\int_C (Pdx + Qdy) = \int_C df$$

$$= \int_{x_a}^{x_b} \nabla f . dx$$

$$= f(x_b) - f(x_a).$$

Thus in $\mathbb{R}^2$ the line integral of an exact differential depends only on the end points and is independent of the path. The same result holds in $\mathbb{R}^m$.

Example: Reconsider example, for which

$$F.\,dx = y\,dx + x\,dy = d(xy).$$

Hence $F.\,dx$ is an exact differential (with $f = xy$), and thus from

$$\int_{(0,0)\to(1,1)} (ydx + xdy) = [xy]_{(0,0)}^{(1,1)} = 1,$$

independent of path chosen for the integral.

Remarks:

1. Similarly in 3D, if $F.\,dx = df$ then $\int_C F.dx$ depends only on the end points.
2. The integral of an exact differential around a closed curve in a simply connected region must vanish because

$$\oint_C df = \int_{C_1} df + \int_{C_2} df$$

$$= \int_{C_1} df - \int_{-C_2} df$$

$$= (f(x_b) - f(x_a)) - (f(x_b) - f(x_a))$$

$$= 0.$$

Conservative Fields

Definitions. Conservative Field. If the line integral

$$\int_{x_a}^{x_b} F.dx$$

is independent of the path taken from x_a to x_b in a simply connected domain, then the vector function $F(x)$ if said to be a conservative field.

Scalar Potential. The scalar potential f of a conservative field $F(x)$ is defined to be

$$f(x) = \int_{x_0}^{x} F.dx,$$

where x_0 is some fixed point. The scalar potential is guaranteed to be a single valued function because, from the definition of a conservative field, the value of the integral is independent of the path.

Property. For the scalar potential

$$\nabla f = F.$$

Verification. The definition of a partial derivative

$$\frac{\partial f}{\partial x}(a) = \lim_{h\to 0} \frac{f(a+he_1)-f(a)}{h}$$

$$= \lim_{h\to 0} \frac{1}{h}\int_a^{a+he_1} F(x).dx.$$

Since F is a conservative field, we may choose the most convenient path for evaluating the integral. In this case we take a straight line defined parametrically by $x(t) = a + hte_1$ for $0 \le t \le 1$. Then, since $\frac{dx}{dt} = he_1$, it follows that

$$\frac{\partial f}{\partial x}(a) = \lim_{h\to 0} \frac{1}{h}\int_a^{a+he_1} F(x).\frac{dx}{dt}dt$$

$$= \lim_{h\to 0} \frac{1}{h}\int_0^1 F_1(a_1+ht, a_2, a_3)hdt$$

$$= \lim_{h\to 0} \int_0^1 (F_1(a_1,a_2,a_3)+o(1))dt \text{ from using Taylor's theorem}$$

$$= F_1(a).$$

Similarly for the other components we deduce that

$$\frac{\partial f}{\partial y} = F_2 \text{ and } \frac{\partial f}{\partial z} = F_3.$$

Putting these results together it follows that

$$\nabla f = F.$$

Property. The relationship between the differential and the gradient of a function,

$$F.\, dx = \nabla f \,.\, dx = df.$$

Thus it follows that the differential form $F\,.\,dx$ of a conservative field F is exact.

Remarks on Work. From the work done on a particle transversing a curve C is

$$W = \int_C F.dx.$$

1. When a force F is a conservative field we refer to it as a conservative force.
2. Friction is not a conservative force because the work done in transversing a closed loop is non-zero (because friction opposes motion), and hence $\int_{x_a}^{x_b} F.dx$ cannot be independent of the path taken.
3. If $F = (0, 0, -g)$ is the gravitational force at the surface of the earth, it follows from the necessary and sufficient conditions for a differential form to be exact, i.e. that $F\,.\,dx$ is exact. Hence from the 3D extension, i.e. the result that the line integral of an exact differential is independent of path, it follows that that the work done in moving from x_a to x_b against gravity is independent of path. The gravitational force at the earth's surface is thus a conservative force.

Non Simply Connected Domains

Suppose that the vector function F is defined by

$$F = \left(-\frac{y}{x^2+y^2}, \frac{x}{x^2+y^2}, 0\right) \text{ for } x \neq 0$$

Then $F.\, dx$ is exact in any simply connected domain excluding the origin since

$$\frac{\partial}{\partial y}\left(-\frac{y}{x^2+y^2}\right) = \frac{y^2-x^2}{(x^2+y^2)^2} = \frac{\partial}{\partial x}\left(\frac{x}{x^2+y^2}\right).$$

Further

$$f = \tan^{-1}\left(\frac{y}{x}\right) \text{ satisfies } df = \nabla f.\, dx = F.\, dx.$$

The question then arises: what if we drop the simply connectedness?

For instance, suppose we consider the domain $R = \mathbb{R}^2 / (0, 0)$ (i.e. the real plane minus the origin ... this is not a simply connected domain since a curve encircling the origin cannot be shrunk to a point in the domain), and the line integral round the path $C : x = (a \cos \phi, a \sin \phi)$ where $0 \leq \phi \leq 2\pi$. Then

$$\oint_C F.dx = \int_0^{2\pi} F.\frac{dx}{d\phi} d\phi$$

$$= \int_0^{2\pi} \left(\left(\frac{-\sin\phi}{a} \right).(-a\sin\phi) + \frac{\cos\phi}{a}, a\cos\phi \right) d\phi$$

$$= 2\pi.$$

We conclude that the simply connectedness is crucial for the integral to depend only on the end points.

Optional Exercise. Show that for any curve C that encircles the origin exactly once

$$\oint_C F.dx = 2\pi.$$

EXERCISE

1. Let $f(t)$ be a differentiable curve such that $f(t) \neq 0$ for all t. Show that

$$\frac{d}{dt}\left(\frac{f(t)}{\| f(t) \|} \right) = \frac{f(t) \times (f'(t) \times f(t))}{\| f(t) \|^3}$$

2. Let $f(t)$ be a smooth curve such that $f'(t) \neq 0$ for all t. Then we can define the *unit tangent vector T* by

$$T(t) = \frac{f'(t)}{\| f'(t) \|}$$

Show that

$$T'(t) = \frac{f'(t) \times (f''(t) \times f'(t))}{\| f'(t) \|^3}$$

3. Let be the three a $= 2i - j + k$, b $= i + 2j - k$ and c $= i + j - 2k$ vectors . Find the vector in the plane of b and c whose projection on a is of magnitude $\sqrt{\frac{2}{3}}$.

4. If the vectors $ai + j + k$, $i + bj + k$ and $i + j + ck$ $(a \neq b \neq c \neq 1)$ are coplanar vectors> find the value of $\frac{1}{1-a} + \frac{1}{1-b} + \frac{1}{1-c}$.

5. If a $= i + j - k$, b $= i - j + k$ and c are unit vectors perpendicular to the vector a and coplanar with a nd b. Find the unit vector d perpendicular to both a and c.
6. If $d = \lambda(a \times b) + \mu(b \times c) + v(c \times a)$ and $[a\ b\ c] = \frac{1}{8}$. Find the value of $\lambda + \mu + v$.
7. Show that $f(x, y) = \frac{1}{2\pi\sigma^2} e^{-(x^2+y^2)/2\sigma^2}$, for $\sigma > 0$, is constant on the circle of radius $r > 0$ centered at the origin. This function is called a *Gaussian blur*, and is used as a filter in image processing software to produce a "blurred" effect.
8. Suppose that $f(x, y) \leq f(y, x)$ for all (x, y) in $\mathbb{R}^2$. Show that $f(x, y) = f(y, x)$ for all (x, y) in $\mathbb{R}^2$.
9. Use the substitution $r = \sqrt{x^2 + y^2}$ to show that
10. Evaluate the line integral $\int_C \left[\left(x^2 + xy\right)dx + \left(x^2 + y^2\right)dy\right]$ where C is the square formed by the lines y = ± 1 and x = ± 1.
11. If F = $(5xy - 6x^2)$I + $(2y - 4x)$J, evaluate $\int_C F.dR$ along the curve C in the xy-plane, $y = x^3$ from the point (1, 1) to (2, 8).
12. A vector field is given by F = (sin y)I + x(1 + cos y)J. Evaluate the line integral over a circular path given by $x^2 + y^2 = a^2$, z = 0.
13. Find T, N, B and κ at each point of the helix $f(t) = (\cos t, \sin t, t)$.
14. Show that the arc length L of a curve whose spherical coordinates are $\rho = \rho(t)$, $\theta = \phi(t)$ and $\phi = \phi(t)$ for t in an interval $[a, b]$ is

$$L = \int_a^b \sqrt{\rho'(t)^2 + (\rho(t)^2 \sin^2 \phi(t))\theta'(t)^2 + \rho(t)^2 \phi'(t)^2}\, dt.$$

Chapter 8

Parallelepidial Product

In geometry, a parallelepiped is a three-dimensional figure formed by six parallelograms. Three equivalent definitions of parallelepiped are

1. A polyhedron with six faces (hexahedron), each of which is a parallelogram,
2. A hexahedron with three pairs of parallel faces.
3. A prism of which the base is a parallelogram,

The cuboid (six rectangular faces), cube (six square faces), and the rhombohedron (six rhombus faces) are all specific cases of parallelepiped. Parallelepipeds are a subclass of the prismatoids.

Any of the three pairs of parallel faces can be viewed as the base planes of the prism. A parallelepiped has three sets of four parallel edges; the edges within each set are of equal length. Parallelepipeds result from linear transformations of a cube (for the non-degenerate cases: the bijective linear transformations). Since each face has point symmetry, a parallelepiped is a zonohedron. Also the whole parallelepiped has point symmetry Ci. Each face is, seen from the outside, the mirror image of the opposite face. The faces are in general chiral, but the parallelepiped is not. A space-filling tessellation is possible with congruent copies of any parallelepiped.

The parallelepidial product, except for the sign,

$$[\vec{a}, \vec{b}, \vec{c}] = \vec{a} \cdot (\vec{b} \times \vec{c}) = a_1(b_2c_3 - b_3c_2) + a_2(b_3c_1 - b_1c_3) + a_3(b_1c_2 - b_2c_1)$$

yields the volume of the parallelepiped spanned by the three vectors $\vec{a}, \vec{b}, \vec{c}$.

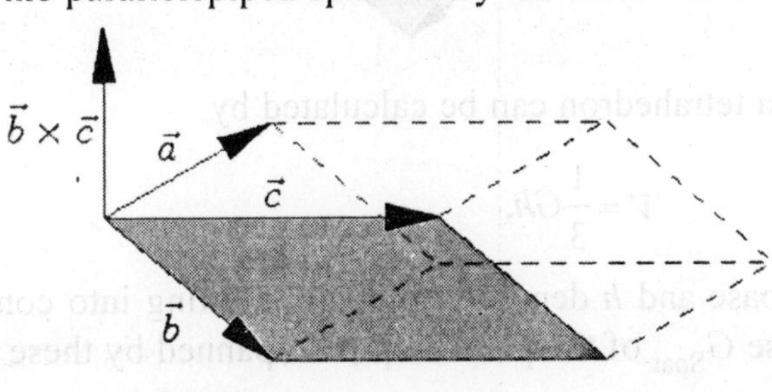

Using the -tensor the parallelepidial product can be expressed as

$$[\vec{a}, \vec{b}, \vec{c}] = \sum_{i,j,k=1}^{3} \varepsilon_{i,j,k,} a_i b_j c_k.$$

With

$$\vec{d} = \frac{\vec{b} \times \vec{c}}{|\vec{b} \times \vec{c}|},$$

the height of the parallelepiped is

$$h = |\vec{a}| \left|\cos(\sphericalangle \vec{b})\right| = |\vec{a} \cdot \vec{d}|.$$

Since the area of the hatched parallelogram is $f = |\vec{b} \times \vec{c}|$, we obtain for the volume

$$hf = \left| \vec{a} \cdot \left(\frac{\vec{b} \times \vec{c}}{|\vec{b} \times \vec{c}|} \right) \right| |\vec{b} \times \vec{c}| = |[\vec{a}, \vec{b}, \vec{c}]|.$$

VOLUME OF A TETRAHEDRON

The volume V of a tetrahedron spanned by the vectors $\vec{a}$, $\vec{b}$, und $\vec{c}$ can be calculated by

$$V = \frac{1}{6} |[\vec{a}, \vec{b}, \vec{c}]|.$$

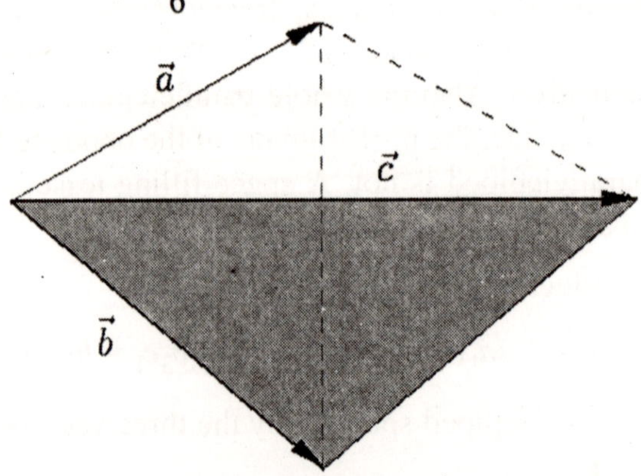

The volume V of a tetrahedron can be calculated by

$$V = \frac{1}{3} Gh,$$

where G denotes the base and h denotes the height. Taking into consideration that G is half as large as the base G_{Spat} of the parallelepiped spanned by these vectors, we obtain

$$V = \frac{1}{6} G_{\text{Spat}} h = \frac{1}{6} V_{\text{Spat}} = \frac{1}{6} |[\vec{a}, \vec{b}, \vec{c}]|$$

by means of the parallelepiped's volume V_{Spat}.

The volume V of the tetrahedron spanned by the vectors

$$\vec{a} = \begin{pmatrix} 0 \\ 1 \\ 1 \end{pmatrix}, \quad \vec{b} = \begin{pmatrix} 1 \\ 1 \\ 0 \end{pmatrix}, \quad \vec{c} \pi \begin{pmatrix} 0 \\ 2 \\ 0 \end{pmatrix}$$

amounts to

$$\begin{aligned} V &= \frac{1}{6} |[\vec{a}, \vec{b}, \vec{c}]| \\ &= \frac{1}{6} \left| \begin{pmatrix} 0 \\ 1 \\ 0 \end{pmatrix} \cdot \left(\begin{pmatrix} 1 \\ 1 \\ 0 \end{pmatrix} \times \begin{pmatrix} 0 \\ 2 \\ 0 \end{pmatrix} \right) \right| \\ &= \frac{1}{6} \left| \begin{pmatrix} 0 \\ 1 \\ 1 \end{pmatrix} \cdot \begin{pmatrix} 0 \\ 0 \\ 2 \end{pmatrix} \right| \\ &= \frac{2}{6} = \frac{1}{3}. \end{aligned}$$

PROPERTIES OF THE PARALLELEPIDIAL PRODUCT

The parallelepidial product is linear in each argument. In addition, it has the following properties:

1. Cyclic Permutation:

$$[\vec{a}, \vec{b}\ \vec{c}] = [\vec{b}, \vec{c}, \vec{a}] = [\vec{c}, \vec{a}, \vec{b}]$$

2. Linear Dependence:

$$[\vec{a}, \vec{b}\ \vec{c}] = 0 \Leftrightarrow \vec{0} = \alpha\vec{a} + \beta\vec{b} + \gamma\vec{c},$$

with at least one scalar α, β, γ beeing unequal to 0.

3. Orientation:

$$[\vec{a}, \vec{b}\ \vec{c}] > 0$$

for every system of three vectors oriented by the Right-Hand-Rule.

PARALLELEPIDIAL PRODUCT-CALCULATION OF COORDINATES

f the vectors $\vec{u}, \vec{v}$ and $\vec{w}$ span a proper parallelepiped, then every vector $\vec{x}$ can be represented by a linear combination

$$\vec{x} = \alpha\vec{u} + \beta\vec{v} + \gamma\vec{\omega}$$

with the coefficients

$$\alpha = \frac{[\vec{x}, \vec{v}, \vec{\omega}]}{[\vec{u}, \vec{v}, \vec{\omega}]}, \qquad \beta = \frac{[\vec{x}, \vec{\omega}, \vec{u}]}{[\vec{v}, \vec{w}, \vec{u}]}, \qquad \gamma\frac{[\vec{x}, \vec{u}, \vec{v}]}{[\vec{w}, \vec{u}, \vec{v}]}.$$

The vectors $\vec{u}, \vec{v}$ and $\vec{w}$ form a proper parallelepiped, that is, $[\vec{u}, \vec{v}, \vec{\omega}] \neq 0$. Thus, computing the scalar product of

$$\vec{x} = \alpha\vec{u} + \beta\vec{v} + \gamma\vec{\omega}$$

and $\vec{v} \times \vec{\omega}$, we obtain

$$\vec{x} \cdot (\vec{v} \times \vec{\omega}) = \alpha\vec{u} \cdot (\vec{v} \times \vec{\omega})$$

and

$$\alpha = \frac{[\vec{x}, \vec{v}, \vec{\omega}]}{[\vec{u}, \vec{v}, \vec{\omega}]},$$

resp. Similarly we obtain the respective results for β and γ.

If you want to represent the vector

$$\vec{x} = \begin{pmatrix} -2 \\ 8 \\ 2 \end{pmatrix}$$

as a linear combination of the vectors

$$\vec{u} = \begin{pmatrix} 0 \\ 1 \\ 1 \end{pmatrix}, \qquad \vec{v} = \pi\begin{pmatrix} 1 \\ 1 \\ 0 \end{pmatrix}, \qquad \vec{w} = \begin{pmatrix} 0 \\ 2 \\ 0 \end{pmatrix},$$

then in a first step you compute

$$[\vec{u}, \vec{v}, \vec{\omega}] = [\vec{v}, \vec{\omega}, \vec{u},] = [\vec{\omega}, \vec{u}, \vec{v}] = 2$$

$$[\vec{u}, \vec{v}, \vec{\omega}] = 4, \; [\vec{v}, \vec{\omega}, \vec{u},] = -4 \; [\vec{\omega}, \vec{u}, \vec{v}] = 8$$

Hence

$$\vec{x} = \frac{4}{2}\vec{u} - \frac{4}{2}\vec{v} + \frac{8}{2}\vec{w} = 2\vec{u} - 2\vec{v} + 4\vec{w}$$

LINES

POINT-SLOPE FORM

The points X on a line through P in direction $\vec{u}$ can be described in parametric form by

$$\overrightarrow{PX} = t\vec{u} \quad t \in \mathbb{R}$$

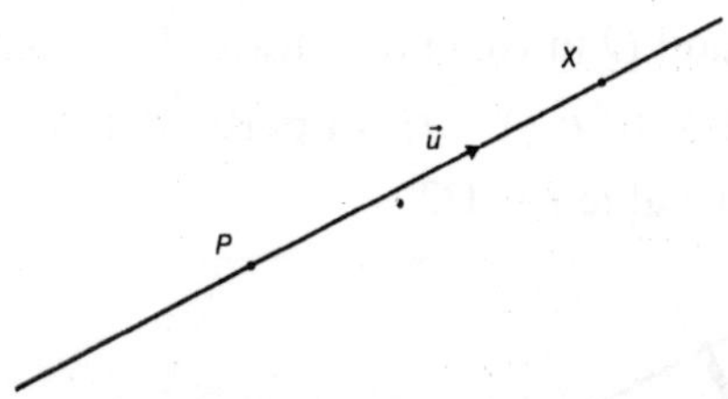

Correspondingly, for the coordinates of the vector $\vec{x}$, we have

$$x_i = p_i + tu_i, \quad i = 1, 2, 3.$$

The figure illustrates the parametric form of a line through point $P = (1, 3)$ in direction $\vec{u} = \begin{pmatrix} 2 \\ -1 \end{pmatrix}$. The point X corresponds to the parameter value $t = 3/2$.

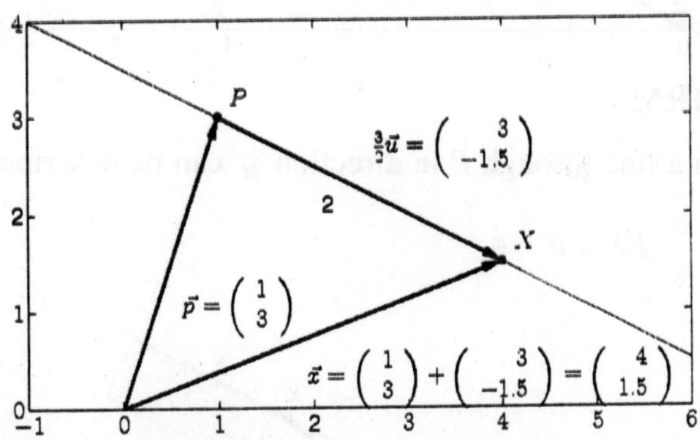

TWO-POINT FORM

The points X on a line passing through two points $P \neq Q$ can be described in the following form:

$$\overrightarrow{PX} = t\overrightarrow{PQ}, \quad t \in \mathbb{R},$$

where parameter values $t \in [0, 1]$ correspond to the points of the line segment $\overline{PQ}$.

Correspondingly, for the coordinates of the vector $\vec{x}$ we have

$$x_i = p_i + t(q_i - p_i), \quad i = 1, 2, 3.$$

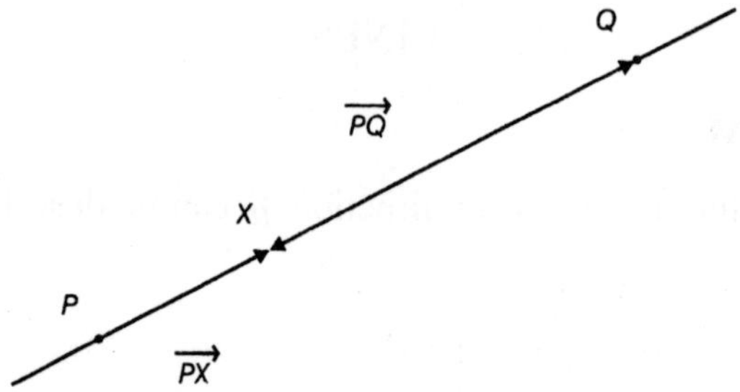

Given a line through P and Q in two point form. For $t \in [0, 1]$ the point X partitions the line segment $\overrightarrow{PQ}$ at a ratio of t: $(1 - t)$. In particular, the midpoint between P and Q corresponds to the parameter value $t = 1/2$.

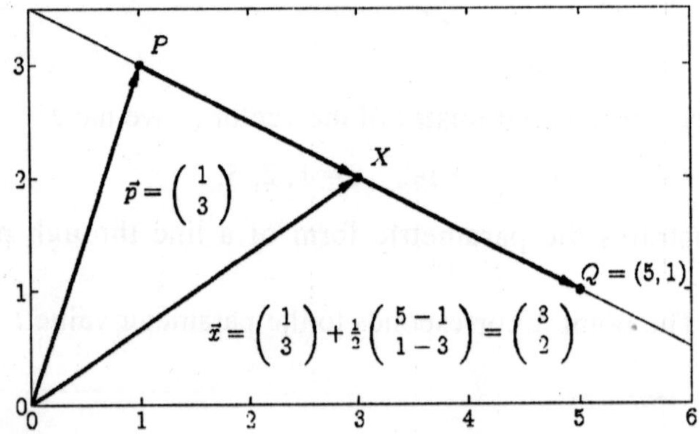

MOMENTUM FORM

The points X on a line through P in direction $\vec{u}$ can be described by

$$\overrightarrow{PX} \times \vec{u} = \vec{0}.$$

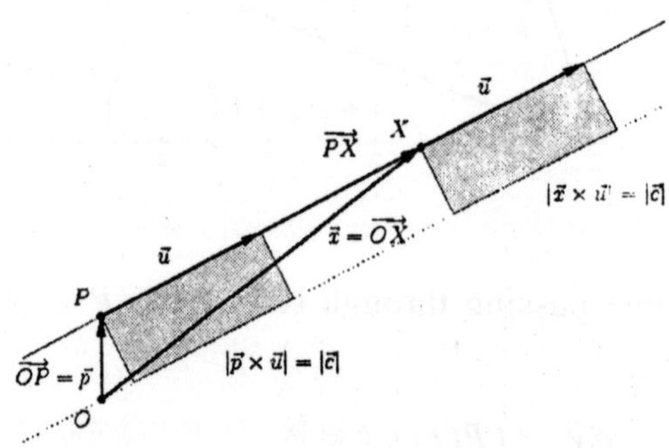

Correspondingly, we have

$$\vec{x} \times \vec{u} = \vec{c}, \quad \vec{c} = \vec{p} \times \vec{u}.$$

The momentum form of a line can be used to verify whether the point X lies on given line. For the depicted figure we obtain for the point X_1

$$\overrightarrow{PX_1} \times \vec{u} = \left(\begin{pmatrix} 5 \\ 1 \\ 0 \end{pmatrix} - \begin{pmatrix} 1 \\ 3 \\ 0 \end{pmatrix}\right) \times \begin{pmatrix} 2 \\ -1 \\ 0 \end{pmatrix}$$

$$= \begin{pmatrix} 4 \\ -2 \\ 0 \end{pmatrix} \times \begin{pmatrix} 2 \\ -1 \\ 0 \end{pmatrix}$$

$$= \begin{pmatrix} 0 \\ 0 \\ 4\cdot(-1) - 2\cdot(-2) \end{pmatrix} = \vec{0}.$$

Whereas we have for the point X_2

$$\overrightarrow{PX_2} \times \vec{u} = \begin{pmatrix} 2\cdot 5 \\ -2 \\ 0 \end{pmatrix} \times \begin{pmatrix} 2 \\ -1 \\ 0 \end{pmatrix}$$

$$= \begin{pmatrix} 0 \\ 0 \\ 2\cdot 5\cdot(-1) - 2\cdot(-2) \end{pmatrix} = \begin{pmatrix} 0 \\ 0 \\ 1.5 \end{pmatrix} \neq \vec{0}.$$

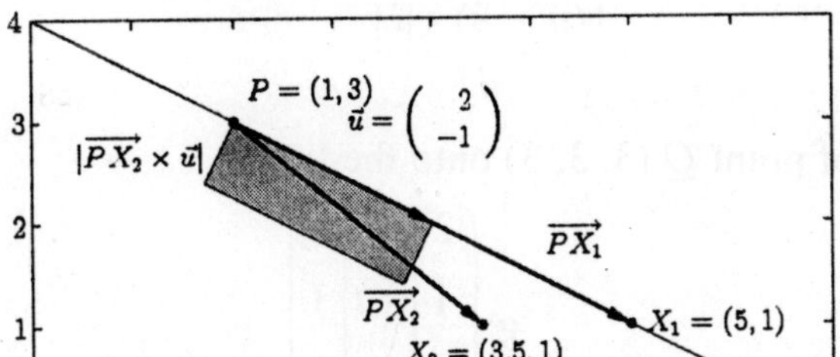

DISTANCE POINT – LINE

The projection X of a point Q on a line through P with direction $\vec{u}$ satisfies

$$\overrightarrow{PX} = t\vec{u}, \quad t = \frac{(\vec{q} - \vec{p})\cdot\vec{u}}{|\vec{u}|^2}.$$

From this we obtain the distance of the point Q from the line by

$$d = \frac{(\vec{q} - \vec{p}) \times \vec{u}}{|\vec{u}|}.$$

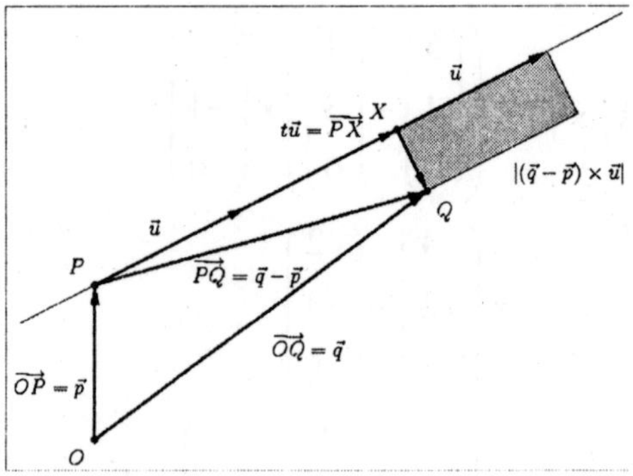

For the projection X of point Q onto the line the vector $\overrightarrow{XQ}$ is orthogonal to $\vec{u}$, what we can verify by computing the scalar product:

$$\overrightarrow{XQ}\cdot\vec{u} = \left(\vec{q}\left(\vec{p}+\frac{(\vec{q}-\vec{p})\cdot\vec{u}}{|\vec{u}|^2}\vec{u}\right)\right).\vec{u} = (\vec{q}-\vec{p})\cdot\vec{u} - \frac{(\vec{q}-\vec{p})\cdot\vec{u}}{|\vec{u}|^2}|\vec{u}|^2 = 0.$$

By the definition of the vector product we have

$$|\vec{a}\times b| = |\vec{a}||\vec{b}|\sin(\sphericalangle(\vec{a},\vec{b})),$$

and so we obtain for the distance:

$$d = \left|\overrightarrow{XQ}\right| = \left|\overrightarrow{PQ}\right|\sin(\sphericalangle(\overrightarrow{PQ},\vec{u})) = \frac{\left|\overrightarrow{PQ}\right||\vec{u}|\sin(\sphericalangle(\overrightarrow{PQ},\vec{u}))}{|\vec{u}|}$$

$$= \frac{|(\vec{q}-\vec{p})\times\vec{u}|}{|\vec{u}|}.$$

The projection of point Q (3, 3, 3) onto the line

$$g: \begin{pmatrix}2\\1\\3\end{pmatrix} + t\begin{pmatrix}1\\1\\1\end{pmatrix}$$

is the point with position vector

$$\vec{x} = \vec{p} + \frac{|(\vec{q}-\vec{p})\times\vec{u}|}{|\vec{u}|^2}\vec{u} = \begin{pmatrix}2\\1\\3\end{pmatrix} + \frac{\left(\begin{pmatrix}3\\3\\3\end{pmatrix}-\begin{pmatrix}2\\1\\3\end{pmatrix}\right)\cdot\begin{pmatrix}1\\1\\1\end{pmatrix}}{1^2+1^2+1^2}\begin{pmatrix}1\\1\\1\end{pmatrix}$$

$$= \begin{pmatrix}2\\1\\3\end{pmatrix} + \frac{1\cdot 1+2\cdot 1+0\cdot 1}{3}\begin{pmatrix}1\\1\\1\end{pmatrix} = \begin{pmatrix}3\\2\\4\end{pmatrix}.$$

The disctance is

$$d = \frac{\left| \begin{pmatrix} 1 \\ 2 \\ 0 \end{pmatrix} \times \begin{pmatrix} 1 \\ 1 \\ 1 \end{pmatrix} \right|}{\sqrt{3}} = \sqrt{\frac{(2+0)^2 + (0-1)^2 + (1-2)^2}{3}} = \sqrt{2}$$

in accordance with

$$\left|\overrightarrow{XQ}\right| = \sqrt{(3+3)^2 + (3-2)^2 + (3-4)^2} = \sqrt{2}\,.$$

DISTANCE OF TWO LINES

The distance between two lines g and h which are not parallel and with parametric equations

$$g : \vec{p} + t\vec{u},\, t \in \mathbb{R}$$

and

$$h : \vec{q} + s\vec{v},\, s \in \mathbb{R}$$

is given by

$$d = \frac{|(\vec{q} - \vec{p}) \cdot (\vec{u} \times \vec{v})|}{|\vec{u} \times \vec{v}|}.$$

This follows from the geometric properties of the dot and the cross product because the distance coincides with the distance of from the plane containing g and the by $p - q$ translated line h.

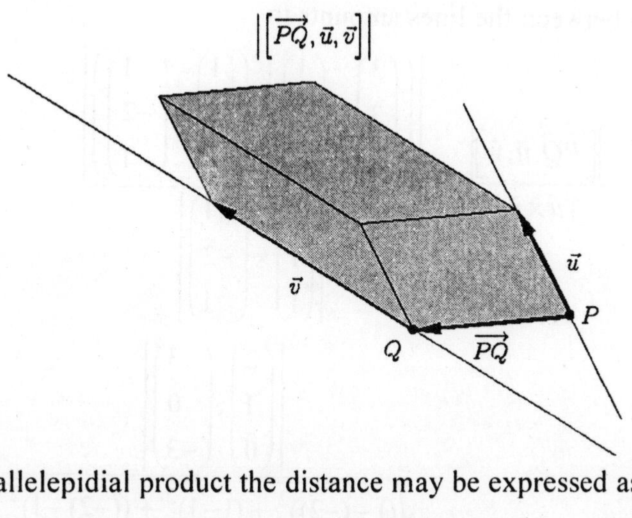

Using the parallelepidial product the distance may be expressed as

$$d = \frac{\left|[\overrightarrow{PQ}, \vec{u}, \vec{v}]\right|}{|\vec{u} \times \vec{v}|},$$

where P and Q are the points with location vector $\vec{p}$ and $\vec{q}$ rsp.

For parallel lines we have

$$d = \frac{\left|\overrightarrow{PQ} \times \vec{u}\right|}{|\vec{u}|}.$$

Two lines are called skew lines if they are not parallel and if the distance between them is positive. Denoting the position vectors of the points of minimal distance by

$$\vec{x} = \vec{p} + s\vec{u}, \quad \vec{y} = \vec{q} + t\vec{v},$$

then

$$\overrightarrow{XY} = \overrightarrow{PQ} + t\vec{v} - s\vec{u}$$

is orthogonal to both direction vectors, and, consequently, is parallel to

$$\vec{c} = \vec{u} \times \vec{v}.$$

Since the magnitude of a vector equals the absolute value of its scalar product with an parallel unit vector, it follows that

$$d = |\overrightarrow{XY}| = \left|(\overrightarrow{PQ} + t\vec{v} - s\vec{u}) \cdot \frac{\vec{c}}{|\vec{c}|}\right| = \left|\overrightarrow{PQ} \cdot \frac{\vec{c}}{|\vec{c}|}\right|,$$

and we obtain the desired formula.

For parallel lines we can use the formula for the distance of a point from a line.

$$g: \begin{pmatrix} 1 \\ 2 \\ 3 \end{pmatrix} + t \begin{pmatrix} 1 \\ 1 \\ 1 \end{pmatrix}, \quad h: \begin{pmatrix} 3 \\ 3 \\ 3 \end{pmatrix} + t \begin{pmatrix} 1 \\ -2 \\ 1 \end{pmatrix}$$

The distance between the lines amounts to

$$d = \frac{\left|\left[\overrightarrow{PQ}, \vec{u}, \vec{v}\right]\right|}{|\vec{u} \times \vec{v}|} = \frac{\left|\left(\begin{pmatrix} 3 \\ 3 \\ 3 \end{pmatrix} - \begin{pmatrix} 1 \\ 2 \\ 3 \end{pmatrix}\right) \cdot \left(\begin{pmatrix} 1 \\ 1 \\ 1 \end{pmatrix} \times \begin{pmatrix} 1 \\ -2 \\ 1 \end{pmatrix}\right)\right|}{\left|\begin{pmatrix} 1 \\ 1 \\ 1 \end{pmatrix} \times \begin{pmatrix} 1 \\ -2 \\ 1 \end{pmatrix}\right|}$$

$$= \frac{\left|\begin{pmatrix} 2 \\ 1 \\ 0 \end{pmatrix} \cdot \begin{pmatrix} 3 \\ 0 \\ -3 \end{pmatrix}\right|}{\sqrt{(1-(-2))^2 + (1-1)^2 + ((-2)-1)^2}} = \frac{6}{3\sqrt{2}} = \sqrt{2}.$$

AIR LANES

Simplifying let us assume that aircrafts fly directly from the starting point to the destination and that the flight path consists in line segments. Then the ascending flight path of a flight from Stuttgart to Kopenhagen is given by

$$g:\overrightarrow{SX} = \begin{pmatrix} 8 \\ 16 \\ 1 \end{pmatrix}, \quad s \in [0,8]$$

and the ascending flight path of a flight from Frankfurt () to Cairo is given by

$$h:\overrightarrow{FX} = \begin{pmatrix} 34 \\ -31 \\ 4 \end{pmatrix}, \quad t \in [0,3]$$

If we attach the origin of our coordinate system to Stuttgart, then Frankfurt has the coordinates $F = (-50, 150, -1/4)$. Hence, the distance between the two flight paths is

$$d = \frac{\left| \left(\begin{pmatrix} -50 \\ 150 \\ -1/4 \end{pmatrix} - \begin{pmatrix} 0 \\ 0 \\ 0 \end{pmatrix} \right) \cdot \left(\begin{pmatrix} 9 \\ 16 \\ 1 \end{pmatrix} \times \begin{pmatrix} 34 \\ -31 \\ 4 \end{pmatrix} \right) \right|}{\left| \begin{pmatrix} 9 \\ 16 \\ 1 \end{pmatrix} \times \begin{pmatrix} 34 \\ -31 \\ 4 \end{pmatrix} \right|}$$

$$= \frac{\left| \begin{pmatrix} -50 \\ 150 \\ -1/4 \end{pmatrix} . \begin{pmatrix} 95 \\ 2 \\ -792 \end{pmatrix} \right|}{\sqrt{95^2 - 2^2 + 792^2}} = \frac{4252}{\sqrt{636293}} \approx 5.33.$$

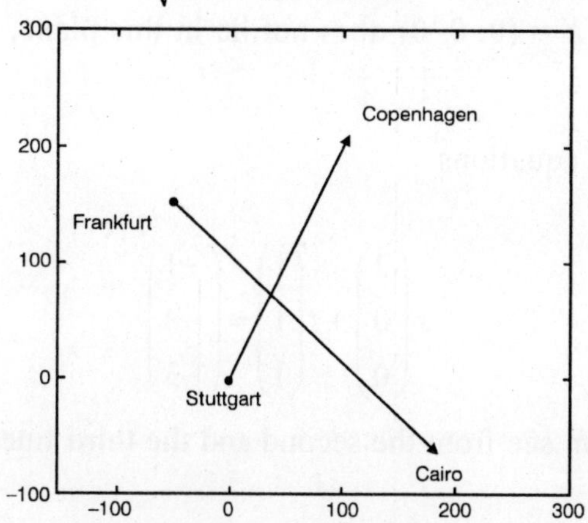

PLANES

PARAMETRIC EQUATION OF A PLANE

The points in a plane through P spanned by two nonparallel directions $\vec{u}$ und $\vec{v}$ satisfy

$$\overrightarrow{PX} = s\vec{u} + t\vec{v}, \quad s, t \in \mathbb{R}.$$

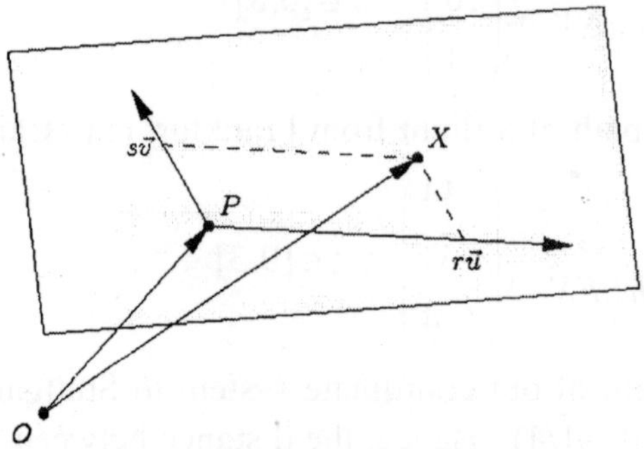

Correspondingly, we have

$$x_i = p_i + su_i + tv_i, \quad i = 1, 2, 3.$$

for the coordinates of the position vector $\vec{x}$.

Given a plane through point $P = (1, 2, 3)$ spanned by the vectors $\vec{u} = \begin{pmatrix} 2 \\ 0 \\ 0 \end{pmatrix}$, $\vec{v} = \begin{pmatrix} 1 \\ 1 \\ 1 \end{pmatrix}$.

Then $X = (1, 1, 2)$ lies in this plane, because we have

$$\overrightarrow{PX} = \begin{pmatrix} 0 \\ -1 \\ -1 \end{pmatrix} = \frac{1}{2}\vec{u} + (-1)\vec{v}$$

On the other hand, $X = (0, 0, 0)$ does not lie in this plane, because we have $\overrightarrow{PX} = \begin{pmatrix} -1 \\ -2 \\ -3 \end{pmatrix}$ and the system of equations

$$s\begin{pmatrix} 2 \\ 0 \\ 0 \end{pmatrix} + t\begin{pmatrix} 1 \\ 1 \\ 1 \end{pmatrix} = \begin{pmatrix} -1 \\ -2 \\ -3 \end{pmatrix}$$

has no solution as we can see from the second and the third line.

THREE-POINT FORM OF A PLANE

Given a plane through three points P, Q, that form a proper triangle. For the points X in this plane the parallelepidial product

$$\left[\overrightarrow{PX}, \overrightarrow{PQ}, \overrightarrow{PR},\right]$$

equals zero.

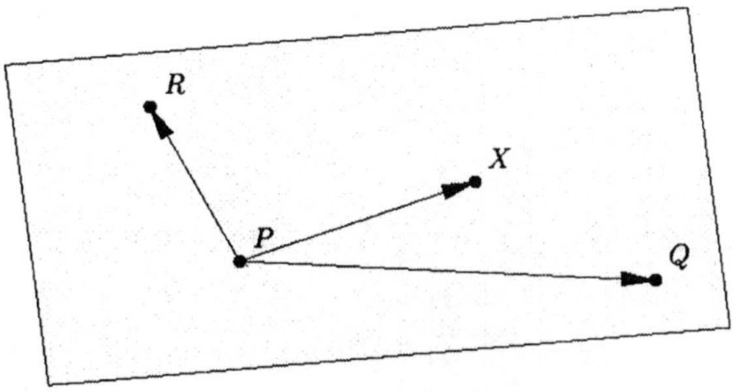

Given a plane through the points $P = (1, 2, 3)$, $Q = (3, 2, 3)$ und $R =$ U(2, 3, 4). Then $X = (1, 1, 2)$ lies in this plane, because we have

$$\left[\overrightarrow{PX}, \overrightarrow{PQ}, \overrightarrow{PR},\right] = \left[\begin{pmatrix} 0 \\ -1 \\ -1 \end{pmatrix}, \begin{pmatrix} 2 \\ 0 \\ 0 \end{pmatrix}, \begin{pmatrix} 1 \\ 1 \\ 1 \end{pmatrix}\right] = \begin{pmatrix} 0 \\ -1 \\ -1 \end{pmatrix} \cdot \begin{pmatrix} 0 \\ -2 \\ 2 \end{pmatrix} = 0$$

On the other hand $X = (0, 0, 0)$ does not lie in this plane, because we have

$$\left[\overrightarrow{PX}, \overrightarrow{PQ}, \overrightarrow{PR},\right] = \left[\begin{pmatrix} -1 \\ -2 \\ -3 \end{pmatrix}, \begin{pmatrix} 2 \\ 0 \\ 0 \end{pmatrix}, \begin{pmatrix} 1 \\ 1 \\ 1 \end{pmatrix}\right] = 4 - 6 = -2 \neq 0$$

HESSE-FORM OF A PLANE

The position vector $\vec{x}$ of a given point X in a plane through point P orthogonal to a normal vector $\vec{n}$ satisfies

$$\vec{x} \cdot \vec{n} = d, \qquad d = \vec{p} \cdot \vec{n}.$$

If $|\vec{n}| = 1$ and $d \geq 0$, then we obtain the so called Hessian normal form and is the distance between the plane and the origin.

Given a plane E with unit normal vector $\vec{n} = \frac{1}{3}\begin{pmatrix} 2 \\ 2 \\ 1 \end{pmatrix}$ throught point $P = (1, 2, 3)$.

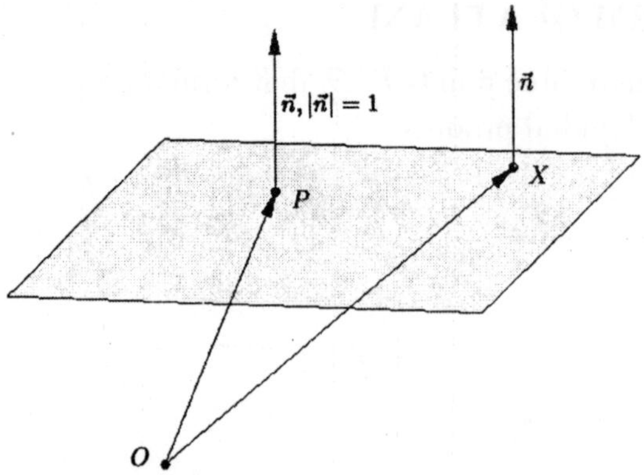

We have

$$d = \begin{pmatrix} 1 \\ 2 \\ 3 \end{pmatrix} \cdot \frac{1}{3} \begin{pmatrix} 2 \\ 2 \\ 1 \end{pmatrix} = 3,$$

i.e. the Hessian normal form of the plane ist given by

$$E : \frac{2}{3}x_1 + \frac{2}{3}x_2 + \frac{1}{3}x_3 = 3.$$

Then $X = (4, 0, 1)$ lies in this plane, because we have

$$\vec{x} \cdot \vec{n} = \frac{1}{3}(8 + 0 + 1) = d.$$

On the other hand, $X = (0, 0, 0)$ does not lie in this plane, because we have

$$\vec{x} \cdot \vec{n} = 0 \neq d.$$

TRANSFORMATION BETWEEN REPRESENTATIONS OF PLANES

1. Given a plane through point P spanned by two nonparallel vectors $\vec{u}$, $\vec{v}$.

We obtain two further points Q and R, lying in this plane but not lying on a common line through P, by

$$\vec{q} = \vec{p} + \vec{u}, \quad \vec{r} = \vec{p} + \vec{v}.$$

We compute the normal vector $\vec{n}$, which is perpendicular to both $\vec{u}$ and $\vec{v}$, and parallel to $\vec{u} \times \vec{v}$. The normalized normal vector for the Hessian normal form ist given by

$$\vec{n} = \sigma \frac{\vec{u} \times \vec{v}}{|\vec{u} \times \vec{v}|},$$

where we must select the sign $\sigma \in \{-1, 1\}$ so that $d = \vec{p} \cdot \vec{n}$ is positive.

2. Given a plane through three noncollinear points P, Q und R.

We obtain two vectors spanning the plane by

$$\vec{u} = \overrightarrow{PQ}, \qquad \vec{v} = \overrightarrow{PR}.$$

As decribed in (i), we can determine the Hessian normal form starting from these two vectors.

3. Given a plane with normal vector $\vec{n}$ through point P.

Finding two linear independent vectors orthogonal to $\vec{n}$ we obtain two vectors spanning the plane. For example

$$\vec{u} = \vec{n} \times \vec{x}, \qquad \vec{v} = \vec{n} \times \vec{u},$$

where $\vec{x}$ is an arbitrary vector $\neq \lambda\vec{n}$.

As in (i), from these vectors we can compute the three point form.

DISTANCE POINT – PLANE

Given a plane with normal vector $\vec{n}$ through point P. Let X be the perpendicular projection of a given point Q onto this plane. The so called perpendicular vector of point Q onto the plane is given by

$$\overrightarrow{XQ} = \frac{\overrightarrow{XQ} \cdot \vec{n}}{|\vec{n}|^2}\vec{n}.$$

Its length

$$d = \frac{\overrightarrow{XQ} \cdot \vec{n}}{|\vec{n}|}$$

is the distance of the plane from Q.

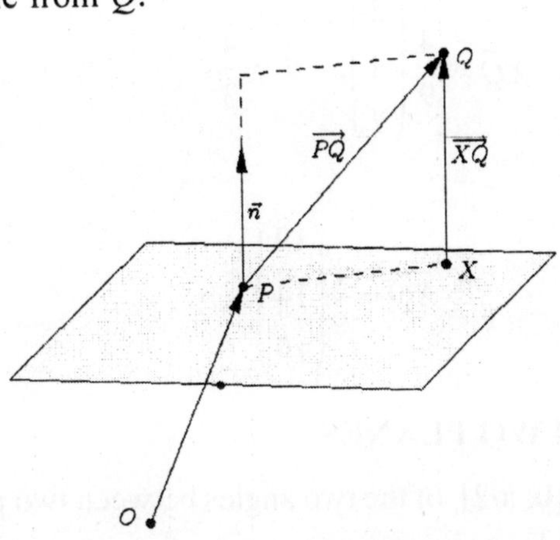

The scalar product

$$\overrightarrow{PQ}\cdot\vec{n} = \cos(\alpha)\,|\overrightarrow{PQ}||\vec{n}| = \pm d\,|\vec{n}|$$

yields the $|\vec{n}|$-fold distance between Q and the plane, where the sign indicates whether $\overrightarrow{PQ}$ lies on the same side of the plane as $\vec{n}$ does.

Thus we have

$$d = \frac{|\overrightarrow{PQ}\cdot\vec{n}|}{|\vec{n}|}$$

and
$$\overrightarrow{XQ} = \pm d\frac{\vec{n}}{|\vec{n}|} = \frac{|\overrightarrow{Q}\cdot\vec{n}}{|\vec{n}|}\frac{\vec{n}}{|\vec{n}|} = \frac{\overrightarrow{PQ}\cdot\vec{n}}{|\vec{n}|^2}\vec{n}.$$

Given the plane with normal vector $\vec{n}\begin{pmatrix}2\\1\\2\end{pmatrix}$ through point $P = (1, 2, 3)$. We want to compute the distance d of point $Q = (3, 2, 2)$ from the plane and the foot X of the perpendicular through this point onto the plane.

First of all we have

$$\overrightarrow{PQ} = \begin{pmatrix}2\\0\\0\end{pmatrix},\quad \overrightarrow{PQ}\cdot\vec{n} = 4, |\vec{n}| = 3.$$

So we obtain

$$\overrightarrow{XQ} = \frac{4}{9}\begin{pmatrix}2\\1\\2\end{pmatrix},\quad d = \frac{4}{3}.$$

and finally we compute

$$\vec{x} = \overrightarrow{XQ} = \frac{1}{9}\begin{pmatrix}19\\14\\19\end{pmatrix}.$$

INTERSECTION OF TWO PLANES

The smaller one, $\varphi\in[0, \pi/2]$, of the two angles between two planes with normal vectors $\vec{n}_i$ is uniquely determined by

$$\cos(\varphi) = \frac{|\vec{n}_1 \cdot \vec{n}_2|}{|\vec{n}_1||\vec{n}_2|}$$

and

$$\vec{n}_1 \times \vec{n}_2$$

is the direction of the line of intersection.

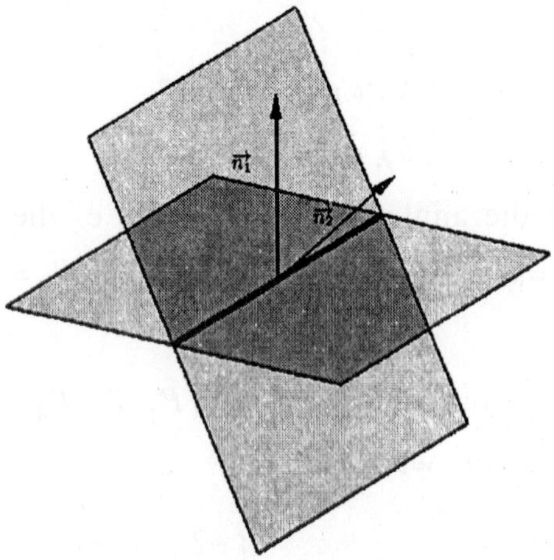

Given two planes with the respective normal vectors

$$\vec{n}_1 = \begin{pmatrix} 1 \\ -1 \\ 0 \end{pmatrix}, \quad \vec{n}_2 = \begin{pmatrix} 0 \\ -1 \\ 1 \end{pmatrix}$$

through the respective points $P_1 = (1, 2, 0)$, P_2 $(0, 1, 3)$. We want to compute the angle φ between these planes and the line of intersection.

We have

$$\cos(\varphi) = \frac{|\vec{n}_1 \cdot \vec{n}_2|}{|\vec{n}_1||\vec{n}_2|} = \frac{1}{2}$$

and therefore

$$\varphi = 60°.$$

The computation

$$\vec{r} = \vec{n}_1 \times \vec{n}_2 = \begin{pmatrix} -1 \\ -1 \\ -1 \end{pmatrix}$$

yields the direction of the line of intersection.

To find a point of intersection we need

$$d_1 = \vec{p}_1 \cdot \vec{n}_1 = -1, \quad d_2 = \vec{p}_2 \cdot \vec{n}_2 = 2$$

and we determine a point S satisfying the equations

$$\vec{s} \cdot \vec{n}_1 = d_1, \quad \vec{s} \cdot \vec{n}_2 = d_2.$$

We find, for example, $S = (2, 3, 5)$ and so we obtain the line of intersection

$$g : \vec{x} = \vec{s} + t\vec{r} = \begin{pmatrix} 2 \\ 3 \\ 5 \end{pmatrix} + t \begin{pmatrix} 1 \\ 1 \\ 1 \end{pmatrix} \quad t \in \mathbb{R}.$$

We want to compute the angle between 2 faces of the regular tetrahedron with vertices $P_1 = (1, 0, 0)$, $P_2 = (-1, 0, 0)$, $P_3 = (0, 1, \sqrt{2})$, $P_4 = (0, -1, \sqrt{2})$.

We can use the planes through P_1, P_2, P_3 and P_1, P_2, P_4, for example.

Then, for the normal vectors we have:

$$\vec{n}_1 = \overrightarrow{P_1P_2} \times \overrightarrow{P_1P_3} = \begin{pmatrix} -2 \\ 0 \\ 0 \end{pmatrix} \times \begin{pmatrix} -1 \\ 1 \\ \sqrt{2} \end{pmatrix} = \begin{pmatrix} 0 \\ 2\sqrt{2} \\ -2 \end{pmatrix}$$

and

$$\vec{n}_2 = \overrightarrow{P_1P_2} \times \overrightarrow{P_1P_4} = \begin{pmatrix} -2 \\ 0 \\ 0 \end{pmatrix} \times \begin{pmatrix} -1 \\ -1 \\ \sqrt{2} \end{pmatrix} = \begin{pmatrix} 0 \\ 2\sqrt{2} \\ 2 \end{pmatrix}$$

resp. It follows

$$\cos(\varphi) = \frac{|\vec{n}_1 \cdot \vec{n}_2|}{|\vec{n}_1||\vec{n}_2|} = \frac{4}{\sqrt{12}\sqrt{12}} = \frac{1}{3}$$

and we obtain $\varphi \approx 70, 59°.$

EXERCISE

1. Evaluate the line integral $\int_C \left[\left(x^2 + xy\right)dx + \left(x^2 + y^2\right)dy\right]$ where C is the square formed by the lines $y = \pm 1$ and $x = \pm 1$.

2. If $F = (5xy - 6x^2)I + (2y - 4x)J$, evaluate $\int_C F.dR$ along the curve C in the xy-plane, $y = x^3$ from the point (1, 1) to (2, 8).

3. A vector field is given by F = (sin y)I + x(1 + cos y)J. Evaluate the line integral over a circular path given by $x^2 + y^2 = a^2$, z = 0.

$$\lim_{(x,y)\to(0,0)} \frac{\sin\sqrt{x^2+y^2}}{\sqrt{x^2+y^2}} = 1.$$

4. If $R = xI + yJ + zK$ and $r \neq 0$, show that

   (a) $\nabla(1/r^2) = -2R/r^4$; $\nabla(R/r^2) = 1/r^2$ (ii) div $(r^n R) = (n+3)r^n$; curl$(r^n R) = 0$.

   (b) $\nabla\left[\nabla.\frac{R}{r}\right] = -\frac{2}{r^3}R.$

5. Show that each of following vectors are solenoidal :

   (a) $(x + 3y)I + (y - 3z)J + (x - 2z)K$ (ii) $3y^4z^2I + 4x^3z^2J + 3x^2y^2K$

   (b) $\nabla f \times \nabla$ y.

6. **Find the points on the circle $x^2 + y^2 = 100$ which are closest to and farthest from the point (2, 3).**

7. **Find the constrained maxima and minima of $f(x, y, z) = x + y^2 + 2z$ given that $4x^2 + 9y^2 - 36z^2 = 36$.**

8. **Find the volume of the largest rectangular parallelepiped that can be inscribed in the ellipsoid**

$$\frac{x^2}{a^2} + \frac{y^2}{b^2} + \frac{z^2}{c^2} = 1.$$

9. **Derive the gradient formula in cylindrical coordinates:** $\nabla F = \frac{\partial F}{\partial r}e_r + \frac{1}{r}\frac{\partial F}{\partial \theta}e_\theta + \frac{\partial F}{\partial z}e_z$

Chapter 9

Vectors and Matrices

A matrix is a rectangular array of numbers. A matrix (plural matrices) is a rectangular table of *elements* (or *entries*), which may be numbers or, more generally, any abstract quantities that can be added and multiplied. Matrices are used to describe linear equations, keep track of the coefficients of linear transformations and to record data that depend on multiple parameters. Matrices are described by the field of matrix theory. They can be added, multiplied, and decomposed in various ways, which also makes them a key concept in the field of linear algebra. All of the following are examples of matrices.

$$P = \begin{bmatrix} -3 & 2 \\ 0 & 4 \end{bmatrix}, \quad Q = \begin{bmatrix} 1 & -2 & 3 \\ 0 & 78 & -5 \\ -5 & -5 & 1 \end{bmatrix}, \quad R = \begin{bmatrix} a & b & c & d \\ e & f & g & h \\ k & l & t & u \end{bmatrix}$$

Matrices are a useful tool for easily representing large amounts of information and solving multivariable problems. Some situations in which matrices are useful include:

1. Representing and solving systems of linear equations
2. Solving systems of differential equations
3. Simplifying formulas involving 2 or more variables or dimensions (examples include multi-variable statistics formulas and many physical formulas in 2 or more dimensions)

We represent the dimensions of a matrix with m rows and n columns with the notation mxn. The matrices above (from left to right) have dimensions of 2x2, 3x3, and 3x4. One can think of a matrix as being a group of row vectors or column vectors. For the matrix R above, the row vectors are [a b c d], [e f g h], and [k l t u] and the column vectors are

$$\begin{bmatrix} a \\ e \\ k \end{bmatrix}, \begin{bmatrix} b \\ f \\ l \end{bmatrix}, \begin{bmatrix} c \\ g \\ t \end{bmatrix}, \begin{bmatrix} d \\ h \\ u \end{bmatrix}$$

The usual two- or three-dimensional vector is typically represented as a single column vector. The elements of a matrix A are frequently labeled as A_{ij} where i is the number of the row the element is in and j is the number of the column the element is in. For the matrix P above, A_{11}=-3, A_{12}=2, A_{21}=0, and A_{22}=4.

TERMINOLOGY AND NOTATION

The horizontal lines in a matrix are called rows and the vertical lines are called columns. A matrix with *m* rows and *n* columns is called an *m*-by-*n* matrix (written $m \times n$) and *m* and *n* are called its dimensions. The dimensions of a matrix are always given with the number of rows first, then the number of columns. It is commonly said that an *m*-by-*n* matrix has an order of $m \times n$ ("order" meaning size). Two matrices of the same order whose corresponding entries are equivalent are considered equal.

The entry that lies in the *i*-th row and the *j*-th column of a matrix is typically referred to as the *i,j*, or *(i,j)*, or *(i,j)*-th, or $(i,j)^{th}$ entry of the matrix. Again, the row is always noted first, then the column.

Almost always upper-case letters denote matrices, while the corresponding lower-case letters, with two subscript indices, represent the entries. For example, the $(i,j)^{th}$ entry of a matrix A is most commonly written as $a_{i,j}$. Alternative notations for that entry are A[*i,j*] or $A_{i,j}$. In addition to using upper-case letters to symbolize matrices, many authors use a special typographical style, commonly boldface upright (non-italic), to further distinguish matrices from other variables. Following this convention, A is a matrix, distinguished from *A*, a scalar. An alternate convention is to annotate matrices with their dimensions in small type underneath the symbol, for example, $A_{m \times n}$ for an *m*-by-*n* matrix.

We often write $A := (a_{i,j})_{i=1,m;\ j=1,\dots,n}$ or $A := (a_{i,j})_{m \times n}$ to define an $m \times n$ matrix A. In this case, the entries $a_{i,j}$ are defined separately for all integers $1 \le i \le m$ and $1 \le j \le n$. In some programming languages, the numbering of rows and columns starts at zero. Texts which use any such language extensively frequently follow that convention, so we have $0 \le i \le m-1$ and $0 \le j \le n-1$.

A matrix where one of the dimensions equals one is often called a *vector,* and interpreted as an element of real coordinate space. An $m \times 1$ matrix (one column and *m* rows) is called a column vector and a $1 \times n$ matrix (one row and *n* columns) is called a row vector.

MATHEMATICAL DEFINITION

An $m \times n$ $(m, n \in \mathbb{N})$ matrix *A* is a function $A : \{1, 2, \dots, m\} \times \{1, 2, \dots, n\} \to S$, where *S* is any non-empty set. ($\{1, 2, \dots, m\} \times \{1, 2, \dots, n\}$ is the Cartesian product of sets $\{1, 2, \dots, m\}$ and ($\{1, 2, \dots, n\}$.)

We say that matrix *A* is a matrix over the set . Important thing to note is that, if we want to have matrix algebra, the set must be a ring and matrix must be a square matrix. Since the set of all square matrices over a ring is also a ring, matrix algebra is usually called matrix ring.

Since this article mainly considers matrices over real numbers, matrices shown here are actually functions $A : (\{1, 2, ..., m\} \times \{1, 2, ..., n\} \to \mathbb{R}$.

Example: The matrix

$$A=\begin{bmatrix} 9 & 8 & 6 \\ 1 & 2 & 7 \\ 4 & 9 & 2 \\ 6 & 0 & 5 \end{bmatrix} \text{ or } A = \begin{pmatrix} 9 & 8 & 6 \\ 1 & 2 & 7 \\ 4 & 9 & 2 \\ 6 & 0 & 5 \end{pmatrix}$$

is a 4×3 matrix. The element $a_{2,3}$ or A [2, 3] is 7. In terms of the mathematical definition given above, this matrix is a function A: $\{1, 2, 3, 4\} \times \{1, 2, 3\} \to \mathbb{R}$ and, for example, $A((2, 3)) = 7$ and $A((3, 1)) = 4$.

The matrix

$$R=\begin{bmatrix} 1 & 2 & 3 & 4 & 5 & 6 & 7 & 8 & 9 \end{bmatrix}$$

is a 1×9 matrix, or 9-element row vector.

BASIC OPERATIONS

SUM

Two or more matrices of identical dimensions m and n can be added. Given m-by-n matrices A and B, their sum A+B is the m-by-n matrix computed by adding corresponding elements:

$$A + B = (a_{i,j})_{1\le i\le m;\, 1\le j\le n} + (b_{i,j})_{1\le i\le m;\, 1\le j\le n}$$

$$= (a_{i,j} + b_{i,j})_{1\le i\le m;\, 1\le j\le n}.$$

For example:

$$\begin{bmatrix} 1 & 3 & 1 \\ 1 & 0 & 0 \\ 1 & 2 & 2 \end{bmatrix} + \begin{bmatrix} 0 & 0 & 5 \\ 7 & 5 & 0 \\ 2 & 1 & 1 \end{bmatrix} = \begin{bmatrix} 1+0 & 3+0 & 1+5 \\ 1+7 & 0+5 & 0+0 \\ 1+2 & 2+1 & 2+1 \end{bmatrix} = \begin{bmatrix} 1 & 3 & 6 \\ 8 & 5 & 0 \\ 3 & 3 & 3 \end{bmatrix}.$$

Another, much less often used notion of matrix addition is the direct sum.

ALGEBRAIC OPERATIONS INVOLVING MATRICES

Matrices can be added, subtracted, and multiplied by scalars in a quite natural manner.

For matrices

$$A = \begin{bmatrix} a & b \\ c & d \end{bmatrix} \text{ and } B = \begin{bmatrix} e & f \\ g & h \end{bmatrix}$$

we have

$$A \pm B = \begin{bmatrix} a & b \\ c & d \end{bmatrix} \pm \begin{bmatrix} e & f \\ g & h \end{bmatrix} = \begin{bmatrix} a \pm e & b \pm f \\ c \pm g & d \pm h \end{bmatrix}$$

$$cB = c\begin{bmatrix} e & f \\ g & h \end{bmatrix} = \begin{bmatrix} ce & cf \\ cg & ch \end{bmatrix}$$

The above examples for 2x2 matrices can be generalized quite easily to larger rectangular matrices. Note that two matrices must have the same dimensions in order to be added or subtracted. Matrix addition is clearly commutative and associative (i.e. A+B=B+A and A+(B+C)=(A+B)+C).

Example

The sum of the matrices

$$P = \begin{bmatrix} -3 & 3 \\ 0 & 1 \end{bmatrix} \text{ and } Q = \begin{bmatrix} -2 & -1 \\ 5 & -4 \end{bmatrix}$$

is

$$P + Q = \begin{bmatrix} -1 & 3 \\ 0 & 1 \end{bmatrix} + \begin{bmatrix} -2 & -1 \\ 5 & -4 \end{bmatrix} = \begin{bmatrix} -3 & 2 \\ 5 & -3 \end{bmatrix}$$

It is also possible to multiply two matrices together. The product of two 2x2 matrices is

$$AB = \begin{bmatrix} a & b \\ c & d \end{bmatrix}\begin{bmatrix} e & f \\ g & h \end{bmatrix} = \begin{bmatrix} ae + bg & af + bh \\ ce + dg & cf + dh \end{bmatrix}$$

The following is an easy way to remember this operation: the element AB_{ij} of the matrix AB is the dot product of the ith row vector of the matrix A (the matrix on the left) with the jth column vector of the matrix B (the matrix on the right). When generalizing to multiply matrices of different sizes, we must note that to perform the multiplication it is necessary for the matrix on the left to have the same number of columns as the matrix on the right has rows (or else we would not be able to carry out the specified dot product). To summarize, a matrix A with m columns can *only* be multiplied (on the right) by a matrix B with m rows.

Example

Suppose we wish to multiply the matrices

$$A = \begin{bmatrix} -1 & -3 & 2 \\ 1 & 0 & 2 \end{bmatrix} \text{ and } B = \begin{bmatrix} 5 & -3 & 1 \\ 4 & -1 & 0 \\ 0 & 0 & -1 \end{bmatrix}$$

The element AB_{11} is the dot product of the row vector [-1 -3 2] of A with the column vector

$$\begin{bmatrix} 5 \\ 4 \\ 0 \end{bmatrix}$$

of B: AB_{11}=-1(5)+(-3)(4)+2(0)=-17. If we continue this process and find the other 5 entries of AB we have

$$AB = \begin{bmatrix} -5-12+0 & -3+3+0 & -1-0-2 \\ 5+0+0 & 3-0+0 & 1+0-2 \end{bmatrix} = \begin{bmatrix} -17 & 0 & -3 \\ 5 & 3 & -1 \end{bmatrix}$$

ERROR NOTICE:

The middle entry in the top row of AB should be (-1)(-3)+(-3)(-1)+(2)(0)= 6.

The middle entry in the bottom row of AB should be (1)(-3)+(0)(-1)+(2)(0)=-3.

If we reverse the order of the matrices A and B to try to find the product BA we find that the multiplication is not possible. We cannot find the dot product of a row vector of B (with length 3) with a column vector of A (with length 2).

We have discovered another one of the algebraic properties of matrices: non-commutativity of multiplication. As our example above illustrates, the product BA might not be defined even if the product AB is defined. For appropriate choices of A, B, and C (i.e. so the multiplication is defined), the following is a list of commonly used *multiplicative* matrix properties:

- Non-commutativity: AB is *not* necessarily equal to BA
- Associativity: A(BC)=(AB)C
- Distributivity: A(B+C)=AB+BC and (A+B)C=AC+BC
- Scalar multiplication: c(AB)=(cA)B=A(cB)

SCALAR MULTIPLICATION

Given a matrix A and a number *c*, the scalar multiplication *cA* is computed by multiplying every element of A by the scalar *c* (i.e.) $(cA)i,j = c \cdot a_{i,j}$). For example:

$$2 \cdot \begin{bmatrix} 1 & 8 & -3 \\ 4 & -2 & 5 \end{bmatrix} = \begin{bmatrix} 2\cdot 1 & 2\cdot 8 & 2\cdot -3 \\ 2\cdot 4 & 2\cdot -2 & 2\cdot 5 \end{bmatrix} = \begin{bmatrix} 2 & 16 & -6 \\ 8 & -4 & 10 \end{bmatrix}$$

Matrix addition and scalar multiplication turn the set $M(m, n, \mathbb{R})$ of all m-by-n matrices with real entries into a real vector space of dimension $m \cdot n$.

MATRIX MULTIPLICATION

Multiplication of two matrices is well-defined only if the number of columns of the left matrix is the same as the number of rows of the right matrix. If A is an m-by-n matrix and B is an n-by-p matrix, then their matrix product AB is the m-by-p matrix given by:

$$(AB)_{i,j} = a_{i,1}b_{1,j} + a_{i,2}b_{2,j} + \ldots + a_{i,n}b_{n,j}$$

for each pair (i,j). For example:

$$\begin{bmatrix} 1 & 0 & 2 \\ -1 & 3 & 1 \end{bmatrix} \times \begin{bmatrix} 3 & 1 \\ 2 & 1 \\ 1 & 0 \end{bmatrix} = \begin{bmatrix} (1\times 3 + 0\times 2 + 2\times 1) & (1\times 1 + 0\times 1 + 2\times 0) \\ (-1\times 3 + 3\times 2 + 1\times 1) & (-1\times 1\times 3 + 1 + 1\times 0) \end{bmatrix} = \begin{bmatrix} 5 & 1 \\ 4 & 2 \end{bmatrix}.$$

Matrix multiplication has the following properties:

1. (AB)C = A(BC) for all k-by-m matrices A, m-by-n matrices B and n-by-p matrices C ("associativity").
2. (A+B)C = AC+BC for all m-by-n matrices A and B and n-by-k matrices C ("right distributivity").
3. C(A+B) = CA+CB for all m-by-n matrices A and B and k-by-m matrices C ("left distributivity").

Matrix multiplication is not commutative; that is, given matrices A and B and their product defined, then generally AB ≠ BA. It may also happen that AB is defined but BA is not defined.

Besides the ordinary matrix multiplication just described, there exist other operations on matrices that can be considered forms of multiplication, such as the Hadamard product and the Kronecker product.

LINEAR TRANSFORMATIONS

Matrices can conveniently represent linear transformations (also known as "linear maps") between finite-dimensional vector spaces. Let R^n be an n-dimensional vector space, and let the vectors in this space be represented in matrix format as column vectors (n-by-1 matrices). For every linear map f: $R^n \to R^m$ there exists a unique m-by-n matrix A such that

$$f(x) = A_x$$

for each vector x in R^n.

We say that the matrix A "represents" the linear map f, or that A is the "transformation matrix" of f.

Matrix multiplication neatly corresponds to the composition of maps. If the k-by-m matrix B represents another linear map $g: R^m \to R^k$, then the composition gof is represented by BA:

$$(g \circ f)(x) = g(f)(x))\ g(Ax) = B(Ax) = (BA)x.$$

This follows from the above-mentioned associativity of matrix multiplication.

More generally, a linear map from an n-dimensional vector space to an m-dimensional vector space is represented by an m-by-n matrix, provided that bases have been chosen for each. This property makes matrices powerful data structures in high-level programming languages.

RANKS

The rank of a matrix A is the dimension of the image of the linear map represented by A; this is the same as the dimension of the space generated by the rows of A, and also the same as the dimension of the space generated by the columns of A.

It can also be defined without reference to linear algebra as follows: the rank of an m-by-n matrix A is the smallest number k such that A can be written as a product BC where B is an m-by-k matrix and C is a k-by-n matrix (although this is not a practical way to compute the rank).

TRANSPOSE

The transpose of an m-by-n matrix A is the n-by-m matrix A^T (also sometimes written as A^{tr}, or ${}^tA$, or A') formed by turning rows into columns and columns into rows, i.e., $A^T[i, j] = A[j, i]$ for all indices i and j.

If A describes a linear map with respect to two bases, then the matrix A^T describes the transpose of the linear map with respect to the dual bases.

We have $(A + B)^T = A^T + B^T$ and $(AB)^T = B^T A^T$.

SQUARE ROOT

Given two bounded matrices T and B, B is a square root of T if T = B*B.

SQUARE MATRICES AND RELATED DEFINITIONS

A square matrix is a matrix which has the same number of rows and columns. The set of all square n-by-n matrices, together with matrix addition and matrix multiplication is a ring. Unless $n = 1$, this ring is not commutative.

M(n, R), the ring of real square matrices, is a real unitary associative algebra. M(n, C), the ring of complex square matrices, is a complex associative algebra.

The unit matrix or identity matrix I_n of size n is the n-by-n matrix in which all the elements on the main diagonal are equal to 1 and all other elements are equal to 0. The identity matrix is named thus because it satisfies the equations $MI_n=M$ and $I_nN=N$ for any m-by-n matrix M and n-by-k matrix N. For example, if $n = 3$:

$$I_3 = \begin{bmatrix} 1 & 0 & 0 \\ 0 & 1 & 0 \\ 0 & 0 & 1 \end{bmatrix}.$$

The identity matrix is the identity element in the ring of square matrices.

Invertible elements in this ring are called invertible matrices or non-singular matrices. An n-by-n matrix A is invertible if and only if there exists a matrix B such that

$$AB = I_n \ (= BA).$$

In this case, B is the inverse matrix of A, denoted by A^{-1}. The set of all invertible n-by-n matrices forms a group (specifically a Lie group) under matrix multiplication, the general linear group.

If λ is a number and v is a non-zero vector such that $Av = \lambda v$, then we call v an eigenvector of A and λ the associated eigenvalue. (*Eigen* means "own" in German and in Dutch.) The number λ is an eigenvalue of A if and only if $A-\lambda I_n$ is not invertible, which happens if and only if $p_A(\lambda) = 0$, where $p_A(x)$ is the characteristic polynomial of A. $p_A(x)$ is a polynomial of degree n and therefore has n complex roots if multiple roots are counted according to their multiplicity. Every square matrix has at most n complex eigenvalues.

The determinant of a square matrix A is the product of its n eigenvalues, but it can also be defined by the *Leibniz formula*. Invertible matrices are precisely those matrices with a nonzero determinant.

The Gaussian elimination algorithm is of central importance: it can be used to compute determinants, ranks and inverses of matrices and to solve systems of linear equations.

The trace of a square matrix is the sum of its diagonal entries, which equals the sum of its n eigenvalues.

The matrix exponential function is defined for square matrices, using power series.

VECTOR AND MATRIX

EUCLIDEAN VECTOR AND EUCLIDEAN SPACE

The n–dimensional (real) euclidean space R^n is the collection of n–tuples of numbers, called vectors. Vectors can be either presented horizontally,

$$x = (x_1, x_2, ..., x_n)$$

or vertically.

$$x = \begin{bmatrix} x_1 \\ x_2 \\ . \\ . \\ x_n \end{bmatrix}$$

The only thing you need to keep in mind is that, when vectors are mixed with matrices, it is often more convenient to use vertical vectors.

The number n is the dimension of the vector x. The numbers $x_1, x_2, ..., x_n$ are called the coordinates of x. Strictly speaking, these are the coordinates with respect to the standard basis vectors

$$e_1 = (1, 0, ..., 0),$$
$$e_2 = (0, 1, ..., 0),$$
$$\vdots$$
$$e_n = (0, 0, ..., 1).$$

We also mention the special zero vector

$$0 = (0, 0, ..., 0).$$

Example: The 1–dimensional euclidean space

$R^1 = \{(x): x \text{ is a real number}\}$

can be identified with all real numbers R and be visualized as a straight line.

The 2–dimensional euclidean space

$R^2 = \{(x_1, x_2): x_1, x_2 \text{ are real numbers}\}$

can be visualized as the coordinate plane.

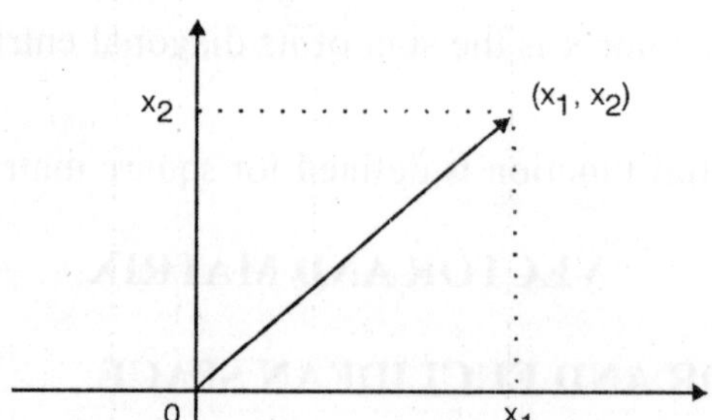

The 3–dimensional euclidean space

$R^3 = \{(x_1, x_2, x_3): x_1, x_2, x_3$ are real numbers$\}$

can be visualized as the world we are living in.

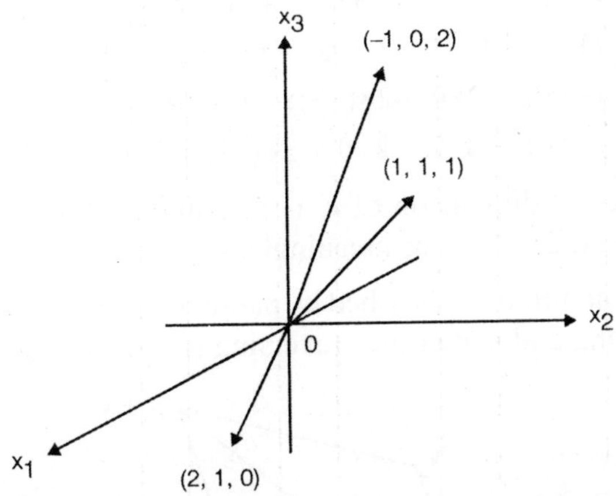

The 0–dimensional euclidean space R^0 consists of a single point – the origin.

The example illustrates that the euclidean vectors can be geometrically presented (in a coordinate system) as an arrow starting from the origin. If an arrow does not start from the origin, we need to parallel shift it so that it begins at the origin.

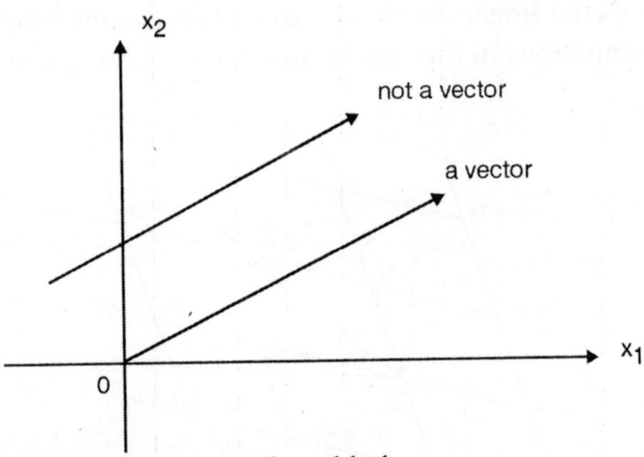

Vectors of the same dimension can be added

$(x_1, x_2,..., x_n) + (y_1, y_2,..., y_n) = (x_1 + y_1, x_2 + y_2,..., x_n + y_n).$

A number (scalar) can also be multiplied to vectors

$c\ (x_1, x_2,..., x_n) = (cx_1, cx_2,..., cx_n).$

The addition and scalar multiplication can also be combined to form linear combinations $c_1x_1 + c_2x_2 +... + c_kx_k$.

Example: Let $u = (1, 2)$, $v = (3, 4)$, $w = (5, 6)$, $x = (1, 0, -3)$, $y = (2, -1, 5)$. Then

$$u + v = (1 + 3, 2 + 4) = (4, 6)$$

$$3u = (3\times1, 3\times2) = (3, 6)$$
$$u + v + w = (1 + 3 + 5, 2 + 4 + 6) = (9, 12)$$
$$3u - 4v + w = (3\times1 - 4\times3 + 5, 3\times2 - 4\times4 + 6) = (-4, -4)$$
$$x + y = (1 + 2, 0 - 1, -3 + 5) = (3, -1, 2)$$
$$-2x = (-2\times1, -2\times0, -2\times(-3)) = (-2, 0, 6)$$
$$-2x + 3y = (-2\times1 + 3\times2, -2\times0 + 3\times(-1), -2\times(-3) + 3\times5) = (4, -3, 21)$$

Moreover, because the dimension of u, v, w is different from the dimension of x, y, expressions such as $u + x$, $2v + 3y$ are meaningless.

Geometrically, the addition is described by parallelogram, and the scalar multiplication is the stretching/shrinking and sometimes reversing (for negative scalars) of vectors.

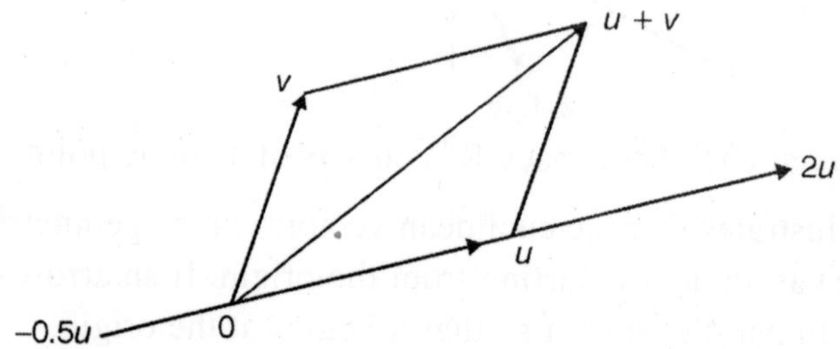

This picture shows the linear combinations of two (nonparallel) vectors. It suggests that all the linear combinations of the two vectors form a plane passing through the origin.

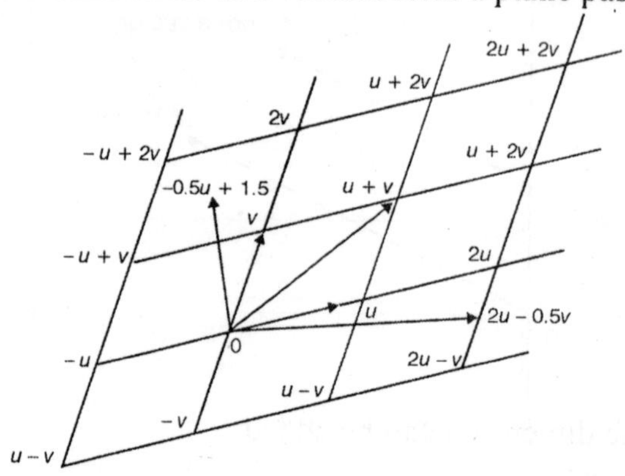

Finally, the operations have the following obvious properties.

commutative

$$u + v = v + u$$

associative

$$(u + v) + w = u + (v + w)$$

zero

$$u + 0 = u = 0 + u$$

distributive

$$r(u + v) = ru + rv$$

distributive

$$(r + s)u = ru + su$$

associative

$$(rs)u = r(su)$$

VECTOR ADDITION IS GIVEN BY PARALLELOGRAM

Let us look at the addition of two vectors (red and blue in the picture) in R^2. The green vector is constructed from the parallelogram.

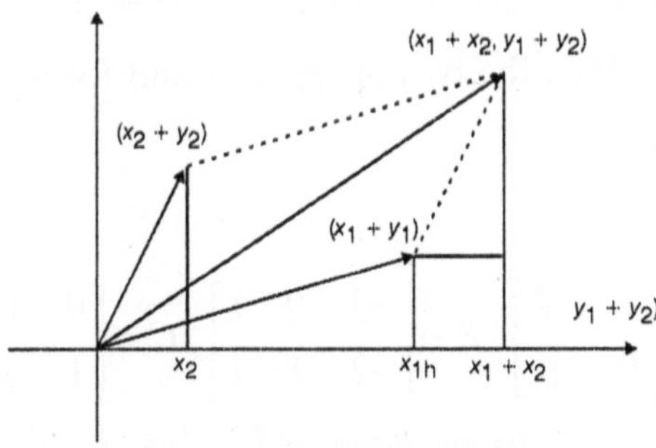

The parallelogram tells us that the two right triangles (in gray shade) are identical. In particular, the two pink segments of the *x*–axis have the same length. This implies that the *x*–coordinate of the green vector is $x_1 + x_2$. By similar argument, the *y*–coordinate of the green vector is $y_1 + y_2$.

MATRIX

Let *M*(*m*, *n*) be the collection of all *m* by *n* matrices. The zero matrix is

$$O = \begin{bmatrix} 0 & 0 & \dots & 0 \\ 0 & 0 & \dots & 0 \\ \vdots & \vdots & & \vdots \\ 0 & 0 & \dots & 0 \end{bmatrix}$$

Like vectors, matrices of the same size may be added.

$$\begin{bmatrix} a_{11} & a_{12} & \dots & a_{1n} \\ a_{21} & a_{22} & \dots & a_{2n} \\ \vdots & \vdots & & \vdots \\ a_{m1} & a_{m2} & \dots & a_{mn} \end{bmatrix} + \begin{bmatrix} b_{11} & b_{12} & \dots & b_{1n} \\ b_{21} & b_{22} & \dots & b_{2n} \\ \vdots & \vdots & & \vdots \\ b_{m1} & b_{m2} & \dots & b_{mn} \end{bmatrix}$$

$$= \begin{bmatrix} a_{11}+b_{11} & a_{12}+b_{12} & \ldots & a_{1n}+b_{1n} \\ a_{21}+b_{21} & a_{22}+b_{22} & \ldots & a_{2n}+b_{2n} \\ \vdots & \vdots & & \vdots \\ a_{m1}+b_{m2} & a_{m2}+b_{m2} & \ldots & a_{mn}+b_{mn} \end{bmatrix}$$

A number (scalar) can also be multiplied to matrices

$$c\begin{bmatrix} a_{11} & a_{12} & \ldots & a_{1n} \\ a_{21} & a_{22} & \ldots & a_{2n} \\ \vdots & \vdots & & \vdots \\ a_{m1} & a_{m2} & \ldots & a_{mn} \end{bmatrix} = \begin{bmatrix} ca_{11} & ca_{12} & \ldots & ca_{1n} \\ ca_{21} & ca_{22} & \ldots & ca_{2n} \\ \vdots & \vdots & & \vdots \\ ca_{m1} & ca_{m2} & \ldots & ca_{mn} \end{bmatrix}$$

Moreover, we may combine the two operations and form the linear combinations of matrices (of the same size).

Example: Let

$$A = \begin{bmatrix} 1 & 1 & -2 \\ 3 & 0 & 1 \end{bmatrix}, B = \begin{bmatrix} -1 & 0 & 1 \\ -2 & 1 & 1 \end{bmatrix}, C = \begin{bmatrix} 0 & 1 & 0 \\ 1 & 0 & 0 \end{bmatrix}, D = \begin{bmatrix} 0 & 1 \\ 1 & 2 \end{bmatrix}$$

Then

$$A + B = \begin{bmatrix} 1+(-1) & 1+0 & (-2)+1 \\ 3+(-2) & 0+1 & 1+1 \end{bmatrix} = \begin{bmatrix} 0 & 1 & -1 \\ 1 & 1 & 2 \end{bmatrix}$$

$$A + B + C = \begin{bmatrix} 1+(-1)+0 & 1+0+1 & (-2)+1+0 \\ 3+(-2)+1 & 0+1+0 & 1+1+0 \end{bmatrix} = \begin{bmatrix} 0 & 2 & -1 \\ 2 & 1 & 2 \end{bmatrix}$$

$$2A = \begin{bmatrix} 2\times 1 & 2\times 1 & 2\times(-2) \\ 2\times 3 & 2\times 0 & 2\times 1 \end{bmatrix} = \begin{bmatrix} 2 & 2 & -4 \\ 6 & 0 & 2 \end{bmatrix}$$

$$-3D = \begin{bmatrix} (-3)\times 0 & (-3)\times 1 \\ (-3)\times 1 & (-3)\times 2 \end{bmatrix} = \begin{bmatrix} 0 & -3 \\ -3 & -6 \end{bmatrix}$$

$$3A - 4B + 2C = \begin{bmatrix} 3+4+0 & 3+0+2 & -6-4+0 \\ 9+8+2 & 0-4+0 & 3-4+0 \end{bmatrix} = \begin{bmatrix} 7 & 5 & -10 \\ 19 & -4 & -1 \end{bmatrix}$$

On the other hand, expressions such as $A + D$, $B + D$, $A + D$ are meaningless.

Euclidean vectors can be considered as matrices of special sizes. Specifically, a vertical n–dimensional euclidean vector is an n by 1 matrix, and the euclidean space R^n is identified with $M(n, 1)$. On the other hand, if we present vectors horizontally, then R^n is identified with $M(1, n)$.

Although euclidean vectors may be presented either vertically or horizontally, you are strongly advised to use vertical vectors when they are mixed with matrices. For example, in expressing a system of linear equation as $Ax = b$, because of the presence of the matrix A, we should consider x and b as vertical vectors.

Conversely, matrices can be considered as an (ordered) collection of vectors. Specifically, an m by n matrix is equivalent to n euclidean column vectors of dimension m.

$$A = [a_1\ a_2 \dots\ a_n] \Leftrightarrow a_1, a_2, \dots, a_n \in R^n.$$

Similarly, we may also consider row vectors, so that an m by n matrix is equivalent to m euclidean row vectors of dimension n.

Example: The matrix

$$A = \begin{bmatrix} 1 & 2 & 3 \\ 4 & 5 & 6 \end{bmatrix}$$

can be considered as three 2–dimensional column vectors

$$A = [a_1\ a_2\ a_3],\ a_1 = \begin{bmatrix} 1 \\ 4 \end{bmatrix},\ a_2 = \begin{bmatrix} 2 \\ 5 \end{bmatrix},\ a_3 = \begin{bmatrix} 3 \\ 6 \end{bmatrix}$$

It can also be considered as two 3–dimensional row vectors.

$$A = \begin{bmatrix} b_1 \\ b_2 \end{bmatrix},\ b_1 = (1, 2, 3),\ b_2 = (4, 5, 6)$$

The relations between the vectors and the matrices are compatible with addition and scalar multiplication. For example, for three column matrices $A = [a_1\ a_2\ a_3]$, $B = [b_1\ b_2\ b_3]$, we have

$$A + B = [a_1 + b_1\ a_2 + b_2\ a_3 + b_3],\ cA = [ca_1\ ca_2\ ca_3].$$

Because of this compatibility, the properties for the addition and scalar multiplication of vectors are also hold for matrices.

SOME MATRIX NOTATIONS AND TERMINOLOGIES

In theoretical discussions about matrices, we sometimes use the notation $A = (a_{ij})$ to denote a matrix, where a_{ij} is the entry at the i–th row and j–th column (the ij–entry). Sometimes we even indicate the size by writing $A = (a_{ij})_{m\times n}$. In terms of this compact notation, we have

$$(a_{ij}) + (b_{ij}) = (a_{ij} + b_{ij}),\ c(a_{ij}) = (ca_{ij}).$$

Example:

$$(i+j)_{3\times3} = \begin{bmatrix} 1+1 & 1+2 & 1+3 \\ 2+1 & 2+2 & 2+3 \\ 3+1 & 3+2 & 3+3 \end{bmatrix} = \begin{bmatrix} 2 & 3 & 4 \\ 3 & 4 & 5 \\ 4 & 5 & 6 \end{bmatrix}$$

$$(ij)_{4\times5} = \begin{bmatrix} 1\times1 & 1\times2 & 1\times3 & 1\times4 & 1\times5 \\ 2\times1 & 2\times2 & 2\times3 & 2\times4 & 2\times5 \\ 3\times1 & 3\times2 & 3\times3 & 3\times4 & 3\times5 \\ 4\times1 & 4\times2 & 4\times3 & 4\times4 & 4\times5 \end{bmatrix} = \begin{bmatrix} 1 & 2 & 3 & 4 & 5 \\ 2 & 4 & 6 & 8 & 10 \\ 3 & 6 & 9 & 12 & 15 \\ 4 & 8 & 12 & 16 & 20 \end{bmatrix}$$

$$(j^2)_{4\times3} = \begin{bmatrix} 1^2 & 2^2 & 3^2 \\ 1^2 & 2^2 & 3^2 \\ 1^2 & 2^2 & 3^2 \\ 1^2 & 2^2 & 3^2 \end{bmatrix} = \begin{bmatrix} 1 & 4 & 9 \\ 1 & 4 & 9 \\ 1 & 4 & 9 \\ 1 & 4 & 9 \end{bmatrix}$$

For an n by n square matrix $A = (a_{ij})$, the terms $a_{11}, a_{22}, \ldots, a_{nn}$ are the diagonal entries of the matrix. A is a diagonal matrix if all the off–diagonal entries are zero. In other words, $a_{ij} = 0$ for $i \neq j$. The following are all the 2 by 2 and 3 by 3 diagonal matrices.

$$\begin{bmatrix} a & 0 \\ 0 & b \end{bmatrix}$$

$$\begin{bmatrix} a & 0 & 0 \\ 0 & b & 0 \\ 0 & 0 & c \end{bmatrix}$$

A is an upper triangular matrix if all the entries below the diagonal are zero. In other words, $a_{ij} = 0$ for $i > j$. The following are all the 2 by 2 and 3 by 3 upper triangular matrices.

$$\begin{bmatrix} a & b \\ 0 & c \end{bmatrix}$$

$$\begin{bmatrix} a & b & c \\ 0 & d & e \\ 0 & 0 & f \end{bmatrix}$$

Similarly, A is a lower triangular matrix if all the entries above the diagonal are zero. In other words, $a_{ij} = 0$ for $i < j$. The following are all the 2 by 2 and 3 by 3 upper triangular matrices.

$$\begin{bmatrix} a & 0 \\ b & c \end{bmatrix}$$

$$\begin{bmatrix} a & 0 & 0 \\ b & c & 0 \\ d & e & f \end{bmatrix}$$

Clearly, A is upper and lower triangular at the same time if and only if it is diagonal.

Exercise: Draw the vectors (1, 3), (0, 1), (0, 0), (–1, –3), (2, –2).

Answer:

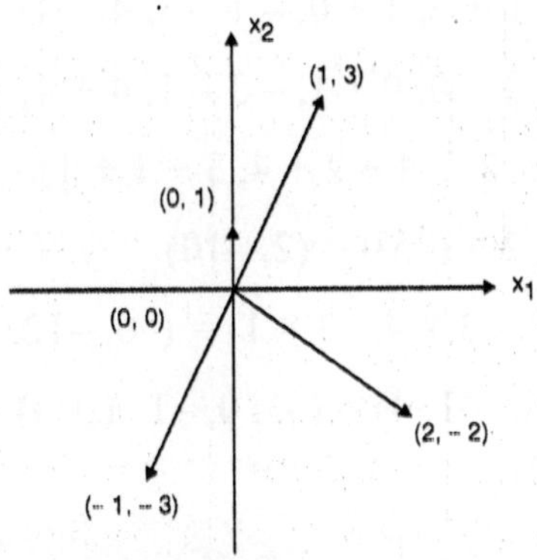

Exercise: Draw the collections of vectors.

$X = \{(0, y): y \text{ is any number}\}$

$Y = \{(x, 0): x \text{ is any number}\}$

$S = \{(x, y): x^2 + y^2 = 1\}$

$D = \{(x, x): x \text{ is any number}\}$

$E = \{(x, y): x + 3y = 0\}$

Answer:

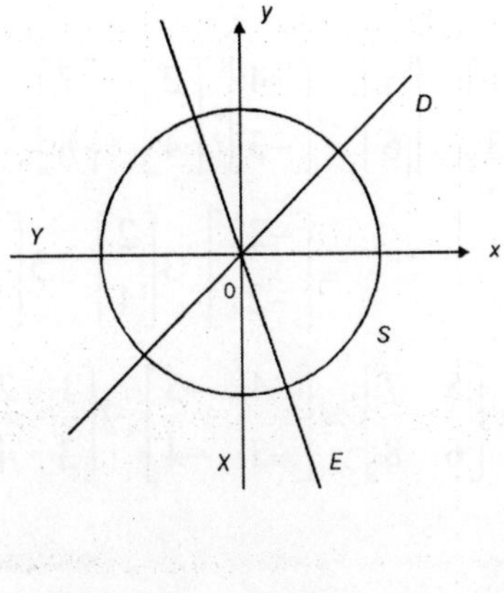

Exercise: Let $u = (1,-5)$, $v = (2, 4,1)$, $w = (3, 0, 1, -1, 4)$, $x = (2, -4, 3)$, $y = (-7, 2, 1)$, $z = (1, 2, 0, -3, 4)$. Compute.

$$u + v,\ v + y,\ y + v,\ w + z,\ z + w,\ x + y + v,$$

$$2u,\ -3v,\ -w,\ 0x,$$

$$2u + 5v,\ -2y + 3v,\ w - 3z,\ 4x + 2y + v.$$

Answer: $u + v$ is meaningless

$$v + y = (2 - 7, 4 + 2, 1 + 1) = (-5, 6, 2)$$

$$y + v = (-7 + 2, 2 + 4, 1 + 1) = (-5, 6, 2)$$

$$w + z = (3 + 1, 0 + 2, 1 + 0, -1 - 3, 4 + 4) = (4, 2, 1, -4, 8)$$

$$z + w = (1 + 3, 2 + 0, 0 + 1, -3 - 1, 4 + 4) = (4, 2, 1, -4, 8)$$

$$x + y + v = (2 - 7 + 2, -4 + 2 + 4, 3 + 1 + 1) = (-3, 2, 5)$$

$$2u = (2 \times 1, 2 \times (-5)) = (2, -10)$$

$$-3v = (-3 \times 2, -3 \times 4, -3 \times 1) = (-6, -12, -3)$$

$$-w = -(3, 0, 1, -1, 4) = (-3, 0, -1, 1, -4)$$

$$0x = (0, 0, 0)$$

$2u + 5v$ is meaningless

$$-2y + 3v = -2(-7, 2, 1) + 3(2, 4,1) = (-2 \times (-7) + 3 \times 2, -2 \times 2 + 3 \times 4, -2 \times 1 + 3\times1) = (20, 8, 1)$$

$$w - 3z = (3, 0, 1, -1, 4) - 3(1, 2, 0, -3, 4) = (3 - 3 \times 1, 0 - 3 \times 2, 1 - 3 \times 0, -1 - 3 \times (-3), 4 - 3 \times 4) = (0, -6, 1, 8, -8)$$

$$4x + 2y + v = 4(2, -4, 3) + 2(-7, 2, 1) + (2, 4,1)$$

$$= (4 \times 2 + 2 \times (-7) + 2, 4 \times (-4) + 2 \times 2 + 4, 4 \times 3 + 2 \times 1+ 1)$$

$$= (-4, 8, 15)$$

Exercise: Compute.

Answer:

$$\begin{bmatrix}1\\3\end{bmatrix} - \begin{bmatrix}5\\6\end{bmatrix} = \begin{bmatrix}-4\\-3\end{bmatrix}, \begin{bmatrix}2\\4\end{bmatrix} - \begin{bmatrix}7\\8\end{bmatrix} = \begin{bmatrix}-5\\-4\end{bmatrix}, 3\begin{bmatrix}1\\3\end{bmatrix} - 5\begin{bmatrix}5\\6\end{bmatrix}$$

$$= \begin{bmatrix}-22\\-21\end{bmatrix}, 3\begin{bmatrix}2\\4\end{bmatrix} - 5\begin{bmatrix}7\\8\end{bmatrix} = \begin{bmatrix}-29\\-28\end{bmatrix}$$

$$\begin{bmatrix}1 & 2\\3 & 4\end{bmatrix} - \begin{bmatrix}5 & 7\\6 & 8\end{bmatrix} = \begin{bmatrix}-4 & -5\\-3 & -4\end{bmatrix}, 3\begin{bmatrix}1 & 2\\3 & 4\end{bmatrix} - 5\begin{bmatrix}5 & 7\\6 & 8\end{bmatrix}$$

$$\begin{bmatrix} 0 & -2 & 3 \\ 1 & 0 & -3 \\ 2 & -2 & 2 \end{bmatrix} + \begin{bmatrix} 2 & 0 & -1 \\ 0 & 4 & -3 \\ 3 & -3 & 0 \end{bmatrix} = \begin{bmatrix} 2 & -2 & 2 \\ 1 & 4 & -6 \\ 5 & -5 & 2 \end{bmatrix}, -2\begin{bmatrix} 0 & -2 & 3 \\ 1 & 0 & -3 \\ 2 & -2 & 2 \end{bmatrix}$$

$$+3\begin{bmatrix} 2 & 0 & -1 \\ 0 & 4 & -3 \\ 3 & -3 & 0 \end{bmatrix} = \begin{bmatrix} 4 & 4 & -9 \\ -2 & 12 & -15 \\ 5 & -5 & -4 \end{bmatrix}$$

$$a\begin{bmatrix} 1 & 0 & 0 \\ 0 & 0 & 0 \\ 0 & 0 & 0 \end{bmatrix} + b\begin{bmatrix} 0 & 0 & 0 \\ 0 & 1 & 0 \\ 0 & 0 & 0 \end{bmatrix} + c\begin{bmatrix} 0 & 0 & 0 \\ 0 & 0 & 0 \\ 0 & 0 & 1 \end{bmatrix} = \begin{bmatrix} a & 0 & 0 \\ 0 & b & 0 \\ 0 & 0 & c \end{bmatrix}$$

$$a_{11}\begin{bmatrix} 1 & 0 \\ 0 & 0 \end{bmatrix} + a_{12}\begin{bmatrix} 0 & 1 \\ 0 & 0 \end{bmatrix} + a_{21}\begin{bmatrix} 0 & 0 \\ 1 & 0 \end{bmatrix} + a_{22}\begin{bmatrix} 0 & 0 \\ 0 & 1 \end{bmatrix} = \begin{bmatrix} a_{11} & a_{12} \\ a_{21} & a_{22} \end{bmatrix}$$

$$\begin{bmatrix} 1 & -3 & 5 & 10 \\ 0 & 5 & 1 & -7 \\ 1 & 0 & 9 & -3 \\ 2 & 0 & 1 & 0 \\ -6 & 3 & 0 & 0 \end{bmatrix} + \begin{bmatrix} 4 & -9 & 8 \\ 4 & 0 & 3 \\ 1 & 4 & 0 \\ 0 & 0 & 4 \\ 0 & 5 & 3 \end{bmatrix} = ?, 12\begin{bmatrix} 0 & 0 & 0 & 0 & 0 \\ 0 & 0 & 0 & 0 & 0 \\ 0 & 0 & 0 & 0 & 0 \end{bmatrix} - 13\begin{bmatrix} 0 & 0 & 0 & 0 & 0 \\ 0 & 0 & 0 & 0 & 0 \\ 0 & 0 & 0 & 0 & 0 \\ 0 & 0 & 0 & 0 & 0 \\ 0 & 0 & 0 & 0 & 0 \end{bmatrix}$$

The last two are meaningless because the sizes do not match.

Exercise: Take the standard basis vectors $e_1 = (1, 0, 0)$, $e_2 = (0, 1, 0)$, $e_3 = (0, 0, 1)$ of R^3. Compute the linear combination $x_1e_1 + x_2e_2 + x_3e_3$. Then generalize your result to other dimensions.

Answer: $x_1e_1 + x_2e_2 + x_3e_3 = x_1(1, 0, 0) + x_2(0, 1, 0) + x_3(0, 0, 1) = (x_1, 0, 0) + (0, x_2, 0) + (0, 0, x_3) = (x_1, x_2, x_3)$. In general, for the standard basis $e_1 = (1, 0,..., 0)$, $e_2 = (0, 1,..., 0),..., e_n = (0, 0,..., 1)$ of R^n, we have $(x_1, x_2,..., x_n) = x_1e_1 + x_2e_2 + ... + x_ne_n$.

Exercise: For $u = (1, 2)$, $v = (3, 4)$, $w = (5, 6)$, compute the linear combination $x_1u + x_2v + x_3w$. Do the same for $a_1 = (1, 2, 3)$, $a_2 = (4, 5, 6)$, $a_3 = (7, 8, 9)$. Express your computation in vertical vectors and draw a general observation.

Answer:

$$x_1\begin{bmatrix} 1 \\ 2 \end{bmatrix} + x_2\begin{bmatrix} 3 \\ 4 \end{bmatrix} + x_3\begin{bmatrix} 5 \\ 6 \end{bmatrix} = \begin{bmatrix} x_1 \\ 2x_1 \end{bmatrix} + \begin{bmatrix} 3x_2 \\ 4x_2 \end{bmatrix} + \begin{bmatrix} 5x_3 \\ 6x_3 \end{bmatrix} = \begin{bmatrix} x_1 + 3x_2 + 5x_3 \\ 2x_1 + 4x_2 + 6x_3 \end{bmatrix}$$

$$x_1\begin{bmatrix} 1 \\ 2 \\ 3 \end{bmatrix} + x_2\begin{bmatrix} 4 \\ 5 \\ 6 \end{bmatrix} + x_3\begin{bmatrix} 7 \\ 8 \\ 9 \end{bmatrix} = \begin{bmatrix} x_1 \\ 2x_1 \\ 3x_1 \end{bmatrix} + \begin{bmatrix} 4x_2 \\ 5x_2 \\ 6x_2 \end{bmatrix} + \begin{bmatrix} 7x_3 \\ 8x_3 \\ 9x_3 \end{bmatrix} = \begin{bmatrix} x_1 + 4x_2 + 7x_3 \\ 2x_1 + 5x_2 + 8x_3 \\ 3x_1 + 6x_2 + 9x_3 \end{bmatrix}$$

The results are of the form Ax, with the matrix A being respectively

$$\begin{bmatrix} 1 & 3 & 5 \\ 2 & 4 & 6 \end{bmatrix} = [u\ v\ w]$$

$$\begin{bmatrix} 1 & 4 & 7 \\ 2 & 5 & 8 \\ 3 & 6 & 9 \end{bmatrix} = [a_1\ a_2\ a_3]$$

In general, we have

$x_1a_1 + x_2a_2 + ... + x_na_n = Ax$

where $A = [a_1\ a_2 ...\ a_n]$ is a matrix with the vectors $a_1, a_2, ..., a_n$ as columns.

Exercise: Express the vectors $b = (4, 6)$, $c = (1, 1)$, $e_1 = (1, 0)$, $e_2 = (0, 1)$ as linear combinations of the vectors $u = (1, 2)$, $v = (3, 4)$, $w = (5, 6)$. Can you express as linear combinations of the vectors u, v only?

***Answer*:** We would like to write b, c, e_1, e_2 in the form $x_1(1, 2) + x_2(3, 4) + x_3(5, 6)$. From the previous exercise, we have

$$x_1\begin{bmatrix} 1 \\ 2 \end{bmatrix} + x_2\begin{bmatrix} 3 \\ 4 \end{bmatrix} + x_3\begin{bmatrix} 5 \\ 6 \end{bmatrix} = \begin{bmatrix} x_1 + 3x_2 + 5x_3 \\ 2x_1 + 4x_2 + 6x_3 \end{bmatrix}$$

Therefore the problem is to solve the system of equations

$$x_1 + 3x_2 + 5x_3 = b_1$$
$$2x_1 + 4x_2 + 6x_3 = b_2$$

For $b_1 = 4$, $b_2 = 6$ on the right side, we find $x_1 = 1$, $x_2 = 1$ is one solution. Therefore $b = u + v$. In fact, the solution is not unique. For example, $x_1 = 2$, $x_2 = -1$, $x_3 = 1$ is another solution, from which we get $b = 2u - v + w$.

Incidentally, we have already expressed b as a linear combination of u and v: $b = u + v$. Such expressions are equivalent to solutions of

$$x_1 + 3x_2 = b_1$$
$$2x_1 + 4x_2 = b_2$$

We do the similar computation for the other vectors and find $c = -1/2\ u + 1/2\ v$, $e_1 = -2u + v$, $e_2 = 3/2\ u - 1/2\ v$.

Exercise: Express the vectors $b = (1, 1, 1)$, $c = (-1, 0, 1)$, $e_1 = (1, 0, 0)$ as linear combinations of the vectors $a_1 = (1, 2, 3)$, $a_2 = (4, 5, 6)$, $a_3 = (7, 8, 9)$. Is the expression (i.e., the coefficients) unique?

Answer: We would like to write b, c, e_1 in the form $x_1(1, 2, 3) + x_2(4, 5, 6) + x_3(7, 8, 9)$. The computation is similar to the previous exercise. From this exercise, we have

$$x_1\begin{bmatrix}1\\2\\3\end{bmatrix}+x_2\begin{bmatrix}4\\5\\6\end{bmatrix}+x_3\begin{bmatrix}7\\8\\9\end{bmatrix}=\begin{bmatrix}x_1+4x_2+7x_3\\2x_1+5x_2+8x_3\\3x_1+6x_2+9x_3\end{bmatrix}$$

Therefore the problem is to solve the system.

$$x_1 + 4x_2 + 7x_3 = b_1$$
$$2x_1 + 5x_2 + 8x_3 = b_2$$
$$3x_1 + 6x_2 + 9x_3 = b_3$$

For $b_1 = b_2 = b_3 = 1$ on the right side, we find $x_1 = -1/3$, $x_2 = 1/3$, $x_3 = 0$ is one solution. Therefore $b = -1/3\ a_1 + 1/3\ a_2$. Since the solution is not unique, the expression is not unique. Another possibility is $b = -1/6\ a_1 + 1/6\ a_3$.

For $b_1 = -1$, $b_2 = 0$, $b_3 = 1$ on the right side, we find $x_1 = 4/3$, $x_2 = 0$, $x_3 = -1/3$ is one solution. Therefore $c = 4/3\ a_1 - 1/3\ a_3$. Again the expression is not unique.

For $b_1 = 1$, $b_2 = 0$, $b_3 = 0$ on the right side, there is no solution. Therefore e_1 is not a linear combination of a_1, a_2, a_3.

Exercise: Express the matrix

$$X=\begin{bmatrix}3&4\\2&3\end{bmatrix}$$

as linear combinations of the matrices.

$$A=\begin{bmatrix}1&2\\0&1\end{bmatrix},\ B=\begin{bmatrix}1&0\\2&1\end{bmatrix},\ C=\begin{bmatrix}0&1\\-1&0\end{bmatrix}$$

Is the expression (i.e., the coefficients) unique? Is it true that any 2 by 2 matrix is a linear combination of A, B and C?

Answer: By comparing the four entries on the both side of the equation $x_1A + x_2B + x_3C = X$,

$$x_1\begin{bmatrix}1&2\\0&1\end{bmatrix}+x_2\begin{bmatrix}1&0\\2&1\end{bmatrix}+x_3\begin{bmatrix}0&1\\-1&0\end{bmatrix}=\begin{bmatrix}3&4\\2&3\end{bmatrix}$$

we have the following system of equations.

$$x_1 + x_2 = 3$$
$$2x_1 + x_3 = 4$$

$$2x_2 - x_3 = 2$$
$$x_1 + x_2 = 3$$

From the row echelon form of the augmented matrix,

$$\begin{bmatrix} 1 & 1 & 0 & 3 \\ 0 & 2 & -1 & 2 \\ 0 & 0 & 0 & 0 \\ 0 & 0 & 0 & 0 \end{bmatrix}$$

we see that the system has non–unique solutions. One solution is $x_1 = 2$, $x_2 = 1$, $x_3 = 0$. Therefore $X = 2A + 1B + 0C$. Moreover, the expression is not unique.

For a general 2 by 2 matrix $X = (b_{ij})$ (i.e., b_{ij} is the entry of X at the i–th row and j–th column), the corresponding system is

$$x_1 + x_2 = b_{11}$$
$$2x_1 + x_3 = b_{12}$$
$$2x_2 - x_3 = b_{21}$$
$$x_1 + x_2 = b_{22}$$

Then from the row echelon form above (or by the fact that there are more equations than variables), we see that solutions do not always exist. In other words, not any X can be expressed as a linear combination of A, B and C.

Remark: In an earlier exercise, we have seen how the problem of expressing a vector as a linear combination of some other vectors is equivalent to solving a system of linear equations. This exercise shows that we have similar relation for matrices.

LINEAR TRANSFORMATION

TRANSFORMATION

Transformations are rules that convert one data into another data. We may denote a transformation by

$T: X \to Y$,

where T is the transformation (also called map, function), X is the source set (or domain), and Y is the target set (or codomain). Taking any element $x \in X$ as input, the transformation produces an element $y = T(x) \in Y$ as output. We also call y the image (or value) of x under the transformation T. We may also include more specific information about the transformation by writing $T: X \to Y, x \to$

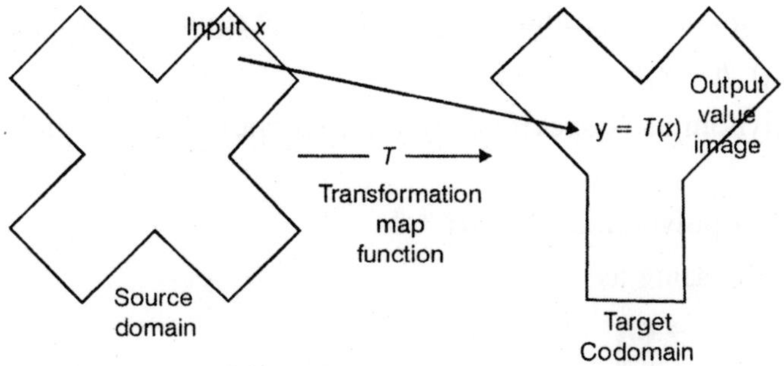

A more rigorous theory of transformations can be found here.

Example: The following are some transformations in everyday life.

Age: people → number

ID_s: student → number

ID_p: professor → number

Instructor: course → professor

Maker: product → manufacturer

Capital City: country → city

Population: city → number

Example In mathematics, transformations are often given by formulae.

Square: $R \to R, x \to x^2$

Reflection in x–axis: $R^2 \to R^2, (x, y) \to (x, -y)$

Polar–Cartesian: $R^2 \to R^2, (r, è) \to (r\cos è, r\sin è)$

Root–Polynomial: $R^2 \to R^2, (x_1, x_2) \to (x_1 + x_2, x_1x_2)$

T: $R^3 \to R^2, (x_1, x_2, x_3) \to (x_1 + 2x_2 + 3x_3, 4x_1 + 5x_2 + 6x_3)$

Here are two examples of mathematical transformations given by descriptions instead of formulae.

Sign: R → R, positive number → 1
negative number → –1
0 → 0

Dirichlet function: R → R, rational number → 1
irrational number → 0

ROOT-POLYNOMIAL TRANSFORMATION

If x_1, x_2 are the roots of a quadratic polynomial $x^2 + ax + b$, then we have $(x - x_1)(x - x_2) = x^2 + ax + b$.

By expanding the left side and compare with the right side, we get

$a = -(x_1 + x_2),\ b = x_1 x_2.$

Since the polynomial is specified by the two coefficients a and b, we see that the transformation

roots $(x_1, x_2) \to$ polynomial $x^2 + ax + b$

is essentially the same as

$R^2 \to R^2,\ (x_1, x_2) \to (x_1 + x_2, x_1 x_2)$

As an exercise for yourself, how would you define the transformation *Root–Polynomial*: $R^3 \to R^3$ for three roots of polynomials of degree 3?

Finally, you may try the general case *Root–Polynomial*: $R^n \to R^n$.

MATRIX TRANSFORMATION

We may think of the left side of a system of linear equations as a transformation. For example, the left side of the system

$$x_1 + 2x_2 + 3x_2 = b_1$$
$$4x_1 + 5x_2 + 6x_4 = b_2$$

can be considered as the linear transformation T in the example above. By expressing vectors in vertical form, we also see that

$$T\begin{bmatrix} x_1 \\ x_2 \\ x_3 \end{bmatrix} = \begin{bmatrix} x_1 + 2x_2 + 3x_3 \\ 4x_1 + 5x_2 + 6x_3 \end{bmatrix} = \begin{bmatrix} 1 & 2 & 3 \\ 4 & 5 & 6 \end{bmatrix}\begin{bmatrix} x_1 \\ x_2 \\ x_3 \end{bmatrix}$$

In particular, the formula for the transformation T is given by a matrix A:

$$T(x) = Ax.$$

In general, for any m by n matrix A, we may construct a transformation T: $R^n \to R^m$ by the above formula. Such transformations are called matrix transformations.

Example: The following matrices

$$\begin{bmatrix} 1 & -1 & 1 \\ 3 & 1 & -1 \\ 2 & 2 & 1 \end{bmatrix}$$

$$\begin{bmatrix} 1 & 3 & 2 & 0 & 1 \\ -1 & -1 & -1 & 1 & 0 \\ 0 & 4 & 2 & 4 & 3 \\ 1 & 3 & 2 & -2 & 0 \end{bmatrix}$$

give us matrix transformations,

$$T_1\begin{bmatrix}x_1\\x_2\\x_3\end{bmatrix}=\begin{bmatrix}x_1-x_2+x_3\\3x_1+x_2-x_3\\2x_1+x_2+x_3\end{bmatrix}:\mathrm{R}^3\to\mathrm{R}^3$$

$$T_1\begin{bmatrix}x_1\\x_2\\x_3\\x_4\\x_5\\x_6\end{bmatrix}=\begin{bmatrix}x_1+3x_2+2x_3+x_5\\-x_1-x_2-x_3+x_4+x_6\\4x_2+2x_3+4x_4+3x_5+3x_6\\x_1+3x_2+2x_3-2x_4\end{bmatrix}:\mathrm{R}^6\to\mathrm{R}^4$$

or written horizontally,

T_1: $\mathrm{R}^3 \to \mathrm{R}^3, (x_1, x_2, x_3) \to (x_1 - x_2 + x_3, 3x_1 + x_2 - x_3, 2x_1 + 2x_2 + x_3)$

T_2: $\mathrm{R}^6 \to \mathrm{R}^4, (x_1, x_2, x_3, x_4, x_5) \to (x_1 + 3x_2 + 2x_3 + x_5, -x_1 - x_2 - x_3 + x_4 + x_6, 4x_2 + 2x_3 + 4x_4 + 3x_5 + 3x_6, x_1 + 3x_2 + 2x_3 - 2x_4)$

Example The transformations

S: $\mathrm{R}^3 \to \mathrm{R}^3, (x_1, x_2, x_3) \to (x_1 - x_2 + x_3, 3x_1 + x_2 - x_3 - 2, 2x_1 + 2x_2 + x_3)$

T: $\mathrm{R}^2 \to \mathrm{R}^2, (x, y) \to (xy, x^2)$

are not matrix transformations because they cannot be expressed as Ax for a *constant* matrix A.

LINEAR TRANSFORMATION

The following example should convince you that the expression Ax is linear:

$A(x + y) = Ax + Ay, A(cx) = cAx.$

A more general proof can be found here.

Example: Let

$$A=\begin{bmatrix}a & b\\c & d\end{bmatrix}$$

The corresponding linear transformation is

$$T\begin{bmatrix}x_1\\x_2\end{bmatrix}=\begin{bmatrix}ax_1+bx_2\\cx_1+dx_2\end{bmatrix}:\mathrm{R}^2\to\mathrm{R}^2$$

We have

$$T\begin{bmatrix} x_1 + y_1 \\ x_2 + y_2 \end{bmatrix} = \begin{bmatrix} a(x_1 + y_1) + b(x_2 + y_2) \\ c(x_1 + y_1) + d(x_2 + y_2) \end{bmatrix} = \begin{bmatrix} ax_1 + bx_2 \\ cx_1 + dx_2 \end{bmatrix} + \begin{bmatrix} ay_1 + by_2 \\ cy_1 + dy_2 \end{bmatrix}$$

$$= T\begin{bmatrix} x_1 \\ x_2 \end{bmatrix} + T\begin{bmatrix} y_1 \\ y_2 \end{bmatrix}$$

and (we use k instead of c as the scalar to avoid confusion in notations)

$$T\begin{bmatrix} kx_1 \\ kx_2 \end{bmatrix} = \begin{bmatrix} akx_1 + bkx_2 \\ ckx_2 + dkx_2 \end{bmatrix} = \begin{bmatrix} k(ax_1 + bkx_2) \\ k(cx_1 + dx_2) \end{bmatrix} = k\begin{bmatrix} ay_1 + by_2 \\ cy_1 + dy_2 \end{bmatrix} kT\begin{bmatrix} x_1 \\ x_2 \end{bmatrix}$$

You are strongly encouraged to verify the linearity for matrices of other sizes.

The discussion suggests us to introduce the following concept.

A transformation T: $R^n \to R^m$ is a linear transformation if it satisfies

$$T(u + v) = T(u) + T(v),\ T(cu) = cT(u).$$

By taking $c = 0$ in the definition, we have

$$T(0) = 0.$$

Moreover, by combining the two properties in the definition, we see that linear transformations preserve linear combinations.

$$T(c_1u_1 + c_2u_2 + ... + c_ku_k) = c_1T(u_1) + c_2T(u_2) + ... + c_kT(u_k).$$

Example Suppose T: $R^2 \to R^2$ is a linear transformation. Suppose we also know (recall $e_1 = (1, 0)$, $e_2 = (0, 1)$)

$$T(e_1) = (1, 2),\ T(e_2) = (3, 4)$$

Then by

$$(5, -3) = 5\,(1, 0) - 3(0,1) = 5e_1 - 3e_2,$$

$$(-2, 4) = -2e_1 + 4e_2,$$

we have

$$T(5, -3) = 5T(e_1) - 3T(e_2) = 5(1, 2) - 3(3, 4) = (-4, -2),$$

$$T(-2, 4) = -2T(e_1) + 4T(e_2) = -2(1, 2) + 4(3, 4) = (10, 12).$$

In general, by

$$(x_1, x_2) = x_1e_1 + x_2e_2,$$

we have

$$T(x_1, x_2) = x_1T(e_1) + x_2T(e_2) = x_1(1, 2) + x_2(3, 4) = (x_1 + 3x_2, 2x_1 + 4x_2).$$

We remark that T is a matrix transformation given by

$$\begin{bmatrix} 1 & 3 \\ 2 & 4 \end{bmatrix} = [T(e_1)\ T(e_2)]$$

Example: The transformation

$$S(x_1, x_2, x_3) = (2x_1 + 3x_2 - 7x_3 + 1,\ 10x_1 - 4x_2 - 8x_3,\ -4x_1 + x_2 - 5)$$

is not a linear transformation because

$S(0) = S(0, 0, 0) = (1, 0, -5)$ “‘ 0

The transformation

$$T(x, y) = (x + y, xy)$$

does satisfy $T(0) = 0$. However, we have

$$T(1, 0) + T(0, 1) = (1 + 0, 1\times 0) + (0 + 1, 0\times 1) = (1, 0) + (1, 0) = (2, 0).$$

This is different from

$$T((1, 0) + (0, 1)) = T(1, 1) = (1 + 1, 1\times 1) = (2, 1).$$

Therefore T is not linear.

Since the transformations are not linear, they are not matrix transformations.

The linearity of Ax helps us to construct solutions of systems of linear equations. The key idea is that, if x and x' are respectively solutions of $Ax = b$ and $Ax = b'$, then $cx + c'x'$ is a solution of $Ax = cb + c'b'$.

Example: The vectors $\mathbf{x} = (1, 1, 2)$ and $x' = (0, 1, -1)$ are the respective solutions of the following systems.

$$\begin{aligned} x_1 - x_2 + x_3 &= 2 \\ 3x_1 + x_2 - x_3 &= 2 \\ 2x_1 + 2x_2 + x_3 &= 6 \end{aligned}$$

$$\begin{aligned} x_1 - x_2 + x_3 &= -2 \\ 3x_1 + x_2 - x_3 &= 2 \\ 2x_1 + 2x_2 + x_3 &= 1 \end{aligned}$$

Note that the two systems have the same left side. The following systems

$$\begin{aligned} x_1 - x_2 + x_3 &= 0 \\ 3x_1 + x_2 - x_3 &= 4 \\ 2x_1 + 2x_2 + x_3 &= 7 \end{aligned}$$

$$x_1 - x_2 + x_3 = 1$$

$$3x_1 + x_2 - x_3 = 1$$
$$2x_1 + 2x_2 + x_3 = 3$$
$$x_1 - x_2 + x_3 = 8$$
$$3x_1 + x_2 - x_3 = -4$$
$$2x_1 + 2x_2 + x_3 = 3$$

still have the same left side but different right sides. By

$$(0, 4, 7) = (2, 2, 6) + (-2, 2\ 1),$$
$$(1, 1, 3) = 1/2\ (2, 2, 6),$$
$$(8, -4, 3) = (2, 2, 6) - 3(-2, 2\ 1),$$

we see that

$$x + x' = (1, 1, 2) + (0, 1, -1) = (1, 2, 1),$$
$$1/2\ x = 1/2\ (1, 1, 2) = (1/2, 1/2, 1),$$
$$x - 3x' = (1, 1, 2) - 3(0, 1, -1) = (1, -2, 5),$$

are the respective solutions of the three new systems.

MATRIX TRANSFORMATIONS ARE LINEAR

Theorem $A(x + y) = Ax + Ay$, $A(cx) = cAx$.

Proof By an earlier exercise, we have

$$x_1a_1 + x_2a_2 + ... + x_na_n = Ax$$

where $A = [a_1\ a_2 ...\ a_n]$ is a matrix with the vectors $a_1, a_2, ..., a_n$ as columns, and $x_1, x_2, ..., x_n$ are the coordinates of x. Thus

$$\begin{aligned} Ax + Ay &= x_1a_1 + x_2a_2 + ... + x_na_n + y_1a_1 + y_2a_2 + ... + y_na_n \\ &= (x_1a_1 + y_1a_1) + (x_2a_2 + y_2a_2) + ... + (x_na_n + y_na_n) \\ &= (x_1 + y_1)a_1 + (x_2 + y_2)a_2 + ... + (x_n + y_n)a_n \end{aligned}$$

Since $x_1 + y_1, x_2 + y_2, ..., x_n + y_n$ are the coordinates of $x + y$, we see that the summation is $A(x + y)$.

The proof for $A(cx) = cAx$ is similar.

LINEARITY IN PICTURE

We can often visualize transformations between euclidean spaces. For example, the name of transformation *Reflection in x–axis*: $R^2 \to R^2$, $(x, y) \to (x, -y)$ is inspired by the picture.

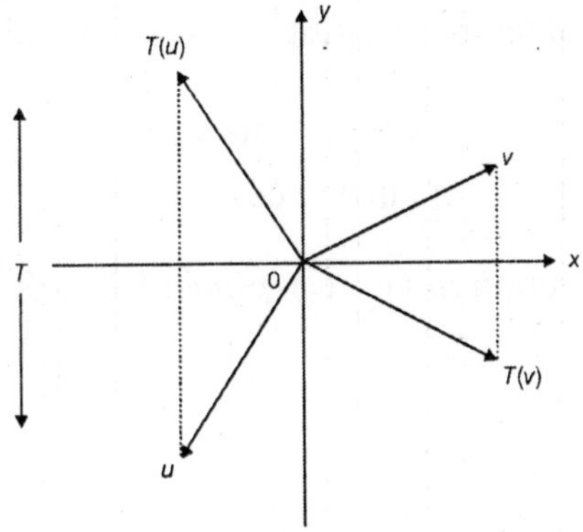

Example: The transformation *Reflection in y–axis*: $R^2 \to R^2$, $(x, y) \to (-x, y)$

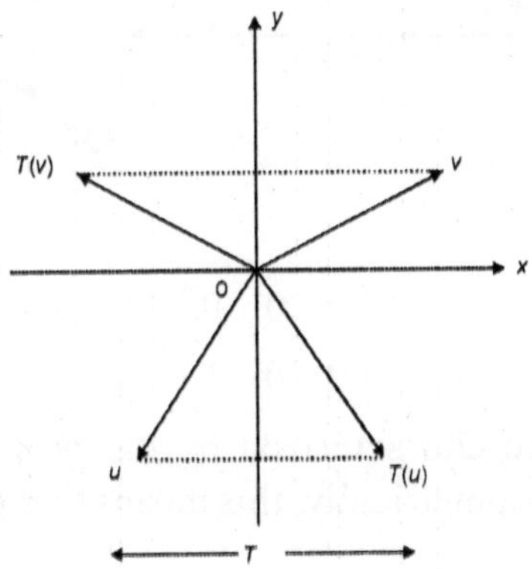

is given by the matrix

$$\begin{bmatrix} -1 & 0 \\ 0 & 1 \end{bmatrix}$$

The transformation *Rotation by &\theta;*: $R^2 \to R^2$, $(x, y) \to$ (*x*cos è + *y*sin è, – *x*sin è + *y*cos è)

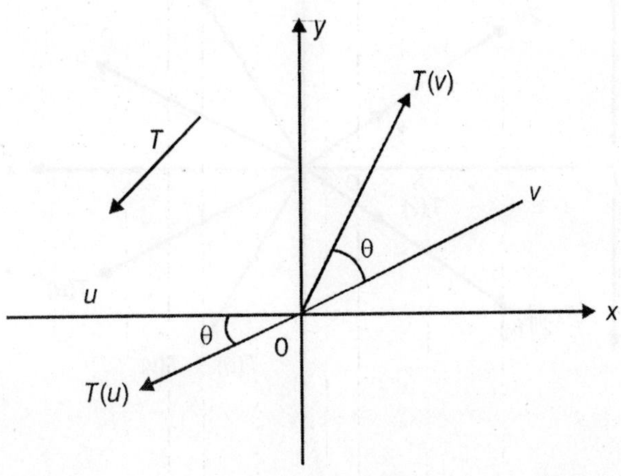

is given by the matrix (remember writing vectors vertically)

$$\begin{bmatrix} \cos\theta & -\sin\theta \\ \sin\theta & \cos\theta \end{bmatrix}$$

The transformation *Projection to* (x_1, x_3)*–plane*: $R^3 \to R^2$, $(x_1, x_2, x_3) \to (x_1, x_3)$

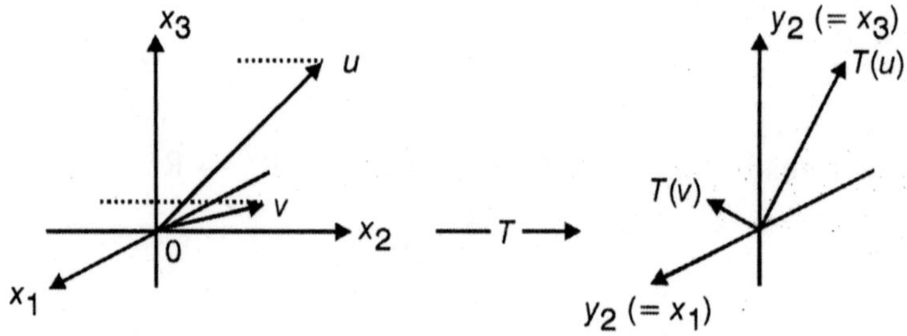

is given by the matrix

$$\begin{bmatrix} 1 & 0 & 0 \\ 0 & 0 & 1 \end{bmatrix}$$

Linear transformations are characterized by the property that addition and scalar multiplication are preserved. Geometrically, this means that parallelograms and stretching/shrinking are preserved.

Example: The transformation *Reflection in x–axis* clearly preserves parallelograms and stretching/shrinking.

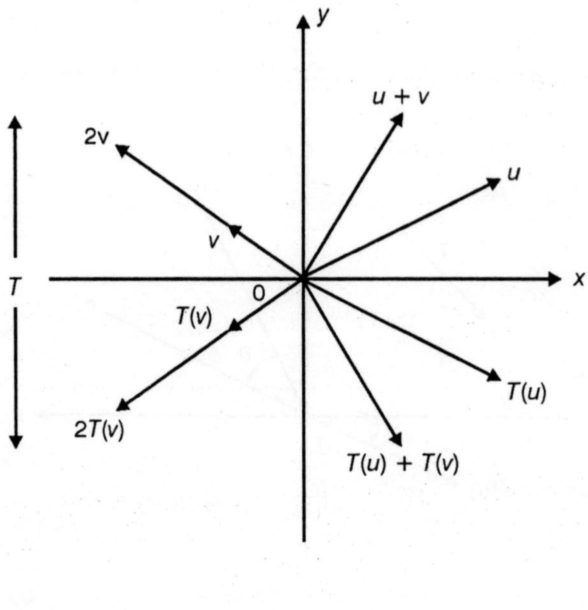

The rotation transformation also preserves the two operations.

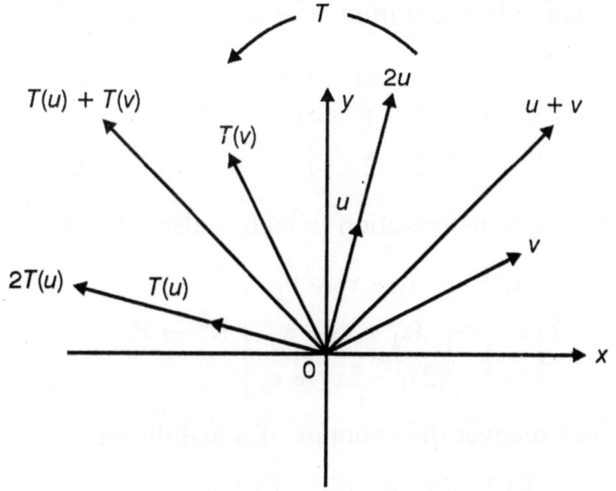

The projection transformation is also a linear transformation.

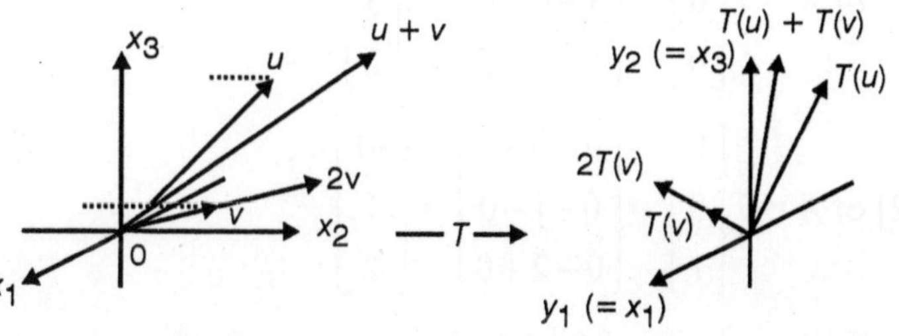

LINEAR TRANSFORMATION = MATRIX TRANSFORMATION

We have seen matrix transformations are linear. Conversely, this example suggests the converse is also true.

Any linear transformation T: $\mathrm{R}^n \to \mathrm{R}^m$ is also a matrix transformation, with the matrix given by

$$A = [T(e_1)\ T(e_2) \ldots\ T(e_n)].$$

In other words, the columns of the matrix A are the values of the linear transformation T at the standard basis. Thus the concepts of matrix transformations and linear transformations (between euclidean spaces) are equivalent.

Example: The identity transformation id: $\mathrm{R}^n \to \mathrm{R}^n$ is clearly linear. The corresponding matrix is

$$[id(e_1)\ id(e_2) \ldots\ id(e_n)] = [e_1\ e_2 \ldots\ e_n] = \begin{bmatrix} 1 & 0 & \ldots & 0 \\ 0 & 1 & \ldots & 0 \\ \cdot & \cdot & & \cdot \\ \vdots & \vdots & & \vdots \\ 0 & 0 & \ldots & 1 \end{bmatrix}$$

called the identity matrix and denoted by I_n.

Example: In this example, we started from a matrix

$$A = \begin{bmatrix} 1 & -1 & 1 \\ 3 & 1 & -1 \\ 2 & 2 & 1 \end{bmatrix}$$

and produced a matrix transformation (which is also a linear transformation).

$$T\begin{bmatrix} x_1 \\ x_2 \\ x_3 \end{bmatrix} = \begin{bmatrix} x_1 - x_2 + x_3 \\ 3x_1 + x_2 - x_3 \\ 2x_1 + 2x_2 + x_3 \end{bmatrix} \quad R^3 \to R^3$$

Conversely, we may recover the columns of A as follows

$$[\text{col } 1] \text{ of } A = T\begin{bmatrix} 1 \\ 0 \\ 0 \end{bmatrix} = \begin{bmatrix} 1-0+0 \\ 3+0-0 \\ 0-1+0 \end{bmatrix} = \begin{bmatrix} 1 \\ 3 \\ 2 \end{bmatrix}$$

$$[\text{col } 2] \text{ of } A = T\begin{bmatrix} 0 \\ 0 \\ 1 \end{bmatrix} = \begin{bmatrix} 0-0-0 \\ 0+1-0 \\ 0-2+0 \end{bmatrix} = \begin{bmatrix} -1 \\ 1 \\ 2 \end{bmatrix}$$

$$[\text{col } 3] \text{ of } A = T\begin{bmatrix} 0 \\ 0 \\ 1 \end{bmatrix} = \begin{bmatrix} 0-0+1 \\ 0+0-1 \\ 0-0+1 \end{bmatrix}$$

Example: The reflections and rotations on R^2 are linear transformations because they preserve parallelogram and stretching/shrinking. To find their matrices, we simply need to apply reflections and rotations to the standard basis.

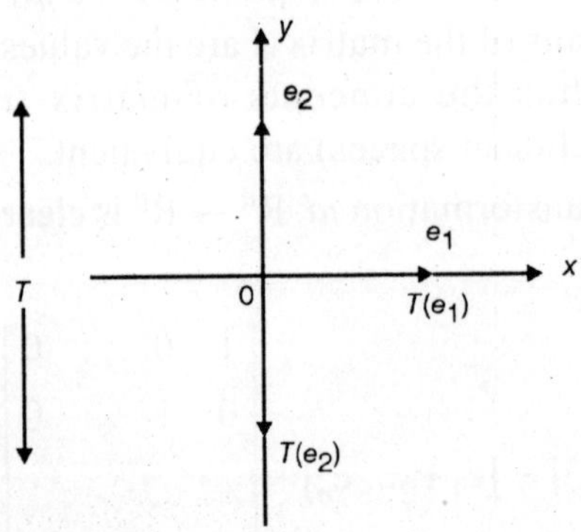

The picture shows that the reflections of e_1, e_2 in x–axis are $e_1 = (1, 0)$, $-e_2 = (0, -1)$.

Therefore the matrix for the reflection is

$$\begin{bmatrix} 1 & 0 \\ 0 & -1 \end{bmatrix}$$

Moreover, we also see from picture that for the rotation by angle è, $T(e_1)$ = (cos è, sin è), $T(e_2)$ = (–sin *è*, cos è).

Therefore the matrix for the rotation is

$$\begin{bmatrix} \cos\theta & -\sin\theta \\ \sin\theta & \cos\theta \end{bmatrix}$$

Exercise: Find the corresponding transformations, and indicate the source and target euclidean spaces.

1. [1 2]

Ans. $T(x_1, x_2) = x_1 + 2x_2$: $R^2 \to R^1$

2. $\begin{bmatrix} 1 \\ 2 \end{bmatrix}$

Ans. $T(x_1) = (x_1, 2x_1)$: $R^1 \to R^2$

3. $\begin{bmatrix} 2 & 0 & 2 & 1 \\ 1 & 1 & -1 & 0 \\ 0 & 1 & -2 & 1 \end{bmatrix}$

Ans. $T(x_1, x_2, x_3, x_4) = (x_1 + 2x_3 + x_4, x_1 + x_2 - x_3, x2 - 2x_3 + x_4)$: $R^4 \to R^3$

4. $\begin{bmatrix} 1 & 2 & 1 \\ 2 & 1 & 1 \\ 0 & 3 & 1 \\ 3 & 0 & 1 \end{bmatrix}$

Ans. $T(x_1, x_2, x_3) = (x_1 + 2x_2 + x_3, 2x_1 + x_2 + x_3, 3x2 + x_3, 3x_1 + x_3)$: $R^3 \to R^4$

5. $\begin{bmatrix} 1 & -1 & 1 & 0 & 2 \\ 2 & -2 & 0 & 2 & 2 \\ -1 & 1 & 2 & -3 & 1 \\ -2 & 2 & 1 & -3 & -1 \end{bmatrix}$

Ans. $T(x_1, x_2, x_3, x_4, x_5) = (x_1 - x_2 + x_3 + 2x_5, 2x_1 - 2x_2 + 2x_4 + 2x_5, - x_1 + x_2 + 2x_3 - 3x_4 + x5, - 2x_1 + 2x_2 + 2x_3 - 3x_4 - x5): R^5 \to R^4$

Exercise: Which of the following are matrix transformations? For the ones that are, find the corresponding matrices. For the ones that are not, provide the reason.

1. $T_1(x_1, x_2, x_3) = (2x_1 + 3x_2 - 7x_3, 10x_1 - 4x_2 - 8x_3, - 4x_1 + x_2)$

Ans. This is a matrix transformation.

$$\begin{bmatrix} 2 & 3 & -7 \\ 10 & -4 & -8 \\ -4 & 1 & 0 \end{bmatrix}$$

2. $T_2(x_1, x_2, x_3) = (2x_1 + 3x_2 - 7x_3 + 1, 10x_1 - 4x_2 - 8x_3, - 4x_1 + x_2 - 5)$

Ans. This is not a matrix/linear transformation because $T_2(0, 0, 0) = (1, 0, - 5)$, whereas any linear transformation must send zero vector to zero vector.

3. $T_3(x_1, x_2, x_3) = 2x_1 + 3x_2 - 7x_3$

Ans. This is a matrix transformation.

$$\begin{bmatrix} 2 & 3 & -7 \end{bmatrix}$$

4. $T_4(x_1, x_2, x_3) = 2x_1 + 3x_2 - 7x_3 + 1$

Ans. This is not a matrix/linear transformation because $T_4(0, 0, 0) = 1$, which is nonzero.

5. $T_5(x_1, x_2, x_3) = (2x_1, 3x_2, - 7x_3)$

Ans. This is a matrix transformation.

$$\begin{bmatrix} 2 & 0 & 0 \\ 0 & 3 & 0 \\ 0 & 0 & -7 \end{bmatrix}$$

6. $T_6(x_1, x_2, x_3) = (2x_2, 3x_3, - 7x_1)$

Ans. This is a matrix transformation.

$$\begin{bmatrix} 2 & 0 & 0 \\ 0 & 0 & 3 \\ -7 & 0 & 0 \end{bmatrix}$$

7. $T_7(x_1, x_2, x_3) = (3x_2, - 7x_3, 2x_1)$

Ans. This is a matrix transformation.

$$\begin{bmatrix} 0 & 3 & 0 \\ 0 & 0 & -7 \\ 2 & 0 & 0 \end{bmatrix}$$

8. $T_8(x) = (2x, 3x, -7x)$

Ans. This is a matrix transformation.

$$\begin{bmatrix} 2 \\ 3 \\ -7 \end{bmatrix}$$

9. $T_9(x, y, z) = (x - y, y - z, z - x)$

Ans. This is a matrix transformation.

$$\begin{bmatrix} 1 & -1 & 0 \\ 0 & 1 & -1 \\ -1 & 0 & 1 \end{bmatrix}$$

10. $\boldsymbol{T}_{10}(x, y) = (x + y, xy)$

Ans. This is not a matrix/linear transformation because $T_{10}(1, 0) = T_{10}(0, 1) = (1,0)$, $T_{10}(1, 1) = (2, 1) \to T_{10}(1, 0) + T_{10}(0, 1)$.

11. $T_{11}(x, y) = (x + y, 2^{1/2}x + 3^{1/2}y)$

Ans. This is a matrix transformation.

$$\begin{bmatrix} 1 & 1 \\ 2^{1/2} & 3^{1/2} \end{bmatrix}$$

12. $\boldsymbol{T}_{12}(u, v) = (u + v, u^2 + v^2, u^3 + v^3)$

Ans. This is not a matrix/linear transformation because $T_{12}(1, 1) = (2, 2, 2)$, $T_{12}(2, 2) = (4, 8, 16)$, so that $T_{12}(2, 2)$ “‘ $2T_{12}(1, 1)$.

Exercise: Convince yourself that the following transformations are linear by drawing the picture and indicating how the parallelogram and scaling by 1/2 are preserved by the transformations. Then find the corresponding matrices.

1. *Reflection in diagonal line*: $R^2 \to R^2$

Ans. The following is the picture for the transformation.

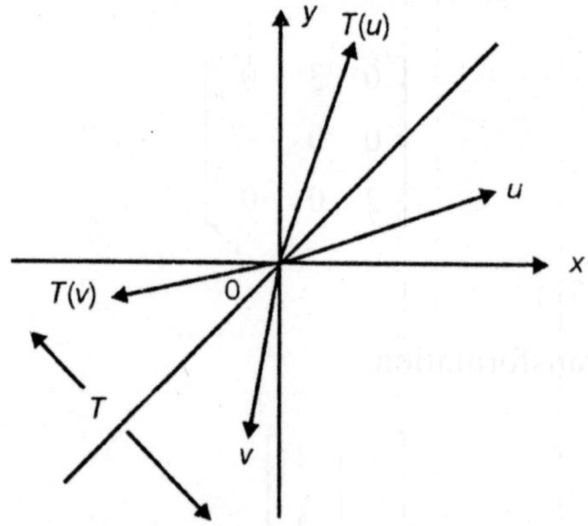

The following illustrates that the transformation is linear.

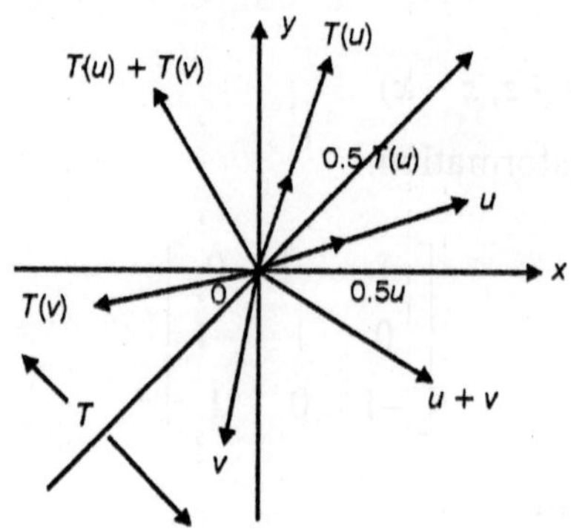

From this picture,

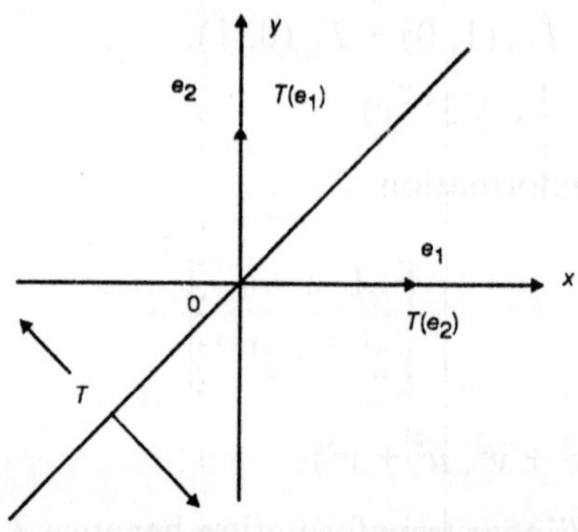

we see $T(e_1) = e_2 = (0, 1)$, $T(e_2) = e_1 = (1, 0)$. Therefore the matrix is

$$\begin{bmatrix} 0 & 1 \\ 1 & 0 \end{bmatrix}$$

2. *Projection to x_1–line*: $R^2 \to R^1$

Ans. The following is the picture for the transformation.

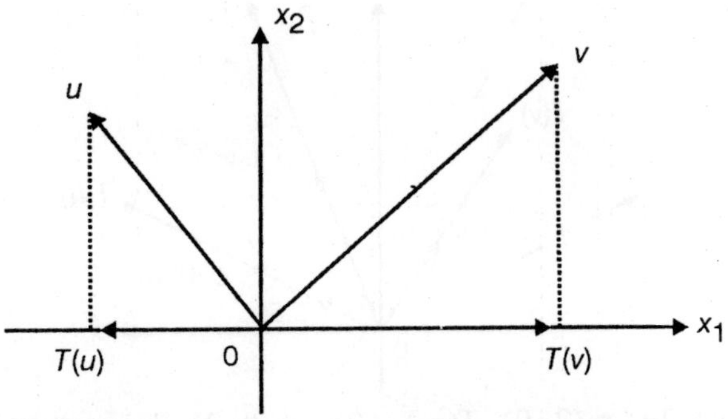

The following illustrates that the transformation is linear.

We clearly have $T(e_1) = e_1 = (1, 0)$, $T(e_2) = (0, 0)$. Therefore the matrix is

$$\begin{bmatrix} 1 & 0 \\ 0 & 0 \end{bmatrix}$$

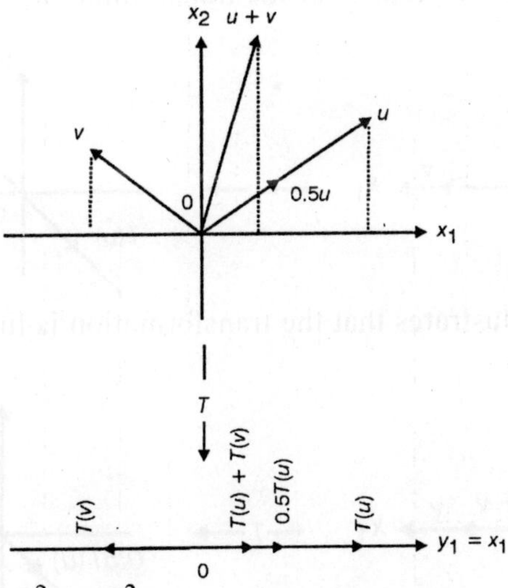

3. *Multiplying by* 2: $R^2 \to R^2$

Ans. The following is the picture for the transformation.

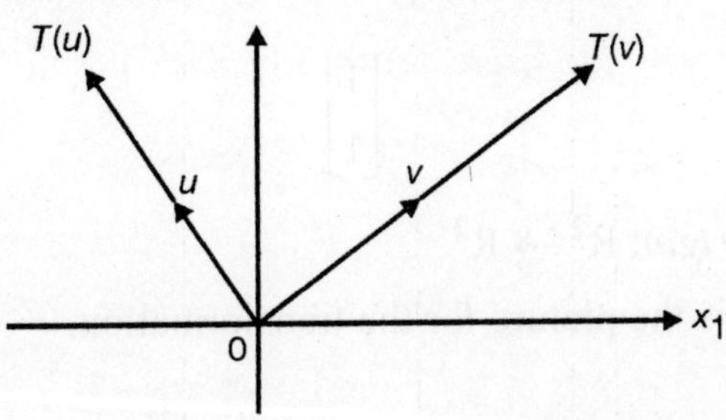

The following illustrates that the transformation is linear.

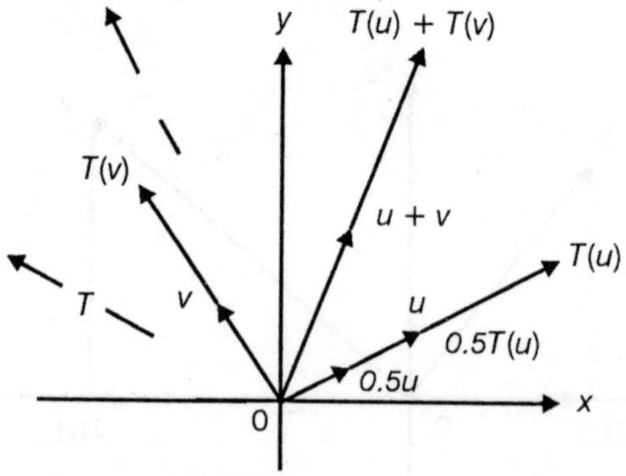

Since $T(e_1) = 2e_1 = (2, 0)$, $T(e_2) = 2e_2 = (0, 2)$, the matrix is

$$\begin{bmatrix} 2 & 0 \\ 0 & 2 \end{bmatrix}$$

4. *Embedding as diagonal*: $R^1 \to R^2$

Ans. The following is the picture for the transformation.

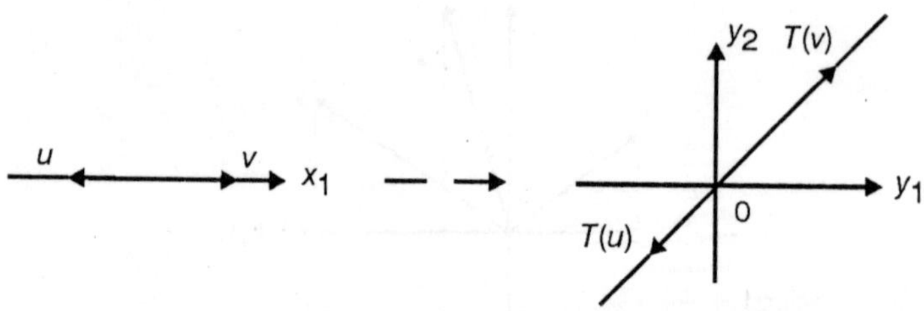

The following illustrates that the transformation is linear.

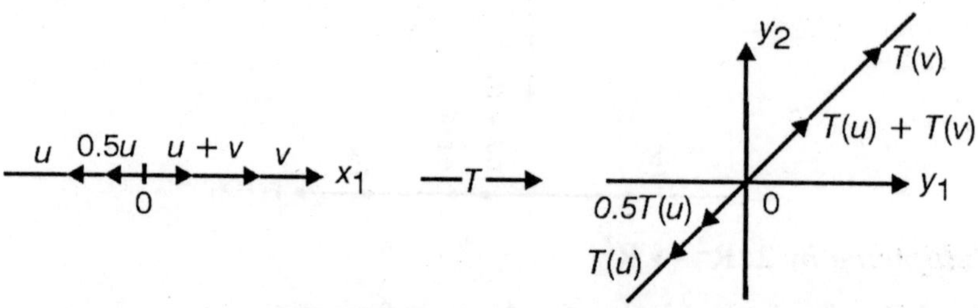

Since $T(e_1) = (1, 1)$ (note that $e_1 = 1$ is a 1–dimensional vector), the matrix is

$$\begin{bmatrix} 1 \\ 1 \end{bmatrix}$$

5. *Reflection in origin*: $R^2 \to R^2$

Ans. The following is the picture for the transformation.

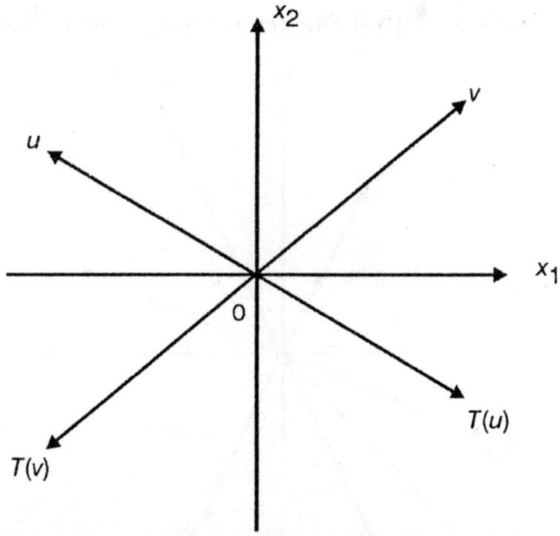

The following illustrates that the transformation is linear.

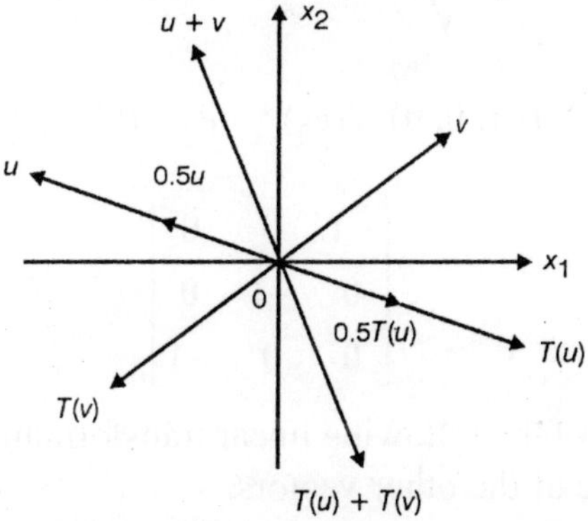

Since $T(e_1) = -e_1 = (-1, 0)$, $T(e_2) = -e_2 = (0, -1)$, the matrix is

$$\begin{bmatrix} -1 & 0 \\ 0 & -1 \end{bmatrix}$$

6. *Reflection in origin*: $R^3 \to R^3$

Ans. The following is the picture for the transformation.

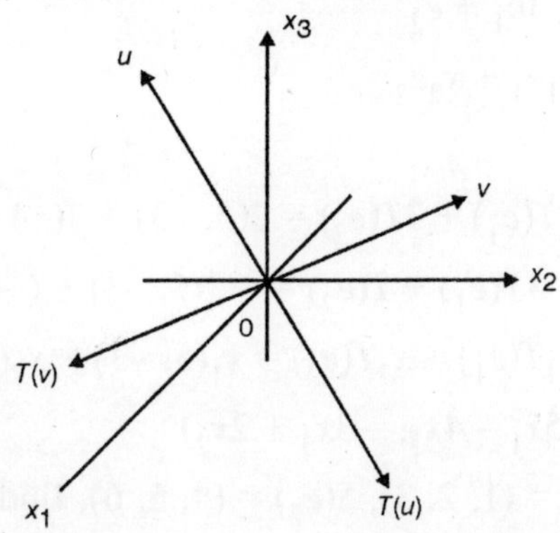

The following illustrates that the transformation is linear.

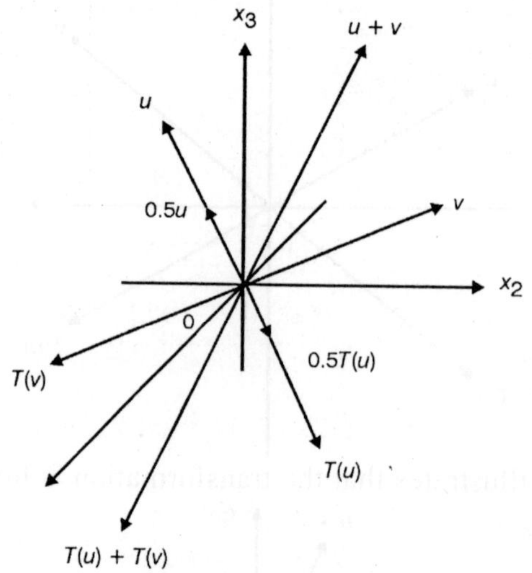

Since $T(e_1) = -e_1 = (-1, 0, 0)$, $T(e_2) = -e_2 = (0, -1, 0)$, $T(e_3) = -e_3 = (0, 0, -1)$, the matrix is

$$\begin{bmatrix} -1 & 0 & 0 \\ 0 & -1 & 0 \\ 0 & 0 & -1 \end{bmatrix}$$

Exercise: The values of the following linear transformations at the standard basis have been given. Find the value at the other vectors.

Ans.

1. $T: R^2 \to R^2$, $T(e_1) = (5, -3)$, $T(e_2) = (-4, 2)$, find $T(2, 3)$, $T(-3, 1)$, $T(x_1, x_2)$.

Ans. By

$$(2, 3) = 2e_1 + 3e_2$$
$$(-3, 1) = -3e_1 + e_2$$
$$(x_1, x_2) = x_1e_1 + x_2e_2$$

We have

$$T(2, 3) = 2T(e_1) + 3T(e_2) = 2(5, -3) + 3(-4, 2) = (-2, 0)$$
$$T(-3, 1) = -3T(e_1) + T(e_2) = -3(5, -3) + (-4, 2) = (-19, 9)$$
$$T(x_1, x_2) = x_1T(e_1) + x_2T(e_2) = x_1(5, -3) + x_2(-4, 2)$$
$$= (5x_1 - 4x_2, -3x_1 + 2x_2)$$

2. $S: R^2 \to R^3$, $S(e_1) = (1, 2, 3)$, $S(e_2) = (4, 5, 6)$, find $S(2, 3)$, $S(-3, 1)$, $S(x_1, x_2)$.

Ans. By

$$(2, 3) = 2e_1 + 3e_2$$
$$(-3, 1) = -3e_1 + e_2$$
$$(x_1, x_2) = x_1e_1 + x_2e_2$$

We have

$$S(2, 3) = 2S(e_1) + 3S(e_2) = 2(1, 2, 3) + 3(4, 5, 6) = (14, 19, 24)$$
$$S(-3, 1) = -3S(e_1) + S(e_2) = -3(1, 2, 3) + (4, 5, 6) = (1, -1, -3)$$
$$S(x_1, x_2) = x_1S(e_1) + x_2S(e_2) = x_1(1, 2, 3) + x_2(4, 5, 6)$$
$$= (x_1 + 4x_2, 2x_1 + 5x_2, 3x_1 + 6x_2)$$

3. R: $R^3 \to R^2$, $R(e_1) = (1, 2)$, $R(e_2) = (3, 4)$, $R(e3) = (5, 6)$, find $R(1, 0, 2)$, $R(-2, 1, 3)$, $R(x_1, x_2, x_3)$.

Ans. By

$$(1, 0, 2) = e_1 + 2e_3$$
$$(-2, 1, 3) = -2e_1 + e_2 + 3e_3$$
$$(x_1, x_2, x_3) = x_1e_1 + x_2e_2 + x_3e_3$$

We have

$$R(1, 0, 2) = R(e_1) + 2S(e_3) = (1, 2) + 2(5, 6) = (11, 14)$$
$$R(-2, 1, 3) = -2R(e_1) + R(e_2) + 3R(e_3)$$
$$= -2(1, 2) + (3, 4) + 3(5, 6) = (16, 18)$$
$$R(x_1, x_2, x_3) = x_1R(e_1) + x_2R(e_2) + x_3R(e3) = x_1(1, 2) + x_2(3, 4) + x_3(5, 6)$$
$$= (x_1 + 3x_2 + 5x_3, 2x_1 + 4x_2 + 6x_3)$$

Exercise: Let T: $R^2 \to R^3$ be a linear transformation. Assume we already know $T(1, 2) = (1, 2, 3)$, $T(3, 4) = (4, 5, 6)$.

- Use this exercise and the linearity of T to compute $T(4, 6)$, $T(1, 1)$, $T(1, 0)$, $T(0, 1)$.
- Use $T(1, 0)$, $T(0, 1)$ to compute $T(x_1, x_2)$ in general.
- Use $3(1, 1, 1) = (4, 5, 6) - (1, 2, 3)$ to find x_1, x_2 such that $T(x_1, x_2) = (1, 1, 1)$

Answer: From the exercise, we have

$(4, 6) = (1, 2) + (3, 4)$

$(1, 1) = -1/2\ (1, 2) + 1/2\ (3, 4)$

$(1, 0) = -2(1, 2) + (3, 4)$

$(0, 1) = 3/2\ (1,2) - 1/2\ (3, 4)$

Thus

$\boldsymbol{T}(4, 6) = \boldsymbol{T}(1, 2) + \boldsymbol{T}(3, 4) = (1, 2, 3) + (4, 5, 6) = (5, 7, 9)$

$\boldsymbol{T}(1, 1) = -1/2\ \boldsymbol{T}(1, 2) + 1/2\ \boldsymbol{T}(3, 4) = -1/2\ (1, 2, 3) + 1/2\ (4, 5, 6) = (3/2, 3/2, 3/2)$

$\boldsymbol{T}(1, 0) = -2\boldsymbol{T}(1, 2) + \boldsymbol{T}(3, 4) = -2(1, 2, 3) + (4, 5, 6) = (2, !, 0)$

$\boldsymbol{T}(0, 1) = 3/2\ \boldsymbol{T}(1,2) - 1/2\ \boldsymbol{T}(3, 4) = 3/2\ (1, 2, 3) - 1/2\ (4, 5, 6) = (-1/2, 1/2, 3/2)$

Moreover,

$T(x_1, x_2) = x_1T(1, 0) + x_2T(0, 1) = x_1(2, 1, 0) + x_2(-1/2, 1/2, 3/2) = (2x_1 - 1/2x_2, x_1 + 1/2x_2, 3/2x_2)$.

Finally, from the equality $3(1, 1, 1) = (4, 5, 6) - (1, 2, 3) = T(3, 4) - T(1, 2) = T(2, 2)$ we have $(1, 1, 1) = T(2/3, 2/3)$. Therefore for the last question, we may take $x_1 = x_2 = 2/3$.

Exercise: Let T: R^3 '! R^2 be a linear transformation. Assume we already know $T(1, 1, 1) = (1, 2)$, $T(1, 1, 0) = (3, 4)$, $T(1, 0, 0) = (5, 6)$. Find the general formula (or the matrix) for T.

Ans: We need to find $T(x_1, x_2, x_3)$. We use the same idea as in the previous exercise. First we solve the three systems $c_1(1, 1, 1) + c_2(1, 1, 0) + c3(1, 0, 0) = e_i$ for $i = 1, 2, 3$ and find

$e_1 = (1, 0, 0)$

$e_2 = (1, 1, 0) - (1, 0, 0)$

$e_3 = (1, 1, 1) - (1, 1, 0)$

Then we use the linearity of $\boldsymbol{T}$ to get

$T(e_1) = T(1, 0, 0) = (5, 6)$

$T(e_2) = T(1, 1, 0) - T(1, 0, 0) = (3, 4) - (5, 6) = (-2, -2)$

$T(e_3) = T(1, 1, 1) - T(1, 1, 0) = (1, 2) - (3, 4) = (-2, -2)$

Finally we use $(x_1, x_2, x_3) = x_1e_1 + x_2e_2 + x_3e_3$ and the linearity of T to get

$T(x_1, x_2, x_3) = x_1T(e_1) + x_2T(e_2) + x_3T(e_3) = x_1(5, 6) + x_2(-2, -2) + x_3(-2, -2) = (5x_1 - 2x_2 - 2x_3, 6x_1 - 2x_2 - 2x_3)$.

Exercise: Solve the following three systems of linear equations.

$x_1 + 2x_2 = 1\ x_1 + 2x_2 = 0$

$2x_1 + 3x_2 = 0\ 2x_1 + 3x_2 = 1$

Then use your results and the linearity to find solutions of the following systems

$$x_1 + 2x_2 = 2 \; x_1 + 2x_2 = 13$$
$$2x_1 + 3x_2 = -3 \; 2x_1 + 3x_2 = 9$$

Finally use the same technique to find the solution of the system

$$x_1 + 2x_2 = b_1$$
$$2x_1 + 3x_2 = b_2$$

Ans: For the two systems $Ax = e_1$ and $Ax = e_2$, we combine the augmented matrices together.

$$[A \; e_1 \; e_2] = \begin{bmatrix} 1 & 2 & 1 & 0 \\ 2 & 3 & 0 & 1 \end{bmatrix}$$

Row operations give us

$$\begin{bmatrix} 1 & 0 & -3 & 2 \\ 0 & 1 & 2 & -1 \end{bmatrix}$$

Thus the solution for the system with $e_1 = (1, 0)$ on the right is $x = (-3, 2)$, and the solution with $e2 = (0, 1)$ on the right is $x' = (2, -1)$.

By $(2, 3) = 2(1, 0) + 3(0, 1)$, the solution of the system with $(2, 3)$ on the right is $2x + 3x' = 2(-3, 2) + 3(2, -1) = (0, 1)$.

By $(13, 9) = 13(1, 0) + 9(0, 1)$, the solution of the system with $(13, 9)$ on the right is $13x + 9x' = 13(-3, 2) + 9(2, -1) = (-21, 17)$.

By $(b_1, b_2) = b_1(1, 0) + b_2(0, 1)$, the solution of the system with (b_1, b_2) on the right is $b_1x + b_2x' = b_1(-3, 2) + b_2(2, -1) = (-3b_1 + 2b_2, 2b_1 - b_2)$.

Exercise: Solve the following three systems of linear equations.

$$x_1 + 2x_3 = 1 \qquad x_1 + 2x_3 = 0 \qquad x_1 + 2x_3 = 0$$
$$2x_1 - x_2 + 3x_2 = 0 \; 2x_1 - x_2 + 3x_2 = 1 \;\; 2x_1 - x_2 + 3x_2 = 0$$
$$4x_1 - x_2 + 8x_2 = 0 \; 4x_1 - x_2 + 8x_2 = 0 \;\; 4x_1 - x_2 + 8x_2 = 1$$

Then use your results and the linearity to find a solution of the system

$$x_1 + 2x_3 = b_1$$
$$2x_1 - x_2 + 3x_2 = b_2$$
$$4x_1 - x_2 + 8x_2 = b_3$$

Ans: For the three systems $Ax = e_1$, $Ax = e_2$, $Ax = e_3$, we combine the augmented matrices together.

$$[A\ e_1\ e_2\ e_3] = \begin{bmatrix} 1 & 0 & 2 & 1 & 0 & 0 \\ 2 & -1 & 3 & 0 & 1 & 0 \\ 4 & -1 & 8 & 0 & 0 & 1 \end{bmatrix}$$

Row operations give us

$$\begin{bmatrix} 1 & 0 & 0 & 5 & 2 & -2 \\ 0 & 1 & 0 & 4 & 0 & -1 \\ 0 & 0 & 1 & -2 & -1 & 1 \end{bmatrix}$$

Thus we conclude

$Ax = e_1$ has solution $x = (5, 4, -2)$

$Ax = e_2$ has solution $x' = (2, 0, -1)$

$Ax = e_3$ has solution $x'' = (-2, -1, 1)$

By $(b_1, b_2, b_3) = b_1e_1 + b_2e_2 + b_3e_3$, the combination

$b_1x + b_2x' + b_3x'' = b_1(5, 4, -2) + b_2(2, 0, -1) + b_3(-2, -1, 1) = (5b_1 + 2b_2 - 2b_3, 4b_1 - b_3, -2b_1 - b_2 + b_3)$

is a solution of the system $Ax = b$.

DETERMINANTS

The determinant of a 2x2 matrix

$$A = \begin{bmatrix} a & b \\ c & d \end{bmatrix}$$

is

$$|A| = \det(A) = \begin{vmatrix} a & b \\ c & d \end{vmatrix} = ad - bc$$

Notice the difference in notation between the matrix and its determinant: matrices are typically enclosed with square brackets whereas determinants of matrices are enclosed by straight lines. The determinant is a scalar quantity. It contains much information about the matrix it came from and is quite useful in many applications. The determinant is most often used to

1. Test whether or not a matrix has an inverse
2. Test for linear dependence of vectors (in certain situations)
3. Test for existence/uniqueness of solutions of linear systems of equations

Example: The determinant of the matrix

$$\begin{bmatrix} 2 & -3 \\ 1 & 12 \end{bmatrix}$$

is given by

$$\begin{bmatrix} 2 & -3 \\ 1 & 12 \end{bmatrix} = 2(12) - (-3)(1) = 27$$

DETERMINANTS OF LARGER MATRICES

The determinant of a 3x3 matrix is a bit more involved:

$$|B| \ \det(B) = \begin{vmatrix} a & b & c \\ d & e & f \\ g & h & k \end{vmatrix} = a\begin{vmatrix} e & f \\ h & k \end{vmatrix} - b\begin{vmatrix} d & f \\ g & k \end{vmatrix} + c\begin{vmatrix} d & e \\ g & h \end{vmatrix}$$

which, after taking the three 2x2 determinants, is equivalent to |B|=a(ek-fh)-b(dk-fg)+c(dh-eg). This is called the expansion of the determinant by its first row. To do this cross out the row and column containing the first element of the first row (the number a). We are left with

$$\begin{vmatrix} e & f \\ h & k \end{vmatrix}$$

Now take this determinant and multiply it by a (the element that was crossed out). Repeat the process for the second and third elements of the first row. These terms are then combined with alternating signs, +, -, +. The result is the quantity shown above. The determinant of any nxn matrix can be found by expanding in this manner on *any* row or column.

Example: The determinant of the matrix

$$\begin{bmatrix} 3 & -5 & 3 \\ 2 & 1 & -1 \\ 1 & 0 & 4 \end{bmatrix}$$

is given by

$$3\begin{vmatrix} 1 & -1 \\ 0 & 4 \end{vmatrix} - (-5)\begin{vmatrix} 2 & -1 \\ 1 & 4 \end{vmatrix} + 3\begin{vmatrix} 2 & 1 \\ 1 & 0 \end{vmatrix} = 54$$

An Alternative Computational Approach

The determinant of a 3x3 matrix

$$\begin{bmatrix} a & b & c \\ d & e & f \\ g & h & k \end{bmatrix}$$

can also be found using the following method: repeat the first two columns of the matrix to the right of the last column:

$$\begin{bmatrix} a & b & c & a & b \\ d & e & f & d & e \\ g & h & k & g & h \end{bmatrix}$$

Now, the determinant is the sum of the products of the upper left to lower right diagonals minus the sum of the product of the upper right to lower left diagonals: |A|=(aek+bfg+cdh)-(bdk+afh+ceg). Compare this result with the one obtained above.

TESTING FOR LINEAR DEPENDENCE OF VECTORS

There are many situations when we might wish to know whether a set of vectors is linearly dependent, that is if one of the vectors is some combination of the others.

Two vectors u and v are linearly independent if the only numbers x and y satisfying xu+yv=0 are x=y=0. If we let

$$\vec{u} = \begin{bmatrix} a \\ b \end{bmatrix} \text{ and } \vec{v} = \begin{bmatrix} c \\ d \end{bmatrix}$$

then xu+yv=0 is equivalent to

$$0 = x\begin{bmatrix} a \\ b \end{bmatrix} + y\begin{bmatrix} c \\ d \end{bmatrix} = \begin{bmatrix} a & c \\ b & d \end{bmatrix}\begin{bmatrix} x \\ y \end{bmatrix}$$

If u and v are linearly independent, then the only solution to this system of equations is the trivial solution, $x = y = 0$. For homogeneous systems this happens precisely when the determinant is non-zero. We have now found a test for determining whether a given set of vectors is linearly independent: A set of n vectors of length n is linearly independent if the matrix with these vectors as columns has a non-zero determinant. The set is of course dependent if the determinant is zero.

Example: The vectors <1,2> and <-5,3> are linearly independent since the matrix

$$\begin{bmatrix} 1 & -5 \\ 2 & 3 \end{bmatrix}$$

has a non-zero determinant.

Example: The vectors u=<2,-1,1>, v=<3,-4,-2>, and w=<5,-10,-8> are dependent since the determinant

$$\begin{vmatrix} 2 & 3 & 5 \\ -1 & -4 & -10 \\ 1 & -2 & -8 \end{vmatrix}$$

is zero. To find the relation between u, v, and w we look for constants x, y, and z such that

$$0 = x\begin{bmatrix} 2 \\ -1 \\ 1 \end{bmatrix} + y\begin{bmatrix} 3 \\ -4 \\ -2 \end{bmatrix} + z\begin{bmatrix} 5 \\ -10 \\ -8 \end{bmatrix} = \begin{bmatrix} 2 & 3 & 5 \\ -1 & -4 & -10 \\ 1 & -2 & -8 \end{bmatrix}\begin{bmatrix} x \\ y \\ z \end{bmatrix}$$

This is a homogeneous system of equations. Using Gaussian Elimination, we see that the matrix

$$\left[\begin{array}{ccc|c} 2 & 3 & 5 & 0 \\ -1 & -4 & -10 & 0 \\ 1 & -2 & -8 & 0 \end{array}\right]$$

in row-reduced form is

$$\left[\begin{array}{ccc|c} 2 & 3 & 5 & 0 \\ 0 & 1 & 3 & 0 \\ 0 & 0 & 0 & 0 \end{array}\right]$$

Thus, y=-3z and 2x=-3y-5z=-3(-3z)-5z=4z which implies 0=xu+yv+zw=2zu-3zv+zw or equivalently w=-2u+3v. A quick arithmetic check verifies that the vector w is indeed equal to -2u+3v.

SYSTEMS OF LINEAR EQUATIONS

Consider the two equations ax+by=c and dx+ey=f. Since these equations represent two lines in the xy-plane, the simultaneous solution of these two equations (i.e. those points (x,y) that satisfy both equations) is merely the intersection of the two lines. The graphs below illustrate the three possible cases: non-parallel lines, parallel (but not identical) lines, and identical lines.

From left to right these cases yield one solution, no solutions, and infinite solutions. The same situation occurs in three dimensions; the solution of 3 equations with 3 unknowns is the intersection of the 3 planes.

There is a simple tool for determining the number of solutions of a square system of equations: the determinant.

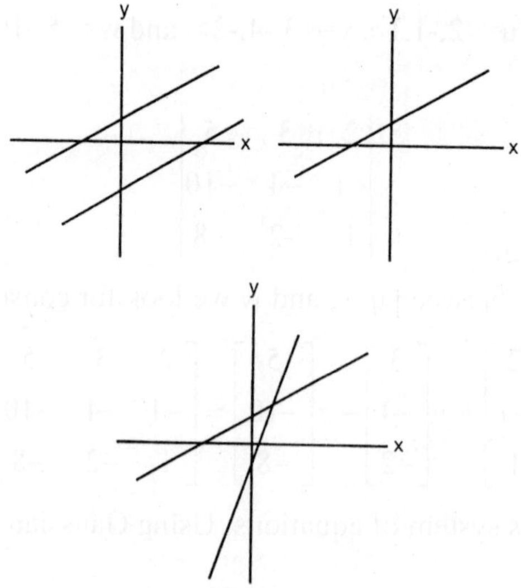

MATRIX NOTATION

It is often convenient to represent a system of equations as a matrix equation or even as a single matrix. Consider the system of equations 2x+3y=-8 and -x+5y=1. Since

$$\begin{bmatrix} 2x+3y \\ -x+5y \end{bmatrix} = \begin{bmatrix} 2 & 3 \\ -1 & 5 \end{bmatrix} \begin{bmatrix} x \\ y \end{bmatrix}$$

we may write the entire system as a matrix equation:

$$\begin{bmatrix} 2 & 3 \\ -1 & 2 \end{bmatrix} \begin{bmatrix} x \\ y \end{bmatrix} = \begin{bmatrix} -8 \\ 1 \end{bmatrix}$$

or as AX=B where

$$A = \begin{bmatrix} 2 & 3 \\ -1 & 5 \end{bmatrix},\ X = \begin{bmatrix} x \\ y \end{bmatrix},\ and\ B = \begin{bmatrix} -8 \\ 1 \end{bmatrix}$$

The 2x2 matrix A is called the matrix of coefficients of the system of equations. Frequently this equation is written as a single augmented matrix:

$$\left[\begin{array}{cc|c} 2 & 3 & -8 \\ -1 & 5 & 1 \end{array}\right]$$

We may now use Gaussian elimination to solve this matrix equation for x and y (as opposed to direct substitution of one equation into the other). This representation can also be done for any number of equations with any number of unknowns. In general, the equation AX=B representing a system of equations is called homogeneous if B is the nx1 (column) vector of zeros. Otherwise, the equation is called nonhomogeneous.

DETERMINING THE NUMBER OF SOLUTIONS OF A NONHOMOGENEOUS SYSTEM OF EQUATIONS

For square systems of equations (i.e. those with an equal number of equations and unknowns), the most powerful tool for determining the number of solutions the system has is the determinant. Suppose we have two equations and two unknowns: ax+by=c and dx+ey=f with b and e non-zero (i.e. the system is nonhomogeneous). These are two lines with slope -a/b and -d/e, respectively. Let's define the determinant of a 2x2 system of linear equations to be the determinant of the matrix of coefficients A of the system. For this system

$$A = \begin{bmatrix} a & b \\ d & e \end{bmatrix} \ and \ |A| = \begin{vmatrix} a & b \\ d & e \end{vmatrix} = ae - bd$$

Suppose this determinant is zero. Then, this last equation implies a/b=d/e; in other words, the slopes are equal. From our discussion above, this means the lines are either identical (there is an infinite number of solutions) or parallel (there are no solutions). If the determinant is non-zero, then the slopes must be different and the lines must intersect in exactly one point. This leads us to the following result:

A nxn *nonhomogeneous* system of linear equations has a unique non-trivial solution if and only if its determinant is non-zero. If this determinant is zero, then the system has either no nontrivial solutions or an infinite number of solutions.

Example: The nonhomogeneous system of equations 2x+3y=-8 and -x+5y=1 has determinant

$$\begin{vmatrix} 2 & 3 \\ -1 & 5 \end{vmatrix} = 2(5) - 3(-1) = 13$$

which implies the system has a unique solution (the point (-43/13,-28/75)). This corresponds to our intuition: the lines have different slopes and therefore must intersect in exactly one point.

Example: The nonhomogeneous system of equations x-4y+6z=3, -2x+8y-12z=-6, and 2x-y+3z=1 does not have a unique solution since its determinant is

$$1\begin{vmatrix} 8 & -12 \\ -1 & 3 \end{vmatrix} + 4\begin{vmatrix} -2 & -12 \\ 2 & 3 \end{vmatrix} + 6\begin{vmatrix} -2 & 8 \\ 2 & -1 \end{vmatrix} = 1(12) + 4(18) + 6(-14) = 0$$

In fact, this system has an infinite number of solutions. To see this, think geometrically. The three equations represent 3 planes. Two of the equations (the first and the second) represent the same plane (why?); so, there are only two distinct non-parallel planes. Since two non-parallel planes intersect in a line, there are an infinite number of points which lie on all three of these planes (i.e. the system has an infinite number of solutions). To actually find the solutions, we must use a method such as Gaussian elimination to solve the system of equations.

DETERMINING THE NUMBER OF SOLUTIONS OF A HOMOGENEOUS SYSTEM OF EQUATIONS

For a homogeneous system of equations ax+by=0 and cx+dy=0, the situation is slightly different. These lines pass through the origin. Thus, there is always at least one solution, the point (0,0). If the slopes -a/b and -c/d are equal then there are an infinite number of solutions since the lines are identical. But as we have seen, the slopes of these lines are equal when the determinant of the coefficient matrix is zero. Thus, for homogeneous systems we have the following result:

A nxn *homogeneous* system of linear equations has a unique solution (the trivial solution) if and only if its determinant is non-zero. If this determinant is zero, then the system has an infinite number of solutions.

Example: The system of equations 2x+y=0 and x-y=0 has exactly one solution since the slopes of the lines are different, i.e. the determinant

$$\begin{vmatrix} 2 & 1 \\ 1 & -1 \end{vmatrix} = 2(-1) - 1(1) = -3$$

is non-zero. The solution is of course (0,0).

GAUSSIAN ELIMINATION

We will now explore a more versatile way than the method of determinants to determine if a system of equations has a solution. We will indeed be able to use the results of this method to find the actual solution(s) of the system (if any).

It should be noted that this method can be applied to systems of equations with an unequal number of equations and unknowns. However, we will only discuss systems with an equal number of equations and unknowns.

It is easiest to illustrate this method with an example. Consider the system of equations

$$\begin{aligned} x - \quad 3y + \quad z &= \quad 4 \\ 2x - \quad 8y + \quad 8z &= \ -2 \\ -6x + \quad 3y - \ 15z &= \quad 9 \end{aligned}$$

To solve for x, y, and z we must eliminate some of the unknowns from some of the equations. Consider adding -2 times the first equation to the second equation and also adding 6 times the first equation to the third equation. The result is

$$\begin{aligned} x - \quad 3y + \quad z &= \quad 4 \\ 0x - \quad 2y + \quad 6z &= -10 \\ 0x - \quad 15y - \quad 9z &= \quad 33 \end{aligned}$$

We have now eliminated the x term from the last two equations. Now simplify the last two equations by dividing by 2 and 3, respectively:

$$\begin{aligned} x - \quad 3y + \quad z &= \quad 4 \\ 0x - \quad 2y + \quad 6z &= -10 \\ 0x - \quad 15y - \quad 9z &= \quad 33 \end{aligned}$$

To eliminate the y term in the last equation, multiply the second equation by -5 and add it to the third equation:

$$\begin{aligned} x - \quad 3y + \quad z &= \quad 4 \\ 0x - \quad y + \quad 3z &= \quad -5 \\ 0x + \quad 0y - \quad 18z &\doteq \quad 36 \end{aligned}$$

The third equation says z=-2. Substituting this into the second equation yields y=-1. Using both of these results in the first equation gives x=3. The process of progressively solving for the unknowns is called back-substitution. This is the essence of Gaussian elimination. However, we may clean up the notation in our work by using matrices.

GAUSSIAN ELIMINATION

Convert the system of equations above into an augmented matrix:

$$\left[\begin{array}{ccc|c} 1 & -3 & 1 & 4 \\ 2 & -8 & 8 & -2 \\ -6 & 3 & -15 & 9 \end{array}\right]$$

This matrix contains all of the information in the system of equations without the x, y, and z labels to carry around. Now carry out the process outlined above. The notation to the right of each matrix describes the row operations that were performed to get the matrix on that line. For example $2R_1+R_2 \rightarrow R_2$ means "replace row 2 with the sum of 2 times row 1 and row 2".

$$\left[\begin{array}{ccc|c} 1 & -3 & 1 & 4 \\ 0 & -2 & 6 & -10 \\ 0 & -15 & -9 & 33 \end{array}\right] \begin{array}{l} -2R_1 + R_2 \rightarrow R_2 \\ -6R_1 + R_3 \rightarrow R_3 \end{array}$$

$$\left[\begin{array}{ccc|c} 1 & -3 & 1 & 4 \\ 0 & -1 & 3 & -5 \\ 0 & -5 & -3 & 11 \end{array}\right] \begin{array}{l} \frac{1}{2}R_2 \rightarrow R_2 \\ \frac{1}{3} + R_3 \rightarrow R_3 \end{array}$$

$$\left[\begin{array}{ccc|c} 1 & -3 & 1 & 4 \\ 0 & -1 & 3 & -5 \\ 0 & 0 & -18 & 36 \end{array}\right] -5R_2 + R_3 \rightarrow R_3$$

If we now reverse the conversion process and turn the augmented matrix into a system of equations we have

$$\begin{aligned} x - 3y + z &= 4 \\ 0x - y + 3z &= -5 \\ 0x + 0y - 18z &= 36 \end{aligned}$$

We can now easily solve for x, y, and z by back-substitution.

Remember: For a system of equations with a 3x3 matrix of coefficients, the goal of the process of Gaussian Elimination is to create (at least) a triangle of zeros in the lower left hand corner of the matrix below the diagonal. Note that you may switch the order of the rows at any time in trying to get to this form.

Example: Let's solve the following system of equations:

$$\begin{aligned} 2x - y + z &= 1 \\ 3x - 2y - 4z &= 4 \\ -6x + 3y - 3z &= 2 \end{aligned}$$

In augmented matrix form we have

$$\left[\begin{array}{ccc|c} 2 & -1 & 1 & 1 \\ 3 & 2 & -4 & 4 \\ -6 & 3 & -3 & 2 \end{array}\right]$$

We now use the method of Gaussian Elimination:

$$\left[\begin{array}{ccc|c} 2 & -1 & 1 & 1 \\ 7 & 0 & -2 & 6 \\ 0 & 0 & 0 & 5 \end{array}\right] \begin{array}{l} 2R_1 + R_2 \rightarrow R_2 \\ 3R_1 + R_3 \rightarrow R_3 \end{array}$$

We could proceed to try and replace the first element of row 2 with a zero, but we can actually stop. To see why, convert back to a system of equations:

$$\begin{aligned} 2x - y + z &= 1 \\ 7x + 0y - 2z &= 6 \\ 0x + 0y + 0z &= 5 \end{aligned}$$

Notice the last equation: 0=5. This is not possible. So the system has no solutions; it is not possible to find values x, y, and z that satisfy all three equations simultaneously.

Example: Solve

$$\begin{aligned} 1x - \quad y + \quad 2z &= -3 \\ 4x + \quad 4y - \quad 2z &= 1 \\ -2x + \quad 2y - \quad 4z &= 6 \end{aligned}$$

In matrix form:

$$\left[\begin{array}{ccc|c} 1 & -1 & 2 & -3 \\ 4 & 4 & -2 & 1 \\ -2 & 2 & -4 & 6 \end{array}\right]$$

Using Gaussian Elimination:

$$\left[\begin{array}{ccc|c} 1 & -1 & 2 & -3 \\ 0 & 8 & -10 & 13 \\ 0 & 0 & 0 & 0 \end{array}\right] \begin{array}{l} 2R_1 + R_3 \to R_3 \\ -4R_1 + R_2 \to R_2 \end{array}$$

Converting back to a system of equations:

$$\begin{aligned} 1x - \quad y + \quad 2z &= -3 \\ 0x + \quad 8y - \quad 10z &= 13 \\ 0x + \quad 0y - \quad 0z &= 0 \end{aligned}$$

Notice the last equation: 0=0 (this resulted from equation 3 being a linear combination of the other two equations). This is always true. And, we can solve the first two equations to get x and y as functions of z alone. Solving the second equation we get

$$y = \frac{5}{4}z + \frac{13}{8}$$

And for the first equation

$$x = y - 2z - 3 = -\frac{3}{4}z - \frac{11}{8}$$

The variable z in this problem is called a parameter since there are no constraints on what values z may take on. Thus, there are an infinite number of solutions - one for each value of z. Examples of solutions are (-11/8,13/8,0) and (-17/8,23/8,1) which come from setting z=0 and z=1, respectively. We may concisely write all solutions as triples of the form

$$\left(-\frac{3}{4}t - \frac{11}{8}, \frac{5}{4}t + \frac{13}{8}, t\right)$$

where t is any real number.

EIGENVALUES AND EIGENVECTORS

Consider multiplying a square 3x3 matrix by a 3x1 (column) vector. The result is a 3x1 (column) vector. The 3x3 matrix can be thought of as an operator - it takes a vector, operates on it, and returns a new vector. There are many instances in mathematics and physics in which we are interested in which vectors are left "essentially unchanged" by the operation of the matrix. Specifically, we are interested in those vectors v for which Av=kv where A is a square matrix and k is a real number.

A vector v for which this equation hold is called an eigenvector of the matrix A and the associated constant k is called the eigenvalue (or characteristic value) of the vector v. If a matrix has more than one eigenvector the associated eigenvalues can be different for the different eigenvectors.

Geometrically, the action of a matrix on one of its eigenvectors causes the vector to stretch (or shrink) and/or reverse direction.

In order to find the eigenvalues of a nxn matrix A (if any), we solve Av=kv for scalar(s) k. Rearranging, we have Av-kv=0. But kv=kIv where I is the nxn identity matrix

$$\begin{bmatrix} 1 & 0 & \cdots & 0 \\ 0 & 1 & \cdots & 0 \\ \vdots & \vdots & \ddots & \vdots \\ 0 & 0 & \cdots & 1 \end{bmatrix}$$

So, 0=Av-kv=Av-kIv=(A-kI)v. This equation is equivalent to a homogeneous system of n equations with n unknowns. This equation has a non-zero solution for v if and only if the determinant det(A-kI) is zero. Thus, by finding the zeros of the polynomial in k determined by the characteristic equation det(A-kI)=0, we will have found the eigenvalues of the matrix A.

Example: To find the eigenvalues of the matrix

$$\begin{bmatrix} 2 & \frac{3}{2} \\ 2 & 0 \end{bmatrix}$$

we substitute A into the equation det(A-kI)=0 and solve for k. The matrix A-kI is given by

$$\begin{bmatrix} 2-k & \frac{3}{2} \\ 2 & 0-k \end{bmatrix}$$

which has determinant k^2-2k-3. Hence, we are looking for values k satisfying k^2-2k-3=0. So, of course, we have k=3 or k=-1. To find the eigenvectors of the eigenvalue k=3 we look for solutions v of the homogeneous system of equations (A-3I)v=0:

$$\left[\begin{array}{cc|c} -1 & \frac{3}{2} & 0 \\ 2 & -3 & 0 \end{array}\right]$$

Since the second equation is a constant multiple of the first, this system of equations reduces to the single equation -x+(3/2)y=0 or equivalently x=1.5y. This system has an infinite number of solutions. So for example, choosing y=2 yeilds the vector <3,2> which is thus an eigenvector that has eigenvalue k=3. In a general form, all eigenvectors with eigenvalue 3 have the form <2t,3t> where t is any real number. It can also be shown (by solving the system (A+I)v=0) that vectors of the form <t,-2t> are eigenvectors with eigenvalue k=-1.

Example: Find the eigenvalues and corresponding eigenvalues for the matrix

$$\begin{bmatrix} 1 & -1 \\ 5 & -3 \end{bmatrix}$$

First, we must find det(A-kI):

$$|A-kI| = \begin{vmatrix} 1-k & -1 \\ 5 & -3-k \end{vmatrix} = (1-k)(-3-k)-(-1)5 = k^2+2k+2$$

This leads to the characteristic equation k^2+2k+2=0 which has complex roots k=-1+i and k=-1-i. To find the eigenvectors for k=-1+i, we solve (A-(-1+i)I)v=0 for v:

$$\left[\begin{array}{cc|c} 1-(-1+i) & -1 & 0 \\ 5 & -3-(-1+i) & 0 \end{array}\right] = \left[\begin{array}{cc|c} 2-i & -1 & 0 \\ 5 & -2-i & 0 \end{array}\right]$$

The second equation is a constant multiple of the first equation so the system reduces to the single equation (2-i)x-y=0 which implies y=(2-i)x.

There are once again an infinite number of eigenvectors of A of the form <t,(2-i)t> with eigenvalue k=-1+i. By examining the system of equations (A-(-1-i)I)v=0 it can also be shown that vectors of the form <t,(2+i)t> are eigenvectors of A with eigenvalue k=-1-i.

From the examples above we can infer a property of eigenvectors and eigenvalues: eigenvectors from distinct eigenvalues are linearly independent. The following examples illustrate that the situation is not so clear cut when the eigenvalues are not distinct.

Example: The matrix

$$\begin{vmatrix} 1 & 0 \\ 0 & 1 \end{vmatrix}$$

has two eigenvalues (1 and 1) but they are obviously not distinct. Since A is the identity matrix, Av=v for any vector v, i.e. any vector is an eigenvector of A. We can thus find two linearly independent eigenvectors (say <-2,1> and <3,-2>) one for each eigenvalue.

Example: The matrix

$$D = \begin{vmatrix} 1 & 1 \\ 0 & 1 \end{vmatrix}$$

also has non-distinct eigenvalues of 1 and 1. All eigenvalues are solutions of (A-I)v=0 and are thus of the form <t,0>. Hence, in this case there do not exist two linearly independent eigenvectors for the two eigenvalues 1 and 1 since <t,0> and <s,0> are not linearly independent for any values of s and t.

SYMMETRIC MATRICES

There is a very important class of matrices called symmetric matrices that have quite nice properties concerning eigenvalues and eigenvectors. A symmetric matrix A is a square matrix with the property that $A_{ij}=A_{ji}$ for all i and j. The matrices

$$\begin{bmatrix} 2 & -3 \\ -3 & 5 \end{bmatrix} \text{ and } \begin{bmatrix} -1 & 3 & 0 \\ 3 & 5 & 2 \\ 0 & 2 & -4 \end{bmatrix}$$

are symmetric matrices. In symmetric matrices the upper right half and the lower left half of the matrix are mirror images of each other about the diagonal.

A nxn symmetric matrix A not only has a nice structure, but it also satisfies the following:

1. A has exactly n (not necessarily distinct) eigenvalues
2. There exists a set of n eigenvectors, one for each eigenvalue, that are mututally orthogonal.

Thus, the situation encountered with the matrix D in the example above cannot happen with a symmetric matrix: A symmetric matrix has n eigenvalues and there exist n linearly independent eigenvectors (because of orthogonality) *even if the eigenvalues are not distinct*.

Example: Find the eigenvalues and a set of mutually orthogonal eigenvectors of the symmetric matrix

$$\begin{bmatrix} 3 & 2 & 4 \\ 2 & 0 & 2 \\ 4 & 2 & 3 \end{bmatrix}$$

First we need det(A-kI):

$$|A-kI|=\begin{vmatrix}3-k & 2 & 4\\ 2 & 0-k & 2\\ 4 & 2 & 3-k\end{vmatrix}=-k^3+6k^2+15k+8$$

Thus, the characteristic equation is $(k-8)(k+1)^2=0$ which has roots k=-1, k=-1, and k=8. Note that we have listed k=-1 twice since it is a double root. We must find two eigenvectors for k=-1 and one for k=8. We now examine (A+I)v=0 to find the eigenvectors for the eigenvalue k=-1:

$$\left[\begin{array}{ccc|c}4 & 2 & 4 & 0\\ 2 & 1 & 2 & 0\\ 4 & 2 & 4 & 0\end{array}\right]$$

It is easily seen that this system reduces to the single equation 2x+y+2z=0 since the other two equations are twice this one. There are two parameters here (x and z); thus, eigenvectors for k=-1 must have the form y=-2x-2z which corresponds to vectors of the form <s,-2s-2t,t>. We must choose values of s and t that yield two orthogonal vectors (the third comes from the eigenvalue k=8). First, choose anything, say s=1 and t=0: <1,-2,0>. Now find a vector <x,-2x-2z,z> such that

$$0=<1,-2,0>\cdot<x,-2x-2z,z>=x+4x+4x+0=5x+4z$$

An easy choice here is x=4 and z=-5. So, we now have two orthogonal vectors <1,-2,0> and <4,2,-5> that correspond to the two instances of the eigenvalue k=-1. It can also be shown that the eigenvectors for k=8 are of the form <2r,r,2r> for any value of r. It is easy to check that this vector is orthogonal to the other two we have for any choice of r. So, let's take r=1. We now have the following: eigenvalues and orthogonal eigenvectors:

for k = –1: <1, – 2, 0 > and <4, 2, – 5>
for k = 8: < 2, 1, 2 >

Note that since this matrix is symmetric we do indeed have 3 eigenvalues and a set of 3 orthogonal (and thus linearly independent) eigenvectors (one for each eigenvalue).

EXERCISE

1. Use $f = u\nabla v$ in the Divergence Theorem to prove:

   (a) *Green's first identity:* $\iiint_S (u\Delta v+(\nabla u).(\nabla v))dV=\iint_\Sigma (u\nabla v).d\sigma$

   (b) *Green's second identity:* $\iiint_S (u\Delta v-v\Delta u)dV=\iint_\Sigma (u\nabla v-v\nabla u).d\sigma$

2. Use the substitution $r=\sqrt{x^2+y^2}$ to show that

3. Evaluate the line integral $\int_C \left[\left(x^2+xy\right)dx+\left(x^2+y^2\right)dy\right]$ where C is the square formed by the lines y = ± 1 and x = ± 1.

4. If F $= (5xy - 6x^2)I + (2y - 4x)J$, evaluate $\int_C F.dR$ along the curve C in the xy-plane, $y = x^3$ from the point (1, 1) to (2, 8).

5. A vector field is given by F = (sin y)I + x(1 + cos y)J. Evaluate the line integral over a circular path given by $x^2 + y^2 = a^2$, z = 0.

$$\lim_{(x,y)\to(0,0)} \frac{\sin\sqrt{x^2+y^2}}{\sqrt{x^2+y^2}} = 1.$$

6. If R = $xI + yJ + zK$ and $r \neq 0$, show that

   (a) $\nabla(1/r^2) = -2R/r^4$; $\nabla(R/r^2) = 1/r^2$ (ii) div $(r^n R) = (n+3)r^n$; curl$(r^n R) = 0$.

   (b) $\nabla\left[\nabla.\frac{R}{r}\right] = -\frac{2}{r^3}R.$

7. Show that each of following vectors are solenoidal :

   (a) $(x + 3y)I + (y - 3z)J + (x - 2z)K$ (ii) $3y^4z^2I + 4x^3z^2J + 3x^2y^2K$

   (b) $\nabla f \times \nabla$ y.

8. Find *T, N, B* and κ at each point of the helix $f(t) = (\cos t, \sin t, t)$.

9. Show that the arc length *L* of a curve whose spherical coordinates are $\rho = \rho(t)$, $\theta = \phi(t)$ and $\phi = \phi(t)$ for *t* in an interval $[a, b]$ is

$$L = \int_a^b \sqrt{\rho'(t)^2 + (\rho(t)^2 \sin^2\phi(t))\theta'(t)^2 + \rho(t)^2\phi'(t)^2}\,dt.$$

10. Let be the three a $=2i - j + k$, b $= i + 2j - k$ and c $= i + j - 2k$ vectors . Find the vector in the plane of b and c whose projection on a is of magnitude $\sqrt{\frac{2}{3}}$.

Chapter 10

Partial Differentiation

In mathematics, a partial derivative of a function of several variables is its derivative with respect to one of those variables with the others held constant (as opposed to the total derivative, in which all variables are allowed to vary). Partial derivatives are useful in vector calculus and differential geometry. The partial derivative of a function f with respect to the variable x is written as f_x, $\partial_x f$, or $\partial f/\partial_x$. The partial-derivative symbol ? is a rounded letter, distinguished from the straight d of total-derivative notation. The notation was introduced by Legendre and gained general acceptance after its reintroduction by Jacobi.

Scalar Functions of One Variable

You should have met most of the ideas, if not the presentation, of this first subsection at school. Suppose $f(x)$ is a real function of $x \in \mathbb{R}$.

Limits

ℓ is the limit of $f(x)$ as $x \to a$ if

$$\lim_{x \to \alpha} f(x) = \ell .$$

Remarks:

1. Limits do not necessarily exist. For instance consider

$$f = 1/x \text{ as } x \to 0.$$

2. The left limit, i.e.

$$\lim_{x \to \alpha -} f(x) = \ell -,$$

need not equal the right limit, i.e.

$$\lim_{x \to \alpha +} f(x) = \ell +,$$

For instance consider $f = \tan^{-1}(1/x)$ as $x \to 0\pm$.

Continuity

$f(x)$ is continuous at a iff

$$\lim_{x \to \alpha} f(x) = f(a).$$

Remark:

1. The limit may exist at $x = a$ without f being continuous. For instance consider

$$f(x) = \begin{cases} 0 & x \neq a \\ 1 & x = a \end{cases}.$$

Dierentiability

$f(x)$ is dierentiable at a with derivative

$$\frac{df}{dx}(a) = f'(a) \text{ if}$$

$$\lim_{x \to \alpha} \frac{f(x) - f(a)}{x - a} = f'(a)$$

Remarks:

1. The derivative measures the rate of change of f with respect to x.
2. A dierentiable function is continuous (because both the numerator and denominator in go to zero in the limit).

Taylor's Theorem

Suppose that f is dierentiable $(N + 1)$ times; then

$$f(a + h) = f(a) + hf'(a) + \frac{h^2}{2!} f''(a) + \cdots + \frac{h^N}{N} f^{(N)}(a) + R_N(h),$$

where for some $0 < \theta < 1$

$$R_N(h) = \frac{h^{N+1}}{(N+1)} f^{(N+1)}(a + \theta h).$$

If $f^{(N+1)}$ is continuous then

$$R_N(h) = \frac{h^{N+1}}{N!} \int_0^1 (1-t)^N f^{(N+1)}(a + th)dt.$$

Notation: If $g(x)/x$ is bounded as $x \to 0$ we say that $g(x) = O(x)$,

If $g(x)/x \to 0$ as $x \to 0$ we say that $g(x) = o(x)$.

Remarks:

1. If f is $(N + 1)$ times dierentiable (on a suitable interval) then

$$f(a+h)-f(a)-\sum_{r=1}^{N}\frac{h^r}{r!}f^{(r)}(a)=O(h^{N+1}),$$

while if f is N times dierentiable then from the definition of dierentiability

$$f(a+h)-f(a)-\sum_{r=1}^{N}\frac{h^r}{r!}f^{(r)}(a)=O((h^N),$$

2. If f is dierentiable

$$\begin{aligned}f(a+h)&=f(a)+hf'(a)+O(h)\\&=f(a)+hf'(a)+...,\end{aligned}$$

where the ellipsis ... is sloppy (but common) notation to indicate terms smaller than those included.

3. Knowledge of the derivative allows us to make a local *linear* approximation of a smooth function.

Functions of Several Variables

We have so far considered scalar functions of a scalar

e.g. $$f(x)=x^2+1 \text{ from } \mathbb{R} \text{ to } \mathbb{R},$$

but it is also possible to have scalar functions of a vector

e.g. $$f(x,y)=x^2+y^2 \text{ from } \mathbb{R}^2 \text{ to } \mathbb{R},$$

e.g. $$f(x)\ f(x_1,...,x_m)=\sum_{j=1}^{m}x_j^2=x_jx_j \text{ (s.c.) from } \mathbb{R}^m \text{ to } \mathbb{R},$$

and vector functions of a scalar or vector

e.g. $$f(x)=\sum_{j=1}^{n}f_j(x)e_j\equiv(f_1(x),\cdots,f_n(x)) \text{ from } \mathbb{R} \text{ to } \mathbb{R}^n,$$

e.g. $$f(x)=\sum_{j=1}^{n}f_j(x)e_j \text{ from } \mathbb{R}^m \text{ to } \mathbb{R}^n,$$

where the e_j $(j = 1, ..., n)$ are orthonormal basis vectors.

Remark:

1. A function may only be defined on a subset of $\mathbb{R}^m$ called its *domain*. For instance the real function f,

$$f(x, y) = (1 - x^2 - y^2)^{\frac{1}{2}},$$

is only defined for $x^2 + y^2 \leq 1$

Limits

For a function $f := \mathbb{R}^m \to \mathbb{R}^n = \ell$, is the limit of $f(x)$ as $x \to a$ if

$$\lim_{x \to a} f(x) = \ell$$

independently of the way that x approaches a.

Remark and Example:

1. The independence of approach route of x to a is crucial (cf. left and right limits being equal for $f: \mathbb{R} \to \mathbb{R}$). To illustrate this, consider the function

$$f(x, y) = \begin{cases} \dfrac{2xy}{x^2 + y^2} & x \neq 0 \\ 0 & x = 0 \end{cases}$$

Suppose that $x = (\rho \cos \phi, \ \rho \sin f) \to 0$ on lines of constant ϕ. Since

$$f(x, y) = \sin 2\phi \quad \rho \neq 0,$$

it follows that lim $\rho \to 0 \, f(x, y)$ can be anything between –1 and 1. Hence no limit exists as $x \to 0$.

Continuity

A function $f: \mathbb{R}^m \to \mathbb{R}^n$ is continuous at a iff

$$\lim_{x \to a} f(x) = f(a)$$

Remark:

1. Continuity in $\mathbb{R}^m$ says more than continuity in x^1 for fixed $x^2, x^3, ..., x^m$, plus continuity in x_2 for fixed $x_1, x_3, ..., x_m$, plus ... plus continuity in x_m for fixed $x_1, x_2, ..., x_m - 1$.

For instance, in the case of $f(x, y)$) no limit exists as $x \to 0$, hence f cannot be continuous at $x = 0$. However $f(x, 0) = 0$ and $f(0, y) = 0$ are continuous functions of x and y respectively.

Derivatives of Vector Functions of One Variable

Suppose $f(x)$ is a real vector function from $\mathbb{R}$ to $\mathbb{R}^n$. The derivative of f (x) at a is defined as

$$f'(a) = \frac{df}{dx}(a) = \lim_{x \to a} \frac{f(x) - f(a)}{x - a}$$

where this limit exists. If we write $f = f_i e_i$ (s.c.), where the e_i ($i = 1, ..., n$) are fixed basis vectors (e.g. i, j, k in $\mathbb{R}^3$), it follows that

$$\begin{aligned} f'(a) &= \frac{df}{dx}(a) = \lim_{x\to a} \frac{f_i(x)e_i - f_i(a)e_i}{x-a} = \frac{df_i}{dx}(a)e_i \\ &= \left(\frac{df_1}{dx}(a), \frac{df_2}{dx}(a), \cdots, \frac{df_n}{dx}(a)\right) \\ &= (f_1'(a), f_2'(a), \cdots, f_n'(a)). \end{aligned}$$

Remark: If the position a particle in $\mathbb{R}^3$ at time t is given by $x(t)$, then we define the velocity $u(t)$ of the particle at time t to be

$$u = \frac{dx}{dt}$$

Example: Consider the function $g(s) = u(s).u(s)$. Then using the summation convention,

$$\frac{dg}{ds} = \frac{d}{ds}(u.u) = \frac{d}{ds}(u_i u_i) = 2u_i \frac{du_i}{ds} = 2u.\frac{di}{ds}.$$

Suppose further that there is a constraint so that $g' = u.u = 1$. It follows that $g' = 0$, and hence from that $u.u' = 0$, i.e. that u and u' are perpendicular. 1/00

Partial Derivatives

How do we dierentiate functions of several variables? The method we will adopt is:

1. Try the obvious;
2. Spot something fishy;
3. Fix it up (but not in this course).

Definition: For a function $f(x, y)$ of two variables define the partial derivative with respect to x at (a, b) as

$$\left(\frac{\partial f}{\partial x}\right) = \lim_{h\to 0} \frac{f(a+h,b) - f(a,b)}{h}.$$

Similarly define the partial derivative with respect to y at (a, b) as

$$\left(\frac{\partial f}{\partial y}\right) = \lim_{k\to 0} \frac{f(a+b,k) - f(a,b)}{k}.$$

More generally for $f: \mathbb{R}^m \to \mathbb{R}^n$, the jth partial derivative of f at a $\in \in \mathbb{R}^m$ is

$$\frac{\partial f}{\partial x_j}(a) = \lim_{h_j \to 0}$$

$$\frac{f(a_1,\cdots a_j-1,h_j,a_j+1,\cdots,a_m)-f(a_1,\cdots a_j-1,a_j,a_j+1,\cdots,a_m)}{h_j}$$

where this limit exists.

Notation: There are a number of alternative notations for partial derivatives:

$$\frac{\partial f}{\partial x_j}(a) = \partial_{xj} f(a) = \partial_j f(a) = f_{xj}(a) = f_{,j}(a) = D_j f(a)$$

$$= \left(\frac{\partial f}{\partial x_j}\right)_{x1,\cdots,xy-1,xj+1,\cdots,x_m}(a) = \frac{\partial f}{\partial x_j}\bigg|_{x1,\cdots xj-1,xj+1,\cdots x_m}(a)$$

Remark: When doing the question on Examples Sheet 1 on Euler's theorem for homogeneous functions you are recommended to use the f, j (or $\delta_j f$ or $D_j f$) notation.

Examples:

1. Sider the function $f(x, y) = x^2 + yx$. Then

$$\left(\frac{\partial f}{\partial x}\right)_y = 2x+y, \left(\frac{\partial f}{\partial y}\right)_x = x.$$

2. Consider the function $f(x, y, z) = r$ where $r = |x| = (x^2+y^2+z^2)^{\frac{1}{2}}$. Then

$$\left(\frac{\partial f}{\partial x}\right)_{y,z} = \left(\frac{\partial}{\partial x}\right)_{y,z}(x^2+(y^2+z^2)^{\frac{1}{2}} = \frac{1}{2(x^2+y^2+z^2)} = 2r\frac{x}{r}$$

and similarly

$$\left(\frac{\partial r}{\partial y}\right)_{x,z} = \frac{y}{r}, \left(\frac{\partial r}{\partial z}\right)_{x,z} = \frac{z}{r}.$$

3. Find $\frac{\partial f}{\partial r_k}$ $f = x_j$, i.e. $\frac{\partial x_j}{\partial x_k}$,First note that for $k = 1$

$$\left(\frac{\partial x_1}{\partial x_2}\right)_{x_2,\cdots,r_m} = 1, \left(\frac{\partial x_2}{\partial x_1}\right)_{x_2,\cdots,r_m} = 0,$$

i.e. $$\left(\frac{\partial x_j}{\partial x_1}\right)_{x_2,\cdots,x_m} = \begin{Bmatrix} 1 & \text{if } j=1 \\ 0 & \text{if } j \neq 1 \end{Bmatrix} = \partial_{j1},$$

where δ_{jk} is the Kronecker delta. Similarly for $k = 2$

$$\left(\frac{\partial x_j}{\partial x_2}\right)_{x_1,x_3,\cdots,x_m} = \partial j2 \;=\; \frac{\partial x_j}{\partial x_k} = \partial_{jk}.$$

and for general k

$$\frac{\partial x_j}{\partial x_k} = \delta_{jk}.$$

We can now use this result to obtain and more cleanly. First we note (invoking the summation convention) that $r^2 = x_j x_j$, and hence that

$$2r\frac{\partial r}{\partial x_k} = \frac{\partial r^2}{\partial x_k} = \frac{\partial}{\partial x_k}(x_j x_j) = 2x_j\frac{\partial x_j}{\partial x_k} = 2x_j\delta_{jk} = 2x_k,$$

i.e. on dividing through by $2r$

$$\frac{\partial r}{\partial xk} = \frac{xk}{r},$$

which, in terms of sux notation.

4. Unlectured example to demonstrate that the partial derivatives of a function can exist at all points in $\mathbb{R}^3$, including the origin, even though the function is not continuous at the origin. The partial derivatives of the function defined in, i.e.

$$f(x.y) = \begin{cases} \dfrac{2xy}{x^2+y^2} & x \neq 0 \\ 0 & x = 0 \end{cases},$$

for $x \neq 0$ are

$$\left(\frac{\partial f}{\partial x}\right)_y = \frac{2y(y^2-x^2)}{(x^2+y^2)^2}, \left(\frac{\partial f}{\partial y}\right)_x = \frac{2x(x^2-y^2)}{(x^2+y^2)^2}$$

and at $x = 0$ are

$$\left(\frac{\partial f}{\partial x}\right)_y = \lim_{h\to 0}\frac{f(h,0)-f(0,0)}{h} = 0$$

$$\left(\frac{\partial f}{\partial y}\right)_x = \lim_{h\to 0}\frac{f(0,h)-f(0,0)}{h} = 0$$

Hence $\frac{\partial f}{\partial x}$ and $\frac{\partial f}{\partial y}$ exist at all points in $\mathbb{R}^2$. However, we concluded that f is not continuous at 0. This is fishy, since for $f : \mathbb{R} \to \mathbb{R}$ dierentiability implies continuity.

Remark: The existence of partial derivatives does not imply dierentiability (although, of course, we have yet to define dierentiability, so this statement is a little presumptive).

Directional Derivatives

This is the generalisation of [first-order] partial derivatives to directions other than parallel to the coordinate vectors e*j* .

Definition: Let $\hat{u} \in \mathbb{R}^m \to \mathbb{R}^m$ be a unit vector. Then for $f : \mathbb{R}^m \to \mathbb{R}^n$, the directional derivative of f at a in direction $\hat{u}$ is

$$f'(a\,;\,\hat{u}) = \lim_{h \to 0} \frac{f(a + h\hat{u}) - f(a)}{h}$$

whenever this limit exists.

Remarks:

1. It follows from and the definition of a partial derivative that
$$\frac{\partial f}{\partial x_j}(a) = f'(a; e_j).$$
2. The directional derivative is a measure of the rate of change of f in the direction $\hat{u}$.
3. The directional derivatives of a function at a can exist for all directions *bu*, without the function necessarily being continuous at a.

Dierentiability

The presence of a margin line indicates material that, in my opinion, does not appear in the schedule, and that, in my opinion, is not examinable.

Remarks:

1. The key idea behind dierentiability is the local approximation of a function by a linear function/map.
2. Note from definition that a function $g(x)$ is dierentiable at a if for some $g'(a)$
$$g(a + h) - g(a) = g'(a)h + (h)|h|\,,$$
where $|\,\varepsilon\,(h)\,| \to 0$ as $|\,h\,| \to 0$.

Scalar Functions of Two Variables

A function $f(x, y)$ is dierentiable at (a, b) if

$$f(a + h,\, b + k) - f(a,\, b) = T_x h + T_y k + \varepsilon(h)\,|\,h\,|$$

where T_x and T_y are independent of $h = (h, k)$, and $\varepsilon|(h)| \to 0$ as $|\,h\,| = (h^2 + k^2)^{\frac{1}{2}} \to 0$.

Remarks:

1. $T_x h + T_y k$ is a linear function of h and k (i.e. we have our linear approximation).
2. Let $z = f(x, y)$ define a surface (cf. $y = g(x)$ defining a curve). Then
$$z = f(a, b) + T_Z(x - a) + T_y(y - b)$$
is the tangent plane touching the surface at the point $(a, b, f(a, b))$ (cf. $y = g(a) + g'(a)(x - a)$ defining the tangent curve at the point $(a, g(a))$).
3. Let $| h | \to 0$ with $k = 0$, then from and the definition of the partial derivative.
$$T_z(x, y) = \left(\frac{\partial}{\partial x}\right)_y f(x, y).$$
Similarly by letting $| h | \to 0$ with $h = 0$
$$T_y(x, y) = \left(\frac{\partial}{\partial y}\right)_x f(x, y).$$
4. Dierentiability implies that the directional derivative exists for all directions $\hat{u} = (\hat{u}_x, \hat{u}_y)$; in particular
$$f'(x, \hat{u}) = T_z \hat{u}_x + T_y \hat{u}_y .$$

It follows that dierentiability implies that partial derivatives exist for all orientations of the coordinate axes. Unlectured example: show that the function $f(x, y)\, xy$ is dierentiable. Starting from the definition we have that

$$f(x + h, y + k) - f(x, y) = xy + yh + xk + hk - xy$$
$$= yh + xk + \frac{hk}{(h^2 + k^2)^{\frac{1}{2}}} | h |.$$

Hence, after writing $h = | h | \cos \phi$ and $k = | h | \sin \phi$ where $| h | = (h^2 + k^2)^{\frac{1}{2}}$,

$$\varepsilon(h) = \frac{hk}{(h^2 + k^2)^{\frac{1}{2}}} = | h | \cos \phi \ \sin \phi \to 0 \text{ as } | h | \to 0.$$

$$\left(\frac{\partial f}{\partial x}\right)_y = T_x = y, \left(\frac{\partial f}{\partial y}\right)_x = T_y = x.$$

Functions of Several Variables

A function $f : \mathbb{R}^m \to \mathbb{R}^n$ is dierentiable at $a \in \mathbb{R}^m$ if for all $h \in \mathbb{R}^m$

$$f(a + h) - f(a) = T\, h + \in(h) | h |,$$

where T is a $n \times m$ matrix that is independent of h, and

$$| \in (h) | \to 0 \text{ as } | h | \to 0.$$

Remarks:

1. *Th* is a linear function of *h*.
2. From the definition of a partial derivative, it follows for the special case $h = (0, ..., 0, h_j, 0, ..., 0)$ that

$$\frac{\partial f_i}{\partial x_j}(a) = \lim_{h_j \to 0} \frac{f(a_1, \cdots a_j + h_j, \cdots, a_m) - f_i(a_1, \cdots a_j, \cdots, a_m)}{h} = T_{ij}.$$

Definition: The $n \times m$ Jacobian matrix *T* (which is a 'representation' of the derivative) is defined by

$$T = \begin{pmatrix} \frac{\partial f_1}{\partial r_1} & \frac{\partial f_1}{\partial r_2} & \cdots & \frac{\partial f_1}{\partial r_2} \\ \frac{\partial f_2}{\partial r_1} & \frac{\partial f_2}{\partial f_2} & \cdots & \frac{\partial r f_2}{\partial r_m} \\ \cdots & \cdots & \cdots & \cdots \\ \frac{\partial f_n}{\partial r_1} & \frac{\partial f_n}{\partial r_2} & \cdots & \frac{\partial f_n}{\partial r_m} \end{pmatrix}$$

Example: For $f(x, y) = x^2 yi + 2xj + 2yk$,

$$T = \begin{pmatrix} \frac{\partial f_1}{\partial x} & \frac{\partial f_1}{\partial y} \\ \frac{\partial f_2}{\partial x} & \frac{\partial f_2}{\partial y} \\ \frac{\partial f_3}{\partial x} & \frac{\partial f_3}{\partial y} \end{pmatrix} = \begin{pmatrix} 2ry & x^2 \\ 2 & 0 \\ 0 & 2 \end{pmatrix}$$

Properties of Dierentiable Functions

We shall just state a couple of these. Analysis both for more properties (many of which are not unexpected), and for proofs.

Continuity. If a function $f(x)$ is dierentiable at a, then *f* is continuous at a.

A sucient condition for dierentiability. If all the partial derivatives of *f* exist 'close to' a, and are continuous at *a*, then *f* is dierentiable at *a*; i.e. the continuity of partial derivatives, rather than just their existence, ensures dierentiability.

Optional Exercises

1. Confirm that for *f* defined, the partial derivatives defined are not continuous at the

origin. Note that if these partial derivatives were continuous at the origin, then from the above two results, f would be both dierentiable and continuous at the origin, which would be in contradiction to the result derived.

2. Give an example of a function $f(x, y)$ that is dierentiable at (0, 0), but for which the partial derivatives are either not continuous at (0, 0), or do not exist at other than at (0, 0).

The Chain Rule and Change of Variables

For dierentiable $x : \mathbb{R} \to \mathbb{R}$ and $f : \mathbb{R} \to \mathbb{R}$ define the composite function $F : \mathbb{R} \to \mathbb{R}$ by $F(t) = f(x(t))$.

Then the chain rule states that

$$F'(t) = f'(x(t))\, x'(\text{t}) \text{ i.e.} \frac{dF}{dt} = \frac{df}{dx}\frac{dx}{dt}.$$

How does this generalise to functions of several variables?

The Chain Rule for $x : \mathbb{R} \to \mathbb{R}^m$ and $f : \mathbb{R}^m \to \mathbb{R}$

Suppose that x and f are dierentiable, and consider

$$F(t) = f(x(t)),$$

and its derivativet

$$\begin{aligned}\frac{dF}{dt} &= \lim_{\delta t \to 0} \frac{F(t+\delta t) - F(t)}{\delta t} \\ &= \lim_{\delta t \to 0} \frac{f(x)(t+\delta t) - f(x(t))}{\delta t} \\ &= \lim_{\delta t \to 0} \frac{f(x)(t+\delta x(t)) - f(x(t))}{\delta t}\end{aligned}$$

where we have written

$$x(t + \delta \text{t}) = x(t) + \delta x(t).$$

Recall also from the definition of a derivative of a vector function of one variable,

$$\lim_{\delta t \to 0} \frac{\delta \text{x}}{\delta t} = \frac{\text{dx}}{\text{dt}}$$

In order to focus ideas first consider $m = \in$ and write $x = (x, y)$. Then from using together with the definitions of a partial derivative, it follows on assuming continuity of the partial derivatives, $\in$ that

$$\frac{\mathrm{d}F}{\mathrm{dt}} = \lim_{\delta t \to 0}\left\{\frac{f(x+\delta x, y+\delta y) + f(x, y+\delta y) - f(x,y)}{\delta t}\right\}$$

$$= \lim_{\delta t \to 0}\left\{\frac{f(x+\delta x, y+\delta y) - f(x, y+\delta y)}{\delta t}\frac{\delta x}{\delta t}\right\}$$

$$+ \lim_{\delta t \to 0}\left\{\frac{f(x, y+\delta y) - f(x,y)}{\delta y}\right\}$$

$$= \left(\frac{\partial f}{\partial x}\right)_y \frac{\mathrm{dx}}{\mathrm{dt}} + \left(\frac{\partial f}{\partial y}\right)_x \frac{du}{dt}$$

Similarly (but much more messily) for general m

$$\frac{\mathrm{dF}}{\mathrm{dt}} = \frac{\mathrm{d}}{\mathrm{dt}} f(x(t)) = \sum^{m} \left(\frac{\partial f}{\partial x_i}\right)_{x_1,\ldots,x_{j-1},x_{j+1},\ldots,x_m} \frac{dx_j}{dt},$$

or, if we invoke the summation convention and do not display the variables that are begin held constant,

$$\frac{dF}{dt} = \frac{d}{dt} f(x(t)) = \frac{\delta f}{\delta x_i}\frac{dx_j}{dt}$$

Example: Suppose that

$$f(x, y) = x^2 + y \,,\; x(t) = (x, y) = (t^2, e^t)\,,$$

and that as above $F(t) = f(x(t))$. We calculate $\frac{dF}{dt}$ in two ways.

1. From the definitions of f and x

$$\left(\frac{\partial f}{\partial x}\right)_y = 2x, \left(\frac{\partial f}{\partial y}\right)_x = 1 \frac{dx}{dt} = 2t,\; \frac{dy}{dt} = e^t.$$

Hence from using the chain rule

$$\frac{dF}{dt} = \frac{\partial F}{\partial x}\frac{dx}{dt} + \frac{\partial f}{\partial y}\frac{dy}{dt} = 2x2t + e^t = 4t^3 + e^t.$$

2. From direct substitution

$$F = t^4 + e^t, \text{ and hence } \frac{dF}{dt} = 4t^3 + e^t.$$

Unlectured example: implicit dierentiation. Suppose that $f \equiv f(x, y)$ and that $y \equiv y(x)$. Write $x = t$ and define $F(t) = f(x(t))$ where $x = (t, y(t))$. Then from

$$\frac{dF}{dt} = \left(\frac{\partial f}{\partial x}\right)_y \frac{dx}{dt} + \left(\frac{\partial f}{\partial y}\right)_x \frac{dy}{dt}$$

$$= \left(\frac{\partial f}{\partial x}\right)_y + \left(\frac{\partial f}{\partial y}\right)_x \frac{dy}{dt}$$

or since $x = t$

$$= \frac{d}{dx} f(x, y(x)) = \left(\frac{\partial f}{\partial x}\right)_y + \left(\frac{\partial f}{\partial y}\right)_x \frac{dy}{dx}.$$

Suppose further $y(x)$ is defined implicitly by the equation

$$f(x, y) = 0.$$

Then $df/dx = 0$, and hence from

$$\frac{dy}{dx} = -\left(\frac{\partial f}{\partial y}\right)_y \Big/ \left(\frac{\partial f}{\partial x}\right)_x$$

The Chain Rule for x : $\mathbb{R}^{\ell} \to: \mathbb{R}^m$ and $f : \mathbb{R}^m \to: \mathbb{R}$

Suppose x and f are dierentiable and consider

$$F(u_1, ..., u_\ell) = f(x(u_1, ... u_\ell)),$$

where the vector $u = (u_1, ..., u_\ell)$ is the independent variable. By keeping $u_2, ..., u_\ell$ constant and dierentiating with respect to u_1, that

$$\left(\frac{\partial F}{\partial u_1}\right)_{u_2,\cdots,u_\ell} = \sum_{j=1}^{m}\left(\frac{\partial f}{\partial x_j}\right)_{x_1,\cdots,x_{j+1},\cdots,x_m} \left(\frac{\partial x}{\partial u_1}\right)_{u_2,\cdots,u_\ell},$$

where we write $\frac{\partial F}{\partial u_1}$ rather than $\frac{\partial F}{\partial u_1}$ since $F(u)$ is a function of two or more variables. Similarly

$$\frac{\partial F}{\partial u_k} = \sum_{j=1}^{m} \frac{\partial f}{\partial x_j}\frac{\partial x_j}{\partial u_k}$$

where we have not displayed the variables that are being held constant — this is a dangerous notational 'simplification', but an often used one. Alternatively if we invoke the s.c.

$$\frac{\partial F}{\partial u_k} = \frac{\partial f}{\partial x_j}\frac{\partial x_j}{\partial u_k}$$

Example: Consider the transformation (actually, as discussed later, a rotation of the co-ordinate axes through an angle θ)

$$x \equiv (x, y) \ (u \cos\theta - v \sin\theta, u \sin\theta + v \cos\theta),$$

where we have written (x, y) for (x_1, x_2) and (u, v) for (u_1, u_2). Then for $F(u) = f(x(u))$, that

$$\left(\frac{\partial F}{\partial u}\right)_v = \left(\frac{\partial f}{\partial x}\right)_y \left(\frac{\partial x}{\partial u}\right)_v + \left(\frac{\partial f}{\partial y}\right)_x \left(\frac{\partial y}{\partial u}\right)_v = \cos\theta\left(\frac{\partial f}{\partial x}\right)_y + \sin\theta\left(\frac{\partial f}{\partial y}\right)_x,$$

$$\left(\frac{\partial F}{\partial v}\right)_u = \left(\frac{\partial f}{\partial x}\right)_y \left(\frac{\partial x}{\partial v}\right)_u + \left(\frac{\partial f}{\partial y}\right)_x \left(\frac{\partial y}{\partial v}\right)_u = -\sin\theta\left(\frac{\partial f}{\partial x}\right)_y + \cos\theta\left(\frac{\partial f}{\partial y}\right)_x$$

Sloppy Notation: Sometimes, indeed often, $F(u, v) = f(x(u, v), y(u, v))$ is written as $f(u, v)$. Admittedly this can be confusing at first sight, but context is all and otherwise it is possible to run out of alphabet! The results then become

$$\left(\frac{\partial f}{\partial u}\right)_v = \cos\theta\left(\frac{\partial f}{\partial x}\right)_y + \sin\theta\left(\frac{\partial f}{\partial y}\right)_x$$

$$\left(\frac{\partial f}{\partial v}\right)_u - \sin = \theta\left(\frac{\partial f}{\partial x}\right)_y + \cos\theta\left(\frac{\partial f}{\partial y}\right)_x,$$

or for more general $f \equiv f(x, y)$ and $x \equiv x(u, v)$

$$\left(\frac{\partial f}{\partial u}\right)_v = \left(\frac{\partial f}{\partial x}\right)_y \left(\frac{\partial x}{\partial u}\right)_v + \left(\frac{\partial f}{\partial y}\right)_x \left(\frac{\partial y}{\partial u}\right)_v,$$

$$\left(\frac{\partial f}{\partial v}\right)_u = \left(\frac{\partial f}{\partial x}\right)_y \left(\frac{\partial x}{\partial v}\right)_u + \left(\frac{\partial f}{\partial y}\right)_x \left(\frac{\partial y}{\partial v}\right)_u,$$

Similarly for $f \equiv f(x)$ and $x \equiv x(u)$, becomes

$$\left(\frac{\partial f}{\partial u_k}\right)_{u1,\ldots,u_{k-1},uk+1,\ldots,u\ell} \quad \left(\frac{\partial f}{\partial x_j}\right)_{x_1,\ldots,x_{y-1},x_y+1,\ldots,x_m} \quad \left(\frac{\partial f}{\partial u_k}\right)_{u_1,\ldots,u_{k-1},uk+1,\ldots,u\ell}$$

or just

$$\frac{\partial f}{\partial u_k} = \frac{\partial f}{\partial x_j}\frac{\partial x_j}{\partial u_k}$$

The Chain Rule for $x : \mathbb{R}^\ell \to f : \mathbb{R}^m$ and $f : \mathbb{R}^m \to : \mathbb{R}^n$

For dierentiable $x(u)$ and $f(x)$, let

$$F(u) = f(x(u)) .$$

Then by applying componentwise and using the s.c., we have that

$$\frac{\partial F_i}{\partial u_k} = \frac{\partial f_i}{\partial x_j}\frac{\partial x_j}{\partial u_k}$$

or in more sloppy notation

$$\frac{\partial f_i}{\partial u_k} = \frac{\partial f_i}{\partial x_j}\frac{\partial x_j}{\partial u_k},$$

The Matrix Form of the Chain Rule

Denote the Jacobian matrices of $F(u)$, $f(x)$ and $x(u)$ by S, T and U respectively.

These matrices are of size $n \times \ell$, $n \times m$ and $m \times \ell$ respectively.

The matrix T, while S and U are given by

$$S = \begin{pmatrix} \frac{\partial F_1}{\partial x_1} & \frac{\partial F_1}{\partial x_2} & \cdots & \frac{\partial F_1}{\partial u_\ell} \\ \frac{\partial F_2}{\partial x_1} & \frac{\partial F_2}{\partial u_2} & \cdots & \frac{\partial F_2}{\partial u_\ell} \\ \cdots & \cdots & \cdots & \cdots \\ \frac{\partial F_n}{\partial x_1} & \frac{\partial F_n}{\partial u_2} & \cdots & \frac{\partial F_n}{\partial u_\ell} \end{pmatrix}, \quad U = \begin{pmatrix} \frac{\partial x_1}{\partial u_1} & \frac{\partial x_1}{\partial u_2} & \cdots & \frac{\partial x_1}{\partial u_\ell} \\ \frac{\partial x_2}{\partial u_1} & \frac{\partial x_2}{\partial u_2} & \cdots & \frac{\partial x_2}{\partial u_\ell} \\ \cdots & \cdots & \cdots & \cdots \\ \frac{\partial x_m}{\partial u_1} & \frac{\partial x_m}{\partial u_2} & \cdots & \frac{\partial x_m}{\partial u_\ell} \end{pmatrix}$$

Using these definitions, can be rewritten (using the s.c.) as

$S = TU$ or, using the s.c., as $S_{ik} = T_{ij}\, U_{jk}$.

This is the matrix form of the chain rule.

Remarks:

1. From the property of determinants

$$|S| = |T|\,|U|.$$

2. If $n = m$ and T is non-singular then

$$U = T^{-1} S \text{ and } |U| = |S| / |T|.$$

Similarly if $m = \ell$ and U is non-singular then

$$T = SU^{-1} \text{ and } |T| = |S| / |U|.$$

Change of Variables

To fix ideas, initially work with a scalar function/field f.

Digression: There are many important physical examples of scalar fields, e.g. the gravitational potential, the electric potential, density, temperature, salinity, the ozone mixing ratio (i.e. the concentration of ozone), the partial pressure of oxygen in blood $\left(P_{O_2}\right)$.

For instance, suppose that $f(x, y, z)$ describes the ozone mixing ratio. Rather than using Cartesian co-ordinates (x, y, z) centred, say, at the centre of the earth, it is more natural to use spherical polar coordinates (r, θ, ϕ), or a modified version of spherical polar co-ordinates that measures distances above the surface of the earth. We need to be able to transform f and its partial derivatives from one co-ordinate system to another.

For simplicity we will consider a two dimensional example, in which instead of using Cartesian coordinates (x, y), we wish to use (u, v), say because of the physical properties of the system. For instance:

RotationPlane polar co-ordinates

$x = u \cos\theta - v \sin\theta$ $\qquad x = u \cos v$

$y = u \sin\theta + v \cos\theta$ $\qquad y = u \sin v$

For U, i.e. $U_{ij} = \dfrac{\partial x_i}{\partial u_j}$,

Rotation $\qquad$ Plane polar co-ordinates

$$U = \begin{pmatrix} \cos\theta & -\sin\theta \\ \sin\theta & \cos\theta \end{pmatrix} \qquad U = \begin{pmatrix} \cos v & -u\sin v \\ \sin v & u\cos v \end{pmatrix}$$

while from respectively

$$S = \left(\frac{\partial F}{\partial u}, \frac{\partial F}{\partial v}\right) \text{ and } T = \left(\frac{\partial f}{\partial x}, \frac{\partial f}{\partial y}\right).$$

Hence

$$\frac{\partial F}{\partial u} = \cos\theta \frac{\partial f}{\partial x} + \sin\theta \frac{\partial f}{\partial y}$$

$$\frac{\partial F}{\partial v} = -\sin\theta \frac{\partial f}{\partial x} + \cos\theta \frac{\partial f}{\partial y}$$

Plane polar co-ordinates

$$\frac{\partial F}{\partial u} = \cos v \frac{\partial f}{\partial x} + \sin v \frac{\partial f}{\partial y}$$

$$\frac{\partial F}{\partial v} = -u \sin v \frac{\partial f}{\partial x} + u \cos v \frac{\partial f}{\partial y}$$

Remarks:

1. For rotations it is more conventional to use (x', y') or (X, Y) rather than (u, v).
2. For plane polar co-ordinates it is more conventional to use (ρ, ϕ) or (r, θ) rather than (u, v).

In the case of (ρ, ϕ) we have after identifying F with f that

$$\rho\left(\frac{\partial f}{\partial \rho}\right)_\phi = \rho\cos\phi\left(\frac{\partial f}{\partial x}\right)_y \; \rho\sin\phi + \left(\frac{\partial f}{\partial y}\right)_x$$

$$= x\left(\frac{\partial f}{\partial x}\right)_y + y\left(\frac{\partial f}{\partial y}\right)_x$$

$$\left(\frac{\partial f}{\partial \phi}\right)_\rho = -\rho\sin\phi\left(\frac{\partial f}{\partial x}\right)_y + \rho\cos\phi\left(\frac{\partial f}{\partial y}\right)_x$$

$$= -y\left(\frac{\partial f}{\partial x}\right)_y + x\left(\frac{\partial f}{\partial y}\right)_x.$$

From solving for f_x and f_y, or equivalently from calculating U^{-1} and using,

$$\left(\frac{\partial f}{\partial x}\right)_y = \cos\phi\left(\frac{\partial f}{\partial \rho}\right)_\phi - \frac{\sin\phi}{\rho}\left(\frac{\partial f}{\partial \phi}\right)_\rho,$$

$$\left(\frac{\partial f}{\partial y}\right)_x = \sin\phi\left(\frac{\partial f}{\partial \rho}\right)_\phi - \frac{\cos\phi}{\rho}\left(\frac{\partial f}{\partial \phi}\right)_\rho$$

A Common Error. An alternative method of calculating would be to use the chain rule in the form

$$\left(\frac{\partial f}{\partial x}\right)_y = \left(\frac{\partial f}{\partial \rho}\right)_y \left(\frac{\partial \rho}{\partial x}\right)_\phi + \left(\frac{\partial f}{\partial \phi}\right)_\rho \left(\frac{\partial \phi}{\partial y}\right)_y,$$

$$\left(\frac{\partial f}{\partial y}\right)_x = \left(\frac{\partial f}{\partial \rho}\right)_\phi \left(\frac{\partial \rho}{\partial y}\right)_x + \left(\frac{\partial f}{\partial \phi}\right)_\rho \left(\frac{\partial \phi}{\partial y}\right)_x.$$

It is then crucial to observe that $\frac{\partial \rho}{\partial x}$ and $\frac{\partial \phi}{\partial x}$ are to be evaluated at constant y, and $\frac{\partial \rho}{\partial y}$ and $\frac{\partial \phi}{\partial y}$ are to be evaluated at constant x. After recalling that $\rho = (x^2 + y^2)^{\frac{1}{2}}$ and $\phi = \tan^{-1}(y/x)$ it is straightforward to show that

$$\left(\frac{\partial \rho}{\partial x}\right)_y = \frac{x}{(x^2+y^2)^{\frac{1}{2}}} = \cos\phi,$$

$$\left(\frac{\partial \phi}{\partial x}\right)_y = -\frac{y}{(x^2+y^2)} = -\frac{\cos\phi}{\rho},$$

A common error is to try to calculate $\left(\frac{\partial \rho}{\partial x}\right)_y$ as follows:

$$x = \rho\cos\phi \Rightarrow \rho = \frac{x}{\cos\phi} \Rightarrow \frac{\partial \rho}{\partial x} = \frac{1}{\cos\phi}$$

The final expression is $\left(\frac{\partial \rho}{\partial x}\right)_\phi$ not the required $\left(\frac{\partial \rho}{\partial x}\right)_y$.

When evaluating a partial derivative remember to hold the right variables constant.

The Gradient of a Scalar Field

Let $f: \mathbb{R}^m \to \mathbb{R}$ be dierentiable at $a \in \mathbb{R}^m$, and let $(e_1, ..., e_m)$ be an orthonormal basis of $\mathbb{R}^m$.

Definition: The gradient of f at a is

$$\text{grad}\, f = \nabla f = \sum_{j=1}^{m} e_j \frac{\partial f}{\partial x_j} \quad \text{(a)}$$

$$= \left(\frac{\partial f}{\partial x_1}(a), \frac{\partial f}{\partial x_2}(a), \ldots, \frac{\partial f}{\partial x_m}(a)\right)$$

Remark: For Cartesian co-ordinates and $f \equiv f(x, y, z)$, the gradient of f can be written

$$\nabla f = i\frac{\partial f}{\partial x} + j\frac{\partial f}{\partial y} + k\frac{\partial f}{\partial z} = \left(\frac{\partial f}{\partial x}, \frac{\partial f}{\partial y}, \frac{\partial f}{\partial z}\right).$$

Example: Suppose that

$$f = \frac{1}{r} = \frac{1}{|x|} = \frac{1}{(x^2+y^2+z^2)^{\frac{1}{2}}} \text{ for } x \neq 0$$

note and remember that the function is singular, indeed not defined, at $x = 0$. Then

$$\nabla f = \left(-\frac{x}{r^3} - \frac{y}{r^3} - \frac{z}{r^3}\right) = -\frac{x}{r^3}$$

and so

$$\frac{\delta f}{\delta x} = -\frac{x}{(x^2+y^2+z^2)^{\frac{3}{2}}} = -\frac{x}{r^3}.$$

Alternatively, using example,

$$\frac{\partial f}{\partial x_j}\frac{\partial f}{\partial x_k} = \frac{\partial f}{\partial x_k}\left(\frac{1}{r}\right) = -\frac{1}{r^2}\frac{\partial f}{\partial x_k} = -\frac{1}{r^2}\frac{xk}{r} = -\frac{xk}{r^3}.$$

Properties of the Gradient. We shall just state a couple of these (they follow in a straightforward manner from, say, use of sux notation). For dierentiable functions $f : \mathbb{R}^m \to \mathbb{R}$ and $g : \mathbb{R}^m \to \mathbb{R}$

$$\nabla(\lambda f + \mu g) = \lambda \nabla f + \mu \nabla g \text{ for } \lambda, \mu \in \mathbb{R}$$

$$\nabla(fg) = g\nabla f + f\nabla g.$$

The Gradient is a Vector

Revision: Orthonormal Basis Vectors.

Suppose that for $\mathbb{R}^m$ $e'_1, \ldots, e_m$ are orthonormal basis vectors with Cartesian co-ordinates x, and that $e'_1, \ldots, e_m$ are orthonormal basis vectors with Cartesian co-ordinates x'.

From Algebra & Geometry, if $v \in \mathbb{R}^m$ then (using the s.c.)

$$v = v_j e_j = (v.e_j)\, e_j\,,$$

since

$$v \,.\, e_j = (v_k e_k) \,.\, e_j = v_k\,(e_k \,.\, e_k) = v_k\,\delta_{kj} = v_j.$$

In particular from for the vector $v = e'_k$, it follows that

$$e'_k = (e'_k.e_j)\, e_j$$

Also, in terms of the e'_k basis

$$v = v'_k e'_k \text{ with } v'_k = v.e'_k = (v_j e_j).e'_k = (e_j.e'_k)\, v_j.$$

Hence for the vector x

$$x'_k = (e_j \,.\, e'_k)\, x_j$$

and so using

$$\frac{\partial x'_k}{\partial y_i} = (ej.e'_k)\frac{\partial x_j}{\partial x_i} = (ej.e'_k)\delta_{ji} = e_i.e'_k$$

The Result. We wish to show that ∇f is independent of its basis vectors, i.e. that

$$\nabla f = e_j \frac{\partial f}{\partial x_j} \text{ and } \nabla f = e'_k \frac{\partial f}{\partial x'_k}$$

are the same vector. To this end we note that

$$e_j \frac{\partial f}{\partial x_j} = e_j \frac{\partial x'_k}{\partial x_j}\frac{\partial f}{\partial x'_k} \text{ using the chain rule}$$

$$= e_j (e_j.e'_k \frac{\partial f}{\partial x'_k}$$

$$= e'_k \frac{\partial f}{\partial x'_k}$$

Remark: Similarly (s.c.)

$$e_j \frac{\partial}{\partial x_j} = e_j \frac{\partial x'_k}{\partial x_j}\frac{\partial}{\partial x'_k} = e_j(e_j.e'_k)\frac{\partial}{\partial x'_k} = e'_k \frac{\partial}{\partial x'_k}$$

Hence the dierential operator

$$\nabla = e_j \frac{\partial}{\partial x_j}$$

is a vector (note the order of e_j and $\frac{\partial}{\partial x_j}$). This operator is referred to as 'del' or 'nabla' (after a harp-like ancient Assyrian/Hebrew musical instrument of similar shape).

Directional Derivatives Revisited

As shown below the directional derivative of a dierentiable scalar function $f : \mathbb{R}^m \to \mathbb{R}$ at a in the direction bu is the component of the gradient vector at a in the direction $\hat{u}$, i.e.

$$f'(a, \hat{u}) = \hat{u}\,\nabla f.$$

Thus the directional derivative has a maximum when bu is parallel to ∇f, i.e. ∇f is in the direction of the maximum rate of change of f.

To derive start from the definition of a directional dierentiability.

$$f'(a;\hat{u}) = \lim_{h\to 0} \frac{f(a+h\hat{u}) - f(a)}{h} = T\hat{u}.$$

For a scalar function f

$$T = \left(\frac{\partial f}{\partial x_1} \quad \frac{\partial f}{\partial x_2} \quad \cdots \quad \frac{\partial f}{\partial x_m} \right), 0$$

and hence (with a slightly cavalier attitude to column and row vectors)

$$T\hat{u} = \frac{\partial f}{\partial x_i}\hat{u}_i = \hat{u}.\nabla f.$$

Partial Derivatives of Higher Order

The partial derivatives of the first-order partial derivatives of a function are called the second-order partial derivatives; e.g. for a function $f: \mathbb{R}^m \to \mathbb{R}$, the jth partial derivative of $\frac{\partial f}{\partial x_i}$ is written

$$\frac{\partial f}{\partial x_j}\left(\frac{\partial f}{\partial x_i}\right) = \frac{\partial^2 f}{\partial x_j \partial x_i} = (fx_i)x_j := fx_ix_j.$$

There is a straightforward generalisation to n^{th} order partial derivatives.

Notation. If $i \neq j$ then a second-order partial derivative is called a mixed second-order partial derivative.

Example: For the scalar function $f(x, y) = xe^y$,

$$f_x = e^y,\ f_{xy} = e^y,\ f_{xx} = 0,$$
$$f_y = xe^y, f_{yx} = e^y, f_{yy} = xe^y.$$

Note that $f_{xy} = f_{yx}$, i.e. that the mixed second-order partial derivatives are equal.

Unlectured example showing that mixed partial derivatives need not be equal. For the scalar function

$$f(x,y) = \begin{cases} xy\left(\dfrac{x^2 - y^2}{x^2 + y^2}\right) & (x,y) \neq 0 \\ 0 & (x,y) = 0 \end{cases},$$

we have from the definitions and of a partial derivative, that

$$\frac{\partial f}{\partial x}(x,y) = \frac{(3x^2 - y^2)y}{x^2 + y^2} - \frac{2x^2 y(x^2 - y^2)}{(x^2 + y^2)^2} = \frac{y(x^4 + 4x^2y^2 - y^4)}{(x^2 + y^2)^2} \quad (x,y) \neq 0,$$

$$\frac{\partial f}{\partial x}(0,0) = \lim_{h\to 0}\frac{f(k,0)-f(0,0)}{h} = \lim_{h\to 0}\frac{0-0)}{h} = 0.$$

$$\frac{\partial}{\partial y}\left(\frac{\partial f}{\partial x}\right)(0,0) = \lim_{k\to 0}\frac{\frac{\partial f}{\partial x}(0,k)-\frac{\partial f}{\partial x}(0,0)}{k}$$

$$= \lim_{k\to 0}\frac{\frac{k(0+0-k^4)}{(0+k^2)^2}-0)}{k} = -1.$$

Similarly

$$\frac{\partial f}{\partial y}(x,y) = -\frac{x(y^4+4y^2x^2-x^4)}{(y^2+x^2)^2}(x,y)\neq 0,$$

$$\frac{\partial f}{\partial y}(0,0) = \lim_{k\to 0}\frac{f(0,k)-f(0,0)}{k}\lim_{k\to 0}\frac{0-0}{k} = 0,$$

$$\frac{\partial}{\partial x}\left(\frac{\partial f}{\partial y}\right)(0,0) = \lim_{k\to 0}\frac{\frac{\partial f}{\partial y}(h,0)-\frac{\partial f}{\partial y}(0,0)}{h} = \lim_{h\to 0}\frac{\frac{-h(0+0-h^4)}{(0+h^2)^2}-0}{h} = +1.$$

A Sucient Condition for the Equality of Mixed Partial Derivatives

For a function $f:\mathbb{R}^m\to\mathbb{R}$ suppose that the partial derivatives $\frac{\partial^2 f}{\partial x_j \partial x_i}$ and $\frac{\partial^2 f}{\partial x_i \partial x_j}$ exist at all points suciently close to a and that they are continuous at a, then

$$\frac{\partial^2 f}{\partial x_j \partial x_i}(\mathrm{a}) = \frac{\partial^2 f}{\partial x_i \partial x_j}$$

Change of Variables

Example: Under a change of variables, it is often necessary to transform second order, or higher order, partial derivatives. We illustrate the procedure with an example, namely the transformation from Cartesian co-ordinates (x, y), to plane polar co-ordinates (ρ, ϕ), where

$$x = \rho\cos\phi \text{ and } y = \rho\sin\phi.$$

In particular suppose for a function f we wish to calculate $\left(\frac{\partial^2 f}{\partial x^2}+\frac{\partial^2 f}{\partial x^2}\right)$ in terms of the partial derivatives in plane polar co-ordinates. First, we recall from that

$$\left(\frac{\partial f}{\partial x}\right)_y = \cos\phi\left(\frac{\partial f}{\partial \rho}\right)_\phi - \frac{\sin\phi}{\rho}\left(\frac{\partial f}{\partial \rho}\right)_\rho,$$

$$\left(\frac{\partial}{\partial x}\right)_y = \cos\phi\left(\frac{\partial}{\partial \rho}\right)_\phi - \frac{\sin\phi}{\rho}\left(\frac{\partial}{\partial \phi}\right)_\rho,$$

Hence

$$\left(\frac{\partial^2 f}{\partial x^2}\right)_y = \left(\frac{\partial}{\partial x}\right)_y\left(\frac{\partial f}{\partial x}\right)_y = \left(\cos\phi\left(\frac{\partial}{\partial \rho}\right)_\phi - \frac{\sin\phi}{\rho}\left(\frac{\partial}{\partial \phi}\right)_\rho\right),\left(\frac{\partial f}{\partial x}\right)_y$$

$$= \cos\phi\left(\frac{\partial}{\partial \rho}\right)_\phi\left(\cos\phi\left(\frac{\partial f}{\partial \rho}\right)_\phi - \frac{\sin\phi}{\rho}\left(\frac{\partial f}{\partial \phi}\right)_\rho\right)$$

$$= \cos^2\phi\left(\frac{\partial^2 f}{\partial \rho^2}\right)_\phi = -\frac{\sin\phi\cos\phi}{\rho^2} = \left(\frac{\partial f}{\partial \phi}\right)_\rho$$

$$-\frac{\sin\phi\cos\phi}{\rho}\left(\frac{\partial}{\partial \rho}\right)_\phi\left(\frac{\partial f}{\partial \rho}\right)_\rho + \frac{\sin^2\phi}{\rho}\left(\frac{\partial f}{\partial \phi}\right)_\rho$$

$$+\frac{\sin\phi\cos\phi}{\rho}\left(\frac{\partial}{\partial \rho}\right)_\rho\left(\frac{\partial f}{\partial \rho}\right)_\phi + \frac{\sin^2\phi\cos\phi}{\rho^2}\left(\frac{\partial f}{\partial \phi}\right)_\rho$$

$$+\frac{\sin^2\phi}{\rho_2}\left(\frac{\delta^2 f}{\delta\phi^2}\right)_\rho$$

On assuming the equality of mixed derivatives we obtain

$$\frac{\partial^2 f}{\partial x^2} = \cos^2\phi\frac{\delta^2 f}{\delta\rho^2} + \frac{\sin^2\phi}{\rho}\frac{\partial f}{\partial \rho} - \frac{2\sin\phi\cos\phi}{\rho}$$

$$\frac{\delta^2 f}{\delta\rho\partial\phi} + \frac{2\sin\phi\cos\phi}{\rho^2}\frac{\partial f}{\partial\phi} + \frac{\sin^2\phi}{\rho^2}\frac{\partial^2 f}{\partial\phi^2}$$

We can calculate $\frac{\partial^2 f}{\partial y^2}$ by observing that $x = \rho\sin\left(\phi + \frac{\pi}{2}\right)$ and $y = -\rho\cos\left(\phi + \frac{\pi}{2}\right)$.

Hence after applying the transformations $x \to -y, y \to x, \phi \to \left(\phi + \frac{\pi}{2}\right)$, we have that

$$\frac{\partial^2 f}{\partial y^2} = \sin^2\phi\frac{\partial^2 f}{\partial \rho^2} + \frac{\cos^2\phi}{\rho}\frac{\partial f}{\partial \rho} + \frac{2\sin\phi\cos\phi}{\rho}$$

$$\frac{\partial^2 f}{\partial \rho \partial \phi} - \frac{2\sin\phi\cos\phi}{\rho^2}\frac{\partial f}{\partial \phi} + \frac{\cos^2\phi}{\rho^2}\frac{\delta^2 f}{\delta \phi^2}.$$

and hence that

$$\frac{\partial^2 f}{\partial x^2} + \frac{\partial^2 f}{\partial y^2} = \frac{\partial^2 f}{\partial \rho^2} + \frac{1}{\rho}\frac{\partial f}{\partial \rho} + \frac{1}{\rho^2}\frac{\partial^2 f}{\partial \phi^2}$$

Taylor's Theorem

Recall from (1.4) that if $f: \mathbb{R} \to \mathbb{R}$ is dierentiable $(N + 1)$ times then Taylor's Theorem states that

$$f(a+h) = f(a) + \sum_{j=1}^{m} \frac{h^r}{r!}\frac{d^r}{dx^r}(a) + R_N(h)$$

where for some $0 < \theta < 1$

$$R_N(h) = \frac{h^{N+1}}{(N+1)!}\frac{d^{N+1} f}{dx^{N+1}}(a+\theta h).$$

There is a generalisation of this result to functions of several variables. Namely, if for a function $\mathbb{R}^m \to \mathbb{R}$ the partial derivatives of orders up to and including the $(N + 1)$th are continuous at a, then there is some $0 < \theta < 1$ such that

$$f(a + h) = f(a) + \sum_{r=1}^{N} \frac{1}{r!} = (h.\nabla)^r (f(a) + R_N(h),$$

where

$$R_N(h) = \frac{1}{(N+1)!}(h.\nabla)^{N+1} f(a+\theta h),$$

and

$$h\nabla = \sum_{j=1}^{m} h_j \frac{\partial}{\partial x_j} = h_j \frac{\partial}{\partial x_j}$$

If for a function $\mathbb{R}^m \to \mathbb{R}$ the partial derivatives of orders up to and including the Nth are continuous at a, then there is the weaker result

$$f(a+h) = f(a) + \sum_{r=i}^{N} = \frac{1}{r!}(h.\nabla)^r f(a) + o(|h|^N)$$

Remarks:

1. Since f is suciently dierentiable, the partial derivatives commute. This allows some simplification in the calculation of

$$(h\,.\,\nabla)^r f \text{ for } r = 1, \ldots N + 1.$$

For instance, for $m = \in$ the binomial theorem gives

$$(h.\nabla)^r f = \left(h_1 \frac{\partial}{\partial y_1} + h_2 \frac{\partial}{\partial y_2}\right)^r f$$

$$\sum_{s=0}^{r} \frac{r! h_1^s h_2^{r-s}}{(r-s)!s!} \frac{\partial^r f}{\partial x_1^s \partial x_2^{r-s};}$$

while the summation convention yields

$$(h.\nabla)^r = \left(h_{j_1} \frac{\partial}{\partial x_{j_1}}\right)\left(h_{j_2} \frac{\partial}{\partial x_{j_2}}\right)\cdots\left(h_{j_r} \frac{\partial}{\partial x_{jr}}\right)$$

$$= h_{j_1} h_{j_2} \cdots h_{jr} \frac{\partial}{\partial x_{j_1}} h_{j_1} \frac{\partial}{\partial x_{j_2}} h_{j_1} \cdots \frac{\partial}{\partial x_{jr}}.$$

2. A form of Taylor's Theorem holds for $\mathbb{R}^m \to \mathbb{R}^n$ by applying the above result componentwise:

$$f_i(a+h) = f_i(a) + \sum_{r=1}^{N} \frac{1}{r!} (h.\nabla)^r (f_i(a) + \ldots.$$

Example: Suppose that $\mathbb{R}^3 \to \mathbb{R}^n$ is given by

$$f(x) = \frac{1}{r} \text{ for } r \neq 0,$$

where $r = |x|$. Then, from using example,

$$(h.\nabla) f(x) = h_j \frac{\partial}{\partial x_j}\left(\frac{1}{r}\right) = -\frac{h_j x_j}{r^3}$$

$$(h.\nabla) f(x) = (h.\nabla)((h.\nabla) f) = h_k \frac{\partial}{\partial x_k}\left(-\frac{h_j x_j}{r^3}\right)$$

$$= -h_j h_k x_j \left(-\frac{3}{r^4}\right).\frac{x_k}{r} - h_j h_k \left(\frac{\delta_{jk}}{r^3}\right) = \frac{h_j h_k}{r^5}(3x_j x_k - r^2 \delta_{jk}).$$

Hence for $|h| \ll |x|$,

$$\frac{1}{|x+h|} = f(x+h) = f(x) + (h.\nabla) f(x) + \frac{1}{2}(h.\nabla)^2 f(x) + \ldots$$

$$= \frac{1}{r} - \frac{h_j x_j}{r^3} + \frac{h_j x_k}{2r^5}(3x_j x_k - r^2 \delta_{jk}) + \ldots..$$

We will make use of this result when we come to discuss monopoles, dipoles, quadrupoles, etc. at the end of the course.

EXERCISE

1. Let $f(t)$ be a smooth curve such that $f'(t) \neq 0$ for all t. Then we can define the *unit tangent vector* T by

$$T(t) = \frac{f'(t)}{\| f'(t) \|}$$

Show that

$$T'(t) = \frac{f'(t) \times (f''(t) \times f'(t))}{\| f'(t) \|^3}$$

2. Find T, N, B and κ at each point of the helix $f(t) = (\cos t, \sin t, t)$.
3. If $R = xI + yJ + zK$ and $r \neq 0$, show that
   (a) $\nabla (1/r^2) = -2R/r^4$; $\nabla (R/r^2) = 1/r^2$ (ii) div $(r^n R) = (n + 3)r^n$; curl$(r^n R) = 0$.
   (b) $\nabla\left[\nabla.\frac{R}{r}\right] = -\frac{2}{r^3}R.$
4. Show that each of following vectors are solenoidal :
   (a) $(x + 3y)I + (y - 3z)J + (x - 2z)K$ (ii) $3y^4z^2I + 4x^3z^2J + 3x^2y^2K$
   (b) $\nabla f \times \nabla y$.
5. Find the points on the circle $x^2 + y^2 = 100$ which are closest to and farthest from the point (2, 3).
6. Find the constrained maxima and minima of $f(x, y, z) = x + y^2 + 2z$ given that $4x^2 + 9y^2 - 36z^2 = 36$.
7. Find the volume of the largest rectangular parallelepiped that can be inscribed in the ellipsoid

$$\frac{x^2}{a^2} + \frac{y^2}{b^2} + \frac{z^2}{c^2} = 1.$$

8. Let be the three a $= 2i - j + k$, b $= i + 2j - k$ and c $= i + j - 2k$ vectors . Find the vector in the plane of b and c whose projection on a is of magnitude $\sqrt{\frac{2}{3}}$ .Exercise

Chapter 11

Poision's Equation and its Application

In mathematics, Poisson's equation is a partial differential equation with broad utility in electrostatics, mechanical engineering and theoretical physics. It is named after the French mathematician, geometer and physicist Siméon-Denis Poisson. The Poisson equation is

$$\Delta\varphi = f$$

where Δ is the Laplace operator, and f and φ are real or complex-valued functions on a manifold. When the manifold is Euclidean space, the Laplace operator is often denoted as Δ^2 and so Poisson's equation is frequently written as

$$\Delta^2\varphi = f$$

In three-dimensional Cartesian coordinates, it takes the form

$$\left(\frac{\partial^2}{\partial x^2}+\frac{\partial^2}{\partial y^2}+\frac{\partial^2}{\partial z^2}\right)\varphi(x,y,z) = f(x,y,z).$$

For vanishing f, this equation becomes Laplace's equation

$$\Delta\varphi = 0$$

The Poisson equation may be solved using a Green's function; a general exposition of the Green's function for the Poisson equation is given in the article on the screened Poisson equation. There are various methods for numerical solution. The relaxation method, an iterative algorithm, is one example.

We have encountered Poision's equation

$$\nabla^2 f = p(x).$$

and Laplace's equation (which is a special case of Poision's equation)

$$\nabla^2 f = 0,$$

where the Laplacian operator is given by

$$\nabla^2 = \begin{cases} \dfrac{\partial^2}{\partial x^2}+\dfrac{\partial^2}{\partial y^2} & \text{2D Cartesians,} \\ \dfrac{\partial^2}{\partial x^2}+\dfrac{\partial^2}{\partial y^2}+\dfrac{\partial^2}{\partial z^2} & \text{3D Cartisians,} \\ \dfrac{1}{p}\dfrac{\partial}{\partial p}\left(p\dfrac{\partial^2}{\partial p}\right)+\dfrac{1}{p^2}\dfrac{\partial^2}{\partial \phi^2}+\dfrac{\partial^2}{\partial z^2} & \text{Cylindrical Polar,} \\ \dfrac{1}{r^2}\dfrac{\partial}{\partial r}\left(r^2\dfrac{\partial}{\partial r}\right)+\dfrac{1}{r^2\sin\theta}\dfrac{\partial^2}{\partial \phi^2} & \text{Spherical Polar,} \end{cases}$$

Equations are examples of partial differential equations.

How do we find solutions to these equations ? What are appropriate boundary conditions ?

LINEARITY OF LAPLACE'S EQUATION

Laplace's equation is linear, in that if f_1 and f_2 are solutions, i.e if

$$\nabla^2 f_1 = 0 \text{ and } \nabla^2 f_2 = 0,$$

then $(\lambda f_1 + \mu f_2)$, where λ *and* μ are real constant, is another solution

$$\nabla^2(\lambda f_1 + \mu f_2) = \frac{\partial^2 f_2}{\partial x_j \partial x_j}(\lambda f_1 + \mu f_2) = \lambda \frac{\partial^2 f_1}{\partial x_j} + \mu \frac{\partial^2 f_2}{\partial x_j}$$

$$= \lambda \nabla^2 f_1 + \mu \nabla^2 f_2 = 0.$$

Hence if we have a number of solutions of Laplace's equations, we can generate further solutions by adding multiples of the known solutions together.

Separable Solutions

you haved already met the general idea of 'separatelity' in IA Differential Equations where you studied separable equations, i.e, the special differential equations that can be written in the form.

$$\underbrace{X(x)dx}_{\text{functions of } x} = \underbrace{Y(y)dy}_{\text{function of } y} = \text{constant.}$$

Sometimes functions can we written inb separable form; for instance,

$$f(x, y) = \cos x \exp y = X(x)\, Y(y), \text{ where } X = \cos x \text{ and } Y = \exp y,$$

is separable in Cartesian co-oridinates, while

$$g(x,y,z) = \frac{1}{(x^2+y^2+z^2)^{\frac{1}{2}}}$$

is not separable in Cartesian co–ordiantes, but is separable in spherical polar co–ordinates since

$$g = R(r)\theta(\theta)\Phi(\phi) \text{ where } R = \frac{1}{r}, \theta = 1 \text{ and } \theta = 1 \text{ and } \Phi = 1.$$

Solutions to Laplace's equation can be found by seeking solutions that can be written in separable from e.g.

2D Cartesians : $f(x, y) = X(x)\, Y(y)$,

3D Cartesians : $f(x, y, z) = X(x)\, Y(y)\, Z(z)$,

Plane Polars : $f(\rho, \phi) = R(\rho)\Phi(\phi)$,

Cylindrical Polars : $f(\rho, \phi, z) = R(\rho)\Phi(\phi)Z(z)$,

Spherical Polars : $f(r, \theta, \phi) = R(r)\theta(\theta)\Phi(\phi)$.

However, we emphasise that not all solutions of Laplace's equation can be written in this form which will be exploited in IB Methods.

Example: Two Dimensional Cartisian Co-ordinates

Seek solutions $f(x, y)$ to Laplace's equation. On substituting into we obtain

$$X''(x)\, Y(y) + X(x) = 0,$$

or after rearrangement

$$\underbrace{\frac{X''(x)}{X(x)}}_{\text{functiono of } x} = \underbrace{\frac{Y''(y)}{Y(y)}}_{\text{function of } y} = C,$$

where C is a constant (the only function of x that equal a function of y).There are three cases to consider.

$C = 0$. In this case

$$X''(x)\, Y''(y) = 0 \Rightarrow X = ax + b \text{ and } Y = cy + d,$$

where a, b, c and d are constants, i.e.

$$f = (ax + b)(cy + d).$$

$$C = K^2 < 0. \text{ In this case}$$

$$X'' - k^2X = 0, \text{ and } Y'' + k^2Y = 0$$

Hence

$$X = ae^{kx} + be^{-kx}, \text{ and } Y = c \sin ky + d \cos ky,$$

where a, b, c and d are constants , i.e.

$$f = (ae^{kx} + be^{-kx})\,(c \sin ky + d \cos ky).$$

$C = -k^2 < 0$. Using the $x \longleftrightarrow y$ symmetry it follows that

$$f = (ac^{ky} + bc^{-ky})\,(C \sin kx + d \cos kx).$$

Remarks:

1. Observe that solutions to Laplace's equation tend to grow or decay smoothly in one direction while oscillating in the other.
2. Without loss of genrality we could also impose, say, $c^2 + d^2 = 1$.

Isotopic Solutions

Letr Laplacian operator ∇^2 is isotropic, i,e. it has no preferred direction, since the operator is unchanged by the rotation of a Cartesian co-ordiante system. This suggests that there may be solutions to Laplace's equation that have no preferred direction.

3D: Spherical Symmetry

On the basis in 3D we seek a spherically symmetric solution to Laplace'sequation, i.e we seek a solution independent of θ and v (this is a special case of $r$the separable solution. Write $f = R(r)$, then from Laplace' equation

$$\nabla f = \frac{1}{r^2}\frac{\partial}{\partial r}\left(r^2\frac{\partial R}{\partial r}\right) + \frac{1}{r^2 \sin\theta}\frac{\partial}{\partial \theta}\left(\sin\frac{\partial R}{\partial \theta}\right) + \frac{1}{r^2 \sin^2 \theta\phi^2} = 0.$$

Hence we require

$$\frac{d}{dr}\left(r^2\frac{dR}{dr}\right) = 0 \;\Rightarrow\; \frac{dR}{dr} = \frac{\beta}{r^2}$$

$$\Rightarrow\; R = \alpha - \frac{\beta}{r},$$

where a and b are constants.

Remarks:

1. There is a 'singularlarity' at $r = 0$ if $\beta \neq 0$; hence the solution is not valid at $r = 0$ if $\beta \neq 0$.
2. We already knew that $\nabla^2\left(\frac{1}{r}\right) = 0$ for $r \neq 0$ from

3. Consider the flux of $u = \nabla f$ out of sphere S of radius r, then using the expression for grad

$$\iint_S \nabla f.dS = \iint_S n.\nabla\left(\alpha - \frac{\beta}{r}\right) dS$$

$$= \iint_S \frac{\partial}{\partial r}\left(\alpha - \frac{\beta}{r}\right) dS$$

$$= \iint_S \frac{\beta}{r^2} dS = 4\pi\beta.$$

Thus flux is thus independent of r, and there is unit flux if $\beta = -\dfrac{1}{4\pi r}$.

Definition: The solution

$$f = -\frac{1}{4\pi r}$$

is refrerred toas to potential of the unit point source, of as the fundamental soution of Poisson's eqution in 3D.

2D: Radial Symmetry

There is an analogous result in 2D. In particular radially symmetric solutions to Laplace's equation exist,

$$f = \alpha + \beta \log \rho,$$

where α and β are constants.

Remarks:

1. There is agian a singularity at the orgin if $\beta \neq 0$; i.e the solution is not valid at $r = 0$ if $\beta \neq 0$.

2. Consider the flux of $u = \nabla f$ out of circle C of radius r, then using the expression for grad

$$\int_C \nabla f.nds = \int_C n.\nabla(\alpha + \beta \log \rho) ds$$

$$= \int_C \frac{\partial}{\partial \rho}(\alpha + \beta \log \rho) ds$$

$$= \int_0^{2\pi} \frac{\partial}{\rho} \rho d\phi = 2\pi\beta.$$

Thus flux is thus independent of r, and there is unit flux if $\beta = \frac{1}{2\pi}$.

Definition: The solution

$$f = \frac{1}{2\pi}\log\rho$$

is referred to as the potintial of the unit line source, or as the fundamental solution of Poisson's equation in 2D.

MONOPOLES AND DIPOLE

We being by nothing that if $f(x)$ is a solution to Laplace's then further solution cane be generated by differenting with respect to Cartesian co-ordinates ' for suppose that

$$g(x) = \frac{\partial f}{\partial x_i},$$

then from

$$\nabla^2 g = \frac{\partial^2}{\partial x_j \partial x_j}\left(\frac{\partial f}{\partial x_i}\right) = \frac{\partial f}{\partial x_i}\left(\frac{\partial^2 f}{\partial x_j \partial x_j}\right) = 0.$$

Definition: The fundamental solution , i.e

$$\varphi_m \equiv -\frac{1}{4\pi r}, r \neq 0,$$

is sometimes referred to as the unit monopole.

Definition: Let d be a s constant vector, then using, it follows that

$$\varphi_d = -d_i\frac{\partial \varphi_m}{\partial x_i} = -\frac{d_i}{4\pi r^2}\frac{\partial r}{\partial x_i} = -\frac{d.x}{4\pi r^3}, r \neq 0,$$

is a solution to Laplace's equation. This is referred to as a dipole of strengh of d.

Remarks: A dipole can be viewed as being generated by two point sources that are of large, but opposite, strength and very close to each other.

HARMONIC FUNCTIONS

Definition: The function $f : \mathbb{R}^m \rightarrow \mathbb{R}(m = 2,3)$ is harmonic iff all its second - order partial derivatives exist and are continuous, and

$$\nabla^2 f = 0.$$

Remark: In IB complex Methods you will see that the real and imaginary parts of a complex differentiable function are both harmonic (this follows in straight forward manner from the Caulchy- Riemann equations). Hence solution of Laplace's equation can be found by taking the real and imaginary parts of any complex differentioble function

Green's Formulae, or Green's Theorem (But A Different One)

Three Dimensions

Suppose $\varphi, \psi, \nabla^2\varphi$ and $\nabla^2\psi$ defined in a region V bounded by a piecewise smooth surface S. Let $\frac{\varphi\psi}{\partial n}$ denote the component $\nabla\psi$ in the direction of the local normal of S, then

Gree's First Formula

$$\iiint_{v} (\varphi\nabla^2\psi + \nabla\varphi.\nabla\psi)dr) = \iint_{S} \varphi\nabla^2\psi.dS = \iint_{S} \varphi\frac{\varphi\psi}{\partial n}dS;$$

Green's second Formula

$$\iiint_{v} (\varphi\nabla^2\psi - \psi\nabla^2\varphi)dr) = \iint_{S} \left(\varphi\frac{\partial\psi}{\partial n} - \psi\frac{\partial\varphi}{\partial n}\right)dS.$$

Remark: Green 's first Formula is analogous to integration by parts : $f(fg'' + f'g')dx = [fg']$.

Outline proof.

Green's First Formula. Put $u = \varphi\nabla\psi$ in the divergence theorem. Then since

$$\nabla.u = \nabla.(\varphi\nabla\psi) = \frac{\partial\psi}{\partial x_j}\left(\varphi\frac{\partial\varphi}{\partial n}\right) = \varphi\frac{\partial^2\psi}{\partial x_j\partial x_j} + \frac{\partial\varphi}{\partial x_j}\frac{\partial\psi}{\partial x_j},$$

it follows that

$$\iiint_{v} (\varphi\nabla^2\psi + \nabla\varphi.\nabla\psi)dr = \iint \varphi\nabla\psi.dS.$$

Two Dimensions

Suppose $\varphi, \psi, \nabla^2\varphi$ and $\nabla^2\psi$ difined on a closed surface A bounded by a simple closed curve C. Then

Green's First Formula

$$\iint_{A} (\varphi\nabla^2\psi + \nabla\psi)dS \oint_{C} \varphi(\nabla\psi.n)dS = \oint_{C} \frac{\partial\psi}{\partial n}ds.$$

Green,s second Formula

$$\iint\limits_A (\varphi\nabla^2\psi - \psi\nabla^2\varphi)dS = \oint\limits_C \left(\varphi\frac{\partial\psi}{\partial n} - \psi\frac{\partial\psi}{\partial n}\right)ds.$$

Outline proof, As above but using the two dimensional divergence theorem

You have considered questions of uniqueness before. For instance, in *I* Algebra & Geometry you saw that if A is a $m \times m$ matrix then the solution to

$$Ax = b \text{ where } x,\ b \in \mathbb{R}^m,$$

is unique iff det $A \neq 0$.

How do we know that a solution of Poisson's/Laplace's equations for a given set of boundary conditions is unique? Indded how can we be sure that the correct number of boundarey conditions have been specified? More generally how do we know that a solution to a specified problem exists? Existence is hard and will not be tackled here - but we can make progress on uniqueness if we already know that a solution exists.

Generic Problems

Suppose that in a volume v boundary by a surface S

$$\Delta^2\varphi = \rho\,.$$

One of three common types of boundary condition is normally specified for this Poisson equation.

Dirichlet Condition: In this case φ is specified on S; i.e.

$$\varphi(x) = g(x) \text{ on } S,$$

where $g(x)$ is a sknown function.

Neumann Condition: In this case $\frac{\partial\varphi}{\partial n}$ is specified on S; i.e.

$$\frac{\partial\varphi}{\partial n} = n.\ \Delta\varphi = h(x) \text{ on } S,$$

where $h(x)$ is a known function.

Mixed Condition: In this case a linear combination of φ and $\frac{\partial\varphi}{\partial n}$ is specified on S; i.e.

$$\alpha(x)\ \frac{\partial\varphi}{\partial n} + \beta(x)\varphi = d(x) \text{ on } S,$$

where $\alpha(x)$, $\beta(x)$ and $d(x)$ are known functions $\alpha(x)$ and $\beta(x)$ are not simultaneously zero.

Uniqueness: Three Dimensions

Dirichlet Condition. Suppose β is a acalar field satisfying $\Delta^2\varphi = p(x)$ in v and $\varphi = g$ (x) on S. Then φ is unique.

Outline Proof. Suppose that there exit two functons, say φ_1 and φ_2, satisfying

$$\Delta^2\varphi_3 = \rho \text{ in } v, \text{ and } \varphi_3 = g \text{ on } S.$$

Let $\Phi = (\varphi_1 - \varphi_2)$, then

$$\Delta^2\Phi = 0 \text{ in } v, \text{ and } \Phi = 0 \text{ on } S$$

From Green's First Formula with $\varphi = \psi = \Phi$,

$$\iiint_V (\Phi\Delta^2\Phi + \Delta\Phi.\Delta\Phi)dr = \iint_S \Phi\frac{\partial\Phi}{\partial n}dS.$$

Hence from

$$\iiint_V |\Delta\Phi|^2\, d\tau = 0.$$

Suppose $\Delta\Phi \neq 0$ at $x_0 \in V$, and assume that the φ_j, and hence Φ, are continuous. Then if

$$|\Delta\Phi(x_0)| = \varepsilon > 0,$$

if follws from the continuity of Φ, that there exists $\delta > 0$ such that

$$|\Delta\Phi(x)| > \frac{1}{2}\varepsilon\forall\, |x - x_0| < \delta. \qquad \forall|x - x_0| < \delta.$$

Hence

$$\iiint_V |\Delta\Phi|^2\, d\tau \geq \iiint_{|x-x_0|<\delta} |\Delta\Phi|^2\, d\tau \geq \frac{4}{3}\pi\delta^3\left(\frac{1}{4}\varepsilon^2\right) > 0$$

It follows that

$$\Delta\Phi = \text{in } V \Rightarrow \Phi = \text{constant in } V.$$

But $\Phi = 0$ on S, hence

$$\Phi = 0 \text{ in } V \Rightarrow \Phi \;\; \varphi_1 = \varphi_2 \text{ in } V.$$

Neumann condition. Suppose instead that.

$$\Delta^2\delta = \rho(x) \text{ in } V \text{ and } \frac{\partial\varphi}{\partial n} = h(x) \text{ on } S.$$

Then φ is unique upto a constant.

Outline Proof: The proof is as above, but with $\frac{\partial \Phi}{\partial n} = 0$ on S, instead of $\Phi = 0$ on S. We again deduce that Φ = constant in V. However, now there is no means of fixing the constant. Hence

$$\varphi_1 = \varphi_2 + k,$$

where k is a constant.

Uniqueness: Two: Dimensions

Analogous uniqueness tresults hold in tow dimensions, and can be proved using instead of. For instance, suppose S is a closed region of the plane bounded by a simple piecewise smooth closed curve C, and suppose φ is a scalar field satisfyuing Poisson's equation with Dirichlet boundarry condition, i. e.

$$\Delta^2\varphi = \rho(x) \text{ on } S, \text{ and } \varphi = g(x) \text{ on } C,$$

then φ is unique.

Boundary Conditions

The uniqueness results imply that solutions to Poisso's equation, $\Delta^2\varphi = p$, are determined by the value of φ or $\frac{\partial \varphi}{\partial n}$ (or a lineaer combination of φ and $\frac{\partial \varphi}{\partial n}$), on the boundary.

Hence appropriate boundary conditions for Poisson's eqation include the specificatrion of φ or $\frac{\partial \varphi}{\partial n}$, on the whoe boundary.

MACIMA AND MINIMA OF HARMONIC FUNCTIONS

Three Dimensions

Suppose $\varphi(x)$ is harmonic, i.e. $\Delta^2\varphi = 0$, in a volume V boundary by a piecewise smooth closed S. Then φ attains its maximum and minimum values on S.

Δ^2 is a smooth operator - no spikes.

Outline Proof: Suppose φ has a maximum at an interior point x_0. Surround x_0 by a sphere S_δ small enough so that $\frac{\partial \varphi}{\partial n} < 0$ on S_δ (always possible if φ differentiable). Also choose S_δ so that it is entirely within V. Denote by V_δ the volume within the S_δ.

Let

$$m = \min_{x \in S_\delta} \varphi(x),$$

and define

$$\psi(x) = \varphi(x) - m + \varepsilon,$$

where $\varepsilon > 0$. Then

(a) ψ is harmonic in V, since

$$\Delta^2 \psi = \Delta^2(\varphi - m + \varepsilon) = 0;$$

(b) $\psi > 0$ on S_δ, since $\varphi - m \geq 0\ S_\delta$;

(c) $\frac{\partial \psi}{\partial n} < 0$ on S_δ, since by construction $\frac{\partial \psi}{\partial n} = \frac{\partial \varphi}{\partial n} < 0$ on S_δ.

Set $\varphi = \psi$ in Green's First Formula (8.12), to obtain

$$\iiint_{V_\delta} (\psi \Delta^2 \psi + |\Delta \psi|^2) dr = \iint_{S_\delta} \psi \frac{\partial \psi}{\partial n} dS.$$

Hence using (a) – (c)

$$0 \leq \iiint_{V_\delta} |\Delta \psi|^2\, dr = \iint_{S_\delta} \psi \frac{\partial \psi}{\partial n} dS < 0.$$

Thus φ cannot have a maximum at an interior point. Similarly for a minimum, or alternatively consider a maximum of $(-\varphi)$.

Remark: The argument does not apply for x_0 on the suface S, because it is then impossible to surrouund x_0 by a sphere S_δ that lies totallly within V.

Two Dimensions

Suppose $u(x, y)$ is harmonic, i. e. $\Delta^2 u = 0$, in a region A bounded by a simple piecewise smooth closed curve C. Then u attains its maximum and minimum values on C.

Outline Proof: Proceed as in three dimensions.

Example: Find the maximum of

$$u = \sin x \cos h\, y$$

on the unit square $0 \leq x \leq 1, 0 \leq y \leq 1$.

Answer: u is *a* special case of separable solution. Thus it is harmonic, and its maximum must be on the boundary. Since both sin and cosh are in creasing on $|0,1|$, it follows that

$$\max u = \sin 1 \cos h\ 1.$$

Unlectured Example: A Mean Value Theorem.

Suppose that the scalar field φ is harmonic in a volume V bounded by a closed surface S. Then, using the diverrgence theorem, we note the useful result that

$$\iint_S \frac{\partial \varphi}{\partial n} dS = \iint_S \nabla \varphi .\ dS = \iiint_v \nabla^2 \varphi dr = 0.$$

Consider the function $f(r)$ defined to be the mean value of φ on a spherical surface S, of radius r and normal n, i. e.

$$f(r) = \frac{1}{4\pi r^2} \iint_S \varphi(x) dS$$
$$= \frac{1}{4\pi} \int_0^{\pi} d\theta \int_0^{2\pi} d\theta \sin\theta \varphi(r,\theta,\phi),$$

where (r,θ,ϕ) are spherical polar co-ordinates. We claim that f is a cnstant, and that it is equal to the value of φ at the origin.

We can show this result either by using Green's Second Formula or directly as follows using:

$$\frac{df}{dr} = \frac{1}{4\pi} \int_0^{\pi} d\theta \int_0^{2\pi} d\theta \sin\theta \frac{\partial \varphi}{\partial r}$$
$$= \frac{1}{4\pi} \int_0^{\pi} d\theta \int_0^{2\pi} d\theta \sin\theta \frac{\partial \varphi}{\partial n}$$
$$= \frac{1}{4\pi r^2} \iint_{S_r} \frac{\partial \varphi}{\partial n} dS = 0.$$

Further, in the limit $r \to 0$ it follows that

$$f(0) = \varphi(0)$$

and hence

$$\varphi(0) = \frac{1}{4\pi r^2} \iint_{S_r} \varphi(x) dS.$$

We conclude the mean value result that: If φ is harmonic, the value of φ at a point is equal to the average of the values of φ on any spherical shell centred at that point.

APPLICATION OF POISSON'S EQUATION

GRAVITY

The Gravitational Potential

The gravitational acceleration, or gravitational filed, g at x due to a particle of mass m_1 at x_1 is given by newton's famous inverse square law

$$g = -Gm_1 \frac{x - x_1}{|x - x_1|^3} \text{ for } x \neq x_1,$$

where G is the universal gravitational constant and space is taken to be unbounded (i.e the whole of $\mathbb{R}^3$).

For the special case $x_1 = 0$ we have seen that curlg = 0; the case $x_1 \neq 0$ follows after a co-ordinate translation, e. g. let $x' = x - x_1$. Hence from the results on irrotational fields in (a) it follows that gravity is a conservative force, and (b) it follows from Poincare's lemma that there exists a scalar gravitational potential φ such that

$$g = \nabla\varphi.$$

Further, we deduce from the evaluation of the gradient of 1/2 that for $x \neq x_1$

$$\varphi = -\frac{Gm_1}{|x - x_1|}.$$

Hence it follows that

$$\nabla . g = -\nabla^2\varphi = Gm_1\nabla^2\left(\frac{1}{|x - x_1|}\right) = 0, \quad x \neq x_1.$$

Remarks:

1. The gravitastional potential for a point mass is a multiple of the fundamental solution for a monopole, i. e. a unit point source.
2. The gravitational potential for a point mass satisfies Laplace's equation except at the position of the point mass $x = x_1$.

Gauss' Flux Theorem

Point masses: Let S be a closed surface, and let ν be the volume interior to S. We start by considering the flux of gravitational field g out of the surface S generated by a single particle of mass m_1 at x_1. There are two cases t consider: $x_1 \notin V$ and $x_1 \in V$(we will side step the case of when the point mass is on the surface).

$x_1 \notin \nu$. Then from the divergence theorem

$$\iint_S g.dS = \iiint_V \nabla.g d\tau = 0.$$

$x_1 \in V$. Before considering a general surface S, suppose that $S = S_\delta$ where S_δ is a sphere of radius δ centred on x_1. Let V_δ be the interior volume to a S_δ, and note that on S_δ the normal is parallel to $(x - x_1)$. Hence proceeding as in the calculation leading,

$$\iint_{S_\delta} g.dS = -\iint_{S_\delta} \frac{Gm_1}{|x - x_1|^3}(x - x_1).ndS$$

$$= -Gm_1 \iint_{S_\delta} \frac{1}{\delta^2} dS = -4\pi Gm_1.$$

Suppose now that S is a general sufrace such that $x_1 \in V$. Then define

$$\hat{V} = V - V_\delta,\ \hat{S} = S - S_\delta,$$

where δ is chosen to be sufficiently small that V_δ lies totally within V and $-S_\delta$ means that the normal is directed towards the origin. Since $x_1 \in \hat{V}$ it follows that

$$0 = \iiint_{\hat{V}} \nabla.g\tau = \iint_{\hat{S}} g.dS = \iint_S g.dS - \iint_{S_\delta} g.dS.$$

hence

$$\iint_S g.dS = \iint_{S_\delta} g.dS = -4\pi Gm_1.$$

We conclude that in both cases that if M is the total mass inside S then

$$\iint_S g.dS = -4\pi GM.$$

Suppose now that there i more than one mass, e.g. $m_1, m_2, ..., m_p$ at pints $x_1, x_2, ..., x_p$ respectively. Then, since the gravitational fields of a number of particles can be superimposed, is modified to

$$\iint_S g.dS = -4\pi G \sum_{S_j \in V} m_j = -4\pi GM.$$

where M is again the total mass inside S.

THE CONTINUUM LIMIT

Supose that $\delta\tau_\xi$ is a valume surronding an arbitrary point ξ, and suppose also that

$\delta\tau_\xi$ contains a mass δm_ξ. The continuum limiut consists in letting $P \to \infty$ and $m \to 0$ ($j = 1, \ldots P$) in such a way that if we subsequently consider the limit in which $\delta\tau_\xi$ contracts ot zero volume, then

$$\rho(\xi) = \lim_{\delta\tau \to 0} \frac{\delta m_\xi}{\delta\tau_\xi}$$

exits for all ξ; the quantity $\rho(\xi)$ is referred ot as the density of mall per unit volume.

We wish to determine the total mass, M, in a volume V. by analogy with the defination of a triple intergral, subdivide V into N sub-domains $V_1 \ldots V_N$, and suppose $\xi_i \in V_1 (i = 1, \ldots N)$. Denote the volume of V_1 $\delta\tau\xi_i$ and the mass containedd within it by δm_{ε_i} where

$$\delta m_{\xi_i} = \sum_{x_j \in V_i} m_j$$

then a little arm-waving yields

$$M = \sum_{x_j \in V} m_j = \sum_{i=1}^{N} \sum_{x_j \in V_i} m_j = \sum_{i=1}^{N} \delta m_{\xi_i}$$

$$= \lim_{\substack{N \to \infty \\ \delta\tau_{\xi_i} \to 0}} \sum_{i=1}^{N} \rho(\xi_i) \delta\tau_{\xi_i} = \iiint_V \rho(\xi) d\tau_\xi .$$

Hence on taking the continuum limit we have Gauss, flux theorem:

$$\iint_S g.dS = -4\pi G \iiint_V \rho d\tau.$$

Poisso's Equation and solution

Poisson's Equation. Applying the divergence theorem to Gauss' flux theorem we have that

$$\iiint_V (\nabla.g + 4\pi G\rho) d\tau = 0.$$

But this true for all vloumes V, hence

$$\nabla.g = -4\pi G\rho.$$

Further, we know that $g = -\nabla\varphi$ for a single point mass. On the assumption that this result continues to hold on taking the continuuum limit, e. g. because of the principle of superposition, it follows that in the continuum limit φ satisfies *Poisson;s* equation

$$\nabla^2\varphi = 4\pi G\rho.$$

The Solution in unbounded Space. We have the principle of superposition, that for more than one mass in unbounded space

$$\varphi(x) = -G\sum_j \frac{m_j}{|x - x_j|}.$$

We again take the continuum limit in a similar way to above, i. e. from using Taylor's theorem and iin particular,

$$\varphi(x) = -G\sum_{i=1}^{N}\sum_{x_j \in V_j} \frac{m_j}{\left|(x-\xi_i) + \underbrace{(\xi_i - x_j)}_{\text{small}}\right|} = -G \lim_{\substack{N\to\infty \\ \delta\tau\xi_i \to 0}} \sum_{i=0}^{N}\left(\frac{\rho(\varepsilon_i)\delta\tau\xi_i}{|(x-\xi_i)|} + \ldots\right),$$

and hence the solution is

$$\varphi(x) = -G\iiint_V \frac{\rho_j}{|x-\xi|} d\tau\xi,$$

where the volume integral extends over all regions where the density is non-zero.

CONDITIONS AT A DISCONTINUITY IN DENSITY

Suppose that the density ρ is discontinuous (but bounded across a surface S. How do φ and ∇_φ very across the discontinuity?)

Let x_0 be a point on the surface S. Surround x_0 with a surface $\hat{S}$ that is a small squashed 'pillbox' f cross- sectional area $\varphi A \ll 1$ and thickness $\delta n \ll (\varphi A)^{\frac{1}{2}}$. Let the enclosing volume be $\hat{\upsilon}$, and suppose that the normal of the cross-sectional faces is aligned with the normal of the surface S at x_0. Then for a very squashed small pill box

$$\iint_{\hat{S}} g.ds \approx (n/\,g(x_0+) - n.g(x_0-))\varphi A,$$

$$\iiint_{\hat{V}} pd\tau(\max_{\hat{V}})\delta n\delta A,$$

where $g(x_0+)$ and $g(x_0-)$ are, respectively, the values of g on the outside and inside of the surface S at x_0 . On applying Gauss' Flux Theorem to $\hat{S}$ and factoring δA we have that

$$(n.g(x_0+) - n.g(x_0-)) = 0(\delta n).$$

Thus, on taking the limit $\delta n \to 0$, we conclude that

$$[n.g]_{x_{0-}}^{x_{0+}} = 0,$$

i.e. that the normal component of the gravitaional field is continuous across a surface of discontinuity in ρ. Moreover, since $g = -\nabla\varphi$; this condition can als be expressed as

$$[n.\nabla\varphi]_{x_0-}^{x_0+} = \left[\frac{\partial\varphi}{\partial n}\right]_{x_0-}^{x_0+} = 0,$$

i.e the normal component of $\nabla\varphi$ is continuous across a surface of discontinuity in ρ.

Further, suppose C is a straight line between x_a and x_b where the two ends are on either side of S. Then suing

$$\varphi(x_a) - \varphi(x_b) = \int_c \nabla\varphi.dx \le |x_b - x_a| \max_c |\nabla\varphi|.$$

If we let $x_a \to x_0 -$ and $x_b \to x_0 +$ it follows, on the assumption that $|\nabla\varphi|$ is bounded, that

$$[\varphi]_{x_0-}^{x_0+} = \varphi(x_0 =) - \varphi(x_0 -) = 0,$$

i.e that φ is continuous across a surface of discontinuity in ρ.

Example: Obtain the gravitational field due to sphere of radius *a*, uniform density ρ_0, and total mass

$$M = \iiint\limits_{\text{sphere}} \rho_0 d\tau = \frac{4}{3}\pi a^3 \rho_0.$$

without loss of generality we assume that the origin of co-ordinates is at the centre of the sphere.

Using Poisso's Equation. From

$$\nabla^2\varphi = \begin{cases} 4\pi G\rho_0 & \text{for } r < a \\ 0 & \text{for } r > a \end{cases},$$

while, φ and $\dfrac{\partial\varphi}{\partial n}$ will be continuous at the discontinuity in density at $r = a$.

Further, from symmetry considrations we anticipate that φ will be a function only of $r = |x|, i.e.\ \varphi \equiv \varphi(r)$. Then from the expression for the laplacilan in spherical polar co-ordinates, for $r < a$ we have that

$$\frac{1}{r^2}\frac{d}{dr}\left(r^2\frac{d\varphi}{dr}\right) = 4\pi G\rho_0,$$

and hence on intergrating

$$r^2\frac{d\varphi}{dr} = \frac{4}{3}rG\rho_0 r^3\beta_i$$

and

$$\varphi = \frac{2}{3}\pi G\rho_0 r^2 + \alpha_i - \frac{\beta_i}{r},$$

where α_i and β_i are constants. Since there is no point mass at the origin, φ is bounded there, and hence $\beta_i = 0$. Similarly, for $r > a$ we deduce that

$$\varphi = \alpha_0 - \frac{\beta_0}{r},$$

where α_0 and β_0 are cnstants. if φ and $\frac{\partial\varphi}{\partial n} = \frac{\partial\varphi}{\partial r}$ are to be continuous at $r = a$ it also follows that

$$\frac{2}{3}\pi G\rho_0 a^2 + \alpha_i = \alpha_0 - \frac{\beta_0}{a}$$

and

$$\frac{4}{3}\pi G\rho_0 a = \frac{\beta_0}{a^2},$$

respectively, i.e. that

$$\beta_0 = \frac{4}{3}\pi G\rho_0 a^3, \quad \alpha_i = \alpha_0 - 2\pi G\rho_0 a^2.$$

It is usual to normalise the gravitational potential so that $\varphi \to 0$ as $r \to \infty$, in which case $a_0 = 0$,

$$\varphi = -\frac{4\pi G\rho_0 a^3}{3r}, \quad g = -\nabla\varphi = -\frac{4\pi G\rho_0 a^3}{3r^2}\hat{x}$$

$$= -\frac{GM}{r^2}\hat{x} \text{ for } r \geq a,$$

and

$$\varphi = \frac{2}{3}\pi G\rho_0(r^2 - 3a^3),$$

$$g = -\nabla\varphi = -\frac{4}{3}\pi G\rho_0 r\hat{x} = -\frac{GMr}{a^3}\hat{x} \ \textit{for}\ r \leq a.$$

Remarks: Outside the sphere, the gravitational acceleration is equal to that of a particle of mass *M*. Within the sphere, the gravitational acceleration increases linearly with distance from the centre.

Using Gauss' Flux Theorem: From symmetry considration we anticipate that *g* will be a function only of $r = |x|$, and will be parallel to $\hat{x}$, i. e. $g = g(r)\hat{x}$. Suppose *S* is a spherical surface of radius *R*. Then

$$\iint_S g.dS = \iint_S (R)\hat{x}.dS = g(R)4\pi R^2.$$

Further,

$$\text{if } R \geq a \text{ then} \iiint_V \rho d\tau = \frac{4}{3}\pi a^3 \rho_0 = M,$$

$$\text{if } R \leq a \text{ then} \iiint_V \rho d\tau = \frac{4}{3}\pi R^3 \rho_0 = MR^3 / a^3.$$

Hence from Gauss' folux theorem, and substituting *r* for *R*,

$$g(r) = \begin{cases} -GM / r^2 & r \geq a \\ -GMr / a^3 & r \leq a \end{cases}.$$

Electrostatics

Similar results hold for the electrostatic force. This is because the *electric field E* due to a point *chage* q_1 at x_1 is also given by an inverse square law,

$$E = \frac{q_1}{4\pi\varepsilon_0} \frac{x - x_1}{|x - x_1|^3},$$

where ε_0 is the permittivity of free space. Thus we can read off the equivalent results for electrostatics by means of the transformations

$$g \rightarrow E,\ m_j \rightarrow q_j\ G \rightarrow -\frac{1}{4\pi\varepsilon_0}.$$

For instance, Gauss' flux theorem becomes

$$\iint_S E.dS = \frac{1}{\varepsilon_0} \iiint_V d\tau,$$

where ρ is the now the charge density. The differential from of this equation is

$$\nabla.E = \frac{\rho}{\varepsilon_0},$$

which is one of maxwell's equations. Further, if we introduce the electric potential φ, where,

$$E = -\nabla\varphi,$$

then

$$\nabla^2\varphi = -\frac{\rho}{\varepsilon_0},$$

with solution in unbounded space

$$\varphi(x) = \frac{1}{4\pi\varepsilon_0}\iiint_{\nu}\frac{\rho(\xi)}{|x-\xi|}d\tau_{\xi}.$$

The Solution to Poisson's Equatin in

We derived the solution ot Poisson's equation in unbounded space, as the continuum limit of a sum over point sources. It is also possible to derive it using Green's second formula. Specifically we aim to find the solution to

$$\nabla^2\varphi = \rho \text{ in } \mathbb{R}^3, \text{with} \varphi \to 0 \text{as} |x| \to \infty,$$

where for simplicity we will assume that $\rho(x) = 0 \text{ for} |x| > R_\rho$ for some real R_ρ.

We begin by supposing that V is the volume between concentric spheres S_δ and S_R of radii δ and $R > R_\rho$, respectively. Suppose taht the spheres are centred on some arbitrary point $\xi \in \mathbb{R}^3$, and let

$$\psi = \frac{1}{|x-\xi|} = \frac{1}{r},$$

where $r = \equiv |x-\xi|$. Then,since $\xi \notin V$, it follows that

$$\nabla^2\psi = 0 \text{ in } V.$$

Also, for x on S_δ and for $r = |x-\xi|$,

$$\psi = \frac{1}{\delta} \text{ and } \frac{\partial\psi}{\partial n} = -\frac{\partial\psi}{\partial r} = \frac{1}{\delta^2},$$

where n is taken to be the normal pointing outward from n similarly, for x on S_R and for the normal pointing outward from n,

$$\psi = \frac{1}{R} \text{ and } \frac{\partial\psi}{\partial n} = \frac{\partial\psi}{\partial r} = -\frac{1}{R}.$$

Then from Green's second formula, we have that

$$-\iiint_{v} \frac{\rho(x)}{[x-\xi]} d\tau_x = \iint_{S_\delta + S_R} \left(\varphi \frac{\partial\psi}{\partial n} - \psi \frac{\partial\psi}{\partial n}\right) dS.$$

Remarks:

1. In an order of magnitude sense it follows from the definitions that

$$Q = O\left(\rho R_\rho^3\right), | d | = O\left(\rho R_\rho^4\right) \text{and } | A_{ij} | = O\left(\rho R_\rho^5\right).$$

   Hence at large distances x from a distributed mass/ charge source, we conclude that the dipole field is less than the monopole field by a factor of $O(Rr/| x |)$, and the quadropole field is less than the dipole field by a factor of $O(Rr/| x |)$,

2. Far from a distributed mass /charge source it follows that the potential is, to leading order, just that of a point mass/charge of the same strength .

3. If the total [mass /]charge of distributed source is zero then, then to leading order, the potential at large distance is that due to a dipole.

4. If r is the [scaled] mass density of a body, then Q is the total mass of the body, d/Q is the centre of gravity of the body, and the quadrupole strength is related to the moment of inertia of the body. Hence if you can accurately measure the gravitional potential of the body (e.g. by satellite) you can back out the moments of inertia of the body.

Heat

What is equation the governs the flow of heat in a saucepan, an engine block, the earth 's core, etc.?

Let $q(x, t)$ denote trhe flux vector for heat flow. Then rthe eneergy in the form of heat (molecular vivbrations) crossing a small surface dS in time δt is.

$$(q.dS)\delta t,$$

and heat flow out of a closed surface S in time.

and the heat flow out of a closed surface S in time δt is

$$\left(\iint_S q.dS\right)\delta t.$$

Let $E(x, t)$ deonte the internal energy per unit mass of the solid, $Q(x, t)$ denote any heat source per unit time per unit per unit volume of the solid, $\rho(x, t)$ denote the mass density of the solid (assumed constant here).

The flow of heat in/out of S must balance the change in internal energy and heat source. Hence in time δt

$$\left(\iint_s q.ds\right)\delta t = -\left(\frac{d}{dt}\iiint_v pEd\gamma\right)\delta t.$$

For 'slow' changes at constant pressure (1st and 2nd law of thermodynamics)

$$E(x,t) = c_p\theta(x,t),$$

where θ is the temperature and c_p is the specific heat (assumed constant here).

Hence using the divergence theorem and exchanging the order of differentiation and integration,

$$\int\left(\nabla.q + \rho c_p \frac{\partial\theta}{\partial t} - Q\right)d\gamma = 0.$$

But this is true for any volume, hence

$$\frac{\partial\theta}{\partial t} = \nu\nabla^2\theta + \frac{Q}{\rho c_p}$$

where $\nu = k/(\rho c_p)$ is the fiffusivity (or coefficient of diffusion).

Special Cases.

Diffusion Equation: If there is no heat source then $Q = 0$, and the governing equation becomes the diffusion equation

$$\frac{\partial\theta}{\partial t} = \nu\nabla^2\theta.$$

Poisson's Equation: If the system has reached a steady state (i.e $\partial_1 = 0$), then with $f(x) = -Q(x)/k$ we recover Poisson's equation

$$\nabla^2\theta = f$$

Laplace's Equation: If the system has reached a steady state and there are no heat sources then the temperature is governed by Laplace's equation

$$\nabla^2\theta = f$$

The maximum and minimum temperatures occu on the boundary, i.e the vartiations of temperature on the boundary are smoothed out in the interior.

EXERCISE

1. Let $f(t)$ be a differentiable curve such that $f(t) \neq 0$ for all t. Show that

$$\frac{d}{dt}\left(\frac{f(t)}{\|f(t)\|}\right) = \frac{f(t)\times(f'(t)\times f(t))}{\|f(t)\|^3}$$

2. Let $f(t)$ be a smooth curve such that $f'(t) \neq 0$ for all t. Then we can define the *unit tangent vector* T by

$$T(t) = \frac{f'(t)}{\|f'(t)\|}$$

Show that

$$T'(t) = \frac{f'(t)\times(f''(t)\times f'(t))}{\|f'(t)\|^3}$$

3. Find T, N, B and κ at each point of the helix $f(t) = (\cos t, \sin t, t)$.
4. Show that the arc length L of a curve whose spherical coordinates are $\rho = \rho(t)$, $\theta = \phi(t)$ and $\phi = \phi(t)$ for t in an interval $[a, b]$ is

$$L = \int_a^b \sqrt{\rho'(t)^2 + (\rho(t)^2 \sin^2 \phi(t))\theta'(t)^2 + \rho(t)^2 \phi'(t)^2}\, dt.$$

5. Let be the three $a = 2i - j + k$, $b = i + 2j - k$ and $c = i + j - 2k$ vectors . Find the vector in the plane of b and c whose projection on a is of magnitude $\sqrt{\frac{2}{3}}$.
6. If the vectors $ai + j + k$, $i + bj + k$ and $i + j + ck$ $(a \neq b \neq c \neq 1)$ are coplanar vectors> find the value of $\frac{1}{1-a} + \frac{1}{1-b} + \frac{1}{1-c}$.
7. If $a = i + j - k$, $b = i - j + k$ and c are unit vectors perpendicular to the vector a and coplanar with a nd b. Find the unit vector d perpendicular to both a and c.
8. If $d = \lambda(a \times b) + \mu(b \times c) + \nu(c \times a)$ and $[a\ b\ c] = \frac{1}{8}$. Find the value of $\lambda + \mu + \nu$.
9. Show that $f(x, y) = \frac{1}{2\pi\sigma^2} e^{-(x^2+y^2)/2\sigma^2}$, for $\sigma > 0$, is constant on the circle of radius $r > 0$ centered at the origin. This function is called a *Gaussian blur*, and is used as a filter in image processing software to produce a "blurred" effect.
10. Suppose that $f(x, y) \leq f(y, x)$ for all (x, y) in $\mathbb{R}^2$. Show that $f(x, y) = f(y, x)$ for all (x, y) in $\mathbb{R}^2$.

12. Use the substitution $r = \sqrt{x^2 + y^2}$ to show that

13. Evaluate the line integral $\int_C \left[\left(x^2 + xy\right)dx + \left(x^2 + y^2\right)dy\right]$ where C is the square formed by the lines y = ± 1 and x = ± 1.

14. If F = (5xy – 6x²)I + (2y – 4x)J, evaluate $\int_C F.dR$ along the curve C in the xy-plane, y = x³ from the point (1, 1) to (2, 8).

15. A vector field is given by F = (sin y)I + x(1 + cos y)J. Evaluate the line integral over a circular path given by $x^2 + y^2 = a^2$, z = 0.

$$\lim_{(x,y)\to(0,0)} \frac{\sin\sqrt{x^2+y^2}}{\sqrt{x^2+y^2}} = 1.$$

16. If $R = xI + yJ + zK$ and $r \neq 0$, show that
   (a) $\nabla(1/r^2) = -2R/r^4$; $\nabla(R/r^2) = 1/r^2$ (ii) div $(r^n R) = (n+3)r^n$; curl$(r^n R) = 0$.
   (b) $\nabla\left[\nabla.\frac{R}{r}\right] = -\frac{2}{r^3}R.$

17. Show that each of following vectors are solenoidal :
   (a) (x + 3y)I + (y – 3z)J + (x – 2z)K (ii) $3y^4z^2I + 4x^3z^2J + 3x^2y^2K$
   (b) $\nabla f \times \nabla$ y.

18. Find the points on the circle $x^2 + y^2 = 100$ which are closest to and farthest from the point (2, 3).

19. Find the volume of the largest rectangular parallelepiped that can be inscribed in the ellipsoid

$$\frac{x^2}{a^2} + \frac{y^2}{b^2} + \frac{z^2}{c^2} = 1.$$

20. Derive the gradient formula in cylindrical coordinates: $\nabla F = \frac{\partial F}{\partial r}e_r + \frac{1}{r}\frac{\partial F}{\partial \theta}e_\theta + \frac{\partial F}{\partial z}e_z$

21. Use $f = u\nabla v$ in the Divergence Theorem to prove:

   (a) *Green's first identity:* $\iiint_S (u\Delta v + (\nabla u).(\nabla v))dV = \iint_\Sigma (u\nabla v).d\sigma$

   (b) *Green's second identity:* $\iiint_S (u\Delta v - v\Delta u)dV = \iint_\Sigma (u\nabla v - v\nabla u).d\sigma$